Central Pacific Drive

HISTORY OF U. S. MARINE CORPS OPERATIONS IN WORLD WAR II

VOLUME III

by

HENRY I. SHAW, JR.

BERNARD C. NALTY

EDWIN T. TURNBLADH

Historical Branch, G-3 Division, Headquarters, U. S. Marine Corps

1966

Other Volumes in this Series

I

LtCol Frank O. Hough, Maj Verle E. Ludwig, and Henry I. Shaw, Jr.

Pearl Harbor to Guadalcanal

II

Henry I. Shaw, Jr. and Maj Douglas T. Kane

Isolation of Rabaul

Library of Congress Catalog Card No. 58-60002

For sale by the Superintendent of Documents, U.S. Government Printing Office
Washington, D.C. 20402 – Price $7.25

Foreword

This book, the third in a projected five-volume series, continues the comprehensive history of Marine Corps operations in World War II. The story of individual campaigns, once told in separate detail in preliminary monographs, has been reevaluated and rewritten to show events in proper proportion to each other and in correct perspective to the war as a whole. New material, particularly from Japanese sources, which has become available since the writing of the monographs, has been included to provide fresh insight into the Marine Corps' contribution to the final victory in the Pacific.

During the period covered in these pages, we learned a great deal about the theory and practice of amphibious warfare. But most of all we confirmed the basic soundness of the doctrine which had been developed in prewar years by a dedicated and farsighted group of Navy and Marine Corps officers. These men, the leaders and workers in the evolution of modern amphibious tactics and techniques, served their country well. Anticipating the demands of a vast naval campaign in the Pacific, they developed requirements and tested prototypes for the landing craft and vehicles which first began to appear in large numbers at the time of the Central Pacific battles. Many of the senior officers among these prewar teachers and planners were the commanders who led the forces afloat and ashore in the Gilberts, Marshalls, and Marianas.

Allied strategy envisioned two converging drives upon the inner core of Japanese defenses, one mounted in the Southwest Pacific under General MacArthur's command, the other in the Central Pacific under Admiral Nimitz. Although Marines fought on land and in the air in the campaign to isolate Rabaul, and played a part significant beyond their numbers, it was in the Central Pacific that the majority of Fleet Marine Force units saw action. Here, a smoothly functioning Navy-Marine Corps team, ably supported by Army ground and air units, took part in a series of amphibious assaults that ranged in complexity from the seizure of tiny and heavily-defended islets, where there was little room for maneuver and no respite from combat, to large islands where two and three divisions could advance in concert.

It was my privilege to take part in much of this campaign, first as operations officer with Tactical Group 1 in the Marshalls and later in the same capacity with the 2d Marine Division in the Marianas. I prize the associations formed then with the officers and men who won the victories, not only those of the Marine Corps but also those of the other services in our joint commands. Sparked by knowledge hard-won at Tarawa, we were able to plan and execute effectively the operations at Kwajalein and move forward, on incredibly short notice, to Eniwetok, seizing there the islands that guarded the lagoon from which most of our ships staged for the Marianas. In the fighting for Saipan, Tinian, and Guam, Marines proved conclusively that their demonstrated effectiveness in short-term amphibious assault carried over to extended combat ashore.

As the narrative in this volume clearly shows, victory against a foe as determined and as competent as the Japanese could not have been won without a high cost in the lives of the men who did the fighting. Our advance from Tarawa to Guam was paid for in the blood of brave men, ordinary Americans whose sacrifice for their country should never be forgotten. Nor will it be by those who were honored to serve with them.

Wallace M. Greene, Jr.

WALLACE M. GREENE, JR.
GENERAL, U.S. MARINE CORPS
COMMANDANT OF THE MARINE CORPS

Reviewed and approved
8 December 1965

Preface

The series of Central Pacific operations that began at Tarawa in November 1943 marked a period of steadily increasing momentum in our drive toward the Japanese home islands. To a great extent, these operations were periods of trial—and occasionally of error—when our amphibious striking force, the Fifth Fleet and the V Amphibious Corps, tested and proved the basic soundness of the doctrine, tactics, and techniques developed by the Navy and Marine Corps in the years before World War II. This was a time of innovation too, when new weapons, improved methods of fire support, and organizational developments all played a large part in our victories. The lessons learned in the Gilberts, reaffirmed and applied with increasing effect in the Marshalls and Marianas, were of priceless value in shortening the war.

Our purpose in publishing this operational history in durable form is to make the Marine Corps record permanently available for study by military personnel and the general public as well as by serious students of military history. We have made a conscious effort to be objective in our treatment of the actions of Marines and of the men of other services who fought at their side. We have tried to write with understanding about our former enemies and in this effort have received invaluable help from the Japanese themselves. Few people so militant and unyielding in war have been as dispassionate and analytical about their actions in peace. We owe a special debt of gratitude to Mr. Susumu Nishiura, Chief of the War History Office, Defense Agency of Japan and to the many researchers and historians of his office that reviewed our draft manuscripts.

This volume was planned and outlined by Mr. Henry I. Shaw, Jr., senior historian assigned to the World War II historical project. Mr. Bernard T. Nalty, originally assigned as the author of this volume, wrote the narrative of the Gilberts and Marshalls campaigns and began the story of Saipan before he left the Marine Corps to become an historian with the Joint Chiefs of Staff. In his work, Mr. Nalty made use of the research material gathered for the monographs prepared by Captain James R. Stockman, *The Battle for Tarawa*, Lieutenant Colonels Robert D. Heinl, Jr. and John A. Crown, *The Marshalls: Increasing the Tempo*, and Major Carl W. Hoffman, *Saipan: The Beginning of the End*. Mr. Edwin T. Turnbladh finished the work on Saipan begun by Mr. Nalty and wrote the part concerning Tinian, using much of the research data amassed by Major Hoffman for his monograph, *The Seizure of Tinian*. Shortly after he had started working on the Guam narrative, Mr. Turnbladh left the Marine Corps to become an Air Force research analyst, and Mr. Shaw completed the book, revising and editing it for publication. In his research on the Guam operation, Mr. Shaw frequently consulted the material assembled for Major Orlan R. Lodge's monograph, *The Recapture of Guam*. The appendices concerning casualties, command and staff, division table of organization, and chronology were prepared by Mr. George W. Garand; the remainder were completed by Mr. Shaw. Successive Heads of the Historical Branch–Colonel William M. Miller, Major John H. Johnstone, Colonel Thomas G. Roe, Colonel Joseph F. Wagner, Jr., and Lieutenant Colonel Richard J. Schening—made the final critical review of portions of the manuscript. The book was completed under the direction of Colonel Frank C. Caldwell, present head of the branch.

A number of leading participants in the actions described have commented on the preliminary drafts of pertinent portions of the book. Their valuable assistance is gratefully acknowledged. Several senior officers, in particular General David M. Shoup, Admiral Harry W. Hill, Lieutenant General Julian C. Smith, and Rear Admiral Charles J. Moore, made valuable additions to their written comments during personal interviews.

Special thanks are due to the historical agencies of the other services for their critical readings of draft chapters of this book. Outstanding among the many official historians who measurably assisted the authors were: Dr. John Miller, Jr., Deputy Chief Historian, Office of the Chief of Military History, Department of the Army; Mr. Dean C. Allard, Head, Operational Archives Branch, Naval History Division, Department of the Navy; and Dr. Robert F. Futrell, Historian, U.S. Air Force Historical Division, Research Studies Institute, Air University, Maxwell Air Force Base.

First Lieutenant John J. Hainsworth, and his predecessors as Historical Branch Administrative Officer, First Lieutenant D'Arcy E. Grisier and Chief Warrant Officer Patrick R. Brewer, ably handled the many exacting duties involved in processing the volume from first drafts through final printed form. Many of the early preliminary typescripts were prepared by Mrs. Miriam R. Smallwood and the remainder were done by Miss Kay P. Sue, who expertly handled the painstaking task of typing the final manuscript for the printer. Much of the meticulous work demanded by the index was done by Miss Sue and Miss Linnea A. Coleman.

The maps were originally drafted by Corporal Robert F. Stibil; later revisions and additions were made by Corporal Thomas L. Russell. Official Department of Defense photographs have been used throughout the text.

W. R. Collins

W. R. COLLINS
MAJOR GENERAL, U.S. MARINE CORPS
ASSISTANT CHIEF OF STAFF, G–3

Contents

PART I LAUNCHING THE CENTRAL PACIFIC OFFENSIVE

PART II THE GILBERTS OPERATION

PART III THE MARSHALLS: QUICKENING THE PACE

PART IV SAIPAN: THE DECISIVE BATTLE

PART V ASSAULT ON TINIAN

PART VI THE RETURN TO GUAM

APPENDICES

ILLUSTRATIONS

ILLUSTRATIONS—Continued

MAPS

MAPS—Continued

PART I

Launching the Central Pacific Offensive

Early Plans for a War with Japan

Between November 1943 and August of the following year, American forces captured a series of key outposts in the Gilbert, Marshall, and Mariana Islands. Under the direction of Admiral Chester W. Nimitz, amphibious forces advanced almost 2,000 nautical miles, thrusting from Tarawa on the outer fringe of the enemy's defenses to within aerial striking distance of the Japanese homeland. Although this Central Pacific campaign saw the introduction of many new weapons, the strategy of attacking directly westward against Japan had for several decades been under study by American war planners. (See Map I, Map Section.)

THE EVOLUTION OF ORANGE PLAN

American acquisition from Spain of Guam and the Philippine Islands was followed within a few years by the emergence of Japan as a world power. The question arose whether the Philippines, at the end of a long and vulnerable line of communications, could be defended against the modern armed forces of Japan. Since both the American Army and Navy would take part in defending these islands, the Joint Board, an agency created to develop plans and policies which would most effectively use the available forces of both services, turned its attention to developing a coordinated plan for a possible war in the Pacific. Defending the Philippines, however, seemed so difficult a task that President Theodore Roosevelt, writing in 1907, termed the islands "our heel of Achilles." [1]

War plans of this era derived their titles from the code name of the probable enemy, and because Japan was designated ORANGE, the plan dealing with a conflict in the Far East was called ORANGE Plan. The earliest drafts required the Army to defend the Philippines until the fleet could shepherd reinforcements across the Pacific. Planners believed that the Japanese Navy would challenge the approach of the American armada and that the ensuing battle would decide not only the fate of the Philippines but the outcome of the war.

Naval strategists realized that before a relief expedition could be dispatched to the Far East, Japan certainly would have seized Guam, thus depriving the United States of its only fleet anchorage between Pearl Harbor and Manila Bay. Either Guam would have to be retaken or some other site occupied as a coaling and repair station. Whichever course of action was adopted, a landing force made up from the various ships' crews could not be used. With Japanese battleships lurking just over the hori-

[1] Henry F. Pringle, *Theodore Roosevelt: A Biography* (New York: Harcourt and Brace, 1931), p. 408.

zon, the Philippine relief expedition could not afford to have any large number of Bluejackets and Marines serving ashore and absent from their battle stations.[2]

Since the recapture of Guam or the taking of some other island—Truk was most frequently designated the prime objective[3]—was an integral part of the war at sea, such missions fell to the naval services and specifically to the Marine Corps, which was especially suited to those operations. During the Spanish-American War, a Marine battalion had landed at Guantanamo Bay to obtain a coaling station for the American ships blockading Cuba. Following the war, Marine units inherited the mission of occupying and defending advanced naval bases, and some naval officers began to urge that specially equipped defense forces be incorporated into each American squadron. Various planners cooperated in applying the lessons learned at Guantanamo Bay to the situation in the Pacific.

Among the first Marines to claim for their Corps an important role in an ORANGE war were Major Dion Williams and Captain Earl H. Ellis. Writing in 1912, Williams offered tables of organization for a brigade to accompany the battle fleet and assist it by occupying poorly defended anchorages, emplacing weapons, and guarding against counterattack.[4] Ellis, whose study appeared a few years later, agreed with the basic theory set forth by Williams, but he prophesied that the day might come when the enemy had fortified those islands suitable as advanced bases. Should this happen, the Marine contingent would be called upon to seize a defended beach. The capture of the objective rather than its subsequent defense would become the primary task of the Marines supporting the battle fleet.[5]

As a result of World War I, during which Japan and the United States had been allies, America's potential enemy gained control over the former German possessions in the Marshalls, Carolines, and Marianas. The Philippines were more vulnerable than before and Guam now was ringed by Japanese outposts. By 1921, the Marine Corps had evaluated recent gains by Japan and developed a realistic framework for its own operational and logistical planning. Staff officers believed that Guam and probably the Philippines would fall to the enemy shortly after the outbreak of war, and that Marine Corps units, in cooperation with Army troops, would face the task of seizing bases in the Marshalls, Carolines, Marianas, and Philippines. In addition, they assumed

[2] Capt William R. Shoemaker, USN, "Strategy of the Pacific: An Exposition of the ORANGE War Plan," dtd Aug14; Admin and StratSecs, ORANGE Plan, dtd Mar16 (NA folder no. 40, OP 29 folder no. 5, OAB, NHD).

[3] RAdm Charles J. Moore cmts on draft MS and interview by HistBr, G–3, dtd 26Nov62, hereafter *Moore comments* (Gilberts Comment File, HistBr, HQMC). Admiral Moore, who served as Admiral Spruance's chief of staff, noted: "This Truk operation became an obsession of the Navy and Marine Corps and was not eliminated as an objective until the raid on Truk on 17 February 1944."

[4] Maj Dion Williams, "The Naval Advanced Base," dtd 26Jul12 (NA folder no. 29, OP 29 folder no. 6, OAB, NHD).

[5] Capt Earl H. Ellis, "The Security of Advanced Bases and Advanced Base Operations," *ca.* 1913 (IntelSec, DivOps and Trng Files, HistBr, HQMC).

that the Marine Corps was to take part in the final advance from the Philippines to Japan itself.[6]

Earl Ellis, now a major, concentrated on one segment of a war against ORANGE and devised Operation Plan 712, which dealt with the systematic reduction of the Marshall Islands. He also outlined the tactics to be used against such objectives as Eniwetok, Wotje, and Maloelap. Although his theories were limited by the equipment then available, he made several sound recommendations, urging among other things that troops fighting ashore have at their disposal the on-call fire of supporting warships. Yet, the amphibious assault depicted by Ellis was understandably crude in comparison to the skilfully coordinated landings of World War II.[7]

Whatever the flaws in his theory, Ellis's plan marked a complete break with tradition. No longer would Marines be used primarily to defend advanced bases. Instead, they would seize these bases from the enemy.

During the 1920s and 1930s, various Marine Corps officers elaborated upon Ellis' amphibious doctrine. Key Pacific islands were studied as potential battlefields, new types of landing craft were tested, and more efficient landing techniques came into use. Thus did the Marine Corps devote its energies to preparing for whatever amphibious missions might be assigned it in an ORANGE war.[8]

Framing the broad strategy for a possible war in the Pacific remained the task of the Joint Army and Navy Board. The ORANGE Plan, actually a preferred course of action rather than a detailed war plan, needed little revision, and the missions first assigned the services before World War I remained much the same as World War II approached. The Army was to deny Manila Bay to the enemy for as long a time as possible, while the Navy, capturing en route as many bases as it might need, steamed westward to defeat the Japanese fleet and break the siege of the Philippines. Although some planners doubted that the Philippine garrison could hold out until help arrived, and in fact believed that the islands were indefensible, the basic concept persisted throughout the 1930s.[9] Finally, on the eve of war with the Axis powers, ORANGE Plan, which had presumed that Japan would

[6] MarCorps War Plan against ORANGE, Resume, 1921 (War Plans File, HistBr, HQMC).

[7] Maj Earl H. Ellis, OPlan 712, 1921 (War Plans File, HistBr, HQMC).

[8] For the story of the Marine Corps role in the development of amphibious doctrine and equipment, see LtCol Frank O. Hough, Maj Verle E. Ludwig, and Henry I. Shaw, Jr., *Pearl Harbor to Guadalcanal—History of U. S. Marine Corps Operations in World War II*, v. I (Washington: HistBr, G-3, HQMC, 1958), pp. 8-34, hereafter Hough, Ludwig, and Shaw, *Pearl Harbor to Guadalcanal*.

[9] Development of Joint Army-Navy War Plan ORANGE (RG 115, WPD 2720-22, WWII RecsDiv, FRC, Alexandria, Va.); Louis Morton, "Strategy and Command: Turning the Tide, 1941-1943—The War in the Pacific—U. S. Army in World War II," MS in OCMH, pt II, pp. 24-31, 38-41, hereafter *Morton MS*. See Louis Morton, *The Fall of the Philippines—The War in the Pacific—U. S. Army in World War II* (Washington: OCMH, DA, 1953), *passim*, for a discussion of the strong body of Army opinion which held that the islands could not be successfully defended with the forces available.

be the only enemy, was incorporated into an overall strategy designed to meet the needs of a two-ocean conflict.

STRATEGY FOR A GLOBAL WAR

As early as 1937, the United States Navy had sent a representative to Great Britain to discuss the employment of the American fleet in the event that these two nations should go to war with Japan. During the conversations, the possibility that Japan might join forces with Italy and Germany was explored. The British Admiralty was satisfied that, in the event of another world war, the United States Navy should concentrate in the Pacific, leaving the effort in the Atlantic to Great Britain and her continental allies.[10] By June 1939, American planners were fully aware that Japan, possibly without the aid of Germany and Italy, might take advantage of the European crisis to seize British, French, or even American holdings in the Orient. Because the potential enemies might either act independently or combine their efforts, the Joint Board in June 1939 ordered that five new war plans be written, the RAINBOW series, each of which might incorporate the features of several "color" plans such as ORANGE. These new plans were designed to meet danger from various sources. Two of them dealt with the defense of the western hemisphere, two others with a war in the Pacific, and still another, RAINBOW 5, with a war in Europe or Africa that pitted the United States, France, and Great Britain against Germany and Italy.

Although a greatly expanded RAINBOW 5 eventually became the basis of America's World War II strategy, work on this particular plan got off to a discouraging start, for France suddenly collapsed, and the Axis nations signed a formal military alliance. A two-ocean war now seemed probable, a conflict in which winning the battle of the Atlantic would be of more consequence than a victory in the Pacific. In the words of Admiral Harold R. Stark, the Chief of Naval Operations, "if Britain wins decisively against Germany we could win everywhere; but . . . if she loses the problem confronting us would be very great; and, while we might not *lose everywhere*, we might, possibly, not win anywhere."[11] During January 1941, President Franklin D. Roosevelt announced a policy that emphasized the greatest possible aid to Britain, and a series of Anglo-American conferences began that same month which saw the two nations agree upon defeating Germany and Italy before turning their full might against Japan.[12]

Since the United States was now committed to assuming a strategic defensive on the outbreak of war in the Pacific, joint planners began rewriting RAINBOW 5 to include the probability that Japan would cooperate with her Axis partners in any future conflict.

[10] *Morton MS*, pt, III, p. 2.

[11] Maurice Matloff and Edwin M. Snell, *Strategic Planning for Coalition Warfare; 1941–1942—The War Department—The U. S. Army in World War II* (Washington: OCMH, DA, 1953), p. 25.

[12] *Morton MS*, pt. III, pp. 34–35, 37; Louis Morton, "American and Allied Strategy in the Far East," *Military Review*, v. 29, no. 9 (Dec49), p. 33, hereafter Morton, "Strategy."

Essentially, this revision consisted in delaying indefinitely the Central Pacific campaign advocated by the ORANGE Plans. Instead of seeking immediately a decisive sea battle in Philippine waters, the Navy would be restricted in its early operations to attacks upon the Marshalls designed to prevent the enemy from concentrating his forces against Singapore. In brief, the naval offensive against Japan, to which the Navy and Marine Corps had devoted so much thought, became but a single element in a global strategy designed primarily to crush Germany and Italy as rapidly as possible.[13]

[13] Morton, "Strategy," pp. 37–38.

214-881 O-67—2

The Central Pacific in Global Strategy

The devastating raid on Pearl Harbor, coupled with the destructive attacks on airfields in the Philippines, enabled Japan to seize the initiative in the Pacific. Since many of its battleships had been disabled by enemy bombs and torpedoes, the American Navy could do nothing to divert the Japanese from Singapore. This bastion fell, the Netherlands Indies were overwhelmed, and the inadequate garrison holding out in the Philippines was encircled. Instead of conducting extensive raids in the Marshalls and preparing for an eventual drive across the Pacific, the United States and her Allies were trying desperately to hold a perimeter that extended from Burma through Australia to Hawaii and the Aleutians. As the triumphant Japanese pushed southward, existing American war plans were abandoned, and the security of Australia became the principal task of the Allies in the Pacific area.

Like the United States, Japan early in the war revised her Pacific strategy. The enemy's change of plans, however, was not caused by unforeseen setbacks but by the ease with which she had gained her primary objectives. Originally, Japan's principal aim was the conquest of the Netherlands Indies, Malaya, the Philippines, and Burma, areas rich in natural resources. Once these regions were taken, the Japanese empire, now self-sufficient in oil, tin, and rubber, would be capable of defending a perimeter that stretched from the Kuriles, along the outer Marshalls, through the Bismarck Archipelago, across the Netherlands Indies, to Malaya and Burma. With Malaya, the Philippines, and the Netherlands colonies firmly in her control, Japan, rather than pausing as planned to consolidate these gains, decided to expand still farther.

These secondary conquests were in a sense defensive, for the enemy wished only to protect his earlier gains. Japan hoped to seize Port Moresby in New Guinea, capture Midway, establish outposts in the Aleutians, and isolate Australia by taking New Caledonia, Fiji, and Samoa. Should these operations succeed, Australia would be useless as an Allied base, and the United States fleet would be confined within a triangle bounded by Alaska, Hawaii, and Panama.

This series of operations ended in disaster for the overextended enemy. The Battle of the Coral Sea stopped the immediate threat to Port Moresby. A month later the Japanese fleet met defeat off Midway, a blow that caused the cancellation of the offensive aimed at the line of communication with Australia. The attempt to gain a foothold in the Aleutians did succeed, but before the summer of 1942 had ended the Americans took the initiative by landing at Guadalcanal in the lower Solomons.[1]

THE CASABLANCA CONFERENCE[2]

Since the war against the Axis powers was a combined effort, both the United States and Great Britain had a major voice in determining Allied strategy. The agency charged with developing the program of Allied military operations and allocating the resources of the two nations was the Combined Chiefs of Staff (CCS), which was composed of the principal military advisers of President Roosevelt and British Prime Minister Winston Churchill. Shortly after the attack on Pearl Harbor, the CCS gave to the United States primary responsibility for the war against Japan, excluding operations in defense of Malaya, Sumatra, and Burma. Thus, planning for the Pacific war fell to the American Joint Chiefs of Staff (JCS).

By the end of 1942, Japan had lost the initiative, and the JCS could begin exploring the most effective ways of damaging the enemy. A limited offensive had already been approved for the South and Southwest Pacific Areas, an operation undertaken to protect the line of communication to Australia. Now the JCS had to determine whether the advance northward and westward from Australia would prove more decisive than a drive across the Central Pacific and then to convince the other members of the CCS of the wisdom of its strategy.

The conference of Allied leaders at Casablanca in January 1943 gave American planners their first opportunity to present to the British their detailed views on Pacific strategy. Admiral Ernest J. King, Commander in Chief, U. S. Fleet and Chief of Naval Operations, expressed his belief that an eventual goal of any Pacific offensive should be the Philippines. These islands lay astride the sea lanes over which rubber, oil, and tin were transported to Japan from the conquered territories. Submarines and aircraft operating from the Philippines could halt the flow of raw materials needed by Japanese industry.

Turning to the problem of selecting of the best route of advance toward the Philippines, King urged a move across the Central Pacific. Both the admiral and General George C. Marshall, Army Chief of Staff, hoped to outflank the defenses that the Japanese were preparing in the area north of Australia. This could be done by striking westward through the Marshalls and Carolines to the Marianas. Since planes flying from the Marianas could strike Japan while submarines based in these islands isolated the Philippines, King looked upon them as important intermediate objectives on the way westward.

British strategists, however, were not

[1] See Hough, Ludwig, and Shaw, *Pearl Harbor to Guadalcanal*, pp. 235–374, for the story of Marine operations in the Guadalcanal campaign.

[2] Unless otherwise noted, the material in this section is derived from: CCS, Minutes of the 55th–60th Meetings 14–18Jan43 (ABC Files, WWII RecsDiv, FRC, Alexandria, Va.); John Miller, Jr., "The Casablanca Conference in Pacific Strategy," *Military Affairs*, v. 13, no. 4 (Winter 49), p. 209.

willing to expand the effort against Japan at the expense of the European war. They agreed that limited offensives were necessary but wanted to undertake only those operations which would help protect India and Australia. British planners believed that the Allies should seize the key enemy base at Rabaul on New Britain, and recover Burma, while remaining on the defensive elsewhere in the Pacific.

Actually, a compromise was easily reached, once the JCS had agreed informally to avoid becoming involved in a series of offensives throughout the Pacific. The Burma operation could not begin immediately, and it appeared that Rabaul might fall in the near future. A Central Pacific campaign, limited in scope, would maintain pressure on the enemy during the period between the capture of Rabaul and the attack from India. After General Marshall stated that the move toward the Carolines would be undertaken with resources available in the Central Pacific Areas, the CCS recommended that the heads of state accept this broadening of the war against Japan. President Roosevelt and Prime Minister Churchill, the final arbiters of Allied strategy, agreed with their advisers.

COORDINATING PACIFIC STRATEGY

Late in February 1943, Admiral King discussed with Admiral Nimitz, Commander in Chief, Pacific Fleet (CinCPac), means of carrying out the policies decided upon at Casablanca. Although General Douglas MacArthur, responsible for directing the war in the Southwest Pacific, had disclosed only the general outline of his proposed operations, King and Nimitz attempted to plot the course of events in the Central Pacific. Nimitz wished to remain temporarily on the defensive, gradually whittling down Japanese strength while augmenting his own. During this build-up, submarines and aircraft would carry the war to the enemy. King agreed that to strike westward with the men, ships, and planes available in 1943 was to take a very considerable calculated risk, but he considered that the American forces must keep the initiative and that they had to be used in order to justify their allocation by the CCS.[3]

As a result of their conference, the two admirals agreed to a limited thrust in the general direction of the Philippines, but neither the objectives nor the timing of the attack were selected. In the meantime, the JCS had been arranging a conference of representatives from the Central Pacific, South Pacific, and Southwest Pacific Areas. At this series of meetings held in Washington, beginning on 12 March 1943, American planners heard additional details concerning the strategic design that had been prepared by General MacArthur.

Delegates to the Pacific Military Conference, as these meetings were termed, did not learn of MacArthur's planned return to the Philippines, for his representatives concerned themselves with immediate operations for the capture of Rabaul. ELKTON, the Rabaul plan,

---

[3] FAdm Ernest J. King and Cdr Walter Muir Whitehill, *Fleet Admiral King: A Naval Record* (New York: W. W. Norton and Co., 1952), pp. 431–432, hereafter King and Whitehill, *Fleet Admiral King*.

called for continuing cooperation between the South and Southwest Pacific Areas in the capture of airfield sites on New Guinea, New Georgia, and Bougainville and in the seizure of two mighty bases–Kavieng on New Ireland and Rabaul itself.

Since a minimum of 79,000 Japanese troops were believed to be stationed in the area through which General MacArthur proposed to advance, his representatives told the conference that five additional divisions and as many air groups would be needed to sustain the offensive. Unfortunately, the JCS could spare neither the men nor the aircraft which MacArthur needed and had no choice but to order ELKTON revised. Instead of seizing Rabaul as quickly as possible, Allied forces in the South and Southwest Pacific would, during 1943, occupy Woodlark and Kiriwina Islands, continue the war in New Guinea, land on New Britain, and advance along the Solomons chain by way of New Georgia to Bougainville. These changes enabled MacArthur and Admiral William F. Halsey, who directed operations in the South Pacific theater, to get along with fewer troops, planes, and ships, some of which might now be employed in the Central Pacific.[4] Although the Pacific Military Conference brought about a tailoring of ELKTON to fit the resources available for the operation, no attempt was made to coordinate this offensive with the attack scheduled for the Central Pacific. The JCS postponed this decision until the next meeting of the CCS which was scheduled to convene at Washington in May 1943.

During the discussions held at Washington the American Strategic Plan for the Defeat of Japan, a general statement of the strategy which would finally force the unconditional surrender of that nation, was presented for British approval. As at Casablanca, Admiral King urged that a powerful blow be struck through the Central Pacific, a more direct route toward the enemy heartland than the carefully guarded approach from the south. In Admiral Nimitz' theater, the American fleet could disrupt Japanese supply lines and hinder any effort to strengthen the defenses in the region. The vast expanse of ocean was dotted with potential naval bases, so the attackers could select their objectives with a view to inflicting the maximum strategic damage with the fewest possible men, ships, and aircraft. In short, the Central Pacific offered an ideal opportunity to use carrier task forces in conjunction with amphibious troops to launch a series of swift thrusts across hundreds of miles of water.

Admiral King, however, did not intend to abandon the Solomons-New Britain-New Guinea offensive. Allied forces fighting in this area could not simply suspend their operations and begin shifting men and equipment to the Central Pacific. Such a maneuver would alert the enemy to the im-

[4] John Miller, Jr., *CARTWHEEL: The Reduction of Rabaul—The War in the Pacific—The U. S. Army in World War II* (Washington: OCMH, DA, 1959), pp. 12–19. For the story of the role of the Marine Corps in operations against Rabaul, see Henry I. Shaw, Jr. and Maj Douglas T. Kane, *Isolation of Rabaul—History of U. S. Marine Corps Operations in World War II*, v. II (Washington: HistBr, G–3, HQMC, 1963), hereafter Shaw and Kane, *Isolation of Rabaul*.

pending offensive and weaken the defenses of Australia. Rather than hurl all Allied resources into a single drive, the American strategist desired two offensives, a major effort through the Central Pacific and a complementary push from the South and Southwest Pacific. Exact timing and the judicious use of available strength would keep the Japanese off balance and prevent their deploying men and ships from the southern defenses to the Carolines and Marshalls.[5]

After listening to Admiral King's arguments, the CCS accepted the American position with one major modification. Originally, the American plan had called for the Allies, while continuing the offensive against Germany, "to maintain and extend unremitting pressure against Japan with the purpose of continually reducing her military power and attaining positions from which her ultimate unconditional surrender can be forced."[6] British members of the CCS considered this statement to be permission to strike at Japan without regard to the war against Germany, and all the American representatives except King agreed that the language was too strong. Finally, the British suggested that the effect of any extension of the Pacific war on overall strategy should be considered by the CCS before actual operations were begun, a sentence to this effect was inserted, and the document approved.[7] Included among the goals proposed for 1943–1944 were the seizure of the Marshalls and Carolines as well as certain operations in Burma and China, the reconquest of the Aleutians,[8] and a continuation of the effort in the South and Southwest Pacific Areas.[9]

FINAL APPROVAL FOR THE CENTRAL PACIFIC OFFENSIVE

As soon as the Washington Conference adjourned, King hurried to San Francisco to discuss with Nimitz the attack against the Marshalls. He repeated his conviction that the Mariana Islands formed the most important intermediate objective on the road to Tokyo and recommended that the conquest of the Marshalls be the first step in the march westward. Because the resources available to Admiral Nimitz would determine the final selection of targets, the two officers also discussed the possibility of striking first at the Gilberts, a group of islands believed to be more vulnerable than the Marshalls.[10]

---

[5] Philip A. Crowl and Edmund G. Love, *Seizure of the Gilberts and Marshalls—The War in the Pacific—The U. S. Army in World War II* (Washington: OCMH, DA, 1955), pp. 12–14, hereafter Crowl and Love, *Gilberts and Marshalls*.

[6] King and Whitehill, *Fleet Admiral King*, p. 441.

[7] *Ibid.*

[8] The invasion of Attu, one of the two Aleutian outposts seized by the Japanese, had begun on 11 May. Since Japan was unable to mount an offensive from this quarter, Attu could have been ignored, but its recapture, along with the planned reconquest of Kiska, would put additional pressure on the enemy, drive him from the fringes of the Western Hemisphere, and release American ships and troops for service elsewhere.

[9] CCS 239/1, Operations in the Pacific and Far East in 1943–1944, dtd 22May43 (ABC Files, WWII RecsDiv, FRC, Alexandria, Va.).

[10] King and Whitehill, *Fleet Admiral King*, pp. 443–444.

In the meantime, General MacArthur was informed of the proposed offensive in Nimitz' theater, an attack which was to be launched in mid-November by the 1st and 2d Marine Divisions. Neither the strategic plan nor its means of execution coincided with MacArthur's views. The general believed that the best avenue of approach to the Philippines was by way of New Guinea. ELKTON was but the first phase of MacArthur's promised return to the Philippines, and even this plan would be jeopardized by the loss of the two Marine divisions, then under his command.

MacArthur objected to the JCS that a campaign in the Central Pacific would be no more than a diversionary effort and that the withdrawal of the Marine units would delay the seizure of Rabaul. American strategists were able to effect a compromise by ordering the release of the 2d Marine Division, then in the South Pacific, while leaving the 1st Marine Division under MacArthur's control. Instead of capturing Rabaul, the South and Southwest Pacific forces were to neutralize and bypass this fortress.[11]

Even before MacArthur had voiced his objections to the proposed Central Pacific operations, appropriate agencies of the JCS were at work selecting objectives for Admiral Nimitz. So little intelligence was available on the Marshalls that planners urged the capture instead of bases in the Gilberts from which American reconnaissance planes could penetrate the neighboring island group. The defenses of the Gilberts appeared to be weaker than those of the Marshalls, and the proposed objective area was near enough to the South Pacific to permit naval forces to support operations in both places. Finally, control of the Gilberts would reduce Japanese threats to American bases in the Ellice Islands and Samoa and shorten as well as protect the line of communication to New Zealand and Australia.[12] On 20 July, the JCS directed Nimitz to begin planning for the capture of Nauru Island and of bases in the Gilberts. Well within American capabilities, this limited offensive would open the way into the Marshalls, which in turn would provide the bases for a later move to the Marianas.[13]

The following month, the CCS, meeting at Quebec, added the Gilbert Islands to a list of proposed Central Pacific operations designed to carry the war to the Marianas by the end of 1944.[14] At last the stage was set for a Central Pacific offensive similar in concept to the campaign outlined in the earliest ORANGE Plans.

---

[11] Miller, *CARTWHEEL*, *op. cit.*, pp. 223–225.

[12] *Moore Comments.*

[13] Crowl and Love, *The Gilberts and Marshalls*, pp. 21–24.

[14] King and Whitehill, *Fleet Admiral King*, p. 489.

The Central Pacific Battleground

In American military terminology, the scene of the proposed offensive was the Central Pacific Area, a subdivision of the Pacific Ocean Areas that extended from the western coast of North America to the shores of China and reached from the equator northward to the 42d parallel. Canton Island, just south of the equator, was included in the Central Pacific, but the Philippines and those parts of the Netherlands Indies that lay in the northern hemisphere were not.[1] Admiral Nimitz, Allied commander-in-chief throughout the Pacific Ocean Areas (CinCPOA), retained immediate control over operations in the Central Pacific. Within this area lay Micronesia, a myriad of islands of varying size and type, the region in which the forthcoming Central Pacific battles would be fought.

THE GEOGRAPHY OF MICRONESIA[2]

That part of Micronesia nearest Japan is the Mariana group of 15 volcanic islands. The Marianas curve southward from the 20th to the 13th parallel north latitude, with the center of the chain lying at about 144 degrees east longitude. The islands themselves, five of which were inhabited at the outbreak of war, are by Micronesian standards vast and mountainous. Guam, the largest, boasts an area of 228 square miles and peaks rising to over 1,000 feet. The highest elevation in the entire group is 3,166 feet on smaller Agrigan. Although the temperature is warm but not unpleasant, rains occur frequently, and there is the threat of typhoons and an occasional earthquake. Saipan, near the center of the Marianas, lies 1,285 nautical miles southeast of Yokohama. (See Map I, Map Section.)

South of the Marianas are the Carolines, a belt of over 500 volcanic islands and coral atolls extending eastward from Babelthuap in the Palaus, 134 degrees east longitude, to Kusaie at 163 degrees. The long axis of this group coincides roughly with the seventh parallel north latitude. At the approximate center of the Carolines, 590

[1] The Pacific Ocean Areas contained three subdivisions: the North Pacific Area north of the 42d parallel, the Central Pacific Area, and south of the equator, the South Pacific Area. Australia, its adjacent islands, most of the Netherlands Indies, and the Philippines formed the Southwest Pacific Area under General MacArthur.

[2] Unless otherwise noted, the material in this section is derived from: *Morton MS*, Introduction, p. 14; R. W. Robson, *The Pacific Islands Handbook, 1944*, North American ed., (New York: Macmillan Co., 1945), pp. 132–175; Fairfield Osborne, ed., *The Pacific World* (New York: W. W. Norton and Co., 1944), pp. 155–159.

miles southeast of Saipan, is Truk, site of a Japanese naval base. The climate in this area is healthful, but the average temperatures are slightly higher than in the Marianas. Some of the larger Caroline Islands are covered with luxuriant vegetation.

Northeast of Kusaie are the Marshalls, a group of 32 flat coral atolls and islands scattered from 5 to 15 degrees north latitude and 162 to 173 degrees east longitude. The highest elevation in the entire group is no more than 40 feet. Kwajalein, the largest atoll in the world, lies near the center of the Marshalls, 955 miles from Truk. The climate is hot and humid.

On a map, the Gilberts appear to be an extension of the Marshalls, an appendix of 16 atolls that terminates three degrees south of the equator. Tarawa, slightly north of the equator, is 540 nautical miles southeast of Kwajalein and 2,085 miles southwest of Pearl Harbor. Heat and humidity are extreme during the rainy season, the soil is poor, and portions of the group are occasionally visited by droughts.

The most striking feature of the Micronesian battlefield is its vastness. An island as big as Guam is little more than a chip of wood afloat in a pond. Although the total expanse of ocean is larger than the continental United States, the numerous islands add up to less than 2,000 square miles, a land area smaller than Delaware.

At one time, all of Micronesia except for the British Gilberts had belonged to Spain. The United States seized Guam during the Spanish-American War, and Spain later sold her remaining Central Pacific holdings to Germany. After World War I, the League of Nations made Japan the mandate power in the Marshalls, Carolines, and Marianas (except Guam). Under the terms of the mandate, a reward for participating in the war against Germany, the Japanese were to govern and develop the islands, but were forbidden to fortify them. This bar to fortification was reinforced by the Washington Naval Treaty of 1922.

In 1935, however, Japan withdrew from the League of Nations without surrendering her authority over the Pacific isles. Because of the strict security regulations enforced throughout the region, Japan succeeded in screening her activities for the six years immediately preceding the attack on Pearl Harbor. As late as 1939, a Japanese scholar assured the English-speaking peoples that his nation was not using the islands of Micronesia for military purposes.[3] The strength of the defenses in this area would not be accurately determined until the Central Pacific campaign was underway.

THE ROLE OF THE MARINE CORPS

Throughout his arguments for an offensive across Micronesia, Admiral King had desired to use Marines as assault troops, for the Marine Corps had pioneered in the development of amphibious doctrine, and its officers and men were schooled in this type of operation.[4] Major General Holland M.

[3] Tadao Yanaihara, *Pacific Islands under Japanese Mandate* (London: Oxford University Press, 1940), p. 305.

[4] King and Whitehill, *Fleet Admiral King*, p. 481.

Smith, who would command the V Amphibious Corps (VAC) for most of the Central Pacific campaign, had directed the amphibious training of Army troops that had participated in the invasion of North Africa. The same staff which would accompany him westward had helped him prepare elements of the 7th Infantry Division for the Attu operation.[5] Experienced leadership would not be lacking, but veteran Marine divisions were at a premium.

Three Marine divisions, two of them proven in combat, were overseas when Admiral Nimitz received the JCS directive to prepare plans for a blow at the Gilberts. The 2d Marine Division, which was recovering from the malarial ravages of the Guadalcanal campaign, continued to train in the temperate climate of New Zealand after its release to Admiral Nimitz and incorporation into General Holland Smith's amphibious corps. The other malaria-riddled veteran unit of the battle for Guadalcanal, the 1st Marine Division, was in Australia. This division was left in General MacArthur's Southwest Pacific Area to execute a part of the revised ELKTON plan, the landing at Cape Gloucester on New Britain. The 3d Marine Division, untested in combat and new to Vice Admiral William F. Halsey's South Pacific Area, was completing its movement from New Zealand to the southern Solomons where it would train for the invasion of Bougainville, an operation that was to begin in November 1943. Although still in training in the United States, the 4th Marine Division was scheduled to be ready by the end of 1943 for service in the Central Pacific.

Also present in the South Pacific were several other Marine combat organizations larger than a battalion in size. The 1st Raider Regiment was committed to the Central Solomons offensive; one battalion was already fighting on New Georgia, and another had just landed on that island. The 1st Parachute Regiment, an airborne unit in name only, was preparing in New Caledonia for possible employment during the advance into the northern Solomons. The 22d Marines, a reinforced regiment that eventually would see action in the Central Pacific, was at this time standing guard over American Samoa.[6]

Like the ground combat units, the bulk of Marine Corps aviation squadrons overseas at the time were stationed in the distant reaches of the South Pacific. An exception was the 4th Marine Base Defense Aircraft Wing. Although the wing, at the time of the JCS directive, was preparing to shift its headquarters from Hawaii to Samoa and most of its squadrons were staging southward, one fighter and one scout-bomber squadron were in the Ellice group near the northern boundary of the South Pacific Area. Since the planes based in the Ellice Islands were short range craft and the pilots unused to carrier operations, neither unit could participate in the Gilberts

[5] Jeter A. Isely and Philip Crowl, *The U. S. Marines and Amphibious War, Its Theory, and Its Practice in the Pacific* (Princeton: Princeton University Press, 1951), pp. 61–63, hereafter Isely and Crowl, *Marines and Amphibious War.*

[6] StaSheet, FMF Gnd, dtd 31Jul43 (HistBr, HQMC).

invasion.[7] At Tarawa, the 2d Marine Division would be supported by Navy carrier squadrons.

TACTICAL ORGANIZATION OF THE MARINE DIVISION

The Marine division that figured in Admiral Nimitz' plans for the Gilberts was organized according to the E series tables of organization adopted in April 1943. With an authorized strength of 19,965 officers and men, the division was constructed in a triangular fashion —three infantry regiments, each of which had three infantry battalions. This arrangement enabled the division commanding general to hold in reserve an entire regiment without impairing the ability of his command to attack or to remain on the defensive. A regimental commander could exercise this same option with his battalions, and the battalion commander with his rifle companies, as well as with a headquarters company and a weapons company. An infantry regiment was authorized a basic strength of 3,242, a battalion 953, and a rifle company 196.

Supporting the divisional infantry components were an engineer regiment of three battalions (engineer, pioneer, and naval construction), an artillery regiment with three battalions of 75-mm pack howitzers and two of 105mm howitzers, Special Troops, and Service Troops. Special Troops, its total strength 2,315, consisted of a light tank battalion which included the division scout company, a special weapons battalion equipped with antitank and antiaircraft guns, and the division headquarters battalion which contained headquarters, signal, and military police companies. Service, motor transport, amphibian tractor, and medical battalions, with a total of 2,200 officers and men, made up Service Troops. The division chaplains, doctors, dentists, hospital corpsmen, and the Seabees of its naval construction battalion were members of the U. S. Navy.

The infantry units, too, had their own support elements. A weapons company armed with heavy machine guns, 37mm antitank guns, and self-propelled 75mm guns was under the direct control of each regimental commander. The battalion commander had his own company of heavy machine guns and 81mm mortars, and a company commander could rely on the light machine guns and 60mm mortars of his weapons platoon.

The basic structure of both division and regiment was altered when necessary. Troops normally under corps control, such as reconnaissance, medium tank, or artillery units, might be used to reinforce the division. For amphibious operations, each regiment was made a combat team by the addition of troops from the artillery and engineer regiments, the amphibian tractor, medical, motor transport, service, tank, and special weapons battalions. These attachments increased the strength of the regiment to as much as 5,393. Some of the additional troops were reassigned to the infantry battalions, so that the combat team

[7] StaSheet, Air, dtd 31Jul43 (HistBr, HQMC); Robert Sherrod, *History of Marine Corps Aviation in World War II* (Washington: Combat Forces Press, 1952), pp. 222, 224, 438–439, hereafter Sherrod, *Marine Air History*.

generally consisted of three landing teams, each with its own engineer, artillery, medical, and tank support.[8]

THE ENEMY'S BASIC TACTICAL ORGANIZATION

During the drive westward, Marine divisions would be opposed by Japanese *Special Naval Landing Forces* as well as by the enemy's infantry divisions. When employed in the attack, a *Special Naval Landing Force* usually consisted of two infantry companies and a heavy weapons unit, plus communications, engineer, medical, supply and transportation elements. With a strength of 1,000–1,500, this organization was comparable in size to a Marine battalion reinforced as a landing team. Artillery support for the landing force was provided by from 8 to 24 guns and howitzers ranging from 70mm to 120mm.

On the defense, however, a *Special Naval Landing Force* could be reinforced to a strength of 2,000, with an appropriate increase in the number of automatic weapons and the addition of antitank guns, mortars, or both. In addition, the Marines might expect to encounter, among the naval units, trained guard forces. Construction or pioneer units, both types composed in part of Korean laborers, were engaged in building airfields and defensive installations throughout Micronesia. The strength of these organizations depended on the particular project assigned them, and their zeal for combat and state of training varied according to the policies of individual island commanders.

Like the Imperial Navy, the Japanese Army habitually altered the strength and composition of its field units to meet the task at hand. The standard infantry division consisted of some 20,000 men organized into a cavalry or reconnaissance regiment, an infantry group of three regiments, and artillery, engineer, and transportation regiments. Signal, hospital, water purification, ordnance, and veterinary units were considered parts of the division headquarters. For the most part, the Japanese adhered to the triangular concept, for each of the three infantry regiments of 3,845 men contained three 1,100-man battalions. If judged necessary, the size of a division could be increased to over 29,000 officers and men.

Usually, these reinforcements were troops not assigned to any division, for the Japanese Army had created a bewildering variety of independent units. Some were larger than the ordinary infantry regiment; others as small as a tank company. By combining independent units or attaching them to divisions, the enemy was able to form task forces to capture or defend a particular place.[9]

These were the forces that would battle for Micronesia. With many potential anchorages in the region, Admiral Nimitz could feint with his carriers before striking with his as-

[8] OrgChart, MarDiv, dtd 15Apr43 (SubjFile: T/Os, HistBr, HQMC). A copy of this table of organization and equipment is contained in Shaw and Kane, *Isolation of Rabaul*, pp. 571–573.

[9] WD, TM-E 30-480, *Handbook on Japanese Military Forces*, dtd 15Sep44, pp. 19–21, 76–81.

sault troops. Yet, the task of destroying the individual enemy would inevitably fall to the infantryman supported by aircraft, naval gunfire, and all the firepower organic to the Marine division.

PART II

The Gilberts Operation

Preparing to Take the Offensive [1]

PLANNING FOR OPERATION GALVANIC

The summer of 1943 saw a revival of the strategy contained in both ORANGE and RAINBOW Plans as well as rigorous training in amphibious techniques, methods that stemmed from further elaboration of the theories of amphibious warfare first advanced by Major Earl Ellis. The Central Pacific offensive, for so many years the keystone of American naval planning, was about to begin with operations against the Gilbert Islands.[2] The attack on this group of atolls would test the Marine Corps concept of the amphibious assault, an idea originated by Ellis and greatly modified by his successors. The major had been confident that a defended beach could be taken by storm, and since his death new equipment and tactics had been perfected to aid the attackers, but the fact remained that such an operation had never been tried against a determined enemy dug in on a small island. Was Ellis' conclusion still valid? Could an army rise out of the sea to overwhelm prepared defenses? These questions soon would be answered. (See Map I, Map Section.)

On 20 July the JCS ordered Admiral Nimitz to begin preparing for the capture, development, and defense of bases in the Gilbert group and on Nauru Island. This directive also provided for the occupation of any other islands that might be needed as air bases or naval facilities for the carrying out of the primary mission. GALVANIC was the code name assigned to the Gilberts-Nauru venture. The operation was intended to be a preliminary step to an attack against the Marshalls.

[1] Unless otherwise noted, the material in this chapter is derived from: CinCPac OPlan 1-43, dtd 5Oct43, hereafter *CinCPac OPlan 1-43*; ComCenPacFor OPlan 1-43, dtd 25Oct 43, hereafter *ComCenPacFor OPlan 1-43*; V PhibFor AR Gilbert Islands, dtd 4Dec43, hereafter *V PhibFor AR*; CTF 54 OPlan A2-43, dtd 23Oct43; CTF 53 OpO A101-43, dtd 17Oct 43, hereafter *CTF 53 OpO A101-43*; VAC AR GALVANIC, dtd 11Jan44, hereafter *VAC AR*; VAC OPlan 1-43, dtd 13Oct43; 2d Mar Div OpO No. 14, dtd 25Oct43, hereafter *2d MarDiv OpO No. 14*; TF 11 AR Baker Island, Sep43 (Baker Island Area OpFile, HistBr, HQMC); Samuel Eliot Morison, *Aleutians, Gilberts, and Marshalls, June 1942–April 1944—History of United States Naval Operations in World War II*, v. VII (Boston: Little, Brown and Company, 1960 ed.), hereafter Morison, *Aleutians, Gilberts, and Marshalls*; Crowl and Love, *Gilberts and Marshalls*; Isely and Crowl, *Marines and Amphibious War*; Capt James R. Stockman, *The Battle for Tarawa* (Washington: HistSec, DivPubInfo, HQMC, 1947), hereafter Stockman, *Tarawa*. Unless otherwise noted, all documents cited in this part are located in the Gilberts Area Op File and the Gilberts CmtFile, HistBr, HQMC.

[2] It is interesting to note that the Gilberts did not appear as an objective in the ORANGE Plans. *Moore Comments*.

214-881 O-67—3

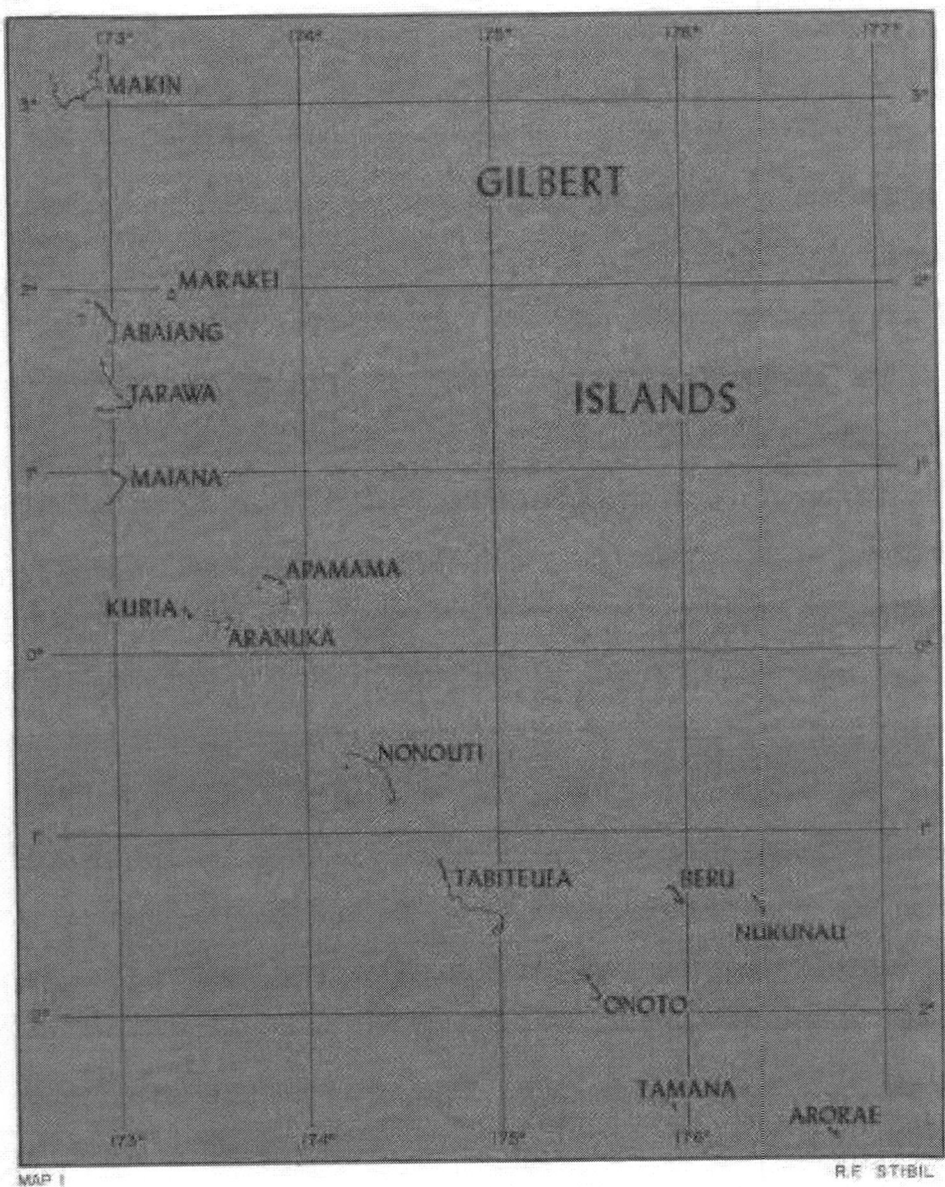
MAKIN
GILBERT
MARAKEI
ABAIANG
TARAWA
ISLANDS
MAIANA
APAMAMA
KURIA
ARANUKA
NONOUTI
TABITEUEA
BERU
NUKUNAU
ONOTO
TAMANA
ARORAE
MAP I
R.F. STIBIL

At the same time, a lodgement in the Gilberts, in addition to bringing the Marshalls within range of land-based bombers, also would insure the safety of Samoa and shorten the line of communication to the South and Southwest Pacific.

The bulk of the Pacific Fleet, the Central Pacific Force under Vice Admiral Raymond A. Spruance, was committed to the Gilberts operation. As commander of the GALVANIC expedition, Spruance was the immediate subordinate of Admiral Nimitz. The actual landings would be conducted by V Amphibious Force, headed by Rear Admiral Richmond K. Turner. This organization, established on 24 August 1943, was similar to the amphibious force that had carried out the Attu invasion. Within Turner's force was V Amphibious Corps (VAC), organized on 4 September and commanded by Marine Major General Holland M. Smith. The general had with him the same staff, with representatives from all services, that had aided in preparing for the Aleutians offensive. Like the original Amphibious Corps, Pacific Fleet, the new VAC was to train and control the troops required for future operations in the Pacific Ocean Areas.

Nimitz had entrusted GALVANIC to a group of experienced Navy and Marine Corps officers. Spruance had commanded a carrier task force at the Battle of Midway; Turner had led the amphibious force that landed the Marines at Guadalcanal and Tulagi and had directed a similar force in operations against New Georgia; and Holland Smith had helped train troops for amphibious operations in Europe, North Africa, and the Aleutians. Preparations for the Gilberts invasion had been placed in capable hands.

The man who would actually command the amphibious phase of the Tarawa operation, Rear Admiral Harry W. Hill, reported to Admiral Turner on 18 September 1943. Hill, a veteran battleship and escort carrier group commander fresh from the South Pacific battles, was designated Commander, Amphibious Group 2. His group would transport, land, and support the assault troops at Tarawa, while a similar group, which Admiral Turner retained under his direct command, would be part of the attack force at Makin. When Hill arrived, the projected D-Day for GALVANIC was 1 November, which was later changed to 20 November, a date which provided only two months to weld a widely scattered force of ships and troops into an effective team.[3]

The major Marine unit available to Turner's amphibious force was the 2d Marine Division, commanded by Major General Julian C. Smith. This division had fought at Guadalcanal and was currently reorganizing in New Zealand. The division commander, a Marine Corps officer since 1909, joined the unit after it had sailed from Guadalcanal; but during his career he had seen action at Vera Cruz, in Haiti, Santo Domingo, and Nicaragua, and had served in Panama and Cuba as well. In the words of the division historian, Julian Smith's "entirely unassuming manner and friendly hazel eyes clothed a determined personality

[3] Adm Harry W. Hill interview with HistBr, G-3, HQMC and cmts on draft MS, dtd 4Oct 62, hereafter *Hill interview/comments.*

that could be forcefully displayed in decisive moments. His concern for his men was deep and genuine."[4]

Because of the scope of the proposed Central Pacific campaign and the need in other theaters for Marine amphibious divisions, Army troops appeared certain to be needed for GALVANIC and later operations. Preliminary training, administration, and logistical support of these Army divisions fell to the Commanding General, Army Forces, Pacific Ocean Areas, Lieutenant General Robert C. Richardson. In carrying on his work, General Richardson was subject to the direction of Admiral Nimitz.

When the JCS first began exploring the possibility of a thrust into the Gilberts, Admiral King had urged that only Marines be used in the operation. General MacArthur's needs for amphibious troops and the shortage of transports to bring a second Marine division to the Central Pacific prevented the carrying out of King's recommendation. General Marshall, on 29 July offered an Army unit instead, the 27th Infantry Division then in Hawaii, close to the scene of future combat. Although the unit had received no amphibious training, this was a deficiency that Holland Smith's VAC could solve.

A part of the New York National Guard, the 27th Division had been inducted into the federal service in the fall of 1940. Upon the outbreak of war, the unit had been ordered from Fort McClellan, Alabama, to various installations in California. After standing guard on the Pacific coast, the division sailed in March 1942 for the Hawaiian Islands. In command of the 27th Division when it was assigned to Operation GALVANIC was Major General Ralph C. Smith, who had studied and later lectured at France's *Ecole de Guerre* and who was considered a keen tactician.[5]

During the planning for and fighting on Guadalcanal, a naval officer, Admiral Turner, had the final responsibility for the conduct of operations, both afloat and ashore. The views of the admiral, a man sure in his opinions and forceful in presenting them, did not always coincide with those of the landing force commander, Major General Alexander A. Vandegrift, when the question was the proper employment of troops. General Vandegrift, who recognized the absolute necessity of naval control of the assault, wanted unquestioned authority over operations ashore. A dispatch to Admiral King proposing this command setup was drafted in late October 1942 by the Commandant, Lieutenant General Thomas Holcomb, who was on an inspection trip to the South Pacific, signed by Admiral Halsey, and endorsed by Admiral Nimitz. On his return to Washington, Holcomb indicated to King that he concurred in the concept that the landing force and naval task force commanders should have equal authority, reporting to a joint superior, once the landing phase of the amphibious assault was com-

[4] Richard W. Johnston, *Follow Me! The Story of the 2d Marine Division in World War II* (New York: Random House, 1948), p. 89, hereafter Johnston, *Follow Me!*

[5] Edmund G. Love, *The 27th Infantry Division in World War II* (Washington: Infantry Journal Press, 1949), pp. 11–22.

pleted.[6] Although such a change in doctrine evolved eventually, in the relatively brief operations against the small atolls of the Central Pacific, the role of the naval commander in tactical operations remained paramount.

In setting up the planning machinery for GALVANIC, Admiral Spruance followed the Guadalcanal precedent and made VAC completely subordinate to Turner. Holland Smith, a man just as forceful and outspoken as Turner, objected vigorously and successfully to this setup. Spruance so revised his system that Holland Smith and Turner faced each other as equals throughout the planning phase. On the surface it would seem that the Navy officer and the Marine had equal voice in shaping plans for GALVANIC, but Spruance naturally looked upon Turner as his principal amphibious planner. The decisions of the force commander, however, depended upon the scheme of maneuver advanced by the corps commander, in this case Holland Smith.[7]

In one respect, their dislike for Nauru as an objective, Spruance, Turner, and Holland Smith were in complete agreement. They did not want to divide the available naval forces and conduct two simultaneous amphibious operations separated by 380 miles, with the enemy naval base at Truk beyond the reach of any land-based reconnaissance planes. Admiral Spruance vigorously protested the selection of Nauru and argued for the substitution of Makin Atoll in its stead. He recalled:

> Kelly Turner and I both discussed this situation with Holland Smith at length. It appeared to me that Nauru had been useful to the Japanese as a position from which to search to the southward the area between the Gilbert and Ellice Groups on the east and the Solomons on the west. Once this area was controlled by us, Nauru was not needed by us, and we could keep it pounded down. On the other hand, Makin was 100 miles closer to the Marshalls where we were going, and it tied in well from the point of view of fleet coverage with an operation against Tarawa. The more we studied the details of capturing Nauru, the tougher the operation appeared to be, and finally it seemed doubtful that the troops assigned for it could take it. The transports available for trooplift were the limiting factor. Makin . . . was an entirely suitable objective, and its capture was well within our capabilities.[8]

The upshot of these discussions was that Holland Smith offered a revised estimate of the situation, endorsed by Turner, that led to a revision of the concept of GALVANIC. On 19 September, the general pointed out that at least one entire division would be required to seize rocky, cave-riddled Nauru. In addition, this island lacked a lagoon as an anchorage for small craft and had only a tiny airfield. This estimate was presented to Nimitz on 24 September, while Admiral King was in Pearl Harbor for a conference with CinCPac, and Spruance recommended to both that the Nauru portion of the plan be scrapped. Proposed sub-

[6] Gen Thomas Holcomb interview by LtCol Robert D. Heinl, Jr., dtd 12Apr49, cited in Hough, Ludwig, and Shaw, *Pearl Harbor to Guadalcanal*, p. 342.

[7] Adm Raymond A. Spruance, "The Victory in the Pacific," *Journal of the Royal United Service Institution*, v. 91, no. 564 (Nov46), p. 544; *Moore comments*.

[8] Adm Raymond A. Spruance ltr to ACofS, G-3, HQMC, dtd 24Jul62.

stitute for this target was Makin Atoll, which boasted a lagoon and ample room for airstrips. The defenses of Makin appeared weaker than those of Nauru, its beaches better, and its location near enough to Tarawa to permit the concentration of American shipping. Convinced by the arguments of the GALVANIC commanders, "Admiral King agreed to recommend to the JCS the substitution of Makin for Nauru."[9]

There was little quarrel with the selection of Tarawa as an objective of the GALVANIC forces. Although this atoll was heavily defended, its capture would cut in half the distance that American bombers would have to fly in raiding the Marshalls. Also, Betio Island in this atoll was the nerve center for the Japanese defense of the Gilberts. The responsible planners believed that the Gilberts could not be neutralized with the American strength then available until Tarawa was overrun.

The capture of Apamama was also thought necessary if the Americans were to consolidate their hold on the Gilberts. Again, the primary consideration was to gain an air base from which to strike the Marshalls. This atoll promised to be the least difficult of the three objectives that Central Pacific planners wished to include in GALVANIC.

The JCS promptly agreed to the substitution of Makin for Nauru, and on 5 October, Admiral Nimitz issued Operation Plan 13-43, containing the revised concept of GALVANIC. This document assigned Admiral Spruance the mission of capturing, developing, and defending bases at Makin, Tarawa, and Apamama. The operation was designed to gain control of the Gilberts and by so doing to smooth the way into the Marshalls, improve the security and shorten the line of communication with Australia, and support operations in the South Pacific, Southwest Pacific, and Burma areas by exerting pressure on the Japanese. (See Map 1.)

[9] *Ibid.*

THE INTELLIGENCE EFFORT[10]

Although the Gilberts group was not included by popular journalists among "Japan's Islands of Mystery," American planners knew very little about the onetime British possession. Charts and tide tables provided by the Navy Hydrographic Office proved unreliable. In fact, the maps prepared by the Wilkes expedition of 1841 were found to be as accurate as some of the more modern efforts. If the assault troops were to get ashore successfully, detailed intelligence had to be obtained on beach conditions, tides, and the depth of water over the reefs that fringed the atolls. Principal sources of such informations were photographs taken from aircraft and submarines as well as interviews with former residents of the islands.

American photo planes, both land-

[10] Additional sources for this section include: VAC G-2 Study, TO Gilbert Islands, Nauru, Ocean, dtd 20Sep43, pp. 21-48, 61-67; 2d MarDiv Est of Sit-Gilberts, dtd 5Oct43; 2d MarDiv SupplEst of Sit, dtd 25Oct43; 2d MarDiv D-2 Study of Makin Island and Little Makin, n.d.; 2d MarDiv D-2 Study of Tarawa, n.d.; IntelEst, Anx D to *2d MarDiv OpO No. 14*.

based Liberators (B-24s) and carrier aircraft, soared over Tarawa on 18-19 September and again on 20 October. Makin was photographed on 23 July and 13 October. Of the two atolls, Tarawa received better coverage, for only vertical aerial photos of Makin reached the joint intelligence center. Without oblique prints, photo interpreters had difficulty in estimating beach conditions and determining the exact nature of shore installations.

In spite of this handicap, and the inability of interpreters to gauge the depth of water from aerial photographs alone, other information was evaluated in conjunction with the photographs and the work done by intelligence officers proved to be extremely accurate. The remarkable ability of the aerial camera to locate enemy positions and the skill with which these photographs were analyzed enabled the interpreters to estimate the size of the enemy garrison from a picture that showed the shoreline latrines on Betio.

In commenting on this impressive bit of detective work, the 2d Division operations officer (D-3),[11] later called the picture they used "the best single aerial photo taken during WWII."[12] Using it, he was able to select the spot where he thought "the headman's CP was, since it was the only place with a baffle and sufficient room to drive a vehicle between the baffle and the door."[13] He also was able to determine which of the latrines were probably used by officers by the difference in type. Figuring that the Japanese would assign more men per latrine than an American force, he was able to present intelligence officers with an interesting problem and method of finding the size of the garrison. Utilizing these factors, the D-2 (intelligence) section came up with a figure that Japanese documents later indicated was within a few men of the actual count. The D-3 commented: "This didn't help much in determining strategy and tactics, but it provided the valuable knowledge of enemy strength. By the laws of chance we happened to strike it right."[14]

Additional and extremely valuable data on reefs, beaches, and currents was obtained by the submarine *Nautilus*. This vessel had been fitted out to take pictures of the atolls through her periscope. None of the cameras issued for this mission could take an intelligible picture, but fortunately one of her officers owned a camera that would work?[15] For 18 days *Nautilus* cruised through the Gilberts, pausing to take panoramic shots of Apamama, Tarawa, and Makin. The negatives did not reach Hawaii until 7 October, but the photos were developed, interpreted, and

[11] During much of World War II, Marine division general staff officers were designated D-1, D-2, D-3, and D-4, and comparable corps staff officers as C-1, etc. Eventually, the Marine Corps adopted the Army system of designating all general staff officers at division and above as G-1, etc.

[12] Gen David M. Shoup interview with Hist Br, G-3, HQMC and cmts on draft MS, dtd 14Aug62, hereafter *Shoup interview/comments.*

[13] *Ibid.*

[14] *Ibid.*

[15] Tests of this camera, made with the cooperation of VAC and 27th Division G-2 officers, "resulted in the development of a holding frame, sequence timing for stereo overlap, etc." Col Cecil W. Shuler comments on draft MS, dtd 12Dec62.

the information contained in them disseminated in time for the invasion.

Intelligence officers also were able to gather together 16 persons familiar with the Gilberts. Travelers, traders, or former residents of the British colony, they were attached to Admiral Turner's staff. Those most familiar with Tarawa were sent to Wellington where the 2d Marine Division was training, while those who knew Makin best were assigned to the 27th Infantry Division in Hawaii. Additional intelligence on Makin came from Lieutenant Colonel James Roosevelt, who had fought there as executive officer of the 2d Marine Raider Battalion during the raid of 17–18 August 1942.

Tarawa, the intelligence officers found, was the sort of objective that Earl Ellis had pictured when he first began his study of the amphibious assault. The target for which the 2d Marine Division had been alerted to prepare was a coral atoll triangular in shape, two legs of the triangle being formed by reef-fringed chains of islands and the third by a barrier reef. The southern chain measured 12 miles, the northeastern 18 miles, and the western or reef side 12½ miles. A mile-wide passage through which warships could enter the lagoon pierced the coral barrier. (See Map 2.)

Key to the defenses of Tarawa was Betio, southwestern-most island in the atoll, just three and one-half miles from the lagoon entrance. On Betio the Japanese had built an airfield, and bases for planes were what the Americans wanted. Like the rest of Tarawa, this island was flat; indeed, the highest point in the entire atoll was but 10 feet above sea level. Betio, completely surrounded by reefs, was only 3 miles long and some 600 yards across at its widest point. The Joint Intelligence Center, Pacific Ocean Areas, after weighing all the evidence estimated that between 2,500 and 3,100 Japanese troops were crammed onto the island. The intelligence officers also reported that the enemy might have 8 or 9 coastal defense guns including 8-inch guns [16], 12 heavy antiaircraft guns, 12 medium antiaircraft guns, and emplacements for 82 antiboat guns and 52 machine guns or light cannon. The fighting on Betio would be bloody, but a difficult problem had to be solved before the Marines could come to grips with the enemy. A way had to be found to cross the reefs that encircled the island. (See Map II, Map Section.)

The best solution would have been to land the division in amphibian tractors (LVTs), for these vehicles, like the legendary river gunboats of the American Civil War, could navigate on the morning dew. Unfortunately, the tractors were in short supply, so that most of the troops would have to come ashore in LCVPs (Landing Craft, Vehicle and Personnel), boats which drew about 3½ feet of water. Because the assault on Betio had been scheduled for 20 November, a day on which the difference between high and low tide would be slight, the attackers could not rely on any flood of water to float them over the troublesome reef. The success of the operation well might de-

[16] The determination of the caliber of these guns was made by relating their size in aerial photographs to the known dimensions of wrecked aircraft on the airfield at Betio. *Shoup interview/comments.*

pend upon an accurate estimate on the depth of the water off Betio.

The first estimate made by amphibious force intelligence officers predicted that during the period of lowest tides no more than two feet of water would cover the reefs off the northern coast of Betio. Turner's staff also was aware that during the lowest period, tides at the island might ebb and flow several times in a single day. There was the remote possibility of a freak "dodging tide," a tide with an eccentric course that could not be foreseen or predicted. Such tides had been reported, but few of the islanders had experienced one. On the other hand, those officers responsible for GALVANIC could take heart from the fact that some of the island traders who had sailed among the Gilberts predicted that there would be five feet of water, more than enough for the landing craft. Some of the Americans chose to be optimistic.

Among those who entertained doubts concerning the depth of water over reef at Betio was Major F. L. G. Holland, a New Zealander and former resident of Tarawa Atoll. Assigned to the staff of General Julian Smith, Holland did not accept the estimate of five feet, but he could not disprove the tide tables prepared by his fellow experts. He could, however, point out that during the neap period tides ebbed and flowed irregularly and warn the Marines to expect as little as three feet of water at high tide.

After listening to the New Zealand major, Julian Smith decided to prepare for the worst. The troops embarked in LCVPs were briefed to be ready to face the possible ordeal of wading ashore in the face of Japanese fire. The best that Julian Smith could foresee was a 50-50 chance that landing craft would clear the reef.

Hydrographic and reef conditions also helped dictate the choice of landing beaches. On the south or ocean side of Betio, the reef lay about 600 yards from the island proper, but heavy swells rolled in from the open sea, a factor which might complicate the landings. To land directly from the west would mean crossing both the barrier and fringing reefs as well as battling strong and unpredictable currents. Aerial photographs showed that the enemy defenses were strongest on the seaward side and that the beaches were heavily mined.[17] The choice, then, narrowed to the lagoon side where the reef, though wide, rose gradually to the surface. In addition, the island itself would serve as a breakwater to ships maneuvering within the lagoon.

Makin, northernmost of the Gilberts, was the objective of a reinforced regiment of the 27th Infantry Division. Like Tarawa, this atoll was shaped like a distorted triangle. Southeast of the spacious lagoon lay the large islands of Butaritari and Kuma. A long reef

[17] VAC and 2d Division planners could plainly see the seaward beaches were mined, but the lagoon side was a different matter. The enemy troops there "were in the business of working on their defenses—unloading steel rails, concrete, etc., besides their regular logistic support within the lagoon. . . . The question was what you would do if you were on the island," General Shoup recalled. "Chances are you would mine everything but the place you use daily—that would be the last place to be sewed up. This conclusion was a very definite factor in our decision to land where we did." *Ibid.*

formed the northern leg of the triangle, but the western portion, made up of scattered islets and reefs, was for the most part open to the sea. Butaritari, some six miles in length, was the principal island in the atoll. Intelligence officers discovered that the western part of the narrow island was swampy and somewhat overgrown. Much of Butaritari, however, had been given over to the cultivation of coconut palms and of the native staple food, taro. (See Map 6.)

Photographs of Butaritari, best clue to Japanese strength, led planners to believe that only 500 to 800 troops were available for the defense of the island. This total included an infantry company, a battery of four heavy antiaircraft guns, and two antiaircraft machine gun batteries. Most of the enemy installations were located in the vicinity of Butaritari Village within an area bounded on east and west by antitank ditches.

Unlike the reef off Betio, the coral outcropping around Butaritari was not considered a particularly difficult obstacle. Along the lagoon shore and off the southern part of the west coast at the island, the reef was considered to be so close to the beaches or so flat that it could be crossed quickly. Even if the LCVPs grounded at the edge of the reef, intelligence officers felt that the soldiers could wade ashore without difficulty.

Apamama, according to intelligence estimates, should cause its attackers no trouble at all. As late as 18 October, the atoll was not occupied by any organized defensive force. The only emplacement that photo interpreters could locate was for a single pedestal-mounted 5-inch naval gun, a weapon that appeared to have been abandoned. It was thought possible, however, that several coastwatchers might lurk among the islands that formed the atoll.[18] (See Map 5.)

TASK ORGANIZATION AND COMMAND STRUCTURE

The basic organization for GALVANIC was established by Operation Plan 1–43, issued by Admiral Spruance on 25 October. The task organization consisted of three major groups: Rear Admiral Charles A. Pownall's Carrier Task Force (TF 50), Admiral Turner's Assault Force (TF 54), and Defense Forces and Land-based Air (TF 57) commanded by Rear Admiral John H. Hoover. The Assault Force was divided into two attack forces. One of these, the Northern Attack Force (TF 52) remained under Turner's command and was assigned to capture Makin. The other, Southern Attack Force (TF 53) under Admiral Hill, was to seize Tarawa and Apamama.

Admiral Pownall's TF 50 was to play an important part in the forthcoming operation. In addition to establishing and maintaining aerial superiority in the area, the carrier pilots were to aid the amphibious assault by neutralizing Japanese defenses, helping to spot the fall of supporting naval gunfire, and flying observation missions over Makin, Tarawa, and Apamama. They also had the mission of searching ahead of the convoys, providing fighter cover for

[18] 2d MarDiv OPlan No. 1, dtd 30Oct43.

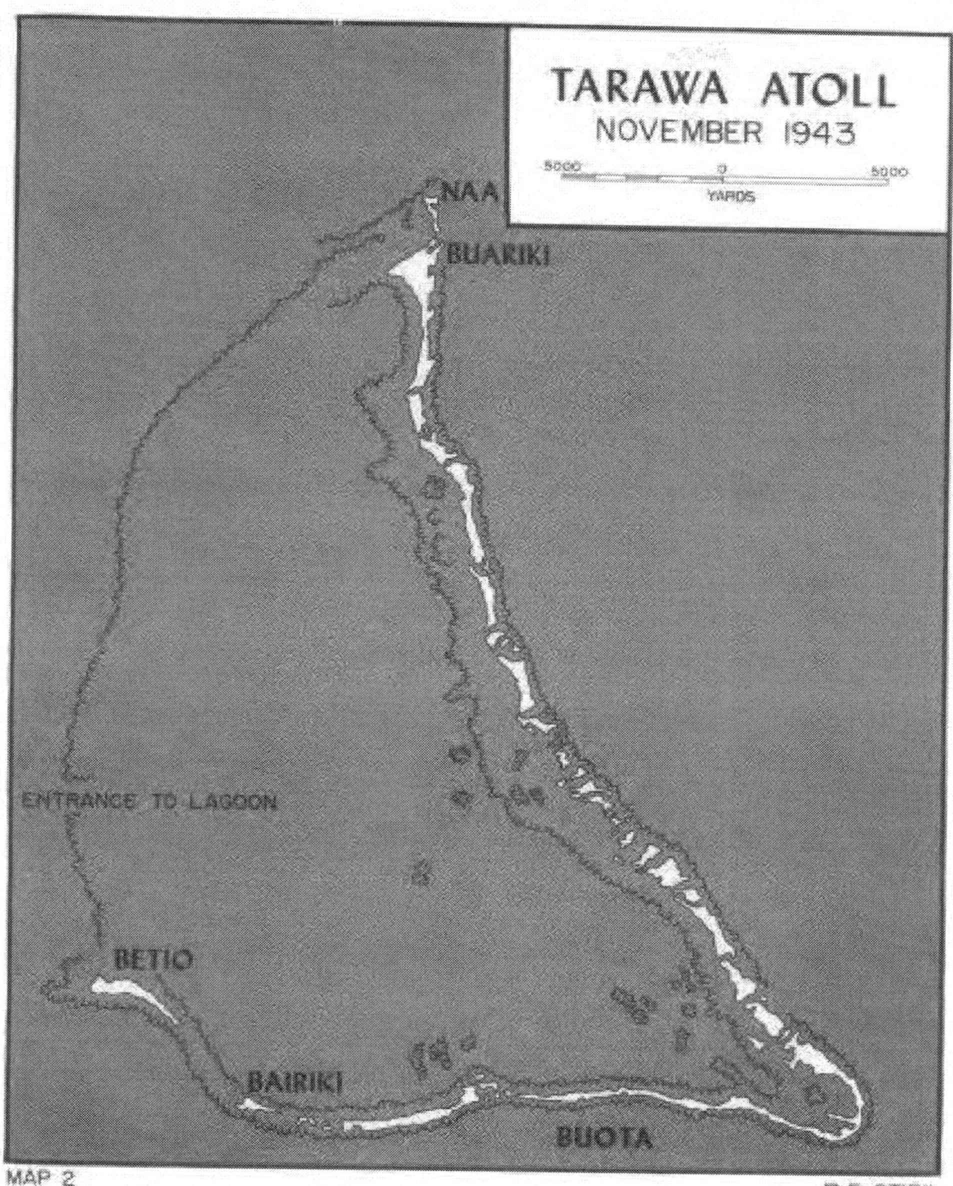
TARAWA ATOLL
NOVEMBER 1943
5000
0
5000
YARDS
NAA
BUARIKI
ENTRANCE TO LAGOON
BETIO
BAIRIKI
BUOTA
R.F. STIBIL

MAP 2

the operation, and guarding against submarine attack.

The land-based planes of TF 57 were to help gain mastery of the skies over the Gilberts. Flying from airstrips on Baker Island and in the Ellice, Phoenix, and Samoan groups, Admiral Hoover's aircraft, including the planes of the Seventh Air Force, were to blast those bases from which the enemy might interfere with GALVANIC. In addition, this force was to bomb the assault objectives and conduct long-range searches.

Before preparing the command relationships paragraph of Admiral Spruance's operation order, his chief of staff, Captain Charles J. Moore, had long discussions with the commanders involved. Continual revisions were made to clarify Holland Smith's position and to satisfy him regarding the role of the landing force commanders at Tarawa and Makin. At each objective, Julian Smith and Ralph Smith were to take independent command of their own forces, once they were established ashore, but "their gunfire support and logistic support and they, themselves, remained under the command of their respective Assault Task Commanders."[19] In the case of Holland Smith as a tactical corps commander, the naval leaders considered—although the Marine general disagreed—that he had no function" in directing the operations of the two independent commanders ashore at Betio and Makin. He could do nothing without the functioning of the Task Force Commander who controlled the ships."[20]

General Holland Smith was to sail in the Assault Force flagship and command the landing force; however, Admiral Spruance made directives issued by the general subject to the approval of Admiral Turner, since the employment of troops was governed by "the capabilities of the surface units to land and support them."[21] The operation plan issued by Turner followed this definition of Holland Smith's duties. The general was to advise the Assault Force commander on the employment of the landing force and the use of reserves, but at both Makin and Tarawa the Attack Force commanders would exercise authority through the commanders ashore. Although Spruance directed that the assault troops would be free of naval control after the beachhead had been secured, his command alignment did not follow the theories advanced by the Marines who had fought at Guadalcanal. Unquestionably, the Central Pacific commander determined that GALVANIC, with two widely separated landings either of which might attract the Japanese battle fleet, was an operation which required naval control throughout all its stages.[22]

As he had done concerning his role in planning, Holland Smith protested his tactical command position to Spruance. The naval officer replied that VAC retained operational control over three garrison units: the 2d and 8th Marine Defense Battalions and the 7th Army

[19] *Moore Comments.*

[20] *Ibid.*

[21] *ComCenPacFor OPlan 1-43*, p. 11.

[22] *Moore Comments.*

Defense Battalion; overall command of assault troops would be exercised by Turner through Holland Smith. "It is considered essential," Spruance continued, "that the responsibility for the assault be placed on the Commander Fifth Amphibious Force. He will require the benefit of your knowledge of amphibious training and operations to ensure the success of the operation with the minimum losses to the troops engaged." [23] In response to a query from General Richardson, who received a copy of this letter, as to whether Admiral Turner or General Holland Smith was the "immediate superior combat commander" of Army troops engaged in GALVANIC, Admiral Nimitz replied that the "immediate superior combat commander of the Commanding General, 27th Infantry Division (Army), is the Commanding General, 5th Amphibious Corps, Major General Holland Smith, USMC." [24]

THE SCHEME OF ATTACK

The overall plan for GALVANIC called for the 2d Marine Division (less one regiment in corps reserve) to storm Tarawa, while the 165th Infantry of the 27th Division took Makin. Elements of the Marine reserve regiment could be employed at either objective, depending upon the enemy's reaction, or used to occupy Apamama. When and where the reserve would be committed was a decision that Admiral Turner alone would make. The force commander, however, might rely upon the advice of Holland Smith.

At both Makin and Tarawa, it was planned that the first few waves would churn ashore in amphibian tractors, vehicles that had been ferried to the objective in tank landing ships (LSTs). Off the atolls, the landing ships would stop, open the huge doors in their bows and disgorge the tractors down a lowered ramp into the water. Since the assault troops would be in transports, it was necessary that they first climb down heavy nets to enter LCVPs from which they later would transfer to LVTs. The tractors would then maneuver to form waves, each one destined for a particular beach. Plans called for minesweepers to sweep the lagoon entrance, anchor buoys to mark the cleared channel, and take position at the line of departure. At this line, the waves were to be guided into lanes leading directly to the assigned beach and at a given signal sent racing across the line toward the island.

The procedure planned for later waves was slightly different, for LVTs had been reserved for the leading assault elements. Since no transfer was necessary, the same LCVPs in which the infantrymen and artillerymen left their transports would carry them to the rendezvous area for the formation of assault waves, to the line of departure where the shoreward movement would be coordinated, and finally to the embattled beachhead. Two Landing

[23] ComdrCenPacFor ltr to CG, VAC, ser 0081, dtd 14Oct43 (S-1 File, Comd Relationships, HistBr, HQMC).

[24] CinCPOA ltr to CGAFPOA, ser 00249, dtd 25Oct43 (OAB, NHD). General Richardson's request for clarification of the tactical command structure was made to insure that it was in accord with Army doctrine and that the corps commander would be the superior officer from whom the 27th Division commander received his combat orders. CGAFPOA ltr to CinCPOA, dtd 17Oct43 (OAB, NHD).

Ships, Dock (LSDs), the USS *Ashland* and *Belle* Grove, had been assigned to carry the medium tanks for Operation GALVANIC. These ships would perform basically the same service for the Landing Craft, Medium (LCMs) and the tanks they carried that the smaller LSTs did for the amphibian tractors. The holds of the LSDs would be flooded to enable the landing craft to float through an opening in the stern. Once afloat, the LCMs would head for the rendezvous area for assignment to the proper boat wave, the first leg in their journey into battle.

D-Day at both objectives was to be ushered in with an aerial attack. From 0545 to 0615 carrier planes would bomb and strafe enemy troops and installations. After the aviators had completed their final runs, the fire support ships would begin a 2½-hour hammering of the objectives. Scheduled to blast Butaritari Island at Makin were four old battleships, four cruisers, and six destroyers. Betio Island, Tarawa Atoll, was destined to shudder under the weight of high explosives thrown into it by three battleships, five cruisers, and nine destroyers. Never before had such powerful seaborne batteries been massed against such small targets. The result, naval gunfire planners optimistically hoped, would be devastating, although few experienced officers looked for total destruction of the enemy defenses.[25] When the naval guns had ceased their thundering, the carrier planes would return for a five-minute attack on the invasion beaches as the assault waves were moving ashore.

Naval gunfire, to be delivered on D-Day and after, was scheduled for both objectives. When the ships opened fire depended upon the enemy's reaction, for any Japanese batteries that threatened the unloading would have to be silenced. Preparatory fires on D-Day, divided into two phases, were to begin after the first air strike. First, the support ships would deliver 75 minutes of pre-arranged neutralization and counterbattery fire, if necessary closing the range to as little as 2,000 yards in order to knock out protected coastal defense guns. The second phase, to last for 45 minutes, called for an increasingly heavy bombardment of assigned areas with the combined purposes of destroying emplacements along the invasion beaches and neutralizing enemy defenses throughout the islands. At Tarawa, the support ships were positioned to fire from the west across Betio, since fire from the south might cause ricochetting shells that could fall into troop assembly areas on the lagoon side of the island.[26] Once the assault troops were ashore, certain warships could be called upon to blast specific targets that impeded the American ad-

[25] The commanding officer of the transport group which landed the Marines at Tarawa, recalled stating his doubts of the efficacy of this fire "very forcibly during a conference at Wellington, N.Z. I had witnessed a similar bombing and bombardment of Gavutu Island, in the Solomons, where I landed a Marine Paratroop outfit. From daylight to noon this little island was subjected to repeated bombing attacks and bombardment by cruisers and destroyers. The results had been most disappointing." RAdm Herbert B. Knowles ltr to ACofS, G-3, HQMC, dtd 1Sep62, hereafter *Knowles ltr.*

[26] *Hill interview/comments.*

vance. During the preparatory shelling, however, all ships were to fire for the most part into areas rather than at selected strongpoints.

Target destruction fires were to be delivered simultaneously with the neutralization of the remainder of the area in which the target was located. The idea of combining neutralization with destruction did not appeal to the 2d Marine Division staff, but naval planners were determined to rain down explosives on the whole of Betio in order to devastate the island in the shortest possible time. The final approved naval gunfire support plan was an amalgam of the desires of the naval and landing force commanders. In commenting on the planning period, General Julian Smith recalled:

> We Marines, all of whom had studied, and in some cases seen in actual combat, the effect of land artillery fire, ships' gunfire, and aerial bombardment, found naval officers unduly optimistic as to the results to be obtained from the bombardment, but never any lack of willingness on their part to listen to our problems and to cooperate most fully in assisting in their solution.[27]

The plan prepared by the 2d Marine Division had its origin early in August, when Admiral Spruance visited General Julian Smith's headquarters at Wellington, New Zealand. At this time the admiral verbally assigned the capture of Tarawa Atoll to the division. During the conference, the problem of the reef at Betio was discussed, and division planners made a tentative decision to land the first three waves in amphibian tractors. The final judgment would depend upon the results of tests of the ability of LVTs to clamber over coral ledges.

Following these talks, the division received its first written directives, documents based on the original Gilberts-Nauru concept. Since the Marines' objectives were Tarawa and Apamama, the later substitution of Makin for Nauru did not disrupt staff planning.

The 2d Marine Division was attached to VAC on 15 September, and on 2 October, Julian Smith and members of his staff flew to Pearl Harbor to coordinate plans with Holland Smith, Turner, and Hill. During the time between his conversations with Spruance and his trip to Pearl Harbor, Julian Smith had been devising a plan for the conquest of Betio. A striking feature of this tentative scheme was the landing of artillery on an island adjacent to Betio prior to the main assault. At Pearl Harbor, Julian Smith learned that the enemy was considered capable of launching a combined air and submarine attack within three days after the American ships arrived off the atoll. Landing howitzers in anticipation of the assault on Betio would forewarn the Japanese, and the enemy might be able to catch the transports before these vessels could be unloaded. Another unpleasant fact that came to light at this time was the decision to hold one regiment of Julian Smith's command in corps reserve. The 2d Marine Division would lack even the strength to make simultaneous assaults against Betio and a secondary objective which

---

[27] LtGen Julian C. Smith, "Tarawa," *U. S. Naval Institute Proceedings*, v. 79, no. 11 (Nov 53), p. 1170, hereafter Smith, "Tarawa"; VAC NGF Spt Plan, dtd 13Oct43, Anx B to VAC OPlan 1-43, dtd 5Oct43; TF 53 NGF Spt n.d., Encl A to TF 53 Rpt of Tarawa Ops, dtd 13Dec43, hereafter *TF 53 AR*.

might serve as the site for artillery. One alternative remained—a direct frontal attack without the prelanding support of division artillery.

Aware that an assault of the type confronting him was the most costly of operations, Julian Smith asked for definite orders. "I discussed the matter fully with the Corps commander," he later recalled, "and when informed that the decision to make the attack directly on Betio was final and must be accomplished by the Second Marine Division less the combat team assigned to Corps reserve, I requested that my orders be so worded as I did not feel that the plan should be my responsibility." [28] Orders were promptly issued by VAC to seize Betio before occupying any of the remaining islands in the atoll.

After the approval of the 2d Division plan, Julian Smith and his party returned to New Zealand. On 19 October, Admiral Hill and key members of his staff followed to go over last-minute details before the issuance of the final plans. Hill brought with him a rough draft of Admiral Spruance's communications plan, whose final version was not available to Task Force 53 until three weeks later, just an hour before the ships left their staging area for the target. When he reached New Zealand, Hill got his first look at his flagship, the USS *Maryland*. The battleship still had 20 yard workmen on board making the necessary alterations for its role as command center for the Tarawa operation.[29]

The 2d Marine Division operations order, completed on 25 October, called for Combat Team 2 (2d Marines, reinforced, with 2/8 attached) to make the assault landings. The remaining two battalions of the 8th Marines, along with the regimental headquarters, were held in division reserve, while the 6th Marines remained under corps control. Elements of the 10th Marines, division artillery, a part of Combat Team 2, would be landed as quickly as possible to support operations ashore. From the 18th Marines, Julian Smith's engineers, came another part of the combat team, demolitions and flamethrower men to assist the infantry battalions, as well as the shore party that had the task of speeding supplies to the front lines.

Combat Team 2 planned to assault Beaches Red 1, Red 2, and Red 3, all on the lagoon side of the island and each the objective of one battalion landing team. As a result of the removal of one regiment from Julian Smith's control, the 2d Marine Division had only an estimated two-to-one numerical edge in infantry over the defending Japanese. Instead of reinforcements, Combat Team 2 would have to rely on the effect of the massive preliminary bombardment in its effort to drive completely across the island, capture the airfield, change direction, and launch a two-battalion thrust down the long axis of the objective. (See Map III, Map Section.)

Minor adjustments had to be made throughout the planning phase. Experiments proved that amphibian tractors could crawl across a coral reef, but these vehicles were in short supply. The 2d Division had 100 tractors, all of

---

[28] Smith, "Tarawa," p. 1167; LtGen Julian C. Smith ltr to ACofS, G-3, HQMC, hereafter *Julian Smith ltr.*

[29] *Hill interview/comments.*

them primitive LVT(1)s which had been designed primarily as cargo carriers and lacked armor protection. Julian Smith's staff obtained sheets of light armor which were fixed to the tractors while the division was in New Zealand. Many of these LVTs, veterans of the Guadalcanal fighting, had outlived their usefulness, but mechanics managed to breathe new life into 75 of them. Each LVT(1) had room for 20 fully equipped men in addition to its crew of 3. Unless the division commander received more LVTs, he did not have enough vehicles for the first three assault waves.[30]

The nearest source of additional tractors was San Diego. Although there was neither time nor shipping to get large numbers of these vehicles to New Zealand, 50 LVT(2)s were shipped to Samoa. Members of the 2d Amphibian Tractor Battalion went to that island to form a new company which would join the division off Tarawa. The new LVT(2) was an improved version of the tractor already in use by the division. Horsepower had been boosted from 146 to 200, a change which enabled the LVT(2) to move slightly faster while carrying 4 more men or 1,500 more pounds of cargo than its predecessor. Also, the new model could cruise for 75 miles in the water, compared to 50 miles for the LVT(1). Tests were ordered in which the new tractors ran 4,000 yards with a full battle load to determine the time it would take the LVT(2), which proved capable of making at least four miles an hour.[31] Each LVT (2) was equipped with portable armor plate for the front, sides, and cab. These plates could be used during assault landings or removed if there was no danger of enemy fire.

Another proposed refinement in the basic plan was a request for additional aerial bombardment. Lieutenant Colonel David M. Shoup, division operations officer, urged that Seventh Air Force planes drop one-ton "daisy-cutters" on and beyond the invasion beaches during the ship-to-shore movement. In addition to killing Japanese, the heavy bombs would shatter buildings that otherwise might provide cover for enemy snipers. This request, although endorsed by the division and listed in the air operations plan, was not carried out.[32] The approach plan prepared by Task Force 53 called for certain of the fire support ships to separate from the main group as the transports neared the transport area. These warships would steam to designated positions to the south, west, and northwest of Betio. Two minesweepers were to lead the destroyer screen into the lagoon. Next to pass through the gap in the barrier reef would be an LSD carrying the medium

---

[30] *Ibid.*, p. 1166. For details concerning various models of the LVT, see ONI, ND, *Supplement No. 1 to ONI 226, Allied Landing Craft and Ships* (Washington, 1945).

[31] *Hill interview/comments.*

[32] *Ibid.* General Shoup was later told that three B-24s with these bombs on board actually took off, but that one crashed on takeoff, one flew into the water, and the third did not reach its destination. *Shoup interview/comments.* A search of Seventh Air Force records in the USAF Historical Archives failed to reveal any mention of this request. Dr. Robert F. Futrell, USAF Historian, ltr to Head, Hist Br, G-3, HQMC, dtd 20Aug62.

214-881 O-67—4

tanks of the division and finally the initial waves of landing craft.

Planning for the employment of the 27th Infantry Division was handicapped by the substitution of Makin Atoll for Nauru Island. In addition, there was a difference of opinion between General Ralph Smith, the division commander, and General Holland Smith of VAC. The corps commander preferred to assault Butaritari from the lagoon side, bringing the maximum strength to bear against a small portion of the coast. On the other hand, the Army general wanted to land two battalions on the west coast of the island and two hours later send a single battalion ashore near the waist of objective. This second blow would be delivered from the lagoon. Ralph Smith's views prevailed, and the Marine general gave rather reluctant approval to the scheme.

Assigned to the Makin operation was the 165th Regimental Combat Team (RCT). With a total of 6,470 men, this heavily reinforced unit outnumbered the estimated defenders of Butaritari by roughly 8-to-1. Three reinforced companies from the 105th Infantry, 582 officers and men in all, had been assigned to the landing force. One of these companies would, if LVTs became available, spearhead each of the assault battalions. This mixing of units was brought about by the shortage of amphibian tractors. Since it seemed for a time that none of these vehicles would be available, Ralph Smith scheduled all assault elements of the 165th Infantry to train with and land from LCVPs, while reserving the tractors for the men of the 105th. Thus, there would be no need to adjust his plans if the promised tractors did not arrive, for the men from the 105th could remain in reserve. On the other hand, if the LVTs did appear, they could be used by the detachments from the 105th Infantry, and again no violence would be done to the basic landing plan.

In Ralph Smith's opinion, the rapid capture of the tank barrier guarding the western limits of the main defenses was of the greatest importance. It was to gain this end that he had proposed two separate landings followed by a pincers movement against the enemy stronghold. Such a maneuver, however, would depend on close coordination between the attacking units and reliable communications with the artillery batteries that had landed over the western beaches. Another solution to the problem posed by the tank barrier would have been to rake the obstacle with naval gunfire. At the time, however, neither Ralph Smith nor his staff were impressed with the effectiveness of seaborne artillery. Instead of seeking aid from sharpshooting destroyers, they preferred a combination of land weapons—infantry, artillery, and armor.

COMMUNICATIONS AND SUPPLY [33]

In general, the overall communica-

[33] Additional sources for this section include: CinCPac Comm Plan, n.d., Anx A to CinCPac OPlan 1-43, dtd 5Oct43; Rpt of GALVANIC Comm, dtd 4Dec43, Encl D to *V PhibFor AR;* TransArea Debarkation and Unloading Plan, n.d., and Unloading and Beach Pty Plan, n.d., Apps 1 and 2 to Anx D to *TF 53 OpO A101-43;* SigRpt, dtd 4Dec43 with Suppl, dtd 3Jan 44, and TQM Report, dtd 30Dec43, Encl F to *VAC AR.*

BETIO ISLAND *as it appeared two weeks after the battle, looking west over GREEN Beach. (USAF B–65141AC)*

ASSAULT TROOPS *cross the log wall behind the RED Beaches and move inland on Betio. (USMC 64032)*

tions plan for GALVANIC was considered adequate, even though it could have been improved. Principal objections to the communications annex issued by Spruance's headquarters were twofold: it was too long, 214 mimeographed pages, and it should have been distributed sooner. The second criticism was justified, but the staff had worked against an impossible deadline; those who objected to the bulk of the document would later admit that an overall plan, huge though it might be, was preferable to several briefer, less detailed, and possibly conflicting plans.

Secrecy was the watchword during the preparation for GALVANIC, and this mood of caution was to prevail during the approach of the expeditionary force. Since strict radio silence was necessary, only VHF (Very High Frequency) and TBS (Talk Between Ships) equipment could be used within the convoys. Visual signals were substituted whenever possible for routine radio messages, but signalmen proved rusty at first. Although speed came with practice, the vast number of visual signals, which reached as many as 80 per day off Tarawa, led to the establishment, en route to the target, of areas of operational responsibility within the task force. Had this practice not been adopted, hours would have been lost in passing messages from ship to ship to insure that every element of the force had got the information. As it was, certain vessels were to pass on information to ships within specified sectors.

Keeping contact between ships and shore was certain to be the most difficult aspect of the GALVANIC communications problem. Neither LSTs, transports, nor the beachmasters were equipped with the SCR–610 radio, and this set turned out to be the best piece of signal equipment ashore on Betio.[34] During the first crucial days, these sets would often provide the only means of radio contact between the beach and the task force. The Marines themselves were saddled with the TBX and TBY, two low power sets whose general worthlessness brought the postoperation comment that: "light weight but powerful and rugged portable equipment having full frequency range and capable of sustained operation does not appear available in any standard type."[35] The TBX lacked the necessary range, and the TBY was not sufficiently waterproof.

Both the Marines and soldiers had wire equipment with which to establish communications within the beachhead area. Unfortunately, the generator armature of the standard EE–8 field telephone and the drop coil of its companion switchboard were not waterproof and therefore unreliable in amphibious operations. Also, to avoid damage by troops and tracked vehicles as well as short circuits caused by dampness, it was desirable to string telephone wires above the ground, something that could not easily be done in the face of enemy fire.

[34] "The SCR–610 and the ship-carried SCR–608 were Army radios 'appropriated' by the Navy transports that served in the Aleutians and were now to take part in GALVANIC. We had to dole them out where most needed and never had anywhere near enough of them during the GALVANIC Operation." *Knowles ltr.*

[35] V PhibFor Rpt of GALVANIC Comm, *op. cit.*, p. 2.

GALVANIC could not succeed unless a steady current of supplies was kept moving from the holds of the transports to the front lines. First step in this process was the rapid unloading of cargo, and to gain speed both the Army and Marine divisions combat loaded their shipping. Cargo was so stowed that items needed early in the fighting were at the top of the holds and close to the hatches. Because vessels were dispatched to the 2d Marine Division piecemeal, as quickly as they were released from other duties, the division staff could not predict how much cargo space would be available. Sometimes the blueprints provided by the arriving ships were outdated and no help to the hard-pressed planners. The Marines, nevertheless, managed to do a creditable job; in fact, the only snag in unloading came as a result of the re-arranging of cargo in ships at anchor off Tarawa.

The vessels carrying the 27th Infantry Division troops also were effectively combat loaded. Lieutenant Colonel Charles B. Ferris, division G-4, organized a transport quartermaster school and sent his students to Pearl Harbor to learn the characteristics of naval transports as well as loading techniques. In addition, the supply section of the Army division resurrected the stowage plans drawn up for the Attu operation, studied them, and used them as guides for plans of their own.

In handling cargo, the Army division had a decided advantage over its Marine counterpart. While the 27th Infantry Division had some 1,800 sled and toboggan type pallets, the 2d Marine Division had almost none, a deficiency caused when the necessary materials to build them failed to reach Wellington in time. Pallets meant easier handling of cargo because several heavy boxes could be lashed to a wooden platform, stowed and unloaded as a single unit, and hauled intact to the using unit.

At Makin, the scheme of maneuver and the relatively small Japanese garrison indicated that cargo could be ferried to the beach without serious enemy interference. Sailors and a part of a company from the 105th Infantry were to act as ship unloading details. On the beachhead, the 102d Engineer Battalion, reinforced by small detachments from the 165th Infantry, would provide shore parties to sort supplies and rush them inland. One engineer company was attached for this purpose to each assault battalion.[36]

Tarawa, however, offered a far greater logistical challenge. The assault waves were to slam directly into the enemy's defenses on Betio, and the craft carrying supplies for the 2d Marine Division also might encounter fierce opposition. Landing craft were certain to be sunk, so extra service would be required of the survivors. Every man was expected to do his duty and more. "Use your brains . . . and guts," urged Captain Herbert B. Knowles, transport group commander; "keep the boats moving, and get the stuff to the Marines." [37]

As soon as the assault waves had hit the beach and landing craft became

[36] SP Ops, Anx 4 to 27th InfDiv AdminO 11, dtd 26Oct43; see also Crowl and Love, *Gilberts and Marshalls*, pp. 48–49, 102.

[37] TF 53 Unloading and Beach Pty Plan, *op. cit.*

available, Marines and sailors would begin unloading cargo from the transports. Supplies were to be loaded into the boats according to a fixed priority, but dispatchers would not send the boats shoreward unless told to do so by the commander of the regiment for whom the cargo was destined. En route to the beach, all supply craft had to report to control officers who made sure that the incoming boats were headed toward the proper sector and that a shore party was on hand to unload them. LVTs, LCVPs, and LCMs all might haul supplies, but the last, with its 30-ton capacity, was considered most valuable.

An orderly logistical effort also required that beach party and shore party units land with the assault battalions. In charge of each beach party was a naval officer, the beachmaster, who assisted the shore party commander, and also supervised marking the beaches, evacuating the wounded, and the other tasks performed by his men. A Marine officer commanded the shore party, which was primarily concerned with unloading the incoming boats, sorting supplies, and storing them or moving them inland. At Betio some of these activities could be concentrated at the long pier near the waist of the island. This structure was accessible to landing craft, for its jutted beyond the reef, and a boat channel had been dredged along its western side.

Since protracted fighting was expected at neither Makin nor Tarawa, both divisions limited the amount of supplies to be carried to the target area. The transports assigned the 165th RCT carried the assault troops, their equipment and weapons, 5 units of fire per weapon, plus 10 days' rations, 2 days' K rations, and such miscellaneous items as medical supplies, ordnance spare parts and cleaning equipment, and fuel enough to last the vehicles on board for 7 days. Stowed in the assault cargo ship assigned to the Makin landing force were 24 days' B rations for the entire command, 15,000 gallons of water, 8 days' motor fuel, and additional ammunition. Three LSTs carried still other supplies.

The 2d Marine Division also attempted to keep a tight rein on its supplies. To be embarked with the convoy carrying the assault and garrison forces were 30 days' B rations, 5 days' C or K rations (later changed to 3 days' K and 10 days' C), 2 days' D rations, and enough water to provide 2 gallons per day to each member of the command for a period of 5 days.[38] Within five days, water distillation equipment would be operating. Enough maintenance supplies, fuels, and lubricants to last 30 days were loaded in the transports. Also on hand were construction, medical, and aviation supplies for 30 days. Although antiaircraft weapons were allotted 10 units of fire, coast defense guns and all other weapons received 5.

---

[38] The usual components of standard rations were: D, an emergency individual ration—a special chocolate bar; C, the individual combat ration—canned hash, stew, or meat and beans, biscuits, sugar, powdered coffee, and candy; K, another emergency or combat ration—breakfast, dinner, and supper units, each consisting of tinned luncheon meat, biscuits, sugar, and gum; B, a rear-area unit ration—canned meats, dried or canned fruit and vegetables, canned bread, or biscuits.

The number of vehicles was to have been reduced to the minimum necessary for operations on an island the size of Betio, but as planning progressed the number of trucks, tanks, half-tracks, LVTs, and trailers thought vital for the attack continued to increase. Eventually, the Marines lifted to the target more vehicles than they could use. The final total, including LVTs, for the assault echelon was 732 wheeled and tracked vehicles plus 205 trailers. The Makin landing force made a similar miscalculation, bringing with it 372 tracked or wheeled vehicles, and 39 trailers.[39]

TRAINING AND PRELIMINARY OPERATIONS

Upon assuming command of the 2d Marine Division on 1 May 1943, Julian Smith inherited a veteran unit but one that still was suffering the effects of the Guadalcanal fighting. The division had arrived in New Zealand with 12,500 diagnosed cases of malaria, many of whom eventually were evacuated to the United States. So serious was the health problem that as late as 10 October malaria victims were being admitted to the hospital at the rate of 40 per day. Even as the new commanding general was taking charge, the first replacements began arriving. More would follow until the organization reached combat strength. Fitting these men into the division team was one of the problems facing the new commander and his staff.

In addition to shattering the health of the division, the Guadalcanal campaign weakened it in a tactical sense. At Guadalcanal, the 2d Marine Division had fought for the most part as a collection of combat teams rather than as a tightly organized unit. The lessons of jungle warfare had to be put aside, and the various elements of the command welded into an effective and well-coordinated striking force capable of seizing a defended atoll.[40]

Late in September, as transports became available, the transition from jungle fighters to amphibious assault troops began in earnest. First the battalion landing teams, then the regiments practiced off Paekakariki, at Hawke Bay, and in Wellington Harbor, while a few LVTs were sent to the Fijis to test their ability against reefs similar to those that guarded Betio. After these preliminary landings, the Marines returned to camp to rest, repair equipment, and prepare for what they thought was going to be a full-scale division exercise.[41]

During this period, the same few ships did most of the work with the Marines, since new arrivals destined for the transport group of the Southern Attack Force needed "to have engineering work done, boats from the scrapped boat pool ashore overhauled and supplied them, and some semblance of communications equipment furnished."[42] The group commander, Captain Knowles, commented:

> Most of these ships arrived lacking full crews, full boat complements and woefully

[39] Details of Loading of GALVANIC Ship, Encl 1 to *V PhibFor AR;* VAC AdminO 4–43, dtd 13Oct43; *2d MarDiv OpO No. 14.*

[40] Smith, "Tarawa," pp. 1165–1166.

[41] *Shoup interview/comments.*

[42] *Knowles ltr.*

> lacking in communications facilities. Some of these ships had been diverted to Wellington while still on 'shakedown' operations. The transport group commander did not know that he was destined for anything except conduct of amphibious training with 2nd Mar. Div. until about the middle of the month; then he had to organize 3 divisions of transports and get them ready for sea by 1 November. His flapship *Monrovia* had been stripped of everything useful in the way of communication facilities except basic commercial ship radios. At Efate we had to install a small command station above the ship's bridge plus sufficient signal yards and signal flags to do the job ahead. Had we not had extra naval personnel and Army SCRs (both 'appropriated' at the end of the Aleutian Operation) we would have been in an even sorrier mess than we were. The few ships that had been in the Aleutians furnished officers and men to give at least a minimum of [experienced] personnel to new arrivals.[43]

Marine Division had a strong leavening of combat experienced men spread combat experienced men spread throughout its units. These veterans gave emphasis to the constant theme in training—keep the attack moving. Should officers fall or units become disorganized, noncommissioned officers would have to assume command, and this would often happen at Betio. Also emphasized were local security and fire discipline during the night, tactics that would forestall Japanese infiltration and local counterattacks. One criticism of the division training program was its failure to spend enough time drilling infantrymen, tank crews, and demolitions men to act as integrated teams in reducing strongpoints. At the time no one realized the tenacity with which the enemy would fight even after the island seemed doomed to fall.[44]

The 27th Infantry Division, untried in combat, was also new to the techniques of amphibious warfare. Preparing this division for Operation GALVANIC was a task shared between General Richardson's headquarters and Holland Smith's VAC. The Army command handled training for ground combat as well as certain phases of pre-amphibious training, while the Marine headquarters concentrated on the ship-to-shore movement. Logistical planning and routine administration for Army troops also lay within the province of General Richardson.

In actual practice, the distinction between ground, pre-amphibious, and amphibious matters tended to disappear. The 27th Infantry Division was first introduced to amphibious warfare in December 1942, when two officers from the unit attended a school offered at San Diego by Amphibious Corps, Pacific Fleet. The information gained at San Diego was passed on to other division officers in a school conducted in Hawaii. After the division had been selected to provide troops for GALVANIC, the tempo of training increased, and those portions of amphibious training which could be carried out ashore were undertaken at Army installations. In addition, the division began organizing liaison parties to direct naval gunfire and drilling its supply personnel in the complexities of combat loading. Ship-to-shore exercises, however, awaited

---

[43] *Ibid.*

[44] *Ibid.*, p. 1168; Maj Arthur J. Rauchle ltr to CMC, dtd 12Jun47; 2d MarDiv, 3/2, and 3/6 WarDs, Oct–Nov43 (Unit Rpt File, Hist Br, HQMC); Johnston, *Follow Me!*, pp. 94–95.

Admiral Turner's ships and General Holland Smith's instructors.

While the training of Army troops for GALVANIC was getting underway, VAC found itself preoccupied with two demanding tasks, organization of the corps itself and planning for the scheduled operation. In the meantime, General Richardson went ahead with his training program, absorbing the necessary amphibious doctrine from War Department manuals, Navy and Marine Corps publications, and the recorded experience of other Army divisions. Of particular value were notes prepared by the 9th Infantry Division during its indoctrination under Holland Smith as well as the original loading plans for the Attu landing. By the time that VAC began to assert itself in the training setup, Richardson had come to look upon the Marine organization as simply another echelon to clutter up the chain of command. The Army general believed that Admiral Turner, who controlled the necessary ships, was the logical person to train troops for the ship-to-shore movement, and felt that there was no need at the time for a corps of any sort.[45]

Thus, on the eve of GALVANIC both Richardson and Holland Smith were complaining about the status of VAC. The Marine general objected because his headquarters had been restricted in its exercise of tactical command, and the Army general urged that the corps be abolished completely. Their respective higher headquarters gave each essentially the same advice: to make the best of the situation. This they did, and preparations for combat continued.[46]

Training of the 27th Infantry Division came to a climax with a series of amphibious exercises held in the Hawaiian Islands. Bad weather and poor beaches hampered the earlier efforts, and the rehearsals were of questionable value. During the first two rehearsals, the troops landed, but no supplies were put ashore. Because of rock-strewn beaches, assault craft did not advance beyond the line of departure in the third or dress rehearsal. Preserving scarce LVTs from possible damage was judged more important than any lessons the troops might learn.

Preparations for the Gilberts invasion included certain preliminary combat operations, some remotely connected with GALVANIC, and others designed specifically to batter the assault objectives. American might first made itself felt in the Gilberts in February 1942 when carrier planes lashed at Makin Atoll. In August of the same year, Marine Raiders startled the Japanese by making a sudden descent on Butaritari Island. In April of the following year, after a series of reconnaissance flights, heavy bombers of the Seventh Air Force, operating from Funafuti and Canton Island, began harassing Nauru and targets in the Gilberts.

These early aerial efforts were sorely handicapped by the lack of bases close to the Gilberts. To remedy this situation, the 7th Marine Defense Battalion occupied Nanomea Atoll in the Ellice

[45] LtGen Robert C. Richardson, Jr., USA, ltr to LtGen Thomas T. Handy, USA, dtd 5Nov 43 (OPD File 384 PTO–Sec II, RG 115, WW II RecsDiv, FRC, Alexandria, Va.).

[46] *Morton MS*, ch. 23, pp. 22–25.

Islands and the 2d Marine Airdrome Battalion established itself at Nukufetau in the same island group. Both landings were made during August 1943. A third air base was established in September at Baker Island, an American possession which had gone unoccupied since the coming of war. The last of these fields to be completed, that at Nukufetau, was ready on 9 October.

The Seventh Air Force began its systematic support of GALVANIC on 13 November by launching 18 Funafuti-based B–24s against Tarawa. On the following day, the hulking bombers divided their attention between Tarawa and Mille in the Marshalls. Gradually the list of targets was expanded to include Makin, Jaluit, Maloelap, and even Kwajalein. Between 13 and 17 November, planes of the Seventh Air Force dropped 173 tons of high explosives on various targets in the Gilberts and Marshalls and destroyed 5 enemy aircraft. Admiral Hoover's land-based naval planes and patrol bombers also began their offensive on 13 November, but limited themselves to night strikes against Nauru, Tarawa, and Makin. (See Map I, Map Section and Map 7.)

The Navy was far from reluctant to risk its carrier planes against Japan's island fortresses. In fact, Admiral Pownall's fast carriers went into action even before the Seventh Air Force had launched its intensive aerial campaign. On 17 and 18 September, planes from three aircraft carriers blasted Makin, Apamama, Tarawa, and Nauru. The naval aviators were assisted by B–24s from Guadalcanal, Canton Island, and Funafuti, aircraft which carried cameras as well as bombs, and other Liberator bombers struck the Gilberts on the following day as the carriers were withdrawing. Next the carriers attacked Wake Island on 5 and 6 November.

The final phase of this campaign of preliminary aerial bombardment took place on 18 and 19 November. Seventh Air Force planes blasted Tarawa and Makin and helped carrier aircraft attack Nauru. After pounding Nauru, Admiral Pownall's fliers on 19 November dropped 130 tons of bombs on Jaluit and Mille. Air power had done its best to isolate the objectives and soften their defenses for the amphibious assault.[47]

THE ENEMY[48]

Japan seized control of the Gilberts on 10 December 1941 in a move designed to gain bases from which to observe American activity in the South Pacific. Since the occupied islands were considered mere observation posts, little was done to fortify them. A handful of men were posted at

[47] Wesley Frank Craven and James Lea Cate, eds., *The Pacific: Guadalcanal to Saipan—The Army Air Forces in World War II*, v. 4, (Chicago: University of Chicago Press, 1950) pp. 290–302 hereafter Craven and Cate, *Guadalcanal to Saipan;* TF 50 AR GALVANIC, 10–27Nov43, dtd 4Jan44; CinCPac–CinCPOA WarD, Nov43, dtd 28Feb44 (CinCPac File, HistBr, HQMC).

[48] Additional sources for this section include: JICPOA Buls 4–44, Study of Japanese Instls on Butaritari Island, Makin Atoll, dtd 14Jan 44, and 8–44, Japanese Fors in the Gilbert Islands, n.d (IntelFile, HistBr, HQMC); VAC G–2 Study and Rpt, Betio, dtd 23Dec43; 2d MarDiv and JICPOA Study of Japanese Def of Betio Island, Tarawa, dtd 20Dec43.

Tarawa, coastwatchers were scattered throughout the island group, and a seaplane base along with some rudimentary defenses were built on Makin. Apparently the Japanese became convinced that geography plus the battering given the American Fleet at Pearl Harbor had made the Gilberts invulnerable, for the small garrisons were shortly reduced. On 17 August 1942, when the hatches of two American submarines eased open and 221 Marines began paddling toward Butaritari Island, only 70 Japanese could be mustered to oppose them.[49]

Although Japanese strategists dismissed the Makin raid as an attempt to pin down troops in the Central Pacific while new operations were launched to the southwest, the vulnerability of the Gilberts certainly shocked them. Unless these outlying islands were garrisoned in some strength, they would fall to the Americans and serve as bases for a thrust into the far more valuable Marshalls. Reinforcements were started toward the Gilberts, fortifications were thrown up throughout the group, and British citizens overlooked since the occupation were hunted down.

While the Gilberts were being reinforced, Japanese strategy was being revised. As early as March 1943, the Imperial Navy was thinking in terms of "interception operations," in which its ships would fall upon and annihilate any American fleet attempting to land troops along the fringes of Micronesia. Operations of this sort would be possible only if the defending garrison were strong enough to hold the attackers at bay until Japanese aircraft, submarines, and surface craft could reach the area.[50]

In May 1943, Japanese naval leaders conferred at Truk, and out of these discussions evolved a plan to counter any American thrust into the Gilberts. Should Nimitz choose to attack, Japanese bombers from the Bismarcks would swoop down on his convoys, land at fields in the Gilberts and Marshalls, refuel, rearm, and return to action. Meanwhile, short-range planes were to be shuttled into the threatened area by way of Truk and other bases. Fleet units would steam eastward from Truk to cooperate with Bismarcks-based submarines in destroying the already battered invasion force.

This scheme for the defense of the Gilberts was but a single aspect of *Z Operation*, an overall plan of defense. This larger concept called for the establishment of an outer perimeter stretching from the Aleutians through the Marshalls and Gilberts to the Bismarcks. Vigorous action by the Imperial Fleet coupled with a stubborn fight by the island garrisons would thwart any American attempt to penetrate the barrier. The type of strategy espoused

[49] Chief, War Hist Off, Def Agency of Japan ltr to Head, HistBr, G-3, HQMC, dtd 19Nov62, hereafter *Japanese Gilberts comments*.

[50] Mil HistSec, Japanese Research Div, HQ, AFFE, Japanese Monographs No. 161, Inner South Seas Islands Area NavOps, Pt. 1: Gilbert Islands (Nov41–Nov43) and No. 173, Inner South Seas Area NavOps, Pt. 2: Marshall Islands (Dec41–Feb44); Takushiro Hattori, *Dai Toa Senso Zenshi* [*The Complete History of the Greater East Asia War*] (Tokyo: Masu Publishing Company, 1953—MS translation in 4 vols. at OCMH), II, pt. 5, p. 43, hereafter Hattori, *Complete History*.

in *Z Operation*, modified because of American successes in the Aleutians, was reviewed at an Imperial conference held during September 1943 and was considered acceptable.[51]

Betio Island, in keeping with defensive theory advanced as part of *Z Operation*, was heavily fortified. The basic defensive pattern selected for the island called for a series of strongpoints with the spaces between them covered by fire. American assault forces were to be cut down at the beach. Should the invaders manage to gain a foothold on the island, determined counterattacks would be launched to hurl them back into the sea.

In command at Tarawa was Rear Admiral Keiji Shibasaki of the *3d Special Base Force*. He had at his disposal 1,122 members of this force and 1,497 men of the *Sasebo 7th Special Naval Landing Force (SNLF)*. In addition to these combat troops, the admiral had a large contingent of laborers, 1,247 from the *111th Construction Unit* and 970 from the *4th Fleet Construction Department Detachment.* Since many of the laborers were Koreans and most were untrained, Shibasaki could rely on no more than about 3,000 effectives.

The defenses of Betio were cleverly integrated, with coast defense guns, automatic weapons, and various kinds of obstacles complementing one another. Upon approaching the island, the invader would have to brave the fire of 20 coastal defense guns, ranging in size from 80mm to 8-inch. Concrete tetrahedrons scattered along the reef would be encountered next; these had been placed to force assault craft to follow routes swept by fire from the smaller coastal defense weapons, automatic cannon, and machine guns. To scourge the incoming waves, the Japanese on Betio had, in addition to the weapons already mentioned, 10 75mm mountain howitzers, 6 70mm guns, 9 37mm field pieces, at least 31 13mm machine guns, and an unknown number of 7.7mm machine guns.[52] The defenders could also press into service dual-purpose antiaircraft weapons and the 37mm guns of seven light tanks. To make the firepower of this arsenal more effective, the Japanese strung double-apron barbed-wire fences between reef and beach and along the beaches themselves. (See Map II, Map Section.)

Admiral Shibasaki planned to destroy the enemy forces as they landed, but he did not overlook the possibility that the attackers might gain a lodgment on the island. A log fence just inland of the beaches, antitank ditches, and other obstacles were arranged to confine the assault force to a tiny strip of coral sand, where it could be wiped out.

If the ring of defenses along the shores of Betio could be penetrated, the attackers would find the inland defenses organized in a more haphazard fashion. The command posts, ammunition dumps, and communications centers were housed in massive bunkers of reinforced concrete, structures that were built to withstand even direct hits by high explosive naval shells or ae-

[51] Hattori, *Complete History*, III, p. 4–5.

[52] So great was the destruction on the island that a postoperation count of light machine guns was impossible.

rial bombs. These positions, however, were not designed primarily for defensive fighting. Although some fitted into patterns of mutual defense, most of them had blind spots, not covered by fire, from which flamethrower or demolition teams could close for the kill.

Far less formidable were the defenses of Butaritari Island. There Lieutenant Junior Grade Seizo Ishikawa commanded no more than 384 combat troops, 100 of them marooned aviation personnel and the remainder members of his *3rd Special Base Force Makin Detachment.* Also present, but of doubtful effectiveness, were 138 men of the *111th Construction Unit* and 276 from the *4th Fleet Construction Department Detachment.*

Japanese defenses on Butaritari were concentrated around King's Wharf, about one-third of the way down the lagoon side of the island from its western foot. At the base of the wharf, the Japanese had built their seaplane base. The perimeter was bounded on the southwest by an antitank ditch linked to an earthen barricade. This obstacle, about 2,000 yards from King's Wharf, stretched almost across the island but was defended by only one antitank gun, a single pillbox, six machine gun emplacements, and numerous rifle pits. A similar ditch-and-barricade combination was located about the same distance from King's in the opposite direction and marked the northeastern limits of the main defenses. Six machine guns, three pillboxes, and a string of rifle pits guarded this barrier. Throughout the principal defensive area, the majority of heavy weapons pointed seaward, so the greatest threat to an assault from the lagoon lay in the trio of 80mm guns emplaced at the base of King's Wharf. (See Map 6.)

ON TO TARAWA

The departure from Wellington of the 2d Marine Division was shrouded in secrecy. Announced destination of the division was Hawkes Bay, the site of most amphibious exercises, and a rumor was planted that the troops would be back in Wellington in time for a scheduled dance. "With regard to the dance," reminisced Julian Smith, "one of the Division wits remarked that maybe we didn't leave many broken hearts in New Zealand but we certainly left a lot of broken dates." [53]

Instead of steaming to Hawkes Bay, the transports joined Admiral Hill's Southern Attack Force at Efate in the New Hebrides, where rehearsals were held. During the first of these, troops landed at Mele Bay while the support ships simulated a bombardment of Pango Point. The second rehearsal saw the division land again at Mele Bay and the warships actually pump shells into Erradaka Island. At this time, the commander of the assault regiment, Colonel William McN. Marshall, fell ill. To replace the stricken leader, Julian Smith selected his operations officer, Lieutenant Colonel David M. Shoup, who was spot promoted to colonel.

On 13 November, Task Force 53 set sail for Tarawa, but not until the following day did Julian Smith announce to his men the name of the island which they were to assault. "I know

[53] Smith, "Tarawa," p. 1169.

. . . ," read his message to the division, "you will decisively defeat and destroy the treacherous enemies of our country; your success will add new laurels to the glorious tradition of our Corps." [54] An intensive briefing for all hands followed this announcement, and the mighty task force bored onward toward its goal.

The first contact with the enemy came on 18 November, when a carrier plane sighted a Japanese aircraft far in the distance. On the following morning, a four-engine patrol bomber was picked up on radar, intercepted, and destroyed. No waves of bombers challenged Hill's ships as they began their final approach to Tarawa. At 2033, 19 November, USS *Ringgold*, the destroyer leading the task force, picked up Maina Atoll, and Hill's ships altered course to close with their objective.

Around midnight the fire support sections began steaming to their assigned stations. Transports crammed with Marines eased into unloading areas. Finally, at 0507 on 20 November, shore batteries on Betio opened fire, and the battle was underway.

General Holland Smith did not accompany the Tarawa expedition, for he had been ordered to embark in Admiral Turner's flagship, and the latter officer had taken personal command of the Makin task force. The admiral reasoned that since Makin was nearer to the Marshalls, Japanese surface units, if they chose to intervene, would probably strike the Northern Attack Force.

Events during the approach of the Northern Attack Force seemed to bear out Turner's theory. On 18 November a Japanese bomber attacked a group of LSTs but was beaten off by antiaircraft fire. Another bomber appeared the following afternoon and fell victim to Navy fighters. A night attack, delivered against the LSTs on 19 November, ended with the destruction of one enemy bomber and the escape of a second.[55] The Japanese, however, did not contest the final maneuvering of the task force, and at first light on 20 November the preliminary bombardment began.

The Japanese were never able to carry out the ambitious program of counterattacks against a Gilberts invasion force envisioned in their *Z Operation* plan. The carrier aircraft that were to have sortied from Truk and the Bismarcks had been severely depleted in a series of air battles over Rabaul in early November, as Admiral Halsey's and General MacArthur's fliers struck the enemy base in covering strikes for the landing at Bougainville.[56] Although it was not known at the time, Admiral Turner's Assault Force was insured against an enemy attack in any significant strength.

---

[54] Quoted in Stockman, *Tarawa*, p. 86.

[55] Four Japanese bombers failed to return from attacks made on the 19th and 10 from attacks mounted on the 20th. *Japanese Gilberts Comments*.

[56] For the story of this significant series of aerial assaults see Shaw and Kane, *Isolation of Rabaul*, pp. 481–486.

The Assault on Betio[1]

PREPARATIONS PRIOR TO H-HOUR

The transports halted at approximately 0355,[2] and the Marines of Combat Team 2 began groping down the sides of their ships toward the LCVPs waiting below. The troopships, victims of an unexpectedly strong current, had halted in the wrong area and masked certain of Admiral Hill's fire-support ships. The task force commander at 0431[3] ordered the transports to stop disgorging the troops and steam northward to their proper positions. As the larger vessels glided off into the night, the landing craft attempted to follow, but some of the LCVPs became separated from their assigned ships. Rounding up these strays further delayed unloading, subsequent transfer of men from landing craft to amphibian tractors, and the final formation of the assault waves.

From 0507, when enemy shore batteries first opened fire, until 0542, American warships attempted to reduce these troublesome batteries and neutralize known enemy positions. The naval guns then fell silent to enable carrier planes to scourge the objective. Admiral Hill ceased firing to prevent possible collisions between shells and planes as well as to allow the dust raised by explosions to settle before the pilots began diving toward their targets. Unexpectedly, the aircraft failed to appear. One explanation for this failure is that the request for a dawn attack may have been misunderstood, with the result that the strike was scheduled in-

[1] Unless otherwise noted, the material in this chapter is derived from: *VAC AR; TF 53 AR;* 2d MarDiv Rpt on GALVANIC, dtd 22Dec43, hereafter *2d MarDiv OpRpt;* 2d MarDiv D-3 Jnl, 19Nov-4Dec43, hereafter *2d MarDiv D-3 Jnl;* 2d MarDiv ADC Jnl, 19-21Dec43; 2d Mar Rpt of Ops, Betio Island (including 1/2, 2/2, 3/2, 2/8, WpnsCo, and H&S Co ARs), dtd 21Dec43, hereafter *2d Mar Op Rpt;* 2d Mar UJnl, 12-24Nov43; 8thMar SAR (including 3/8 AR), dtd 1Dec43, hereafter *8th Mar SAR;* 18th Mar CbtRpt (including 1/18, 2/18, and 3/18 CbtRpts), dtd 23Dec43, hereafter *18th Mar CbtRpt;* 10th Mar Rpt of Ops, Tarawa, (including 1/10, 2/10, 3/10, 4/10, and 5/10 Notes on or Rpts of Ops), dtd 22Dec43, hereafter *10th Mar OpRpt;* 2d PhibTracBn SAR, dtd 23Dec43, hereafter *2d PhibTracBn SAR;* 2d TkBn SAR, dtd 14Dec43, hereafter *2d TkBn SAR;* BGen Merritt A. Edson, "Tarawa Operation," lecture delivered at MCS, Quantico, Va., 6Jan44, cited hereafter as *Edson Lecture;* Stockman, *Tarawa;* Johnston, *Follow Me;* Isley and Crowl, *Marines and Amphibious War;* Morison, *Aleutians, Gilberts, and Marshalls.*

[2] *TF 53 AR,* Anx A, p. 18. The times for various actions of the task force contained in this document have been accepted as accurate. Stockman, *Tarawa,* p. 11, maintains that disembarkation had begun by 0320.

[3] Crowl and Love, *Gilberts and Marshalls,* holds that this maneuver took place at 0550. Other sources indicate that the transports actually shifted position sometime between 0430 and 0510.

stead for sunrise. This seems logical, for the fast carriers, from which the planes were to be launched, had been excused from the Efate rehearsals where any misinterpretation of orders would have come to light. Another version maintains that the principal commanders of Task Force 52 had agreed to a strike at 0610 because pilots diving from sun-filled skies toward the darkened earth could not locate their targets. This change, the account continues, was incorporated into the overall plan for both task forces, the information was passed to the carrier pilots, but somehow word did not reach Admiral Hill at Efate. Since the planes materialized over both Makin and Tarawa within a few minutes of sunrise, this too seems plausible.[4] Whatever the reason, the enemy was granted a brief respite from the storm of high explosives that was breaking around him.

Earlier that morning, when the *Maryland* had opened fire against Betio, the concussion from her main batteries had damaged her radio equipment, leaving Admiral Hill without any means of contacting the tardy planes. While Hill waited, Japanese gunners took advantage of the lull to hurl shells at the transports. The admiral scanned the skies until 0605, at which time he again turned his guns on the island. After the supporting warships had resumed firing, the transports, which had unloaded all troops in the initial assault waves, steamed out of range of the determined enemy gunners. At 0613, the aircraft finally appeared over Betio, naval gunfire again ceased, and for about 10 minutes the planes swept low over the objective, raking it with bombs and machine gun fire. Because the Japanese had taken cover in concrete or log emplacements, neither bomb fragments nor bullets did them much harm. Yet, the blossoming explosions looked deadly, and as the pilots winged seaward, the warships returned to their grim task of battering the island. This resumption of naval gunfire marked the beginning of the prelanding bombardment.

A few minutes after sunrise, the minesweeper USS *Pursuit*, carrying on board a pilot familiar with the atoll, began clearing the entrance to Tarawa lagoon. Astern of this vessel was another minesweeper, USS *Requisite*. Smoke pots laid by LCVPs were used to screen the sweeping operation.[5] Two destroyers, the USS *Ringgold* and *Dashiell*, waited off the entrance until a passage had been cleared. Fortunately for both *Pursuit* and *Requisite*, the pair of destroyers were in position to silence, at least temporarily, the shore batteries that had opened fire on the minesweepers. Once a path had been cleared, the *Pursuit*, assisted by an observation plane, began marking the line of departure, assault lanes, and those shoals which might cripple ships or small craft. In the meantime, the other minesweeper steamed out to sea to pick up the destroyers and lead them into the lagoon.

The enemy batteries, so recently silenced, again began firing as the destroyers came through the passage. A

[4] *Cf.* Crowl and Love, *Gilberts and Marshalls*, p. 219*n*; Morison, *Aleutians, Gilberts, and Marshalls*, p. 156.

[5] *Hill interview/comments.*

shell sliced through the thin skin of the *Ringgold*, penetrated to the after engine room, but failed to explode. Another dud glanced off a torpedo tube, whistled through the sick bay, and thudded to a stop in the emergency radio room. Moments after the guns of the ship had been unleashed at the supposed artillery position, a vivid explosion rocked the area. One of the destroyer shells must have touched off the enemy ammunition supply.

At 0715, the *Pursuit*, which had taken position astride the line of departure, switched on her searchlight to guide the waves of LVTs through the curtain of dust and smoke that hung between the minesweeper and the assembly area. While the *Ringgold* was fighting her duel with Japanese cannoneers, the *Pursuit* tracked the approaching waves on radar. The minesweeper reported to Admiral Hill that the assault waves were 24 minutes behind schedule and could not possibly reach the beaches by 0830, the time designated as H-Hour.[6] Lieutenant Commander Robert A. Macpherson, flying spotter plane off the *Maryland*, also reported that the LVTs could not meet the schedule, so Hill, in the *Maryland*, radioed instructions to postpone H-Hour until 0845.[7]

When the task force commander issued this order the erratic radios of the *Maryland* were still misbehaving. Though in contact with surface craft, Hill could not raise the aircraft that were scheduled to attack Betio just before H-Hour.[8] While his communications men were struggling with the balky radio sets, the admiral at 0823 received a report from Lieutenant Commander Macpherson that the amphibian tractors had just crossed the line of departure.[9] Since tests had indicated the lead LVT(2)s could make 4–4½ knots, he granted them an additional 40 minutes in which to reach the beach and announced that H-Hour would be 0900.

At this point, carrier planes reappeared over Betio and began strafing the assault beaches, delivering what was supposed to have been a last-minute attack. The cessation of main battery fire on the *Maryland* enabled its support air control radio to reach the planes so that Hill could call off the premature strike. The aviators finally made their runs between 0855 and 0900.

While the fliers were waiting their turn, the task force continued blasting the island. Five minutes before H-Hour, Hill's support ships shifted their fires inland, the planes strafed the beaches, and at 0900 the bombardment, except for the shells fired by the two destroyers in the lagoon, was stopped.

Awesome as it had been, the preliminary bombardment did not knock out all the defenses. The coast defense guns had been silenced, many of the dual purpose antiaircraft weapons and antiboat guns had been put out of action, but most of the concrete pillboxes and emplacements protected by coconut logs and sand survived both bombs and

[6] USS *Pursuit, Requisite*, and *Ringgold* ARs, dtd 6, 13, and 1Dec43.

[7] *Hill interview/comments.*

[8] In regard to the communication setup on his flagship, Admiral Hill reported: "not only are the transmitters, receivers, and antenna so close to each other as to cause mutual interference, but several of the installations, particularly SAC [Support Air Control] equipment were made entirely inoperative during main battery gunfire." *TF 53 AR*, Anx A, p. 62.

[9] *Hill interview/comments.*

214-881 O-67—5

shells. Of major importance, however, was the effect of the preliminary bombardment on Japanese communications. According to some of the prisoners taken during the battle, the preinvasion shelling had ripped up the enemy's wire and forced him to rely on messengers. Since these runners often were killed or pinned down by bursting shells, few messages got through.

Betio has been compared in shape to a bird, whose legs were formed by the 500-yard pier that passed just beyond the fringing reef. On 20 November, as H-Hour drew near, the bird appeared lifeless, the plumage on its carcass badly charred. Colonel Shoup had decided to use three of his landing teams in the assault and hold one in reserve. Major Henry P. Crowe's 2/8, attached for this operation to the 2d Marines, was given the job of storming Beach Red 3. This objective was the bird's belly, the invasion beach that lay east of the long pier. Ordered to land on Crowe's right was Lieutenant Colonel Herbert Amey's 2/2. Amey was to attack Beach Red 2, the breast of the bird, which included the base of the pier and stretched 500 yards westward to an indentation in the shoreline. Major John F. Schoettel's 3/2 would land to the right of Amey's battalion, assaulting Beach Red 1, a crescent-shaped portion of the coast that measured about 500 yards in width and served as throat and lower bill for the bird of Betio. The legs, or long pier, were reserved for the 2d Scout-Sniper Platoon, which was to secure its objective immediately before the assault waves landed. In regimental reserve was 1/2 commanded by Major Wood B. Kyle. (See Map III, Map Section.)

THE LANDINGS

The precisely arranged waves of amphibian tractors that roared across the line of departure had difficulty in making headway toward the island. At the time, the slowness of the assault waves was blamed upon "overloading, wind, sea, and an ebb tide, together with poor mechanical condition of a number of the leading LVTs." [10] Students of the operation, as well as the men who fought at Betio, have since absolved the wind, sea, and tide of some of the responsibility for the tardiness of the assault waves. The time lost earlier in the morning when the transports had first shifted their anchorage could not be made up. Because they had missed the rehearsals, the drivers of the new LVT(2)s were not familiar with signals, speeds, and load limitations, a factor which slowed both the transfer of men from the LCVPs and the forming of assault waves. The waves had to dress on the slowest tractors, and fully loaded LVT(1)s could not keep up with the LVT(2)s. The older vehicles were not in sound enough mechanical condition to maintain even 4 knots during their long journey from assembly area to the assault beaches.[11]

While the assault waves were moving from the line of departure toward the Betio reef, Japanese shells first began bursting over the Marines huddled inside the amphibian tractors. These air bursts proved ineffectual, as did the

[10] CinCPac Monthly Rpt, Nov43, Anx E (CinCPac File, HistBr, HQMC).

[11] Isley and Crowl, *Marines and Amphibious War*, pp. 227–228; MajGen Thomas E. Watson ltr to CMC, dtd 17Jun47, hereafter *Watson ltr; Hill interview/comments.*

long-range fire of machine guns on Betio, and none of the tractors was damaged. Upon crossing the reef, the LVTs swam into a hail of machine gun and antiboat fire, but even so casualties among the troops were relatively light. Few of the LVTs failed to reach the beach.

The first unit to land on Betio was First Lieutenant William D. Hawkins' 2d Scout-Sniper Platoon,[12] a part of which gained the end of the pier at 0855. Hawkins, with engineer Second Lieutenant Alan G. Leslie, Jr., and four men, secured the ramp that sloped downward from the pier to the edge of the reef. Next, the platoon leader ordered the men who had remained in the boat to scramble up the ramp. When enemy fire began crackling around the gasoline drums that the Japanese had stored at the end of the pier, Hawkins waved the men back into their LCVP. With his four scouts and Leslie, who was carrying a flamethrower, he began advancing shoreward along the pier, methodically destroying or clearing anything that might shelter enemy snipers. Blazing gasoline from Leslie's weapon splattered against two shacks that were thought to be serving as machinegun nests, the flimsy structures ignited like twin torches, but unfortunately the flames spread to the pier itself. Although the gap burned in the pier by this fire would later handicap the movement of supplies, this difficulty was a small price to pay for driving the enemy from a position that gave him the opportunity of pouring enfilade fire into the assault waves.

After the Japanese on the pier had been killed, Hawkins and his handful of men rejoined the rest of the section in the LCVP and moved along the boat channel toward the island. Beyond the end of the pier, Hawkins tried unsuccessfully to commandeer an LVT to carry his men to the beach. In the meantime, the second boatload of scout-snipers was being held off the reef on order of a control officer. The platoon leader finally made contact with them, got hold of three LVTs, and started the entire platoon toward shore. Two tractor loads, Hawkins among them, landed in the proper place and reported to the regimental command post, but the third group came ashore on the boundary between Red 1 and Red 2 to join in the fighting there. The difficulties in getting ashore experienced by the 2d Scout-Sniper Platoon were typical of the Betio operation.[13]

Two of the assault battalions hurled against Betio made the last part of their shoreward journey unaided by naval gunfire. Dust and smoke screened the movement of the LVTs, so that the distance yet to be traveled could not be accurately gauged. At 0855, according to plan, all but two ships in Hill's task force lifted their fires to avoid striking either the advancing tractors or the planes which were beginning their final 5-minute strafing of the beaches. Out in the lagoon, however, the destroyers *Ringgold* and *Dashiell* continued to loft 5-inch shells into Red 3. These ships, whose officers

[12] A scout-sniper platoon from the 8th Marines also saw action at Betio. There is no record of the employment of a similar unit by the 6th Marines.

[13] H&S Co OpRpt of Sct-Sniper Plat, dtd 15Dec43, Encl J to *2d MarDiv OpRpt*.

DAMAGED LVTs *and the bodies of Marines killed during the landing on Betio are grim witnesses to the fury of D-Day. (USMC 63578)*

RUBBER RAFT *is used to float wounded men across the lagoon reef to LCVPs waiting offshore at Betio. (USMC 63454)*

were able to follow the progress of the tractor waves, did not cease firing until 0910.

The first assault battalion to reach its assigned beach was Major Schoettel's 3/2, which landed on Red 1 at 0910. On the right half of that beach, and at the extreme right of Combat Team 2, the Marines of Company I leaped from their LVTs, clambered over the log beach barricade, and began advancing inland. On the left of the beach, astride the boundary between Red 1 and Red 2, was a Japanese strongpoint, which raked Company K with flanking fire before that unit could gain the shelter of the barricade. After getting ashore, Company K was to have tied in with the troops on neighboring Red 2. Since the company commander could see no Marines in that direction, he made contact with Company I and did not attempt to advance toward his left.[14] During the next two hours, these assault companies of 3/2 would lose over half their men.

Little better was the lot of Company L and the battalion mortar platoon. These units, boated in LCVPs, grounded on the reef 500 yards offshore. While wading toward Red 1, Company L suffered 35 percent casualties.

The next battalion to touch down on Betio was Major Crowe's 2/8, which reached Red 3 at 0917, just seven minutes after the destroyers in the lagoon had ceased firing. The fire of these warships had kept the Japanese underground as the Marines neared the beach, and the enemy did not have time to recover from the effects of the barrage before the incoming troops were upon him. Two LVTs found a gap in the beach barricade and were able to churn as far inland as the airstrip before unloading their men. The other amphibians halted before the log obstacle, discharged their troops, and turned about to report to the control boats cruising in the lagoon. Of the 552 men in the first three waves that struck Red 3, fewer than 25 became casualties during the landing.

The most violently opposed landing was that made by Lieutenant Colonel Amey's 2/2. Company F and most of Company E gained Red 2 at 0922, but one platoon of Company E was driven off course by machine gun and antiboat fire and forced to land on Red 1. Although Company G arrived only three minutes behind the other companies to lend its weight to the attack, the battalion could do no more than carve out a beachhead about 50 yards in depth. Losses were heavy, with about half the men of Company F becoming casualties.

NOTHING LEFT TO LAND

Behind the first three waves of amphibian tractors, came two waves of LCVPs and LCMs carrying additional infantrymen, tanks, and artillery. When the leading waves had crawled across the reef, they discovered that the depth of water over this obstacle varied from three feet to a few inches. Since standard landing craft had drafts close to four feet, they were barred from approaching the beach. Infantrymen and pack howitzer crews had to transfer to LVTs or wade ashore with their weapons and equipment. The tanks

[14] Rpt of Capt James W. Crain, n.d., in Rpts of 2d MarDiv BnComdrs, dtd 22Dec43, hereafter *Rpts of 2d MarDiv BnComdrs.*

were forced to leave the LCMs at the edge of the reef and try to reach Betio under their own power.

The men who attempted to wade to the island suffered the heaviest casualties on D-Day. Japanese riflemen and machine gunners caught the reserve elements as they struggled through the water.[15] The only cover available was that provided by the long pier, and a great many men died before they reached this structure. During the movement to the beach, platoons and sections became separated from their parent companies, but junior officers and noncommissioned officers met the challenge by pushing their men forward on their own initiative. On D-Day, few reserve units reached Betio organized in their normal combat teams.

Like the reserve and supporting units, the battalion command groups were unable to move directly to their proper beaches. All the battalion commanders, each with a part of his staff, had embarked in landing craft which took position between the third and fourth waves as their units started toward the island. As was true with the assault waves, the least difficulty was encountered at Red 3, but establishing command posts on Red 2 and Red 1 proved extremely hazardous.

The first landing craft to slam onto the reef off Red 3 was that carrying Major Crowe, a part of his communications section, and other members of his battalion headquarters. On the way in, one of Crowe's officers told him that if things did not go well on the first day, the staff could swim back to the transport, brew some coffee, and decide what to do next. "Jim Crowe," recalled First Lieutenant Kenneth J. Fagan, "let out with his bull bellow of a laugh and said it was today, and damn soon, or not at all." [16] When his boat grounded on the reef, Crowe fitted action to these words, ordered his men to spread out and start immediately for the island, and reached Betio about four minutes after his assault companies. Such speed was impossible on Red 2 and Red 1, beaches that had not received a last-minute shelling from the pair of destroyers in the lagoon.

Off Red 2, Lieutenant Colonel Amey's LCM also failed to float over the reef, but the commander of 2/2 was fortunate enough to flag down two empty LVTs that were headed back to the transports. These amphibian tractors became separated during the trip toward the island, and the one carrying Amey halted before a barbed wire entanglement. The battalion commander then attempted to wade the rest of the distance, but after he had taken a few steps he was killed by a burst from a machine gun. Lieutenant Colonel Walter I. Jordan, an observer from the 4th Marine Division and the senior officer present, was ordered by Colonel Shoup to take command until Major Howard J. Rice, the battalion executive officer, could get ashore.[17] Although Rice was a mere 13 minutes behind the first assault waves, he was in no posi-

[15] Capt James R. Stockman, Notes on an Interview with Col David M. Shoup, dtd 26 May 1947, hereafter *Shoup-Stockman interview.*

[16] Quoted in *Watson ltr.*

[17] *Shoup interview/comments;* Rpt of LtCol Walter I. Jordan, dtd 27Oct43, in Rpts by SplObservers on GALVANIC, Encl G to *VAC AR,* hererafter *Jordan Rpt.*

tion to relieve Jordan of responsibility for 2/2. The executive officer was pinned down and out of contact with his unit, so Jordan retained command until he was relieved of this task by Shoup.

Though the beachhead held by Jordan's men was admittedly precarious, the most disturbing news came from neighboring Red 1, where 3/2 was in action. There the battalion commander, Major Schoettel, was unable to get ashore until late in the afternoon. At 0959, Schoettel informed Colonel Shoup that the situation on Red 1 was in doubt. "Boats held up on reef of right flank Red 1," said his next message, "troops receiving heavy fire in water." Shoup then ordered the battalion commander to land his reserve over Red 2 and attack westward. To this the major replied, "We have nothing left to land." [18]

Colonel Shoup and his regimental headquarters experienced difficulties similar to those that had plagued the battalion commanders. At the reef, Shoup happened upon an LVT which was carrying wounded out to the transports. The colonel had these casualties transferred to his LCVP, commandeered the amphibian tractor, and started toward the left half of Red 2. As the tractor neared the island, it entered a maelstrom of fire and a hail of shell fragments "started coming down out of the air. It was strong enough to go through your dungarees and cut you," Shoup recalled. Then, as the command group continued its way shoreward, he said, "a kid named White was shot, the LVT was holed, and the driver went into the water. At that point I said, 'let's get out of here,' moved my staff over the side and waded to the pier. From then on it was a matter of getting from the pier on down. You could say my CP was in the boat, then in the LVT, and then on the pier on the way in, but there was very little business conducted." [19] After determining what portion of Red 2 was in the hands of 2/2, Shoup established his command post on that beach at approximately 1200.

Even before he reached the island, Colonel Shoup kept a close rein on the operations of his command. At 0958, in the midst of his exchange of messages with Major Schoettel, he directed his reserve battalion, 1/2, to land on Red 2 and attack westward toward the embattled Marines on Red 1. Shoup's plan, however, was slow of execution, for only enough LVTs could be rounded up to carry Companies A and B. Company C had to wait until noon for transportation. While Major Kyle's battalion was moving toward Red 2, the leading waves of amphibian tractors drew heavy fire from the right hand portion of the beach. As a result, some of the vehicles veered from course to touch down on Red 1, and the 4 officers and 110 men that they carried joined in the fighting there. The remainder of the LVTs bored onward to the left half of Red 2, where the bulk of Kyle's command aided in expanding the beachhead. Not until the morning of the second day was the entire battalion ashore on Betio.

---

[18] LT 3/2 to CO CT 2, ser no. 27, CO CT 2 to LT 3/2, ser no. 28, and LT 3/2 to CO CT 2, ser no. 30, in *2d MarDiv D-3 Jnl.*

[19] *Shoup interview/comments.*

3/8 IS COMMITTED

At his command post in the *Maryland*, General Julian Smith was convinced that a foothold had been won, but he realized how vital it was that the attack be kept moving. Because the 6th Marines had been placed under the control of Turner, Julian Smith had but two battalion landing teams as his own reserve. These two units were Major Robert H. Ruud's 3/8 and 1/8 commanded by Major Lawrence C. Hays, Jr. The commanding general could select the regimental headquarters of the parent 8th Marines to control either or both of its battalions.

At 1018 on the morning of D-Day, Julian Smith radioed Colonel Elmer E. Hall, commanding officer of the 8th Marines, to send 3/8 to the line of departure where it would come under the tactical control of Colonel Shoup. Since Shoup was more familiar with the situation on Betio, it was logical that he, rather than Hall, should determine where this portion of the reserve would be landed. Ruud's battalion became a part of Shoup's command at 1103 and was promptly ordered to land on Red 3 in support of Crowe's 2/8.

Since Shoup and his party were moving alongside the pier at this time, he could watch what was happening to the incoming Marines. As soon as their boats grounded on the reef and the ramps were lowered, Ruud's men started wading toward the island. Landward of the reef, the water proved deep, in places well over a man's head. Some Marines, weighted down by the equipment, plunged into deep water and drowned; others were killed by enemy bullets and shell fragments. Only 100 men from Ruud's first wave, approximately 30 percent of the total, survived the ordeal to set foot on Betio.

From the pier, Shoup and his staff signalled frantically to the men of the second wave, directing these troops to seek the shelter of the pier. This structure, however, offered little protection, so the toll claimed by Japanese gunners continued to mount. "Third wave landed on Beach Red 3 were practically wiped out," reported Ruud—who had lost radio contact with Shoup—to Hall. "Fourth wave landed on Beach Red 3," he continued, "but only a few men got ashore and the remainder pulled away under heavy MG and 37mm fire."[20] Shortly after the fourth wave landed, the battalion commander received a message to "Land no further troops until directed." The remainder of the battalion gathered off the end of the pier and was finally ordered in about 1500. By 1730, all of 3/8 was ashore, and, on Shoup's orders, Rudd deployed one of his companies to plug a gap directly inland from the pier between 2/8 and 1/2.[21] Company K, which had landed in the first waves, was already attached to Crowe's battalion and continued to serve with 2/8 through the rest of the battle.

SUSTAINING MOMENTUM

In spite of the light losses suffered by the LVTs that carried the assault waves, the number of amphibian tractors available to the division dwindled

[20] LT 3/8 to CO CT 8, ser no. 88, *2d MarDiv D-3 Jnl.*

[21] Col Robert H. Ruud comments on draft MS, dtd *ca.* 10Aug62.

rapidly as the day progressed. Some were destroyed while bringing supplies or reinforcements to the island, others were so badly damaged that they sank upon reaching the deep water of the lagoon, and a few either broke down or ran out of gas. To reduce losses to a minimum, the LVTs had to be restricted to the boat channel that paralleled the long pier, but even so Japanese gunners still managed to cripple some of the incoming amphibians.

General Julian Smith realized that he could not afford a stalemate at the beaches. Strength would have to be built up rapidly and the attack pressed vigorously if Betio were to be taken with a minimum of losses. Yet, after the assault waves had gained a foothold, the operation bogged down, for the reef effectively barred landing boats, and the number of LVTs available for duty was fast diminishing. Since the battle was raging only a few yards inland, the reserve units that attempted to wade ashore were under fire from the moment they stepped into the water. Those who survived the trek from reef to beach found themselves in the thick of the fight as soon as they set foot on the island.

Reserve units became so disorganized while wading toward shore that battalion and even company control was virtually impossible. The resultant confusion was offset by the grim determination of the individual Marines, who simply kept coming in spite of all the enemy could hurl at them. For a distance of 400 yards, Japanese machine gunners or riflemen grazed the water with streams of bullets. The only cover available to the Marines was that afforded by the pier, and even from here they had no opportunity to fight back. All the attackers could do was take their punishment and keep moving. Many Marines were hit in the water, but the survivors waded onward, moving doggedly to join their comrades ashore.

During the early morning, the situation on Betio was literally cloudy, for the explosions of shells and bombs had sent a column of dust and smoke towering above the island. As the morning wore on, the smoke from burning emplacements and buildings continued to cloak parts of the island so that it was impossible, even from the air, to see much of the island at one time. Neither Julian Smith nor Colonel Shoup could observe much of the action ashore. The general had remained in the *Maryland*, the best place, given adequate communications, from which to control his division.[22] The movements of the commander ashore were restricted and his communications, especially with the unit on Red 1, unreliable. Colonel Shoup, however, was by no means pinned down. "Once ashore," he recalled, "I was never off my feet for over 50 hours, standing for the most time protected by an enemy pillbox with 26 live Japs therein." [23]

By noon, the situation ashore began to come into sharper focus. Colonel Shoup made contact with his subordinates, and requests for medical supplies, ammunition, and air support

[22] The transport group commander offered the opinion that Julian Smith "could have had much more ready communication means (radio, boats, etc.) than were available to him on the *Maryland*" on board the comparatively close-in transport flagship. *Knowles ltr.*

[23] *Shoup-Stockman interview.*

began trickling back to the *Maryland*. Julian Smith also profited from observation flights made during the afternoon by naval pilots. Little information, however, could be had concerning the battle on Red 1.

THE FIGHTING ASHORE

A source of grave concern throughout the morning was the fate of 3/2 on Beach Red 1. Actually, Major Schoettel's men, though isolated from the other Marines on Betio, had fared better than Julian Smith suspected. The assault companies had received a severe scourging as they started moving inland, but Major Michael P. Ryan, commander of Company L, managed to organize an effective fighting force from remnants of several units. By midafternoon, his contingent could boast portions of every company of 3/2, four platoons and part of the headquarters of 2/2, as well as the 113 officers and men of 1/2 who had been driven off course during their attempt to reach Red 2. Among the members of 2/2 who ended up on Red 1 was Major Rice, the battalion executive officer, who had with him a usable radio. This set provided Ryan his only link with Colonel Shoup's command post.

During the afternoon, Ryan's Marines consolidated their beachhead on Betio's beak, clearing an area 500 yards deep and 150 yards wide. The farthest penetration made by this conglomerate command was to the antitank ditch 300 yards from the south coast of the island, but this advanced position could not be held with the number of men at Ryan's disposal. For this reason he pulled back to within 300 yards of the tip of the beak and dug in for the night.

The key to Ryan's success was the pair of medium tanks that reached Red 1 about 1130. All Shermans employed at Betio were from the Company C, I Marine Amphibious Corps (IMAC) Tank Battalion, the entire company having been attached to the 2d Marine Division for GALVANIC. Company C joined the division at Efate and made the voyage to Tarawa in the USS *Ashland*, a new LSD.

On D-Day morning, a total of six medium tanks started toward Red 1 in LCMs, but the coxswains of these craft could find no place to unload. Fortunately, Major Schoettel happened upon these boats as they were circling off the reef and ordered them to run up on the reef and lower their ramps. The Shermans then nosed into the water to begin a 1,200-yard journey to the island. Reconnaissance parties waded in front of the tanks, carefully marking potholes with flags, so none of the tanks drowned out before reaching the beach. As the lumbering vehicles approached a gap already blown in the log barricade, the platoon leader saw to his horror that the coral sands were littered with wounded and dead. Rather than risk crushing those Marines who were still alive, he led the platoon back into the water, drove to a position off Green Beach, and waited for engineers to pierce the barrier. During this second move, four tanks wandered into potholes, drowning out their engines.[24] Both the surviving Shermans were hit during Ryan's advance. One was gutted by flames, but the other, with only its bow machine gun still in work-

[24] *Ibid.*

ing order, was used to protect the flank of the beachhead during the night.

Ryan's men held the beak of the Betio bird, but the head, throat, back, and most of the breast were controlled by the enemy. The nearest American troops were elements of 1/2 and 2/2 fighting on that part of Red 2 near the pier, an area some 600 yards from the Red 1 perimeter. That part of the line manned by 1/2, originally Shoup's reserve, extended from a point about 350 yards inland from the base of the pier along the triangular plot formed by the runway and west taxiway then veered toward the beach. The area between 1/2 and the edge of the water was the zone of 2/2, the battalion that had stormed Red 2.

Both medium tanks and artillery reached Red 2 before D-Day had ended. Three Shermans that had landed on Red 3 crossed the boundary and halted in a previously selected assembly area. This trio of tanks supported 2/2 in its advance toward the runway by rolling up to pillboxes and firing at point blank range through the openings in these structures. Two of the Shermans were knocked out, but one of these was retrieved on the following morning.

The artillery that arrived on Red 2, 1/10 commanded by Lieutenant Colonel Presley M. Rixey, had first been destined for Red 1. A member of Colonel Shoup's command post group, Rixey had landed before noon. In the meantime, his 75mm pack howitzers and their crews were boated at the line of departure awaiting further orders. During the afternoon it became apparent to both Shoup and Rixey that Red 1 was no place to land artillery, and they finally decided to bring the battalion ashore over Red 2. Since the boats carrying the unit could not cross the reef, LVTs had to be found. Two gun sections, one from Battery A and one from Battery B, were transferred to amphibian tractors and at dusk ordered ashore. Also ordered to land were three howitzers and crews of Battery C, elements that were believed to be in LVTs but which had not yet actually been shifted from their original landing craft. The two sections in the tractors moved rapidly to Red 2. The other three were boated to the edge of the pier where the artillerymen plunged into the water and began wading ashore carrying their dismantled pack howitzers. None of the guns reached the beach until after dark, and the crews could do little more than wait, ready to move into position at dawn.[25]

Inland of the pier, at the dividing line between Red 2 and Red 3, responsibility passed to elements of 3/8 and Crowe's 2/8. Its initial blow at the midsection of Betio had carried a part of Crowe's battalion into the triangle formed by the runway and taxistrip, but on the left flank his men collided with a powerful strongpoint near the base of the Burns-Philp pier. During the afternoon, some 70 Marines from 3/8 were sent into the triangle to hold that sector of the line. A group of men, survivors of various battalions whose weapons had been lost or ruined by water, were found crouching under the Burns-Philp pier, led ashore, rearmed, and fed into the battle being waged on Crowe's left flank.[26]

Throughout D-Day, the Marines of

[25] *Shoup interview/comments.*

[26] *Watson ltr.*

2/8 attempted to batter their way through the fortifications inland of Burns-Philp pier in order to advance eastward along Betio's tail. Four medium tanks from the Company C, IMAC Tank Battalion, threw their weight and firepower into this effort but to no avail. One Sherman was destroyed by a friendly dive bomber, a second bulled its way into an excavation used by the enemy as a fuel dump and was burned when an American plane set the gasoline aflame, and a third was disabled by Japanese gunners. Although damaged by an enemy shell, the fourth tank continued to fight.

REBUILDING A RESERVE

General Julian Smith's decision to land 3/8 left him with but a single landing team in division reserve. Early in the afternoon, it began to appear as though it might be necessary to land 1/8, the last of the reserve, to help the five battalions already in the fight. If this were done, the general would be left with no reserve except for his support group, made up of elements of the 10th Marines (artillery), the 18th Marines (engineers), Special Troops, and Service Troops. In short, he would be forced to rely upon an assortment of specialists in case of an emergency.

There was present, however, an organized unit which might spell the difference between victory and defeat. This was the 6th Marines, designated as corps reserve and under the control of Admiral Turner. Having informed Holland Smith of the situation at Betio, the division commander at 1331 requested the release of the 6th Marines to his control. Admiral Hill seconded Julian Smith's request and within 50 minutes Turner's message of approval arrived. "Meanwhile," commented Julian Smith, "consideration was being given to a plan to organize the support group into provisional battalions." [27]

Once the 6th Marines had been released to him, Julian Smith felt it safe to land 1/8. At 1343, Colonel Hall's regimental headquarters and his remaining landing team, commanded by Major Hays, was ordered to proceed to the line of departure and wait there for further orders. The division commander then asked Colonel Shoup to recommend the best site for a night landing by this battalion.

This message concerning Hall's unit never reached Colonel Shoup, another of the communications failures so typical of the Tarawa operation. The radios of the *Maryland* had proved balky, and the portable sets carried ashore by the assault troops were little better. Water, shell fragments, bullets, and rough handling played havoc with communications equipment, but some radios were repaired with parts pirated from other damaged sets. Both the TBYs and the MUs, the latter light-weight hand sets, were exceptionally vulnerable to water damage, and the TBX the more durable and somewhat waterproof battalion radio, was so heavy that it could hardly be called portable.

Colonel Hall's headquarters and 1/8, "cramped, wet, hungry, tired, and a large number . . . seasick," [28] waited throughout the afternoon at the line of departure. At 1625, Julian Smith sent

[27] Smith, "Tarawa," p. 1173.

[28] LtCol Rathvon McC. Tompkins ltr to CMC, dtd 13Jun47.

a message ordering Hall to land on the north shore of the extreme eastern end of the island. These last uncommitted elements of the 8th Marines were to have gone ashore at 1745 and to have attacked to the northwest, but the orders failed to reach the regimental commander.

To observe the general progress of the battle, a scout plane was launched from the *Maryland* at 1548. Colonel Merritt A. Edson, division chief of staff, and Lieutenant Colonel Arnold F. Johnston, the operations officer, contacted the plane and asked the fliers to report any movements in the area where 1/8 was waiting. As the observation craft circled overhead, an artillery battery from 1/10 started toward Red 2. Since Hall was believed to have received his orders, the artillerymen were mistaken for a portion of 1/8. The thought that Hall was landing on the wrong beach caused consternation at division headquarters, but his supposed position was duly plotted on the situation map. Not until midnight did the division staff discover that Hall's command was still waiting on the line of departure.

THE FLOW OF SUPPLIES AND INFORMATION [29]

In assessing the work of his shore party, Lieutenant Colonel Chester J. Salazar admitted that carefully prepared and basically sound standing operating procedure had to be abandoned during the Tarawa operation. Elements of the shore party had difficulty in finding the combat units to which they were assigned. Salazar's demolitions men and bulldozer operators were needed to blast or bury enemy positions, and the assault battalions could not spare riflemen to serve as stevedores on the crowded, hard-won beachheads. Finally, there were not enough LVTs to move supplies directly to the battalions from the ships offshore.

With Colonel Shoup throughout the morning of D-Day was Lieutenant Colonel Evans F. Carlson, leader of the previous year's Makin raid, who had been assigned to GALVANIC as an observer. Because of the continuing difficulty in keeping radio contact with division, Shoup at 1230 asked Carlson to make his way to the *Maryland* and sketch for Julian Smith an accurate picture of what was happening ashore. The commander of Combat Team 2 could then be certain that higher headquarters knew his basic plan for the conquest of Betio—to expand southward and to unite the beachheads before attempting a final thrust. The division could best help by landing reserves on Red 2. As the two men parted, Shoup told Carlson, "You tell the general and the admiral that we are going to stick and fight it out." [30]

Before starting toward the lagoon, Carlson noticed some Marines from Ruud's 3/8 clinging to the pier and unable to get into the fight. With

[29] Additional source for this section include: Rpts of LtCol Chester J. Salazar and Maj George L. H. Cooper, dtd 22Dec43, in *Rpts of 2d MarDiv BnComdrs;* Rpt of LtCol Evans F. Carlson, dtd 27Oct43, in Rpts by SplObservers on GALVANIC, Encl G to *VAC AR*, hereafter *Carlson Rpt;* MajGen Leo D. Hermle ltr to CMC, dtd 9Jun47; Maj Ben K. Weatherwax ltr to CMC, dtd 17Jun47.

[30] *Shoup-Stockman interview.*

Shoup's permission, Carlson interrupted his journey to bring several LVT-loads of able-bodied infantrymen to the island, returning each time with wounded men whom he had transferred to boats at the reef. This done, he left his tractor at the reef, embarked in an LCVP, and at 1800 reported to Julian Smith in the *Maryland*.

Early in the afternoon the division commander ordered Brigadier General Leo D. Hermle, assistant division commander, to prepare to land his command post group on order. General Hermle was told at 1343 to go to the end of the pier, form an estimate of the situation, and report his findings to General Julian Smith. On the way to the pier he attempted to learn the location of Shoup's command post but could not contact the regimental commander by radio. At 1740, Hermle reported that he had reached the pier and was under fire. He tried a short time later to radio to the *Maryland* details of the action ashore, but again he was victim of a communications failure. He then entrusted the information to a messenger.

While he was on the pier, Hermle managed to establish intermittent radio contact with Shoup and Crowe, who informed him that ammunition and water were desperately needed ashore. Since many Marines from 3/8 had taken cover beneath the pier, Hermle had enough men available to organize carrying parties to bring these vital items to the island. Supplies, which kept arriving by boat throughout the night, were unloaded by the carrying parties and manhandled to the beach. En route to Betio, the Marines doing this important job had to wade through a 50-yard area that was exposed to Japanese fire.

In addition to the able-bodied men who were formed into carrying parties, a number of wounded Marines had gained protection of the pier. Captain French R. Moore (MC), USN, assistant division surgeon and a member of Hermle's party, had the wounded collected, given first aid, and evacuated in landing craft that had finished unloading supplies. The captain later returned to the transport area with a boatload of seriously wounded men.

General Hermle's radio link with Shoup and Crowe was severed early in the evening. About 1930, the assistant division commander sent Major Rathvon McC. Tompkins and Captain Thomas C. Dutton to find Shoup's command post and learn where and when the regimental commander wanted the reserves to land. The two officers, after working their way across a 600-yard strip of coral swept by enemy machine gun fire, reached their goal. They found out the needed information, but it was 0345 before they could report back to Hermle.

Although he had obtained answers to Julian Smith's questions, Hermle lacked a rapid means of communicating this intelligence to the division commander. For this reason, the assistant division commander and his party ventured into the lagoon to use the radio on the destroyer *Ringgold*. Word that Shoup wanted 1/8 to land near the pier on Red 2 was dispatched to the *Maryland* at 0445.

General Hermle next was ordered to report to Julian Smith in the battleship. Here he learned that at 1750 an order had been issued giving him com-

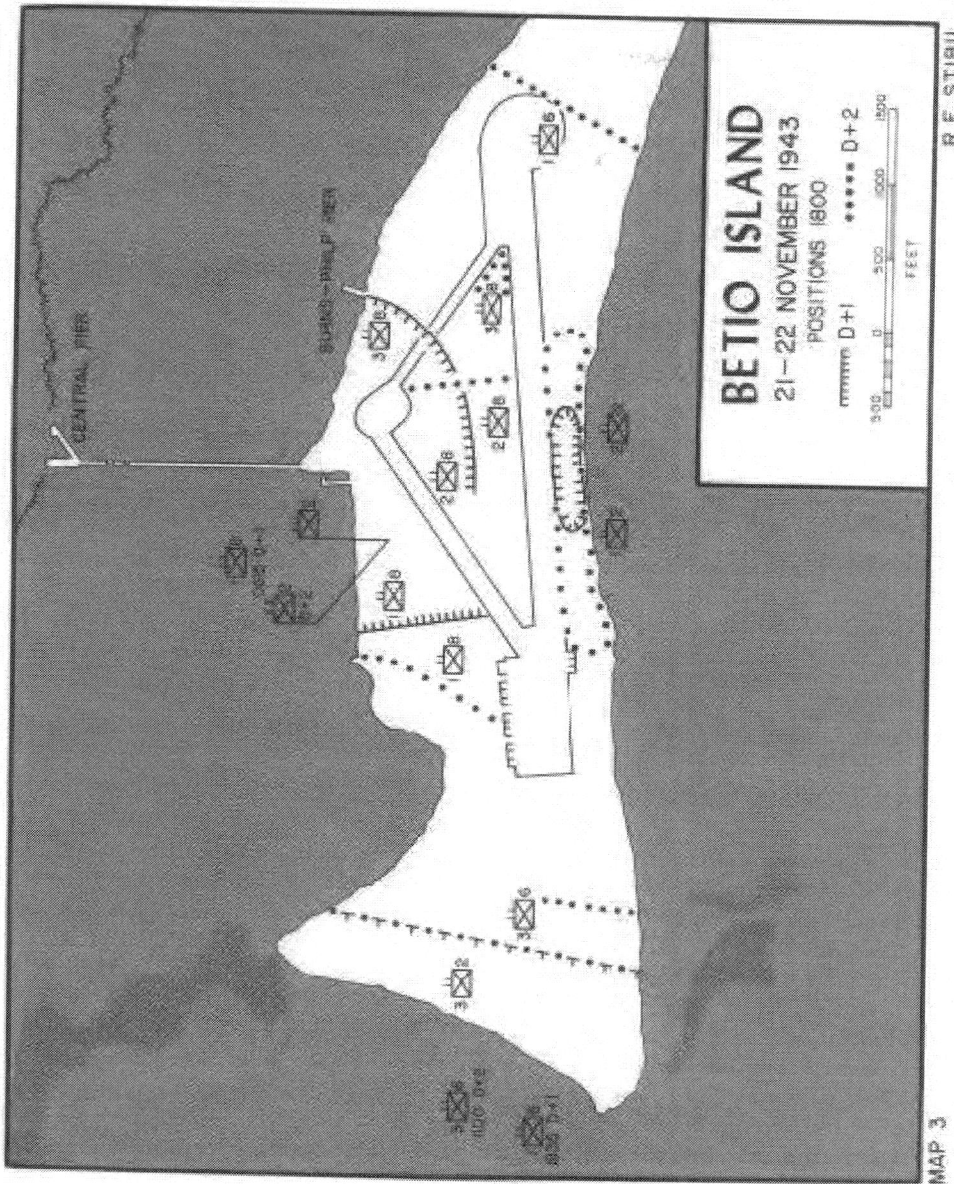
BETIO ISLAND
21–22 NOVEMBER 1943
POSITIONS 1800
D+1
D+2
FEET
CENTRAL PIER
BURNS-PHILIP PIER
R.F. STIBIL

MAP 3

mand of the troops ashore. Because of a communications failure, the message had gone astray. Command ashore was to remain the responsibility of Colonel Shoup.

The transports had been unloading water, plasma, ammunition, and other supplies throughout the day, but judging from the requests that continued to pour in from the island, few of these articles were finding their way to the front lines. Captain Knowles, commander of the transport group, who shared with the division supply section responsibility for coordinating the logistical effort, directed the Assistant D-4, Major Ben K. Weatherwax, to go to Betio and find out what had gone wrong. The major was to contact either General Hermle or Colonel Shoup.

Weatherwax left the transport USS *Monrovia* at 2100 and went to the *Pursuit*, where he obtained directions for landing. He approached Betio by way of the pier, climbing out of his boat on the beach side of the gap burned in the structure by Hawkins' men. Had the major landed at the end of the pier instead of following the boat channel, he would have met General Hermle and learned the details of the logistical situation. As it was, he reached the beachhead, made his way to Shoup's command post, and there learned that the troops ashore needed still more of the types of supplies that already had been sent them. Weatherwax then encountered the same problem that had plagued Hermle—inability to reach the *Monrovia* by radio. He finally went along the pier until he found a boat and arrived at the transport just before dawn. There, he informed Captain Knowles of the supply situation and of the need for getting additional tanks ashore. The transport group commander gave Weatherwax authority to order in any boat with a tank aboard to any beach where there was a good chance for the armor to land.[31]

THE EVENING OF D-DAY

As daylight waned on 20 November, the position of the Marines on Betio seemed precarious. The front lines were perilously close to the beach, and the enemy had effectively dammed the torrent of supplies that was to have sustained the embattled riflemen. Small boats dashed to the end of the pier and unloaded. Carrying parties managed to keep a trickle of supplies moving toward the island, an effort that was supplemented by the work of the surviving LVTs. During the afternoon these tractors had carried water, ammunition, and medical supplies directly to the beaches. In the meantime, the transports were unloading as rapidly as possible. Soon the waters around the line of departure were dotted with landing craft waiting for an opportunity to dart toward the pier and unload their cargoes.

The picture ashore seemed equally confused, with the assault battalions confined to small, crowded areas. Forward progress had been slow, a matter of a few feet at a time. It was worth a man's life to raise his head a few inches. Yet, a Marine could not fire his weapon unless he exposed himself, however briefly. Shoup's men did this

---

[31] *Knowles ltr.*

and even more. "A surprising number . . . ," the colonel would recall, "displayed a fearless eagerness to go to the extreme for their country and fellow men." [32]

At dusk, the Marines held two separate portions of Betio Island. On the right, Major Ryan's composite unit, isolated from the remainder of Colonel Shoup's command, had withdrawn to a compact perimeter on the island beak. Another perimeter fanned out from the base of the long pier. The segment nearest Ryan's lines was manned by troops from 1/2 and 2/2 and curved from the water into the triangle formed by runway and taxiway. Within this triangle, the left-hand portion of the line was held by 3/8, while 2/8 had responsibility for the sector facing the strongpoint at the base of Burns-Philp pier. The larger perimeter was not a continuous line, for this beachhead was defended by small groups of Marines who had taken advantage of whatever cover they could find.

Most of the Marines on Betio prepared for the night with the uneasy feeling that a Japanese counterattack was inevitable. On the control vessels and transports there was a restless feeling that at any moment reports would come flooding in telling of a Japanese attempt to hurl the invaders into the sea. In the *Maryland*, the division staff strained to pick up the sounds of rifle fire that would herald the enemy attack. Silence reigned. Marine fire discipline was superb; few shots were wasted on imagined targets. Enemy weapons too were quiet, for the expected attack never came.

According to Julian Smith, Admiral Shibasaki lost the battle by failing to counterattack on that first night, for never again would the beachhead be so vulnerable. Shibasaki's failure was probably due to a collapse of his communications. The fact that few field message blanks were captured during the course of the battle seems to indicate a reliance on wire communication. Naval gunfire ripped out the carefully strung wire, and the Japanese command post was isolated from troops it was to direct.[33] Important as this lapse in control may have been, it was the combat effectiveness of Shoup's Marines, men who overcame incredible obstacles to maintain cohesive fighting teams, that promised failure to any enemy assault.

[32] Gen David M. Shoup, "Some of My Thoughts" (Gen David M. Shoup Personal Folder, HistBr, HQMC). This folder contains notes, impressions, and reminiscences dating from early in the general's career until his appointment as CMC.

[33] Smith, "Tarawa," pp. 1173–1174; Major Eugene P. Boardman ltr to CMC, dtd 16Jun47.

214-801 O-67—6

Amphibious Victory[1]

1/8 LANDS

Colonel Hall, his regimental headquarters group, and the men of 1/8 had spent the night at the line of departure, waiting in vain for orders to land on Betio. Although division had issued such an order, on the afternoon of D-Day, the message had not reached the regimental commander. Finally, at 0200 on the morning of 21 November, Hall was contacted and told to report the position of 1/8 and the condition of its men. He replied that his Marines, in boats near the control vessel, were "resting easy," a surprisingly cheerful description of men that had spent over 12 hours in bobbing landing craft.[2] In 2 1/2 hours, Hall's radio again came to life, as division headquarters directed him to arrange with the *Pursuit* for a new line of departure and to land his troops at 0900 on D plus 1.

While Hall was preparing to make this move, General Hermle radioed division headquarters that Colonel Shoup wanted 1/8 to land on Beach Red 2. The general's message was sent at 0513, and a few minutes later, Hall was told to start at once toward Red 2. Once ashore, 1/8 was to attack westward toward Red 1.

The LCVPs carrying the first waves of 1/8 grounded on the reef at 0615, and the men began wading the 500 yards to shore. En route, the troops were hit from both flanks by machine gun fire. Casualties were severe, and the survivors were badly disorganized. But by 0800, Major Hays, commander of 1/8, had over half his men ashore and under cover. He then reported to Shoup, who told him to reorganize the battalion for an attack westward toward Ryan's beachhead. This action would have to be fought with machine guns, rifles, and grenades, for Hays' battalion had lost its demolitions and flamethrowers during the passage from reef to beach.

Of great assistance to the landing of 1/8 were Rixey's pack howitzers, which had been organized into a five-gun composite battery. During the night, a bulldozer had thrown up an earthen embankment on the exposed sides of the artillery position to protect the cannoneers from small arms fire. Early on the morning of D plus 1, two guns were moved from this makeshift

[1] Unless otherwise noted, the material in this chapter is derived from: *2d MarDiv Op Rpt; 2d MarDiv D-3 Jnl; 2d Mar OpRpt;* 2d Mar Unit Jnl, 12–24Nov43; 6th Mar SAR (including 1/6 and 3/6 SARs), dtd 20Dec43, hereafter *6th Mar SAR; 8th Mar SAR; 18th Mar CbtRpt; 10 Mar OpRpt; 2d PhibTracBn SAR; 2d TkBn SAR;* 2/6 Narrative Account of Ops, 21–29Nov43, n.d.; *Edson Lecture;* Stockman, *Tarawa.*

[2] *8th Mar SAR,* p. 1.

cover to fire directly into a pair of blockhouses located on the boundary between Red 1 and Red 2. Using high explosive ammunition with delay fuses from a range of about 125 yards, the Marine howitzers succeeded in penetrating these log and coral structures and in temporarily silencing machine guns that were sited to cut down Hays' incoming troops.[3] By 0800, after this mission had been fired, all five howitzer sections were in position to support Shoup's renewed offensive.

Colonel Hall, commanding officer of the 8th Marines, reached Shoup's command post about 1400. Although he was now the senior officer ashore on Betio, Hall did not assume command. In his opinion, nothing would have been gained from such an action, for Shoup, "who was doing very well and was the division's selected commander, was in a position to know more about what was going on ashore. . . ."[4] The senior colonel aided his junior by placing his own communications equipment at Shoup's disposal.

EXPANDING THE BEACHHEAD

Colonel Shoup's plan for the second day called for 1/8 to fight its way westward toward the Red 1 beachhead, while 1/2 and 2/2 drove across the island. Crowe's 2/8 was to reduce the enemy pocket at the base of Burns-Philp pier, and 3/2 was given the task of securing Green Beach. By 1200, Hays had his battalion ready to launch its thrust. His unit relieved the composite force that was holding the western segment of the beachhead and, with the aid of a medium tank, attempted to forge ahead. This lone tank could not shatter all the fortifications in the path of 1/8, and the attackers were unable to gain momentum. The battalion made little progress, killed few Japanese, and suffered light casualties, for the troops lacked the tools with which to destroy Japanese positions. "Hays had no flamethrowers, the most important weapon we had on Tarawa," commented his regimental commander, "and without them a unit had little chance to advance."[5] At dark the Marines paused, ready to continue the attack on the following morning.

The task assigned Major Kyle's 1/2 on the morning of the second day was to strike across the airstrip to the ocean shore. Two of Kyle's companies were located in the triangular area formed by the runway and taxiways, and the third was on their right facing to the west. Early in the day, the battalion commander had reinforced Company C, the unit on the right, with a platoon of water-cooled .30 caliber machine guns. In the meantime, members of Weapons Company headquarters had found a pair of .50 caliber machine guns on the beach. These weapons, manned by volunteer crews, joined the .30 caliber guns to give Company C

[3] The commander of 1/10 recalls that he "personally sited the two pieces in view of their being masked by disabled LVTs. We, of course, desired to use those LVTs as partial cover for personnel manning the sections as small-arms fire was sweeping across the area at the time." BGen Presley M. Rixey ltr to ACofS, G-3, dtd 10Aug62.

[4] Capt James R. Stockman, Notes on an Interview with Col Elmer E. Hall, dtd 10Jun47.

[5] *Ibid.*

still greater firepower. Neither reinforcements nor supplies could be sent to Companies A and B in the triangle, for enemy machine guns had been moved into position to graze the taxiway behind those units, thus isolating them from Kyle and the remainder of his command.

When 1/2 began its advance, Company C and Kyle's headquarters were prevented by the fire of the Japanese machine guns from crossing the airstrip, so the other two companies had to attack on their own. Assisted by elements of neighboring 2/2, the battalion on Kyle's left, Companies A and B reached the south coast. The Marines occupied an abandoned position about 200 yards long, a series of entrenchments that lay between two Japanese strongpoints. No sooner had the attackers gained the cover of the trenches than the Japanese attacked from the east, a blow that was beaten back at the cost of heavy Marine casualties.

Lieutenant Colonel Jordan, the observer who had taken command of 2/2 when Lieutenant Colonel Amey was killed, was not in contact with that portion of his command farthest from the lagoon shore. After runners had failed to return with news of these isolated units, Jordan reported this breakdown of communications to Shoup and was instructed to move his command post to the south coast of the island. Upon reaching the recently established perimeter, Jordan took command over remnants of three of his own companies, 50 to 75 men in all, plus Companies A and B of Kyle's battalion, an additional 135 Marines. Now that he had regained control, the commanding officer of 2/2 hoped to carry out Shoup's latest instructions—to join up with Crowe's 2/8 and form a continuous line facing eastward and stretching from the vicinity of Burns-Philp pier to the perimeter now held by Jordan's men.

Jordan soon realized that he lacked both ammunition and men to fight his way to the lines held by 2/8. With Shoup's permission, he postponed the effort until the following morning. Amphibian tractors carrying food, ammunition, and supplies reached the coastal perimeter during the afternoon, unloaded, and evacuated 30 wounded Marines.

In the meantime, Company C of Kyle's command had been trying to push across the island. The pair of .50 caliber machine guns managed to kill or discourage the Japanese gunners firing along the taxiway, the rifle platoons fell back from their positions on the right of Kyle's line, and, aided by the fires of the battalion machine gun platoon, Company C crossed the airstrip. By dusk, a stronger perimeter had been established along the south coast, with Company B and most of the recently arrived machine guns on the west, Company A and portions of 2/2 in the center, and Company C on the east. On both east and west, formidable Japanese positions lay within 25 yards of the Marine line. Since Major Kyle had arrived with Company C, Jordan, at Shoup's direction, attached the men from 2/2—who represented less than one sixth of the forces holding the position—to the 1st Battalion. Relinquishing his command to Kyle, Jordan reverted to his original role as observer, "having done," in Shoup's words, "a fine job in the task he was assigned"[6] without warning or preparation in the hail of fire on Red Beach 2.[7]

At the eastern end of the main beachhead, Major Crowe's 2/8 fought hard but was unable to gain much ground. To the left of the battalion position was the Burns-Philp pier, inland of which lay several ruined buildings, a steel reinforced pillbox, a log and coral emplacement, and a large bombproof shelter. The eastward drive made no headway against these defenses, nor did the Marines attacking to the south in the vicinity of the airfield have much success. Company E did reach the main runway, but the unit had to fall back to avoid being isolated from the rest of the battalion. At dusk, a patrol reoccupied the Burns-Philp pier, a structure which served as a sort of no man's land, and by the following dawn these Marines had killed 15 Japanese infiltrators at the cost of 2 casualties. In brief, the best that 2/8 could do was to strengthen its position and maintain pressure on the weary enemy.

The main beachhead, by dusk of the second day, extended from 400 to 500 yards along the lagoon coast on either side of the control pier. To the right of that structure, 1/8 manned a line running from the beach directly inland to the west taxiway. Neither this taxi strip nor the main runway were occupied by American troops, so a gap almost 250 yards wide separated 1/8 from the perimeter on the south coast. This stretch, sandwiched between two Japanese strongpoints, encompassed a 300-yard portion of the coastline directly across the island from the base of the pier. The right flank of Crowe's 2/8 was near the middle of the airfield triangle, some 200 yards behind the left flank of the south coast position. From the triangle, Crowe's line curved in a quarter circle toward positions held by elements of 3/8 near the Burns-Philp pier. The Red 2 - Red 3 beachhead had been enlarged, but its defensive positions were marked by gaps which might be the target of enemy counterattacks and infiltration. (See Map 3.)

SUCCESS IN THE WEST

While the fight was raging for the central portion of Betio, Ryan's men, members of 3/2 aided by Marines from 1/2 and 2/2, were securing Green Beach, code designation for the entire western side of the island. Since the Japanese had a dozen anti-boat guns and a pair of 5-inch naval guns emplaced at the southwestern corner of this beach, positions that were protected by a maze of rifle pits, Ryan felt that he needed the help of naval gunfire in overrunning his portion of the island. A naval gunfire spotter contacted a destroyer lying off the coast and coached her guns onto Japanese strongpoints. Another destroyer joined the action, and at 1100 Ryan decided that the enemy was sufficiently battered to enable him to begin his attack. The Marines of 3/2, supported by two medium tanks, encountered only slight resistance. According to Ryan, his troops "got another medium tank,

[6] *Shoup interview/comments.*

[7] *Jordan Rpt;* Rpt of Capt Maxie R. Williams, n.d., in *Rpts of 2d MarDiv BnComdrs.*

and naval gunfire came from somewhere. . . . There was little opposition."[8] Late in the afternoon, the Marines organized a defensive line that stretched across the island about 200 yards inland from Green Beach.

Ryan's success, called by Julian Smith "the most cheering news of D plus 1,"[9] enabled the division commander to land elements of the 6th Marines without exposing them to enemy machine gunners. Although the 6th Marines had been released to his control, Julian Smith refused to commit this force until he had a clear picture of the situation ashore. On the morning of D plus 1, the general conferred with Colonel Maurice G. Holmes, the regimental commander, and outlined for him several possible missions which the 6th Marines might be called upon to perform. Holmes left the conference with the understanding that he was to prepare for any of these eventualities and await further orders from division. At 1230, while Holmes was passing on to his battalion commanders the instructions he had received from Julian Smith, he was told by division to land one of his battalions immediately. After reaching the southern part of Green Beach in rubber boats, this unit was to pass through the lines of 3/2 and attack to the east. In addition, Holmes was to have a second battalion ready to land in close support of the first. He selected 1/6, under Major William K. Jones, to make the landing and placed Lieutenant Colonel Raymond L. Murray's 2/6 in close support. Circumstances, however, upset Holmes' plans. The landing could not be executed immediately, nor did 2/6 remain available for close support.

At 1303, a message that was to alter Holmes' plans reached division headquarters. An unidentified observer, using the TBS circuit, claimed to have seen Japanese troops attempting to wade eastward from Betio toward the island of Bairiki. Half an hour later, Colonel Shoup sent a similar report to division and requested naval gunfire to prevent an enemy withdrawal. Before receiving this second message, Julian Smith ordered Holmes to land a battalion on Bairiki. The commander of the 6th Marines elected to use Murray's 2/6. Now the colonel had one battalion preparing to land on Betio and another getting ready to capture Bairiki. The third landing team, Lieutenant Colonel Kenneth F. McLeod's 3/6, was ordered to embark in boats in the event it might be needed at either objective.

At 1655, while 1/6 and its supporting tanks were preparing to land on Green Beach, Murray's 2/6 gained the coral sands of Bairiki. During the approach to the island, the 15 Japanese ashore opened fire with a pair of machine guns. Neither naval gunfire nor preliminary air strikes had silenced these guns, but division, upon learning that 2/6 was under fire, ordered the planes to try once again. This last minute strafing run proved a spectacular success, for a bullet struck a can of gasoline in the pillbox that housed the entire enemy contingent and turned the structure into an oven. The Marines landed against no resistance,

---

[8] Rpt of Maj Michael P. Ryan, n.d., in *Rpts of 2d MarDiv BnComdrs.*

[9] Smith, "Tarawa," p. 1174.

carefully searched the island, and found no live Japanese.

Getting ashore at Green Beach, however, proved a complex undertaking. Company B, 2d Tank Battalion, had the mission of supporting with its light tanks the operations of 1/6. The vehicles belonging to each of the tank platoons had been loaded in a different transport. The tanks had been stowed in bottom holds, and while the transports were unloading off Betio, the cargo within the vessels was rearranged in such a way that the armored vehicles were buried under a mass of supplies. Several hours were lost in digging out the tanks and lowering them into the waiting LCMs.

Nor was the experience of Jones' unit, the only battalion landed in rubber boats, much less frustrating. Just as the transport carrying 1/6, the USS *Feland*, was ready to lower the boats, it was ordered away from the reef into deeper water. Later this ship came in closer again, but it was still 12,000 yards from the beach when it launched the rubber boats. LCVPs towed the boats toward the beach, six of them strung behind each landing craft; outboard motors, which would have permitted the rubber craft to proceed independently, had proved unreliable. After the first wave was safely ashore on the southern part of Green Beach, Major Jones learned that the area was heavily mined and rerouted the rest of 1/6 to the northern portion of the beach. One of the two LVTs carrying food, water, and medical supplies for the battalion struck a mine en route to the island and was destroyed; only one man survived. By 1835, 1/6 was ashore on Betio.[10]

Jones then conferred with Ryan, made a reconnaissance of his zone of action, and decided to attack at 2000 that night. In the midst of his preparations for this thrust, he received a message originated by Shoup and approved by division that directed 1/6 to hold fast until daylight and then strike inland. Upon receiving these orders, Jones organized his companies for the night and coordinated with Ryan concerning the next day's operations.

Unloading the light tanks from the transports had been difficult; getting them beyond the reef proved almost impossible. In fact, only one platoon managed to reach the island in time to support the attack originally scheduled for 2000. Potholes, treacherous currents, and a steep drop-off on the inland side of the reef caused the tank company commander to request permission to land the rest of his unit on Red 2. Division agreed, instructing the remaining two platoons to follow the west side of the long pier. The company commander complied, but it was the morning of the third day before all the tanks were ashore.

THE SUPPLY SITUATION

Lieutenant Colonel Carlson, who had acted as Colonel Shoup's liaison officer on D-Day, was to serve in the same capacity on D plus 1. On the morning of 21 November, Carlson left the divi-

[10] Rpt of Maj William K. Jones, in *Rpts of 2d MarDiv BnComdrs;* Col William K. Jones interview with HistBr, G-3, HQMC and cmts on draft MS, dtd 20Aug62, hereafter *Jones interview/comments.*

sion command post in the *Maryland*, obtained an LVT, and made his way to the central part of the island. He found Shoup at the latter's command post, located in the shadow of an enemy bunker some 30 yards inland on Red 2. Since Japanese troops still lurked in the interior of the structure, guards had been posted at each exit. Shoup informed Carlson that ammunition and water remained in short supply and asked Carlson to return to division headquarters with news of the fighting ashore. At this point, Carlson volunteered to help organize the handling of supplies, an offer which Shoup promptly accepted.

About noon, Carlson met Lieutenant Colonel Salazar, the shore party commander, who had first reached the pier late the previous day. Salazar and Major George L. H. Cooper, operations officer of 2/18, had been supervising beachhead logistics and trying to keep a steady flow of supplies to the front lines. Carlson now urged that LVTs instead of carrying parties be given the job of shuttling supplies inland. Standard landing craft would continue to unload at the pier, which was being repaired by division engineers, and the amphibian tractors would be used for runs from the pier to the units on the island. Each vehicle could bring out wounded on its return trip. Now Carlson faced the problem of finding the necessary LVTs.

From the pier, Carlson journeyed to the minesweeper *Pursuit*, where he explained to Captain John B. McGovern, USN, who was coordinating the movement of landing craft, the need for additional amphibian tractors. The captain responded by making 18 LVTs available to the shore party. Thanks to the presence of these vehicles, the pier could be used as an artificial beach for the unloading and sorting of supplies. The items in greatest demand were loaded in LVTs and rushed to the embattled Marines, while the remaining articles were stacked on the pier.

Another important development was the return to the island of Captain Moore, the assistant division surgeon, who helped speed the evacuation of casualties. With Moore was Major Homer W. Sharpenberg, an engineer assigned the task of locating the water reservoir known to be on the island. Sharpenberg found this source of water on Red 2, not far from Shoup's command post. Later, water purification equipment was installed and by D plus 3, the reservoir had become the principal source of drinking water for the Marines ashore.

After his conversation with McGovern, Carlson reported to Colonel Edson, division chief of staff, to add details to the picture which the officers in the *Maryland* had of operations ashore. Edson felt that progress had been good, and he sketched Julian Smith's plans for crushing the Japanese garrison. Two battalion landing teams from the 6th Marines were scheduled to land over Green Beach, while the third was to seize neighboring Bairiki Island. In addition, the division command post was to move ashore during the night.[11]

[11] Rpts of LtCol Chester J. Salazar and Maj George L. H. Cooper, n.d., in *Rpts of 2d Mar Div BnComdrs; Carlson Rpt; Weatherwax ltr.*

WE ARE WINNING: THE SECOND DAY

At dawn on 21 November, sharp bursts of small-arms fire had served notice that the bitter action of the previous day was to continue with unabated vigor. Because the island had no terrain features big enough to mask preparations for a large-scale attack, ground was gained at Betio by small groups of Marines fighting from the cover of shell craters, ruined buildings, fallen coconut trees, or piles of debris. Often a unit was enabled to advance by the determination of two or three men who worked their way forward by fire and movement to a position from which they could hurl grenades into a bunker or deliver a sudden burst of fire into an enemy position. Engineers attached to the rifle companies tied together blocks of TNT and threw these makeshift charges into pillboxes. The men handling the flamethrowers slipped close to an enemy blockhouse and, while covered by riflemen, suddenly jumped up, ran to the entrance, and sprayed the interior with liquid fire. The riflemen then surged forward to mop up the position, and the flamethrower operator sought the nearest cover to get ready for his next mission. At Betio ground was gained a few yards at a time.

Such had been the fighting on the second day, but even as the Marines were battling across the island, officers of the 2d Marine Division began to sense victory. "At about 1230," recalled Colonel Edson, "things broke rapidly for us."[12] The messages sent and received by Colonel Shoup on 21 November accurately trace the shifting tide of battle. At 1022, division asked Shoup if he had troops enough to complete the conquest of Betio, to which the colonel replied that the situation did not look good. Julian Smith's headquarters radioed for a clarification of this statement, and again Shoup was less than optimistic. "Situation ashore uncertain," was his evaluation of the battle. During the afternoon, however, the picture began coming into sharper focus. At 1345, the best that Shoup could offer was a hopeful "Doing our best," but at 1706 he radioed: "Casualties many. Percentage dead not known. Combat efficiency—we are winning."[13]

Colonel Edson reached Shoup's command post at 2030, obtained an estimate of the situation from the leader of Combat Team 2, and assumed the burden of overall command ashore. Edson's arrival meant some measure of relief for Shoup, who had been responsible thus far for all the troops on Betio. Almost isolated from division headquarters, handicapped by unreliable communications with his battalions, he had succeeded in coordinating the efforts of his combat team. Neither enemy opposition, failures in communications, nor the slow delivery of supplies could stop the Marines who struggled ashore on Betio, for Shoup

[12] *Edson Lecture*, p. 28.

[13] CG 2d MarDiv msg to CO CT 2, ser no. 118, CO CT 2 msg to CT 2 (Rear), ser no. 134, CG 2d MarDiv msg to CT 2, ser no. 139, CO CT 2 msg to ADC, ser no. 148, CO CT 2 msg to LT 1/6, ser no. 169, CO CT 2 msg to 2d MarDiv, ser. no 198 in *2d MarDiv D-3 Jnl.*

had infused them with the spirit of victory.

PLANS FOR THE THIRD DAY

Immediately after his arrival, Colonel Edson enlisted Shoup's aid in laying plans for the next day's attack. Their first task was coordinating air support and naval gunfire. They requested naval gunfire to work over the eastern end of Betio, starting on a line across the island east of the turning circle and the end of the main airstrip. The supporting ships were to slam their shells into the eastern third of the island, keeping 500 yards forward of friendly troops. Aircraft were to bomb and strafe the same general areas assigned to supporting warships. At 0700, ships and planes were to bombard their assigned target areas for 20 minutes, a pounding which was to be repeated at 0830, 0930, and 1030.

Edson next turned his attention to Lieutenant Colonel Kenneth McLeod's 3/6 which had been waiting in landing boats at the line of departure since 1600 of the second afternoon. Shoup's command post was in contact with division, but neither Combat Team 2 nor division was in direct contact with McLeod's battalion or with Jones' 1/6. To reach either landing team it was necessary to contact the headquarters of the 6th Marines and have the message relayed. Edson set to work establishing communications with 1/6 so that it could fight under Shoup's control, at least until more elements of the 6th Marines were ashore on Betio. Edson also recommended that McLeod's men land over Green Beach at 0800. Division accepted Edson's advice, 3/6 was ordered to a rendezvous area, but Julian Smith's headquarters finally decided that the battalion would land on Shoup's order.

After consulting with Colonel Shoup, Edson issued verbal orders for the attack of 22 November. Since Edson could not communicate directly with Jones at this time, he was forced to rely on a messenger. His choice to deliver the message to 1/6 was Major Tompkins, assistant division operations officer.

Plans for the third day of the Betio operation had to take into account the disposition of the troops ashore on the island. On Green Beach, at the western end of the battlefield, were Jones' 1/6 and the composite unit which had secured this area the day before. These troops were in position to advance eastward toward the central beachhead, where three battalions of the 8th Marines and two battalions of the 2d Marines manned the American line. Nearest the troops on Green Beach was Hays' 1/8, which faced westward from positions that extended from the lagoon coast to the western taxiway. No Marines were posted on the runway to the left of 1/8, but 1/2 and a fragment of 2/2 occupied a perimeter on the south coast inland of the pier. Another gap lay between the left flank of this perimeter and the line held by 2/8 and 3/8. The final portion of the beachhead curved from the center of the airfield triangle to the Burns-Philp pier. Strong Japanese fortifications lay at the juncture of Red 1 and Red 2 between Ryan's troops and those led by Hays, on either end of the south coast perimeter, and inland from the Burns-Philp pier.

The attack order for the morning of 22 November called for 1/6 to pass through 3/2 and strike eastward from Green Beach along the south shore in order to establish contact with Kyle's command. What time this assault got underway would depend upon the speed with which Tompkins could reach the battalion command post. At daylight, Hays' 1/8 was to attack to the west along the lagoon shore to eliminate the pocket of resistance astride the boundary between Red 1 and Red 2. Meanwhile, Colonel Hall, with the other two battalions of the 8th Marines, was to continue the drive toward the east, shattering in the process the defenses that had contained Major Crowe's battalion.

In addition to ordering air and naval gunfire support for the morning of 22 November, the 2d Marine Division prepared to augment the fires of its artillery already on Betio by emplacing batteries on neighboring Bairiki. As soon as Murray reported that 2/6 had landed successfully, division headquarters instructed Colonel Holmes to send the artillery element of his combat team, 2/10 (less Battery D), ashore on Bairiki early the next morning. Holmes then directed the artillery battalion commander, Lieutenant Colonel George R. E. Shell, to start transferring his men and equipment to boats by 0330, 22 November, and to land on Bairiki.

At 0300, 2/10 began loading into LCVPs, but this work was suddenly interrupted by an air alert which sent the transports steaming out to sea. At the time of the alert, Battery E was fully loaded, and a portion of Headquarters and Service Battery was boated. Leaving the troops in the transports to come ashore when those vessels returned, the boats already loaded moved directly to Bairiki, arriving off the designated beach at about 0630. Immediately after landing, Battery E dropped trails and, with the aid of the fire direction center on Betio, began registering on the larger island.[14] The transports returned to their unloading area at 0700, but it was noon before Battery F and the rest of the Headquarters and Service Battery reached Bairiki. Battery D, which had been ordered to land on Green Beach at Betio rather than at Bairiki, found that hydrographic conditions off its assigned beach were unfavorable to the landing of artillery. In order to get the howitzers of the battery into action as quickly as possible, division ordered it to land instead on Bairiki.

THE ATTACK OF HAYS' BATTALION

At 0700 on the morning of the third day, 1/8 attacked to the west in an attempt to drive the enemy from strong positions on the boundary between Red 1 and Red 2. Company B advanced along the shore on the battalion right, Company A in the center, and Company C on the left or inland flank. During the morning the attack was supported by three light tanks which maneuvered into positions from which they could deliver almost point-blank

[14] The 1/10 commander noted that this procedure required his forward observer to adjust fires of 2/10 "while looking into their muzzles," an unusual situation that "had been foreseen, planned, and rehearsed in New Zealand during regimental exercises." Rixey ltr, *op. cit.*

SQUAD LEADER *points out the enemy ahead as Marines crawl inland under fire at Betio. (USMC 63575)*

REINFORCEMENTS, *seen through the framework of a Japanese pier, wade ashore at Betio. (USMC 63515)*

fire into the opening of the Japanese pillboxes that were holding up the Marine advance. The enemy positions, however, were far too rugged to be badly damaged by the 37mm guns mounted on these tanks, and the steel and concrete structures had to be reduced by hand-placed bangalore torpedoes and shaped charges.

The light tanks proved unable to perform their assigned task, and at 1130, after one of them had been put out of action by what was thought to be a magnetic mine, they were withdrawn.[15] Two self-propelled 75mm guns from Weapons Company, 2d Marines, were ordered forward to support the attack. One of these half-tracks had its radiator holed by a bullet and was forced to retire before it could contribute anything to the success of Hays' battalion.[16]

Although little ground was gained, the men of 1/8 succeeded in destroying several cores of Japanese resistance. The best progress was made by Companies A and C, which outflanked the enemy position, while along the beach Company B kept unremitting pressure on the Japanese. Late in the afternoon the defenders of the Red 1 - Red 2 strongpoint attempted a minor and futile counterattack which was easily beaten back. As a result of the day's action, the enemy in this area were effectively isolated from their comrades. When Hays' men dug in for the night, they held a semi-circular line reaching from the beach to the airfield dispersal area. (See Map 3.)

THE ATTACK OF 1/6

As late as 0505 on 22 November, Colonel Edson had no contact with Major Jones of 1/6. An hour had passed since Edson had issued his attack orders, and the colonel was anxious that Jones learn of his mission in ample time to make the necessary preparations. Edson asked division to notify the battalion of his plan, but contact between the two headquarters had been lost temporarily. Shortly after 0600, however, Major Jones was able to contact Colonel Shoup by radio and was told the details of the operation plan. Consequently, the battalion commander reported, his men were ready to go at first light.[17]

Colonel Shoup, under whose control 1/6 was operating, ordered Jones to attack at 0800 in order to clear the south side of the island and make contact with 1/2 and 2/2. Once this was done, Jones was to pass through these battalions and prepare to continue the attack on order. With three light tanks in the lead, Jones launched his attack on time, driving forward on a one-platoon front in a zone of action only 100 yards wide. Company C, the assault company, had its lead platoon about 50 yards behind the trio of tanks. Thus the infantrymen were able to protect the tanks from suicidal Japanese who might attempt to destroy them, while the tanks carried out their mission of blasting

[15] According to Japanese sources, no magnetic mines were used but some 3,000 small contact mines were planted, mainly on west and south coast beaches. *Japanese Gilberts comments.*

[16] Maj Robert J. Oddy ltr to CMC, dtd 11 Jun47.

[17] *Jones interview/comments.*

enemy strongpoints. Jones kept his flamethrowers up front with Company C, where they proved useful in knocking out covered emplacements.[18] Resistance, however, proved minor as 1/6 swept along the coast, and the battalion made contact with 1/2 at 1100. During this thrust, Jones' men killed about 250 Japanese, but suffered only light casualties themselves. Losses in the assault company were kept to a minimum by the effective employment of tanks and flamethrowers and rapid movement of the infantry. Given adequate infantry protection, the light tanks proved unusually effective in closing with and destroying enemy installations that might have delayed the battalion for several hours.

Just before 1/6 made contact with the Marines manning the coastal perimeter, Jones was ordered to report as quickly as possible to Shoup's command post. There he received orders to continue the attack to the east at 1300. With the exception of the Red 1 - Red 2 pocket, Japanese resistance on the western half of the island had been crushed. Since this surviving strongpoint was under pressure from 3/2 on the west and 1/8 on the east, it no longer posed a serious threat to the 2d Marine Division. For his afternoon attack, Jones was to have one medium and seven light tanks, the support of naval gunfire, and aid from field artillery on both Betio and Bairiki. In addition, the 8th Marines, except for Hays' 1/8, would attack in conjunction with Jones' battalion. Colonel Hall, commander of the 8th Marines, pointed out that 2/8 and elements of 3/8 had been fighting for two days to reduce the stubborn positions between the east taxiway and the Burns-Philp pier. He felt that these men were capable of just one more effort, and Shoup agreed. As soon as these two battalions had shattered the defenses that had so far contained them, they would rest while 1/6 and 3/6 assumed the entire burden of conquering the eastern half of Betio.

[18] *Ibid.*

THE ATTACK OF 2/8

Early in the morning of 22 November, Major Crowe reorganized his troops for the day's attack. The strongpoints had to be reduced before substantial progress could be made. One was a steel pillbox to the left front, near the Burns-Philp pier and in the zone of action of Company F. To the front of Company K, a 3d Battalion unit temporarily under the command of Major William C. Chamberlain, Crowe's executive officer, was a coconut log emplacement from which Japanese machine gunners kept the company pinned down. The third position that impeded the advance was a large bombproof shelter, inland and to the south of the steel pillbox.

The aid of mortars and tanks was needed because the three positions were mutually supporting. None of them could be attacked unless the assault troops exposed themselves to fire from the other two. Crowe's entire battalion was to be involved in the attack. Company F was to strike first at the steel pillbox, then the next company would move forward, and the advance would be taken up all along the battalion front.

At 0930 the mortars supporting Company K made a direct hit on the roof of the log emplacement, and fortune took a hand in the operation. The detonation of the mortar round touched off a supply of ammunition and the bunker exploded. In the meantime, a medium tank slammed several 75mm shells against the steel pillbox and Company F also was able to advance. The bombproof shelter, though, was a job for an infantry-engineer team. While riflemen kept the defenders busy, flamethrower operators and demolitions men darted forward. After an hour of bitter fighting, the Marines were clinging to the top of the structure.

As soon as the Marines had overrun the bombproof shelter, the Japanese counterattacked. The enemy's effort was smashed largely through the work of one man, First Lieutenant Alexander Bonnyman, who turned his flamethrower on the charging Japanese and drove back those not burned to death. The lieutenant lost his life as a result of this action, but he helped make possible the advance of 2/8.[19]

After this counterattack had failed, the defenders began fleeing the interior of the shelter, running out the east and south entrances. Those trying to escape from the eastern side were mowed down by machine gun and rifle bullets and grenade fragments. The Japanese who broke to the south ran into a hail of fire from machine guns and 37mm cannon.

As the attack progressed, Companies E and G moved around the north end of the bombproof shelter, while Company K moved up on the south, pausing in the process to touch off demolitions at the southern entrance. Company K made contact with Company E east of the shelter, and the enemy defenses were broken. For the time being, riflemen were left to guard the entrances to the bombproof, but in a short time a bulldozer arrived to push sand and dirt into the openings, thus sealing the doom of any Japanese still lurking within the structure.

The thrust of 2/8 carried it to a point at the east end of the airfield, where the battalion paused for fear of being fired upon by 1/6 operating along the south coast. In order to be completely safe from misdirected friendly fire, Crowe ordered his men to fall back about 150 yards. Even though forward progress had stopped, the men of 2/8 still faced the dangerous task of killing the Japanese that infested the area just overrun. At the east end of the airfield triangle, Companies I and L remained in place for the night, even though 1/6 already had advanced beyond that point. At dusk Company C of Jones' command took over the segment of the line along the north shore at the east end of the airfield, and Crowe's Company K was pulled back to form a secondary defensive line.

3/6 LANDS

Lieutenant Colonel McLeod's 3/6 spent the night of 21–22 November in its boats near the line of departure. Twice during the early morning of 22

[19] Maj William C. Chamberlain ltr to OinC, HistSec, DivPubInfo, HQMC, dtd 18June46; Capt James R. Stockman memo to telcon with Maj William C. Chamberlain, dtd 17Jun47.

November the battalion received a change of orders. Originally McLeod was to have landed over Green Beach at 0800, but he was later told to rendezvous off the beach at 0800 and wait there for further instructions. By 0730 on the third morning, 3/6 was in position off Green Beach standing by for word from division. At 0850 orders were issued for McLeod to land his team on the north part of Green Beach, reorganize, and prepare to attack eastward. Although the reef made the landing difficult, by 1100 the entire battalion was ashore. McLeod immediately formed a line with Companies L and I, while holding Company K in reserve. At 1700, 3/6 began moving along the south coast, following the same route taken by Jones' battalion earlier in the day. McLeod halted about 600 yards to the rear of the lines held by 1/6 and remained there in close support of Jones' command. Early that evening, Company I was ordered forward to strengthen the forward position.[20] Much later that night, about 0340, four enemy planes arrived over Tarawa to bomb the island. One whole stick of bombs fell along the southern part of Green Beach, destroying the remaining LVT belonging to 1/6, but fortunately "the flamethrower supplies it contained were saved."[21]

In the morning of 22 November, General Julian Smith decided to establish his command post ashore. With the general and his 10-man command post group were Brigadier General Thomas E. Bourke, commanding the 10th Marines, and Brigadier General James L. Underhill, an observer from VAC. After landing on Green Beach at 1155 and inspecting the troops in the area, the commanding general became convinced that he could best control operations from Red 2, where Colonel Shoup had set up his own command post.

The best route from Green Beach to Red 2 was by water via amphibian tractor, but even at its best the journey was far from easy. The Japanese holding out on the boundary between Red 1 and Red 2 fired on the general's LVT, wounding the driver, disabling the vehicle, and forcing the command post group to transfer to another tractor. Not until 1335 did Julian Smith arrive at Shoup's command post to be briefed on the situation.[22]

THE THIRD AFTERNOON

Orders for the afternoon's effort, issued by Colonel Edson at 1117, called for 1/6, "with all available attached," to pass through the lines of 1/2 and at 1330 attack toward the tank trap at the eastern end of the airfield. In the meantime, 2/8 and 3/8 were to continue their efforts to destroy the enemy to their front, while the 2d Marines, with 1/8 attached, was maintaining pressure on the strongpoint between Red 1 and Red 2.[23] During the afternoon of D plus 2, 2/8 and 3/8 bulled their way to the east end of the air-

[20] Rpt of LtCol Kenneth F. McLeod, n.d., in *Rpts of 2d MarDiv BnComdrs.*

[21] *Jones interview/comments.*

[22] Rpt of BGen James L. Underhill, n.d., in Rpts by SplObservers on GALVANIC, Encl G to *VAC AR;* MajGen Thomas E. Bourke ltr to CMC, dtd 8Jun47.

[23] 2d MarDiv D-3 Jnl, no ser no.

field, and 1/8 helped isolate the Red 1 - Red 2 strongpoint, but the most spectacular gains were punched out by Jones' 1/6.

In order to reach Kyle's coastal perimeter, Jones' Marines had battled for 800 yards against a disorganized enemy. The men of the battalion were hot, weary, and thirsty, but fresh water was so scarce that only a small fraction of the command could fill canteens before the afternoon attack got underway. Although 1/6 had one medium and seven light tanks attached, only six of the light tanks saw action, for the seventh had to be held at the battalion command post so that Jones would have radio contact with the other armored vehicles. The enemy fought with his characteristic determination, finally stalling the Marines after gains of from 300 to 400 yards. At 1500, Company C was ordered to the northern coast to relieve a portion of 2/8, while Companies A and B dug in to the right of the runway.[24]

By dusk on 22 November, the Japanese, except for those manning the stronghold on the boundary between Red 1 and Red 2, had been driven back to the tail of the island. The troublesome redoubt along the beach boundary was effectively isolated. Marines from 1/6, backed up elements of 2/8 and 3/8, manned a line that stretched across the island at the eastern end of the airfield. As before, no troops had dug in on the airstrip itself, but the gap, in this case between Companies A and C of 1/6, was covered by fire. In spite of the day's progress, division headquarters was far from optimistic concerning the possibility of a rapid conquest of the island. "Progress slow and extremely costly," reported the commanding general, "complete occupation will take at least five days more." [25] (See Map 3.)

COUNTERATTACK

The staff of the 2d Marine Division believed at this time that the entire 6th Marines would be needed to dig out and destroy the Japanese holed up on Betio. Colonel Holmes, commanding officer of the regiment, had established his command post ashore. According to the division plan, 2/6 would be brought to Betio from Bairiki and used to support a morning attack by 3/6. Both 2/8 and 3/8 were scheduled to be evacuated to Bairiki. The plan could not be carried out, for most of the available landing craft were being used to carry supplies, and the first elements of 2/6 did not reach Betio until the following morning.

While plans were being made for the attack of the 6th Marines, the Japanese, as Colonel Edson phrased it, "gave us very able assistance by trying to counterattack." [26] The first blow fell at about 1930, when some 50 Japanese, taking advantage of the thick vegetation east of the airfield, infiltrated the outpost line and opened a gap between Companies A and B of 1/6. Within an hour, the battalion reserve, a force made up of Marines from Headquarters and Weapons Companies had mopped up the infiltrators and sealed the gap.

---

[24] Rpt of Maj William K. Jones in *Rpts of 2d MarDiv BnComdrs.*

[25] *2d MarDiv D-3 Jnl*, ser no. 206.

[26] *Edson Lecture.*

214-881 O-67—7

MACHINE GUN AMMUNITION BEARERS *race forward to the front lines at the height of the battle for Tarawa. (USMC 64013)*

FLAMETHROWER SMOKE *rises above the top of an enemy bombproof shelter on Betio as Marines cautiously advance up its slopes. (USMC 63458)*

To contain and destroy any future penetration, Major Jones asked Kyle, commander of 1/2, to set up a one-company secondary line 100 yards to the rear of the main line of resistance. Company I of McLeod's 3/6 later relieved Kyle's men, and additional ammunition, grenades, and water were rushed to 1/6. To disorganize the enemy and disrupt his communications, Jones arranged for a destroyer to shell the tail of the island to within 500 yards of Marine lines. The 10th Marines delivered harassing fire in the area from 75-500 yards in front of the battalion position.[27]

At 2300 the enemy struck again. About 50 Japanese created a disturbance in front of Company A in order to screen an attack by 50 enemy soldiers on the position held by Company B. The defenders used machine guns, grenades, and mortars to blunt this thrust, but the attackers did succeed in learning the location of Marine automatic weapons.

About 0300, Japanese machine gunners opened fire from some wrecked trucks that lay about 50 yards in front of the Marine positions. Although 1/6 silenced some of the enemy weapons with its own heavy machine guns, three of the Japanese guns had to be destroyed by Marines who crawled through the darkness to throw grenades into the ruined vehicles. An hour after the firing had begun, an estimated 300 Japanese hit Company B from the front and Company A from the right front. Artillery fire from 1/10 was pulled back to within 75 yards of the front lines, destroyers opened fire from the lagoon, and the infantrymen caught the enemy silhouetted against the sky. By 0500 of 23 November the attack had been shattered. Within 50 yards of the Marine foxholes lay the bodies of almost 200 Japanese, and sprawled throughout the naval gunfire and artillery impact area were still other corpses.[28]

[27] *Jones interview/comments.*

BETIO SECURED

On the morning of the fourth day, 23 November, the 2d Marine Division faced two difficult tasks, the elimination of the Red 1 - Red 2 strongpoint and the capture of the tail of Betio Island. The coastal redoubt was to be attacked from two sides, with Schoettel's 3/2, of which Ryan's group was a part, advancing toward the northeast, while Hays' 1/8 pushed westward into the heart of the enemy defenses. The final stage of the drive along the length of Betio was made the responsibility of Holmes' 6th Marines. Supported by medium and light tanks, including those which had fought with 1/6, as well as the flamethrowers from Jones' battalion,[29] 3/6 was ordered to destroy the 500 Japanese believed to be on the easternmost portion of the island. In the event McLeod's Marines were fought to a standstill, Holmes would have to call upon 1/6, because 2/6, originally slated to support the assault

[28] Regarding this supporting fire, the 2d Marines commander noted: "The destroyers really cut these people to pieces. . . . They really laid it in there . . . cutting across their flanks and putting the rounds where they count." *Shoup interview/comments.*

[29] *Jones interview/comments.*

battalion, had not yet arrived from Bairiki.

McLeod's battalion began its attack at 0800, and for the first 200 yards of the advance, it met relatively light resistance. Company I, however, was stalled in front of a group of bombproofs and pillboxes located along the lagoon shore. The battalion commander, noticing a great deal of cover in the zone of action of Company L, ordered that unit to bypass the troublesome position and then spread out across the width of the island before continuing its advance. Company I was to remain behind to reduce the enemy pocket, while Company K followed in the path of Company L. (See Map IV, Map Section)

All in all, the attacking Marines had a comparatively easy time. Commented Major McLeod:

> At no time was there any determined defensive. I did not use artillery at all and called for naval gunfire for only about five minutes, which was all the support used by me. We used flamethrowers and could have used more. Medium tanks were excellent. My light tanks didn't fire a shot.[30]

At 1310, 23 November, 3/6 reached the eastern tip of the island, and Betio was secured. During this final drive, the battalion killed 475 Japanese and captured 14 at the cost of 9 Marines killed and 25 wounded. The enemy, though willing to fight to the death, was too tired, thirsty, and disorganized to put up a coordinated defense. Courage and determination proved no substitute for cohesive action; the Japanese were overwhelmed by Marines who displayed teamwork as well as personal bravery.

While McLeod's command was overrunning the tip of the island, Hays' 1/8 and Schoettel's 3/2 were wiping out the pocket of resistance on the northern shore. Since 2/8 was no longer in action, Colonel Hall, in command of the 8th Marines, directed that the flamethrowers formerly attached to the idle unit be released to 1/8. Once these weapons arrived, Hays' battalion made good progress.[31] Also supporting 1/8 were demolitions teams from the 18th Marines and half-tracks mounting 75mm guns. At 1000, 1/8 made physical contact with 3/2, the two units increased their pressure on the trapped Japanese, and by 1305 the western part of Betio was secured. Of the estimated 4,836 Japanese troops and Korean laborers who defended Betio, only 146 were taken prisoner, and a mere 17 of these were Japanese.

The men of the *3d Special Base Force* had died fighting to hold Betio. So great was the destruction wrought by the battle that few enemy documents of any significance survived. Most intelligence of the conduct of the defense by the *Sasebo 7th SNLF* was derived from combat observations and post-combat examination of the shattered and flame-charred remnants of the enemy installations. Somewhere in the ruins lay the bodies of Admiral Shibasaki and his principal commanders, silent forever on their part in the brief, furious struggle. The last word that Tokyo received from the island, a radio message sent early on 22 November read: "Our weapons have been destroyed and from now on everyone is

[30] Rpt of Maj Kenneth F. McLeod, n.d., in *Rpts of 2d MarDiv BnComdrs.*

[31] Hall interview, *op. cit.*

attempting a final charge."[32] The enemy *SNLF* troops, so often called Japanese Marines, met their American counterparts head on in a bitter, close-quarter clash that was never surpassed for its ferocity on any Pacific battleground.

Die-hard survivors of the garrison continued to crop up even though the island was secured. Mopping up continued on the 23d and 24th. The dead were buried, and the weary Marine battalions organized a systematic beach defense in the event that the Japanese should attempt a counterlanding. The island was in shambles. "The stench," wrote a Marine artilleryman, "the dead bodies, the twisted, torn, and destroyed guns of Betio are things which I shall long remember."[33] Later the Marine dead were buried in a military cemetery on the island where they had fallen. On this plot of sacred ground was placed a plaque which read:

> So let them rest on their sun-scoured atoll,
> The wind for their watcher, the wave for their shroud,
> Where palm and pandanaus shall whisper forever
> A requiem fitting for Heroes so proud.[34]

[32] Quoted from a Japanese report, Military Action in the Gilbert Islands, dtd 3May44, cited in Morison, *Aleutians, Gilberts, and Marshalls*, p. 173.

[33] *Watson Ltr.*

[34] Smith, "Tarawa," p. 1175. General Smith recalled that the author of these lines was Captain Donald L. Jackson. *Julian Smith ltr.*

Completing the Capture

JAPANESE RAIDS ON TARAWA

The beach defenses manned by the victorious Marines at Betio were never tested. Although the Japanese had hoped to respond to an American invasion of the Gilberts with an interception operation—a combined attack by aircraft, submarines, and surface ships—the swift capture of the key atolls made the enemy plans obsolete before they could be activated. The most important element of the counterattack force, carrier air, was rendered impotent by losses of Rabaul early in November. Little choice was left the Japanese but to turn their attention from the Gilberts to the Marshalls.[1] All that the enemy could send to aid the defenders of Betio and Makin was a few planes and submarines to stage harassing attacks against the task forces of Admirals Turner and Hill.

On the afternoon of D-Day at Betio, after the invasion ships had been alerted to expect an aerial attack, a single plane winged toward the fleet at an altitude of about 300 feet. Antiaircraft gunners opened fire, and the target settled toward the surface of the sea. The plane was not hit, fortunately in this instance, for it was an observation craft from the *Maryland*, carrying Lieutenant Colonel Jesse S. Cook, Jr., D-4 of the 2d Marine Division. After drifting throughout the night, Cook's pilot managed to taxi his plane back to the battleship.[2]

Just before dawn on 21 November, approximately eight enemy aircraft soared over Betio, dropped a few bombs, and returned to their base. Four planes bombed the island on the following morning, but again the Japanese aviators ignored Admiral Hill's shipping. The task force, however, remained on the alert for a major aerial attack from either Mille or Maloelap in the southern Marshalls. (See Map 7.)

The gravest threat to the transports came not from planes but from submarines. At noon on 22 November, the destroyer USS *Gansevoort* reported a contact with a submarine to the west of the transports. Other contacts with this marauder were made during the afternoon, but not until 1627 was its position fixed. The destroyers *Meade* and *Frazier* took over from the *Gansevoort*, dropping depth charges until the enemy was forced to surface. Shells from the *Meade* and *Frazier* burst around the damaged raider. Finally, the impatient *Frazier* rammed the submarine, sending her plummeting to the bottom. The few Japanese who

[1] Hattori, *Complete History*, vol. 3, pp. 59, 68.

[2] Col Jesse S. Cook, Jr., ltr to CMC, dtd 11 Jun47.

survived to be taken prisoner identified the doomed vessel as the *I–35*.[3]

Aside from these few incidents of support, the enemy left the defense of Tarawa to the hopelessly cut-off garrison. What success the aerial and undersea counterattacks had was gained at Makin.

MAKIN TAKEN[4]

At Makin, the fire support ships began launching spotter planes at 0540 on 20 November, while Admiral Turner's attack force was about three miles from the island. The transports carrying the 6,472 assault troops of the 165th RCT slipped into their designated area at 0601 and began lowering their LCVPs and debarking soldiers within a few minutes. While the landing craft were rapidly filling with men and weapons in the gathering light, carrier aircraft struck targets on Butaritari Island. At 0640, the naval support ships began firing a preparation that lasted until 0824, by which time the island was hidden in a haze of dust and smoke.

As the waves of assault troops of the 165th Infantry were forming off the western end of Butaritari, the only Marine unit to fight at Makin was making a preliminary landing. The 4th Platoon, VAC Reconnaissance Company, along with a rifle platoon and a machine gun squad from the 165th, occupied Kotabu, a reef-fringed islet that guarded the entrance to the atoll lagoon. Although this effort was unopposed, the Marines later saw action on 21 and 22 November when they assisted in the mop up on Butaritari.[5] (See Map 4.)

General Ralph Smith had been able to obtain the LVTs he wanted to spearhead the assault landing at Makin, although they arrived in the Hawaiian Islands only 13 days before their LST transports were due to sail for the target. Forty-eight tractors, manned by a provisional company from the 193d Tank Battalion, and loaded with men of the 3d Battalion, 105th Infantry, made up the first wave. Off the western beaches, 32 LVTs, formed in two inverted Vs, led the landing craft carrying the 1st and 3d Battalions of the 165th Infantry into the silent shore. The preliminary bombardment by air and naval guns had ripped apart the vegetation in the landing area and discouraged any attempt to meet the Americans on the beaches. Instead, the Japanese commander chose to remain in the area around King's Wharf, guarded on each flank by a cross-island tank trap and barricade.

It was fortunate that the Japanese did not contest the landing. Although the LVTs carrying the assault detachments of 3/105 landed without undue difficulty, the following LCVPs and LCMs were often unable to reach the shore across the reef, which was studded with coral rocks and potholes. Many men had to wade to the beach, and

---

[3] CinCPac WarD, Nov43, Anx E, pp. 13–14 (CinCPac File, HistBr, HQMC); USS *Gansevoort* AR, 17–26Nov43, dtd 3Dec43.

[4] Unless otherwise noted, the material for this section was derived from: Crowl and Love, *Gilberts and Marshalls;* Morison, *Aleutians, Gilberts, and Marshalls.*

[5] Rpt of 1stLt Harvey C. Weeks, n.d., Encl D to VAC G–2 Rpt, dtd 8Dec43, Encl C to *VAC AR.*

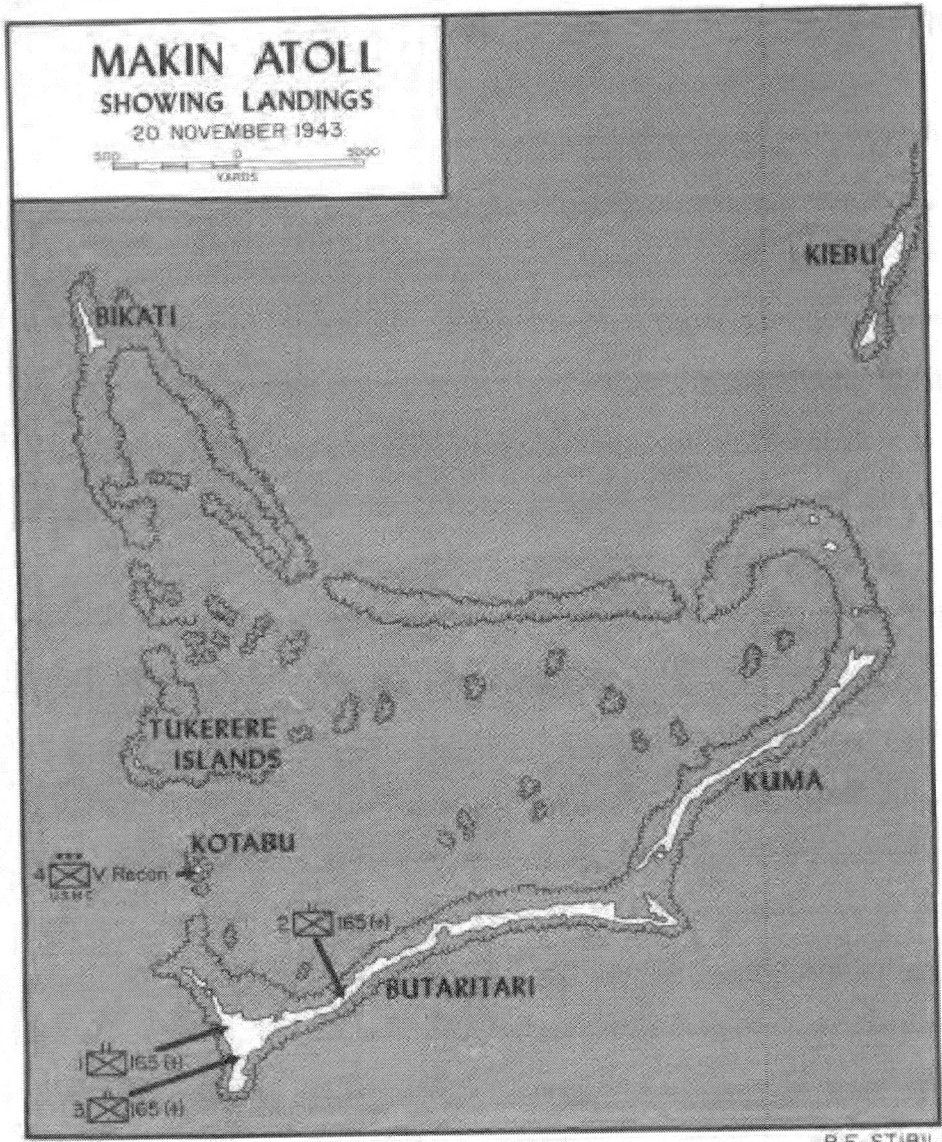
MAKIN ATOLL
SHOWING LANDINGS
20 NOVEMBER 1943
YARDS
KIEBU
BIKATI
TUKERERE ISLANDS
KUMA
KOTABU
4 V Recon
USMC
2 165 (-)
BUTARITARI
1 165 (-)
3 165 (-)
MAP 4
R.F. STIBIL

the waterproofed tanks which accompanied the infantry were taxed to their limit to reach land through the surging swells.

An innovation, rocket fire by some of the assault LVTs, was tried at Makin. Designed to neutralize beach defenses, the launchers mounted on the tractors were far from successful. Most of the rockets fell short, exploding harmlessly in the surf, and many never fired at all because of defects in their firing mechanisms caused by sea water. Machine guns on the tractors provided close-in suppressive fire as the first waves came ashore at 0832.[6]

Only scattered sniper fire met the first assault troops, and the confusion arising from the late arrival of infantrymen and tanks negotiating the rock-strewn reef and beaches had little effect on the course of operations. By 1000, the western end of Butaritari was secured, and the 1st battalion of the 165th advanced toward the western tank trap encountered increasing resistance as it approached the main Japanese position.

While 1/165 was moving toward the enemy, the second landing planned by Ralph Smith was taking place. Moving through the lagoon entrance in columns of LVTs and landing craft, 2/165, reinforced by a 3/105 assault detachment, turned and headed in for beaches which gave entrance to the heart of the Japanese prepared positions. Naval gunfire and carrier air hit the island where the 2d Battalion was headed, and the 16 tractors made for shore with machine guns firing and landed at 1040. Behind them, the tanks and infantry in following waves found the water too shallow to proceed in boats and had to wade across the reef shelf about 300 yards to the beaches.

For the rest of the day, the troops which had landed on the lagoon side tangled at close quarters with small, determined enemy groups that held up any appreciable advance. Across the western tank trap, the men of the 1st Battalion of the 165th, whose regimental commander was killed in the fighting, were stalled by the fire of enemy machine gunners and riflemen. At nightfall, the troops dug in in close contact with the Japanese. What followed in the darkness was a harrowing experience for the American soldiers, as infiltrators and trigger-happy green troops filled the night with grenade explosions and rifle and machine gun fire.

Ruefully wiser in the ways of the Japanese in night combat, the men of the 165th spent most of 21 November consolidating their beachheads, all the while so closely enmeshed with the Japanese defenders that naval gunfire, air, and supporting 105s, which had landed on D-Day, could furnish little effective support. Much of the day's fighting, during which the two elements of the RCT joined forces, was concentrated in efforts to reduce Japanese pillboxes and machine gun nests, much like those on Tarawa. Tank-infantry teams, working with flamethrowers and demolitions, cleaned out the stubbornly resisting Japanese naval

[6] LtCol S. L. A. Marshall, USA, "Supplementary Notes on Makin Operation," dtd 2Jun 44 (File supporting *The Capture of Makin*, OCMH), pp. 3-4; VAC G-3 Rpt, dtd 11Jan44, p. 12, Encl B to *VAC AR*.

LIGHT TANKS *and artillery wait their turn to leave Betio two weeks after the battle, while Marine transports and fighters use the rebuilt airstrip. (USAF 68045AC)*

SOLDIERS *of 2/165 wade toward the lagoon beach at Butaritari, as smoke rises from oil dumps hit by naval gunfire. (USA SC183574)*

troops, but progress was slow. General Holland Smith, who landed to inspect the progress of the battle during the day, reported to Admiral Turner: "Enemy losses very heavy, own light. Consider situation in hand." [7] The Japanese, however, were still full of fight and those few that survived the first two days now withdrew across the eastern tank trap to take up positions in the heavy vegetation on the long, narrow tail of the island.

With the situation ashore rapidly clarifying and the destruction of the enemy garrison in sight, Turner gave Ralph Smith permission to land troops on Kuma Island north of Butaritari to cut off any Japanese attempt to retreat. On 19 November, the 27th Division commander had requested permission to make a landing on Kuma on D-Day, but on Holland Smith's advice, the admiral denied the request for the last-minute change in the tactical plan. Neither of the senior officers wanted to make the subsidiary effort until the destruction of the Butaritari garrison was certain. On the morning of the 22d, a detachment of the 105th Infantry rode LVTs ashore on the neighboring island and found no Japanese, only a welcoming party of natives.

On Butaritari, on 23 November, the main effort was made by 3/165, in reserve since D-Day. Crossing the eastern tank trap without opposition, the battalion moved quickly into the brush, flushing a few Japanese and maintaining a steady, continuous rate of advance that brought them up to a prepared line of enemy defenses by dusk. All night long, the Japanese tried to infiltrate the American positions and attacked in small groups to no avail. When morning came, there were 51 enemy dead in front of the lines of 3/165; the battalion had lost 3 men killed and 25 wounded. Sweeping on to the tip of the island on the 23d, the soldiers encountered no further organized resistance. At 1130, General Ralph Smith signalled Admiral Turner: "Makin taken. Recommend command pass to commander garrison force." [8]

Makin Atoll was captured at the cost of 218 Army casualties, 66 of whom were killed in action or died of wounds. Only one member of the *3d Special Base Force Makin Detachment* was made prisoner, but 104 Korean laborers surrendered to the assault force. The total number of enemy killed was estimated to be about 445, 300 of them Japanese combat troops.

The United States Navy suffered a far greater number of casualties in supporting the operation than did the Army units fighting ashore. On D-Day, an accidental explosion in a turret of the battleship *Mississippi* killed 43 sailors and wounded 19 others. Some 20 miles southeast of Butaritari on the morning of 24 November, the Japanese submarine *I–175* torpedoed the escort carrier *Liscome Bay*. Bombs stowed in the vessel exploded, and in 23 minutes the carrier had perished, claiming the lives of 53 officers and 591 enlisted men.

---

[7] TF 52 NarrativeRpt, dtd 4Dec43, p. 21, Encl A to *V PhibFor AR*.

[8] CG, 27th InfDiv msg to CTF 52, dtd 23Nov 43, in 27th InfDiv G–3 Rpt, Encl 5 to 27th InfDiv Rpt of Participation in GALVANIC Op, dtd 11Dec43, hereafter *27th InfDiv OpRpt*.

The battle for Butaritari made veterans of the men of the 165th RCT. They had met the test of combat. Thanks to the experience gained at Makin, these troops could be expected to make a greater contribution to future victories in the Central Pacific.[9]

OTHER TARAWA ISLANDS

The occupation of the less important islands in Tarawa Atoll began while the battle for Betio still was raging. On 21 November, elements of Company D, 2d Tank Battalion, the division scout company, landed on Eita, west of Bairiki, and Buota, near the southeast corner of the atoll, where an estimated 100 Japanese were discovered. Another part of the scout company went ashore on an unnamed island that lay about one-fourth the distance from Buota to the northern apex of Tarawa. Two days later, while 3/10 was setting up its weapons on Eita to support the Marines on Betio, the Japanese on Buota escaped unopposed to the north. By 25 November, elements of Company D had scouted the southern half of Tarawa's eastern side, but at this point the entire unit was recalled to Eita to prepare for a reconnaissance of three nearby atolls—Abaiang, Marakei, and Maiana.[10] (See Map 5.)

In the meantime, Lieutenant Colonel Murray's 2/6, which had overrun Bairiki, had undertaken the mission of clearing the enemy from the outlying islands. Murray's men encountered no resistance until the late afternoon of 26 November, when the battalion reached Buariki, the northernmost of the larger islands of the atoll. At sunset on that day, a Marine patrol engaged in a fire fight with a small Japanese force. In spite of enemy harassment, the Marines held their fire throughout the night.

Murray resumed his advance on the following morning and soon located the enemy defenses, a haphazard arrangement of rifle pits and log barricades concealed in dense undergrowth. Because Company E had taken several casualties, the battalion commander ordered Company F to continue the attack while Company G maneuvered to strike the enemy on his eastern flank. Although Murray had a battery of pack howitzers at his disposal, poor visibility and the short range at which the infantrymen were fighting prevented the cannoneers from firing more than a single concentration. In spite of this absence of artillery support, Murray's troops crushed the position, killing 175 and taking 2 prisoners. The Marines lost 32 killed and 59 wounded as a result of this fight.

Naa, a tiny island north of Buariki and the final objective of the battalion, was found on 28 November to be free of Japanese. The men of 2/6 then returned to Eita to rest from their mission. By boat and on foot, these Marines had covered a distance of about 35 miles in moving from Betio to Naa.[11]

[9] For a detailed analysis of the operation to capture Makin, see Crowl and Love, *Gilberts and Marshalls*, pp. 75–126.

[10] Co D, 2d TkBn SAR, dtd 20Dec43, Encl to *2d TkBn SAR*.

[11] 2/6 Narrative Account of Ops, 21–29Nov 43, n.d.

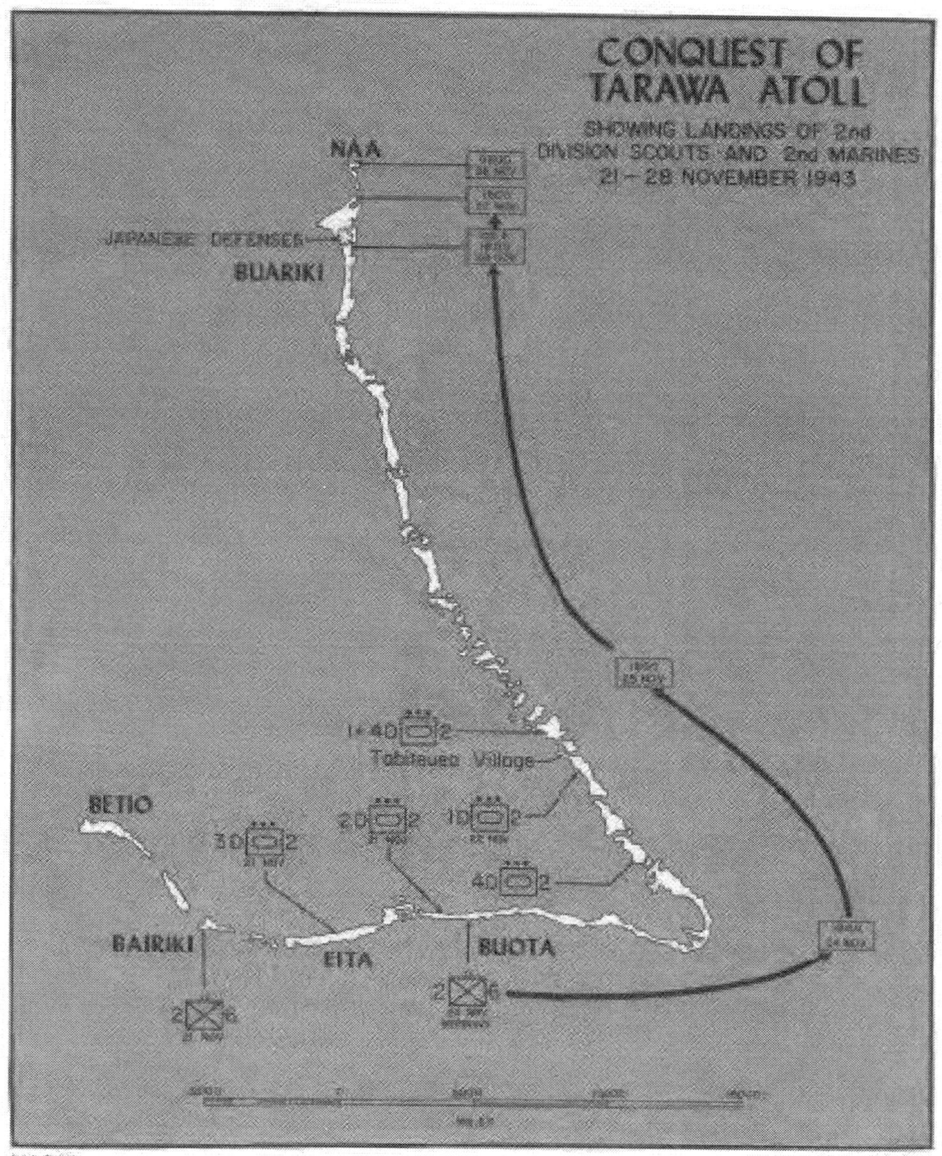
CONQUEST OF
TARAWA ATOLL
SHOWING LANDINGS OF 2nd
DIVISION SCOUTS AND 2nd MARINES
21 – 28 NOVEMBER 1943
NAA
JAPANESE DEFENSES
BUARIKI
Tabiteuea Village
BETIO
BAIRIKI
EITA
BUOTA
MAP 5
R.F. STIBIL

APAMAMA [12]

In operations to seize the third major objective of GALVANIC, plans called for VAC Reconnaissance Company, commanded by Captain James L. Jones, to sail to Apamama in the submarine USS *Nautilus*, land on the night of 19–20 November, and determine whether the atoll was as lightly held as aerial photographs had indicated. Should Jones' Marines encounter any large number of the enemy, they were to break off the action and make their way back to the *Nautilus*. The objective which the company was to explore was a large atoll shaped like a partially inflated football and measuring 12 miles long by 5 miles wide. Abatiku Island to the southwest, which might serve as the laces of the football, was bordered on either side by a passage into the lagoon. Beyond South Passage lay Entrance Island, then an expanse of reef, and finally Apamama Island, which curved to the northwest then swung southward toward Western Passage and Abatiku. Apamama Island was broken by shallow water into six segments, each of which had been given a name by the planners of the operation. Those of concern to Jones' men were, in order of their separation from Entrance Island, JOE, JOHN, ORSON, and OTTO. (See Map 6.)

The *Nautilus* was to pick up the reconnaissance unit at Pearl Harbor, cruise for a time off Tarawa to observe Japanese ship movements, and then sail the 76 miles that separated Tarawa from Apamama. On 19 November, after departing from Tarawa, the submarine surfaced in order to increase her speed. The destroyer *Ringgold*, assigned to protect Task Force 53—which had received no word that the *Nautilus* was in the area—[13] spotted the submarine, scored a direct hit with a 5-inch shell, and forced her to crash dive. Luckily, the shell failed to explode and the minor damage it caused could be repaired once it was safe to come to the surface. The submarine arrived off Apamama on the afternoon of 20 November.

Before dawn of the following day, Jones and his men embarked in rubber boats for JOHN Island. An unexpectedly strong current forced them to land on JOE, just to the west, but they later crossed to their original objective. On JOHN, the Marines collided with a 3-man enemy patrol, killed one of the Japanese, and prepared to move to the next portion of Apamama Island, the part called ORSON. After this crossing had been made, the troops encountered a group of natives who informed Jones that about 25 Japanese were entrenched at the southern tip of neighboring OTTO.

The company attempted on 23 November to move across the sandspit connecting ORSON and OTTO Islands. The fire of enemy rifles and light machine guns proved so intense that Jones decided to break off the action and attempt to outflank the defenders. On the following morning, while the *Nauti-*

[12] Unless otherwise noted, the material in this section is derived from: TF 53 Rpt on Apamama Op, dtd 15Dec43; VAC ReconCo WarD, 8–27Nov43, dtd 12Dec43; Sgt Frank X. Tolbert, "Apamama: A Model Operation in Miniature," *Leatherneck*, v. 28, no. 2 (Feb45), pp. 26–27.

[13] *Hill interview/comments.*

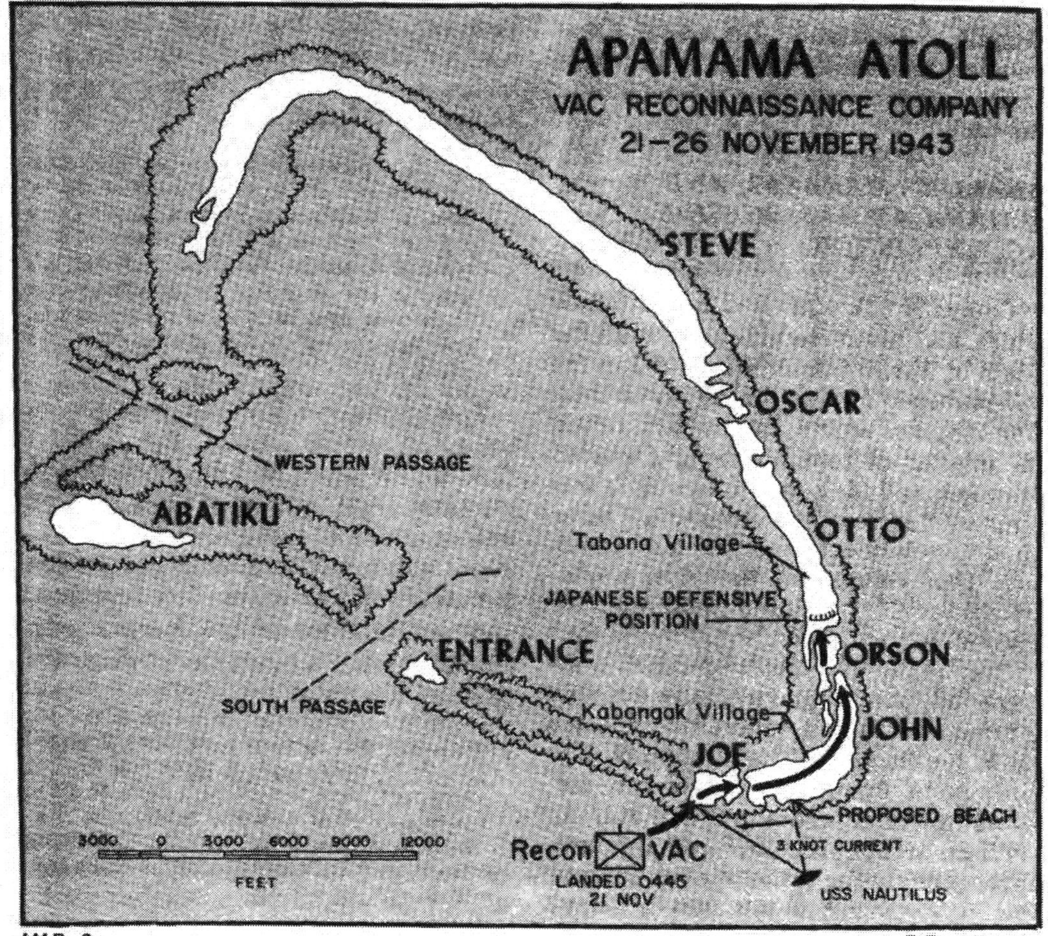

MAP 6 R.F. STIBIL

lus shelled the Japanese position, the company commander tried to disengage so that his troops could enter the rubber boats, bypass the strongpoint, and attack it from the rear. The plan, however, could not be executed, for the Japanese kept firing steadily in spite of the bursting shells. Late in the day, a friendly destroyer arrived off the island to slam a few additional rounds into the stubborn emplacements.

A native reached Jones on the next morning with the startling news that all the Japanese were dead. Patrols soon discovered that this report was true, for 18 of the enemy had killed themselves after the other 4 had perished in the bombardment. Marine losses in the Apamama action were two killed, two wounded, and one injured.[14] That afternoon, Brigadier General

[14] LtCol Merwin H. Silverthorn, Jr. ltr to Head, HistBr, G-3, HQMC, dtd 9Sep62.

Hermle, Assistant Division Commander, reached Apamama with 3/6 to assume responsibility for the defense of the atoll.

ABAIANG, MARAKEI, AND MAIANA ATOLLS: MISSION ACCOMPLISHED

Company D, 2d Tank Battalion, which had begun securing the outlying parts of Tarawa Atoll, embarked in the minesweeper *Pursuit* on 29 November. The division scouts had been assigned the mission of reconnoitering Abaiang, Marakei, and Maiana Atolls. By 1 December, this unit had checked all three objectives without encountering opposition, thus completing the Marine contribution to Operation GALVANIC. (See Map 1.)

Even while these last operations were going on, the 2d Marine Division was moving to Hawaii, there to prepare again for combat. The 2d and 8th Marines went first on 24 November, followed in short order by the 6th, as garrison troops responsible to the area naval commander took over the defense and development of the islands seized by the Marines. On 4 December 1943, General Julian Smith turned over command of Tarawa to the Navy and headed for the new division base, Camp Tarawa, high in the hills of the big island of Hawaii. As the division historian recalled the 2,000-mile voyage, he described it as:

> . . . a postscript to horror. The transports reeked of the awful smell of the island, of disinfectant, and of blood. There were no fresh clothes for unwounded Marines, and almost everyone had lost his gear in the shuffle of battle. Every day there were funerals aboard the transports, and flag-covered bodies slipping into the silent seas.[15]

The 2d Marine Division casualty reports show the battle for Tarawa claimed the lives of 984 Marines and attached Navy personnel and that an additional 2,072 men were wounded.[16] These harsh statistics serve as a lasting tribute to the courage, determination, and self sacrifice of the Marines who fought there. As a national magazine phrased it:

> Last week some two to three thousand U. S. Marines, most of them now dead or wounded, gave the nation a name to stand beside those of Concord Bridge, the *Bonhomme Richard*, the Alamo, Little Big Horn, and Belleau Wood. The name was Tarawa.[17]

---

[15] Johnston, *Follow Me!*, p. 166.

[16] *Julian Smith ltr.* See Appendix H for the final official compilation of Marine casualties.

[17] *Time*, v. 42, no. 23 (6Dec43), p. 15.

The Importance of GALVANIC

THE GILBERTS IN AMERICAN STRATEGY

To American planners, the capture of bases in the Gilberts marked the beginning of a major effort against Japan, the type of offensive outlined in the ORANGE Plans. The loss of Tarawa, Apamama, Makin, Abaiang, Marakei, and Maiana Atolls did not cripple the enemy, for GALVANIC had not been designed to do so. Although Admirals King and Nimitz believed that a victory in the Marshalls would be more damaging to the enemy than the conquest of the Gilberts, geographical considerations plus slender military resources forced them to strike first at the Gilberts. Both Nimitz, who was willing to undertake any operation that had "a reasonable prospect of success," and King, who was willing to accept "very considerable calculated risks," refused to plunge blindly into the mandated islands.[1]

Not until bombers and photographic planes had penetrated the Marshalls were the American naval leaders willing to risk the ships and men necessary for amphibious operations in that area. The capture of airfield sites in the Gilberts brought the Marshalls within more effective range of land-based planes and enabled the Navy to launch its westward drive. "This operation," commented Nimitz, "is considered to have been highly successful. Island bases essential to our advance across the Pacific were captured from the enemy with the complete destruction of all his defending forces."[2]

As a result of GALVANIC, the Army Air Forces gained four new airfields from which to launch strikes at targets in the Marshalls. At Tarawa, a 6,000-foot runway was built on Betio, while 7,000 and 4,000-foot runways were constructed on Buota. On 15 December, the first bombers, twin-engine B–25s (North American Mitchells), arrived at Tarawa, but neither of the two atoll bases was then ready to handle its full complement of planes. As the year 1944 arrived, heavy B–24s began flying bombing and reconnaissance missions from Tarawa.

[1] King and Whitehall, *Fleet Admiral King*, p. 432.

[2] CinCPac WarD, Nov43, Anx F, p. 10. In regard to the decision to attack fortified islands, like Tarawa and targets in the Marshalls, rather than undefended objectives, Admiral Hill commented that prevailing trade winds required an island air base to have fields running in the wind direction and that the Japanese, recognizing this fact, had built their bases on the relatively few islands that were situated to take best advantage of the winds. Many of the undefended atoll islands were too small for airfields or would require too much construction work to be usable. *Hill interview/comments.*

214-881 O-67—8

In spite of swamps and soft ground, a 7,000-foot flight strip was finished at Makin early in January. Because this runway was built on sand rather than coral, it could not support the weight of B–24s. The Apamama facility, completed by 15 January, boasted 7,000 feet of hard coral surface ideally suited to heavy bombers.[3]

Although the bases gained as a result of GALVANIC were in themselves important, far more valuable was the experience amassed by American Army, Navy, and Marine Corps commanders. By capturing Betio Island, the men of the 2d Marine Division had proven that Marine Corps amphibious doctrine was essentially sound. Although the casualty list shocked the American public, the operation was nonetheless a success, for the capture of Makin, Tarawa, and four lesser atolls had neutralized the entire Gilberts group and advanced American might across some 700 miles of ocean. Because the loss of life was confined to so short a period, the impact on civilian morale was especially severe. Almost unnoticed was the possibility that a land campaign over a similar distance, even though comparatively few men fell each day, might in the end prove more costly than a violent but brief assault from the sea. GALVANIC, moreover, did show means by which losses could be reduced in future amphibious operations.

PERSONNEL PROBLEMS[4]

Because Betio was to be taken by a previously specified number of men in what was expected to be a brief but furious battle, corps personnel officers were not concerned with a replacement system, which would be needed in a campaign of longer duration. Routine administration, however, had to be carried on as usual, and at Betio personnel accounting proved a difficult task. Breakdowns in communications, plus the hectic tempo of the fighting prevented the 2d Marine Division from checking each day on the number of able-bodied men in its ranks. Summaries of casualties and prisoners taken were submitted to VAC immediately after the action, but the confusion of reembarkation resulted in incomplete and inaccurate returns. To provide more thorough statistics in future operations, VAC urged that periodic G–1 reports be submitted as promptly as possible.

At Butaritari, where 27th Infantry Division headquarters was rapidly established ashore, two periodic reports of losses were prepared. Details of the Apamama venture, however, remained unknown to Holland Smith's G–1 section. No reports were submitted by VAC Reconnaissance Company until the unit returned to Hawaii.

Although GALVANIC represented a greater concentration of naval might than any previous effort against the Japanese, the size of the expeditionary force was limited by the number of transports and trained men available in the Central Pacific area. The 2d Ma-

[3] Craven and Cate, *Guadalcanal to Saipan*, pp. 303–304; Morison, *The Aleutians, Gilberts, and Marshalls*, pp. 221–212.

[4] Unless otherwise noted, the material in this section is derived from: VAC G–1 Rpt GALVANIC, dtd 6Jan44, Encl E to *VAC AR;* Isely and Crowl, *Marines and Amphibious War*, pp. 203–205.

rine Division, the only experienced amphibious division that could be assigned to the Gilberts expedition, was given the mission of capturing Tarawa. In order to lessen the risk of loss of valuable ships, the Marines had to assault Betio before landing elsewhere in that atoll. Had the American fleet been strong enough to accept the possible loss of several transports or warships, the assault force could have risked prolonging the action by first seizing islands near Betio and emplacing field artillery to support the storming of the principal objective.

In addition, one Marine regiment had to be retained in corps reserve, thus leaving only two regiments at the disposal of the commanding general of the division. In brief, circumstances forced upon Julian Smith a plan that called for the direct assault by an understrength division against a heavily fortified objective. Although he later was able to employ his third regiment, he could not count on its use and had to rely on aerial and naval bombardment to make up for what he lacked in numbers. The lesson was clear. In assessing operations to seize the atoll, Admiral Nimitz wrote: "Under present conditions, it is necessary to plan for the employment of not less than one division for the capture of an enemy position comparable in strength to Tarawa." [5]

Compared to the defending garrison, the force assigned to capture Butaritari was of overwhelming strength. Yet, in the opinion of both Admiral Turner and General Holland Smith, the might of this reinforced regiment could not be dissipated in secondary landings on neighboring islands until Butaritari had been won. After the fall of the major objective was certain, the possible need for further reinforcements at Betio kept General Ralph Smith from employing his reserve battalion as he desired. Not until victory at Betio was assured, could the Army general carry out his plan to trap the remnants of the Makin garrison by landing troops on Kuma Island.

The operations against the Gilberts were the most damaging blows that could be struck against the enemy with the resources then available to Admiral Nimitz. The expedition, no more than equal to its task, was the largest that could have been mounted in the fall of 1943. The GALVANIC force, in comparison to expeditions sent forth later in the war, was small, but these few troops were able to shatter Japanese power in the Gilberts and open the way into the Marshalls.

INTELLIGENCE

American intelligence officers, working from photographs taken by submarine and aircraft, were able to locate almost all of the enemy's defensive installations before the operation got underway. If anything, the interpretation of these photos was too cautious, for several dummy gun emplacements on Butaritari were listed as containing actual weapons. Intelligence specialists, however, failed to foresee the adverse conditions off the beaches at Butaritari. Although the unexpected boulders and coral outcroppings there, together with unforeseen tides, compli-

[5] CinCPac 1stEnd to ComCenPacFor ltr to Cominch, dtd 10Dec43 (OAB, NHD).

cated the unloading of men and supplies, these conditions had little effect on the assault landings.

A greater number of oblique photos, taken at irregular hours over a period of several days, might have given a clearer indication of Betio's regular tides, but no available information could have plotted the freak dodging-tide that occurred on D-Day. When traders and British colonial administrators familiar with the Gilberts failed to agree on tidal conditions, American officers were forced to use a concensus estimate in order to prepare their carefully drawn landing plans. While General Julian Smith felt there was one chance in two that standard landing craft would be able to cross the reef at Betio, he approved a plan that envisioned "a tide that would not float our boats across the reef." [6]

More thorough photographic coverage would be needed in future amphibious undertakings, but only prolonged observation could give a hint of the course of eccentric tides. Another partial solution to the problem lay in the use of the Naval Combat Demolitions Units that had been organized prior to the invasion of Kiska. Although the six-man team destined for the Aleutians was inadvertently left behind at San Francisco, Admiral Turner felt that a similar team would have been valuable in destroying underwater obstacles off Betio. In the course of the Pacific war, these units, designated Underwater Demolitions Teams (UDTs), also were employed to collect last-minute information on the depth of water, approaches, and gradients off various objectives.[7]

During the 76-hour battle for Betio, there was little opportunity to collect, evaluate, and disseminate intelligence information. Most of the Japanese and Koreans preferred death to surrender, and the information provided by the few prisoners had no effect on the conduct of the fighting. At Butaritari, friendly natives confirmed preinvasion estimates of the size of the enemy garrison and its location. In addition, villagers on Kuma Island provided an accurate count of the Japanese in the area, information which helped General Ralph Smith prevent the enemy from retreating along the atoll. Most of the intelligence gathered in the Gilberts, however, was applicable to future operations rather than to the situation at hand.

After Betio was secured, Japanese language officers of the 2d Marine Division scoured the island in search of enemy documents. The most important find was a set of plans and specifications for some of the defenses encountered on the island. This document and the examination by engineers of shattered emplacements enabled the Americans to build sample blockhouses and test their durability. Such experiments led to improvements in naval gunfire techniques and infantry tactics

---

[6] LtGen Julian C. Smith interview with HistBr, G-3, HQMC, dtd 4Oct62. For an interesting discussion of the development of tide information at Tarawa see: Patrick L. McKiernan, "Tarawa: The Tide that Failed," *U. S. Naval Institute Proceedings*, v. 88, no. 2 (Feb62).

[7] Cdr Francis D. Fane and Don Moore, *The Naked Warriors* (New York: Appleton-Century-Crofts, Inc., 1956), pp. 24, 30–31.

in time for the Marshalls operation.[8]

Of the prisoners questioned by 2d Division intelligence specialists, only Ensign Kiyoshi Ota was able to provide valuable information. He testified that although air strikes had destroyed two or three protected installations, the men in shelters or covered fortifications were safe from both bombs and naval shells. In his opinion, naval gunfire had devastated antiaircraft emplacements, shattered communications, but failed to destroy concrete structures. The ensign, however, was impressed by the effective on-call fire delivered by destroyers posted in the lagoon. Since the building where Ota was stationed was destroyed by a tank-infantry team, his testimony indicated both the importance of coordination between infantry and armor as well as the need for a more accurate and powerful preparatory naval bombardment.[9]

COMMAND AND COMMUNICATIONS

The command relationships decided upon for GALVANIC satisfied neither Holland Smith nor Ralph Smith. Before the expedition sailed, the Marine general had pointed out that, although nominally a corps commander, he had no troops under his tactical control. The Army general felt that, since he had not been free to alter his tactical plans without Admiral Turner's approval until after the naval officer had directed him to assume command ashore, the commander of the landing force was for the most part "a conduit for the issue of orders" [10] by the assault force commander.

Ralph Smith, however, was quick to admit that an amphibious operation was a type of combat in which the concern of the Navy for its ships might conflict with the scheme of maneuver ashore. Obviously, some sort of compromise was necessary. In the general's opinion, "the successful execution of an amphibious operation is dependent not on who or what component of the armed forces commands, but on the mutual confidence between all commanders and a comprehensive understanding of the problems faced by each." [11] Apparently, there was no lack of confidence and understanding, for Turner's system of command was adequate to the situation at both major objectives.

The difficulty in transmitting orders and information rather than any weakness in the command structure caused confusion and needless delay at both Makin and Tarawa. Because there was little opposition at the beaches of Butaritari, communications failures did not jeopardize the success of the 27th Infantry Division assault troops. Radio contact between ship and shore was reliable enough, but elements of the assault battalions at times had difficulty in exchanging messages.[12]

As a command ship for the Tarawa operation, the battleship *Maryland*

---

[8] Maj Eugene P. Boardman ltr to CMC, dtd 16Jun47.

[9] 2d MarDiv PrelimIntelRpt of Tarawa Op, dtd 7Dec43, Encl N to VAC G-2 Rpt, dtd 8Dec43, Encl C to *VAC AR*.

[10] *27th InfDiv OpRpt*, p. 2.

[11] *Ibid.*

[12] *Ibid.*; Rpt of 27th InfDiv SigO, dtd 4Dec 43, Encl no. 2 to *27th InfDiv OpRpt*.

proved an unhappy choice. Transmitters, receivers, and antennas, installed in a compact area, interfered with each other, severely hampering communication efficiency. In addition, the concussion from the 16-inch guns of the vessel ruined some of the more delicate pieces of radio equipment. Admiral Hill recommended that, as a temporary expedient, the number of radio channels in use be drastically reduced, but the problem of ship-to-shore communications could not be solved until specially designed command ships were introduced in the Central Pacific.

The communications difficulties extended to transports and landing control craft as well. There was an evident need for better facilities, better trained control personnel, and a more systematic command setup. As a result of lessons learned, the transport group, and later the transport squadron commander, "was given a greatly enlarged staff and made responsible for 'traffic control' off the beaches. With the better communications facilities made available to him, he was," in the words of the transport commander at Tarawa, "the logical one to be charged with this duty." [13]

Once ashore on Betio, the Marines continued to have communication troubles. Batteries in the MU radio, the handset carried by platoon leaders, wore out too quickly to suit the men who depended on these sets. Officers in the division complained that the TBX radios were susceptible to water damage, but VAC analysts held that the case containing the radio was watertight if assembled properly. Neither division nor corps, however, had a kind word for the TBY, and Admiral Hill's headquarters recommended that this piece of equipment be replaced by its Army equivalent. Waterproof bags or cartons also were needed to protect telephones and switchboards during the ship to shore movement. Because each used a different type of radio, contact between infantrymen and tank commanders was uncertain.[14]

TACTICAL LESSONS

The most important feature of the assaults upon Betio and Butaritari was the role of the amphibian tractor as an assault craft. Prior to GALVANIC, LVTs had been used to haul supplies from transports to dumps inland of the beaches, but the conquest of the Gilberts marked the first time that these vehicles had carried the first wave of troops. The tractors proved so successful in their new role that Holland Smith came to believe that LVTs were vital to the amphibious assault.

As valuable as the tractors had been, they were not perfect. Greater speed, additional armor protection, and a ramp for discharging troops were the improvements suggested by VAC.[15] At

[13] *Knowles ltr.*

[14] V PhibFor CommRecoms and Cmts, Encl A to *V PhibFor AR*, p. 62; *VAC AR*, p. 17; VAC Analysis of CommRpts, dtd 3Jan44, Encl 5 to SpStfRpts, n.d., Encl F to *VAC AR*.

[15] Of the 125 LVTs used at Tarawa, 35 were sunk at sea, 26 were filled with water on the reef, 9 were burned on the beach as gas tanks ignited, and 2 were destroyed by mines on the beach. Eight tractors were put out of action by mechanical failures. Of the 500 men in the 2d Amphibian Tractor Batallion, 323 were killed, wounded, or missing in action, including the battalion commander, Major Henry C.

the time of GALVANIC, an armored amphibian tractor mounting a 37mm gun, the LVT(A), and an amphibious 2½-ton truck, the DUKW, were in production. Even though neither of these types had undergone an adequate combat test,[16] corps recommended that a battalion of armored tractors, two companies of the new DUKWs, and two battalions of ordinary LVTs be assigned each division in future assault landings.[17]

The fighting on Betio centered around pillboxes and shelters built of either steel and concrete or log and coral. Many of the flamethrowers which the Marines used so effectively against these installations had been made available by the Army Chemical Warfare Service detachment in Hawaii.[18] In spite of the help of the Army, there were not enough flamethrowers at Betio, so VAC recommended that in the future one such weapon be assigned to each rifle platoon. In addition, the Army, Navy, and Marine Corps agencies in Hawaii began cooperating in the development of a flame-throwing tank.

The Sherman tanks and the halftracks, which also mounted high-velocity 75mm guns, proved effective against the lighter Japanese installations. Because the 37mm guns of the light tanks could do little damage to prepared fortifications, Holland Smith's headquarters recommended that these vehicles be replaced by the heavier Shermans. Pack howitzers, which had to be wheeled into position by their crews, were not as effective in delivering direct fire against pillboxes as were the more maneuverable tanks and self-propelled guns.

Demolitions had proved so deadly that corps recommended the issue of one demolitions kit to each rifle squad. Flamethrowers, demolitions, and armor had enabled Marine infantrymen to close with and kill the enemy by means of grenades and rifle fire. Grenades, in fact, were so valuable that VAC urged still greater emphasis on the offensive or concussion type. Perhaps the most important lesson learned was that the destruction of a Japanese garrison as skilfully entrenched as the defenders of Betio was a task that required teamwork as well as courage.

Because units tended to become intermingled during the amphibious assault, individual Marines might find themselves commanded by a stranger. Under these adverse conditions, the riflemen had to fight as part of a hastily organized team. In the opinion of corps operations officers, this kind of teamwork could only result from the self-discipline, resourcefulness, and initiative of every unit leader. Leadership, then, would continue to be stressed in future training.[19]

ARTILLERY AND NAVAL GUNFIRE

Had circumstances not forced him to do otherwise, General Julian Smith would have seized the islands adjacent to Betio, emplaced artillery on them,

Drewes, killed on D-Day. Information supplied by LtGen Julian C. Smith, dtd 15Oct62.

[16] Two of the DUKWs were used at Makin Atoll.

[17] *VAC AR*, p. 12.

[18] Col George F. Unmacht, USA, "Flame Throwing Seabees," *U. S. Naval Institute Proceedings*, v. 74, no. 4 (Apr48), pp. 425–426.

[19] *VAC AR*, p. 20.

and shelled the main objective before attempting to storm it. The need to capture Betio as quickly as possible prevented him from landing elsewhere in the atoll prior to the principal assault, but a study of the operation indicated the soundness of the original idea. Although Holland Smith's headquarters had no choice but to veto such tactics at Tarawa, the corps headquarters now urged that every effort be made in future operations to land artillery on lightly defended islands within range of the major objective.[20]

At Betio in particular, great things had been expected of the preparatory naval bombardment. Representatives of V Amphibious Force, V Amphibious Corps, and the 2d Marine Division, had contributed their knowledge to the drafting of a naval gunfire plan. As a result of their combined efforts, a greater weight of metal was hurled into each square foot of Betio than had rained down on any previous amphibious objective, but the bombardment, awesome as it seemed, did not kill enough Japanese. The VAC commander noted, however, in his report on the operation that "without naval gunfire the landing could not have been made." [21]

GALVANIC taught naval gunfire officers that, when the requirements of a surprise attack did not preclude it, adequate preparation required days rather than hours of precision bombardment. To fire for two or three hours, much of the time shifting from one sector to another, was not enough. The sturdiest Japanese installations, many of them dug into the coral sands, could be penetrated only by a base-fused, armor-piercing shell plunging at a steep angle. Instead of the armor-piercing type, comparatively ineffective point-detonating, high-capacity ammunition was used at Betio. Although the training and rehearsals for GALVANIC had helped, especially in the accurate delivery of on-call fire, still more training was thought necessary. A simpler and more effective target designation system needed to be developed.[22] In the future, the officers of every supporting ship should know just what was expected of their guns. The ideal solution to the problems posed by the fortifications at Tarawa appeared to be the early arrival of the objective of thoroughly trained fire support units stocked with the proper ammunition, a deliberate bombardment designed to shatter possible strongpoints, additional shelling by destroyers and landing craft during the assault, and finally the accurate delivery of whatever fires the troops ashore might request.[23]

---

[20] *Ibid.*, p. 20.

[21] NGF SptRpt, n.d., p. 49, Encl A to *TF 53 AR.*

[22] The naval gunfire grid and target designation system used in the Gilberts proved to be "cumbersome and inaccurate" at times. In future Central Pacific operations, the Tactical Area Designation system, developed by a group of Army, Navy, and Marine intelligence and mapping officers at Pearl Harbor, was standard. The new system, based on a 1,000-yard grid broken down into 200-yard lettered squares was readily usable by all fire support agencies. Col Cecil W. Shuler comments on draft MS, dtd 12Dec62.

[23] *VAC AR*, pp. 16–17; Rpt of NavShoBomb, dtd 4Dec43, Encl H, and Important Recoms, dtd 4Dec43, Encl J, to *V PhibFor AR;* NGF Rpt, dtd 7Jan44, Encl 2 to SplStfRpts, n.d., Encl F to *VAC AR.*

Because the preliminary hammering of Betio had not achieved the spectacular results hoped for, the importance of naval gunfire to the success of the operation could easily be underestimated. The 3,000 tons of explosives that blasted the island caused many casualties, disrupted Japanese communications, and enabled the first three assault waves to gain the beaches without meeting organized resistance.[24] Once these Marines were ashore, the enemy rallied to inflict serious casualties on succeeding waves. This seemingly remarkable recovery was due in part to the lifting of naval gunfire where the LVTs were some distance from shore. Out of the entire task force, only the pair of destroyers in the lagoon could see the progress of the amphibian tractors and time their fires accordingly. The other fire support ships halted their bombardment according to a prearranged schedule that did not take into account the distance yet to be traveled by the assault waves. To prevent the premature lifting of preliminary fires, Admiral Hill's staff recommended that destroyers take up positions from whch they could track the incoming waves and thus keep firing as long as the friendly troops were not endangered.[25]

LOGISTICS

The original logistical plan for the Betio operation, though carefully drafted and based on previous amphibious experience, proved unrealistic. A beachhead was needed for the unloading of supplies and evacuation of casualties, but at Betio the Marines fought through the first day with their backs against the sea. Not until the long pier was pressed into service as a transfer point was there room to store or sort cargo. Even if space had been available ashore, landing craft could not have crossed the reef to reach the island. The carrying of supplies from the end of the pier, a point accessible to LCVPs and LCMs, to the front lines was best done by LVTs. Casualties were evacuated in the same vehicles that brought food, water, and ammunition to the embattled units. Wounded Marines were placed in the tractors and carried to the end of the pier where they were given emergency treatment and transferred to landing craft for the journey out to the transports. A naval officer in a minesweeper at the line of departure was given control over boat traffic, and the improvised system worked quite well. The Navy and Marine Corps officers responsible for beachhead logistics, when confronted with an unforeseen difficulty, had responded to the challenge.

The waters off the pier were usually dotted with landing craft waiting their turn to unload. This congestion was due to the conflicting needs of the Navy and Marine Corps as well as to the fact that supplies had to be funneled along the pier. The Marines wanted items landed in the order of their importance, but the Navy had to unload the transports as rapidly as possible regardless of the value of the cargo to the attack-

[24] General Shoup, noting the few casualties in the leading assault waves, commented: "I always attributed this to a destroyer on the flank which kept firing and kept the Japanese in their holes." *Shoup interview/comments.*

[25] TF 53 NGF SptRpt, n.d., p. 49, Encl A to *TF 53 AR.*

ing troops.[26] The longer the transports remained at the objective, the greater the danger to these valuable ships from Japanese planes and submarines.

In their eagerness to aid the assault troops, the ship crews were often too cooperative. As the commander ashore on D-Day noted:

> In their enthusiasm, they did not load what I wanted, they just loaded. By the time they got a message from me requesting certain items the boats were already filled with other material. Tarawa made SOP [Standing Operating Procedure] that the Navy would not unload supplies except as requested by the landing force commander ashore. . . . Items that come ashore must be in accordance with the requirements of those ashore.[27]

Although pallets, a few of which were used at Tarawa, were recommended for adoption, Marine planners pointed out that not all bulk supplies could be lashed to wooden frames. During the early hours of the assault, or when the beachhead was narrow and under fire, supplies would have to be landed rapidly and in comparatively small quantities. Once the beachhead had been won, these platforms appeared to be one of the best means of speeding the movement of cargo from the transports, across the beaches, and to inland dumps.

At Betio, supplies piled up on the beaches, for enemy opposition and a shortage of manpower prevented the shore party from functioning as planned. A single pioneer battalion from the engineer regiment was not equal to the task, and the Marines from certain of the rifle companies, men who were supposed to be sorting and moving supplies, had joined in the fighting, leaving their work to be done by whomever the shore party officers could press into service. In the opinion of the corps G–4 section, the shore party machinery was in need of overhaul, for the pioneer unit was not large enough to do its work without reinforcement. Until the table of organization could be revised, Marines from service rather than combat units should be detailed to aid in the logistical effort.[28]

The evacuation of casualties became increasingly efficient as the beachhead was expanded. On the first day, the wounded were placed in rubber boats and towed by hand to the edge of the reef where they were transferred to landing craft for the journey out to the transport area. Later, LVTs became available to evacuate the wounded to boats waiting at the end of the pier. Although the supplies of morphine, sulpha drugs, splints, dressings, and plasma proved adequate, the collecting platoons did run short of litters. At Tarawa, the doctors and corpsmen did a heroic job. The major recommendation to result from the operation was that hospital ships be assigned to task forces charged with seizing heavily defended atolls.[29]

---

[26] LtCol Robert D. Heinl, Jr., notes of an interview with BGen Merritt A. Edson, dtd 26May47.

[27] *Shoup interview/comments.*

[28] G–4 Rpt, dtd 4Jan44, pp. 3–4, Encl D, and Rpt of TQM, dtd 30Dec43, Encl 8 to SplStfRpts, n.d., pp. 13–14, Encl F to *VAC AR*.

[29] Rpt of MedObserver, FwdEch, GALVANIC, dtd 1Dec43, Encl 6 to SplStfRpts, n.d. Encl F to *VAC AR*.

THE ROLE OF AVIATION

Both defense against air attack and the close support of ground troops were entrusted to carrier planes flown by Navy pilots. One force of carriers stood ready to intercept Japanese planes striking from the Marshalls. Other carriers protected the Makin task force and attacked targets ashore, while a third group performed the same tasks at Tarawa. Except for unsuccessful night attacks on the retiring task forces, the enemy offered no serious challenge to American air defenses.

Air support at Tarawa, in the opinion of both Navy and Marine Corps officers, suffered many of the ills that had plagued naval gunfire support. The strikes made prior to the assault accomplished little, for not enough bombs were delivered, and those dropped were not heavy enough to damage Japanese emplacements. On D-Day, because of severe communication difficulties, there had been poor coordination of the aerial effort with the progress of the assault waves toward the beaches. The planes scheduled to attack at dawn arrived late, and those which were to support the landing were early, their pilots unaware of a postponement of H-Hour. Although the beaches were strafed just before H-Hour, the attack was not effective, for the Japanese, who had taken cover in air raid shelters and pillboxes, were immune to harm from machine gun bullets. Later in the operation, while supporting units ashore, the aviators had difficulty in locating their targets.

These shortcomings indicated, among other things, that effective air support was impossible unless the pilots and ground troops had trained as a team. Marine Colonel Peter P. Schrider, VAC air officer, was convinced that the division and its supporting carriers should train together for two or three days—the longest possible time at this stage of the war.[30] Holland Smith recommended that Marine aviators thoroughly schooled in the principles of direct air support should be assigned to escort carriers and included in any future amphibious operation undertaken by a Marine division. If this request could not be granted, he continued, the Navy airmen selected for the task should be carefully indoctrinated in the tactics they would employ.[31]

Air operations at Tarawa led to certain changes in doctrine, which would benefit soldiers and Marines storming other beaches. Unless dive bombers were attacking a particular target which might be obscured by smoke and dust, there was no need to suspend naval gunfire while the planes executed the strike. No danger of shells striking aircraft existed as long as the pilots pulled out of their dives at an altitude higher than the maximum ordinate of the naval guns. In addition, the practice of scheduling the final aerial attack in relation to H-Hour was judged to be unrealistic. Those planes assigned to maintain the neutralization of the beaches just prior to the assault should begin their attack when the landing craft were about 1,000 yards from the objective and continue the bombing and strafing until the assault waves were approximately 100 yards from shore.

---

[30] Rpt of AirO, dtd 6Jan44, Encl 1 to SplStfRpts, n.d., p. 2, Encl F to *VAC AR*.

[31] *VAC AR*, p. 16.

Finally, since machine gun fire had proved ineffective against beach defenses, some sort of gasoline bomb was needed, a device which would insure that the defenses remained silenced during the last few minutes of the ship-to-shore movement.[32]

Not only were important lessons learned from GALVANIC, but many necessary changes in amphibious tactics and techniques were made almost immediately. By the time of the Marshalls operation, for example, naval gunfire would improve in both accuracy and volume. On the other hand, the war had reached the Marianas before effective coordination was achieved between air strikes and naval gunfire. As one study of amphibious warfare has phrased it, "Tactically, Betio became the textbook for future amphibious landings and assaults." [33] In the flames of Tarawa was tempered the sword that would cut to the heart of the Japanese Pacific empire.

[32] Rpt of AirSpt, n.d., Encl A to *TF 53 AR*, pp. 58–61.

[33] Isely and Crowl, *Marines and Amphibious War*, p. 251.

PART III

The Marshalls: Quickening the Pace

FLINTLOCK Plans and Preparations[1]

GETTING ON WITH THE WAR

During the series of Allied conferences that resulted in approval for the Central Pacific campaign, the first proposed objective was the Marshalls. Because of the lack of information concerning these islands and the shortage of men and materiel, the initial blow struck the Gilberts instead. After the capture of Apamama, Makin, and Tarawa, planes based at these atolls gathered the needed intelligence. As this information was processed, American planners prepared and revised several concepts for an offensive against the Marshalls.

Like GALVANIC, the invasion of the Marshalls was the responsibility of the Commander in Chief, Pacific Ocean Areas, Admiral Nimitz. His principal subordinate planner was Admiral Spruance, Commander, Fifth Fleet and Central Pacific Force.[2] Admiral Turner, Commander, V Amphibious Force, and General Holland Smith, Commanding General, V Amphibious Corps, were the officers upon whom Spruance relied for advice throughout the planning of the operation.

EARLY PLANS FOR THE MARSHALLS

The Marshalls consist of two island chains, Ratak (Sunrise) in the east and Ralik (Sunset) in the west. Some 32 atolls of varying size form the Marshalls group. Those of the greatest military importance by late 1943 were Mille, Maloelap, and Wotje in the Ratak chain, and in the Ralik chain, Jaluit, Kwajalein, and Eniwetok. Except for

[1] Unless otherwise noted, the material in this chapter is derived from: US PacFleet OPlan 16–43, rev, dtd 14Dec43; CenPacFor OPlan Cen 1–44, dtd 6Jan44; USAFPOA, Participation in the Kwajalein and Eniwetok Ops, dtd 30Nov4 (OAB, NHD); TF 51 OPlan A6–43, dtd 3Jan44, hereafter *TF 51 OPlan A6–43;* VAC Rpt on FLINTLOCK, dtd 6Mar44, hereafter *VAC AR FLINTLOCK;* VAC AdminO 1–44, dtd 5Jan44; TF 52 AtO A1–44, dtd 14Jan44; TF 53 OpO A157–44, dtd 8Jan44; TF 53 Rpt of PhibOps for the Capture of Roi and Namur Islands, dtd 23Feb44, hereafter *TF 53 AR Roi-Namur;* 4th MarDiv OPlan 3–44 (rev), dtd 10Jan44, hereafter *4th MarDiv OPlan 3–44;* 4th MarDiv Final Rpt of FLINTLOCK Op, dtd 28Mar44, hereafter *4th MarDiv AR;* Crowl and Love, *The Gilberts and Marshalls;* LtCol Robert D. Heinl, Jr. and LtCol John A. Crown, *The Marshalls: Increasing the Tempo* (Washington: HistBr, G–3, HQMC, 1954), hereafter Heinl and Crown, *The Marshalls.* Unless otherwise noted, all documents cited are located in the Marshalls Area OpFile and Marshalls Cmt File, HistBr, HQMC.

[2] The Central Pacific Force was, at this stage of the war, also known as the Fifth Fleet. After the Marshalls operation, the latter title was habitually used.

Jaluit, which was a seaplane base,[3] all of these atolls were the sites of enemy airfields, and those in the Ralik chain were suitable as naval anchorages.[4] (See Map 7.)

In May 1943, at the Washington Conference, the CCS recommended to the Allied heads of state that an offensive be launched into the Marshalls. At this time, American planners believed that the services of two amphibious divisions and three months' time would be needed to neutralize or occupy all of the major atolls in the group and Wake Island, as well. The JCS considered the 1st, 2d, and 3d Marine Divisions available for immediate service and was certain that the 4th Marine Division, then training in the United States, would be ready for combat by the end of the year.[5]

After the Washington Conference had adjourned, the JCS directed Admiral Nimitz to submit a plan for operations against the Marshalls, and the admiral responded with a preliminary proposal,[6] necessarily vague because he lacked adequate information on the area. Within three weeks after receiving Nimitz' views, on 20 July the JCS directed him to plan for an attack against the Gilberts, a move to be made prior to the Marshalls offensive. Thus, early planning for the Marshalls coincided with preparations for GALVANIC.

By the end of August, Nimitz and his staff had carefully evaluated the proposed Marshalls operation. In their opinion, the United States was strong enough to undertake an offensive that would strengthen the security of Allied lines of communications, win bases for the American fleet, force the enemy to redeploy men and ships, and possibly result in a stinging defeat for the *Imperial Navy*. The attackers, however, would need to gain aerial superiority in the area and obtain accurate intelligence. A solution was required for the logistical problem of sustaining the fleet in extended operations some 2,000 miles west of Pearl Harbor. Finally, VAC would have to speed the training of the 35,000 amphibious troops required for the campaign. The proposed objectives were key islands in Kwajalein, Wotje, and Maloelap Atolls. Central Pacific amphibious forces were to seize all of these simultaneously while ships and aircraft neutralized Jaluit and Mille. Nimitz now requested specific authorization to seize control of the Marshalls, urging that "thus we get on with the war." [7]

At the Quebec Conference of August 1943, Allied leaders agreed that an effort against the Marshalls should follow the successful conquest of the Gilberts. Accordingly, the JCS on 1 September issued Nimitz a directive to undertake the operations he had recently proposed and, upon their completion, to seize or neutralize Wake Is-

[3] RAdm Charles J. Moore cmts on draft MS, dtd 25Jan63, hereafter *Moore comments Marshalls*.

[4] VAC G-2 Study of the Theater of Ops; Marshall Islands, dtd 26Nov43, pp. 1–2.

[5] JCS 304, Ops in the Pac and Far East in 1943–1944, dtd 12May44 (OPD File, ABC Pac, WWII RecsDiv, FRC, Alexandria, Va.)

[6] CinCPac disp to CominCh, ser no. 0096, dtd 1Jul43, referred to in CinCPac disp to ComincCh, ser no. 00151, dtd 20Aug43 (OPlan File, OAB, NHD).

[7] CinCPac disp to CominCh, ser no. 00151, dtd 20Aug43.

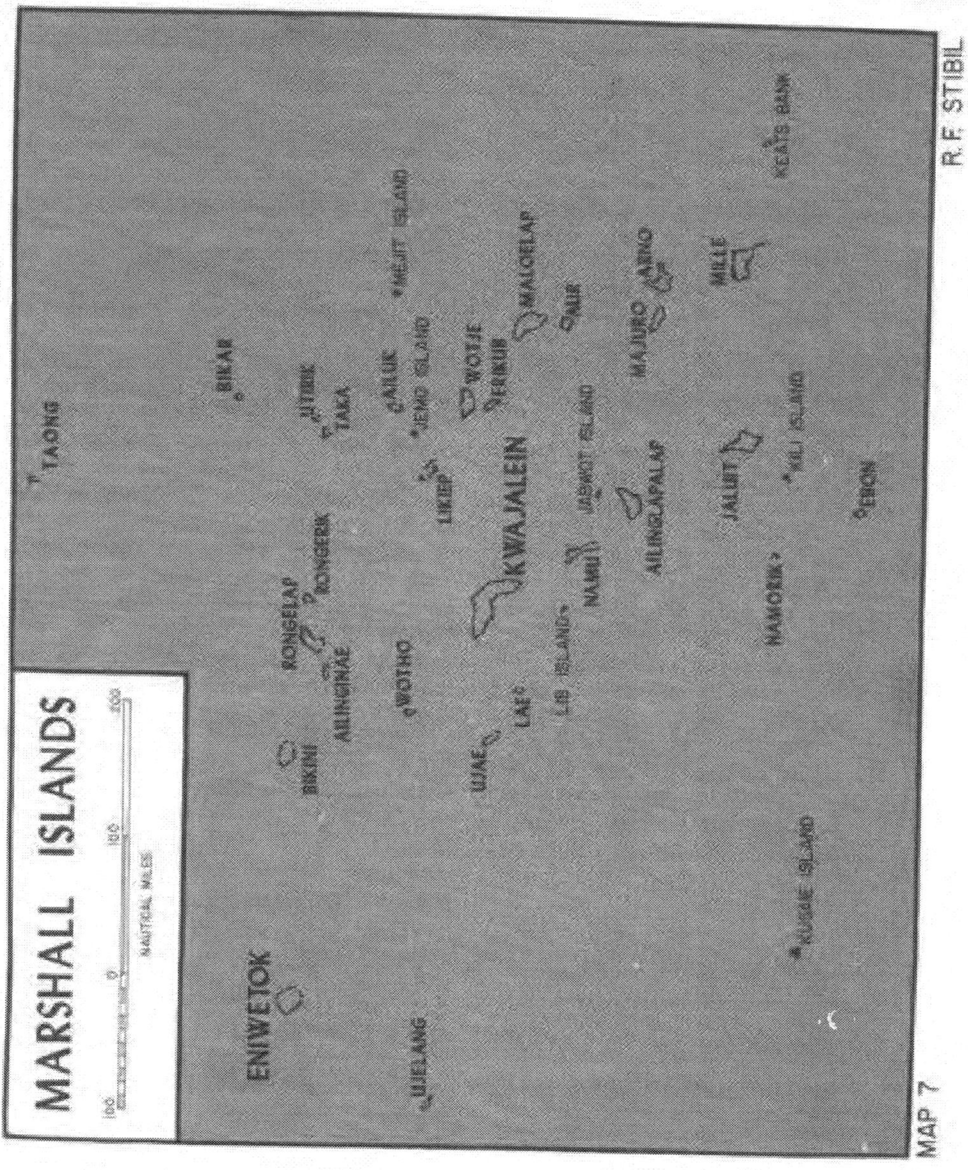
MARSHALL ISLANDS
NAUTICAL MILES
ENIWETOK
UJELANG
BIKINI
AILINGINAE
RONGELAP
RONGERIK
WOTHO
UJAE
LAE
KWAJALEIN
LIKIEP
WOTJE
ERIKUB
AILUK
MEJIT ISLAND
MALOELAP
AUR
MAJURO
ARNO
MILLE
JALUIT
KILI ISLAND
EBON
NAMORIK
NAMU
JABWOT ISLAND
AILINGLAPALAP
BIKAR
TAONG
KEATS BANK
KUSAIE ISLAND
MAP 7
R. F. STIBIL

214-881 O-67—9

land and Eniwetok, as well as Kusaie in the Carolines. By this time, the 2d Marine Division was committed to GALVANIC, the 1st to the New Britain operation, and the 3d to the Solomons offensive. As assault troops for the Marshalls, the JCS made available, pending the completion of its training, the 4th Marine Division and also selected the 7th Infantry Division, which had seen action in the Aleutians, and the 22d Marines, then guarding Samoa.[8] See Map I, Map Section.)

THE SHAPING OF FLINTLOCK [9]

On 22 September, Nimitz handed Spruance a preliminary study in support of the course of action he had proposed to the JCS and directed him to prepare to assault the Marshalls on 1 January 1944. The study itself was not considered complete, so the objectives might be altered as additional intelligence became available.[10] Because of this lack of information on the Marshalls area, Spruance began studying two alternatives to Nimitz' suggested course of action. All of these proposals called for simultaneous assaults, at sometime in the operation, upon three atolls, Maloelap, Wotje, and either Mille or Kwajalein.[11]

Although Nimitz on 12 October issued an operation plan for FLINTLOCK, the Marshalls Operation, he avoided selecting specific objectives. Within two days, however, he decided to employ the 7th Infantry Division against both Wotje and Maloelap and to attack Kwajalein with the 4th Marine Division and 22d Marines. He fixed 1 January 1944 as target date for the storming on Wotje and Maloelap and proposed to attack Kwajalein on the following day.

General Holland Smith's VAC staff now prepared an estimate of the situation based on the preliminary plans advanced by Admirals Nimitz and Spruance. The likeliest course of action, according to the VAC paper, was to strike simultaneously at Wotje and Maloelap, with the Kwajalein assault troops serving as reserve. On the following day, or as soon as the need for reinforcements had passed, the conquest of the third objective would begin. Smith's headquarters drew up a tentative operation plan for such a campaign, but at this point the attack against the Gilberts temporarily halted work on FLINTLOCK.

Prior to the GALVANIC operation, Admiral Turner had done little more than gather information concerning the proposed Marshalls offensive. Immediately following the conquest of the Gilberts, Turner's staff carefully examined the FLINTLOCK concept and concluded that Maloelap and Wotje should be secured before Kwajalein was attacked. Meanwhile, every planning agency in the Central Pacific Area was digesting the lessons of GALVANIC. Among other things, the theories regarding naval gunfire were revised.

---

[8] JCS disp to CinCPac, dtd 1Sep43, Encl A to CinCPac disp to ComCenPac, ser no. 00190, dtd 22Sep43 (OPlan File, OAB, NHD).

[9] Originally, the Marshalls operation had been given the code name BANKRATE, but this title was abandoned early in the planning phase.

[10] CinCPac disp to ComCenPac ser no. 01900, dtd 22Sept43 (OPlan File, OAB, NHD).

[11] ComCenPac disp to Com VPhibFor and CG VAC ser no. 0053, dtd 10Oct43 (OPlan File, OAB, NHD).

As an Army officer assigned to General Smith's staff phrased it, "Instead of shooting at geography, the ships learned to shoot at definite targets." [12] After they had evaluated events in the Gilberts and assessed their own strength, Turner and Smith agreed that with the forces available Kwajalein could not be taken immediately after the landings on Wotje and Maloelap. Nimitz, acting on the same information available to his subordinates, also desired to alter FLINTLOCK, but in an entirely different manner.

On 7 December, CinCPac proposed an amphibious thrust at Kwajalein in the western Marshalls, coupled with the neutralization of the surrounding Japanese bases. In a series of conferences of senior commanders that followed, General Smith joined Admirals Turner and Spruance in objecting to this bold stroke.[13] Spruance, the most determined of the three, pointed out that immediately after the capture of Kwajalein units of his Central Pacific Force were scheduled to depart for the South Pacific. Once the fast carriers had steamed southward, he could no longer maintain the neutralization of Wotje, Maloelap, Mille, and Jaluit, and the enemy would be able to ferry planes to these Marshalls bases in order to attack the line of communications between the Gilberts and Kwajalein. Spruance also desired to ease the logistical strain by seizing an additional fleet anchorage in the Marshalls. To meet the last objective Nimitz included in FLINTLOCK the capture of a second atoll, one that was weakly defended. To cripple Japanese air power, he approved a more thorough pounding of the enemy bases that ringed Kwajalein.[14]

After informing the JCS of his change of plans, Nimitz on 14 December directed Spruance and his other subordinates to devise a plan for the assault on Roi and Kwajalein Islands in Kwajalein Atoll. The alternative objectives were Maloelap and Wotje, but whichever objectives were attacked, D-Day was fixed as 17 January 1944.[15] On 18 December, Nimitz informed King that he had set back D-Day to 31 January in view of the need for additional time for training and the need to make repairs to the carriers USS *Saratoga*, *Princeton*, and *Intrepid*.[16]

The assignment of another reinforced regiment, the 106th RCT of the 27th Infantry Division, to the FLINTLOCK force increased the number of men available for the expanded plan, but Turner continued to worry about the readiness of the various units. On 20 December, he requested that D-Day be postponed until 10 February to allow the two divisions to receive the proper equipment and to enable the 4th Marine Division to hold rehearsals.[17] No further delays were authorized, however, as the JCS had directed that

[12] Col Joseph C. Anderson, USA, ltr to CMC, dtd 23Jan53, hereafter *Anderson ltr.*

[13] FAdm Chester W. Nimitz ltr to CMC, dtd 27Feb53; Adm Richmond K. Turner ltr to CMC, dtd 27Feb53, hereafter *Turner ltr I.*

[14] *Turner ltr I;* Adm Raymond A. Spruance ltr to CMC, dtd 12Jan53.

[15] CinCPac disp to ComCenPac, ser no. 001689, dtd 14Dec43 (OPlan File, OAB, NHD).

[16] CinCPac disp to ComInCh, ser no. 0236, dtd 18Dec43 (OPlan File, OAB, NHD).

[17] *Turner ltr I.*

the operation get under way "not later than 31 January 1944." [18]

Nimitz' headquarters on 20 December issued FLINTLOCK II, a joint staff study which incorporated the results of his recent conversations with Spruance. Carrier aircraft, land-based bombers, and surface ships were to blast the Japanese bases at Wotje and Maloelap. If necessary, the carriers would launch strikes to aid land-based planes in neutralizing Mille, Jaluit, Kusaie, and Eniwetok. The primary objectives remained Roi and Kwajalein Islands, but a secondary target, Majuro Atoll, was also included.

Admiral Spruance, in reviewing the reasons that he recommended Majuro as an objective, stated:

> Airfields on Majuro would enable us to help cover shipping moving in for the buildup of Kwajalein, and it would give us a fire protected anchorage at an early date for fleet use, if the capture of Kwajalein were a protracted operation. We had been fortunate during the Gilberts operation in being able to fuel fleet forces at sea without having them attacked by submarines. This we did by shifting the fueling areas daily. There were too many islands through the Marshalls for that area to lend itself to this procedure.[19]

With the final selection on 26 December of an assault force for Majuro, the FLINTLOCK plan was completed. For a time, General Smith had considered using most of Tactical Group I, the 22d Marines and the 106th Infantry, against Majuro. A staff officer of Tactical Group I, who was present during the discussions of this phase of the operation, recalled that "General Holland Smith paced the floor of the little planning room, cigar butt in mouth or hand —thinking out loud." Thanks to additional intelligence, the choice by this time lay between employing an entire regiment or a smaller force. After weighing the evidence, Smith announced he was "convinced that there can't be more than a squad or two on those islands today . . . let's use only one battalion for the Majuro job." [20] As a result, 2/106 was given the task of seizing Majuro, while the remainder of that regiment and the 22d Marines were designated the reserve for FLINTLOCK.

ORGANIZATION AND COMMAND

Task Force 50, commanded by Admiral Spruance, included all the forces assigned to the FLINTLOCK operation. Its major components were: Task Force 58, Rear Admiral Marc A. Mitscher's fast carriers and modern battleships; Task Force 57, Defense Forces and Land-Based Air, commanded by Rear Admiral John H. Hoover; Task Group 50.15, the Neutralization Group under Rear Admiral Ernest G. Small; and Admiral Turner's Task Force 51, the Joint Expeditionary Force. Admiral Spruance decided to accompany the expedition to the Marshalls, but he would not assume tactical command unless the *Imperial Japanese Navy* chose to contest the operation.

Admiral Turner, as commander of

[18] CominCh memo to CinCPac, ser no. 002415, dtd 4Nov44 (OPlan File, OAB, NHD).

[19] Adm Raymond A. Spruance ltr to ACofs, G–3, HQMC, dtd 10Sep62, hereafter *Spruance 62 ltr.*

[20] Col Wallace M. Greene, Jr., ltr to CMC, dtd 23Nov52, hereafter *Greene ltr I.*

the Joint Expeditionary Force, was primarily concerned with conveying the assault troops to the objective and getting them safely ashore. Within his command were: the Southern Attack Force, over which he retained personal command; the Northern Attack Force, entrusted to Rear Admiral Richard L. Conolly, a veteran of the Sicily landings; the Majuro Attack Group under Rear Admiral Hill, commander at Tarawa; Captain Harold B. Sallada's Headquarters, Supporting Aircraft, the agency through which Admirals Turner, Conolly, and Hill would direct aerial support of the landings; and General Smith's Expeditionary Troops. Among the 297 vessels assigned to Turner for FLINTLOCK were two new AGC command ships, 7 old battleships, 11 carriers of various classes, 12 cruisers, 75 destroyers and destroyer escorts, 46 transports, 27 cargo vessels, 5 LSDs, and 45 LSTs.[21]

As far as General Smith's status was concerned, Spruance's command structure for FLINTLOCK fit the situation and continued the primary responsibility of Admiral Turner for the success of the operation.[22] Until the amphibious phase was completed and the troops were ashore, Admiral Turner would, through the attack force commanders, exercise tactical control. After the 7th Infantry Division had landed on Kwajalein Island and the 4th Marine Division on Roi-Namur, General Smith was to assume the authority of corps commander and retain it until Admiral Spruance declared the capture and occupation of the objectives to be completed. The authority of the Marine general, however, was as limited as it had been in the Gilberts operation, for he could not make major changes in the tactical plan nor order unscheduled major landings without the approval of Admiral Turner. Included in Expeditionary Troops with the two assault divisions were the 106th Infantry, 22d Marines, the 1st and 15th Marine Defense Battalions, Marine Headquarters and Service Squadron 31, and several Army and Navy units which would help garrison and develop the captured atolls.

At Roi-Namur, objective of the Northern Attack Force, and at Kwajalein Island, where the Southern Attack Force would strike, Admirals Conolly and Turner were initially to command the assault forces through the appropriate landing force commander. As soon as the landing force commander knew that his troops had made a lodgement, he was to assume command ashore. The Majuro operation was an exception, for Admiral Hill, in command of the attack group, was in control from the time his ships arrived, throughout the fighting ashore, until Admiral Spruance proclaimed the atoll captured.

APPLYING THE LESSONS OF TARAWA

Everyone who took part in planning FLINTLOCK profited from the recent GALVANIC operation. To prevent a repetition of the sort of communications failures that had happened off

[21] ComInCh, *Amphibious Operations: The Marshall Islands, January-February 1944*, dtd 20May44, p. 1:5, hereafter ComInCh, *Marshall Islands*.

[22] *Moore comments Marshalls.*

Betio, the commander of each attack force was to sail in a ship especially designed to serve as a floating headquarters during an amphibious assault. The AGC *Rocky Mount* would carry Turner to Kwajalein, while Conolly would command the Roi-Namur assault from the AGC *Appalachian*. The *Cambria*, a transport equipped with additional communications equipment, was assigned to Admiral Hill for use at Majuro.

Prior to the attack on Tarawa, Marine planners had requested permission to land first on the islands near Betio to gain artillery positions from which to support the main assault. The loss of surprise and the consequent risk to valuable shipping were judged to outweigh the tactical benefits to be gained from these preparatory landings, and the 2d Marine Division was directed to strike directly at the principal objective. Such was not the case in FLINTLOCK. Plans called for both the 7th Infantry Division and the 4th Marine Division to occupy four lesser islands before launching their main attacks.

In addition to providing for artillery support of the major landings, planners sought to increase the effectiveness of naval gunfire. On D minus 1, while cruisers and destroyers of Task Force 51 bombarded Maloelap and Wotje, Admiral Mitscher's fast battleships were to hammer Roi-Namur and Kwajalein Islands. At dawn, elements of Task Force 58 would begin the task of destroying Japanese aircraft, making the flight strips temporarily useless, and shattering coastal defense guns. After pausing for an air strike, the ships were to resume firing, primarily against shore defenses. On D-Day, the landing forces would seize certain small islands adjacent to the main objectives. These operations were to be supported by naval gunfire and aerial bombardment in a manner similar to that planned for the assaults on Kwajalein Island and Roi-Namur. Plans also called for the American warships to maintain the neutralization of principal objectives while supporting the secondary landings elsewhere in the atoll.

About 25 minutes before H-Hour for the main landings, cruisers, destroyers, and LCI(G)s were to begin firing into the assault beaches, distributing high explosives throughout an area bounded by lines 100 yards seaward of the edge of the water, 200 yards inshore, and 300 yards beyond both flanks. Admiral Turner directed that cruisers continue their bombardment until the landing craft were 1,000 yards from shore, destroyers until the assault waves were 500 yards or less from the island, and LCI(G)s until the troops were even closer to their assigned beaches. Since the plan depended upon the progress of the assault rather than on a fixed schedule, the defenders would not be given the sort of respite gained by the Betio garrison.[28]

[28] The executive officer of the 106th Infantry recalled that, during a briefing of principal commanders and staff officers at Pearl Harbor in January, Admiral Turner said, in effect: "I say to you commanders of ships—your mission is to put the troops ashore and support their attack to the limit of your capabilities. We expect to lose some ships! If your mission demands it, risk *your* ship!" Col Joseph J. Farley, AUS, ltr to Head, HistBr, G-3, HQMC, dtd 20Oct62.

The LCI(G)s which figured so prominently in Admiral Turner's plans were infantry landing craft converted into shallow-draft gunboats. These vessels mounted .50 caliber machine guns, 40mm and 20mm guns, as well as 4.5-inch rockets. Another means of neutralizing the beach defenses was provided by the armored amphibian, LVT(A)(1), which boasted a 37mm gun and five .30 caliber machine guns. One machine gun was located atop the turret, one was mounted coaxially with the cannon, a third was located in a ball and socket mount in the forepart of the hull, and the other two were placed on ring mounts to the rear of the turret.[24] Protection for the crew of six was provided by ¼ to ½ inch of armor plate and by small shields fixed to the exposed machine guns. Neither the LCI(G)s nor the LVT(A)(1)s were troop carriers.[25] A few LVT(2)s with troops embarked were equipped with multiple rocket launchers to assist in the last-minute pounding of Japanese shore defenses.

Admiral Turner and General Smith also attempted to increase the effectiveness of supporting aircraft. The strikes delivered to cover the approaching assault waves were scheduled according to the progress of the LVTs. When the amphibian tractors reached a specified distance from the beaches, the planes would begin their attacks, diving parallel to the course of the landing craft and at a steep angle to lessen the danger of accidentally hitting friendly troops. During these pre-assault aerial attacks, both naval guns and artillery were ordered to suspend firing.

Throughout the operation, carrier planes assigned to support ground troops were subject to control by both the Commander, Support Aircraft, and the airborne coordinator. The coordinator, whose plane remained on station during daylight hours, could initiate strikes against targets of opportunity, but the other officer, who received his information from the attack force commanders, was better able to arrange for attacks that involved close cooperation with artillery or naval gunfire. During GALVANIC, the airborne coordinator had performed the additional task of relaying information on the progress of the battle. This extra burden now fell to a ground officer, trained as an aerial observer, who would report from dawn to dusk on the location of friendly units, enemy strongpoints, and hostile activities.[26]

THE LANDING FORCE PLANS

The objectives finally selected for FLINTLOCK were Majuro and Kwajalein Atolls. Measuring about 24 miles from east to west and 5 miles from north to south, Majuro was located 220 nautical miles southeast of Kwajalein. Admiral Hill, in command of the Majuro force, decided to await the results of a final reconnaissance before choosing his course of action. Elements of the VAC Reconnaissance Company

[24] Col Louis A. Metzger ltr to Head, HistBr, G-3, HQMC, dtd 24Oct62, hereafter *Mtezger ltr*.

[25] ONI, ND, *Allied Landing Craft and Ships, Supplement No. 1 to ONI 226* (Washington, 1945), PhibVehsSec.

[26] CominCh, *Marshall Islands*, p. 2:7.

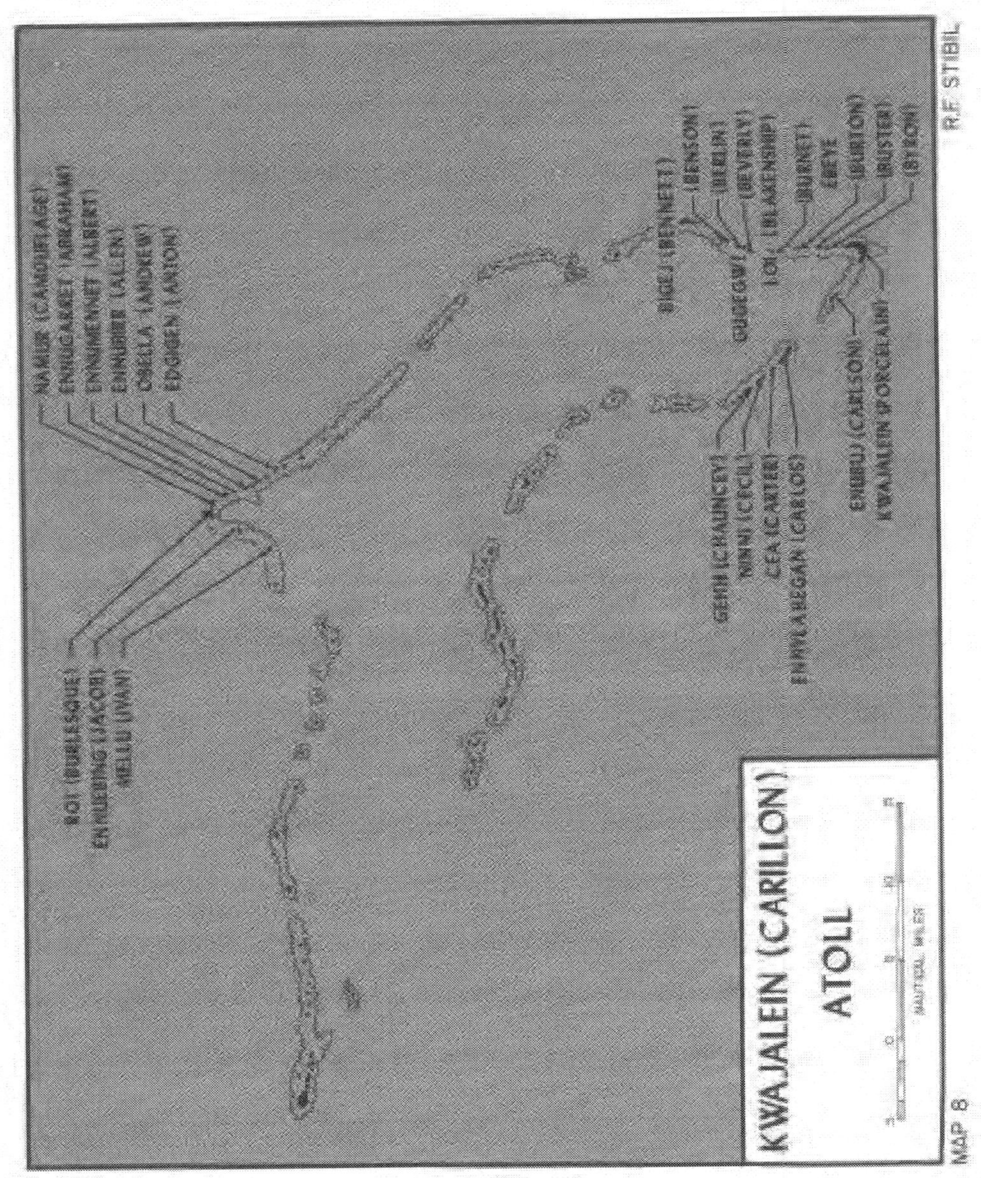

MAP 8

R.F. STIBIL

would land on Eroj and Calalin, the islands that guarded the entrance to Majuro lagoon, then scout the remaining islands. Once Japanese strength and dispositions had been determined, the landing force, 2/106, could make its assault.

Kwajalein Atoll, 540 miles northwest of Tarawa, is a triangular grouping of 93 small reef-encircled islands. The enclosed lagoon covers 655 square miles. Because of the vast size of the atoll, Admiral Turner had divided the Expeditionary Force into Northern and Southern Landing Forces. In the north, at the apex of the triangle, the recently activated 4th Marine Division, commanded by Major General Harry Schmidt, a veteran of the Nicaraguan campaign, was to seize Roi-Namur, twin islands joined by a causeway and a narrow strip of beach. The site of a Japanese airfield, Roi had been stripped of vegetation, but Namur, where the enemy had constructed numerous concrete buildings, was covered with palms, breadfruit trees, and brush. The code names chosen for the islands were CAMOUFLAGE for wooded Namur and for Roi, because so little of it was concealed, BURLESQUE.[27] (See Map 8.)

Crescent-shaped Kwajalein Island, objective of the Southern Landing Force, lay at the southeastern corner of the atoll, some 44 nautical miles from Roi-Namur. Major General Charles H. Corlett, who had led the Kiska landing force, would hurl his 7th Infantry Division against the largest island in the atoll. Here the enemy had constructed an airfield and over 100 large buildings. Although portions of the seaward coastline were heavily wooded, an extensive road net covered most of the island.

Throughout the planning of the Marshalls operation, General Schmidt and his staff were located at Camp Pendleton, California, some 2,200 miles from General Smith's headquarters at Pearl Harbor. The problem posed by this distance was solved by shuttling staff officers back and forth across the Pacific, but division planners continued to work under two disadvantages, a shortage of time and a lack of information. These twin difficulties stemmed from Admiral Nimitz' sudden decision to attack Kwajalein Atoll, bypassing Wotje and Maloelap. The division staff, however, proved adequate to the challenge, and by the end of December its basic plan had been approved by VAC. The timing of approval and issue was so tight, however, that some units sailed for Hawaii without seeing a copy.[28]

The Northern Landing Force plan consisted of three phases: the capture of four offshore islands, the seizure of Roi-Namur, and the securing of 11 small islands along the northeastern rim of Kwajalein Atoll. The first phase was entrusted to the IVAN Landing Group, the 25th Marines, Reinforced, commanded by Brigadier General James L. Underhill, the Assistant Division Commander. These troops were to seize ALBERT (Ennumennet), ALLEN (Ennubirr), JACOB (Ennuebing), and IVAN (Mellu) Islands as firing positions for

[27] BGen Homer L. Litzenberg ltr to CMC, dtd 31Jan53.

[28] *Metzger ltr.*

the 14th Marines, the division artillery regiment. The troops involved in this operation would land from LVTs provided by the 10th Amphibian Tractor Battalion. Company A, 11th Amphibian Tractor Battalion, which reinforced the 10th, along with Companies B and D, 1st Armored Amphibian Battalion, were chosen to spearhead the assaults. When this phase was completed, the LVT and artillery units would revert to division control, and the 25th Marines would become the division reserve for the next phase.

The 23d Marines received the assignment of storming Roi while the 24th Marines simultaneously attacked Namur. Both regiments were to land from the lagoon, the 23d Marines over Red Beaches 2 and 3 and the 24th Marines on Green 1 and 2. In the meantime, the 25th Marines could be called upon to capture ABRAHAM (Ennugarret) Island.[29] Detailed plans for the final phase were not issued at this time.

General Schmidt organized his assault waves to obtain the most devastating effect from his armored amphibians and LCI gunboats. The LCI(G)s were to lead the way until they were about 1,000 yards from the beach. Here they were to halt, fire their rockets, and continue to support the assault with their automatic weapons. Then the LVT(A)s would pass through the line of gunboats, open fire with 37mm cannon and machine guns, and continue their barrage "from most advantageous positions."[30] The troop carriers were directed to follow the armored vehicles, passing through the line of supporting amphibians if it was stopped short of the beach. The few LVT(2)s armed with rockets were to discharge these missiles as they drew abreast of the LCIs.

The 7th Infantry Division faced fewer difficulties in planning for the capture of Kwajalein Island. General Corlett was experienced in large-scale amphibious operations, and two of his regiments, the 17th and 32d Infantry, had fought at Attu, while the third, the 184th Infantry, had landed without opposition at Kiska. The Army division easily kept pace with the changes in the FLINTLOCK concept, for its headquarters was not far from General Smith's corps headquarters.

Like the Marine division in the north, General Corlett's Southern Landing Force faced an operation divided into several phases. The first of these was the capture of CARLSON (Enubuj), CARLOS (Ennylabegan), CECIL (Ninni), and CARTER (Gea) Islands by the 17th Infantry and its attached units. Once these objectives were secured and artillery emplaced on CARLSON, the 17th Infantry would revert to landing force reserve. Next, the 184th and 32d Infantry would land at the western end of Kwajalein Island and attack down the long axis of the island. The third phase, the seizing of BURTON (Ebeye), BURNET (unnamed), BLAKENSHIP (Loi), BUSTER (unnamed), and BYRON (unnamed), as well as the final oper-

[29] The attack on ABRAHAM eventually was scheduled to precede the Roi-Namur landings.

[30] 4th MarDiv LdgSked, dtd 10Jan44, Anx M to *4th MarDiv OPlan 3-44*.

ations, the landings on BEVERLY (South Gugegwe), BERLIN (North Gugegwe), BENSON (unnamed), and BENNETT (Bigej), were tentatively arranged, but the assault troops were not yet designated.[31] (See Map 8.)

The assault formations devised by Corlett's staff differed very little from those in the 4th Marine Division plan. Instead of preceding the first assault wave, the armored LVTs, amphibian tanks in Army terminology, were to take station on its flanks. Also, the Army plan called for the LVT(A)s to land regardless of Japanese opposition and support the advance from positions ashore. After the infantry had moved 100 yards inland, the amphibians might withdraw.[32]

INTELLIGENCE [33]

When Admiral Nimitz first began planning his Marshalls offensive, he had little information on the defenses of those islands. Because the enemy had held the area for almost a quarter-century, the Americans assumed that the atolls would be even more formidable than Tarawa. The first photographs of the probable objectives in the western Marshalls were not available to General Smith's staff until after GALVANIC was completed. The corps, however, managed to complete its preliminary area study on 26 November. Copies of this document were then sent to both assault divisions. Throughout these weeks of planning, the 7th Infantry Division G–2 was a frequent visitor to General Smith's headquarters, and this close liaison aided General Corlett in drafting his landing force plan. Unfortunately, close personal contact with the 4th Marine Division staff was impossible, but corps headquarters did exchange representatives with General Schmidt's command.

Carrier planes photographed Kwajalein Atoll during a raid on 4 December, but the pictures they made gave only limited coverage of this objective. Interviews with the pilots provided many missing details. Additional aerial photos of the atoll were taken during December and January. Reconnaissance planes took pictures of Majuro on 10 December. A final photographic mission was scheduled for Kwajalein atoll just two days before D-Day.

Submarines also contributed valuable intelligence on reefs, beaches, tides, and currents of Kwajalein. The *Seal* photographed the atoll in December, and the *Tarpon* carried out a similar mission the following month. Plans called for Underwater Demolition Teams, making their first appearance in combat, to finish the work begun by the undersea craft. These units were to scout the beaches of Kwajalein and Roi-Namur Islands on the night of 31 January-1 February. After obtaining up-to-date hydrographic data, the swimmers would return to destroy mines and antiboat obstacles.

---

[31] 7th InfDiv FO No. 1, dtd 6Jan44, FO Phase II, dtd 6Jan44, FO Phase III, dtd 12Jan44, FO Phase IV, dtd 12Jan44, FO Phase V, dtd 12Jan44, and FO No. 5, dtd 4Feb44.

[32] LVT Anx, dtd 8Jan44, Anx 8 to 7th InfDiv FO No. 1, dtd 6Jan44, hereafter *7th InfDiv FO 1*.

[33] Additional sources for this section include: 4th MarDiv Est of Sit for Kwajalein Island, n.d.; IntelPlan, n.d., Anx 3 to *7th InfDiv FO 1*.

By mid-January, VAC intelligence officers had concluded that Kwajalein Atoll, headquarters of the *6th Base Force* and, temporarily, of the *Fourth Fleet*, was the cornerstone of the Marshalls fortress. Originally, most of the weapons emplaced on the larger islands of the atoll had been sited to protect the ocean beaches, but since the Tarawa operation, in which the Marines had attacked from the lagoon, the garrisons were strengthening and rearranging their defenses. Except for Kwajalein Island, where photographs indicated a cross-island line, the Japanese had concentrated their heaviest installations along the beaches. In general, the assault forces could expect a bitter fight at the beaches as the enemy attempted to thwart the landing. Once this outer perimeter was breached, the defenders would fight to the death from shell holes, ruined buildings, and other improvised positions.

The atoll garrison was believed to be composed of the *6th Base Force, 61st Naval Guard Force,* a portion of the *122d Infantry Regiment*, and a detachment of the *4th Civil Engineers*. Intelligence specialists believed that reinforcements, elements of the *52d Division*, were being transferred from the Carolines to various sites in the Marshalls. The enemy's total strength throughout Kwajalein Atoll was estimated to be 8,000–9,600 men, 6,150–7,100 of them combat troops.

General Smith's intelligence section predicted that the 7th Infantry Division would face 2,300–2,600 combat troops and 1,200–1,600 laborers. The enemy appeared to have built a defensive line across Kwajalein Island just east of the airfield, works designed to supplement the pillboxes, trenches, and gun emplacements that fringed the island. Photographs of Roi-Namur disclosed coastal perimeters that featured strongpoints at each corner of both islands. Very few weapons positions were discovered in the interior of either island. Namur, however, because of its many buildings and heavy undergrowth, offered the enemy an excellent chance to improvise a defense in depth. At both Kwajalein and Roi-Namur Islands, the installations along the ocean coasts were stronger than those facing the lagoon. No integrated defenses and only a small outpost detachment were observed on Majuro. (See Map V, Map Section.)

Corps also had the task of preparing and distributing the charts and maps used by the assault troops, naval gunfire teams, defense battalions, and other elements of FLINTLOCK Expeditionary Troops. Each division received 1,000 copies of charts (on a scale of one inch to one nautical mile) and of special terrain maps (1:20,000), and as many as 2,000 copies of another type of special terrain map (1:3,000). On the 1:3,000 maps, the particular island was divided into north, east, west, and south zones. Within each zone, known gun positions were numbered in clockwise order, each number prefixed by N, E, W, or S to indicate the proper zone. All crossroads and road junctions also were given numbers. Besides the customary grid system, these maps also showed the number and outline of all naval gunfire sectors. By compressing so much information onto a single sheet, the corps devised a map that suited a variety of units.

The information gathered, evaluated, and distributed by Admiral Nimitz' Joint Intelligence Center, General Smith's amphibious corps, and Admiral Turner's amphibious force was both accurate and timely. Sound intelligence enabled Nimitz to alter his plans and strike directly at Kwajalein. A knowledge of the enemy defenses made possible an accurate destructive bombardment and, together with hydrographic information, guided attack force and landing force commanders in the selection of assault beaches.

COMMUNICATIONS AND CONTROL[34]

Generals Corlett and Schmidt planned to destroy the enemy garrison in a series of carefully coordinated amphibious landings. For this reason, success depended to a great extent upon reliable communications and accurate timing. Although the introduction of command vessels had given attack force and landing force commanders a better means of controlling the different phases of the operation, not every communications problem had been solved.

The Marine assault troops assigned to FLINTLOCK used much of the same communications equipment that had proved inadequate in the Gilberts. The radios in the LVTs were not waterproofed, a fact which would greatly reduce communication effectiveness during the landing.[35] Both the TBX and TBY radios, neither type adequately waterproof, had to be used again in the Marshalls. Eventually, it was hoped, these sets could be replaced, the TBX by some new, lighter, and more reliable piece of equipment and the TBY by the portable SCR 300 and mobile SCR 610 used by the Army. Although intended for infantrymen rather than communications men, the hand-carried MU radios were too fragile to survive the rugged treatment given them in rifle units. The SCR 610 worked well, but it too was vulnerable to water damage. No waterproof bags were available for either spare radio batteries or telephone equipment.

In an attempt to insure unbroken communications, both the 4th Marine Division and the 7th Infantry Division were assigned Joint Assault Signal Companies (JASCOs). The Marine 1st JASCO was activated on 20 October 1943 at Camp Pendleton, California. The primary mission of this unit was to coordinate all supporting fires available to a Marine division during an amphibious operation. In order to carry out this function, the company was divided into Shore and Beach Party Communications Teams, Air Liaison Parties, and Shore Fire Control Parties. Early in December, the company joined VAC and was promptly attached to the 4th Marine Division. During training, the various teams were attached to the regiments and battalions of the division. Thus each assault battalion could become familiar with its shore and beach party, air liaison, and fire control teams. The Army 75th JASCO was attached in the same manner to the battalions of the 7th Infantry Division.

Communications equipment, how-

[34] An additional source for this section is ComInCh, *Marshall Islands*, pp. 6:1–6:4, 8:1–8:13.

[35] *Metzger ltr.*

ever, was but a means of control. If the landings were to succeed, they would have to be precisely organized and accurately timed. Unit commanders and control officers would have to be located where they could see what was happening and influence the conduct of the battle. For FLINTLOCK, the movement of assault troops from the transports to the beaches was carefully planned, and an adequate system of control was devised.

Instead of transferring from transports to landing craft and finally to LVTs, as had been done at Tarawa, the first waves of assault troops were to move from the transports directly to the LSTs that carried their assigned tractors.[36] The men would climb into the assault craft as the LSTs steamed to a position near the line of departure from which the ships would launch the amphibians. Next, the LVTs were to form waves, each one guided by a boat commander. At the line of departure, the commander of each wave reported to the control officer, a member of the V Amphibious Force staff.

Among other vessels, each control officer had at his disposal two LCCs (Landing Craft Control vessels), steel-hulled craft similar in appearance to motor torpedo boats. These carried radar and other navigational aids and were designated as flank guides for the leading assault waves. After the first four waves had crossed the reef, the LCCs, which were incapable of beaching and retracting, would take up station in a designated area 2,000 yards from shore. Since reserve units were to follow a transfer scheme similar to that planned for Tarawa, officers in the LCCs now had to supervise the shifting of men from landing craft to returning LVTs, as well as the formation of waves, and the dispatch of tractors to the beach.

A submarine chaser was assigned the control officer to enable him to move wherever he might be needed in the immediate vicinity of the line of departure. A representative of the landing force commander, the commander of the amphibian tractor battalion, a representative of the division supply officer, and a medical officer were embarked in the same craft. These men were given power to make decisions concerning the ship-to-shore movement, the landing of supplies, and the evacuation of wounded.

A second submarine chaser, this one stationed continuously at the line of departure, carried a representative of the transport group commander. This officer saw to it that the waves crossed the line either according to the prearranged schedule, as the control officer directed, or in the case of later waves as the regimental commander requested.

Off the beach his troops were assaulting, the regimental commander was to establish a temporary floating command post in a submarine chaser. While in this vessel, he would be able to contact by radio or visual signals the landing force commander, the various boat waves, and his battalions already ashore. As soon as the regimental commanders had established command posts ashore, the submarine chasers

[36] This method was to be used on D plus 1. The troops bound for the outlying island were scheduled to transfer at sea from LCVPs to LVTs.

could be used by the division headquarters.

LOGISTICS [37]

The geographical separation of the units assigned to FLINTLOCK affected logistical planning as well as tactical training. The 4th Marine Division trained at Camp Pendleton and prepared to sail from San Diego, the 7th Infantry Division and 106th Infantry trained on Oahu, and the 22d Marines made ready in Samoa prior to its movement to the Hawaiian Islands. In spite of the distance involved, General Smith later reported that in the field of logistics "no major difficulties were encountered." [38]

There were, however, several minor problems. The 22d Marines, for example, was unable to obtain from Marine sources either 2.36-inch rocket launchers and ammunition for them, or shaped demolitions charges, but a last-minute request to Army agencies was successful.[39] The 4th Marine Division had to revise its logistical plans in the midst of combat loading. Originally, Admiral Nimitz had prescribed that each division carry to the objective five units of fire for each of its weapons except antiaircraft guns. Officers of the 7th Infantry Division requested additional ammunition, but the admiral was reluctant to accept their recommendations. Not until 5 January did he approve 10 units of fire for 105mm howitzers and 8 for all other ground weapons. Nor was the 7th Infantry Division without its troubles, for the water containers provided by Army sources proved useless, and drums had to be obtained from the Navy.

A total of 42 days' rations was scheduled to be carried to Kwajalein Atoll. Each Marine or soldier was to land with 2 days' emergency rations. A 4-day supply of the same type of food was loaded in LSTs, and an additional 6-day amount was lashed to pallets for storage in the transports. The cargo ships assigned to the expedition carried enough dried, canned, and processed food to last the assault and garrison troops for 30 days. Five day's water, in 5-gallon cans and 55-gallon drums, was stowed in the LSTs and transports. Logistical plans also called for a 30-day quantity of maintenance, medical, and aviation supplies, as well as fuels and lubricants. The assault divisions and the garrison units also brought with them large amounts of barbed wire, sandbags, and light construction material.

Not all of this mountain of supplies and ammunition was combat loaded. Those items likely to be needed early in the operation were stowed in easily accessible places according to probable order of use. The remaining supplies were loaded deep within the cargo vessels in a manner calculated to conserve space. Some emergency supplies, including ammunition, water, and rations, were placed in LSTs.

Admiral Conolly divided his trans-

[37] Additional sources for this section include: VAC Rpt of LogAspects of FLINTLOCK Op, dtd 23Mar44 hereafter *VAC Rpt of LogAspects;* LtCol S. L. A. Marshall, USA, "General and Miscellaneous Notes on Central Pacific: Supply." (Hist MS File, OCMH); CominCh, *Marshall Islands,* pp. 5:1–5:25, 6:13–6:16.

[38] *VAC Rpt of LogAspects.*

[39] *Greene ltr I.*

ports into three groups, one per infantry regiment, each with four transports and a cargo vessel. The 105mm howitzers of the 4th Marine Division were loaded into LCMs, landing craft that would be ferried to Roi-Namur in an LSD. The 75mm pack howitzers were placed in LVTs, and these tractors embarked in LSTs. A second LSD carried the LCMs in which the 15 Shermans of the division medium tank company were loaded. All 36 light tanks of the 4th Tank Battalion were stowed in the transports. Admiral Turner, who had retained responsibility for conducting the 7th Infantry Division to Kwajalein Island, organized his shipping in much the same way.

At Tarawa, the flow of supplies to the assault units had been slow and uncertain. Admiral Turner, in an effort to prevent a similar disruption, directed that beach party and shore party units sail in the same transports, draw up joint plans, and land rapidly. Skeleton beach parties and elements of shore parties were assigned to the fourth wave at each beach, and the remainder of the units were ordered to follow as quickly as possible.

The corps directed the 7th Infantry Division to form shore parties from its 50th Engineer Battalion and elements of the Kwajalein Island garrison force, while the 4th Marine Division was to rely upon men from the 20th Marines, its engineer regiment. One shore party, reinforced by medical, quartermaster, ordnance, and other special troops, was attached to each infantry battalion. The principal weakness in this phase of the supply plan was the use of men from reserve combat units to bring the shore parties to their authorized strength of approximately 400. Pontoon causeways, broken into sections and loaded in LSTs, were made available for use at Roi-Namur, Kwajalein, and CARLSON Islands, and at Majuro Atoll. The pontoons could be joined together to serve as piers for the unloading of heavy equipment.

Enough emergency supplies were loaded in LSTs to sustain the battle until the beaches were secured. At Roi-Namur, LVTs, the only amphibious cargo vehicles available to the Marine division, were to serve as the link between the LSTs and the battalions advancing inland. After the beaches had been secured, the transports would begin unloading.

The 7th Infantry Division had, in addition to its amphibian tractors, 100 DUKWs. These 2½-ton amphibian trucks were called upon to perform at Kwajalein Island much of the work expected of LVTs at Roi-Namur. Sixty DUKWs were assigned to land the division artillery, and 40 of them, also stowed in LSTs, were to give logistical support to infantry units by bringing ashore emergency supplies. Some of the critical items were loaded in the trucks before the parent LSTs sailed from Hawaii.

Admiral Turner's medical plan gave beachmasters authority over the evacuation of wounded. Theirs was the task of selecting the boats or amphibious vehicles that would carry away casualties. The medical section of the beach party was responsible for distributing the wounded among the cargo ships and transports. All of these vessels could receive the injured, but by D plus 3 all casualties would be collected in specified vessels or transferred to the hospital ships scheduled to arrive on that day.

TRAINING FOR FLINTLOCK

The 4th Marine Division was able to undergo amphibious training in conjunction with Admiral Conolly's support ships and transports. A division exercise was held on 14–15 December, before either the admiral or General Schmidt were certain what course FLINTLOCK would follow. Another exercise took place at San Clemente Island off the California coast on 2–3 January 1944. This second landing was in effect a rehearsal, for all amphibious shipping joined many of Conolly's warships and carriers in the exercise.

The January landing also gave the division a chance to test its aerial observers. These were the ground officers who would be flown over Roi-Namur to report throughout the day on the progress of the battle. This aspect of the exercise was a complete success, but the work of the LVTs and LSTs was far less impressive.

On 5 December, the division's 4th Amphibian Tractor Battalion was broken up, and four/seventh's of its men were used to form the cadre of the 10th Amphibian Tractor Battalion, reinforced by Company A, 11th Amphibian Tractor Battalion.[40] The two units were then brought up to authorized strength by the addition of recruits and the transfer of trained crews from the 1st Armored Amphibian Battalion. By the time these changes had been made, less than a month remained in which to check the tractors, install armor plate, waterproof radios, train the new crews, lay plans for the landings, take part in the San Clemente rehearsal, load the vehicles into amphibious shipping, and make a final check to determine that the LVTs were fit for combat. These varied tasks had to be carried out simultaneously with the obtaining of supplies, processing of men, and the other duties routine to a unit preparing for action.

Unfortunately, many of the LSTs were manned by sailors as inexperienced as the Marine tractor crews. Admiral Conolly recalled:

> A number of these ships were rushed from their Ohio River building yards straight to the West Coast. They had inadequate basic training, little or no time to work with their embarked troops, and, in some cases, arrived in San Diego a matter of a few days before final departure for the Marshalls.[41]

Although the San Clemente exercise was staged to promote close cooperation between the LSTs and LVTs, the sailors and Marines gained little confidence in one another. Some of the ships refused to recover any tractors except those they had launched; as a result several tractors ran out of gas and were lost. There also was one collision between an LST and an LVT. "All this," one participant drily ob-

---

[40] LtCol Victor J. Croizat ltr to Drs. Jeter A. Isely and Philip A. Crowl, dtd 30Apr51, encl to Col Victor J. Croizat ltr to Head, HistBr, G–3, HQMC, dtd 13Sep62.

[41] VAdm Richard L. Conolly ltr to Dr. Jeter A. Isely, dtd 31Aug49, encl to Gen Harry Schmidt ltr to CMC, dtd 22Oct62.

214-881 O-67—10

served, "was very poor for morale just before combat." [42]

The LSTs, loaded with amphibian tractors, sailed from San Diego on 6 January, to be followed a week later by the remainder of Admiral Conolly's attack force. At the time of its departure, the first convoy had not yet received copies of the final operations plans issued by Admirals Spruance, Turner, and Conolly. These documents did not arrive until 18 January, two days before the LSTs set sail for the Marshalls and two days prior to the arrival of the rest of Conolly's ships in Hawaiian waters. Since the two groups shaped different courses toward the objective, there was no opportunity for last-minute coordination.[43]

General Corlett, like General Schmidt, had carefully studied the lessons of Tarawa, so the 7th Infantry Division also was thoroughly trained for atoll warfare. The Army unit, however, had its share of problems in finding crews for its amphibian tractors. On 25 November, the division established a school to train members of the regimental antitank companies as LVT drivers and mechanics. The graduates of this course were selected to man the tractors that would carry the assault waves. The landings would be supported by the 708th Amphibian Tank Battalion which was attached to the division early in December. For FLINTLOCK the amphibian tractors were incorporated into the Army tank battalion and the resultant organization called the 708th Provisional Amphibian Tractor Battalion.[44]

By the time of its attachment to VAC for operational control, the 7th Infantry Division was well grounded in tank-infantry-engineer teamwork. The amphibious training of General Corlett's troops took place in December and January, with the most attention devoted to the comparatively inexperienced 184th Infantry. The division and the 22d Marines conducted their final rehearsals between 12 and 17 January. The troops landed at Maui's Maalaea Bay and made a simulated assault on Kahoolawe Island. The Majuro landing force, 2/106, made a practice landing on the shores of Oahu on 14 January.

PRELIMINARY OPERATIONS [45]

Aircraft of all services joined surface ships in a series of raids planned to batter Kwajalein Atoll, neutralize the Japanese bases that surrounded it, and gain information on the enemy's defenses. Mille, Jaluit, and Maloelap were the principal targets hit during November and December by Army and Navy planes of Admiral Hoover's command. During January, after the Gilberts fields had been completed, the heaviest tonnage fell on Kwajalein and Wotje. Land-based planes in De-

[42] LtCol Louis Metzger ltr to CMC, dtd 13Nov52.

[43] *Ibid.*; Col William R. Wendt ltr to CMC, dtd 19Feb53; LtCol Victor J. Croizat ltr to CMC, dtd 10Nov52, hereafter *Croizat ltr.*

[44] Marshall notes, *op. cit.*, pp. 41–42.

[45] Additional sources for this section include: CinCPac–CinCPOA WarDs, Nov43–Feb44 (CinCPac File, HistBr, HQMC); ComCenPac Rpt on FLINTLOCK Op, n.d.; ComInCh, *Marshall Islands*, pp. 1:1–1:4; Craven and Cate, *Guadalcanal to Saipan;* Morison, *Aleutians, Gilberts, and Marshalls.*

cember and January dropped 326 tons of explosives on targets in Maloelap Atoll, 313 on Kwajalein, 256 on Jaluit, 415 on Mille, and 367 on Wotje. The Japanese retaliated by loosing a total of 193 tons of bombs on Makin, Tarawa, and Apamama. In the meantime, patrol bombers from Midway were active over Wake Island.

On 4 December, while Army bombers were raiding Nauru Island and Mille, carrier task groups commanded by Rear Admirals Charles A. Pownall and Alfred E. Montgomery launched 246 planes against Kwajalein and Wotje Atolls. The aviators sank 4 cargo ships, damaged 2 old light cruisers, shot down 19 enemy fighters, and destroyed many other planes on the ground. Japanese fliers, stung by this blow, caught the retiring carriers, and in a night torpedo attack damaged the USS *Lexington.*

Except for an attack by carrier aircraft and surface ships against Nauru on 8 December, land-based planes swung the cudgel until 29 January. On that day, carriers and fast battleships returned to the Marshalls, attacking the Japanese bases in an unexpected thrust from the westward.[46] Rear Admiral Samuel P. Ginder's carriers hit Maloelap, and Rear Admiral John W. Reeves sent his aircraft against Wotje, while carrier task groups commanded by Rear Admiral Frederick C. Sherman and Admiral Montgomery attacked Kwajalein and Roi-Namur Islands. Surface ships bombarded the targets in conjunction with the air raids.

On 30 January, Reeves took over the preparatory attack against Kwajalein Island, while Sherman began a 3-day effort against Eniwetok Atoll.[47] Ginder maintained the neutralization of Wotje, refueled, and on 3 February replaced Sherman. The task groups under Reeves and Montgomery continued to support operations at Kwajalein Atoll until 3 February.

As these preparations mounted in intensity, the Northern and Southern Attack Forces drew near to their objectives. On 30 January, fire support ships of these forces paused to hammer Wotje and Maloelap before continuing onward to Roi-Namur and Kwajalein. Meanwhile, the supporting escort carriers (CVEs) joined in the preparatory aerial bombardment of the objectives. On 31 January, the 4th Marine Division and 7th Infantry Division would begin operations against island fortresses believed to be stronger than Betio.

[46] A feature of Admiral Spruance's plan was that the fast battleships and carriers would form up at Funafuti in the Ellice Islands well to the southeast of the Marshalls. Battleships arriving from the Atlantic anchored there in time to join the carriers and launch the pre-invasion attack. Japanese searches were conducted to the eastward. *Moore comments Marshalls.*

[47] The original CinPac plan for air support had called for the fast carriers to make a 2-day strike and then withdraw for several days before returning to cover the landings. Admiral Spruance objected to this plan and substituted his own, which insured that Japanese air was "taken out on all positions except Eniwetok on the first day," and that the airfields on Wotje, Taroa [Maloelap], and Kwajalein were "kept immobilized thereafter by naval gunfire on the runways." He sent Sherman's group to hit Eniwetok and "keep the air pipeline . . . inoperative while we captured Kwajalein." *Spruance 62 ltr.*

ROI-NAMUR, *under bombing attack by Seventh Air Force planes, appears in an intelligence photo taken just prior to the pre-landing bombardment (USAF B50003AC)*

THE DEFENSES OF KWAJALEIN ATOLL [48]

Just as he had startled his subordinates by proposing an immediate attack on Kwajalein, Nimitz also surprised his adversaries. "There was divided opinion as to whether you would land at Jaluit or Mille," a Japanese naval officer confessed after the war. "Some thought you would land on Wotje, but there were few who thought you would go right to the heart of the Marshalls and take Kwajalein." [49]

Unlike their leaders, the defenders of Kwajalein Atoll, dazed by a succession of air raids, quickly became convinced that their atoll ranked high on Nimitz' list of objectives. "I welcome the New Year at my ready station beside the gun," commented a squad leader in the *61st Guard Force.* "This will be a year of decisive battles. I suppose the enemy, after taking Tarawa and Makin, will continue on to the Marshalls, but the Kwajalein defenses are very strong." [50]

Actually the Japanese high command had been slow to grasp the importance of the Marshalls. Prewar plans called principally for extensive mine-laying to deny the atolls to United States forces, but the effectiveness of medium bombers during the war against the Chinese had indicated that similar planes based on atolls could be a grave threat to shipping. A survey showed that the best sites for air bases were Wotje, Maloelap, Majuro, Mille, and Kwajalein. This last atoll, now the target of the American expeditionary force, was selected as administrative and communications center for the Marshalls area.

During 1941, the *6th Base Force* and the *24th Air Squadron* of the *Fourth Fleet* [51] were made responsible for defending the islands. The base force immediately set to work building gun emplacements and other structures at Kwajalein, Wotje, Maloelap, and Jaluit. By December 1941, the various projects were nearly complete, and the Japanese forces employed against the Gilberts and Wake Island were able to operate from the Marshalls.[52]

The number of troops assigned to the Marshalls grew throughout 1942, but the islands themselves began to diminish in strategic importance. Japanese planners came to regard the Marshalls, like the Gilberts, as outposts to protect the more important Carolines and Marianas. Although the *Imperial Navy* began, in the fall of 1943, to speed work on the defenses of the Carolines and Marianas, the Marshalls were not neglected. If attacked, the outlying atolls were to hold out long enough for naval forces and aircraft to arrive on

---

[48] Additional sources for this section include: JICPOA Bul 48–44, Japanese Defs, Kwajalein Island, dtd 10Apr44; 4th MarDiv IntelRpt on FLINTLOCK Op, n.d.; USSBS, *Campaigns of the Pacific War* (Washington, 1946), hereafter USSBS, *Campaigns of the Pacific War.*

[49] USSBS (Pac), NavAnalysis Div, *Interrogations of Japanese Officials*, 2 vols (Washington, 1946), Interrogation Nav No. 34, Cdr Chikataka Nakajima, IJN, dtd 21Oct45, I, p. 144, hereafter *USSBS Interrogation* with relevant number and name.

[50] JICPOA Item No. 5913, Diary of Mimori.

[51] Chief, WarHistOff, DefAgency of Japan, ltr to Head, HistBr, G–3, HQMC, dtd 14Jan63.

[52] MilHistSec, Japanese RschDiv, HqAFFE, Japanese Monograph No. 173, *Inner South Seas Islands Area Naval Operations, Part II: Marshall Islands Operations (Dec44–Feb44).*

the scene and destroy the American warships and transports. These were the same tactics that had failed in the Gilberts.[53]

Late in 1943, large numbers of Army troops began arriving in the Marshalls, and by the end of that year 13,721 men of the *1st South Seas Detachment*, *1st Amphibious Brigade*, *2d South Seas Detachment*, and *3d South Seas Detachment* were stationed on atolls in the group, on nearby Wake Island, and at Kusaie. Of these units, only the *1st South Seas Detachment* had seen combat. Its men had been incorporated into the *122d Infantry Regiment* and had fought for three months on Bataan Peninsula during the Japanese conquest of the Philippines.

The enemy also sent the *24th Air Flotilla* to the threatened area. This fresh unit served briefly under the *22d Air Flotilla* already in the area, but at the time of the first preparatory carrier strikes, the remaining veteran pilots of the *22d* were withdrawn and their mission of defending the Marshalls handed over to the newcomers.[54] As the Kwajalein operation drew nearer, progressively fewer Japanese planes were able to oppose the aerial attacks. By 31 January, American pilots had won mastery of the Marshalls skies.

At Roi-Namur, principal objective of the 4th Marine Division, was the headquarters of the *24th Air Flotilla*, commanded by Vice Admiral Michiyuki Yamada, who had charge of all aerial forces in the Marshalls. The enemy garrison was composed mainly of pilots, mechanics, and aviation support troops, 1,500-2,000 in all. Also, there were between 300 and 600 members of the *61st Guard Force*, and possibly more than 1,000 laborers, naval service troops, and stragglers.[55] Only the men of the naval guard force were fully trained for ground combat.

In preparing the defenses of Roi-Namur, the enemy concentrated his weapons to cover probable landing areas, an arrangement in keeping with his goal of destroying the Americans in the water and on the beaches. The defenders, however, failed to take full advantage of the promontories on the lagoon shores of both Roi and Namur, sites from which deadly flanking fire might have been placed on the incoming landing craft. Both beach and antitank obstacles were comparatively few in number, although a series of antitank ditches and trenches extended across the lagoon side of Namur Island.[56] Ten pillboxes mounting 7.7mm machine guns, a 37mm rapid-fire gun, a pair of 13mm machine guns, and two 20mm cannon were scattered along the beaches over which General Schmidt intended to land. Most of these positions were connected by trenches. Although two pair of twin-mounted 127mm guns were emplaced on Namur, these weapons covered the

[53] Hattori, *Complete History*, v. 3, pp. 50–51; Sako Tanemura, *Confidential Diary of the Imperial General Staff Headquarters*, tr by 165th MIS Co, 1952, hereafter Tanemura, *Confidential Diary*.

[54] *USSBS Interrogation* Nav No. 30, Cdr Goro Matsuura, IJN, dtd 20Oct45, I, p. 132.

[55] The 4th Marine Division counted 3,472 enemy dead on the various islands in the northern part of the atoll. Since other bodies lay sealed in bunkers, it was impossible to reconstruct the exact strength of the various components of the Roi-Namur garrison.

[56] *Metzger ltr.*

ocean approaches to the island. The enemy had no integrated defenses within the coastal perimeter, but he could fight, on Namur at least, from a myriad of concrete shelters and storage buildings. (See Map V, Map Section.)

Kwajalein Island was the headquarters of Rear Admiral Monzo Akiyama's *6th Base Force*,[57] and its garrison was stronger in ground combat troops than that at Roi-Namur. About 1,000 soldiers, most of them from the Army *1st Amphibious Brigade*, fewer than 500 men of the Navy *61st Guard Force*, and a portion of a 250-man detachment from the *4th Special Naval Landing Force* were the most effective elements of the defense force. A few members of the base force headquarters and a thousand or more laborers also were available. In the southern part of the atoll, the enemy had some 5,000 men, fewer than 2,000 of them skilled combat troops.

The defenses on Kwajalein Island, like those on Roi-Namur, lacked depth and were strongest along the ocean coast. The western end of Kwajalein Island, where General Corlett planned to land, was guarded by 4 twin-mounted 127-mm guns (weapons emplaced to protect the northwest corner of the island), 10 pillboxes, 9 machine gun emplacements, and a few yards of trenches. The cross-island defenses noted in aerial photographs actually consisted of an antitank ditch, a trench system, and seven machine gun positions. The trenches, though, began near a trio of 80mm guns that were aimed seaward. Although he had few prepared positions in the interior of the island, there were hundreds of buildings from which the enemy might harry the attackers.

Both of the principal objectives were weak in comparison to Betio Island. Few obstacles protected the assault beaches, and work on many installations was not yet finished. In spite of these deficiencies, the soldiers and Marines could expect bitter fighting. "When the last moment comes," vowed one of the atoll's defenders, "I shall die bravely and honorably."[58] In happy contrast to Kwajalein Atoll was Majuro, where a Navy warrant officer and a few civilians had been left behind when the Japanese garrison was withdrawn.[59]

[57] For a brief time just prior to the American attack, Vice Admiral Mashashi Kobayashi had maintained on the island temporary headquarters for his *Fourth Fleet*.

[58] JICPOA Item No. 5913, *op. cit.*

[59] Aerial photographs of Majuro showed a fair-sized barracks area. Since the atoll seemed to be abandoned, Admiral Spruance's chief of staff suggested to Admiral Hill that these buildings not be bombarded. They were found in excellent shape and were useful to U. S. Forces. *Moore comments Marshalls*.

D-Day in the Marshalls [1]

The final version of the FLINTLOCK plan called for three distinct operations, each of which required several amphibious landings. The capture of Majuro Atoll, correctly judged to be the simplest of the three, was entrusted to the VAC Reconnaissance Company and 2/106. Each of the others was believed to require an entire division.

In the northern part of Kwajalein Atoll, the 4th Marine Division had the mission of seizing on 31 January IVAN (Mellu), JACOB (Ennuebing), ALBERT (Ennumennet), ALLEN (Ennubir), and ABRAHAM (Ennugarret). On the following day, D plus 1, this division was scheduled to storm Roi-Namur. In the southern sector, the 7th Infantry Division was to attack CARTER (Gea), CECIL (Ninni), CARLSON (Ennubuj), and CARLOS (Ennylabegan) on D-Day, then assault the beaches of Kwajalein Island on 1 February. Once these principal objectives were secured, the assault divisions were to overcome enemy resistance throughout the remainder of the atoll. (See Map 8.)

MAJURO: BLOODLESS VICTORY [2]

An irregularly shaped collection of islands and partially submerged reefs, Majuro lies approximately 265 nautical miles southeast of Kwajalein Atoll. Majuro lagoon, 24 miles long by 5 miles wide, was a tempting prize, and Dalap Island, at the easternmost point of the atoll, seemed suitable for an airfield. Other large islands thought useful for military installations were Majuro, to the south, as well as Uliga and Darrit, just north of Dalap. Calalin and Eroj, midway along the northern rim of the atoll, were important, for they guarded the two entrances to the lagoon. (See Map 7.)

In planning the operation, Hill faced the problem of employing deep draft ships in an area for which he had only a small segment of a hydrographic chart. He ordered high angle vertical

[1] Unless otherwise noted, the material in this chapter is derived from: TF 51 Rpt of FLINTLOCK and CATCHPOLE Ops, dtd 25Feb44, hereafter *TF 51 AR; TF 53 AR Roi-Namur; 4th MarDiv AR* (which includes rpts of IVAN LdgGru, 23d, 24th, 25th, 14th, and 20th Mar); 4th MarDiv Jnl, 13Jan–2eb44, hereafter *4th MarDiv Jnl;* 10th AmTracBn Rpt on FLINTLOCK, dtd 12Apr44, hereafter *10th AmTracBn Rpt;* 1/25 Rpt of Activities, D-Day and D plus 1, dtd 16Feb44; 2/25 Rpt of Activities, dtd 20Feb44; 3/25 Hist, 11Jan44–8Mar44, n.d.; DesRon 1 AR, dtd 9Feb44; CominCh, *Marshall Islands;* Crowl and Love, *Gilberts and Marshalls;* Morison, *Aleutians, Gilberts, and Marshalls;* Heinl and Crown, *The Marshalls.*

[2] Additional sources for this section include: TG 51.2 Majuro AR, dtd 15Feb44; VAC ReconCo WarD, Majuro, dtd 16Mar44, Encl I to *VAC AR FLINTLOCK.*

aerial photographs made of the lagoon for use by a Coast and Geodetic Survey team attached to his staff and with its help prepared a detailed chart. With this as a navigation guide, he was able to move into the lagoon, once the operation was underway, without difficulty.[3]

To overwhelm what was known to be a small garrison, Admiral Hill could employ 2/106, commanded by Lieutenant Colonel Frederick B. Sheldon, USA, and carried in the task group command ship, *Cambria*. This battalion had been reinforced by the VAC Reconnaissance Company led by Captain James L. Jones. To transport, protect, and defend his landing force, Hill had a heavy cruiser, four destroyers, two escort carriers, two destroyer transports, three minesweepers, and an LST.

One of the transports, the converted destroyer (APD) USS *Kane* left the convoy on 30 January to steam directly to the objective. That night, the ship reached the twin entrances to Majuro lagoon and by 2300 had landed a small detachment from the reconnaissance company. This group found both Eroj and Calalin to be unoccupied. A native told the Marines that 300–400 Japanese were located on Darrit, and this information was relayed to Admiral Hill at 0608. Other inhabitants of Calalin, however, had noted the withdrawal of the enemy troops. They reported that a lone warrant officer and a few civilians were the only Japanese in the atoll.

The *Kane* next landed the remainder of Jones' company on Dalap. Patrols fanned out over the island but discovered no Japanese. At Uliga, an English-speaking native confirmed the earlier reports that the enemy garrison had been evacuated.

At this time, the reconnaissance company lost radio contact with the task force. Unaware that the enemy had abandoned Darrit, Admiral Hill ordered the USS *Portland* to shell the island at 0634. Within 20 minutes, contact was regained, the bombardment was stopped, and a scheduled air strike was cancelled. The troops then occupied Darrit, raising the American flag for the first time over prewar Japanese territory at 0955.[4]

On the night of 31 January, a platoon from Jones' company landed on Majuro Island and captured the naval warrant officer who was responsible for Japanese property left behind on the atoll. The civilians who assisted him in caring for the equipment escaped into the jungle. Thus ended the only action at Majuro Atoll.

About midnight on 1 February, a detachment of VAC Reconnaissance Company, investigating reports of a downed American plane, landed from the *Kane* on Arno Atoll, about 10 miles east of Majuro. The Marines found no Japanese, and natives told them that the plane crew had been removed to Maloelap. Reembarking their APD, the men returned to Majuro on the 2d.

NORTHERN KWAJALEIN: IVAN AND JACOB

During darkness on the morning of 31 January, ships of the Northern Attack Force steamed into position in the

[3] *Hill comments/interview Marshalls.*

[4] *Ibid.*

vicinity of Roi-Namur. The schedule for D-Day called first for the capture of IVAN and JACOB, two islands southwest of Roi-Namur, between which lay a deep-water passage into Kwajalein lagoon. Elements of Lieutenant Colonel Clarence J. O'Donnell's 1/25 were to land at 0900 on both objectives. For the day's action, the battalion had been reinforced with Company D, 4th Tank Battalion, the division's scout company. (See Map 8.)

Because they commanded the lagoon entrance, both IVAN and JACOB had to be attacked from the seaward side. Company B of O'Donnell's battalion was to assault Beach Blue 1 on JACOB, while Company C and the attached scout company struck Blue 2 on neighboring IVAN. Once these landings had been made, the ships supporting the IVAN force, led by mine sweepers, could enter the lagoon to carry out the remaining parts of the D-Day plan. In the meantime, artillery batteries from the 14th Marines would begin arriving on the Blue Beaches to move into positions from which to assist the next day's operation.

Unlike the men who were to make the main landings, the Marines of General Underhill's IVAN group had to transfer at sea from LCVPs to LVTs. O'Donnell's troops entered the landing craft at 0530 and began their journey to the transfer area where they would meet the LVTs of Company B, 10th Amphibian Tractor Battalion. The wind was brisk and the sea rough as the LCVPs plowed toward their rendezvous. By the time the boats reached the tractors, many of the assault troops were soaked by the spray.

The preparatory bombardment of northern Kwajalein Atoll got underway at 0651. In addition to shelling IVAN and JACOB, supporting warships pounded Roi-Namur and stood ready to blast ABRAHAM if necessary. Naval gunfire was lifted at 0715 to permit an 8-minute strike by carrier planes and then resumed.

During the battering of the northern islands, the remainder of Colonel Samuel C. Cumming's 25th Marines was preparing for action later in the day. Both 2/25, under Lieutenant Colonel Lewis C. Hudson, and 3/25, commanded by Lieutenant Colonel Justice M. Chambers, were scheduled to load into LCVPs. Hudson's battalion was to transfer to the LVTs of Company C, 10th Amphibian Tractor Battalion, and seize ALLEN. Tractors released by the IVAN and JACOB forces were to land Chambers' men on ALBERT. After overrunning ALBERT, 3/25 was to prepare to attack on order across the shallow strait separating that island from ABRAHAM.

At 0800, while 1/25 was forming to assault IVAN and JACOB, Admiral Conolly confirmed 0900 as H-Hour. He selected 1130 as A-Hour, the time of the landings on ALBERT and ALLEN, and designated 1600 as B-Hour, when Chambers' battalion would storm ABRAHAM. Adhering to this timetable, the supporting warships ceased firing at 0825 to permit a second aerial attack. At this point, the effects of choppy seas and makeshift rehearsals made themselves felt, and it soon became obvious to Admiral Conolly that the assault waves could not meet his deadline.

The postponement of H-Hour was

partially the fault of the elements. Swells, aided by a 14-knot wind, complicated the transfer of troops, cut the speed of the LVTs almost in half, and raised spray that drowned the radios carried by the tractors. Yet many of the misfortunes that hounded Company B, 10th Amphibian Tractor Battalion, could be traced to the improvised rehearsals that had been held off the California coast.

"A rehearsal with complete plans and orders," the company commander later suggested, "would be of much value prior to D-Day landing."[5] Unfortunately the tractor battalion had received the revised plans long after its final exercise.

For these reasons, the transfer area soon became the site of an amphibious traffic jam. Tractors were slow in leaving the LSTs, landing craft had difficulty in finding the proper amphibians, and rumored changes of plan could not be verified because of the drenched radios.[6] Order, however, eventually prevailed, and the troop-laden LVTs were directed into formation.

Conolly, alerted by a destroyer astride the line of departure that the troop carriers were late, at 0903 issued orders delaying H-Hour until 0930. Within a few minutes of this change, the LCI gunboats and LVT(A)s that were to spearhead the assault crossed the line of departure. Now the aerial observers and air coordinator undertook the task of timing the final strikes according to the progress of the approaching tractors.

"Will hold up attack until boats are in proper position," radioed the air coordinator at 0854 after he had noted that the approaching landing craft were 5,000–6,000 yards from IVAN and JACOB.[7] At 0917, when the LVT(A)s and LCI gunboats were about 3,000 yards from shore, the coordinator ordered the waiting planes to begin their attack. The bombing attacks pinned down the defenders of both objectives until the LCIs were in position to launch their rockets. No strikes were made against the beaches while the rocket bombardment was being delivered, but when the LCIs had accomplished this task, the coordinator directed fighter planes to strafe the islands. Air observers kept close watch over the approaching troops and carefully reported the distance that remained to be covered. Since the force bound for JACOB made better speed, the final strafing of that island was halted shortly after 0940, while the final strike against IVAN continued past 1000.

As the bombing attacks were beginning, a 127mm battery on Roi rashly opened fire on warships supporting the preliminary landings. A cruiser silenced the enemy position but did not destroy the twin-mounted guns. For the time being, though, this threat was removed.

[5] Rpt of CO, Co B, 10th AmTracBn, dtd 17Mar44, Encl A to *10th AmTracBn Rpt.*

[6] The transport division commander noted that the "short delay in How hour was the result of inability of certain LCVPs loaded with troops to locate the LVTs into which they were to transfer. Indications are that all LSTs were not in designated areas and that some LVTs wandered away from the launching LST." ComTransDiv 26 AR, ser no. 0013, dtd 18Feb44 (OAB, NHD).

[7] *4th MarDiv Jnl*, msg dtd 0854, 31Jan44.

105MM AMMUNITION *is unloaded from landing craft at Mellu Island for the bombardment of Roi-Namur. (USA SC324729)*

24TH MARINES *assault troops on the beach at Namur await the word to move inland. (USMC 70450)*

"Good luck to the first Marines to land on Japanese soil," radioed Colonel Cumming to the elements of his regiment that were approaching JACOB and IVAN.[8] The gunboats halted, the armored LVTs passed through the line of LCIs to take up positions just off the beach at JACOB, and at 0952 the tractors carrying Company B, 1/25, rumbled onto the island. Off neighboring IVAN, Company C and the attached scouts were encountering serious difficulties.

A rugged segment of reef, brisk winds, and adverse seas had slowed to a crawl the speed of the LVTs carrying Company C. Continuous strafing attacks prevented the defenders of IVAN from taking advantage of the delay, but the persistent battery on Roi resumed firing until silenced a second time. Finally, Colonel Cumming was able to hasten the landing by diverting the tractors carrying the scout company around the island and onto its lagoon beaches.[9]

While Company C was struggling in vain to reach the ocean shore, the scouts at 0955 landed on southeastern beaches of IVAN and set up a skirmish line facing toward the north. A few minutes later, a regimental staff officer reached Company C and directed it to land in the wake of the scout company. At 1015, the tardy company landed to support the scouts.[10]

The fighting on JACOB and IVAN was brief and not especially violent. JACOB was overrun within a quarter of an hour. After the mop up that followed, a total of 17 enemy dead, 8 of them apparent suicides, were found on the island. Two prisoners were taken. IVAN yielded 13 dead and 3 prisoners.[11]

As soon as the two islands were secured, LVTs from Company A, 10th Amphibian Tractor Battalion began landing elements of the 14th Marines. The 75mm howitzers of 3/14 were carried to JACOB in the tractors and a few LCVPs. Equipped with 105mm howitzers, weapons too bulky to be carried by LVTs, 4/14 landed from LCMs on IVAN. IVAN was selected for the command post of the Colonel Louis G. DeHaven's artillery regiment, but sites on JACOB were chosen by the commanding officers of both the 25th Marines and 1/25.

ALLEN, ALBERT, AND ABRAHAM

Once 1/25 had seized IVAN and JACOB, the scene of action shifted across the lagoon to ALLEN, ALBERT, and ABRAHAM, three islands that ascend the northeastern rim of the atoll toward Roi-Namur. This trio of islands was needed to serve as artillery positions and to secure the flank of the boat waves that would assault the main objective. General Underhill's IVAN Landing Group, the conqueror of IVAN and JACOB, had also been assigned to make these later landings. When landing on ALLEN and ALBERT, the assault forces were to strike from the lagoon, but 3/25 was to approach

[8] *4th MarDiv Jnl*, msg dtd 0920, 31Jan44. Actually, the VAC Reconnaissance Company had already begun the conquest of Majuro.

[9] MajGen Samuel C. Cumming interview with HistBr, G-3, HQMC, dtd 24Nov52.

[10] *Ibid.*

[11] *Ibid.*; LtCol Arthur E. Buck, Jr., ltr to CMC, dtd 21Jan53; LtCol Michael J. Davidowitch ltr to CMC, dtd 26Nov52.

ABRAHAM by moving parallel to and just inside the reef.

A-Hour, the time of the landings on ALLEN and ALBERT, had been tentatively set by Admiral Conolly for 1130. The manner of execution was similar to that used during the earlier D-Day landings. One destroyer was assigned to support each of the two assault battalions, while rockets from LCIs, automatic weapons from LVT-(A)s, and the strafing by aircraft insured the neutralization of the beaches.

The Marines of 2/25, chosen as reserve for the IVAN-JACOB phase of the operation, had loaded into LCVPs at 0530 and within two hours had completed their transfer to LVTs. They spent the remainder of the morning being rocked ceaselessly by the pitching waves. More fortunate were Chambers' men, for they did not begin loading in LCVPs until after dawn. Almost two hours were lost when the transport carrying the battalion was twice forced to get underway in order to maintain station in the buffeting seas. During the morning, the landing craft carrying 3/25 plowed through the swells to the vicinity of the transfer area where they were to meet tractors returning from the IVAN and JACOB landings.

Once they had embarked in the landing craft, the men of 3/25 were as roughly treated by the sea as their fellow Marines of Hudson's battalion. "The sea was not too calm," reads the report of 3/25, "and as a result, many of the Marines found themselves wishing the boats would head for the beach instead of circling in the transport area." [12] The men did not get their wish, but at midday the LCVPs moved the short distance to the transfer area. Here the unit encountered still other misadventures.

In spite of the morning's delays, Admiral Conolly believed that 1430, three hours later than his earlier estimate, was an attainable A-Hour. The passes into the lagoon and the boat lanes were cleared by minesweepers, and supporting ships continued the bombardment of ALLEN, ALBERT, and Roi-Namur. Namur in particular rocked under the hammering of naval guns, but Roi was not slighted. Admiral Conolly signaled to the warships blasting Roi: "Desire *Maryland* move in really close this afternoon for counterbattery and counter blockhouse fire. . . ." [13] This message earned the admiral his nickname of "Close-in" Conolly.

As the minesweepers were clearing JACOB Pass, they had discovered that it was too shallow to permit the entry of the destroyer *Phelps*, the control vessel for all the D-Day landings. As a result, the ship was routed through IVAN Pass, for it was thought necessary to have the *Phelps* inside the lagoon in time to protect the ships sweeping mines from the boat lanes leading to the objectives. The *LCC 33*, a specially equipped shallow-draft vessel, had been selected to shepherd the assault craft in the absence of the destroyer, but the alternate control craft failed to learn of the change in plans. As a result, responsibility for control temporarily passed to *SC 997*,

[12] 3/25 Hist, *op. cit.*, p. 4.

[13] *4th MarDiv Jnl*, msg dtd 1210, 31Jan44. At one point on the afternoon of 31 January, the *Maryland* moved to within 1,000 yards of Roi.

the submarine chaser in which General Underhill and his staff had embarked.

Although the general had neither copies of the control plan nor adequate radio channels to coordinate the movement of the waves, he attempted to restore order. First, the *SC 997* rounded up the tractors carrying 2/25, which had mistakenly attempted to follow the *Phelps*. These strays, as well as some LVTs carrying 3/25 that had wandered from their proper station, were herded back to the transfer area.

The naval officers assigned to guide the various waves cooperated to the best of their ability in reorganizing the assault force, and Admiral Conolly soon steamed onto the scene to supervise. "This was to prove," the admiral commented, "the only case in my experience before or later where I had any difficulty controlling the craft making the landing." [14]

While the waves were being reformed in the transfer area, a few additional LVTs arrived, and these were used to carry Marines of 3/25. There were, however, enough tractors for less than half of Chambers' battalion.[15] By now, JACOB Pass was known to be free of mines, so, rather than wait for additional LVTs, General Underhill ordered both battalions to follow the submarine chaser through the passage toward the line of departure.

The *Phelps*, which had finished her support mission, was now nearing the line of departure within the lagoon where she would again take over as control vessel. Observers in the destroyer viewed the progress of the approaching tractors and reported to Admiral Conolly that A-Hour could not be met.[16] He then postponed the time of the landings to 1500.

The delay imposed a strain on the system of aerial control, for the planes assigned to attack just prior to the landings could not be held on station for the additional 30 minutes. Such a decision would have disrupted the schedule worked out for the carriers and possibly have prevented later flights from arriving on time. To insure complete coverage throughout the day, the Commander Support Aircraft directed the planes then on station to attack targets of opportunity. The relieving flight of bombers was employed to support the landings, but it seemed that no fighters would be on hand to deliver the final strafing. The combat air patrol on station over the northern part of the atoll lacked enough fuel for the attack. Fortunately, another group of fighters arrived as the landings were about to begin. Since these relief pilots were familiar with the air support plan and the radar screen was free of hostile aircraft, they were able to sweep low over the islands and keep the enemy pinned down during the crucial moments just prior to the assault.

At 1432, the assault waves began

[14] VAdm Richard L. Conolly ltr to CMC, dtd 26Nov52.

[15] Rpt of CG, IVAN LdgGru, dtd 29Feb44, Encl C to *4th MarDiv AR*, p. 4, states that 1½ waves were in LVTs, but 3/25 History, *op. cit.*, p. 4, says three or four.

[16] Admiral Conolly reported that the slow progress of the tractors was "due to the low speed of the LVTs proceeding against the wind and the inexperienced LVT drivers permitting their vehicles to drift down wind while waiting for waves to form up." *TF 53 AR Roi–Namur*, p. 5.

crossing the line of departure along which the *Phelps* had taken station. The LCI(G)s led the way, followed by armored amphibians and finally by the troop-carrying LVTs. The gunboats discharged their rockets, raked the beaches with cannon fire, and got clear of the boat lanes. Company D, 1st Armored Amphibian Battalion, plunged past the LCIs to maintain the neutralization of the islands with fire from cannon and machine guns. The supporting destroyers ceased their shelling to permit planes to execute the revised schedule of aerial strikes, and at 1510 3/25 reached ALBERT. The Marines of 2/25 landed on ALLEN just five minutes later.

Both objectives were quickly taken. Chambers' 3d Battalion secured ALBERT by 1542, killing 10 Japanese in the process at a cost of 1 Marine killed and 7 wounded. Hudson's men, their progress impeded by the dense undergrowth in the northern part of ALLEN, needed help from a platoon of tanks to wipe out the Japanese platoon defending the island. By 1628, ALLEN too had been captured.

When it became apparent that Company G, Hudson's reserve, would not be needed at ALLEN, that unit was dispatched to ANDREW (Obella). The unit landed at 1545 and found the island unoccupied. Although opposition had so far been light, the operation had moved slowly. Before darkness, ABRAHAM had to be seized and additional artillery landed.

After a prolonged stay in the rough seas of the lagoon, 1/14 and 2/14 with 75mm pack howitzers came ashore on ALBERT and ALLEN in time to move into firing positions just before dark. Registration, however, was postponed until the morning of D-Day. Although the weapons were emplaced promptly, Colonel William W. Rogers, division chief of staff, was not entirely pleased with the conduct of this phase of the operation. He felt that not enough ammunition was on hand at ALBERT and ALLEN. Forced to buck heavy seas all the way from the transport area to the islands, many of the LVTs that were loaded with ammunition had run out of gas short of their destination. Tractors, however, labored throughout the night to ferry an adequate number of artillery rounds to ALBERT and ALLEN.[17]

While the howitzer battalions were preparing to land, Chambers was readying 3/25 for the seizure of ABRAHAM, the last of the day's objectives. Although this island was not to be a site for howitzer batteries, its capture was important, for Japanese guns emplaced there could fire into the flanks of the assault waves bound for Roi-Namur.[18] Chambers, however, had a difficult time mounting the attack.

The battalion commander suddenly found himself desperately short of assault craft. Because the amphibian tractor unit had received no orders concerning the ABRAHAM landing, its vehicles withdrew to refuel immediately after ALBERT had fallen.[19] The only tractors that remained behind were the two that carried Chambers and his headquarters.

---

[17] MajGen William W. Rogers ltr to Dir DivPubInfo, HQMC, dtd 3Feb48.

[18] *Ibid.*

[19] *Croizat ltr.*

Admiral Conolly had directed that the attack upon ABRAHAM be launched at 1600 or as soon thereafter as practicable. B-Hour had already passed when Colonel Cumming landed on ALBERT to confer with the battalion commander on the quickest method of completing the D-Day operations. Chambers decided to attack at 1800 if landing craft were available by then. Three self-propelled 75mm guns from the Regimental Weapons Company, the battalion's attached 37mm guns, and its organic mortars were to support the landing.

Prior to the advance against ABRAHAM, 3/25 occupied ALBERT JUNIOR, a tiny island 200 yards north of ALBERT. Although no Japanese were posted on ALBERT JUNIOR, the ABRAHAM garrison opened fire on the occupation force. Machine guns were then mounted on the island to support the scheduled landing.

A patrol waded toward ABRAHAM and returned with information concerning the route over which Chambers would attack. In the meantime, the battalion commander had gained the services of two additional tractors that wandered near the ALBERT beaches. He decided to load 120 of his Marines into the four amphibians, dispatch them in a single wave to seize a small beachhead, and then use the same vehicles to shuttle the remainder of his troops across the shallow strait.

The assault began on schedule. A smoke screen laid by the battalion 81-mm mortars concealed the approaching LVTs, and the enemy chose not to defend the southern beaches. By 1830 two companies had reached the island and carved out a beachhead 250 yards deep. In 45 minutes, the island was under American control, but mopping-up continued into the night. Six Japanese were killed on ABRAHAM; one Marine was wounded during a misdirected strafing attack by a friendly plane.[20]

Since this last objective was a scant 400 yards from the southeast shoreline of Namur, it could provide a base of fire for the next morning's attack. During the night, as many weapons as possible were rushed into position. By morning, 5 self-propelled 75mm guns, 17 37mm antitank guns, 4 81mm mortars, 9 60mm mortars, and 61 machine guns stood ready to assist planes, ships, and field pieces in their deadly work.

General Underhill's IVAN Landing Group had executed all its D-Day assignments, but the operation had not been without its flaws. Writing some years after FLINTLOCK, an officer of the 10th Amphibian Tractor Battalion attempted to analyze the work of his battalion in northern Kwajalein. During World War II, he served at Guadalcanal and Saipan as well as in the Marshalls and as a result felt "somewhat qualified to appreciate confusion." He maintained that the period from the organization of his unit through the securing of Roi-Namur was the most exhausting both physically and mentally of any operation in which he took part.[21]

The numerous landings scheduled for

---

[20] This account of the ABRAHAM landing is based on an interview with Colonel Justice M. Chambers, dtd 6May48, cited in Heinl and Crown, *The Marshalls*, pp. 50–52. No transcript of the interview is available.

[21] *Croizat ltr.*

214-881 O-67—11

D-Day placed a grave burden on the LVTs, their crews, and the officers who were to control their employment. The control system, moreover, depended upon reliable communications, and the radios carried in the tractors were vulnerable to water damage. In a heavy sea, such damage was unavoidable.

In commenting on the employment of LVTs on 31 January, the 4th Division chief of staff observed that problems were anticipated and tentative plans were made to insure the success of the operation. He wrote:

> The Commanding General and Staff of the Northern Landing Force were well aware that things might not go as planned on D-Day. In fact the 4th Amphibian Tractor Battalion was withheld entirely on D-Day in spite of urgent requests from subordinate units, in order that we would be sure to be able to land the 23d Marines on Roi on D+1, either from the outside or from the inside of the lagoon. In other words, it was considered that the mission could have been accomplished by the capture of IVAN and JACOB and the subsequent landing on Roi by the 23d Marines utilizing the 4th Amphibian Tractor Battalion, even if the landings on the east side of the lagoon had not been possible on D-Day. This would have involved the subsequent capture of Namur by assault from Roi, with or without a landing from the lagoon. Possession of the Eastern Islands naturally made the entire operation easier.[22]

In the words of Admiral Conolly, the plan for D-Day, "under the sea conditions prevailing, was . . . too complicated and beyond the state of training and discipline of the LVT units to execute *smoothly*, especially when the unexpected complications were imposed. However, the plans were made to work and that is the final test of a command and its organization." [23] In spite of the unfavorable seas, the difficult reefs, and the lapse in control that occurred while the *Phelps* entered the lagoon, the Marines had taken all their objectives. More reliable radios, closer cooperation between LSTs and LVTs, and a tighter rein by control officers would have resulted in a less hectic operation, but these facts were of no consolation to the Japanese killed on the outlying islands.

THE ARMY IN SOUTHERN KWAJALEIN [24]

On D-Day, while the Marines were seizing the islands near Roi-Namur, General Corlett's Army troops were to make a similar series of landings in the immediate vicinity of Kwajalein Island. The 7th Reconnaissance Troop, reinforced by part of the garrison force, men from Company B, 111th Infantry, was scheduled to occupy CECIL (Ninni) and CARTER (Gea), two small islands believed to be undefended. When this task was done, the troop might be called upon to reconnoiter CHAUNCEY (Gehh) not far from CECIL. CARLSON (Enubuj), which was thought to be defended by a force of 250–300, and less formidable CARLOS (Ennylabegan) were the objectives of the 17th Infantry. (See Map 8.)

[22] BGen William W. Rogers ltr to CMC, dtd 1Dec52.

[23] Conolly ltr, *op.cit.*

[24] A detailed account of Army operations on D-Day may be found in Crowl and Love, *The Marshalls and Gilberts*, pp. 219–229.

Although artillery was to be emplaced on CARLSON only, all of these islands figured in General Corlett's plans. CECIL and CARTER were important because they bounded a passage into the lagoon, while a wider deep-water channel lay between CARLOS and CARLSON. In addition, CARLOS was considered a suitable site for the 7th Infantry Division supply dumps.

The invasion of southern Kwajalein, like the operation in the north, was not without its moments of frustration. Attempting to land from rubber boats on a moonless night, that portion of the reconnaissance troop destined for CARTER started off toward neighboring CECIL. The error was detected, the men landed on the correct island, and after a brief fire fight they secured the objective.

While the APD USS *Manley* was launching the boats bound for CARTER, her sister ship USS *Overton* was attempting to locate CECIL. "Intelligence received gave a good picture of both Gea and Ninni Islands," reported the skipper of the *Overton*, "but little of Gehh, the contour of which was, in a way, similar to Ninni."[25] In the darkness, the attackers mistook CHAUNCEY (Gehh) for CECIL (Ninni) and landed there instead. A brief skirmish followed, but before the island had been secured General Corlett learned of the error and ordered the reconnaissance troop to move to the proper island.

Leaving a small force to contain the Japanese on CHAUNCEY, the soldiers re-embarked and occupied CECIL shortly after noon. The group left behind soon encountered a larger number of the enemy and had to be withdrawn. Taking CHAUNCEY was postponed until an adequate force was available.

Off the two objectives assigned to the 17th Infantry, the assault waves began forming in the morning darkness. Poor visibility resulted in confusion, and the attack had to be postponed from 0830 to 0910. At CARLOS, 1/17 landed without opposition and rapidly overran the mile-long island. The few defenders, who lacked prepared positions from which to fight, were either killed or captured. The Americans suffered no casualties.

The CARLSON landing force, 2/17, expected to meet skillfully organized resistance. The LVTs carrying the assault waves reached the island at 0912, and the soldiers promptly began moving inland. Contrary to intelligence estimates, not a single Japanese was found on the island, although 24 Korean laborers were taken prisoner. The most serious opposition came from artillery on Kwajalein Island, but these pieces were silenced by naval gunfire before they could do the attackers any harm.

Army artillery, four battalions of 105mm howitzers and a battalion of 155mm howitzers, promptly landed on CARLSON, moved into position, and began registering. Some of the lighter pieces fired for effect during the night, but not all of the 155mm howitzers were emplaced when darkness fell. Meanwhile, a medical collecting station and LVT maintenance shop were being set up on CARLOS.

In spite of the numerous delays, the D-Day landings in both the north and

[25] USS *Overton* AR, dtd 8Feb44, p. 4.

south had been successful. Roi-Namur and Kwajalein Islands had been isolated, battered, and brought within range of field artillery. The enemy garrisons were given no rest during the night, for warships continued shelling the objectives. Although Army 105s joined in the shelling of Kwajalein Island, the Marine howitzers, scheduled to register at dawn on Roi-Namur, were temporarily silent. Under cover of darkness, Underwater Demolition Teams examined the assault beaches at both islands. Neither mines nor other artificial obstacles were found. The way was clear for the next day's operations.[26]

[26] Com V PhibFor msg to CinCPac, ser no. 00334, dtd 13Mar44 (AR File, OAB, NHD).

FLINTLOCK: Completing the Conquest[1]

On D plus 1, after the capture of the outlying islands, General Schmidt's 4th Marine Division was to storm Roi-Namur. At Roi, where the enemy had built an airfield, Colonel Louis R. Jones would land two reinforced battalions of the 23d Marines on the Red Beaches along the lagoon coast of the island. Namur, to the east of the sandspit that joined the twin islands, was the objective of another reinforced regiment, the 24th Marines, commanded by Colonel Franklin A. Hart. There two assault battalions were to strike northward across the island after landing on the Green Beaches. LVTs of the 4th Amphibian Tractor Battalion were to carry the Roi battalions, and the Marines destined for Namur would rely on the 10th Amphibian Tractor Battalion, veteran of the D-Day landings. (See Maps 9 and 10.)

[1] Unless otherwise noted, the material in this chapter is derived from: *TF 51 AR; TF 53 AR Roi-Namur; VAC AR FLINTLOCK; 4th MarDiv Ar; 4th MarDiv Jnl;* 4th MarDiv CommOpsRpts, dtd 29Mar44; 4th MarDiv D–3 Rpts, 31Jan–12Feb44; 7th InfDiv Rpt of Participation in FLINTLOCK Op, dtd 8Feb44; 7th InfDiv SAR, Kwajalein Island, dtd 27Mar44; 20th Mar Rpt on FLINTLOCK Op, dtd 16Mar44; 23d Mar Rpt on FLINTLOCK Op, dtd 4Mar44; 23d Mar Jnl, 31Jan–4Feb44; 24th Mar Prelim Rpts on Roi-Namur Op, dtd 10Feb44 (including rpts of 1/24, 2/24, and 3/24); 1/23 OpRpt, dtd 10Feb44; 2/23 OpRpt, dtd 14Feb44; 3/23 Rec of Events, 31Jan–5Feb44, dtd 12Feb44; 1st Armd PhibBn, Cmts on LVT(A)(1)s during FLINTLOCK Ops, dtd 3Feb44; 4th TkBn Rpt on FLINTLOCK Op, dtd 20Apr44; *10th AmTracBn Rpt;* Heinl and Crown, *The Marshalls.*

LAND THE LANDING FORCE

Admiral Conolly and his staff were quick to profit from the mistakes of D-Day. The long journey through heavy seas from the transfer area to the beaches had been too much for the short-legged LVTs.[2] The original plan for D plus 1 called for the landing force to transfer to LSTs and there load in the tractors. When the Marines had entered the assault craft, the parent LSTs were to lower their ramps and launch the tractors. The LVTs would then battle the waves to enter the lagoon, move to a position off the objective, and form for the assault. Although this plan spared the troops the discomfort of transferring at sea from one type of landing craft to another, it did not reduce the distance which the tractors had to travel. To avoid the delays of D-Day and move the LVTs closer to their line of departure, Conolly invoked his rough weather plan. The troop transfer arrangement was left unchanged, but the LSTs were

[2] The LVT(2)s had only power-driven bilge pumps. When the gasoline supply was exhausted, these failed, and the unfortunate vehicle usually foundered.

directed to enter the lagoon before launching their tractors.[3]

This change, however, could not prevent a repetition of many of the difficulties that had marred the D-Day landings. The principal offenders were the LVTs and LSTs, for the two types did not cooperate as well as they should have. The troubles of the 10th Amphibian Tractor Battalion began on the night of 31 January as its vehicles were returning from ALBERT and ALLEN.

Some of the parent LSTs failed to display the pre-arranged lights, so that many tractors became lost in the gathering darkness. The boats that were to guide the LVTs fared no better, and the battalion soon became disorganized. Since the tractors did not carry identifying pennants, the LST crews could not easily determine which vehicles had been entrusted to their care. Concerned that they would be unable to refuel their own LVTs, the captains of a few landing ships refused to give gasoline to strangers. The commander of the tractor battalion felt that the trouble stemmed from the feeling, apparently shared by many of the LST sailors, that the LVTs were boats rather than amphibious vehicles. "They should be made to appreciate the fact that LVTs are not boats," he admonished, "cannot maneuver or operate in the manner of boats, nor are they tactically organized in the manner of boat units." [4]

Although the bulk of the battalion vehicles either reached the haven of the LSTs or remained for the night on one of the captured islands, seven tractors were not yet accounted for when FLINTLOCK ended.[5] As dawn approached, the battalion commander realized that the LSTs had not retrieved enough tractors to execute the morning's operations. He notified Admiral Conolly who put into effect a replacement scheme. The company commander, Company A, 11th Amphibian Tractor Battalion was ordered to send a specific number of LVTs to certain of the landing ships to make up the shortage.

The ordeal of the 10th Amphibian Tractor Battalion did not affect the preparations of the Roi-Namur landing force. As soon as there was daylight enough for safe navigation, the LSTs carrying the 4th Amphibian Tractor Battalion began threading their way into the lagoon. At 0650, the old battleship USS *Tennessee* opened fire against a blockhouse on the sandspit that linked Roi with Namur, while other vessels commenced hammering Namur. The bombardment of Roi, delayed by the passage of LSTs between the support units and the island, began at 0710. Carrier planes arrived over the twin islands, and howitzers of the 14th Marines joined in the shelling. W-Hour, the hour of the landings, was set for 1000.

Meanwhile, the LSTs had arrived in

[3] Rough Weather AltnPlan, dtd 26Jan44, Anx V to TF 53 OpO A157–44, dtd 8Jan44.

[4] *10th AmTracBn Rpt*, p. 2. At the conclusion of the operation, the action reports of the transport division commander and Admiral Conolly both voiced the view the LVTs used for assault waves should be regarded and organized as boats and manned by carefully-trained Navy crews. *TF 53 AR Roi-Namur*, p. 10; ComTransDiv 26 AR, *op. cit.*

[5] 10th AmphTracBn Rpt of LVT(2) Activities in Kwajalein Op, dtd 17Feb44, p. 2.

position to disgorge the tractors assigned to the 23d Marines. Like those LVTs used on D-Day, the tractors loaded on the weather decks of the ships had to be lowered by elevators to join the vehicles stored on the tank decks and then be sent churning toward the beaches. Before the convoy sailed, tests had shown that the LVT(2)s were too long for the elevators. As a result, an inclined wooden plane was built on the elevator platform. If the tractor was driven up this ramp, it was sufficiently tilted to pass down the opening with a few inches to spare. Maneuvering the vehicles into position was a time-consuming job, an impossible task unless clutch and transmission were working perfectly. Yet, this was the only method of getting these LVTs into the water.

The elevator in one LST broke down midway through the launching, leaving nine tractors stranded on the weather deck. The Marines assigned to these vehicles were sent to the tank deck and placed, a few at a time, in the LVTs loading there. On another LST, the ramp was so steep that few vehicles could negotiate it. Drivers pulled as far up the incline as they could, then stopped, while a crew of men with a cutting torch trimmed the splash fenders at the rear of the tractors until clearance was obtained.[6]

At 0825, all fire-support ships had acknowledged Conolly's message confirming 1000 as W-Hour, but within a few minutes General Schmidt was sending Colonel Hart some disquieting news. "We are short 48 LVTs as of 0630," the commanding officer of the 24th Marines had reported. The commanding general now replied: "Every effort being made to get LVTs. Use LCVPs for rear waves and transfer when LVTs are available."[7] A two-hour search for amphibian tractors proved fruitless. Because of the night's confusion, the necessary number of LVTs was not at hand.

Both regiments were falling behind schedule, although sailors and Marines alike were trying desperately to get the assault craft into formation. When Admiral Conolly asked the commander of the transport group if a postponement was necessary, he immediately received the reply: "Relative to your last transmission, affirmative."[8] At 0853, the time of the attack was delayed until 1100.

The schedule of fires was adjusted to meet this new deadline, and the task of destruction continued. At 0925, another crisis arose. A salvage boat sent to ABRAHAM by the transport USS *Biddle* reported: "Japs are counterattacking from CAMOUFLAGE. Send support immediately."[9] This message was instantly relayed to Conolly, and even though aerial observers could not locate the enemy troop concentration, the admiral took no chances. Torpedo bombers, warships, and artillery batteries hurled high explosives into the southern part of Namur, but by

[6] *Croizat ltr;* Maj Theodore M. Garhart ltr to CMC, dtd 14Nov52.

[7] CO 24th Mar msg to CG 4th MarDiv, dtd 0630, 1Feb44 and CG 4th MarDiv msg to CO 24th Mar, dtd 0830, 1Feb44, 4th *MarDiv Jnl.*

[8] ComTransGru msg to CTF 53, dtd 0841, 1Feb44, *4th MarDiv Jnl.*

[9] SalvBoat 8 msg to *Biddle,* dtd 0925, 1Feb44, *4th MarDiv Jnl.*

1000 it was clear that the report of a counterattack had been incorrect.

When this sudden flurry of action ended, support ships returned to their tasks, firing deliberately and accurately until 1026 when the shelling was stopped to permit an airstrike. A glide-bombing attack followed by strafing runs kept the enemy occupied. As the planes were departing, the naval bombardment resumed.

Colonel Jones arrived at the line of departure 15 minutes before W-Hour. Although he had ample time to transfer with his staff to the pair of LVTs that had been assigned him, the tractors could not be found. He eventually would land from an LCVP.[10]

Lieutenant Colonel Edward J. Dillon's 2/23, the force destined for Red 3, loaded into LVTs, left the LSTs, and then moved to the line of departure without waiting for the other assault battalion. Within a few minutes, 1/23, commanded by Lieutenant Colonel Hewin O. Hammond, had reached the line and begun the final adjustment of its formation prior to the storming of Red 2. Somehow, Hammond's battalion had failed to learn of the postponement of W-Hour, and the men of the unit felt that they "failed miserably" to meet the deadline.[11] Actually they were a few minutes ahead of schedule. (See Map 9.)

W-Hour came, then passed and still the 23d Marines remained at the line of departure. Although Jones' troops were ready, Hart's 24th Marines was not. Since 0630, control officers had been trying without success to round up enough LVTs to carry the two assault battalions of the regiment. The transport group commander began releasing LCVPs to Hart, but contacting the boats and directing a sufficient number to the proper LSTs were difficult tasks. In spite of Admiral Conolly's decision to delay the attack, the Namur landing force needed still more time.

Hart soon became convinced that his assault waves could not possibly cross the line of departure in time to complete the 33-minute run to the Green Beaches by 1100. He requested another postponement and received word that "W-hour would be delayed until the combat team could make an orderly attack." This message led him to assume that "he was to report when his waves were in position and ready to move."[12] Satisfied that his schedule had been made more flexible, the regimental commander began making last-minute changes in the composition of his assault waves.

Because of the shortage of amphibian tractors, neither 3/24, the battalion destined for Green 1, nor 2/24, which was to attack Green 2, had enough LVTs for all its rifle companies. Lieutenant Colonel Francis H. Brink, commanding 2/24, noted that the company scheduled to remain in reserve had its full quota of vehicles, so he designated it as an assault company and placed the unit with the fewest tractors in reserve. Lieutenant Colonel Austin R. Brunelli of 3/24 ordered the tractors assigned to his

[10] BGen Louis R. Jones ltr to Dir, DivPubInfo, HQMC, dtd 11Apr49.

[11] 1/23 OpRpt, *op. cit.*, p. 3.

[12] 24th Mar Rpt of FLINTLOCK Op, p. 7, dtd 10Mar44, Encl D to *4th MarDiv AR.*

reserve to be divided between the assault companies. (See Map 10.)

When the two battalions reached the line of departure, each was but two-companies strong. Control officers assigned to work with 2/24 found the situation especially confusing, for Company E, the unit originally designated battalion reserve and consigned to the fourth wave, was now the left element on the second and third waves. Additional time was lost as the company commander attempted in vain to explain the change, but his unit finally was formed in a single wave as the discarded plan had directed.[13]

To replace the absent reserves, Colonel Hart turned to Lieutenant Colonel Aquilla J. Dyess, commanding officer of 1/24, the regimental reserve, and ordered him to release one company to each of the assault battalions. While the LCVPs carrying these two units were moving into position, the third rifle company rejoined 2/24. The arrival of this unit, embarked in seven LVTs and two LCVPs,[14] brought Brink's battalion up to full strength. As a result, one of Dyess' companies was returned.

While the composition of the Namur assault force was thus being altered, Colonel Jones' Marines were waiting impatiently at the line of departure. At 1107, the colonel asked the control vessel *Phelps* why the attack was being delayed. Five minutes later, the red flag dropped from the yardarm of the destroyer, the signal which was to send both regiments toward their objective.[15] LCI gunboats, armored amphibians, and finally the LVTs carrying the assault battalions charged toward Roi. At 1150, naval gunfire was lifted from the Red Beaches, the gunboats and armored amphibians fired as long as the safety of the incoming troops permitted, and at 1157 the 23d Marines was reported to have reached Roi.

The signal to launch the attack came as a surprise to Colonel Hart, for he was under the impression that his regiment would not make its assault until all its elements were in position. He attempted to intercept Brunelli's 3/24, which had responded to the control ship signal, but when he saw that the regiment on his left was moving toward Roi, he realized that such an

[13] LtCol John F. Ross, Jr. ltr to Head, HistBr, HQMC, dtd 21Jan53. Commenting on the differing solutions to this problem of the shortage of tractors, the commander of 2/24 believed in retrospect, that Brunelli's procedure "was probably better than mine. At the time," he noted, "I considered shifting tractors between scattered LSTs a time-consuming project in which I could lose control of some I already had." BGen Francis H. Brink ltr to ACofs, G-3, HQMC, dtd 20Oct62, hereafter *Brink ltr*.

[14] *Brink ltr*.

[15] Admiral Conolly noted that the order to execute this signal "was a command decision made by me after consultation with General Schmidt and with his full concurrence. With the information on hand that the 24th RCT had two battalions formed, and considering the already delayed How Hour and other factors such as gasoline consumption in the waiting tractors, and the waning effects of the bombardment, the Landing Force commander and I had to reach a decision to wait further or go ahead. The decision to go ahead was a calculated risk of the kind responsible commanders must make in time of war." VAdm Richard L. Connolly ltr to Dr. Jeter A. Isely, dtd 31Aug49, encl to Gen Harry Schmidt ltr to CMC, dtd 22Oct62.

action would only add to the confusion. Preceded by LCI(G)s and LVT(A)s, the first waves reached Namur at 1155. The weapons emplaced on ABRAHAM supported the landing of the 24th Marines.

The four battalions that stormed Roi-Namur benefited from an experiment in air support directed by the air coordinator. Bomber pilots who were to participate in the strikes just prior to W-Hour were warned to remain above 2,000 feet. At this altitude, above the maximum ordinate of artillery, naval gunfire, and rockets, they could attack while the other supporting weapons were firing.

Just as the carefully arranged bombing attack was to begin, a rain squall blanketed the area east of the islands where the aircraft were on station. For a time, it seemed that the strike would have to be cancelled, but an opening in the clouds was spotted from the bridge of the *Appalachian*. The Commander, Support Aircraft was notified, and the planes were directed to the rift in the clouds west of Roi-Namur. The bombers were able to change station and complete their runs by the time the first wave was 750 yards from the beaches.

This technique assured the assault troops of a "thorough, accurate, and continuous bombing attack . . . during the critical approach phase." [16] Since the naval bombardment was not lifted during the bombing attack, air support on 1 February was more effective than that given on the previous day. When the war had reached the Marianas, coordinated attacks such as this one would become commonplace.

"THIS IS A PIP:" THE CONQUEST OF ROI

Red Beach 2, the objective of Lieutenant Colonel Hammond's 1/23, seemed to be a stoutly defended strip of coral. The battalion zone of action was bounded on the left by Wendy Point, the westernmost tip of the island, and extended on the right to a point within 200 yards of Tokyo pier. The enemy appeared to have built heavy blockhouses on the point and scattered pillboxes along the beach. What was believed to be another blockhouse had been erected not far from the right limit of the zone. (See Map 9 and Map V, Map Section.)

Since flanking fire could be delivered from Wendy Point, that portion of the beachhead had to be secured as quickly as possible. Once the fangs of the blockhouses had been drawn, Hammond's battalion was to attack in the eastern part of its zone to aid the advance of the adjacent 2d Battalion. Armored amphibians played a spectacular role in executing this plan.

Admiral Conolly had not specified whether the armored amphibian bat-

[16] AirSpt: FLINTLOCK, n.d., Encl C to *TF 53 AR Roi-Namur*. The commander of 2/14, located with his forward observer party on ABRAHAM, witnessed one success of this bombing attack. Planes hit a blockhouse on the eastern end of Namur, which had been "barely visible because of the surrounding jungle. After the bomb drop, it was completely denuded of trees and Japanese military personnel rushed from the blockhouse in an apparent daze. These men were picked off by the Marines stationed on the forward part of ABRAHAM." BGen John B. Wilson, Jr. ltr to ACofS, G-3, HQMC, dtd 15Oct62.

talion would support the landings from positions off the beaches or from the island itself. The officer in command of the assault regiment could decide how these vehicles might give the more effective support and place them accordingly.[17] At Red 2 the tractors thundered ashore at 1133, several minutes ahead of the first wave of LVT(2)s, moved inland to seek hull defilade, and turned their 37mm cannon against the Wendy Point fortifications. Companies A and B of Hammond's command were both ashore by 1158. While Company A pushed toward the point, Company B began its advance toward the farthest edge of the runway to its front.

The battalion landed slightly out of position, with the companies somewhat bunched toward the left of the zone. This misalignment was caused when the tractors carrying the adjacent battalion had veered westward from the proper boat lanes. The Marines, however, met only scant fire at Red 2 and advanced with ease to their first objective, the 0-1 Line.

Armored amphibians fired across the island into Norbert Circle to protect the flank of Company A as that unit probed Wendy Point. Instead of the concrete blockhouses they expected, these Marines found a single pillbox that had been blown to shreds by bombs and shellfire. Company B encountered no manned enemy positions between the beach and the 0-1 Line. At 1145, Company C, 4th Tank Battalion, began landing its medium tanks and flamethrower-equipped light tanks. These armored vehicles overtook the infantry on the runway and prepared to race across the remainder of the island.

Upon crossing the line of departure, Lieutenant Colonel Dillon's 2/23 found its destination, Red 3, to be covered with a pall of dense smoke. The tractor drivers, unable to orient themselves, tended to drift from their assigned lanes. The LVT(A)s had the most difficult time. A total of 18 of these vehicles, in contrast to the 12 that led the way to Red 2, were crowded into a single wave. One participant recalled that "there was a good deal of 'accordion action,' with the result that several were squeezed out of line from time to time, and there were a number of collisions. . . ." [18] Worse yet, rockets launched by some of the LVT(2)s fell short and exploded in the water close to the armored amphibians.[19] The LVT(A)s overcame these difficulties and took positions just off the island in order to support the advance of the infantry.

Red 3, objective of 2/23, embraced all of the lagoon coast that lay between the battalion boundary west of Tokyo Pier to the base of the sandspit that linked Roi to neighboring Namur. The sandspit itself lay in the zone of action of the 24th Marines. At approximately 1150, the assault waves began passing through the LVT(A)s and landing on the island. Some tractors rumbled ashore outside the proper zone, a few on either flank. The troops that landed out of place were shepherded

[17] PhibAtkO (Main Ldgs), p. 4, dtd 8Jan44, Anx B to TF 53 OpO A157-44, dtd 8Jan44.

[18] Maj Ellis N. Livingston ltr to CMC, dtd 8Nov52.

[19] LtCol Louis Metzger ltr to CMC, dtd 13Nov52.

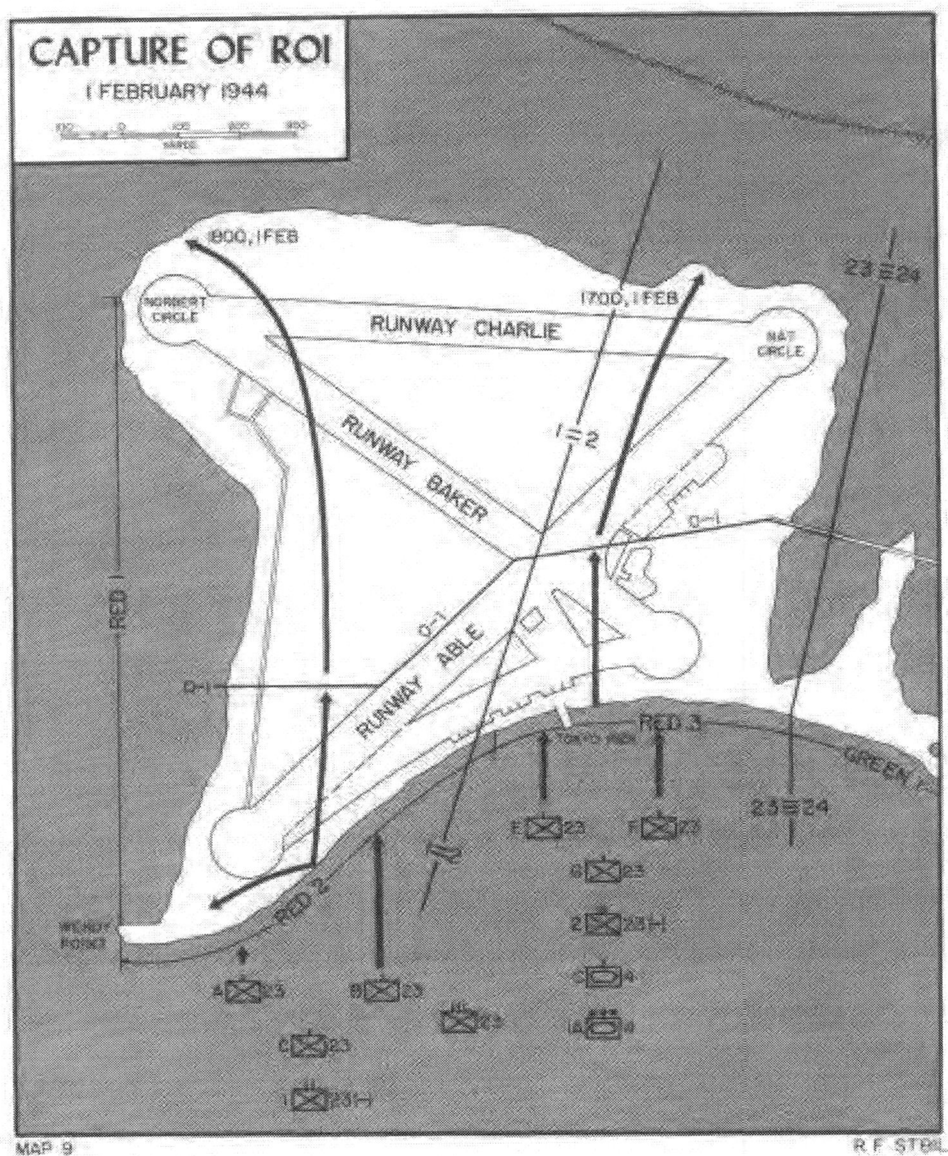

MAP 9 R.F. STEIL

onto Red 3 by alert noncommissioned officers, but those who landed too far to the right had to destroy some Japanese positions before they could cross the regimental boundary.

Resistance on the eastern part of Red 3 was ineffectual. Most of the Japanese seemed dazed by the fierce bombardment which had shattered their prepared defenses. "We received very little enemy fire," recalled an officer who landed there with Company G, "and what fire we did receive came from the northeast corner of Roi." [20] To the west, a few defenders had survived both bombs and shells. "Although these enemy troops were few and dazed from the bombardment," stated an officer of Company E, "they were determined to give their all, as evidenced by the two who left their entrenchment to rush the landing troops." [21]

The surviving Japanese did not lack courage, but they were too few and too stunned to offer serious opposition to Dillon's Marines. Tanks started landing shortly after noon, and by 1215 the battalion commander had set up his command post on the island. Companies E and F had reached the 0–1 Line, which extended from the causeway leading toward Namur to the junction of runways Able and Baker, while Company G was busy ferreting out the Japanese who had taken cover in ruined buildings or in the culverts along the runways.

To an aerial observer circling over Roi, the actions of the assault troops were startling. "Can observe along southwest tip of island;" came one report, "troops seen not to be taking advantage of cover." [22] Colonel Jones, who landed at 1204, soon clarified the situation. "This is a pip," crackled General Schmidt's radio. "No opposition near the beach. . . ." [23] Fifteen minutes later, the commanding officer of the 23d Marines had additional heartening news. "0–1 ours," he reported. "Give us the word and we will take the rest of the island." [24]

In thrusting across the beach, the assault troops had gained such momentum that they approached the 0–1 Line like so many greyhounds in pursuit of a rabbit. Naval gunfire had drastically altered the landmarks which were to designate the line, and this contributed to a breakdown in control. The individual Marines, moreover, were inspired by their incredibly successful landing to finish off the Japanese as quickly as possible. With a confidence that bordered on recklessness, squads, platoons, and even companies launched an uncoordinated, and completely unauthorized, attack toward the northern shore.

If zest for combat can be considered a crime, the worst offenders were the tank and armored amphibian units. The crews of these vehicles, protected by armor plate, were indifferent to the .256 caliber rifle bullets that were cracking across the island. Upon

[20] Maj John J. Padley ltr to Dir, DivPubInfo, HQMC, dtd 16Apr49.

[21] Maj Carl O. Grussendorf ltr to Dir, DivPubInfo, HQMC, dtd 30Mar49.

[22] Air Observer 2 msg, dtd 1210, 1Feb44, *4th MarDiv Jnl.*

[23] CO, 23d Mar msg to CG, 4th MarDiv, dtd 1311, 1Feb44, *4th MarDiv Jnl.*

[24] CO, 23d Mar msg to CG, 4th MarDiv, dtd 1326, 1Feb44, *4th MarDiv Jnl.*

MARINE LIGHT TANK *moves past the radios of a beachhead CP on Namur to lead the attack. (USMC 70203)*

JAPANESE PRISONERS *surrender to Marines near a concrete blockhouse at Roi-Namur. (USMC 70241)*

reaching the 0–1 Line, the commander of Company C, 4th Tank Battalion, radioed for permission to continue the attack, but interference prevented his message from getting through. He then decided to advance rather than wait at the edge of the runway for further orders.

The company commander later justified his action by pointing out that: "If the enemy had had anti-tank guns in his blockhouses on the northern edge of the airfield, he would have been able to seriously damage any tanks remaining for long on the exposed runways."[25] Whatever the danger to the medium tanks might be, plans had called for the assault troops to pause at the 0–1 Line. As it turned out, the menacing blockhouses had been leveled by naval gunfire, and the company commander's aggressiveness prevented Colonel Jones from coordinating the efforts of his regiment.

The tanks roared northward firing cannon and machine guns at every ditch or heap of rubble that might harbor Japanese troops. The Marine infantrymen, trained to protect the tanks and as eager as anyone to advance, also crossed the line, firing frequently and sometimes wildly. A platoon of armored LVTs promptly joined the hunt. The amphibians moved northward along the western coast, some of them in the water, others on land, but all of them firing into trenches and other enemy positions.[26]

Although this impromptu attack killed numerous Japanese and sent most of the survivors scurrying toward the north, it imposed a hardship on the officers directing the campaign. As the tanks were approaching the northeastern corner of Roi, General Schmidt advised Colonel Jones to "await orders for further attack." "Can you," he continued, "control tanks and bring them back to 0–1 Line for coordinated attack?"[27] The tank company commander, in the meantime, was trying to raise Colonel Jones' command post to obtain additional infantry support. Again there was interference on the tank-infantry radio net, and the request was not received. After ranging over the island for about an hour, the Shermans pulled back to the 0–1 Line. Once the tanks began to withdraw, the infantry units followed their example, and by 1445 the colonel was reorganizing his command for a coordinated attack.

This drive was scheduled for 1515, with the two assault battalions advancing along the east and west coast. Once the shoreline had been captured, reserve units could mop up the stragglers who still lurked along the runways. At 1510, 2/23 called for a naval gunfire concentration to be fired against Nat Circle at the northeastern corner of the island. By 1530, the attack was underway.

Supported by the fire of half-tracks mounting 75mm guns, Dillon's Marines pushed resolutely toward Nat Circle. The enemy troops, with little time to

[25] Co C, 4th TkBn AR, dtd 25Mar44, p. 1, Encl C to 4th TkBn Rpt of Activities in the FLINTLOCK Op, dtd 31Mar44, hereafter *4th TkBn Rpt*.

[26] Maj James S. Scales ltr to Dir, DivPubInfo, HQMC, dtd 16Mar49.

[27] Co C, 4th TkBn AR, dtd 25Mar44, p. 1, dtd 1325, 1Feb44, *4th MarDiv Jnl*.

recover and reorganize after the earlier impromptu tank-infantry attack, were readily overcome. Tanks fought in co-operation with the infantry, and by 1600 organized resistance in the battalion zone was confined to the rubble-strewn tip of Roi. Behind 2/23 moved a company from 3/23, the battalion commanded by Lieutenant Colonel John J. Cosgrove, Jr. Because of the speed with which the assault units were moving, this company could not carry out its mission of supporting the advance by fire and had to content itself with mopping up.

Dillon's troops were approaching Nat Circle by the time Hammond's 1/23 launched its attack. From 1530 to 1600, the 1st Battalion supported by fire the thrust of its adjacent unit, then Hammond ordered his infantrymen and their supporting tanks and half-tracks to strike northward along the west coast. Within 45 minutes, all organized resistance in the zone of action had been crushed. During the advance by 1/23, two of Cosgrove's companies stood ready along the beach to thwart any Japanese attempt to attack across the sandspit from Namur.

By 1800, 1/23, in complete control along the western coast, was preparing defenses in the event of an enemy counterlanding. Tanks, riflemen, 37mm guns, a 75mm self-propelled gun, and demolitions teams combined their efforts to destroy the Japanese defending Nat Circle. At 1802, Colonel Jones was able to report that the coastline was secured and that his men were "mopping up, working toward center from both sides." [28] Three minutes later, Roi was declared secured.

Once the situation on Roi was in hand, General Schmidt was able to concentrate on Namur, where the 24th Marines were facing determined resistance. The tanks supporting 3/23 were withdrawn even before the island was secured and sent across the sandspit. Although the defenders had been destroyed, quiet did not immediately descend upon Roi, for even as the last Japanese were being hunted down, an epidemic of "trigger-happiness" swept the island. Near Nat Circle, 3/23 extended between 3,000 and 5,000 rounds against a nonexistent sniper. Only a handful of these Marines actually knew why they were firing, but those who joined in had a sufficient motive. As members of the reserve battalion, they had played a minor role in a spectacularly successful assault, and, as their commanding officer discovered, "they wanted to be able to say they had fired at a Jap." [29] Three Marines were wounded as a result of this outburst.

On the west coast, men from 1/23 opened fire on a group of coral heads in the mistaken belief that these were Japanese troops swimming toward Roi. Observed through binoculars, the coral formations bore no resemblance to human beings, but, as one officer admitted, "to the unaided eye, those coral heads did look like swimmers." [30] No

---

[28] CO, 23d Mar msg to CG, 4th MarDiv, dtd 1802, 1Feb44, *4th MarDiv Jnl.*

[29] 3/23 Rpt of Firing in Vic of Southern Hangar on BURLESQUE, dtd 12Feb44, Encl D to 3/23 Rec of Events, *op. cit.*

[30] Scales ltr, *op. cit.*

one was injured as a result of this incident.

Colonel Jones had been absolutely correct when he called the Roi landings a "pip." Supporting weapons, especially naval gunfire, had done their work so well that the Japanese were incapable of putting up a coordinated defense. The level terrain enabled Marine tanks to roam the island at will. The fight for Roi had been an easy one. Such was not the case on neighboring Namur.

THE STORMING OF NAMUR [31]

The signal to launch the assault on Namur came before the two assault battalions were fully organized. Both Brink's 2/24 and Brunelli's 3/24 had difficulty in getting enough tractors for their commands, and some last-minute arrivals were being fitted into the formation when the destroyer *Phelps* signaled the LVTs to start shoreward. The firepower of supporting weapons helped compensate for the lack of organization. The weapons massed by Lieutenant Colonel Chambers on the northern coast of ABRAHAM added their metal to that delivered by naval guns, artillery pieces, and aircraft. LCI gunboats and LVT(A)s led the assault troops toward the Green Beaches. (See Map 10.)

Unlike the troops who were seizing Roi, the men of the 24th Marines got little benefit from the support of the armored amphibians. These vehicles stopped at the beaches and attempted to support by fire the advance inland. The actions of the LVT(A)s confounded Colonel Hart, the regimental commander, for he had planned that the armored amphibians would precede the assault waves to positions 100 yards inland of the Green Beaches. On the evening prior to the Namur landings, after he discovered that LVT(A)s had supported the landings on the outlying islands from positions offshore, the colonel sent a reminder to his attached armored amphibian unit. To guard any error, he told the unit commander: "You will precede assault waves to beach and land, repeat land, at W-Hour, repeat W-Hour, as ordered." [32] Explicit as these orders were, the LVT(A)s nonetheless could not carry them out. The antitank ditches backing the lagoon beaches and the cut-up jumble of trenches and debris proved to be an impassable barrier for the LVT(A)s in the short time that elapsed between the touchdown of the armored amphibians and the landing of the first waves of Marines.[33] As the infantry moved inland, the LVT(A)s furnished support with all guns blazing until their fire was masked by the advance of the assault troops.

The lagoon coast of Namur was divided into Beaches Green 1, the objective of 3/24, and Green 2, where 2/24 was to land. The boundary between

[31] In addition to the sources already cited, two manuscripts, both of them monographs prepared for the Amphibious Warfare School, MCS, Quantico, Va., have been valuable. They are: LtCol Richard Rothwell, "A Study of an Amphibious Operation: The Battle of Namur, 31Jan–2Feb44," and LtCol Austin R. Brunelli, "Historical Tactical Study: The Capture of Namur Island, February 1–2, 1944."

[32] 2/24 CbtRpt, dtd 7Feb44, p. 2, in 24th Mar PrelimRpts, *op. cit.*

[33] *Metzger ltr.*

214-881 O-67—12

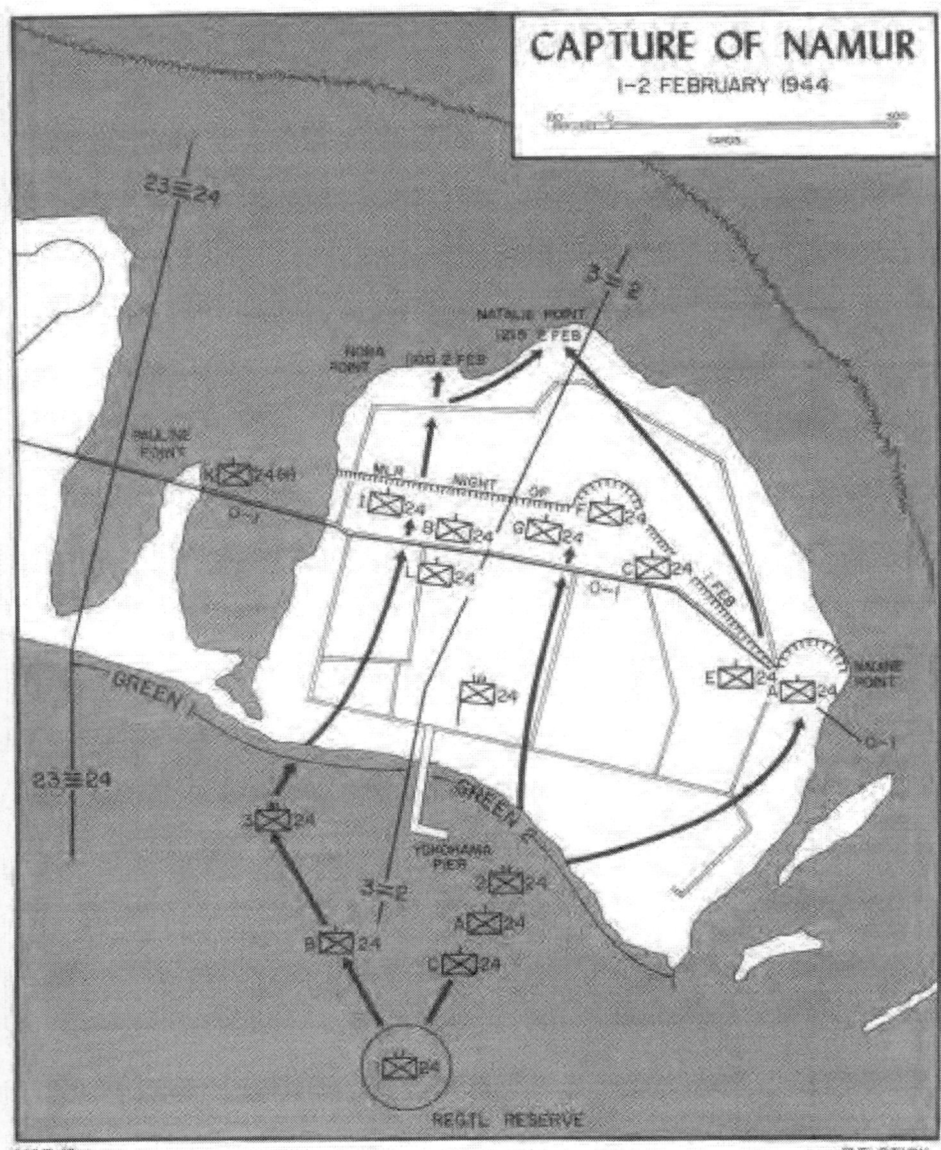

MAP 10

R.F. STIBIL

the two beaches was a line drawn just west of Yokohoma pier. Green 2 encompassed the eastern two-thirds of the coast, while the remainder of the southern shore and the entire sandspit was designated Green 1. Brink's battalion was slated to land two companies abreast on Green 2, but the first unit ashore landed in the middle of the zone. Part of the other assault company scrambled from its LVTs directly behind the leading company. The rest of these riflemen began advancing inland in the left-hand sector on the battalion zone of action.

On the right, 2/24 landed one company, arranged in a single wave, at 1155. The two waves into which the other assault company had been organized began landing on Green 2 about five minutes later. Smoke and dust, which bedeviled the amphibian tractors bound for Roi, also caused the Namur assault elements to stray from their proper boat lanes.

The Marines of 2/24 had been instructed to leave their tractors, thrust immediately toward their first objective, the 0–1 Line and there reorganize. As the various rifle platoons landed, each sent ahead an assault team to deal with any fortifications that had survived the preliminary bombardment. The remainder of the platoon, divided into two groups, followed in the path of the assault element.[34] At the 0–1 Line, which ran along the road that extended from the causeway to within a few yards of the eastern shore, the platoons were to pause and reorganize. Here, too, company commanders would regroup their units for the drive across the island.

As was true on Roi, naval gunfire had so devastated Namur that many of the features designated to mark boundaries and phase lines were eradicated. Thick underbrush also made control difficult, for in places visibility was no more than a few feet. The 2/24 assault companies, nevertheless, continued to advance inland, but because they had landed out of position, a gap soon opened between their left flank and the battalion boundary. As landing craft became available, additional elements of the battalion reserve were landed, and Brink ordered these into the opening.

Within two hours after landing, the assault units, Companies E and F, were intermingled along the 0–1 Line. A contingent from Company G and a part of Company E had overcome a knot of resistance and advanced some 175 yards inland along the battalion left flank. The farther the reserve unit moved, however, the more intense grew the opposition. The effort on the left came to a halt, pinned down by fire from a thicket near the battalion boundary and north of the 0–1 Line. As soon as it became available, the remainder of Company G also was committed to aid in securing the open flank, but this group was stopped by a com-

[34] The assault companies of both 2/24 and of 3/24 were organized into boat teams of 18–20 men, each led by a lieutenant or senior NCO. In the 2d Battalion, Lieutenant Colonel Brink decided to fight his men as boat teams until they reached 0–1, about 300 yards from the beach, where they would reorganize into platoons. *Brink ltr.*

bination of impenetrable undergrowth and Japanese fire.[35]

The first wave of LVT(2)s carrying elements of Lieutenant Colonel Brunelli's 3/24 reached Green 1 at about 1155, and within five minutes, Companies I and K were beginning the advance toward 0–1. The final dash to the beach had been hampered by low-hanging clouds of smoke, and units had strayed from formation. In effect, the assault companies simply exchanged platoons, for, as one officer recalled, "the major part of one platoon from Company I landed in the K/24 zone and approximately the same number from K in the I/24 zone."[36] These units advanced directly inland, remaining with their "adopted" companies until the 0–1 Line was gained.

The volume of fire that greeted 3/24 was somewhat heavier than that which had been encountered by the battalion on its right. Small groups of Japanese, most of them still groggy from the bombardment, fought from the ruins of their emplacements, but there was no organized defense. The communications center on Namur, from which the defense of the twin islands was to have been directed, had been destroyed. Although the enemy would, as expected, fight to the death, he was no longer capable of launching a coordinated counterattack against the rapidly expanding beachhead.

Company B of Dyess' battalion, which had been assigned as Brunelli's reserve, shore party units, and self-propelled 75mm guns landed on Green 1, while the assault companies drove inland through the underbrush and debris. Riflemen and demolition teams worked together to destroy the scattered enemy pillboxes and covered emplacements and keep the attack moving. Many Japanese, hidden in the underbrush and shattered rubble, were bypassed by the assault units and left to reserve forces to mop up.

At 1300, three light tanks from Company B, 4th Tank Battalion, arrived on Green 1. Two of them bogged down in soft sand along the beach, and the other vehicle roared some 30 yards inland, tumbled into a shell crater, and threw a tread. Twice, groups of from 15 to 20 Japanese leaped from the shelter of pillboxes to attack the stranded tanks, but the Marines beat off both groups and cleared the structures where the enemy had hidden. Two Japanese were captured and 30 killed as a result of these forays. Later in the day, the remaining two lights of the supporting tank platoon landed and helped get the disabled vehicles back into the fight.

By 1400, 3/24 was reorganizing along the 0–1 Line. Company I had advanced about 150 yards beyond the control line, but Brunelli promptly ordered the unit to withdraw.

Although the enemy resisted the advance of 3/24 with greater vigor, the other battalion of Hart's regiment suffered a higher number of casualties, losses caused only indirectly by the defenders. At 1305, assault teams of 2/24 were attacking a massive concrete building in the vicinity of 0–1. As the Marines were placing shaped charges against the wall, the Japanese in the immediate vicinity took to their heels.

[35] Maj Charles T. Ireland, Jr. ltr to CMC, dtd 3Feb53, hereafter *Ireland ltr.*

[36] LtCol Albert Arsenault ltr to CMC, dtd 10Feb53.

Once the wall had been breached, the demolitions detail began hurling satchel charges inside. Suddenly, the structure vanished in a pillar of smoke.

At this moment the regimental command post group, in the process of moving ashore, was approximately 300 yards off Namur. While Lieutenant Colonel Homer L. Litzenberg, Jr., the executive officer, watched, "the whole of Namur Island disappeared from sight in a tremendous brown cloud of dust and sand raised by the explosion." [37] Two other blasts occurred in rapid succession, and within seconds large chunks of concrete and other debris began raining down on Colonel Hart's command post, causing some injuries.[38]

The devastation ashore was awesome. An officer who was standing on the beach at the time of the first explosion recalled that "trunks of palm trees and chunks of concrete as large as packing crates were flying through the air like match sticks. . . . The hole left where the blockhouse stood was as large as a fair sized swimming pool." [39] This series of blasts killed 20 members of 2/24 and wounded 100 others. Among the injured was Lieutenant Colonel Brink, who refused to be evacuated.

At first, the tragedy was believed to have been caused by a fluke hit by a 16-inch shell on a warehouse filled with explosives. Investigation proved that the satchel charges thrown into the bunker had detonated row upon row of torpedo warheads. This violent blast could have touched off two smaller magazines nearby, or the enemy may possibly have caused the later explosions in the hope of inflicting additional casualties.[40]

The three explosions, which caused about one-half of its casualties on Namur, were a severe blow to 2/24. Colonel Hart attached Company A of Dyess' command to the battered unit, and a delay ensued as Brink's organization was restored to effectiveness. In the meantime, 3/24 was poised to attack toward the northern coast.

From the undergrowth across the 0–1 Line, a trio of Japanese emplacements were holding Brunelli's Marines at bay. The commanding officer of 3/24 planned to attack at 1630 in conjunction with Brink's unit. In preparation for this effort, light tanks and armored amphibians rumbled inland to fire into the enemy strongpoints. Two of these positions were silenced, but the third, a pillbox near the eastern shore, continued to enfilade the ground along the 0–1 Line.

Company L finally landed at 1531, an unavoidable delay since, as its commander pointed out, the unit "had no means of getting ashore earlier other than swimming." [41] This company relieved Company B as 3d Battalion reserve, assumed responsibility for mopping up, and sent men to strengthen Company I. Company B then moved

[37] BGen Homer L. Litzenberg, Jr., ltr to CMC, dtd 31Jan53.

[38] *Ibid.*

[39] 1stLt Samuel H. Zutty ltr to CMC, dtd 28Jan53.

[40] Capt Joseph E. LoPrete, "The Battle of Roi-Namur," monograph prepared for the Amphibious Warfare School, MCS, Quantico, Va. A platoon leader on Namur, Captain LoPrete commanded one of the two assault teams that attacked the explosives-laden bunker.

[41] LtCol Houston Stiff ltr to CMC, dtd 26Jan53.

into line in place of Company K, which was sent to the sandspit. Company K was to consolidate control over Pauline Point, which extended beyond the front lines, and support by fire the advance on Namur proper.

At 1630, as the advance division command post was being established on Namur, 3/24 launched its drive. Because of the tragic blast, 2/24 was not yet ready to advance. Brunelli's Marines found that the Japanese had recovered from the effects of the bombardment. Although resistance was not coordinated, dense thickets and the enemy's willingness to die fighting combined to slow the offensive.

While 3/24 was attacking, Lieutenant Colonel Brink was busy shuffling his units in an effort to restore 2/24 to fighting trim. Company A moved to the right-hand portion of the battalion zone. To its left was another attached organization, Company C, along with fragments of Companies E and F and approximately half of Company G.[42] Light tanks of the Headquarters Section and 1st Platoon, Company B, 4th Tank Battalion added their weight and firepower, and at 1730 2/24 joined 3/24 in plunging northward.

Tanks, protected insofar as the foliage permitted by infantrymen, spearheaded both battalions. These vehicles fired 37mm canister rounds which shredded the stubborn undergrowth in addition to killing Japanese. Whenever the riflemen encountered an especially difficult thicket, they temporarily lost sight of the tanks they were to protect, and the vehicles to the rear had to defend those in front of them. If enemy soldiers attempted to clamber aboard the leading tanks in an attempt to disable them with grenades, 37mm guns in the covering wave would unleash a hail of canister that swept the enemy to oblivion.

Without this sort of protection, a light tank was all but helpless, as proved by an incident in the 3/24 zone. One vehicle from Company B struck a log, veered out of position, and stopped to orient itself. A squad of Japanese swarmed onto the tank, and a grenade tumbled through a signal port which had been left open to allow engine fumes to escape. The blast killed two of the four Marines inside and wounded the others. Another tank and its accompanying rifle squad arrived in time to cut down the fleeing enemy.

Elements of 2/24 managed to make deep penetrations during the afternoon action. On the left, a few riflemen and some tanks reached a position within 35 yards of the north coast. This position, however, could not be maintained, and the men and machines were ordered to rejoin the rest of the battalion about 100 yards to the south. On the right, the elements of 2/24 that were probing Nadine Point encountered vicious machine gun fire. Although these Marines were able to beat off a local counterattack, they could not advance far beyond 0–1.

[42] The remainder of Company G was having troubles of its own. "No orders for a concerted attack during the afternoon ever reached me," recalled the executive officer. "The situation for my portion of G during the rest of the daylight hours was one of no contact with 2/24, no visible elements of 3/24 on my left, visual contact with a unit of 1/24 on my right, and heavy fire from the front." *Ireland ltr.*

Near 1700, General Schmidt landed and conferred with Colonel Hart. Within an hour, the general had opened his command post on Namur and was shifting his troops to assist the 24th Marines. He ordered Jones' reserve battalion (3/23) and the medium tanks of the combat team to move at once to Namur.[43] The Shermans lumbered across the sandspit in time to take part in the afternoon's fighting.

A platoon of these tanks reported to Lieutenant Colonel Brunelli at 1830, when 3/24 had advanced some 175 yards beyond the 0–1 Line. Rather than waste time feeding the Shermans into the battalion skirmish line, Brunelli used them to spearhead a sweep along the west coast. The tanks, a 75mm self-propelled gun, and several squads of infantry brushed aside enemy resistance to secure the abandoned emplacements on Natalie Point, northernmost part of the island. Isolated and low on ammunition, the task force had to withdraw before darkness.

At 1930, Colonel Hart ordered his Marines to halt and defend the ground they already had gained. Except for two bulges, the regimental main line of resistance ran diagonally from a point roughly 100 yards south of Nora Point to the intersection of 0–1 and the eastern coast. Toward the left of Brink's sector, the line curved to include the group of light tanks and riflemen that had been ordered back from near the north shore.[44] On the far right, the line again veered northward to encompass the elements of 2/24 that had overrun a part of Nadine Point. As Brink's Marines were digging in, the missing portion of Company G rejoined its parent unit along the battalion boundary.[45]

NAMUR SECURED

The night of 1–2 February was somewhat confusing but not particularly dangerous to the embattled Marines. From the front, the Japanese attempted to harass the assault troops, while to the rear by-passed defenders would pop out of piles of debris, fire their weapons, and quickly disappear. In addition, Colonel Hart's men had to put up with the "eerie noise of the star shell as it flew through the air," a sound which they at first found disturbing.[46] Since this was their first night of combat, the Marines did engage in some needless shooting at imagined snipers. When some machine gunners along the beach opened fire into the treetops to their front, General Schmidt himself emerged from his command post to calm them.[47] The troops, however, conducted themselves well enough, and the enemy, although able to launch local attacks, was incapable of making a serious effort to hurl the invaders into the sea.

Darkness found the medium tanks that had crossed over from Roi in difficult straits. The armored unit was located inland from Green 1, but its gasoline and ammunition were on Red 3. Boats could not be found to ferry the needed supplies from Roi, and the tank

[43] Gen Harry Schmidt ltr to ACofS, G–3, HQMC, dtd 22Oct62.

[44] LtCol Frank E. Garretson interview by HistBr, G–3, HQMC, dtd 12Jan53.

[45] *Ireland ltr.*

[46] Zutty ltr, *op. cit.*

[47] Gen Harry Schmidt ltr to CMC, dtd 10Nov52.

crews did not have pumps with which to transfer gasoline from one vehicle to another. They had no choice but to pool all the remaining 75mm shells and divide them among the four Shermans that had the most fuel.

The coming of light proved the wisdom of this arrangement, for the tanks were able to assist Companies I and B in shattering a counterattack. During the night, contact between the two units had been lost, and the enemy was now trying to exploit the gap. While the tanks charged forward, Company L moved into position to contain any breakthrough, and Company K began withdrawing from the sandspit to the island proper.

The Japanese counterattack failed, though the fighting raged for 25 minutes. When Company L arrived to seal the gap, it found that the medium tanks and the men of Companies I and B had broken the enemy spearhead and advanced about 50 yards. All that remained was the task of pushing to the north shore.

Colonel Hart planned to attack at 0900 with two battalions abreast. Enough medium tanks were now available to provide assistance to the riflemen of both battalions. Lieutenant Colonel Brink, injured on the previous day when the blockhouse exploded, yielded command of 2/24 to Lieutenant Colonel Dyess of 1/24. Two rifle companies from 1/24 were to take part in the morning attack of the 2d Battalion, while the third served as reserve for 3/24. Mopping-up was to be carried out concurrent with the advance.

Brunelli's Marines, aided by medium tanks, launched their blow exactly on schedule. The Shermans concentrated on pillboxes and other concrete structures, firing armor-piercing rounds to penetrate the walls and then pumping high explosives shells into the interior. Nora Point was taken within two hours, and by 1215, 3/24 was in control of Natalie Point on the northern coast.

The medium tanks destined for 2/24 were late in arriving, so the attack by the battalion was delayed until 1006. On the left, a blockhouse had to be destroyed by tanks and self-propelled guns, but elsewhere the Marines moved steadily northward. The final enemy strongpoint proved to be an antitank ditch, part of the defenses along the ocean shore, from which the Japanese were firing at the advancing troops. Light tanks wiped out these defenders by moving to the flank of the ditch and raking it with canister and machine gun fire. Lieutenant Colonel Dyess, who had repeatedly risked his life throughout the morning to keep the attack moving, was killed as he urged his men toward Natalie Point. At 1215, the two battalions met at Natalie Point; Namur had been overrun. The island was declared secured at 1418.

Because of the more determined resistance on Namur, Navy corpsmen assigned to the 24th Marines had a more difficult job than those who served with the 23d Marines on Roi. A corpsman accompanied every assault platoon, "and wherever and whenever a man was hit, he went unhesitatingly to his assistance, often . . . coming directly into an enemy line of fire."[48] Shell craters became aid stations, as

[48] 1stLt John C. Chapin memo to Capt William G. Wendell, dtd 8Jan45.

corpsmen struggled to save the lives of wounded Marines. Once again, these sailors had performed their vital work skillfully and courageously.

Colonel Hart's 24th Marines had conquered Namur in spite of serious obstacles. The most spectacular of these was the tragic explosion of the blockhouse, but the shortage of tractors, the incompletely formed assault waves, poor communications, and tangled undergrowth also conspired against the regiment. Colonel Hart remained convinced that "had LVT(2)s and/or LCVPs been available as originally planned, or had the departure . . . been delayed until 1200," the island would have been taken more quickly and with fewer casualties.[49]

The men of both regiments were brave and aggressive, if somewhat lacking in fire discipline. Their primary mission accomplished, the men of the 4th Marine Division could allow their guns to cool, absorb the lessons of the past few days, and prepare for the final phase of the FLINTLOCK operation. To the south, however, the fight for Kwajalein Island still was raging. As the Marines rested, soldiers of the 7th Infantry Division continued to press an attack of their own.

THE CONQUEST OF KWAJALEIN ISLAND [50]

At 0930, 1 February, the 32d and 184th Infantry Regiments of General Corlett's 7th Infantry Division landed at the western end of Kwajalein Island. The preliminary bombardment by field artillery and naval guns, as well as the aerial strikes, had been extremely effective. Admiral Turner, at the request of General Corlett, had ordered two of his battleships to close to 2,000 yards, an extremely short range for these big ships, and level a wall inland of the assault beaches. The captains involved did not believe the figure was correct and asked for clarification, so Turner subtracted 500 yards from his original order, and had them open fire.[51] (See Map 11.)

Aided by this kind of fire support, the well-rehearsed assault proceeded relatively smoothly. The formation headed for each of the two landing beaches was shaped somewhat like the letter U. On either flank, extending forward at an angle of about 45 degrees from the base, was a line of LVT(A)s. These vehicles joined the LCI gunboats in neutralizing the beaches and then crawled ashore to protect the flanks of the beachhead. At the base of the U were the troop-carrying LVTs, with both rifle and engineer platoons in the first wave.

The landings were executed as planned. The only difficulty, telescoping toward the right of the assault waves, stemmed from a mechanical characteristic of the tractors used at Kwajalein Island. These vehicles tended to pull toward the left. The drivers attempted to compensate by inclining toward the right, and in their

[49] 24th MarRpt of FLINTLOCK Op, p. 8, dtd 10Mar44, Encl E to *4th MarDiv AR*, p. 8.

[50] The official Army account of this action is contained in Crowl and Love, *The Gilberts and Marshalls*, pp. 230–282.

[51] MajGen Charles H. Corlett, USA, ltr to CMC, dtd 14Jan53. The closest range reported by the bombardment battleships on the morning of 1 February is 1,800 yards. Dir NHD cmts on draft MS, dtd 27Nov62.

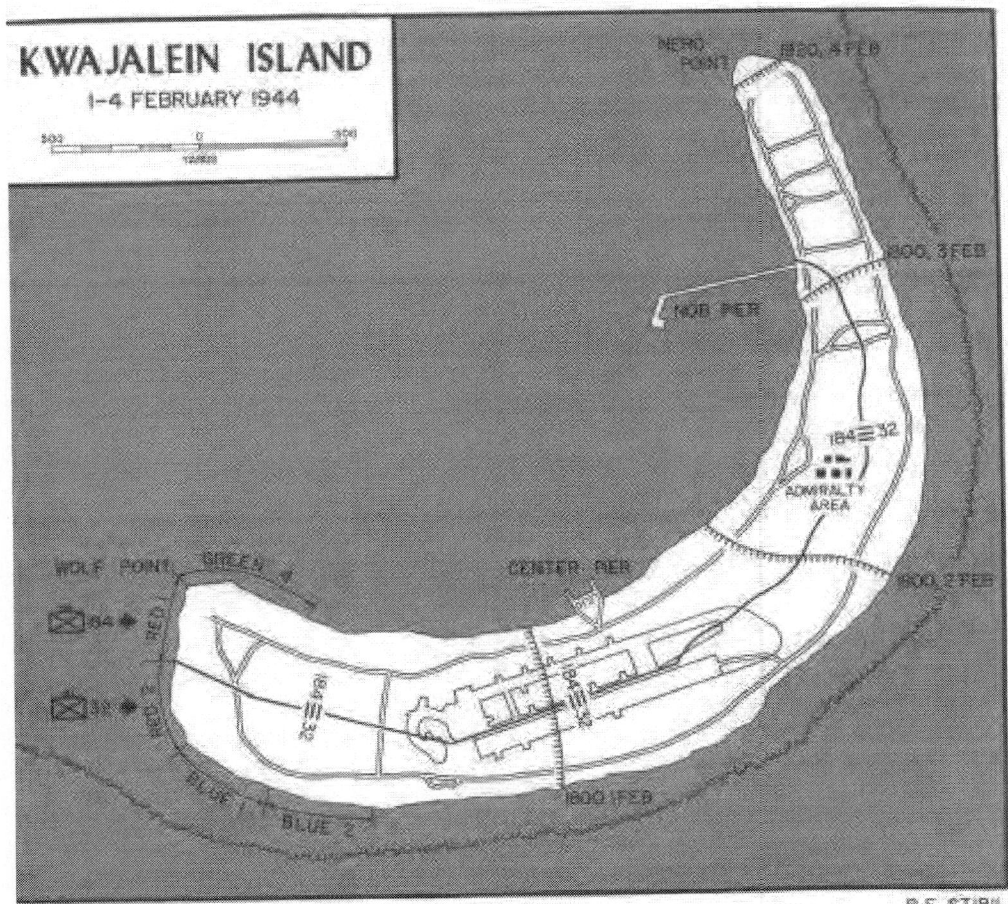
KWAJALEIN ISLAND
1-4 FEBRUARY 1944
YARDS
NERO POINT
1920, 4 FEB
1800, 3 FEB
NOB PIER
184 32
ADMIRALTY AREA
CENTER PIER
1800, 2 FEB
WOLF POINT
GREEN 4
RED 1
RED 2
184
32
BLUE 1
BLUE 2
1800 1 FEB
MAP II
R.F. STIBIL

eagerness to remain in the proper lanes they veered too far.[52]

On the first day, the infantry-engineer teams quickly secured the beaches. No serious opposition was encountered until the attackers had overrun the western third of the airfield. At this point, however, the nature of the battle changed. By the end of the first day, the swift amphibious thrust had become a systematic and thorough offensive designed to destroy a Japanese garrison that was fighting from ruined buildings, shattered pillboxes, and piles of debris. Massive artillery concentrations and close coordination between tanks and riflemen characterized the advance which ended at 1920 on 4 February with the securing of the farthest tip of the island.

As far as Marines were concerned, the most interesting feature of this operation was the logistical plan devised by General Corlett and his staff. Instead of LVTs, the 7th Infantry Division used DUKWs as supply vehicles. Amphibious trucks, filled with items certain to be needed early in the operation, were loaded in LSTs before the convoy left the Hawaiian Islands. These vehicles were sent ashore as needed. As soon as they had unloaded, they reported to the beachmaster. That officer placed the wounded in some of the trucks, but whether or not they carried casualties, all DUKWs next reported to a control officer off the beach. Here a representative of the division medical officer directed the wounded to vessels equipped to care for them, while the control officer saw to it that the DUKWs maintained an uninterrupted flow of supplies from the LSTs to the assault units.

In general, the so-called "hot cargo" system worked well, for by noon of D-Day DUKWs were already arriving on the island. The only serious breakdown, which occurred that night, was caused by a flaw in the basic plan. As evening approached, two of three LSTs that were feeding cargo to trucks destined for the 184th Infantry were recalled from their unloading area. The remaining ship carried no 75mm ammunition for the tanks assigned to support the next day's advance. As a result, the Shermans were late in getting into action.[53]

The logistical plan, however, cannot account for the comparative ease with which the assault waves gained Kwajalein's beaches. The tractor and LVT(A) units assigned to the 7th Infantry Division benefited from rehearsals held in Hawaii prior to departure for the target area. By the time these exercises were held, the plan of attack had been completed. Not so fortunate were the tractor units that landed the 4th Marine Division, for their final rehearsal was held even before the landing force scheme of maneuver had been decided upon. The lack of a last-minute rehearsal gravely hampered the Marines.

The D-Day operations also had a more serious effect on the Marine LVT

[52] LtCol S. L. A. Marshall, USA, "Notes Prepared in the Central and Middle Pacific: The Assault on PORCELAIN" (Hist MS File, OCMH), pp. 1–4.

[53] *Ibid.*, p. 95; LtCol S. L. A. Marshall, USA, "General and Miscellaneous Materials on Central Pacific: Supply" (Hist MS File, OCMH), pp. 44–45.

ARMY 37MM ANTITANK GUN *fires at enemy pillbox on Kwajalein during 7th Infantry Division advance. (USA SC18555)*

MARINE 105MM HOWITZERS, *transported by DUKWs, set up on an off-lying island to cover the landings on Engebi, Eniwetok Atoll. (USN 80-G-233228)*

units than on the Army tractor battalion. On 31 January, General Corlett had employed tractor groups against two islands and held back two other groups, one per assault battalion, for the following day's operation. More complicated was the task facing General Schmidt, whose troops had to seize five small islands. Although he did maintain a reserve for Roi-Namur, these idle tractors had to be augmented by vehicles that took part in the D-Day landings. This was necessary since four battalions were to storm the twin islands. Because of the series of delays and other misfortunes, not enough tractors could be retrieved before nightfall. Thus, the number of landings scheduled for D-Day, the width of the beachheads the 4th Marine Division was to seize, and the lack of rehearsals combined to complicate the Roi-Namur landings.

General Corlett could well be satisfied with the conduct of his veteran division at Kwajalein Island. "I think the Navy did a marvelous job as did the Marines," he later observed, "and I think the Army did as well as either of them." [54] With the capture of Kwajalein Island on 4 February, the last of FLINTLOCK's principal objectives was secured, but several lesser islands remained to be taken.

THE FINAL PHASE [55]

On Roi-Namur the work of burying the enemy dead, repairing battle damage, and emplacing defensive weapons was begun as promptly as possible. Antiaircraft guns of the 15th Defense Battalion were being landed even as the fighting raged. Once the battle had ended, the 20th Marines began clearing Roi airstrip, but on D plus 5, these engineers were relieved of the task by a naval construction battalion. During this same period, various elements of the 4th Marine Division got ready to depart from Kwajalein Atoll.

Badly pummelled by American carrier planes, Japanese air power had been unable to contest the Roi-Namur operation, but early in the morning of 12 February, 12–14 enemy seaplanes struck at Roi. The raiders dropped strips of metal foil to confuse American radar and managed to catch the defenders by surprise. From the Japanese point of view, the attack was a complete success. An ammunition dump, 85 percent of the supplies stockpiled on the island, and roughly one-third of the heavy construction equipment were destroyed. Thirty Americans were killed and an estimated 400 wounded.

The raid on Roi, however, had no effect on the final phase of the 4th Marine Division overall plan. By the time of the aerial attack, Company A, 10th Amphibian Tractor Battalion, and the 25th Marines had investigated the remaining islands in the northern part of Kwajalein Atoll. On 2 February, Lieutenant Colonel Hudson's 2/25 seized eight islands. No resistance was met,

[54] Corlett ltr, *op. cit.*

[55] Additional sources for this section include: 20th Mar Rpt on FLINTLOCK Op, dtd 16Mar44, pp. 2–4; Col Peter J. Negri ltr to CMC, dtd 5Feb53; Carl W. Proehl, ed., *The Fourth Marine Division in World War II* (Washington: Infantry Journal Press, 1946), p. 34, hereafter Proehl, *4th MarDiv History.*

and after the first two landings, the planned artillery preparations were cancelled. Lieutenant Colonel O'Donnell led 1/25 against three islands, and Lieutenant Colonel Chambers' 3/25 secured 39 others within four days. Once the final landings were completed, the regiment served for a time as part of the atoll garrison force. (See Map 8.)

While Colonel Cumming's regiment was occupying the lesser islands in the north, the 17th Infantry and the 7th Reconnaissance Troop were performing a similar mission in the southern part of Kwajalein Atoll. Unlike the Marines, the soldiers encountered vicious fighting on some of the objectives. At CHAUNCEY, where the unit had landed by mistake on D-Day, the reconnaissance troop killed 135 Japanese. BURTON required the services of two battalions of the 17th Infantry, but within two days, 450 of the defenders were dead and the 7 survivors taken prisoner. In spite of the frequent opposition, the last of the southern islands was captured on 5 February.

Both assault divisions could look back on a job well done. According to intelligence estimates, the Northern Landing Force had defeated enemy garrisons numbering 3,563, while the Southern Landing Force accounted for 4,823 Japanese and Koreans. Thus, each division had overwhelmed in a series of landings a total force approximately the same size as the Betio garrison. Yet, American losses in FLINTLOCK were far fewer than the casualties suffered at Betio. The 4th Marine Division had 313 killed and 502 wounded, while the 7th Infantry Division lost 173 killed and 793 wounded.[56]

While the combat troops might pause to congratulate themselves, Admiral Nimitz and his staff continued to look to the future. Planners had to determine how best to capitalize on the stunning victory at Kwajalein Atoll. Should the blow at Eniwetok Atoll, tentatively scheduled for May 1944, be launched immediately?

[56] A breakdown by unit of casualty figures for the 4th Marine Division and 7th Infantry Division is contained in Heinl and Crown, *The Marshalls*, pp. 169–171. Final official Marine Corps casualty totals for the Kwajalein Operation are listed in Appendix H.

Westward to Eniwetok[1]

"Will the enemy attack Eniwetok?" asked Norio Miyada, one of the defenders of the atoll. To him the answer was obvious. "He will strike this island after attacking Roi." The only problem lay in deciding when the Americans would enter the lagoon. This noncommissioned officer, confident of Japanese aerial superiority in the Marshalls, expected a slow advance. "How will the enemy be able to attack us?" he continued. "Will it be by his hackneyed method of island hopping?" [2]

REVISING THE TIMETABLE

Actually, Admiral Nimitz looked forward to leapfrogging the central part of the Marshalls group. He planned to vault from Kwajalein to Eniwetok, neutralizing the Japanese bastions in between. Even before FLINTLOCK was launched, troops were preparing for CATCHPOLE, as the Eniwetok operation was called.

On 1 January 1944, the 2d Marine Division began intensive training for the assault upon Eniwetok Atoll. Within two weeks, the 27th Infantry Division was alerted to ready itself for an attack on Kusaie Island in the eastern Carolines, the second objective in the current CATCHPOLE plan. The target date for Eniwetok had been fixed at 1 May to enable units of the Fifth Fleet to assist in the seizure of Kavieng, New Ireland, an operation that eventually was cancelled.[3]

Admiral Spruance, however, felt that his warships should strike at Eniwetok before steaming southward to Kavieng. This opinion was shared by Admiral Turner, whose staff prepared a tentative plan to advance promptly to Eniwetok if the FLINTLOCK operation was executed smoothly. General Holland Smith's VAC planners also looked ahead to the rapid capture of Eniwetok, but theirs, too, was a tentative concept.[4]

Execution of the Eniwetok proposals depended upon the intelligence that

[1] Unless otherwise noted, the material in this chapter is derived from: CinCPac CATCHPOLE Plan, dtd 29Nov43; TF 51 OPlan A9-44, dtd 7Feb44; TG 51.11 OpO A105-44, dtd 13Feb44; TG 51.11 AtkO A106-44, dtd 9Feb44; TG 51.11 Rpt of Eniwetok Opns, dtd 7Mar44, hereafter *TG 51.11 OpRpt;* VAC MiscOs and Rpts File, Eniwetok; TG 1 OpO 2-44, dtd 10Feb44; TG 1 AdminO 2-44, dtd 10 Feb44; TG 1 SAR CATCHPOLE Op, dtd 10Mar44, hereafter *TG 1 SAR;* CominCh, *Marshall Islands*; Heinl and Crown, *The Marshalls.*

[2] JICPOA Item No. 8200, Extracts from the Diary of Norio Miyada.

[3] VAC WarD, Jan44, p. 11. For the story of the Kavieng venture and of the intended part of Marine units in its capture see Shaw and Kane, *Isolation of Rabaul*, p. 501*ff*.

[4] VAC G-3 Rpt on FLINTLOCK, dtd 12Feb44, Encl B to *VAC AR, FLINTLOCK; TF 51 AR*, p. 6; Isely and Crowl, *Marines and Amphibious War*, pp. 291-292.

could be obtained concerning the objective and on the cost in lives and time of the Kwajalein campaign. During an aerial reconnaissance on 28 December 1943, the first successful penetration of Eniwetok during the war, cameras were trained only on Engebi Island, site of an airstrip. Within a month, however, the Joint Intelligence Center, Pacific Ocean Areas, had amassed enough data to issue a bulletin describing the atoll and its defenses. Last-minute details were provided by the carrier planes which photographed the atoll on 30 January. FLINTLOCK itself brought a windfall of captured documents, among them navigational charts of Eniwetok Atoll. The various parts of the puzzle were assembled, and the task of fitting them together was begun. (See Map 12.)

The fighting within Kwajalein Atoll also was progressing rapidly at a reasonable cost to the attackers. On 2 February, Admiral Turner recommended to Admiral Spruance that the CATCHPOLE operation begin immediately. Turner offered a plan to strike with the 22d Marines and two battalions of the 106th Infantry as soon after 10 February as the necessary ships had taken on fuel and ammunition and the carrier air groups had been brought up to full strength.

Admiral Nimitz, who had received copies of Turner's dispatches, now asked Spruance's views on an amphibious assault upon Eniwetok to be preceded immediately by a carrier strike against Truk. The Fifth Fleet commander favored such a course of action,[5] and on 5 February, Admiral Nimitz arrived at Kwajalein to discuss the proposed operation with his principal subordinates. The Commander in Chief, Pacific Ocean Areas, approved the concept set forth by Admiral Turner. Originally, 15 February was selected as the target date, but D-Day had to be postponed until 17 February to give the fast carriers more time to prepare for their concurrent attack on Truk.[6]

TASK ORGANIZATION

Admiral Hill, commander of the Majuro attack force, reported to Admiral Turner at Kwajalein on 3 February. "I had no forewarning of the possibilities of my being put in command of the Eniwetok operation," Hill recalled, but warning or none, he was given overall command of Task Group 51.11, the Eniwetok Expeditionary Group.[7] In organizing his force, he followed the pattern he had used for the Majuro landing.

With only seven days for planning, and again only a small segment of a larger hydrographic chart to work

[5] On receiving Admiral Nimitz' request, Admiral Spruance related: "I at once went over to see Admiral Kelly Turner and General Holland Smith about it. They were both favorable. When I asked about the time needed to prepare the plans, Holland Smith said he had already prepared a plan while they were coming out from Pearl. This set the operation up, and we covered it with a strike on Truk on 16 and 17 February by Task Force 58." *Spruance 62 ltr.*

[6] Adm Richmond K. Turner ltr to CMC, dtd 13Apr53, hereafter *Turner ltr II;* Isely and Crowl, *Marines and Amphibious War*, pp. 291–292.

[7] VAdm Harry W. Hill ltr to CMC, dtd 24Feb53.

from, Admiral Hill's "first request was for high and low angle photographs taken at high and low tide and particularly in the early morning with its usually still waters." [8] Using the facilities of Admiral Turner's AGC, a photo-based map was reproduced in quantity for the use of the task group. Right after this map was run off, Admiral Hill was presented with a captured Japanese chart taken from a ship wrecked on the shore of one of the islands of Kwajalein. The enemy map, which was used during the operation, showed the area clear of mines and the preferred channel into the lagoon at Eniwetok.[9]

The Eniwetok Expeditionary Group consisted of: Headquarters, Support Aircraft (Captain Richard F. Whitehead, USN); Expeditionary Troops, commanded by Marine Brigadier General Thomas E. Watson; Carrier Task Group 4, under Rear Admiral Samuel P. Ginder; plus the Eniwetok Attack Force and the Eniwetok Garrison Group, these last two commanded by Admiral Hill. The flagship was the USS *Cambria*, which had served Hill during the conquest of Majuro, but the total number of ships assigned to him was far greater than he had commanded during FLINTLOCK.

The assault troops required eight transports of various types, two attack cargo ships, one cargo ship, an LSD, nine LSTs, and six LCIs. Ten destroyers were assigned to screen the transports and cargo vessels, while three battleships, three heavy cruisers, and another seven destroyers formed the fire support group. An escort carrier group, three carriers and three destroyers, joined a fast carrier group, three larger carriers and their screen, in providing aerial support for the operation.

General Watson was to have operational control over expeditionary troops once the landing force was established ashore. Since General Smith would not be present at Eniwetok, Admiral Turner charged the Commander, Expeditionary Troops, with duties similar to those carried out by the corps commander at Kwajalein Atoll. "General Watson," Turner has explained, "was in over-all command of all [troop units], but did not exercise detailed tactical command on shore of any one of them." [10] Like Smith during FLINTLOCK, Watson could issue no orders "as to major landings or as to major changes in tactical plans" without the naval commander's approval.[11]

Because Turner's Eniwetok operation plan did not require Watson to report to the attack force commander when he was ready to assume command ashore, a Marine officer on Admiral King's staff interpreted the command arrangement as a modification of the structure used during FLINTLOCK. "Previous orders," he noted, "did not give this command to the ground force commander until he stated he was ready to assume it. In other words, it formerly required positive action on the ground force com-

[8] *Hill interview/comments Marshalls.*

[9] *Ibid.* Admiral Hill noted that when the fleet anchorage was established at Eniwetok, it was in the area picked by his staff from the original chart based on aerial photographs.

[10] *Turner ltr II.*

[11] *TF 51 OPlan A6–43*, p. 9.

214-881 O-67—13

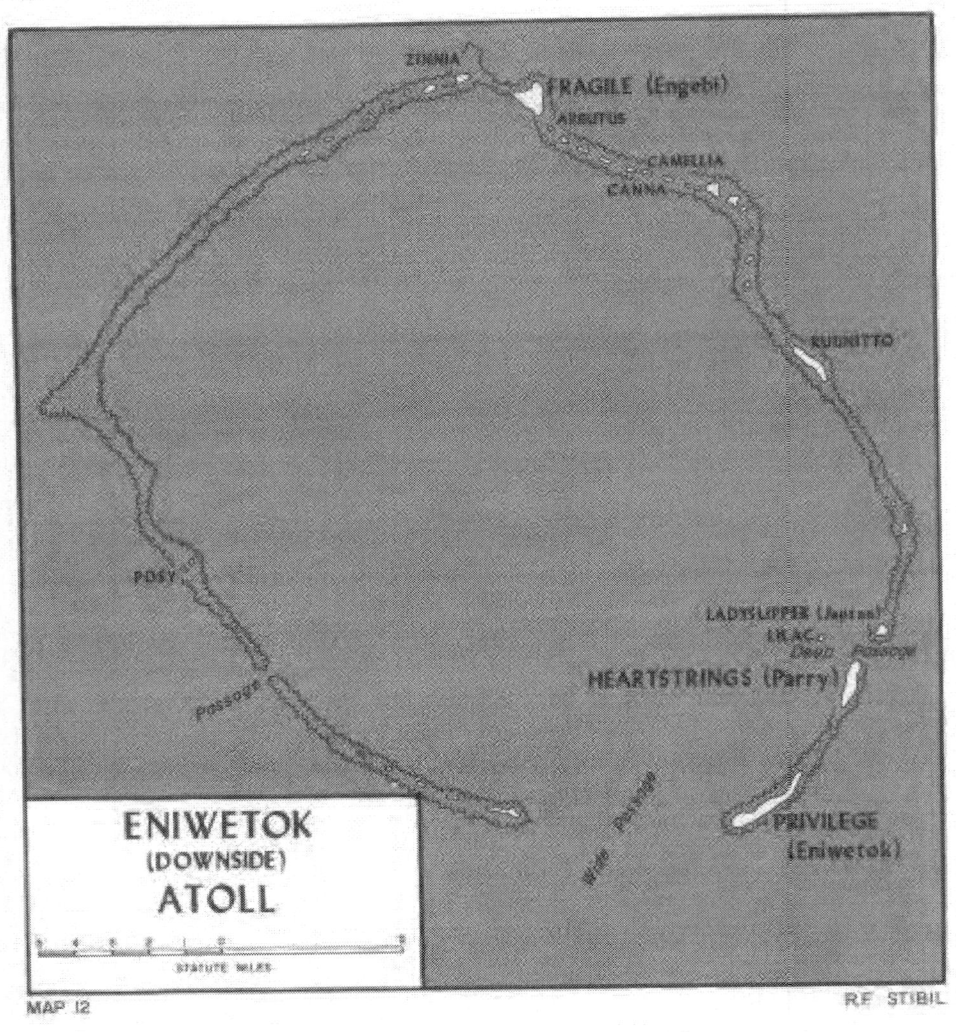
ZINNIA
FRAGILE (Engebi)
ARBUTUS
CAMELLIA
CANNA
LADYSLIPPER (Japtan)
Deep Passage
HEARTSTRINGS (Parry)
PRIVILEGE
(Eniwetok)
Wide Passage
Passage
POSY
ENIWETOK
(DOWNSIDE)
ATOLL
STATUTE MILES
MAP 12
R.F. STIBIL

mander's part. Now it is established before the operation began." [12]

In fact, no change had actually been made, for Watson was, according to Turner, the commander of a small-scale corps. The Marine general in command of the Eniwetok forces was holding a position comparable to that held by Smith at Kwajalein. Both were to "command all landing and garrison forces when ashore." [13] The command structure remained substantially the same, for as one student of amphibious warfare has pointed out, "there was a distinction without a difference." [14]

The Eniwetok landing force was to be provided by Watson's Tactical Group 1, the FLINTLOCK reserve, which had cruised eastward of Kwajalein Atoll while the Northern and Southern Landing Forces had effected their landings. Tactical Group 1 was composed of the 22d Marines, 106th Infantry (less 2/106, assigned to the Majuro operation), the Army 104th Field Artillery Battalion, the Marine 2d Separate Pack Howitzer Battalion, 2d Separate Tank Company, 2d Separate Motor Transport Company, and 2d Separate Medical Company, plus shore party and JASCO units. After 3 February, when the group entered Kwajalein Atoll, further attachments were made to strengthen Watson's command for the CATCHPOLE Operation. The additions were: VAC Reconnaissance Company, Company D, 4th Tank Battalion (a scout unit), 102 LVTs and 17 LVT(A)s from the 708th Provisional Amphibian Tractor Battalion, and a provisional DUKW unit, provided by the 7th Infantry Division, with 30 amphibious trucks and 4 LVTs. By the time CATCHPOLE began, General Watson had command over some 10,000 assault troops, more than 4,000 of them soldiers.[15]

The tactical group was prepared to handle only such administrative chores as might be incident to combat operations. General Watson's staff was small in size and suited only to brief periods of combat. This so-called "streamlined" staff, partly an experiment and partly the result of a shortage of officers with staff experience, was not adequate to the strain imposed by CATCHPOLE. "I can personally attest," stated the group G–3, "that I and all members of the staff came out of the Eniwetok operation utterly exhausted by day and night effort. The streamlined staff idea died a rapid and just death as the staff itself was about to expire." [16]

Colonel John T. Walker's reinforced 22d Marines, the largest single component of Tactical Group 1, had spent almost 18 months as part of the Samoa garrison force prior to its transfer to Hawaii. The regiment had undergone rudimentary amphibious training in preparation for FLINTLOCK. Late in December, the 106th Infantry, two battalions strong, was detached from

[12] BGen Omar T. Pfeiffer memo to BGen Gerald C. Thomas, dtd 23Feb44.

[13] *Cf.* TF 51 OPlan A9–44, dtd 7Feb44, p. 9 and TF 51 OPlan A6–43, dtd 17Jan44, p. 15.

[14] Cmt of Dr. Philip A. Crowl quoted in Maj Leonard O. Friesz, USA, ltr to CMC, dtd 16Mar53.

[15] The size of Tactical Group 1 varies according to the source consulted. Admiral Turner reported a total of 10,269, 5,760 of them Marines. *TF 51 AR*, p. 3.

[16] Col Wallace M. Greene, Jr., ltr to CMC, dtd 4Mar53.

the 27th Infantry Division and assigned to Watson's group. The Army unit also received a brief refresher course in amphibious warfare. These exercises, according to the commanding general of the group, were far from realistic. "We were sent to attack a coral atoll," complained General Watson, but "we rehearsed on the large island of Maui on terrain and approaches totally unlike those of the target." Neither artillery shells, naval gunfire, nor aerial bombs added realism to the exercise. The group's artillery battalions did not land from DUKWs, few of the infantrymen landed from LVTs, and the assault teams did not practice moving inland from the beach. "In the attack on Eniwetok," the general concluded, "the infantry, amphibian tractors, amphibian tanks, tanks, aircraft, supporting naval ships, and most of the staffs concerned had never worked together before." [17]

As far as the 22d Marines was concerned, thorough training in infantry tactics offset the effect of haphazard rehearsals. Colonel Walker's regiment, in the opinion of the group G–3, was "at its peak in small unit training—training which was anchored firmly around a basic fire team organization (three or four-man teams, depending on the battalion)." [18] Since each rifle squad could be divided into teams, the squad leader's problems of control were greatly eased. In jungle or amid ruined buildings, the teams were capable of fighting independent actions against an enemy pillbox or machine guns. The intense training which it had received in Samoa had made the 22d Marines a spirited, competent unit, one which would distinguish itself in the forthcoming operation.

The 106th Infantry, however, had not received the kind of training that the Marine regiment had undergone in Samoa. An Army officer who was serving in General Smith's VAC planning section, observed that the Army regiment was "far from being in an ideal state of combat readiness." Yet, "many fine and highly trained individuals and small units . . . collectively made up the 106th Infantry." [19]

During CATCHPOLE, moreover, Colonel Russell G. Ayers, commanding officer of the 106th Infantry, would labor under still another handicap. He had only two battalions, and if these were committed to an attack, his reserve would have to be provided by the 22d Marines. Thus, the colonel might find himself commanding a hastily combined organization, one third of which was differently trained and unfamiliar to him. "Effective combat units," a member of the VAC staff has pointed out, "are achieved by effective unit training, and can never be replaced by assorted combinations of component units, however highly trained." [20]

Tactical Group 1, then, had its shortcomings. Its staff was designed to assist the commanding general during brief operations rather than in an involved campaign against a large atoll. The infantry components were not of

[17] TG–1 SplRpt of FLINTLOCK and CATCHPOLE Ops, dtd 1Mar44, p. 7, hereafter *TG 1 SplRpt*.

[18] Greene ltr of 4Mar53, *op. cit.*

[19] *Anderson ltr.*

[20] *Ibid.*

equal quality, nor had they received amphibious training beyond what was necessary for them as FLINTLOCK reserve. Still, this group was available at once, and American planners were determined to sustain the momentum of the Marshalls offensive.

In addition to Tactical Group 1, General Watson, as Commander, Expeditionary Troops, had operational control over the Eniwetok Garrison Forces. Although Hill retained control of the landing force until it was set up ashore, Watson was in overall command of the garrison contingent during the landings. Finding the necessary defense forces proved difficult. No extra occupation units had been included in the FLINTLOCK force, for neither the men nor the transports were available. "When the decision was made to capture Eniwetok without waiting for additional forces," Admiral Turner remarked, "we had to rob both Kwajalein Island and Roi-Namur of considerable proportions of their garrisons and carry them forward in order to start the more urgent development of the new base."[21] The Marine 10th Defense Battalion, the Army 3d Defense Battalion, and the Army 47th Engineers formed the nucleus of the hastily formed Eniwetok garrison.

INTELLIGENCE

"Before departure from Hawaii our information concerning Eniwetok was scanty," commented Admiral Turner, "we had only a few high altitude photographs . . . and our maritime charts were of small scale made from ancient surveys."[22] Navigational charts, current enough to be considered secret by the Japanese, were captured during the Kwajalein operation. Gradually the photographic coverage was expanded, and the enemy order of battle began to emerge.

A complete aerial mosaic of Eniwetok Atoll would disclose a vast lagoon, which measures 17 by 21 miles, enclosed by a ring of islands and reefs. Both principal entrances to this lagoon, Wide Passage just west of Eniwetok Island and Deep Passage between Parry and Japtan, lie along its southern rim. The largest of the 30-odd islands in the atoll are Eniwetok, Parry, Japtan, and Engebi farther to the north. (See Map 12.)

In its study dated 20 January 1944, JICPOA reported an airstrip, fortifications, and large buildings on Engebi. An installation believed to be a radio direction finder was plotted on the map of Parry, and the stretch of Eniwetok Island bore the legend "no known defenses."[23] Within a few weeks, Admiral Nimitz' intelligence officers were offering more disturbing news.

Late in January, JICPOA noted that a mobile unit of the Japanese Army, some 4,000 men, had sailed eastward from Truk. The strength of the Eniwetok garrison, once reported as 700 men concentrated on Engebi Island, was revised drastically upward. By 10 February, enemy strength throughout the atoll was placed at 2,900–4,000 men. "These estimates are made without the advantage of late photo-

[21] *Turner ltr II.*

[22] *Ibid.*

[23] JICPOA InfoBul No. 3–44, Eniwetok, dtd 20Jan44.

graphs," JICPOA explained. "Good photographs should be able to settle the question of the presence of such a large body of troops and furnish a more reliable basis of estimation." [24]

The photographs for which JICPOA awaited were taken while FLINTLOCK was in progress. Photo interpreters examined every shadow but discovered few signs of enemy activity. On Engebi, already considered the hub of the enemy defenses, the garrison had improved and extended its network of trenches and foxholes. A few foxholes dug near a collection of storehouses were the only indication that Parry was defended. Eniwetok Island bore the scars of about 50 freshly dug foxholes, and other signs indicated that a small number of Japanese occupied its southwestern tip. In short, the last-minute intelligence belied the presence of a large concentration of troops. Both Parry and Eniwetok Islands appeared weakly held. Whatever strength the enemy had seemed to be massed on Engebi.

TACTICAL AND LOGISTICAL PLANS

The CATCHPOLE plan, prepared in the light of the intelligence available to Admiral Hill and General Watson, bore certain similarities to FLINTLOCK. The operation was divided into four phases, the first of which was the capture of three islands adjacent to Engebi. On 17 February, D-Day, the VAC Reconnaissance Company was to seize CAMELLIA (Aitsu) and CANNA (Rujioru) Islands southeast of Engebi, while the scouts of Company D, 4th Tank Battalion, took ZINNIA (Bogon) northwest of Engebi. Army and Marine artillery batteries would then land at CANNA and CAMELLIA to support the next phase. On 18 February, the second phase of CATCHPOLE was to begin as the 22d Marines stormed lagoon beaches of Engebi. Two objectives, Eniwetok Island and Parry, were included in phase III. As soon as it was certain that additional troops would not be needed at Engebi, the 106th Infantry was to assault Eniwetok Island.[25] The Army objective was considered so poorly defended that the understrength regiment was directed to prepare to move on to Parry within two hours after the Eniwetok landing. Both regiments were scheduled to see action during phase IV, the securing of the remaining islands of the atoll.

Naval gunfire also was to follow a schedule similar to that employed in the Kwajalein landings. On D-Day, destroyers and LCIs were to support operations against the lesser islands, while battleships, cruisers, and other destroyers shelled Engebi from positions outside the lagoon. During the afternoon, two of the battleships would enter the lagoon and assist in shattering the enemy's defenses. On D plus 1, the supporting warships were to de-

[24] *TF 51.11 OpRpt*, p. 1.

[25] The 106th Infantry commander wanted to land artillery on the island opposite Eniwetok, across Wide Passage, but "Hill and Watson said 'no,' as they did not feel that it was warranted since it would take a day out of the schedule to get set up and that there were not enough signs of a garrison on the island to warrant the move." *Hill interview/comments Marshalls.*

stroy beach defenses and other targets. Pausing only when aerial strikes were being executed, battleships, destroyers, and LCIs would hammer the beaches until the assault craft were 300 yards from Engebi's shore, then shift their fire to bombard for five minutes more the area on the left flank of the landing force. The plan called for a heavy cruiser to interdict enemy movement in the northern part of the island for an hour after the landing.

On the morning of the attack upon Eniwetok Island, which would take place as soon as possible after the securing of Engebi, Admiral Hill's cruisers and destroyers were to deliver some 80 minutes of preparatory fire. The heavy cruisers would rain down both destructive and interdictory fires for 30 minutes, then pause if the scheduled aerial attack was delivered. When the planes had departed, destroyers were to join the cruisers in shelling the island. In 25 minutes, after a second air attack, the close support phase was scheduled to begin, with the cruisers ceasing fire when the LVTs were 1,000 yards from the beach and the destroyers shifting to targets on the flanks when the assault waves were 300 yards from shore. The schedule for the Parry landing differed in that the bombardment would last 100 minutes and that a destroyer would join two cruisers in the 50-minute shelling that preceded the first air strike. At all the objectives, LCI(G)s and LVT(A)s were to assist in neutralizing the beach defenses.

Aerial support of the CATCHPOLE operation was scheduled to begin on D minus 1, when carrier planes attacked and also photographed the principal islands in Eniwetok Atoll. On the following day, fighters, dive bombers, and torpedo bombers were to attack specified targets on Engebi no earlier than 0800 and no later than 0830. Naval gunfire would be lifted during the strike. Planes were to remain on station over the atoll in the event they were needed to support the day's operations. Any unscheduled strikes would be directed by the airborne coordinator and the Commander, Support Aircraft.

The schedule of strikes in support of the Engebi landing called for the planes to attack perpendicular to the beach 35 minutes before H-Hour. The strike had to be completed within 10 minutes, for naval guns and artillery pieces would resume firing at H minus 25. No definite timetable was prepared for the Eniwetok and Parry Island operations, but Admiral Hill indicated his intention of scheduling similar aerial attacks 50 and 25 minutes before the troops reached shore.

The 22d Marines, assault force for the Engebi landing, was directed to load its assigned LVTs with ammunition and water before the convoy sailed from Kwajalein Atoll. The tractors were to be carried to the target area in LSTs. Off the objective, the Marines would load in LCVPs, move to the LSTs, and there embark in the amphibious vehicles. LVT(A)s, manned like the troop carriers by Army crews, had the mission of helping neutralize the beaches and then supporting the advance inland by landing on the flanks of the assault battalions. The group reserve, provided for Engebi by the 106th Infantry, was to remain in its transports and, if needed, transfer at sea from LCVPs to LVTs.

General Watson, faced with a series of landings, expected a great deal from his amphibian tractor unit. The 708th Provisional Amphibian Tractor Battalion, a composite Army command which included both armored and unarmored tractors, had a total of 119 vehicles. Of these, 17 were LVT(A)(1)s, 46 LVT(2)s, and 56 LVT(A) (2)s, which were simply LVT(2)s with improved armor. Since 8 to 10 tractors were assigned to each of the four or five waves required by each battalion, the 708th would be required to brave enemy fire time and again.

The method of control prescribed for CATCHPOLE differed little from the system used during the conquest of Kwajalein Atoll. Because of the shortage of LVTs, General Watson directed the amphibian tractor battalion commander to embark in the control vessel. After they had landed the assault troops, all LVTs were to report to this vessel. If reserves were needed ashore, the tractors could be routed to a transfer area behind the line of departure where the troops would load from LCVPs. The evacuation of the wounded was left to the supervision of the beachmaster, and a control officer embarked in an LCI was charged with directing empty supply craft to the proper ships.

DUKWs on loan from the 7th Infantry Division helped ease General Watson's supply problems. Firing batteries of both the 104th Field Artillery and 2d Separate Pack Howitzer Battalion were to land on D-Day in amphibian trucks. When this task was finished, the DUKWs were to report to certain of the LSTs to assist in unloading. Two pontoon causeways brought to the objective by Admiral Hill's transport group could be counted upon to speed the unloading of heavier equipment.

In comparison to the huge FLINTLOCK expedition, Tactical Group 1 carried few supplies, but enough ammunition, water, food, and fuel were loaded to sustain the men, weapons, and machines for five days. The rations carried in the convoy included a two-day supply of types C and K along with one day's D rations. Once the atoll was secured, the stockpile of food was to be increased until there was a minimum of 60 days' B, 8 days' C or K, and 2 days' D rations on hand.

Since ammunition and water were loaded in the tractors of the assault waves, the build-up of supplies would begin at the moment the troops landed. General Watson also directed that boats, each one loaded with a different kind of item, begin collecting at the line of departure as the fourth wave was moving toward the island under attack. Every boat was to fly a particular flag to indicate whether it carried ammunition, rations, fuel, water, or medical supplies. Requests from shore were to be routed through the beachmaster's radio net to the group logistical control officer who would then directed the appropriate landing craft to the proper beach.

The shore party organization appeared to be the weakest part of the logistical scheme. Major John F. Schoettel, the Betio veteran who commanded the composite shore party unit, would have to rely on "low priority combat personnel" to augment his organization.[26] The additional men were to be provided by the battalion upon whose beach the shore party component was working.

Plans and preparations for CATCHPOLE had been completed in a remarkably brief time. Only five days elapsed between Admiral Nimitz' arrival at Kwajalein and General Watson's issuance of his basic operation order; two days later, Admiral Hill's order was dispatched. The enemy too had been busy, trying frantically to convert Eniwetok into a series of fortified islands.

THE JAPANESE PREPARE[27]

The Japanese were slow to begin fortifying Eniwetok Atoll. The war against the United States had been raging for 11 months before 300 construction workers landed at Engebi Island to begin work on an airstrip. In December, 500 Korean laborers joined this original detachment. The runway was completed in mid-1943, and most of the men who built it promptly sailed for Kwajalein. Meanwhile, the first troops, a few sailors of the *61st Guard Force* who arrived from Kwajalein in January 1943, had established lookout stations on Eniwetok and Engebi Islands. By October 1943, a detachment from the Kwajalein guard force had come ashore to garrison the atoll.

Aerial photographs taken late in December showed little activity at Eniwetok Atoll. At the time, Warrant Officer Masimori Osano, in command at the atoll guard detachment, had no more than 61 men at his disposal. He assigned 10 of these to man a picket boat, sent 5 to the lookout station on Eniwetok Island, and retained the rest on Engebi, where a total of three lookout posts had been established. To defend Engebi, he had a pair of 120mm guns with about 87 rounds of ammunition, machine guns, rifles, pistols, and hand grenades.

The airstrip itself lay idle until November, when it was pressed into service as a maintenance stop for planes being ferried westward. Accommodations had been built for more than 300 aviation officers and men, but fewer than 50 mechanics or other specialists were on hand by the end of 1943. Although the atoll appeared quiet to the intelligence officers who studied the earliest American photographs, Eniwetok soon would become the scene of hectic activity.

On 4 January 1944, the convoy carrying the *1st Amphibious Brigade* dropped anchor in Eniwetok lagoon to land 2,586 troops and 95 civilian employees of the unit. The brigade boasted three infantry battalions, each with its own mortar, artillery, and engineer components, plus automatic cannon, tank, engineer, signal, and medical units. One battalion reinforced by elements of the brigade signal, engineer, and medical detachments had been detached for service elsewhere in the Marshalls.

Under strict secrecy, this Army amphibious organization had been detached from the *Kwantung Army* in

---

[26] TG 1 SP Plan, dtd 10Feb44, Anx L to TG-1 OpO 2-44, *op. cit.*

[27] Additional sources for this section include: JICPOA Buls 89-44, Japanese Def of Eniwetok Atoll, dtd 12Jun44, and 88-44, 1st PhibBrig, Japanese Army, dtd 5Jun44.

Manchuria and routed to the Marshalls by way of Fusan in Korea, Saeki in the home islands, and finally Truk.[28] JICPOA had noted the arrival of the convoy at Truk but had been unable to track it farther eastward. Documents captured at Kwajalein and prisoner of war interrogations placed the bulk of the brigade at Eniwetok.

Major General Yoshima Nishida, the brigade commander, found the atoll almost defenseless, a condition he immediately began correcting. "We have been working without sleep or rest on the unloading," [29] complained one of Nishida's men on the day after his arrival, but this back-breaking labor marked only the beginning of a hectic period of construction. The Japanese general inspected the atoll, dispatched garrisons to the various islands, and put the troops to work throwing up fortifications.

Additional help came on 13 January, when 200 Okinawan laborers, probably destined for Kwajalein, paused at Eniwetok. Nishida promptly put these men to work alongside the soldiers and the 200 or more Koreans who had remained behind when the airfield construction detachment was transferred. The mechanics and other aviation technicians on Engebi were probably of little help, for these men were in the process of being withdrawn. Evacuees from Kwajalein, however, temporarily swelled the numbers of the aviation unit, and when the American warships entered the lagoon many were trapped on the atoll.[30]

The general selected Parry as the site of his command post. Here he concentrated 1,115 combat troops and 232 civilians, aviation mechanics, laborers, and members of a naval survey party. The island garrison force was a 197-man rifle company supported by mortar crews, artillerymen, and engineers—305 men in all. Also present on the island was the brigade reserve, with which he could reinforce the other islands. Since an engineer and an antiaircraft unit were deployed to Eniwetok Island prior to the American attack, the reserve numbered only 810 by D-Day. The two reserve rifle companies at Nishida's disposal were reinforced by tank, signal, medical, engineer, and automatic cannon units.

Although the enemy expended a tremendous effort to fortify Parry, he accomplished comparatively little. The Japanese, who lacked both time and heavy equipment, suffered from the effects of short rations and an unfamiliar climate. "In all units," wrote a Japanese who visited Parry, "there are many men suffering from exhaustion. The infirmary is full." [31] The foxholes and trenches which the troops hastily gouged out of the island soil were not lined with rocks or logs, as were the few positions dug before the arrival of the brigade. Often a series of emplacements were linked to form a "spider web." The enemy would con-

---

[28] JICPOA Item No. 7811, Diary of 2dLt Kakino, hereafter *JICPOA Item 7811*.

[29] JICPOA Item No. 7603, Excerpts from a Diary of a Member of the 1st PhibBrig, hereafter *JICPOA Item 7603*.

[30] On D-Day over 200 such troops were stranded on Parry, Engebi, and Eniwetok Islands. All pilots en route westward from the outer Marshalls had already been evacuated.

[31] *JICPOA Item 7603*.

struct a large log-protected bunker sunk close to the surface of the earth. Tunnels led from the shelter to an outlying ring of foxholes, and these holes also were connected by other tunnels. The log-roofed bunker itself and the open foxholes were concealed by strips of corrugated iron covered in turn with a layer of sand. The defenders might take refuge in the large shelters during a shelling, then deploy to the foxholes, lift the roofs and open fire. A spider web, carefully camouflaged, was scarcely visible from the ground, let alone from high-flying photographic planes. Although fortifications on Parry were weaker than the steel-and-concrete pillboxes found on other atolls, they were the best Nishida's men could prepare.

The next largest garrison was that on Engebi, where the general had stationed a rifle company and support elements, which included mortars, tanks, and artillery. In addition to the 692 soldiers from the brigade, Engebi boasted 44 members of *61st Guard Force*, and 540 laborers, civilians, and support troops. The existing 120mm guns were incorporated into the defensive scheme, as were the few poor-quality pillboxes. The garrison, however, dug new trenches and foxholes along the lagoon coast.

The Eniwetok Island force consisted of a 779-man composite unit provided by General Nishida plus 24 civilians and 5 lookouts from the naval guard force. This smallest of the atoll garrisons had dug the most durable entrenchments. Mines were planted, and work on a system of concrete pillboxes was begun but never completed.

General Nishida clung to the Japanese tactical doctrine of destroying the invader at the beaches. "If the enemy lands, make use of the positions you are occupying during the daytime," he directed. "Endeavor to reduce losses, and at night strike terror into the enemy's heart by charges and destroy his will to fight." [32] Colonel Toshio Yano, in command at Engebi, was convinced the Americans would enter the lagoon, seize islands adjacent to the one he was charged with defending, and then storm the lagoon beaches. In keeping with Nishida's overall plan, the colonel ordered his garrison to "lure the enemy to the water's edge and then annihilate him with withering fire power and continuous attacks." [33]

The Japanese, most of their defensive installations undetected by American cameras, awaited Watson's soldiers and Marines. Including the crews of stranded vessels, Nishida's force totaled approximately 3,500 men. Not all were trained for combat, but each of them, with the possible exception of the Korean and Okinawan laborers, would fight to the death.

PRELIMINARY OPERATIONS [34]

The first carrier strikes against Eniwetok Atoll were delivered in conjunction with the FLINTLOCK operation, for the Engebi field had to be

[32] JICPOA Item No. 6637, 1st PhibBrig Plans for Defending Eniwetok Atoll, dtd 28Dec43.

[33] JICPOA Item No. 7539, 3d PhibBn OpO A-38, dtd 10Feb44.

[34] Additional sources for this section include: Craven and Cate, *Guadalcanal to Saipan;* Morison, *Aleutians, Gilberts, and Marshalls;* Crowl and Love, *The Gilberts and Marshalls.*

neutralized to prevent enemy planes from refueling there to bomb the Kwajalein task force. On 30 January, the Americans destroyed the 15 medium bombers based at Engebi and sunk several small craft in Eniwetok lagoon. Between 1 and 7 February, additional raids battered the objective, and the planes returned on the 11th and 13th. These earlier attacks leveled most of the structures built above ground on the various islands. The preparatory strikes began on 16 February, as Admiral Hill's ships neared the lagoon.

Life for the enemy garrison was hell on earth. "When such a small island as Engebi is hit by about 130 bombs a day, and, having lost its ammunition and provisions, lies helpless, it is no wonder that some soldiers have gone out of their minds." The island defenders, this same diarist admitted, were surviving on a single ball of rice each day, for their food had to be sent from Parry in outriggers under cover of darkness. The soldier thought of his family seated at dinner somewhere in Japan: ". . . my family's joy helps me to bear these hardships, when I realize that it is because of just such hardships as these I am now suffering that they are able to eat their rice cakes in peace." [35]

Conditions were not quite so desperate on Parry, where an occasional issue of rice wine spiced the reduced rations, nor on Eniwetok Island, but the Japanese knew that death was fast approaching. Lieutenant Kakino of the Parry garrison read somber portents in a raid of 12 February. "We celebrated the anniversary of the coronation of the Emperor Jimmu, this fourth year of our holy war, under enemy air attack. There must be some meaning for us in that." [36]

While the defenders of Eniwetok Atoll dug and wondered, American forces were attempting to isolate the objective. Army bombers attacked the eastern Carolines, concentrating on Ponape and Kusaie. The most dangerous of the Japanese bases, however, appeared to be Truk, 669 miles southwest of Eniwetok. This enemy Gibraltar of the Pacific was to be neutralized by Task Force 58, commanded during this action by Admiral Spruance himself.

On 17 February, D-Day at Eniwetok, carrier planes began a 2-day hammering of Truk. The aviators sank 2 auxiliary cruisers, a destroyer, 2 submarine tenders, an aircraft ferry, 6 tankers, and 17 merchantmen, a total of about 200,000 tons of shipping. Over 200 Japanese aircraft were damaged or destroyed. The blow at Truk also included a series of one-sided surface actions in which Spruance's battleships, cruisers, and destroyers sank a light cruiser, a destroyer, a trawler, and a submarine chaser. The larger units of the Imperial Navy had left Truk prior to the raid.

Tactical Group 1, without the benefit of last-minute rehearsals, boarded the transports of Admiral Hill's task group and on 15 February set sail for Eniwetok. The voyage proved uneventful; neither enemy planes nor submarines tried to contest Hill's approach. In the morning darkness of

[35] *JICPOA Item 7603.*

[36] *JICPOA Item 7811.*

17 February, a soldier on Parry Island looked up and "from the sea toward the east . . . saw a light and heard something like airplane motors." As daybreak approached, warships began shelling the atoll. "I thought to myself," he wrote in his diary, "that finally what must come has come." [37] The battle for Eniwetok had begun.

[37] JICPOA Item No. 7005, Diary of WO Shionaya, hereafter *JICPOA Item 7005*.

The CATCHPOLE Operation

CATCHPOLE AND THE LESSER MARSHALLS[1]

During the early morning of 17 February, an air alarm sounded on Parry Island, and as the defenders sought cover, naval shells came screaming into the island. The bombardment reached its height shortly after sunrise, or so it seemed to the enemy. At 0915, the Japanese watched Hill's bombardment group steam boldly into the lagoon to continue their firing. American planes joined in the action, so that the enemy garrison had no respite from the deluge of explosives. "There were one man killed and four wounded in our unit during today's fighting," noted one of the defenders. "There were some who were buried by shells from the ships, but we survived by taking care in the light of past experience. How many times must we bury ourselves in the sand?"[2]

To American eyes, the hammering of Parry, Engebi, Japtan, and Eniwetok Islands was equally impressive. After 28 moored mines, the first encountered in Central Pacific operations, had been cleared from Wide Passage, Hill's landing ships entered the lagoon. At the same time, the heavy vessels of the task group were passing through Deep Passage, whose current, Hill's staff had judged, was too swift to allow moored mines.[3] "To see the force enter this lagoon in column through a narrow entrance and between the shores of islands on either flank, and steam something over 20 miles through the enemy lagoon was," in the words of Admiral Hill's operations officer, "one of the most thrilling episodes which I witnessed during the entire war."[4]

The expedition had gained entry to Eniwetok lagoon without opposition from hostile batteries, although the big ships passed within 200 yards of Parry. Once the maneuver had been completed, a veil of tension was lifted from the task group. Ashore, however, the enemy realized that the decisive

[1] Unless otherwise noted, the material in this chapter is derived from: *TG 51.11 OpRpt;* VAC Cmts on *TG 1 SAR*, dtd 1Apr44; VAC NGF Rpt on CATCHPOLE Op, dtd 17Mar44; *TG 1 SAR*, including TG 1 Rpt of Atk on Eniwetok Atoll, dtd 27Feb44; *TG 1 SplRpt*, including VAC ReconCo AR, n.d., Co D (Sct), 4th TkBn AR, n.d., and SplRpt of Casualties and Prisoners, dtd 23Feb44; TG 1 URpts, 1800, 16Feb44–1800, 23Feb44; TG 1 Jnl, 4Nov43–23Feb44, hereafter *TG 1 Jnl;* 22d Mar WarD 1Jan–29Feb44; 1/22 Rpt on FLINTLOCK [actually CATCHPOLE] Op, n.d.; 2/22 Rec of Events, Feb44; 3/22 Rpt on CATCHPOLE Op, dtd 10Apr44, Encl to Capt Buenos A. W. Young ltr to CMC, dtd 9Mar53; Heinl and Crown, *The Marshalls;* Crowl and Love, *Gilberts and Marshalls.*

[2] *JICPOA Item 7005.*

[3] *Hill interview/comments Marshalls.*

[4] Capt Claude V. Ricketts, USN, ltr to CMC, dtd 9Mar53.

moment had come. General Nishida reported the entry of the task group to Tokyo and futilely requested reinforcements.[5]

THE PRELIMINARY LANDINGS

While the battleships were concentrating their fire against the larger islands, the destroyers *Heermann* and *McCord* bombarded CAMELLIA and CANNA. As the Marines of the VAC Reconnaissance Company were preparing to transfer from the APD *Kane* to the LST that carried their six amphibian tractors, word came of a change in plans. The boats carrying the company followed these latest instructions and shaped courses toward *LST 29*. There the unit learned that the original scheme was still in effect, so off the company went to *LST 272*, the ship first prescribed in its orders. (See Map 12.)

Two LCIs supported Captain Jones' command as it headed toward CAMELLIA and CANNA. Jones, with two other officers and 57 men, landed on CAMELLIA at 1320 and promptly reported that neither Japanese nor natives were on the island. The only difficulty was that encountered by two of the Army-manned LVTs, which had a hard time plowing through the sand beyond the beach. A similar group, 4 officers and 57 men, landed 10 minutes later on CANNA. This second island yielded 25 natives but no Japanese.

Captain Jones found the villagers to be friendly, cooperative, and in possession of what proved to be fairly accurate information. He forwarded to General Watson the natives' estimate that 1,000 combat troops guarded each of three major objectives of Tactical Group 1. An additional 1,000 laborers were believed located on Engebi.

The reconnaissance company now began investigating other islands southeast of Engebi. Five landings were made, but no Japanese were encountered. While Jones' Marines were patrolling, artillery units began landing on CAMELLIA and CANNA.

The reconnaissance and survey party sent by the 2d Separate Pack Howitzer Battalion to CAMELLIA found the island covered with undergrowth but lost little time in selecting firing positions. General Watson, however, feared that the battalion would land too late to register before dark, for the 104th Artillery, bound for CANNA, was making better progress. Actually, both battalions were ashore in time to complete registration by 1902.[6]

While the artillery units were selecting base points and check points as well as plotting harassing fires for the evening of D-Day, underwater demolition teams, screened by naval gunfire, were examining the beaches off Engebi. Leaping from LVTs when the tractors were about 100 yards from the lagoon coast, the Navy men swam to within 50 yards of the shoreline. They located no artificial obstacles on either Blue or White Beach. (See Map 13.)

In the last operation planned for D-Day, Tactical Group 1 ordered the 4th Marine Division scout company to seize ZINNIA, just west of Engebi.

[5] *Ibid.*, citing a message intercept.

[6] LtGen Thomas E. Watson ltr to CMC, dtd 1Mar53; MajGen John T. Walker ltr to CMC, dtd 3Apr53, hereafter *Walker ltr*.

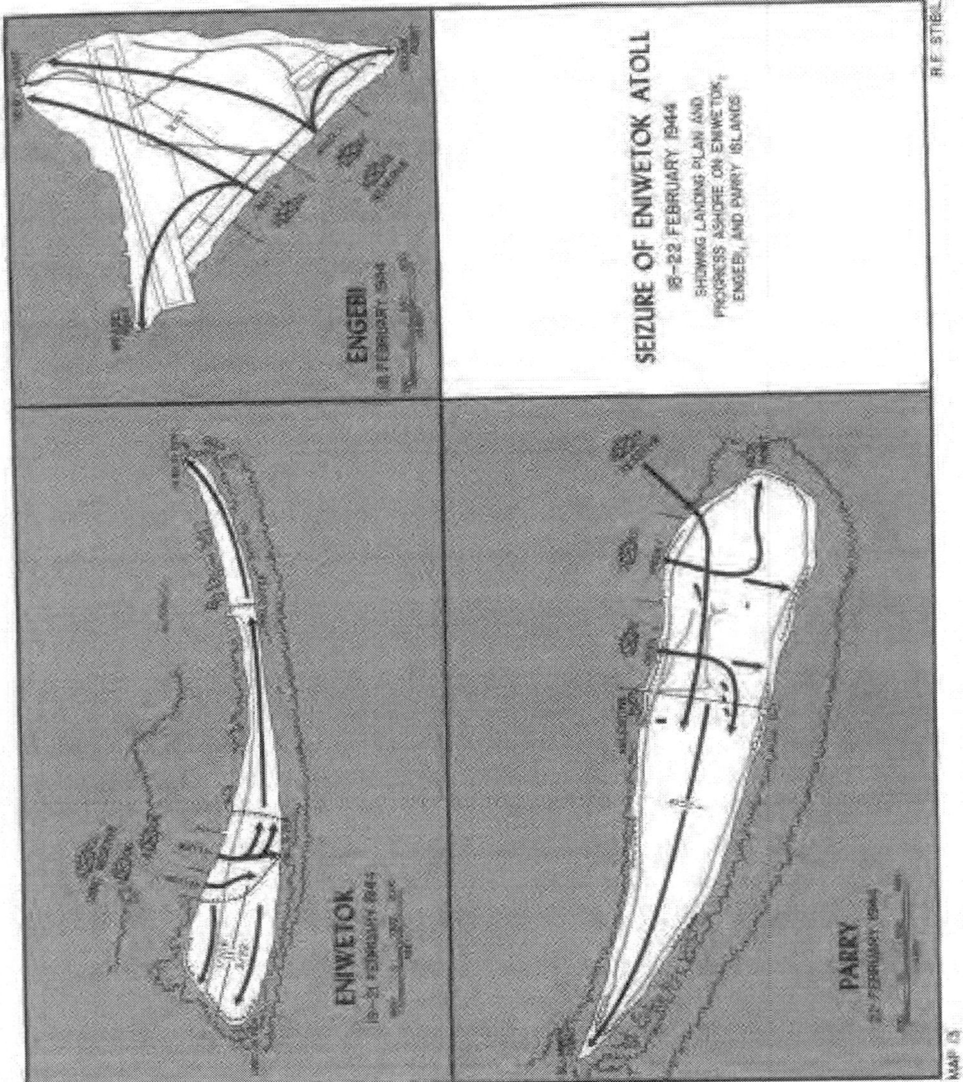
SEIZURE OF ENIWETOK ATOLL
18–22 FEBRUARY 1944
SHOWING LANDING PLAN AND PROGRESS ASHORE ON ENIWETOK, ENGEBI, AND PARRY ISLANDS
ENGEBI
18 FEBRUARY 1944
ENIWETOK
19–21 FEBRUARY 1944
PARRY
22 FEBRUARY 1944
R.F. STIBIL

MAP 13

In the darkness, the scouts, buffeted by a 25-mile wind and 8-foot waves, were unable to paddle to the island for which they were aiming.[7] A landing on a neighboring islet gave them an opportunity to reach their original objective, and at 0327 ZINNIA was in American hands. Now the Japanese could not prolong the battle by retreating from island to island. The first phase of CATCHPOLE came to a successful close. The invaders had suffered no losses and encountered no opposition.

THE ATTACK ON ENGEBI

General Watson's plan for 18 February designated 1/22, commanded by Lieutenant Colonel Walfried H. Fromhold, and 2/22, under Lieutenant Colonel Donn C. Hart, to storm Engebi, while Major Clair W. Shisler's 3/22 remained in reserve. The 1st Battalion was to seize Beach White 1, immediately to the right of a pier that jutted into the lagoon from the southern shore of the triangular-shaped island. Blue 3, objective of the 2d Battalion, lay just to the left of the pier. The boundary between battalions followed the trail that led northward from the pier to a junction with a second trail. The boundary followed the left fork to the vicinity of the airstrip and then curved slightly toward Newt Point, the terrain feature farthest from the landing beaches. (See Map 13.)

Both assault battalions were reinforced for the operation. The 2d Separate Tank Company[8] and an Army platoon of two self-propelled 105mm guns were kept under regimental control. Also, the Marine pack howitzers and the Army 105mm howitzers were to support Colonel Walker's regiment.

Since most of the runway lay within Hart's zone of action, his unit would advance across comparatively open terrain. Shattered buildings just inland of the beaches and a stand of coconut trees near Newt Point were the principal obstacles in the path of 2/22. Most of the enemy troops were located in the 1st Battalion zone in the large coconut grove near the center of the island. Permanent pillboxes had been built at the three corners of the objective, and entrenchments were scattered all along the coastline.

On D-Day, the Engebi garrison had received a severe battering from bombs and naval shells. "One of our ammunition dumps was hit and went up with a terrifying explosion," read the final diary entry of one of the island defenders. "At 1300 the ammunition depot of the artillery in the palm forest caught fire and exploded, and a conflagration started in the vicinity of the western positions."[9] During the night, Army and Marine field artillery batteries harassed the weary Japanese, and at 0655 on 18 February the fire-

[7] Col Edward L. Katzenbach, Jr., ltr to Head, HistBr, G-3, HQMC, dtd 25Sep62.

[8] This unit was equipped with medium tanks obtained from the Army. In six weeks, the Marine crews painted, water-proofed, and otherwise modified the vehicles—in addition to undergoing training. Maj Robert S. Reinhardt, Jr. ltr to CMC, dtd 18Mar53, hereafter *Reinhardt ltr*.

[9] JICPOA Item No. 8200, Extracts from the Diary of Norio Miyada.

214-881 O-67—14

support ships began their final bombardment. W-Hour was scheduled for 0845.

Reveille sounded for the 22d Marines at 0400. As the naval guns were flaying Engebi, the two artillery battalions fired their planned concentrations, and Colonel Walker's Marines transferred to vehicles manned by the provisional amphibian tractor battalion. Planes from the escort carrier group delivered their 10-minute strike at about 0800. Midway through the aerial attack, the first assault waves crossed the line of departure, and when the planes had made their final run over the objective, the warships resumed firing.

The boat landing plan called for the tractors to employ a formation similar to that used at Kwajalein Island. The LVTs formed a W in the wake of the LCIs, with the troop carriers at its base and five of the LVT(A)s on each of the projecting wings. At the base of the formation, between the two battalions, were seven additional armored amphibians.

As the 22d Marines bored through the waves, the LCIs hammered the beaches with 40mm shells. Unfortunately, the rockets launched from the gunboats to neutralize the coastal defenses fell short and exploded in the surf. Smoke and dust obscured the vision of tractor drivers and caused some vehicles to wander from course. Mechanical failures, not surprising since the same LVTs had taken part in the several phases of the southern Kwajalein operation, also slowed the first wave of 1/22. In spite of these difficulties, the first of the tractors reached Engebi some two minutes before the scheduled W-Hour.[10]

The armored amphibians and the first wave of tractors were to have advanced inland to a road that ran parallel to the lagoon coast, but fallen coconut trees and other debris stalled the vehicles. They were able, however, to support the advance by firing from positions just across the beaches. Although the attack lost some of its initial momentum, the enemy remained too dazed from the effects of the preliminary bombardment to contain the Marines.[11]

On the right of the pier, Fromhold's 1st Battalion advanced with Company B on the left and Company A on the right. Behind them, Company C thrust toward Skunk Point in an effort to secure the southeastern corner of the island. During the movement to White 1, the LVTs carrying one platoon of Company A had broken down, so that unit was late in landing. The platoon hurried into position, but as the battalion swept inland, a gap opened between Companies A and C, an opening which the Japanese discovered by accident.

Under relentless pressure from Company C, many of the defenders of Skunk Point sought to escape northward. In doing so, they found themselves within the gap and in position to fire into the exposed flank of Company A. Since this unit was just entering the tangled undergrowth of the coconut grove, an area riddled with spider web

[10] *Walker ltr;* Capt William G. Wendell, Notes of Interview with LtCol Walfried H. Fromhold, n.d., hereafter *Fromhold interview.*

[11] *Ibid.*

defenses, it could do little to protect its dangling flank. The battalion commander checked the advance of Companies A and B until a platoon of tanks could be dispatched to the scene.[12]

On the opposite side of the pier, Hart's 2/22 thrust rapidly inland after a somewhat confused landing. Although a few tractors had landed as much as 200 yards out of position, junior officers and noncommissioned officers quickly led their men to the proper zones. As soon as order had been restored, the advance got underway.

Companies E and F spearheaded the attack of 2/22. These Marines paused at the nearer edge of the runway while artillery concentrations blanketed the area in front of them. Elements of the 2d Separate Tank Company joined, and when the barrage lifted, both armor and infantry plunged forward. The arrival of the Marine armor proved fortunate, for the Japanese were using light tanks as makeshift pillboxes in this area. Although the enemy tanks were firing from earthen emplacements, they could not survive American shells.

Once the Japanese armor had been silenced, the Marines dashed rapidly toward the opposite shore, bypassing many knots of enemy resistance. Fifteen Japanese were killed attempting to flee across the level ground. A handful of men of Company F reached the coast approximately 300 yards east of Weasel Point, the southwestern tip of Engebi. When this was reported to Major Robert P. Felker, the battalion executive officer, he exclaimed: "My gosh, Fox Company is trying to take the whole island."[13] By 1030, when Colonel Walker came ashore, resistance in the 2d Battalion zone was confined to Weasel and Newt Points.

The news which the regimental commander received from the other battalion was not quite so encouraging. Company A, receiving severe fire from the wooded area to its front and from the Japanese within the gap on its right, lacked enough men to push onward. When contact with Company C had been broken, one platoon of Company A had become separated from its parent unit.[14] In addition, the company had been suffering light casualties throughout the morning, and this steady attrition gradually sapped its strength. Since 3/22 was already ashore on White 1, one of its rifle companies was attached to 1/22. Fromhold ordered the fresh unit, Company I, to prepare to pass through Company A.[15]

While tanks were assisting a part of Company C in sealing the gap, Company B continued its advance. To keep the attack moving, Fromhold ordered Company I forward and directed the remnants of Company A to mop up. The task facing Company I was grim, for the ground to its front was covered by dense underbrush and fallen trees which screened a series of open trenches and spider web emplacements.

These positions were extremely difficult to locate, for the defenders had

[12] *Fromhold interview.*

[13] *Reinhardt ltr.*

[14] LtCol Glenn E. Martin memo to Head, HistBr, G–3, HQMC, dtd 12Mar53, Subj: Eniwetok.

[15] *Fromhold interview.*

THREE WAVES *of 22d Marines assault troops approach Engebi as smoke and dust from the preliminary bombardment drifts across landing lanes. (USMC A411182)*

NAVY FIGHTER PLANE *attacks enemy positions ahead of Marines during the fighting at Eniwetok Atoll. (USCG 26-WA-6G-8)*

placed pieces of driftwood over their foxholes and the firing ports of their bunkers. The Japanese could push open these "doors" without exposing themselves to Marine riflemen. If a sniper was spotted, he would retreat into the underground maze connecting the several holes that made up a spider web. The attacking Marines soon hit upon a method of destroying completely these underground defenses. When the bunker at the center of the web had been located, a member of the assault team would hurl a smoke grenade inside. Although this type of missile did no harm to the Japanese within, it released a cloud of vapor which rolled through the tunnels and escaped around the loose-fitting covers of the foxholes. Once the outline of the web was known, the bunker and all its satellite positions could be shattered with demolitions.[16] In this way, Fromhold's command punched its way forward.

Near Skunk Point, however, 1/22 encountered concrete pillboxes which were the key to the enemy defenses around the southeastern corner of the island. Two self-propelled 105mm guns from the 106th Infantry Cannon Company, weapons originally sent to the aid of Company A, were withdrawn upon request of Company C and sent to deal with the heavy fortifications. The self-propelled guns fired almost their entire day's allowance of ammunition, 80 rounds, in order to destroy two pillboxes, one of which sheltered 25 to 30 Japanese.

Effective as they were, the self-propelled 105s were not especially popular with the infantry units which they supported. "They made a very inviting target for mortar and other small arms," commented a veteran of the Skunk Point fighting, "and, consequently, were not a very comforting thing to have around."[17] The 105s, though, had revenge on their Japanese tormentors, for during the afternoon they wiped out an enemy mortar crew.

While 1/22 was battling to secure its portion of Engebi Island, 2/22 overran Weasel Point and that part of Newt Point to the left of the battalion boundary. Both were reported in American hands at 1310, but the fight for the northern corner of the island had been bitter. Woods and undergrowth concealed a maze of underground positions from which enemy riflemen claimed many victims.

General Watson landed at 1400 and 50 minutes later declared that Engebi was secured. Six minutes later, 1/22 reported that Skunk Point had been overrun; only the right-hand portion of Newt Point remained in Japanese hands. The general then ordered 3/22 and the 2d Separate Tank Company to reembark for the Eniwetok Island operation. These units began loading at 1700. Although Company I was thus withdrawn from his control, Fromhold's Companies A and B secured the remainder of Newt Point by 1830.

While the battle for Engebi was raging, the reconnaissance and scout companies continued their exploration of the nearby islands. The two units made a total of eight landings, captur-

[16] Martin memo, *op. cit.*; *Walker ltr.*

[17] Maj Thomas D. Scott ltr to CMC, n.d., hereafter *Scott ltr.*

ing in the process one Japanese soldier. Machine gun fire wounded three members of the contingent sent to ARBUTUS (Muzinbaarikku),[18] but the bullets were proved to be "overs" aimed at the enemy on Skunk Point.

As night descended upon Eniwetok atoll, General Watson and his staff issued necessary modifications to the group operation order. After analyzing the latest information, stories told by natives that indicated the presence of 1,000 men and a captured brigade document listing a 600-man garrison, the group commander alerted Colonel Ayers' 106th Infantry to prepare for the following day's attack against Eniwetok Island. The regiment would land two battalions abreast instead of in a column of battalions as originally planned. Neither component of the 106th was to be withdrawn until the objective was secured. The general attached 3/22 and the group tank company to reinforce the Army unit. Once this objective had been captured, the composite regiment was to get ready for the Parry operation. Although casualty figures were not complete, fragmentary reports indicated that the 22d Marines had lost 64 killed, 158 wounded, and 81 missing during the Engebi battle. Since the 3d Battalion had committed only one company on 18 February, its losses would not prevent its subsequent use at Eniwetok Island.

While General Watson looked ahead to phase III, Lieutenant Colonels Fromhold and Hart could view the afternoon's action with satisfaction. The swift overrunning of the island did not leave "the enemy an opportunity to reorganize resistance."[19] There were plenty of individual Japanese who had been bypassed, but they had no semblance of organization. According to the Task Group 1 report, "isolated snipers and pillboxes and Japanese who were in underground shelters" survived the day's fighting. "These positions," the report continued, "were methodically reduced."[20] To those who remained on Engebi on the night of 18–19 February, the surviving enemy seemed far from isolated.

Under cover of night, the Japanese crept from their underground shelters and opened fire with grenade dischargers, rifles, and automatic weapons. Those who had lost their weapons helped themselves to the Japanese and American rifles, grenades, and ammunition that lay scattered throughout the island.[21] As one Marine officer phrased it, "snipers and by-passed areas made movement to and from the front lines exceedingly difficult and resulted in many enemy contacts and a generally uncomfortable first night. . . ."[22] On 19 February, after a formal flag raising, demolition teams and flamethrower operators from the group engineer unit joined the infantry in the systematic destruction of the stubborn Japanese. Over 1,200 Japanese, Okinawans, and Koreans

[18] One of these men died as a result of his wounds en route from the beach to the hospital ship. LtCol Merwin H. Silverthorn, Jr., ltr to Head, HistBr, G–3, HQMC, dtd 12Sep62.

[19] BGen Donn C. Hart ltr to ACofS, G–3, HQMC, dtd 13Nov62.

[20] TG 1 Rpt, p. 5, in *TG 1 SAR*.

[21] *Fromhold interview.*

[22] LtCol John P. Lanigan memo to LtCol John A. Crown, n.d., Subj: Eniwetok.

were on Engebi when the battle began; only 19 of them surrendered.

PHASE III: ENIWETOK ISLAND

The objective selected for 19 February, Eniwetok Island, is shaped like a war club, its heavy end resting against Wide Passage. The long axis of the island extends from the lagoon entrance northeastward toward neighboring Parry. At Y-Hour, 0900, Colonel Ayers' soldiers were to storm Beaches Yellow 1 and Yellow 2 on the lagoon coast at the thicker end of the club. On the right, 1/106, commanded by Lieutenant Colonel Winslow Cornett, was to cross to the farther shore, then secure that portion of the island between the beachhead and Wide Passage. Lieutenant Colonel Harold I. Mizony's 3/106 would thrust to the ocean coast and then use one company to defend the left flank of the beachhead. The remainder of Mizony's command was to remain inland of the Yellow Beaches, ready to assist Cornett's battalion. When the wider portion of the island had been overrun, Cornett faced the task of attacking through Mizony's blocking position with one rifle company and securing the remainder of the objective. The regimental reserve, Shisler's 3/22, was directed to remain afloat off the Yellow Beaches in the event it was needed. The 104th Field Artillery was to land as promptly as possible, move into position, and register to support the Parry landing, an operation tentatively scheduled to begin as soon as Eniwetok Island had come under American control. (See Map 13.)

During phases I and II, Eniwetok Island was battered by bombs and shells. As early as D-Day, one of the garrison soldiers had admitted that "If such raids keep up, they will intimidate us. . . ."[28] The ordeal continued, but the defenders of Eniwetok were not forced to endure as heavy a weight of high explosives as was showered on Engebi and Parry. None of the battleships turned their guns upon Eniwetok; the cruisers and destroyers fired 1,094 8-inch rounds and 4,348 of 5-inch. These shells struck in an area of approximately 130 acres. In contrast, Engebi, some 220 acres, had shuddered under 497 16-inch, 954 14-inch, 673 8-inch, and 4,641 5-inch projectiles. Parry would receive an even heavier bombardment. Although the pounding which the defenders of Eniwetok Island dreaded so intensely was continued into the morning of D plus 2, the garrison, as events would prove, was far from intimidated.

The timetable which Tactical Group 1 was striving to follow imposed a special hardship on the 2d Separate Tank Company, the LCMs assigned to it, and the LSD *Ashland* from which the landing craft operated. Because maneuvering the *Ashland* in the darkness was believed too dangerous, the LCMs, guided by a submarine chaser, were directed to carry the rearmed and refueled tanks from Engebi to Eniwetok Island. Admiral Hill twice postponed Y-Hour to give the armor ample time in which to complete the 25-mile voyage, and the LCMs arrived by 0900,

[28] JICPOA Item No. 6808, excerpts from the Diary of Cpl Masamichi Kitama.

the time originally fixed for the landings.[24]

The bombardment and aerial strikes were executed according to schedule, and at 0909 Hill ordered the two battalions of the 106th Infantry to cross the line of departure. The soldiers gained the Yellow Beaches at 0916. Armored amphibians, ordered to move 100 yards inland, thundered forward a short distance and were halted by a 9-foot embankment. The infantrymen scrambled from their LVTs, some of which had landed out of position, and found themselves confronted by an intricate network of spider webs similar to that which 1/22 had encountered on Engebi.

Mizony's 3/106 had the easier going, for by 1030 a part of one company had reached the ocean shore. In spite of admonitions from General Watson to "push your attack" and "clear beaches,"[25] Colonel Ayers' troops could make little progress elsewhere in the regimental zone. By noon, the 106th Infantry had won a beachhead that extended from the left flank of Yellow 1 directly across the island, included some 250 yards of the ocean coastline, and then meandered in an irregular fashion toward the right flank of Yellow 2.

A stubborn enemy, a series of defenses that emerged undamaged from the preliminary bombardment, plus the inadequate rehearsal and lack of amphibious experience combined to rob the regimental attack of its momentum. The Japanese were quick to seize this opportunity to strike back. Lieutenant Colonel Masahiro Hashida withdrew about half of his command into the defensive network dug near the southwestern corner of the island and sent the other half creeping through the underbrush toward Cornett's battalion. Early in the afternoon, 300–400 Japanese leaped up and hit 1/106 on both sides of the trail that ran parallel to the lagoon shoreline.

Although the enemy had the twin advantages of surprise and accurate mortar fire, his penetrations of the American line were short-lived. The fighting was bitter and brief; by 1245 the soldiers had beaten off the Japanese. Cornett's troops, however, reported 63 casualties. Hashida's thrust had been vigorous enough to convince Colonel Ayers that a single battalion could not overrun the southwestern end of the island. Since he had already ordered Mizony's battalion to attack instead of remaining on the defensive as planned, the regimental commander had no choice but to commit his reserve. Major Shisler was directed to land 3/22 during the afternoon, relieve a portion of Cornett's unit, and assume responsibility for the left half of the island. The Marine unit also was charged with maintaining lateral contact during the attack. At approximately 1515, the two battalions began advancing toward Hashida's redoubt.

As daylight waned, Shisler's troops collided with a series of log emplacements carefully hidden in the dense underbrush. These positions had survived naval shells and were impervious to damage from mortar rounds. Marine infantrymen and engineers finally killed or at least stunned the

[24] USS *Ashland* Eniwetok AR, dtd 3May44.

[25] *TG 1 Jnl*, msgs nos. 6 and 8, 18–19Feb44.

defenders with bursts from flamethrowers and with satchel charges. Shisler continued to press the attack, but progress was slow.

Across the battalion boundary, which was a line drawn on a map rather than a recognizable terrain feature, 1/106 plodded onward. Cornett's troops lagged a short distance behind the Marines, so a gap soon was opened between the units. To the rear, elements of the 104th Artillery were beginning to come ashore.

Ayers, having experienced a counterattack earlier in the day, was determined to prevent the Japanese from striking again under cover of darkness. At 1850 the regimental commander informed all battalions that they were to "advance until you have reached the end of island. Call for illumination when necessary." [26] At dusk, as the artillery was beginning to register from positions to the rear, a few of the six light tanks attached to the 106th Infantry that had landed at 1745, reported for orders at Shisler's command post.[27]

When the commander of 3/22 told the officer in charge of the tanks that the attack was to continue throughout the night, the leader of the armored unit, knowing his vehicles were ineffective in the dark, ordered them to the rear.[28] Since the tanks would be of no help and the illumination of this portion of the battlefield had not yet begun, Shisler ordered his battalion to defend from its present positions. On the right, however, 1/106 kept plodding along, advancing through an area pockmarked with covered foxholes.

At 1945, Cornett reported that his battalion was in contact with Shisler's Marines, but as the Army unit moved forward, contact once again was broken. By 0333, Cornett's command occupied a 3-company perimeter at the tip of the island along the lagoon coast. The commander of 1/106 indicated that his unit was on line with the adjacent battalion.[29] Actually the Marine flank lay over 100 yards to the left rear of the Army position.

Throughout the night, the Marine battalion fought off sporadic attempts at infiltration. When the skies grew light enough, they looked to their right and were startled to discover that the Army battalion had vanished. The soldiers had executed an order which one of Holland Smith's planners, an Army colonel, later branded as "absurd." [30] Darkness prevented 1/106

---

[26] 106th Inf UJnl, msg dtd 1850, 19Feb44 (WWII RecsDiv, FRC, Alexandria, Va.).

[27] Some of the information concerning the actions of 3/22 on the night of 19 February comes from an interview with LtCol Clair W. Shisler, extracts from which are printed in Heinl and Crown, *The Marshalls*, pp. 140–141. The present location of the transcript of this interview is not known.

[28] Capt Donald J. Myers ltr to CMC, dtd 28Feb53.

[29] 106th Inf UJnl, msgs dtd 1837 and 1945, 19Feb44.

[30] *Anderson ltr.* Colonel Anderson felt that the night attack had been launched as part of a "race" between the 106th Infantry and 22d Marines. But there are at least two views of every military decision. Colonel Ayers' executive officer pointed out that the night attack was intended to keep pressure on the Japanese and to give the enemy no "respite in which to reorganize and counterattack in force (the invariable Japanese reaction in every previous

from mopping up. Thus Hashida's remnants were able to enter the gap between battalions and, screened by the underbrush, deliver another blow. Thirty to 40 Japanese attacked Shisler's command post at about 0900 and for a time it looked as though its defenses would be shattered.

The enemy struck without warning and pressed his thrust with desperate fury. "In my opinion," wrote an eyewitness to the fight, "what would have been a complete rout for us was prevented by the initiative and resourcefulness of two men—Sergeant Major John L. Nagazyna and Captain Leighton Clark . . . who . . . got many men back on line by threatening, encouraging, cajoling, and dragging them back into position." Under this kind of leadership, the clerks, radiomen, and other specialists hurled back the Japanese. A detachment of riflemen, which arrived just as the enemy struck, helped stiffen the command post defenses. Marine losses in this furious action were 4 killed and 8 wounded among the command group. Since other units in the immediate vicinity also suffered casualties, the attackers may have killed as many as 10 Americans.[31]

Most of Hashida's troops now were fighting with their backs against Wide Passage. No longer was there any threat of a coordinated counterattack, but the task of locating and killing these last stubborn defenders occupied most of the day. Again, self-propelled 105mm guns from the Army regiment proved their worth. By 1445, Marines and soldiers had overwhelmed this final pocket of resistance, and the southwestern end of the island was secured.

While 1/106 and 3/22 were fighting their way toward Wide Passage, Mizony's battalion was advancing toward the opposite end of the island. Like Cornett's unit, 3/106 attempted to continue its attack after darkness. The effort was soon frustrated, for "it was impossible to see the camouflaged holes, contact was poor, and the troops as a whole did not seem to have the confidence in themselves that was so apparent during the day."[32] The soldiers, halted approximately one-quarter of the way to the narrower tip of the island, resumed their offensive after daybreak, and at dark were still short of their final objective. Not until the afternoon of 21 February was this portion of Eniwetok Island overrun.

During the fight for Eniwetok Island, the 4th Marine Division scout company and the VAC Reconnaissance Company continued operations elsewhere in the atoll. Among the islands seized was a coral outcropping just north of Parry, Japtan, which was occupied on 19 February. On the next day, the 2d Separate Pack Howitzer Battalion emplaced its 75mm weapons here to support the attack upon Parry. LILAC, between Parry and Japtan, also was occupied, and to the southwest across the lagoon, a party of scouts

engagement)." In his opinion, had the attack not continued "the counterattack might well have been more effective," and that Colonel Ayers' "order was a logical application of available means (illumination and continued pressure) to counter a relatively certain Japanese reaction." Col Joseph J. Farley, AUS, ltr to Head, HistBr, G-3, HQMC, dtd 20Oct62.

[31] Young ltr, *op. cit.*

[32] 3/106 S-3 Sum (WWII RecsDiv, FRC, Alexandria, Va.).

killed nine Japanese on POSY (Rigili). By the afternoon of 21 February, only Parry remained in Japanese hands. (See Map 12.).

PARRY: THE LAST OBJECTIVE

General Watson had hoped that Eniwetok Island could be overrun quickly, but the assault force had become bogged down. In spite of the disappointing results of the first day's fighting, he alerted the components of the 22d Marines on Engebi to embark for an attack on Parry tentatively scheduled for 0900, 21 February. The two battalions reported for further orders on the afternoon of 20 February. The commanding general, however, decided to strike at Parry on 22 February, after reembarking 3/22 as landing force reserve.

By this time, the effect of the continuous series of landings was beginning to make itself felt. On the evening of 20 February, Watson estimated, on the basis of recent reports, that the 22d Marines had suffered 116 dead, 254 wounded, and 6 missing, and the 106th Infantry, 18 killed, 60 wounded, and 14 missing. Although the effectiveness of Tactical Group 1 remained "very satisfactory," the commanding general admitted "some reduction of combat efficiency." [33]

To insure that the Parry landing force would be at peak combat efficiency, he attached to it the 2d Separate Tank Company, VAC Reconnaissance Company, and 4th Marine Division scout unit. He also alerted the light tank company attached to the 106th Infantry to be available for action on the morning of the Parry assault. Mizony's 3/106 was designated as group reserve, available for employment upon 4-hours' notice. Finally, the 10th Marine Defense Battalion was ordered to organize five 100-man rifle companies for possible use at Parry.

Tactical Group 1 also was experiencing shortages in certain types of munitions. All transports were searched for any demolitions that might have been overlooked during the earlier operations. The shortage was relieved to some extent by an aerial shipment of 775 offensive grenades and 1,500 demolition caps that arrived from Kwajalein before noon of 22 February. General Watson also limited the amount of artillery ammunition to be fired in preparation for the Parry landing. The 105s and 75s were to expend no more than one and one-half units of fire during daylight of 21 February and no more than 2,400 rounds per battalion on the following night. In addition, the 22d Marines lacked the prescribed number of rifles and automatic rifles. Before the operation began, many Marines had discarded these weapons in favor of carbines, but the bullets fired from the lighter weapons lacked the killing power of rifle ammunition. "The BARmen and riflemen," recalled an officer of the 22d Marines, "seemed very happy to discard their carbines and take up their former weapons prior to the Parry landing." [34] Rifles and automatic rifles from other

[33] TG 1 URpt, 19–20Feb44, Encl G to *TG 1 SAR.*

[34] *Scott ltr.*

units were redistributed until the Marine infantry units had their full complement of weapons.

The resistance encountered at Engebi and Eniwetok Islands brought about two changes of plan. Admiral Hill had decided to increase the tonnage of naval shells fired into Parry's defensive installations. On 20 February, while Watson was conferring with his unit commanders, the battleships *Tennessee* and *Pennsylvania*, ships which had not figured in the original bombardment scheme, closed to within 850 yards of the objective to batter suspected defensive installations. On the following day, these vessels increased the range to deliver high trajectory fire, a type judged more effective against underground emplacements.

General Watson now re-examined the frontages assigned the assault battalions in his tentative plan. The landing beaches, designated Green 2 and Green 3, were separated in the original plan by a small pier that jutted 20 yards into the lagoon. Green 2 stretched from this structure northward to within a short distance of the fringing reef. Green 3 extended southward to another longer pier, a structure called Valentine Pier. Together the proposed assault beaches encompassed most of the northern third of the wooded island. The commanding general, deciding that the battalion frontages were too large, compressed the beachhead toward the north, so that the landing area, divided equally between the assault battalions, was flanked by the reef on one side and on the other by an arbitrary line 300 yards south of the shorter pier.[35] (See Map 13.)

At Z hour, finally fixed at 0900 on 22 February, Lieutenant Colonel Fromhold's 1/22 would land on Green 3, seize that portion of the island directly to its front, and then reorganize at a phase line just south of Valentine Pier before attacking southward to the narrow tail of the torpedo-shaped objective. On the left, Lieutenant Colonel Hart's 2/22 had the mission of striking across the island, passing into regimental reserve, and then mopping up the northern sector of Parry. Major Shisler's 3/22 was to follow 1/22 ashore and move into position to the right of Fromhold's battalion on the 0–1 Line for the 2-battalion advance southward, with the units separated by a line drawn down center of the long axis of the island.

Tallying the number of casualties, designating reserve units, distributing weapons and ammunition, and revising tactical plans, all of them difficult tasks, were accomplished rapidly. Colonel Walker, commanding officer of the 22d Marines, at 2100 reported: "Assault battalions are embarked on LSTs; all preparations complete."[36] Meanwhile, the 104th Field Artillery on Eniwetok Island and the 2d Separate Pack Howitzer Battalion on Japtan joined four destroyers in a night harassment of Parry. On the following morning, Walker's men would storm

[35] *Fromhold interview*. The account of this change of plans contained in Heinl and Crown, *The Marshalls*, p. 144, is based in part on an interview with Col Floyd R. Moore, dtd 1Mar 53. Apparently, no notes of this conversation have been retained.

[36] *TG 1 Jnl*, msg no. 25, 21–23Feb44.

the last enemy bastion in the entire atoll.

Like their comrades elsewhere in the atoll, the members of the Parry garrison were determined to resist to the last man. On the morning of 18 February, the Japanese, though suffering from the effects of reduced rations, bombing, and naval gunfire, were ready for the impending battle. "We thought they would land this morning," wrote one of Parry's defenders in his diary, "but there was only a continuation of their bombardment and no landing. As this was contrary to our expectations, we were rather disappointed." On the following day, Hill's old battleships began hammering the island, driving the Japanese into underground shelters which soon became "unendurably hot." This same Japanese soldier had glanced over the waters of the lagoon and seen "boats . . . entering and leaving . . . at will, making fools out of us." Now as death drew near, he consoled himself with the thought that "When they land, we will pay them back for what they have given us. . . ."[37]

Determined as most of them were, the enemy troops on Parry staggered under the blows of American warships, planes, and howitzer batteries. Fire-support ships slammed 143 16-inch, 751 14-inch, 896 8-inch, and 9,950 5-inch shells into an area of 200 acres. Although the battleships had fired more rounds at Engebi, smaller ships more than made up the difference, so that Parry Island rocked under the heaviest weight of metal delivered during the Eniwetok campaign.

On 22 February, both artillery battalions began an intense bombardment at 0600, to be joined an hour later by supporting warships. Clouds of dust and smoke rose from the battered island and began rolling out over the lagoon. By 0845, when the first wave crossed the line of departure, the line of lighted buoys that marked the boundary between beaches was no longer visible, nor could the LVT drivers make out many landmarks along the lagoon coast. A "terrific set" in the current off the beaches,[38] combined with poor visibility and the fact that not all control officers were aware of the recent revision in plans, insured a confused landing.

During the ship-to-shore movement, three LCI(G)s, supporting the tractors on the right flank, were hit by 5-inch shells from destroyers, which were firing by radar because of the smoke.[39] Moments later, an observation plane dived too low, was struck by shells fired toward Parry, and crashed in flames. In spite of these accidents, the assault battalions landed at 0900, but not on their assigned beaches. Both units were some 300 yards south of where they should have been.

On the left, 2/22 landed out of position, but the unit met light resistance along the beach. The heaviest fire came from the vicinity of Valentine Pier in the zone of 1/22. Some of Hart's Marines, however, were killed by enemy mines, and others lost their lives trying to eliminate individual Japanese who fought viciously from foxholes inland of Green 2.

---

[37] *JICPOA Item 7005.*

[38] *Hill interview/comments Marshalls.*

[39] *Ibid.*

MARINE HEAVY MACHINE GUNS *fire at Japanese defenses 400 yards ahead on the beach at Eniwetok Island. (USN 80–G–216019)*

NAVY CORPSMEN *administer blood plasma to a wounded Marine on the beach at Eniwetok Island. (USN 216030)*

During the day's fighting at the northern end of the island, bulldozers were used to bury many Japanese in their underground spider holes. Army light tanks landed at 1100 to support the Marines, and two 55-man detachments from VAC Reconnaissance Company also took part in the fighting. By 1400, Hart's men had secured their portion of the objective, but mopping up was not yet completed.[40] On the adjacent beach, 1/22 faced sterner resistance.

On the right flank, the first wave of Fromhold's battalion landed just south of Valentine Pier, one of the few landmarks visible to the tractor crews. The next wave went ashore 200 yards north of that pier, and the third touched down between the first two. As Marine officers and noncommissioned officers attempted to sort out the intermingled units and lead the men inland, the Japanese cut loose with machine guns and mortars.

Because of the confusion and the devastation caused by the preliminary bombardment, Fromhold was unable to check his exact position. Yet, he had no reason to doubt that the tractors had landed his battalion in the proper place so he proceeded to execute the attack as planned. Hand-to-hand fighting raged along the shoreline, as the Marines wiped out the Japanese who manned the beach defenses. The fighting centered around a sand dune separated from the water by a narrow strip of coral. Interlocking bands of machine gun fire grazed the face of the dune to cut down any Marines who sought its protection. Once the automatic weapons had been located and destroyed, 1/22 was able to advance inland. The battalion commander described his losses as "fairly heavy."[41]

While the bulk of 1/22 was engaged in its fight for the beaches, assault elements of two companies somehow shouldered their way through the melee to thrust across the island. These Marines reached the ocean shore by 1000 and established a defensive perimeter.[42] In the meantime, the remainder of the battalion had cleared the enemy from the immediate vicinity of Green 3. Supported by Marine medium tanks that had just landed, Fromhold's command plunged forward.

Just inland of the dune, General Nishida had prepared another unpleasant surprise for the advancing Marines. He had emplaced three of his light tanks in this area. Although the vehicles were hidden in pits, he had no intention of using them as improvised pillboxes. Ramps enabled the armor to crawl from their protected positions and thunder down on the beachhead. Fortunately for the Americans, the enemy delayed his tank attack until elements of the 2d Separate Tank Company were ashore. "If they had attacked the infantry before tank support arrived," commented one of Fromhold's officers, "the battle for Parry Island would have been very bloody, indeed."[43]

[40] *Reinhardt ltr.*

[41] *Fromhold interview.*

[42] LtCol Charles F. Widdecke ltr to CMC, dtd 10Mar53.

[43] *Scott ltr.* The battalion commander later pointed out that his unit antitank weapons, 37mm cannon and 2.36-inch rocket launchers, had been ineffective during CATCHPOLE, but he did believe that the rocket launcher could

Even though Shermans were at hand to destroy the lighter vehicles with 75mm shells, the fight was far from easy. Deadly fire from enemy cannon emplaced on the right flank raked the battalion, so Fromhold requested the aid of naval gunfire. In doing so, he located the Japanese weapons in relation to where 1/22 should have been, rather than where the unit actually was. The only landmark that the battalion commander had recognized so far was the shattered pier near which he had landed. He had no way of knowing that this was Valentine Pier instead of the shorter structure that lay, also in ruins, farther to the north. Advised by aerial observers of the actual progress of the battalion, higher headquarters rejected his request, but a shore fire control party managed to get through to the supporting warships, and soon American shells began screaming toward the supposed location of the enemy guns.

The first of five salvos struck just as the Marine armor was finishing off the enemy light tanks. Some of the Shermans were hit, one by two shells, yet only one crewman was killed and three wounded. The rounds also fell among the infantry, but these Marines suffered fewer than 10 casualties. The misdirected concentration of naval gunfire took a far heavier toll of Japanese men and equipment. "Our troops were stunned and shocked momentarily," Fromhold has admitted, "but so were the Japs." [44]

be developed into an important implement of war. 1/22 Suggestions for Future Atoll Ops Based on CbtExpc on Engebi and Parry Islands, Eniwetok Island, n.d.

[44] *Fromhold interview.*

The Marines recovered more quickly than their adversaries and by noon had reached the ocean coast. As the battalion was reorganizing after its advance across the island, a group of 150–200 Japanese were seen marching northward in single file along the shoreline. These defenders may have taken refuge on the reef during the preliminary bombardment and were just now entering the fight. Although the enemy troops carried rifles, carbines, and automatic rifles, they had little chance to use them. As the Marine battalion commander phrased it: "The Japs were caught like rats in a trap and exterminated." [45] Once its zone of action had been secured, 1/22 was ready for the drive southward, an offensive that would begin when 3/22 relieved that portion of Fromhold's command which was holding the right flank of the beachhead.

The 3d Battalion had been scheduled to land behind 1/22 on Beach Green 3, but because Fromhold's men had landed out of position, the unit was diverted to the extreme left flank of Green 2. Shisler's men landed at 1000, opposed by Japanese small-arms fire, mortar concentrations, and by mines scattered along the shore. Within an hour, 3/22 had advanced southward, neutralizing en route bypassed Japanese defenses, to reach its assigned position along the right half of the 0–1 Line. Colonel Walker, followed by his regimental command post group, landed during the morning. By 1300, General Watson had committed VAC Reconnaissance Company to assist 2/22 and the scout company to aid

[45] *Ibid.*

1/22 in mopping up the captured portion of the island.

During the afternoon, 1/22 on the left and 3/22 on the right were to thrust toward the southern tip of the island, an attack that would be supported by the 2d Separate Tank Company. After a 15-minute artillery bombardment, the Marines started forward at 1330. The defenders fought as viciously as ever, resisting to the death from spider webs and other concealed positions, but close cooperation among armor, infantry, and artillery enabled the attackers to push steadily forward.

Medium tanks neutralized enemy strongpoints with 75mm weapons and machine guns, while the assault teams darted close enough to employ demolitions or flamethrowers. When the armored vehicles had expended their ammunition, they retired a short distance to replenish their magazines. During the interim, a 60mm mortar barrage was called down close to the forward infantry units, while 75mm howitzers pounded the area just beyond that covered by the mortars. Screened by this fire, half-tracks evacuated the wounded and DUKWs delivered supplies to the rifle companies. When the tanks were ready, the advance resumed.[46]

By nightfall, the two battalions were approximately 450 yards from the southern end of the island. Since operations during darkness on so narrow a front might result in firefights between friendly units, the Marines halted for the night. Although an unknown number of Japanese had survived the day's fighting, Colonel Walker was confident that the battle had been won. At 1930 he announced that Parry was secured.

Marine losses during the night of 22–23 February were few. The enemy sniped continually at the Marines, but illumination provided by the supporting warships deprived the enemy of any moral advantage. Abandoned weapons and ammunition had been carefully collected by the units assigned to mop up the island, so the infiltrators had no ready-made "arsenal" such as they had enjoyed at Engebi. Also, the fire discipline of the regiment had improved once the men became used to night combat.

All that remained for 23 February was the elimination of the defenders trapped at the point of the island. Elements of 3/22 and 1/22 overran the remaining territory by 0900, but some mopping up remained. This task was undertaken by 3/106. On 25 February, the 22d Marines and its attached units sailed from the atoll, leaving behind Colonel Ayers' troops as part of the Eniwetok garrison. Tactical Group 1 was disbanded on 22 March.

CATCHPOLE had proved a complete success. The 22d Marines had been exhausted from its "repeated landing operations," but the "loss in combat efficiency due to fatigue and casualties was compensated for by the outstanding fighting spirit of all hands."[47] Marine casualties during the entire operation were reported as 254 killed and

[46] *Scott ltr.*

[47] TG 1 URpt, 22–23Feb44, Encl G to *TG 1 SAR.*

555 wounded.[48] The 106th Infantry, which lost 94 killed and 311 wounded, proved equal to its first test of combat despite the handicaps under which it operated. Perhaps the best indication of the viciousness of the fighting is that only 66 of the enemy surrendered.

Inadequate rehearsals, General Watson maintained, caused most of the difficulties encountered by the 106th Infantry. At Eniwetok Island, he continued, "the assault troops did not move forward rapidly from the beaches . . . did not operate in close cooperation with tanks, and failed to realize the capabilities of and to use to the fullest extent naval gunfire and close support aviation." [49] Similar sentiments were expressed by Colonel Joseph C. Anderson, USA, of the VAC planning staff. "The comments of General Watson relative to the training of troops for this operation are certainly valid," the Army officer has commented, "as the execution by the 106th Infantry (less 2d Battalion) so clearly demonstrated." [50]

The Eniwetok victory brought American forces to within 1,100 miles of the Marianas. An objective tentatively scheduled for May was secured almost three months ahead of time. In addition, a related operation had showed Truk to be far less formidable than anticipated. Now Admiral Nimitz could devote his energies to preparing for a blow against the Marianas while consolidating his grip on the Marshalls.

[48] Final official Marine casualty totals for the Eniwetok Operation are listed in Appendix H.

[49] *TG 1 SplRpt*, p. 7.

[50] *Anderson ltr.*

FLINTLOCK, JUNIOR [51]

The 22d Marines returned to Kwajalein Atoll on 26 February to begin relieving the 25th Marines as the garrison force. The Eniwetok veterans manned defensive positions on several different islands. In the north, 2/22 went ashore on Roi-Namur, 3/22 on ANTON (Edgigen), regimental headquarters and some support units on ALLEN, and the remaining support units on ANDREW. To the south, 1/22 assumed responsibility for BENNETT. If Colonel Walker's troops anticipated a lengthy period of rest, they were mistaken, for Rear Admiral Alva D. Bernhard, the atoll commander, had received orders to neutralize those of the Marshall Islands which were undefended or lightly held.

Admiral Bernhard was charged with a four-part task. Under his direction, the garrison force was to: destroy Japanese installations or materials which might aid enemy air, surface, or submarine forces; capture Japanese or natives sympathetic to the enemy

[51] Additional sources for the following sections include: Atoll Cdr, Rpt of Occ of Ailuk, Mejit, Jemo, and Likiep, dtd 12Apr44; Atoll Cdr, Rpt of Occ of Bikini, Ailinginae, Rongelap, Rongerik, Utirik, Bikar, and Taka, n.d.; Atoll Cdr, Rpt of Occ of Lib, dtd 28Mar44; Atoll Cdr, Rpt of Occ of Wotho, Ujae, and Lae, dtd 28Mar44; Atoll Cdr, Rpt of Occ of Ailinglapalap and Namu, n.d.; 22d Mar Rpt of Ops into Lesser Marshalls, dtd 6Apr44; 22d Mar, Atoll Hopping: the lesser Marshalls, n.d.; TU 57.10.9 Rpt of Recon of Ailinglapalap, Kili, Ebon, and Namorik, dtd 30Mar44; 2/22 Lesser Marshalls Ops, n.d.; 3/22 Rpt of Ops against Ailinglapalap, Ebon, Namorik, and Kili, dtd 29Mar44; Civil Affairs Rpt of Recon of Ebon, Namorik, and Kili, dtd 6Apr 44; Sherrod, *Marine Air History*.

cause; inform the islanders that American forces were in control of the region; and, establish friendly relations with the natives by assisting them as much as practicable. The admiral and Colonel Walker formed a joint staff to plan and direct the series of landings.

Operation FLINTLOCK, JR., dealt with five areas. Included in the West Group were Wotho, Ujae, and Lae Atolls. The South Group embraced Namu, Ailinglapalap, Namorik, and Ebon Atolls, as well as Kili Island. Bikini, Rongelap, Ailinginae, and Rongerik Atolls formed the North Group, while Bikar, Utirik, Taka, Ailuk, and Likiep Atolls and Jemo and Mejit Islands were assigned to the Northeast Group. Lib Island, due south of Kwajalein Atoll, was designated as a separate area. Before any landing was made, a patrol plane would take photographs of the particular objective. If the defenses did appear weak, Admiral Bernhard could then dispatch a force adequate to the task. A typical expedition might consist of two or more companies from Colonel Walker's 22d Marines, an LST carrying as many as nine LVTs, two LCIs, a destroyer or destroyer escort, and a minesweeper. Marine scout bombers based at Roi had the mission of providing air support for the landings. To each of the forces that he dispatched, the admiral assigned civil affairs and medical specialists as well as interpreters and native guides. (See Map 7.)

FLINTLOCK, JR. began on 8 March, when two reinforced rifle companies from 1/22 arrived off Wotho Atoll. Major Crawford B. Lawton, in command of the force, learned from natives that only 12 Japanese, survivors of a plane crash, were present. The Marines landed unopposed on the following morning, suffered one casualty from the accidental explosion of a grenade, and cornered the enemy fliers, who committed suicide rather than surrender. Five of the six Japanese weather observers at Ujae Atoll, where the Marines landed on the 10th, killed themselves, but one man elected to become a prisoner of war.

During the securing of Wotho and Ujae, the natives had been friendly, but such was not the case at Lae Atoll. For some time the inhabitants of this third objective remained aloof, though they were not hostile. A short time before the Americans arrived, a box containing a hand grenade had drifted ashore. When the grenade exploded, a child was injured, and the natives for a time blamed the Marines, who had the misfortune of landing while memories of the tragedy were fresh.[52]

On 14 March, the conquest of the West Group by now completed, Lawton's men returned to Kwajalein. While they were absent, Colonel Walker had been reassigned to VAC headquarters. His successor as regimental commander was his executive officer, Colonel Merlin F. Schneider. During this period, on 11 March, a reinforced platoon from the 1st Battalion had raised the American flag on Lib Island, which was not occupied by the enemy.

Operations against the South Group,

[52] This story of the Lae occupation was based on comments by LtCol Crawford B. Lawton, n.d., in Heinl and Crown, *The Marshalls*, p. 154. No record of his remarks has been discovered.

delayed when one of the landing craft began shipping water, finally got underway on 19 March. On that day, two separate forces, some 650 Marines in all, set sail from Kwajalein for Ailinglapalap Atoll. On the morning of 21 March, Major William E. Sperling, III landed his portion of 3/22, to be followed ashore by the group commanded by Major Shisler. The defenders of the atoll manned a defensive line that crossed Ailinglapalap Island at its narrowest point. Marine infantrymen attacked behind an 81mm mortar barrage to crush the position. Two Marines were wounded on Ailinglapalap Island and 39 Japanese killed. Two of the defenders escaped the American onslaught, but both were captured elsewhere in the atoll.

Once this objective was secured, the two forces parted company as planned. Shisler's men landed at Ebon Atoll on the morning of 23 March and, on the following day, killed 17 Japanese in a vicious fight that cost the lives of two Marines. Six enemy noncombatants were taken into custody. From Ebon, Shisler's command proceeded to Namorik Atoll and Kili Island, neither of which had been garrisoned. Meanwhile, Sperling's Marines investigated Namu Atoll, where seven Japanese civilians willingly surrendered.

On 30 March, two days after Shisler returned to Kwajalein, Major Earl J. Cook led a reinforced rifle company toward the Northeast Group. Once again, the troops were provided by 3/22. The number of objectives was reduced to three—Mejit, Ailuk, and Likiep—for aerial photographs and reports by friendly natives indicated that these were the only inhabited places in the area. No Japanese were found at either Ailuk or Likiep Atoll, but every member of the six-man weather station on Mejit died resisting the landing.

While 3/22 was carrying out its assignments, the 2d Battalion dispatched a reinforced rifle company to secure the North Group and as much of the Northeast Group as remained under enemy control. Aerial reconnaissance indicated that Ailinginae, Rongerik, and Bikar Atolls were not inhabited, so Major Robert P. Felker, the force commander, was free to concentrate upon Bikini, Rongelap, and Utirik Atolls.

Felker's Marines landed on 28 March at Bikini, but no fighting occurred, for the five Japanese on the atoll killed themselves. The 6-man detachment reported at Rongelap apparently had been withdrawn; at any rate, the Americans found no trace of it. On 5 April, the company landed at Utirik, killed 14 Japanese, and then re-embarked for Kwajalein.

During FLINTLOCK, JR. elements of the 22d Marines had made 29 separate landings in order to secure 12 atolls and 3 islands. This campaign, which lasted from 7 March to 5 April, brought an estimated 60,000 square miles of ocean under American control. Once the mission had been accomplished, the 22d Marines embarked for Guadalcanal to prepare for further operations.

FINAL OPERATIONS

The final landings of the Marshalls campaign took place shortly after the departure of the 22d Marines. On 17 April, a detachment from the 1st Marine Defense Battalion set sail for

Erikub Atoll, some 5 miles from Wotje, and Aur Atoll, located 10 miles from Maloelap. No Japanese were found at either objective and the nearby enemy garrisons did not attempt to interfere. In February, Marines from this battalion had left their camp at Majuro to occupy Arno Atoll east of Majuro and north of the Japanese stronghold of Mille. Soldiers from the 111th Infantry landed at Ujelang Atoll on 21 April to begin a 2-day reconnaissance that resulted in the killing of 18 Japanese. No further landings would be made until hostilities had ended.

By the end of April, the enemy retained control over only Wotje, Mille, Maloelap, and Jaluit Atolls in The Marshalls group. The task of keeping these bases neutralized fell to the 4th Marine Base Defense Aircraft Wing, later redesignated the 4th Marine Aircraft Wing. In February, Marine fighters arrived at Roi and Engebi, but the systematic battering of the bypassed atolls did not begin until 4 March, when scout bombers from Majuro braved dense antiaircraft fire to attack Jaluit.

By the time of this first Marine raid, Navy and Army aviators had destroyed the enemy aircraft assigned to defend the Marshalls. The only aerial opposition encountered by these Marine pilots occurred during a strike launched on 28 March against Ponape in the Carolines. Six F4U (Corsair) fighters, escorting four Army bombers, shot down eight Japanese fighters and destroyed another on the ground. Although Ponape was visited several times during the months that followed, no Japanese planes attempted to intercept the raiders.

From 4 March 1944 until the end of hostilities in August of the following year, the Marines continued to bomb and strafe Mille, Maloelap, Wotje, and Jaluit. They unleashed 6,920 tons of bombs and rockets, approximately half the total tonnage employed against the four atolls during the entire war. These missiles, along with 2,340 tons of naval shells, killed 2,564 Japanese out of garrisons that totalled over 13,000. In carrying out their part of the Marshalls mop-up, Marine airmen learned lessons in fighter-bomber techniques applicable elsewhere in the Pacific.[53]

The FLINTLOCK and CATCHPOLE operations resulted in the rapid capture of bases for further Pacific operations. During FLINTLOCK, JR. and the landings that followed, American control over the Marshalls was confirmed. Then, while the assault troops advanced into the Marianas, Marine aviators assumed the mission of maintaining the neutralization of the bypassed strongholds in the group. So well did the flyers succeed that those Japanese who survived the rain of bombs and rockets either starved to death or became so weak from hunger that they were no longer even a remote threat to American forces.

[53] A more complete treatment of the activities of Marine aviation in the Marshalls will be contained in the fourth volume of this series.

CHAPTER 6

The Fruits of Victory

The conquest of the Marshalls was a far more significant victory than the previous success in the Gilberts. FLINTLOCK and CATCHPOLE represented a shattering of the line of outposts that protected the inner defenses of the enemy homeland. As far as the Japanese were concerned, the Marshalls themselves were not indispensable, but the speed with which the American forces moved robbed the enemy of the time he needed to prepare for the defense of the more vital islands that lay nearer to Japan.

THE ENEMY SITUATION [1]

Among the victims of FLINTLOCK was the Japanese naval base at Truk. On 10 February, immediately after the loss of key islands in Kwajalein Atoll, the enemy decided to withdraw his major fleet units to prevent their destruction by American air power. The carrier raid that preceded CATCHPOLE made Truk a rattlesnake without fangs. Nimitz concluded that no amphibious assault would be necessary and abandoned a plan that called for the employment of five divisions and one additional regiment. Once Truk had joined Rabaul and Kavieng in the backwash of World War II, the 1st, 3d, and 4th Marine Divisions, the 4th Marines, and the 7th and 77th Infantry Divisions were freed for service elsewhere in the Pacific. American planners could now look forward to the Marianas.

The loss of the Marshalls and the resultant neutralization of Truk caused the enemy to revise his strategy. Early in March, Admiral Mineichi Koga, commander in chief of the Japanese *Combined Fleet*, established still another zone in which interception operations could be carried out against the American fleet. Patrol planes, submarines, and picket boats were charged with detecting any attempt by Nimitz to penetrate the Central Pacific Front, an area stretching from the Kuriles past Honshu, through the Bonins, Marianas, and Carolines, and terminating in New Guinea. Should the United States fleet venture into the area, land-based planes would blast the carriers to enable surface ships to close with and sink the troop transports. The Japanese admiral urged his men to destroy as many of the invaders as possible while the expedition was at sea. The survivors, in keeping with current tactical doctrine, were to be annihilated at the beaches. This was the gist of Koga's proposed *Z Operation*.

Admiral Koga was killed in a plane crash before his plan could be executed,

[1] Unless otherwise noted, the material in this section is derived from: Hattori, *Complete History*, v. 3, p. 5; Isely and Crowl, *Marines and Amphibious War;* USSBS, *Campaigns of the Pacific War*.

but his successor in command of the *Combined Fleet*, Admiral Soemu Toyoda, had ample time to modify this basic strategy before the Americans struck. This revised plan, dubbed *A-GO*, also called for strengthening the island defenses along the Central Pacific Front. Toyoda, however, established two "decisive battle areas," the Palau Islands and the western Carolines. If amphibious forces should attempt to seize outposts in either the western Carolines, the neighboring Marianas, or the Palaus, the defense forces already posted in the threatened sector were to hold fast. The bulk of the Imperial fleet, now based at Tawi Tawi off Borneo, would stream northeastward to crush the Americans.

CATCHPOLE, following so closely after FLINTLOCK, made it inevitable that Japan would have extreme difficulty in completing the preparations necessary for *A-GO*. The decision to bypass Truk left Nimitz with enough well-trained troops, many of them combat veterans, to deliver a sudden blow at the Marianas. The enemy wanted to reinforce these islands before they were attacked, but in order to move the necessary men and their supplies, the Japanese had to place a heavy strain on an already weakened merchant marine. By the beginning of March 1944, the available enemy merchant shipping, almost 6½ million tons at the outbreak of the war, had been reduced to about 4 million. American submarines had wrought most of this destruction. Now, thanks to the Marshalls operations, these undersea raiders would be able to operate from a base 1,200 miles closer to the Marianas.

AMERICAN GAINS

This second part of the Central Pacific campaign had come to a close with the capture of bases some 800 miles within enemy territory. The Japanese had been driven back to their inner defenses. American amphibious forces demonstrated beyond question that they had absorbed the lessons offered by the assault upon Betio.

Besides securing bases from which to mount further operations, the Pacific Fleet, by seizing certain key objectives, had succeeded in neutralizing the more powerful Japanese bastions in and near the Marshalls. The "unsinkable aircraft carriers," in which the enemy had placed so much confidence, remained ready to receive planes, but none could be sent them. American carrier task forces had driven the Japanese from the skies over the Marshalls, and land-based planes from the recently captured atolls stood ready to maintain this mastery of the air. The careful selection of targets coupled with a skillful use of the available air, ground, and sea forces meant a saving in lives as well as time.

In addition to these strategic gains, the Americans amassed additional experience in atoll warfare. Although the Marshalls operations represented some improvements in tactics and techniques over previous efforts, planners as well as troop leaders were well aware of errors that thus far escaped correction. Only by coldly assessing the campaigns just completed, noting weaknesses, and making the necessary corrections could Nimitz' amphibious forces lay the groundwork for future victories.

LESSONS LEARNED[2]

Like the Gilberts expeditionary force, the organizations formed for FLINTLOCK and CATCHPOLE did not include replacement pools. Planners had decided that casualties in the brief but intense Marshalls actions would not be severe enough to impair the effectiveness of the landing forces. At Kwajalein Atoll, no serious difficulties were encountered by the assault divisions, but the Roi garrison had to call for an emergency draft of 27 men to replace casualties suffered during the 12 February Japanese air raid. At Eniwetok, however, a series of relatively brief fights gradually wore down Tactical Group 1 to such an extent that General Watson ordered rifle units formed from among elements of his garrison forces.

The staff of Tactical Group 1 had been hard pressed to keep an accurate tally of casualties. What was needed, General Watson decided, was a single center for compiling such data, a clearing house that would be located in the flagship of the attack force commander. VAC pointed out that directives then in force called for just such a system of accounting. The fault lay in the various commanders, who seldom reported accurately or on time.

American intelligence concerning Kwajalein Atoll was accurate, but General Smith's staff desired more extensive coverage by oblique aerial photographs and wanted the pictures delivered at least 90 days prior to D-Day. General Watson, whose Parry Island assault troops had been confused by a redesignation of the beaches, noted that the arbitrary designation of "color" beaches by higher headquarters was not always advisable. He believed that the attack force commander should have additional leeway in designating the area to be stormed. Looking back on the Eniwetok operation, Watson also called for the marking of known defenses on maps as small in scale as 1:20,000, a more careful delineation of the coastline and of all landmarks upon which the tractor waves might guide, and photographic coverage of every island within the atoll to be attacked. JICPOA had done an accurate job in placing the bulk of the *1st Amphibious Brigade* at Eniwetok Atoll, but photographs taken prior to D-Day did not indicate the type of defenses that the enemy had prepared. By the time of the main landings, General Watson was aware of the enemy's strength and probable dispositions. The extent of the Japanese underground defenses, however, was not known until the Americans actually encountered them.

Neither of the Marshalls operations represented any departure from the established command structure for amphibious operations. Although the position of the corps commander was clarified for the Kwajalein landings, the nature of the undertaking prevented General Smith from exercising close tactical supervision. Essentially, FLINTLOCK consisted of two distinct series of

---

[2] Unless otherwise noted, the material in this section is derived from: *TF 51 AR; TF 53 AR Roi-Namur; VAC AR FLINTLOCK* (including rpts of staff sections, Encls C–H); *VAC Rpt of LogAspects;* VAC Cmts on TG 1 SAR, dtd 1Apr44; *TG 51.11 OpRpt; 4th Mar Div AR; TG 1 SAR;* ComInCh, *Marshall Islands;* Sherrod, *Marine Air History.*

RIFLEMEN *of the 22d Marines advance toward the last Japanese-held area on Parry. (USMC 74488)*

MARIANAS INVASION FORCE *assembled in the lagoon at Eniwetok Atoll on 9 June 1944. (USN 80-G-248207)*

landings by widely separated divisions. Admiral Turner, however, later noted that "all Central Pacific amphibious operations in which I was concerned needed at least a corps command of expeditionary troops." [3] The principle, then, was already accepted. As soon as VAC attacked a suitable land mass, its commanding general would direct more closely the actions of its assigned divisions.

As far as planning was concerned, the staffs of both VAC and Tactical Group 1 voiced the same complaint —not enough time. Indeed, every agency involved in the planning of FLINTLOCK and CATCHPOLE was working against a rapidly approaching deadline. For this reason, final versions of certain annexes of the basic plans were late in reaching the assault units. The most conspicuous victims of this situation were the LVT and LVT(A) battalions, which had not received their orders for the Roi-Namur landings in time for rehearsals. In particular, this fact hampered their communications, since radio frequencies had to be set en route to the line of departure.[4]

If nothing else, the Marshalls fighting proved the value of sound training climaxed by realistic rehearsals. The shortcomings of Marine amphibian tractor crews at Roi-Namur and of Army infantrymen on Eniwetok Island were blamed on a lack of indoctrination and practice. VAC headquarters could account for the poor quality of the amphibious rehearsals staged for Tactical Group 1. Watson's command completed its training while the FLINTLOCK expedition was being mounted, so the assault units rather than the reserve had first call for the limited number of DUKWs and LVTs then available. The 10th Amphibian Tractor Battalion had completed its final exercise before the plan of attack was ready for distribution. As General Smith went on to point out, there would be times when speed was so essential that rehearsals were certain to be inadequate.

In the case of the Marshalls operations, speed denied the enemy time to convert the objectives into fortresses as powerful as Betio Island had been. Firepower helped the American landing forces to succeed in spite of the minor defects in their training and employment. The prolonged naval bombardment of targets in Kwajalein Atoll was supplemented by the effective fires of artillery units emplaced on islands off the principal objectives. The quality of naval gunfire, and of air support as well, had improved since GALVANIC. General Schmidt, for example, estimated that between 50 and 75 percent of the Roi-Namur garrison was killed by either naval shelling or aerial attack. At Eniwetok Atoll, neither planes nor warships were as deadly, for the planned bombardment was based on an incomplete knowledge of the nature of the Japanese defenses. The troops fighting ashore at Eniwetok, however, benefited from the first heavy use of night illumination shells by supporting naval vessels.

The landings, screened by the fires of

---

[3] *Turner ltr I.*

[4] The radio frequencies received in the plan could not be set up within the LSTs, consequently units had to struggle to establish them once they were launched for the landings. *Metzger ltr.*

LVT(A)s, LCI gunboats, and warships, were hampered by poor communications. In Kwajalein Atoll, where the sea was rough, many of the radios carried in LVTs were drowned out by spray, thus insuring a confused advance toward the beaches. Once again, Marine officers renewed their appeal for communications equipment that was adequately waterproofed.

Ashore the Marine troops fought well. What mistakes they did make were those expected of men entering combat for the first time. The unauthorized sprint across Roi, for example, upset the prearranged scheme of maneuver, although it undoubtedly kept the enemy off balance. This advance was traceable to the Marines' desire to excel in their first battle. Although the 22d Marines had trained ceaselessly during its stay in Samoa, this unit, too, needed the experience of actual warfare. During CATCHPOLE, unnecessary firing decreased in volume as the troops became used to fighting at night. In addition, the men of the regiment learned to avoid leaving weapons scattered about the battlefield where Japanese infiltrators could find them. In attacking enemy emplacements, whether concrete bunkers or underground spider webs, flamethrowers, demolitions, and hand grenades proved most deadly. The division of rifle squads into fire teams, as practiced by the 22d Marines, was a successful innovation, for these elements were especially effective in dealing with enemy positions that were located in wooded or overgrown areas.

Most aspects of the logistical plans for FLINTLOCK and CATCHPOLE represented improvements over GALVANIC. The DUKW justified the confidence that General Corlett had placed in it, and the "hot cargo" system, as practiced by the 7th Infantry Division, proved a reliable method of getting priority cargo ashore during the early hours of an amphibious operation. Generals Schmidt and Watson also had critical items of supply preloaded in amphibious vehicles, in their case LVTs, ready to be landed at the request of the units ashore. Corps observers were convinced that the amphibian truck was better suited for carrying supplies than the tractor, for the DUKW had a larger cargo compartment and was easier to repair.

During the Marshalls fighting, the LST performed several important duties. Except for those units which seized the islands adjacent to Roi-Namur, all the Marine assault forces boarded their assigned tractors before the LVTs were launched by their parent landing ships. Thus, the troops were spared the ordeal of transferring in the open sea. Besides carrying LVTs and providing enclosed transfer areas, this same type of ship participated in the logistical plan. Certain LSTs carried food, water, and ammunition, others served as hospital wards, and still others carried tools and spare parts with which to repair damaged tractors.

The amount of supplies carried to Kwajalein Atoll proved, in some instances, more than sufficient, but the troops at Eniwetok Atoll endured shortages in concussion grenades and demolitions fuzes. Fortunately, the men had enough ammunition. One item that was habitually discarded as soon as the troops landed was the gas mask, which General Watson considered a "distinct nuisance." [5]

The limited area available prevented the proper dispersal of supply dumps, but otherwise the movement of cargo to the troops inland was well executed. Pallets permitted the rapid landing of bulk cargo, and a permanent beach party organization assumed responsibility for controlling boat traffic and the evacuation of the wounded. After observing these beach parties in operation, Admiral Turner's headquarters recommended that a permanent shore party similar to that used in the Southwest Pacific be organized. A well-trained nucleus could be reinforced as necessary by labor contingents and garrison units, so that the handling of supplies no longer would depend on men borrowed from the assault battalions.

The role of Marine aviation in the Marshalls was little changed from the previous operation, for General Smith's recommendation that Marine pilots based on carriers support future landings had not been accepted. The performance of Navy airmen, however, was improved, thanks to better planning and careful briefing. During FLINTLOCK, aircraft had attacked in conjunction with the preliminary naval bombardment. Since the experiment had proved successful, VAC recommended that similar aerial attacks be carried out in forthcoming landings. General Watson's command had not benefited from this kind of coordination. Rather than suspend naval gunfire to enable the planes to make a final strafing run, he urged that this last-minute strike be omitted in future landings.

Immediately after the capture of bases in Kwajalein and Eniwetok Atolls, Marine fighter planes arrived to help defend these conquests. Between 15 and 23 February, elements of two Marine fighter squadrons (VMFs–224 and –532) began flying combat air patrols from Kwajalein. VMF(N)–532, using radar-equipped F4Us, was responsible for patrolling the night skies. Although ground crews landed on Engebi while that island was being mopped up, Marine fighter craft did not make their appearance there until 27 February. VMF–113 operated during daylight, and a detachment from VMF(N)–532 took over after dark. On 14 April, the Engebi-based night fighters made their first kills of the war, destroying two Japanese planes and probably shooting down a third.

During the critical hours after the landings, the antiaircraft units from defense battalions were employed to protect the beachheads. Scout bombers also assisted indirectly in the aerial defense of the Marshalls bases by helping neutralize the bypassed atolls. In short, Marine aviators played a slightly larger role in FLINTLOCK and CATCHPOLE than they had in GALVANIC, but support of the landings remained the responsibility of the Navy. In addition, naval aviators operating from carriers prevented the Japanese from launching aerial attacks against the expanding beachhead, a task which they shared with Marine and Army antiaircraft units.

In summing up the FLINTLOCK operation, General Smith noted that the lessons learned in the Gilberts had

[5] *TG 1 SplRpt*, p. 15.

proved invaluable. "In the attack of coral atolls," read his report, "very few recommendations can be made to improve upon the basic techniques previously recommended and utilized in FLINTLOCK. However, there is still much to be desired to improve planning, improve coordination of efforts, and prepare for the attack of more difficult objectives." [6] As the Central Pacific drive moved westward, the enemy's island defenses seemed certain to improve.

---

[6] *VAC AR FLINTLOCK*, p. 11.

PART IV

Saipan: The Decisive Battle

Background to FORAGER

STRATEGIC AND TACTICAL PLANS

While the Japanese bolstered their defenses along the Central Pacific Front, American strategists were concluding their lengthy debate concerning the future course of the Pacific war. At the Casablanca Conference in January 1943, the CCS had accepted in principle a Central Pacific offensive aimed toward the general area of the Philippines but proceeding by way of the Marshalls, Carolines, and Marianas. In spite of objections by General MacArthur, this proposed offensive was finally incorporated in the Strategic Plan for the Defeat of Japan, with the seizure of the Marshalls and Carolines listed among the Allied goals for 1943–1944. Overall strategy against Japan called for two coordinated drives, one westward across the Central Pacific and the other, by MacArthur's forces, northward from New Guinea.

THE IMPORTANCE OF THE MARIANAS [2]

The staunchest advocate of operations against the Marianas was

[1] Unless otherwise noted, the material in this chapter is derived from: FifthFlt OPlan Cen 10–44, dtd 12May44 (with changes); TF 51 Rpt of PhibOps for the Capture of the Marianas Islands, dtd 25Aug44, hereafter *TF 51 OpRpt;* TF 56 OPlan 3–44, dtd 26Apr44 (with changes); TF 56 Rpt of FORAGER Op (with encls covering Planning, Ops, Intel, Log, Pers, and StfRpts), dtd 25Oct44, hereafter *TF 56 OpRpt;* TF 52 AtkO A11–44, dtd 21May 44 (with changes); TG 52.2 Rpt of Saipan Op, dtd 23Aug44, hereafter *TG 52.2 OpRpt;* NTLF OPlan 3–44 (with changes), dtd 1May 44; NTLF Rpt of Marianas Op, Phase I (Saipan) (with encls containing Op and AdminOs, Daily DispSums, Stf and SpecRpts), dtd 12Aug44, hereafter *NTLF OpRpt;* 27th InfDiv Rpt of Ops, Saipan (with Narrative, Rpts of StfSecs and of SuborUs), dtd 24Oct44, hereafter *27th InfDiv OpRpt;* CominCh, *Amphibious Operations: Invasion of the Marianas, June to August 1944* dtd 30Dec44, hereafter CominCh, *The Marianas;* Craven and Cate, *Guadalcanal to Saipan;* Philip A. Crowl, *Campaign in the Marianas—The War in the Pacific—U. S. Army in World War II* (Washington: OCMH, DA, 1960), hereafter Crowl, *Marianas Campaign;* Maj Carl W. Hoffman, *Saipan: The Beginning of the End* (Washington: HistDiv, HQMC, 1950), hereafter Hoffman, *Saipan;* Samuel Eliot Morison, *New Guinea and the Marianas, March 1944–August 1944—History of U. S. Naval Operations in World War II*, v. VIII (Boston: Little, Brown, and Co., 1953), hereafter Morison, *New Guinea and the Marianas.* Unless otherwise noted, all documents cited are located in the Marianas Area OpFile and Marianas CmtFile, HistBr, HQMC.

[2] Additional sources for this section include: CCS 397 (Rev), SpecificOps for the Defeat of Japan, dtd 3Dec43, CCS 417, 417/1, 417/2, Overall Plan for the Defeat of Japan, dtd Dec43, JCS 581, 581/1, 581/2, SpecificOps for the Defeat of Japan, dtd Nov–Dec43; JPS 264, Outline Plan for the Seizure of the Marianas, Incl Guam, dtd 6Sep43 (OPD–ABC Files,

214-881 O-67—16

Admiral King, who believed that the capture of these islands would sever the enemy's lines of supply to Truk and Rabaul and provide bases for operations against targets farther west. During the Quebec meeting of Anglo-American planners, a conference that lasted from 14 to 24 August 1943, the admiral again stressed the importance of the Marianas. British representatives asked King if it might not be wise to restrict operations in MacArthur's theater so that the Allies might divert to Europe some of the men and material destined for the Southwest Pacific. The admiral answered that "if forces were so released they should be concentrated on the island thrust through the Central Pacific."[3] He added, however, that he considered the two offensives against the Japanese to be complementary. General Marshall then pointed out that the troops scheduled for the New Guinea operations were either en route to or already stationed in the Southwest Pacific.

At Quebec the CCS approved the forthcoming operations against the Gilberts and Marshalls but merely listed the Marianas as a possible objective to be attacked, if necessary, when American forces had advanced to within striking distance. The American Joint Planning Staff, acting upon this tentative commitment, began preparing an outline plan for the conquest of the Marianas. When Admiral Nimitz turned his attention to the Central Pacific drive approved at Quebec, he noted that the Marianas might serve as an alternate objective to the Palaus. In brief, amphibious forces might thrust to the Philippines by way of the Carolines and Palaus or strike directly toward the heart of the Japanese empire after seizing bases in the Marianas and Bonins. The agreements reached at Quebec also affected General MacArthur's plans, for the Allies gave final acceptance to the JCS recommendation that Rabaul should be bypassed. This decision, although it changed the general's plans, actually enabled him to speed his own advance toward the Philippines. (See Map I, Map Section.)

As the next meeting of the Anglo-American Chiefs of Staff, scheduled for November 1943, drew nearer, the JCS began preparing its proposals for the future conduct of the Pacific war. Among the items under discussion was the employment of a new long-range bomber, the B–29, against Japanese industry. This plane, according to General Henry H. Arnold, Commanding General, Army Air Forces, "would have an immediate and marked effect upon the Japanese and if delivered in sufficient quantities, would undoubtedly go far to shorten the war."[4]

At this time, Arnold was planning to strike from bases on the Chinese main-

WWII RecsDiv, FRC, Alexandria, Va.); CinCPOA Campaign GRANITE, prelim draft, dtd 27Dec43; CinCPOA Outline Campaign Plan GRANITE, dtd 13Jan44; CinCPOA Outline Campaign Plan GRANITE II, dtd 3Jun44; CinCPOA JntStfStudy FORAGER, dtd 20Mar 44; CinCPac-CinCPOA memo to ComInCh, dtd 30Sep43, subj: GarRequirements for CenPacArea, with encls A–C (OPlan File, OAB, NHD).

[3] King and Whitehill, *Fleet Admiral King*, p. 485.

[4] JCS, Minutes of the 122d Meeting, 9Nov43, p. 2 (OPD–ABC Files, WWII RecsDiv, FRC, Alexandria, Va.).

land, an undertaking which required new flying fields, a vast amount of fuel and supplies, numerous American flight crews, mechanics, and technicians, and a strengthening of the Chinese Nationalist armies defending the bases. The large airfield nearest Japan was at Chengtu, 1,600 miles from any worthwhile target. If necessary, the B–29's, loaded with extra gasoline instead of high explosives, could take off from India, fly to advanced airfields in China where the emergency fuel tanks could be replaced with bombs, then continue to the Japanese home islands. Unfortunately, the Chinese might prove incapable of holding these way-stations on the aerial road to Japan. What was needed were bases secure from enemy pressure but within range of the Home Islands. The solution lay in the Marianas, some 1,200 miles from the Japanese homeland, but this group was in the hands of the enemy. Army Air Force planners urged that the Marianas be captured and developed as B–29 bases, but they also desired to begin the strategic bombing of Japan as quickly as possible, using the India-China route.[5]

General Arnold was confident that masses of B–29s could destroy Japan's "steel, airplane, and other factories, oil reserves, and refineries," which were concentrated in and around "extremely inflammable cities."[6] His colleagues, already looking ahead to the invasion of Japan, apparently shared his conviction, for they accepted as a basis for planning the assumption, set forth by Vice Admiral Russell Willson, that: "If we can isolate Japan by a sea and air blockade, whittle down her fleet, and wipe out her vulnerable cities by air bombardment, I feel that there may be no need for invading Japan—except possibly by an occupying force against little or no opposition—to take advantage of her disintegration."[7]

The importance attached to strategic bombardment and naval blockade caused the Marianas to assume an increasing significance in American plans, since submarines as well as aircraft might operate from the island group. Evidence of the value of the Marianas was the recommendation by the Strategy Section to the Strategy and Policy Group of the Army Operations Division that the island bases, once they were ready for operations, should have priority over the mainland fields in the allotment of aircraft. "It is self-evident," Army strategists remarked, "that these aircraft should operate from bases within striking range of Japan proper, if that is possible, rather than from a more distant base such as Chengtu."[8] Throughout SEXTANT, as the latest international meeting was called, the United States emphasized the need for air bases in the western Pacific.

The SEXTANT conference, 22

[5] Gen of the AF Henry H. Arnold, USAF, *Global Mission* (New York: Harper and Brothers, 1949), pp. 477–480.

[6] JCS, Minutes of the 123d Meeting, 15Nov 43, p. 9 (OPD–ABC Files, WWII RecsDiv, FRC, Alexandria, Va.).

[7] VAdm Russell Willson memo to Adm Ernest J. King, dtd 11Nov43, subj: Plan for Defeat of Japan (OPD–ABC Files, WWII RecsDiv, FRC, Alexandria, Va.).

[8] Col J. J. Billo, USA, memo to BGen George A. Lincoln, USA, dtd 7Dec43, subj: Specific Ops for the Defeat of Japan (CCS 397) (OPD–AGC Files, WWII RecsDiv, FRC, Alexandria, Va.).

November–7 December 1943, actually was a series of discussions among Allied leaders. After conversations with Generalissimo Chiang Kai-shek at Cairo, President Roosevelt, Prime Minister Churchill, and their advisors journeyed to Teheran, Iran, where they met a Soviet staff led by Marshal Joseph Stalin. The Anglo-American contingent then returned to Cairo so that the combined staffs might revise their world-wide strategy to include commitments made to either Nationalist China or the Soviet Union.

Out of SEXTANT came a schedule, drafted for planning purposes, which called for the invasion of the Marianas on 1 October 1944 and the subsequent bombing by planes based in the islands of targets in and near the Japanese home islands. The date of the Marianas operation, however, might be advanced if the Japanese fleet were destroyed, if the enemy began abandoning his island outposts, if Germany suddenly collapsed, or if Russia entered the Pacific war. The strategy behind this timetable called for two series of mutually supporting operations, one by MacArthur's troops, and the other by Nimitz' Central Pacific forces. Since the advance across the Central Pacific promised the more rapid capture of airfields from which to attack Japan and could result in a crushing defeat for the Japanese navy, Nimitz would have priority in men and equipment. The timing of MacArthur's blows would depend upon progress in the Central Pacific. Planners believed that by the spring of 1945 both prongs of the American offensive would have penetrated deeply enough into the enemy's defenses to permit an attack in the Luzon-Formosa-China area.

On 27 December, area planning began as Nimitz issued his GRANITE campaign plan, a tentative schedule of Central Pacific operations which also helped to establish target dates for landings in the Southwest Pacific that would require support by the Pacific Fleet. First would come FLINTLOCK, scheduled for 31 January 1944, then the assault on Kavieng, 20 March, which would coincide with an aerial attack on Truk. On 20 April, MacArthur's troops, supported by Nimitz' warships, would swarm ashore at Manus Island. The fighting would then shift to the Central Pacific for the Eniwetok assault, then set for 1 May, the landing at Mortlock (Nomoi) 1 July, and the conquest of Truk to begin on 15 August. The tentative target date for the Marianas operation, which included the capture of Saipan, Tinian, and Guam, was 15 November 1944.

As if to prove that his GRANITE plan was more flexible than the mineral for which it was named, the admiral on 13 January advanced the capture of Mortlock and Truk in the Carolines, to 1 August. If these two landings should prove unnecessary, the Palau Islands to the west could serve as an alternate objective. From the Palaus, the offensive would veer northeastward to the Marianas, where the assault troops were to land on 1 November. Late in January 1944, Nimitz summoned representatives from the South Pacific and invited others from the Southwest Pacific to confer with his own staff officers on means of further speeding the war against Japan.

Nimitz, informed of the recent decisions concerning B–29 bases, offered the conference a choice between storming Truk on 15 June, attacking the Marianas in September, and then seizing the Palaus in November or bypassing Truk, striking at the Marianas on 15 June, and then landing in the Palaus during October. Some of those present, however, were interested in neither alternative. The leader of these dissenters was General George C. Kenney, commander of Allied air forces in General MacArthur's theater, who managed to convince various Army and Navy officers that the Central Pacific campaign be halted in favor of a drive northward from New Guinea to the Philippines. As Kenney recalled these sessions, he remarked that "we had a regular love feast. [Rear Admiral Charles H.] McMorris, Nimitz' Chief of Staff, argued for the importance of capturing the Carolines and the Marshalls [FLINTLOCK was about to begin], but everyone else was for pooling everything along the New Guinea-Philippines axis."[9] Although fewer than Kenney's estimated majority were willing to back a single offensive under MacArthur's leadership, a sizeable number of delegates wanted to by-pass the Marianas along with Truk. Nimitz, however, brought the assembled officers back to earth by pointing out that the fate of the Marianas was not under discussion. When reminded that the choice lay between neutralizing or seizing Truk before the advance into the Marianas, they chose to bypass the Carolines fortress.

General MacArthur also saw no strategic value in an American conquest of the Marianas. He dispatched an envoy to Washington to urge that the major effort against Japan be directed by way of New Guinea and the Philippines. Like those who dissented during Nimitz' recent conference, the general's representative accomplished nothing, for the JCS had reached its decision.

On 12 March, the JCS issued a directive that embodied the decisions made during the recent Allied conferences. General MacArthur's proposed assault on Kavieng was cancelled, and the New Ireland fortress joined Rabaul on the growing list of bypassed strongholds. Southwest Pacific forces were to seize Hollandia, New Guinea, in April and then undertake those additional landings along the northern coast of the island which were judged necessary for future operations against the Palaus or Mindanao. This revision in the tasks to be undertaken in the South and Southwest Pacific enabled the Army general to return to Nimitz the fleet units borrowed for the Kavieng undertaking.

In the Central Pacific, where amphibious forces had seized Kwajalein and Eniwetok Atolls and carrier task groups had raided Truk, Nimitz was to concentrate upon targets in the Carolines, Palaus, and Marianas. His troops were scheduled to attack the Marianas on 15 June, while aircraft continued to pound the bypassed defenders of Truk. In addition, the admiral had the responsibility of protecting General MacArthur's flank during the attack upon Hollandia and sub-

[9] Gen George C. Kenney, USAF, *General Kenney Reports: A Personal History of the Pacific War* (New York: Duell, Sloan, and Pearce, 1949), pp. 347–348.

sequent landings. Throughout these operations, the two area commanders would coordinate their efforts to provide mutual support.

Although the Marianas lacked protected anchorages, a fact which Nimitz had pointed out to the JCS, these islands were selected as the next objective in the Central Pacific campaign. The major factor that influenced American planners was the need for bases from which B–29s could bomb the Japanese homeland. Instead of seizing advance bases for the fleet, the mission which the Marine Corps had claimed at the turn of the century, Leathernecks would be employed to capture airfield sites for the Army Air Forces.

After receiving the JCS directive, Nimitz ordered his subordinates to concentrate upon plans for the Marianas enterprise and to abandon the staff work that had been started in preparation for an assault on Truk. On 20 March, the admiral issued a joint staff study for FORAGER, the invasion of the Marianas. The purpose of this operation was to capture bases from which to sever Japanese lines of communication, support the neutralization of Truk, begin the strategic bombing against the Palaus, Philippines, Formosa, and China. Target date for FORAGER was 15 June.

The decision to bypass Truk and Kavieng enabled Admiral Nimitz to alter the established schedule for the Central Pacific offensive. The revised campaign plan, GRANITE II, called for the capture of Saipan, Guam, and Tinian in the Marianas, to be followed on 8 September by landings at Palau. Southwest Pacific Area forces were to invade Mindanao on 15 November.

MAP 14 R.F. STIBIL

The final Central Pacific objective, with a tentative target date of 15 February 1945, would be either southern Formosa and Amoy or the island of Luzon. Not until October 1944 did the JCS officially cancel the Formosa-Amoy scheme, an operation that would have required five of six Marine divisions, in favor of the reconquest of Luzon.

The first of the Marianas Islands scheduled for conquest was Saipan. This objective was, in a military as well as a geographic sense, the center of the group. Ocean traffic destined for the Marianas bases generally was channeled through Saipan. There, also, were the administrative headquarters for the entire chain, a large airfield and supplementary flight strip, as well as ample room for the construction of maintenance shops and supply depots. Finally, Saipan could serve as the base from which to attack Tinian, only three miles to the southwest, the island which had the finest airfields in the area. From Saipan, artillery could dominate portions of Tinian, but the western beaches of the northern island were beyond the range of batteries on Tinian. Thus, to strike first at Saipan was less risky than an initial blow at the neighboring island. Once the Americans had captured Saipan, Tinian, and Guam, the Japanese base at Rota would be isolated and subject to incessant aerial attack. (See Map 14.)

SAIPAN: THE FIRST OBJECTIVE [10]

The Mariana group is composed of 15 islands scattered along the 145th meridian, east longitude. The distance from Farallon de Pajaros at the northern extremity of the chain to Guam at its southern end is approximately 425 miles. Since the northern islands are little more than volcanic peaks that have burst through the surface of the Pacific, only the larger of the southern Marianas are of military value. Those islands that figured in American and Japanese plans were Saipan, some 1,250 miles from Tokyo, Tinian, Rota, and Guam.

Ferdinand Magellan, a Portugese explorer sailing for Spain, discovered the Marianas in 1521. The sight of Chamorros manning their small craft so impressed the dauntless navigator that he christened the group *Islas de las Velas Latinas*, Islands of the Lateen Sails, in tribute to native seamanship. His sailors, equally impressed but for a different reason, chose the more widely accepted name *Islas de los Ladrones*, Islands of the Thieves. Possibly moved by this latter title to reform the Chamorros, Queen Maria Anna dispatched missionaries and soldiers to the group, which was retitled in her honor the Marianas.

All of these islands were Spanish possessions at the outbreak of war with the United States in 1898. During the summer of that year, an American warship accepted the surrender of Guam, a conquest that was affirmed by the treaty that ended the conflict. In 1899, the remaining islands were sold to Germany as Spain disposed of her Pacific empire. Japan seized the German Marianas during World War I. After the war, the League of Nations appointed Japan as trustee over all the group except American-ruled Guam. When Japan withdrew from the League of Nations in 1935, she retained her portion of the Marianas as well as the Marshalls and Carolines. In the years that followed, the Japanese government

[10] Additional sources for this section include: JICPOA InfoBul 7–44, The Marianas, dtd 25Jan44, pp. 50–65; VAC G–2 Study of Southern Marianas, dtd 5Apr44; Tadao Yanaihara, *Pacific Islands under Japanese Mandate* (London: Oxford University Press, 1940); R. W. Robson, *The Pacific Islands Handbook* (New York: Macmillan Co., 1945, North American ed.).

kept its activities in the group cloaked in secrecy.

No single adjective can glibly describe the irregularly shaped island of Saipan. Three outcroppings, Agingan Point, Cape Obiam, and Nafutan Point, mar the profile of the southern coast. The western shoreline of Saipan extends almost due north from Agingan Point past the town of Charan Kanoa, past Afetna Point and the city of Garapan to Mutcho Point. Here, midway along the island, the coastline veers to the northeast, curving slightly to embrace Tanapag Harbor and finally terminating at rugged Marpi Point. The eastern shore wends its sinuous way southward from Marpi Point, beyond the Kagman Peninsula and Magicienne Bay, to the rocks of Nafutan Point. Cliffs guard most of the eastern and southern beaches from Marpi Point to Cape Obiam. There is a gap in this barrier inland of Magicienne Bay, but a reef, located close inshore, serves to hinder small craft. Although the western beaches are comparatively level, a reef extends from the vicinity of Marpi Point to an opening off Tanapag Harbor, then continues, though broken by several gaps, to Agingan Point. (See Map 15.)

Saipan encompasses some 72 square miles. The terrain varies from the swamps inland of Charan Kanoa to the mountains along the spine of the island and includes a relatively level plain. The most formidable height is 1,554-foot Mount Tapotchau near the center of the island. From this peak, a ridge, broken by other mountain heights, runs northward to 833-foot Mount Marpi. To the south and southeast of Mount Tapotchau, the ground tapers downward to form a plateau, but the surface of this plain is broken by scattered peaks. Both Mounts Kagman and Nafutan, for example, rise over 400 feet above sea level, while Mount Fina Susu, inland of Charan Kanoa, reaches almost 300 feet. The most level regions—the southern part of the island and the narrow coastal plain—were under intense cultivation at the time of the American landings. The principal crop was sugar cane, which grew in thickets dense enough to halt anyone not armed with a machete. Refineries had been built at Charan Kanoa and Garapan, and rail lines connected these processing centers with the sugar plantations.

Saipan weather promised to be both warm, 75 to 85 degrees, and damp, for the invasion was scheduled to take place in the midst of the rainy season. Planners, however, believed that the operation would end before August, usually the wettest month of the year. Typhoons, which originate in the Marianas, posed little danger to the expedition for such storms generally pass beyond the group before reaching their full fury.

As American strategists realized, Saipan offered no harbor that compared favorably with the atoll anchorages captured in previous operations. The Japanese had improved Tanapag Harbor on the west coast, but there the reef offered scant protection to anchored vessels. Ships which chose to unload off Garapan, just to the south, were at the mercy of westerly winds. The deep waters of Magicienne Bay, on the opposite shore, were protected on the north and west but exposed to winds from the southeast.

The geography of the objective influenced both planning and training. The size of the island, the reefs and cliffs that guarded its coasts, its cane fields and mountains, and the disadvantages of its harbors had to be considered by both tactical and logistical planners. Whatever their schemes of maneuver and supply, the attackers would encounter dense cane fields, jungles, mountains, cities or towns, and possibly swamps. The Marines would have to prepare to wage a lengthy battle for ground far different from the coral atolls of the Gilberts and Marshalls.

COMMAND RELATIONSHIPS

Since FORAGER contemplated the eventual employment of three Marine divisions, a Marine brigade, and two Army divisions against three distinct objectives within the Mariana group, the command structure was bound to be somewhat complex. Once again, Admiral Nimitz, who bore overall responsibility for the operation, entrusted command of the forces involved to Admiral Spruance. As Commander, Central Pacific Task Forces, Spruance held military command of all units involved in FORAGER and was responsible for coordinating and supervising their performance.[11] He was to select the times of the landings at Tinian, Guam, and any lesser islands not mentioned in the operation plan and to determine when the capture and occupation of each objective had been completed. As Commander, Fifth Fleet, he also had the task of thwarting any effort by the *Combined Fleet* to contest the invasion of the Marianas.

Vice Admiral Turner, Commander, Joint Amphibious Forces (Task Force 51), would exercise command over the amphibious task organizations scheduled to take part in FORAGER. The admiral, under the title of Commander, Northern Attack Force, reserved for himself tactical command over the Saipan landings. As his second-in-command, and commander of the Western Landing Group, which comprised the main assault forces for Saipan, Turner had the veteran Admiral Hill.[12] At both Tinian and Guam, Turner would exercise his authority through the appropriate attack force commander.

In command of all garrison troops as well as the landing forces was Holland M. Smith, now a lieutenant general. Smith, Commanding General, Expeditionary Troops, also served as Commanding General, Northern Troops and Landing Force (NTLF) at Saipan. As commander of the expeditionary troops, he exercised authority through the landing force commander at a given objective from the time that the amphibious phase ended until the capture and occupation phase was completed. Thanks to his dual capacity at Saipan, the general would establish his command post ashore when he believed the beachhead to be secured, report this move to the attack force commander, and begin directing the battle for the island. Since Saipan was a large enough land mass to require a 2-divi-

[11] RAdm Charles J. Moore cmts on draft MS, dtd 1Feb63, hereafter *Moore comments Saipan.*

[12] Adm Harry W. Hill cmts on draft MS, dtd 6Feb63, hereafter *Hill comments Saipan.*

sion landing force, Smith would be the equivalent of a corps commander.

Faced with the burdens of twin commands, the Marine general reorganized his VAC staff as soon as the preliminary planning for the Marianas operation had been completed. For detailed planning, he could rely on a Red Staff, which was to assist him in exercising command over Northern Troops and Landing Force, and a Blue Staff, which would advise him in making decisions as Commander, Expeditionary Troops.

Apart from his role in FORAGER, Smith was charged, in addition, with "complete administrative control and logistical responsibility for all Fleet Marine Force units employed in the Central Pacific Area." [13] Since all Marine divisions in the Pacific were destined for eventual service in Nimitz' theater, the general was empowered to establish an administrative command which included a supply service. Once the Marianas campaign was completed, Nimitz intended to install Smith as Commanding General, Fleet Marine Force, Pacific, with control over the administrative command and two amphibious corps.[14]

Northern Troops and Landing Force was composed of two veteran divisions led by experienced commanders. The 2d Marine Division, which had earned battle honors at Guadalcanal and Tarawa, was now commanded by Major General Thomas E. Watson, whose Tactical Group 1 had seized Eniwetok Atoll. Major General Harry Schmidt's 4th Marine Division had received its introduction to combat during FLINTLOCK. The second major portion of Expeditionary Troops, Southern Troops and Landing Force, was under the command of Major General Roy S. Geiger, a naval aviator, who had directed an amphibious corps during the Bougainville fighting. Geiger's force consisted of the 3d Marine Division, tested at Bougainville, and the 1st Provisional Marine Brigade. The brigade boasted the 22d Marines, a unit that had fought valiantly at Eniwetok Atoll, and the 4th Marines. Although the 4th Marines, organized in the South Pacific, had engaged only in the occupation of Emirau Island, most of its men were former raiders experienced in jungle warfare.

During the interval between the Kwajalein and Saipan campaigns, the Marine Corps approved revised tables of organizations for its divisions and their components, a decision which affected both the 2d and 4th Marine Divisions. Aggregate strength of the new model division was 17,465, instead of the previous 19,965. The principal components now were a headquarters battalion, a tank battalion, service troops, a pioneer battalion, an engineer battalion, an artillery regiment, and three infantry regiments. Service troops included service, motor transport, and medical battalions; the component once designated "special troops" existed no longer. The new tables called for the elimination of the naval construction battalion that had been part of the discarded engineer regiment and the transfer of the scout company,

[13] AdminHist, FMFPac, 1944–1946, dtd 13 May46 (AdminHist File, HistBr, HQMC).

[14] A provisional Headquarters, FMFPac was established on 24 August 1944. A detailed account of the formation of FMFPac along with its administrative and supply components will be included in the fourth volume of this series.

now reconnaissance company, from the tank battalion to headquarters battalion. The artillery regiment was deprived of one of its 75mm pack howitzer battalions, leaving two 75mm and two 105mm howitzer battalions. The infantry regiments continued to consist of three infantry battalions and a weapons company. The old 12-man rifle squad was increased to a strength of 13 and divided into three 4-man fire teams. Finally, the special weapons battalion had been disbanded and its antitank duties handed over to the regimental weapons companies, while the amphibian tractor battalion was made a part of corps troops.

Except for the absence of LVTs, the most striking change in the revised division's equipment was the substitution of 46 medium tanks for 54 light tanks within the tank battalion. The authorized number of flamethrowers had been gradually increased from 24 portables to 243 of this variety plus 24 of a new type that could be mounted in tanks, thus giving official approval to the common practice of issuing prior to combat as many flamethrowers as a division could lay hands upon. The artillery regiment lost 12 75mm pack howitzers, but the number of mortars available to infantry commanders was increased from 81 60mm and 36 81mm to 117 60mm and 36 81mm. Since each of the newly authorized fire teams contained an automatic rifle, the new division boasted 853 of these weapons and 5,436 M1 rifles instead of 558 automatic rifles and 8,030 M1s. Although it would seem that the reorganized division could extract a greater volume of fire from fewer men, such a unit also would require reinforcements, notably a 535-man amphibian tractor battalion, before attempting amphibious operations.[15]

Both Marine divisions scheduled for employment at Saipan were almost completely reorganized before their departure for the objective. Neither had disbanded its engineer regiment although the organic naval construction battalions were now attached and would revert to corps control after the landing.[16] The two surviving Marine battalions were originally formed according to discarded tables of organization as pioneer and engineer units. Thus, they could perform their usual functions even though they remained components of a regiment rather than separate battalions. Reinforced for the Saipan landings, its infantry battalions organized as landing teams and its infantry regiments as combat teams, each of the two divisions numbered approximately 22,000 men.[17] In contrast, the 27th Infantry Division, serving as FORAGER reserve, could muster only 16,404 officers and men when fully reinforced.

During the battle for Saipan, the attacking Marines would be supported by heavier artillery weapons than the 75mm and 105mm howitzers that had aided them in previous Central Pacific operations. Two Army 155mm how-

[15] TO F-100, MarDiv, dtd 5May44; F-30, ArtyRegt, dtd 21Feb44; F-80 TkBn, dtd 4 Apr44; F-89, ReconCo, HqBn, dtd 4Apr44; F-70, ServTrps, dtd 12Apr44 (TO File, HistBr, HQMC). A summary of TO F-100, Marine Division, is included as Appendix F.

[16] BGen Ewart S. Laue ltr to ACofS, G-3, HQMC, dtd 29Jun63, hereafter *Laue ltr*.

[17] 4th MarDiv and 2d MarDiv WarDs, Feb–May44 (Unit File, HistBr, HQMC).

itzer battalions joined a pair of Army 155mm gun battalions to form XXIV Corps Artillery under command of Brigadier General Arthur M. Harper, USA. In addition, a Marine 155mm howitzer battalion was attached by VAC to the 10th Marines, artillery regiment for the 2d Marine Division. The 4th Marine Division, however, had to be content with an additional 105mm howitzer battalion. The remainder of VAC artillery was retained under corps control in Hawaii.[18]

Another division which might see action at Saipan was the FORAGER reserve, the 27th Infantry Division, an organization that had yet to fight as a unit. During GALVANIC, the division commanding general, Major General Ralph C. Smith, had led the 165th Infantry and 3/105 against enemy-held Makin Atoll. As part of Tactical Group 1, 1/106 and 2/106 had fought at Eniwetok Island. The remaining battalion of the 106th Infantry landed at Majuro where there was no opposition, and the other two battalions of the 105th Infantry lacked combat experience of any sort. Also in reserve was the inexperienced 77th Infantry Division, but this unit would remain in Hawaii as a strategic reserve until enough ships had returned from Saipan to carry it to the Marianas. Not until 20 days after the Saipan landings would the 77th Division become available to Expeditionary Troops for employment in the embattled islands.

The effort against Saipan, then, rested in capable hands. The team of Spruance, Turner, and Holland Smith had worked together in the Gilberts and Marshalls. Both assault divisions were experienced and commanded by generals who had seen previous action in the Pacific war. Only the Expeditionary Force reserve, which might be employed at Saipan, was an unknown factor, for the various components of the 27th Infantry Division had not fought together as a team, and there was considerable difference in experience among its battalions.

LOGISTICAL AND ADMINISTRATIVE PLANNING[19]

In attacking Saipan, Nimitz' amphibious forces would encounter an objective unlike any they seized in previous Central Pacific operations. The mountainous island, with a total land area of some 72 square miles, was a far different battleground from the small, low-lying coral atolls of the Gilberts and Marshalls. The capture of this limited land mass could not be accomplished at a single stroke, a fact that was reflected in the plan of resupply adopted at the urging of General Holland Smith. The assault forces were directed to carry a 32-day supply of rations, enough fuel, lubricants,

[18] As a consequence of the assignment of XXIV Corps Artillery to FORAGER, VAC Artillery served as part of the XXIV Corps in the assault on Leyte. The role of Marine artillery and air units in the Philippines campaign will be covered in the fourth volume of this series.

[19] Additional sources for this section include: NTLF AdminO 3-44, dtd 1May44; 2d MarDiv SplCmts, Phase I, FORAGER, n.d., p. 23; 4th MarDiv OpRpt Saipan, 15Jun–9Jul44, (incl Narrative, StfRpts, and Rpts of SuborUs), dtd 18Sep44, hereafter *4th MarDiv OpRpt*.

chemical, ordnance, engineer, and individual supplies to last for 20 days, a 30-day quantity of medical supplies, 7 days' ammunition for ground weapons, and a 10-day amount for antiaircraft guns.

Vast as this mountain of supplies might be, the Commanding General, Expeditionary Troops, wanted still more. The Navy accepted his recommendations that an ammunition ship anchor off Saipan within five days after the landings and that supply vessels sailing from the continental United States be "block loaded." In other words, those ships that would arrive with general supplies after the campaign had begun should carry items common to all troop units in a sufficient quantity to last 3,000 men for 30 days. The portion of the plan dealing with ammunition resupply worked well enough, but block loading proved inefficient. Since the blocks had been loaded in successive increments, each particular item had to be completely unloaded before working parties could reach the next type of supplies. Admiral Turner later urged a return to the practice of loading resupply vessels so that the various kinds of cargo could be landed as needed. He saw no need in forcing many ships to carry a little bit of everything, when, by concentrating certain items in different ships, selective unloading was possible.

As usual, hold space was at a premium, so Expeditionary Troops kept close watch on the amount of equipment carried by assault and garrison units. The three divisions that figured in the Saipan plan adhered to the principles of combat loading, but only one, the 27th Infantry Division, made extensive use of pallets. In fact, the Army unit exceeded the VAC dictum that from 25 to 50 percent of embarked division supplies be placed on pallets. The 2d Marine Division lashed about 25 percent of its bulk cargo to these wooden frames, while the 4th Marine Division placed no more than 15 percent of its supplies on pallets. General Schmidt's unit lacked the wood, waterproof paper, and skilled laborers necessary to comply with the wishes of corps. To complicate the 4th Marine Division loading, G–4 officers found that certain vessels assigned to carry cargo for Schmidt's troops were also to serve other organizations. In addition, the transports finally made available had less cargo space than anticipated. Under these circumstances, division planners elected to use every available cubic foot for supplies, vehicles, and equipment. Even if material had been available, there would have been room for few pallets.

Applying the lessons of previous amphibious operations, VAC addressed itself to the problems of moving supplies from the transports to the units fighting ashore. In April 1944, a Corps Provisional Engineer Group was formed, primarily to provide shore party units for future landings. The two Marine Divisions assigned to VAC for FORAGER had already established slightly different shore party organizations, but since both were trained in beachhead logistics, the engineer group did not demand that they be remodeled to fit a standard pattern. Backbone of the shore parties for both divisions were the pioneer battalions and the attached naval construction battalions.

The 2d Marine Division assigned pioneer troops as well as Seabees to each shore party team, while the 4th Marine Division concentrated its naval construction specialists in support of a single regiment. If this construction battalion should be needed for road building or similar tasks, the 4th Marine Division would be forced to reorganize its shore party teams in the midst of the operation. Neither Marine division used combat troops to assist in the beachhead supply effort.

To support both the garrison and assault units assigned to FORAGER, the Marine Supply Service organized the 5th and 7th Field Depots.[20] Marines trained to perform extensive repairs on weapons, fire control equipment, and vehicles accompanied the landing forces, while technicians capable of making even more thorough repairs embarked with the garrison troops. The 7th Field Depot was chosen to store and issue supplies, distribute ammunition, and salvage and repair equipment on both Saipan and Tinian. The 5th Field Depot would perform similar duties on Guam. At the conclusion of FORAGER, the two depots were to assist in re-equipping the 2d and 3d Marine Divisions by accepting, repairing, and re-issuing items turned in prior to their departure from the Marianas by the 4th Marine Division and 1st Provisional Marine Brigade. Since plans called for Saipan to be garrisoned primarily by Army troops, the 7th Field Depot eventually would move its facilities to nearby Tinian, although it would continue to serve Marine units on the other island.

Authority to determine which boats were to evacuate the wounded from Saipan rested in the beachmasters. During the early hours of the operation, casualties were to be collected in three specially equipped LSTs, treated, and then transferred to wards installed in certain of the transports. One of the hospital LSTs would take station off the beaches assigned to each of the Marine divisions. The third such vessel was to relieve whichever of the other two was first to receive 100 casualties. Each of the trio of landing ships had a permanent medical staff of one doctor and eight corpsmen. An additional 2 doctors and 16 corpsmen would be reassigned from the transports to each of the LSTs before the fighting began. Plans also called for hospital ships to arrive in the target area by D plus three or when ordered forward from Eniwetok by Admiral Turner. Detailed plans were also formulated for the air evacuation of severely wounded men from the Marianas. Planes of the Air Transport Command, Army Air Forces, would load casualties at Aslito airfield and fly them to Oahu via Kwajalein.[21]

In spite of the scope of the Saipan undertaking and the possibility of numerous casualties, no replacement drafts were included in the expedition, for G–1 planners believed that men transferred from one unit could replace those lost by another. During the Saipan fighting, the 2d Marine Division

[20] The story of the development of the Marine Supply Service as part of the overall picture of the formation of Fleet Marine Force, Pacific, will be covered in the fourth volume of this series.

[21] Dr. Robert F. Futrell, USAF HistDiv, ltr to Head, HistBr, G–3, dtd 29Jan63, hereafter *USAF Comments.*

was to be kept at peak effectiveness by the reassignment of troops from the 4th Marine Division. This plan, however, had to be abandoned, for the mass transfers required under such an arrangement would have crippled General Schmidt's division. Instead, replacement drafts were dispatched to Saipan during June and July.

INTELLIGENCE FOR SAIPAN

Until carrier planes attacked Saipan on 22–23 February 1944, American intelligence officers had no accurate information concerning the island defenses. As a result of these strikes, planners received aerial photographs of certain portions of the island. Ideal coverage, General Holland Smith's G–2 section believed, could be obtained if photographic missions were flown 90, 60, 30, and 15 days before the Saipan landings. Unfortunately, Navy carrier groups were too busy blasting other objectives to honor such a request, but additional pictures were taken by long-range Navy photo planes. Between 17 April and 6 June, Seventh Air Force B–24s escorted their Navy counterpart PB4Ys from Eniwetok to the Marianas on seven joint reconnaissance missions.[22] Although the final set of photographs reached Expeditionary Troops headquarters at Eniwetok, where the expedition had paused en route to the objective, the assault elements had already set sail for Saipan. As a result, the troops that landed on 15 June did not benefit from the final aerial reconnaissance. Equally useless to the attacking divisions were the photographs of the island beaches taken by the submarine *Greenling*, for these did not cover the preferred landing areas.

The aerial photographs taken by carrier aviators were not of the best quality, for the taking of pictures was more or less a sideline, and a dangerous one at that. First in the order of importance was the killing of Japanese, but the most profitable target for American bombs was not always the island or area which the intelligence experts wanted photographed. Admiral Spruance did for a time contemplate a second carrier strike against Saipan, a raid which would have netted additional photographs to supplement those taken in February by carrier aircraft and in April and May by Eniwetok-based photographic planes. In order to avoid disclosing the Marianas as the next American objective, the Admiral decided against the raid.

The photos obtained during the February raid along with charts captured in the Marshalls provided the information upon which Expeditionary Troops based its map of Saipan. Since the sources used did not give an accurate idea of ground contours, map makers had to assume that slopes were uniform unless shadows in the pictures indicated a sudden rise or sharp depression. Clouds, trees, and the angle at which the photos were taken helped hide the true nature of the terrain, so that many a cliff was interpreted on the map as a gentle slope. Fortunately, accurate Japanese maps were to be captured during the first week of fighting.

The strength, disposition, and armament of the Saipan garrison was difficult to determine. Documents cap-

[22] *Ibid.*

tured in previous campaigns, reports of shipping activity, and aerial photographs provided information on the basic strength, probable reinforcement, and fixed defenses of the garrison. As D-Day approached, Admiral Turner and General Smith obtained additional fragments of the Saipan jigsaw puzzle, but full details, such as the complete enemy order of battle, would not be known until prisoners, captured messages, and reports from frontline Marine units became available.

On 9 May, Expeditionary Troops estimated that no more than 10,000 Japanese were stationed at Saipan, but by the eve of the invasion, this figure had soared to 15,000–17,600. This final estimate included 9,100–11,000 combat troops, 900–1,200 aviation personnel, 1,600–1,900 Japanese laborers plus 400–500 Koreans, and 3,000 "home guards," recent recruits who were believed to be the scrapings from the bottom of the manpower barrel. The actual number of Japanese was approximately 30,000 soldiers and sailors plus hundreds of civilians.

Although aerial photographs gave the landing force an accurate count of the enemy's defensive installations, these pictures did not disclose the number of troops poised inland of the beaches. The number and type of emplacements, however, did indicate that reinforcements were pouring into the island. By comparing photos taken on 18 April with those taken on 29 May, intelligence experts discovered an increase of 30 medium antiaircraft guns, 71 light antiaircraft cannon or machine guns, 16 pillboxes, a dozen heavy antiaircraft guns, and other miscellaneous weapons.

Intelligence concerning Saipan was not as accurate as the information previously gathered for the Kwajalein campaign. The 1,000-mile distance of the objective from the nearest American base, the clouds which gathered over the Marianas at this time of year, and the fear of disclosing future plans by striking too often at Saipan were contributing factors. The lack of usable submarine photographs was offset by the possession of hydrographic charts seized in the Marshalls and by the boldness of underwater demolition teams. Under cover of naval gunfire, these units scouted the invasion beaches during daylight on D minus 1 to locate underwater obstacles.

TACTICAL PLANS

Northern Troops and Landing Force was assigned the capture of both Saipan and adjacent Tinian. For these operations, service and administrative elements of the command were banded together in Corps Troops, while the combat elements were the 2d and 4th Marine Divisions, supported by XXIV Corps Artillery. One Marine infantry battalion, 1/2, was withdrawn from the 2d Marine Division and placed under corps control for a special operation in connection with the Saipan landing. To replace this unit, General Watson was subsequently given the 1st Battalion, 29th Marines. This outfit, made up of drafts from the 2d Division, was located at Hilo, Hawaii. After the campaign, 1/29 was destined to join the rest of its regiment at Guadalcanal and form part of the

6th Marine Division.[23] In addition to the combat troops, NTLF also controlled two garrison forces, composed mainly of Army units for Saipan and Marine units for Tinian. The 27th Infantry Division, as Expeditionary Troops reserve, might be employed to reinforce Northern Troops and Landing Force at Saipan or Tinian, or to assist Southern Troops and Landing Force at Guam. As a result, the division G-3 section prepared 21 operation plans, 16 of them dealing with possible employment at Saipan.

The basic scheme of maneuver for the Saipan attack called for the 23d and 25th Marines, 4th Marine Division, to land on the morning of 15 June over the Blue Beaches off the town of Charan Kanoa and across the Yellow Beaches, which extended southward from that town toward Agingan Point. At the same time, the 6th and 8th Marines, 2d Marine Division, were to land on the Red and Green Beaches just north of Charan Kanoa. To deceive the enemy, General Smith decided to make a feint toward the coastline north of Tanapag Harbor, a maneuver which he assigned to the 2d Marines, including 1/29, and the 24th Marines. (See Map 16.)

Another portion of the plan, one that eventually was canceled, would have sent 1/2 ashore near the east coast village of Laulau on the night of 14-15 June. This reinforced battalion was to have pushed inland to occupy the crest of Mount Tapotchau and hold that position until relieved by troops from the western beachhead. After this part of the plan had been abandoned, 1/2 remained ready to land on order at Magicienne Bay, or, if the tactical situation demanded, elsewhere on the island.

Striking inland, the 2d and 4th Marine Divisions were to seize the high ground that stretched southward from Hill 410 through Mount Fina Susu to Agingan Point. Since this high ground dominating the beaches had to be seized as rapidly as possible, the LVTs and their escorting LVT(A)s were to thrust toward the ridge line, bypassing pockets of resistance along the shore. From this terrain feature, General Schmidt's division was to push eastward beyond Aslito Airfield to Nafutan Point, while General Watson's Marines secured the shores of Magicienne Bay and attacked northward toward Marpi Point. Among the intermediate objectives of the 2d Marine Division during this final advance were Mount Tipo Pale, Mount Tapotchau, and the city of Garapan.

The ship-to-shore movement that would trigger the battle for Saipan was patterned after earlier amphibious operations in the Marshalls. Because of the reef that guarded the landing sites, LVTs were required by the attacking Marines. Northern Troops and Landing Force had a total of six amphibian tractor battalions, three of them, the 2d, 4th, and 10th, Marine units and the others, the 534th, 715th, and 773d, Army organizations. The tractors assigned to the assault infantry battalions, as well as those assigned to one reserve battalion in each division, were ferried to Saipan in LSTs. Since the tank landing ships also carried the Marines assigned to land in these LVTs, relatively few assault

[23] BGen Rathvon McC. Tompkins ltr to ACofS, G-3, HQMC, dtd 4Jan63.

214-881 O-67—17

troops would be forced to transfer from one type of craft to another. All reserve infantry elements, except for the two battalions assigned to LSTs, were scheduled to proceed in LCVPs from their transports to a designated area where they would change to LVTs. The organic field artillery regiments of both divisions embarked their battalions in LSTs. The 75mm howitzers and crews were to land in LVTs, and the 105s in DUKWs. Both types of weapons were placed in the appropriate vehicles before the expedition sailed. Tanks once again were preloaded in LCMs, and these craft embarked in LSDs.

The assault on Saipan would be led by rocket-firing LCI gunboats which were followed by armored amphibian tractors. The LVT(A)4s, manned by the Marine 2d Armored Amphibian Tractor Battalion, were modifications of the type used in the Marshalls. Instead of a 37mm gun and three .30 caliber machine guns, the new vehicles boasted a snub-nosed 75mm howitzer mounted in a turret and a .50 caliber machine gun. The other unit assigned to the Saipan operation, the Army's 708th Amphibian Tank Battalion, was equipped with older LVT(A)1s and a few LVT(A)4s.

To control the Saipan landings, Admiral Hill selected officers experienced in amphibious warfare. At the apex of the control pyramid was the force control officer, who had overall responsibility for controlling all landing craft involved in getting two divisions ashore on a frontage of some 6,000 yards.[24] A group control officer was assigned each division, and a transport division control officer was in charge of each regiment in the landing force. On D-Day, the force control officer would, by means of visual signals and radio messages, summon the leading waves to the line of departure and dispatch them toward the island. Transport division control officers had the tasks of sending in the later waves according to a fixed schedule and of landing reserves as requested by the regimental commander or his representatives.

One LCC was stationed on either flank of the first wave formed by each assault regiment. These vessels were to set the pace for the amphibian tractors in addition to keeping those vehicles from wandering from course. When the initial wave crossed the reef, a barrier which the control craft could not cross, the LCCs would take up station seaward of that obstacle to supervise the transfer of reserve units from LCVPs to LVTs. Later assault waves would rely on designated LCVPs to guide them as far as the reef.

Since communications had been the key to control in previous operations, Admiral Turner decided to employ at Saipan 14 communications teams, each one made up of an officer, four radiomen, and two signalmen. In addition to placing these teams where he thought them necessary, the admiral had additional radio equipment installed in the patrol craft, submarine chasers, and LCCs that were serving as control vessels. In this way, adequate radio channels were available to everyone involved in controlling the landings, the supply effort, and the evacuation of casualties.

[24] *Hill comments Saipan.*

AIR AND NAVAL GUNFIRE SUPPORT

After Navy pilots based on fast carriers had destroyed Japanese air power in the Marianas, other aviators could begin operations designed to aid the amphibious striking force. Because of its size, Saipan imposed new demands upon supporting aircraft. Pilots assisting an attack against an atoll could concentrate on a relatively small area, but in their strikes against a comparatively large volcanic island, the aviators would have to range far inland to destroy enemy artillery and mortars which could not be reached by naval guns and to thwart efforts to reinforce coastal defense units. The neutralization of the beach fortifications was to follow a flexible schedule, while strikes against defiladed gun positions or road traffic could be launched as required by planes on station over the island.

The first D-Day attack against the beach defenses was a 30-minute bombing raid scheduled to begin 90 minutes before H-Hour. Naval gunfire would be halted while the planes made their runs. This strike was intended to demoralize enemy troops posted along the beaches as well as to destroy particular installations.

To make up for the absence of field artillery support, such as had been enjoyed in the Marshalls, aircraft were ordered to strafe the beaches while the incoming LVTs were between 800 and 100 yards of the island. This aerial attack would coincide in part with the planned bombardment by warships of this same area, for naval gunfire would not be shifted until the troops were 300 yards from the objective. Pilots were informed of the maximum ordinate of the naval guns, and since their shells followed a rather flat trajectory, the approach of the planes would not be seriously hindered. When the leading wave was 100 yards from its objective, the aviators were to shift their point of aim 100 yards inland and continue strafing until the Marines landed.

Prior to H-Hour, all buildings, suspected weapons emplacements, and possible assembly areas more than 1,000 yards from the coastline were left to the attention of naval airmen. Planes armed with bombs or rockets had the assignment of patrolling specific portions of Saipan to attack both previously located installations and targets of opportunity. After the landings, aircraft would cooperate with naval gunfire and artillery in destroying enemy strongpoints and hindering Japanese road traffic.

The air support plan also provided for the execution of strikes at the request of ground units. A Landing Force Commander Support Aircraft was appointed primarily to insure coordination between artillery and support aviation. A requested strike might be directed by any of four individuals: the Airborne Coordinator, aloft over the battlefield; the leader of the flight on station over the target area; the Landing Force Commander Support Aircraft with headquarters ashore; or the Commander Support Aircraft, located in the command ship and aware of the naval gunfire plan.

The decision whether to handle the strike himself or delegate it to another was left to the Commander Support Aircraft. He would select the person best informed on the ground situation to direct a particular attack. He also had the responsibility of insuring that his subordinates were fully informed concerning troop dispositions and any plans to employ other supporting weapons.

The preliminary naval bombardment of Saipan was to begin on D minus 2 with the arrival off the objective of fast battleships and destroyers from Task Force 58. The seven battleships, directed to remain beyond the range of shore batteries and away from possible minefields, would fire from distances in excess of 10,000 yards. The nocturnal harassment of the enemy was left to the destroyers. On the following day, the fire support ships, cruiser, destroyers, and old battleships were scheduled to begin hammering Saipan from close range.

The plan for D-Day called for the main batteries of the supporting battleships and cruisers to pound the beaches until the first wave was about 1,000 yards from shore. The big guns would then shift to targets beyond the 0-1 Line, which stretched from the northern extremity of Red 1 through Hill 410 and Mount Fina Susu to the vicinity of Agingan Point. Five-inch guns, however, were to continue slamming shells into the beaches until the troops were 300 yards from shore, when these weapons also would shift to other targets. The final neutralization of the coastal defenses was left to the low-flying planes which had begun their strafing runs when the LVTs were 800 yards out to sea.

During the fighting ashore, on-call naval gunfire was planned for infantry units. To speed the response to calls for fire support, each shore fire control party was assigned the same radio frequency as the ship scheduled to deliver the fires and the plane that observed the fall of the salvos. A Landing Force Naval Gunfire Officer was selected to go ashore and work with the Landing Force Commander Support Aircraft and the Corps Artillery Officer in guaranteeing cooperation among the supporting arms.

American and Japanese Preparations[1]

As the tactical plans were taking shape, the divisions slated for the Saipan operation began training for the impending battle. Ships were summoned to Hawaii to carry the invasion force to its destination. While the Americans gathered strength for the massive effort to seize the Marianas, the enemy looked to the defenses of the Central Pacific. In Hawaii, Marines and Army infantrymen practiced landing from LVTs in preparation for the Saipan assault. At the objective, Japanese troops were working just as hard to perfect their defenses.

TRAINING AND REHEARSALS

The Marine and Army units selected to conquer Saipan underwent training in the Hawaiian Islands designed to prepare them for combat in the jungle, cane fields, and mountains of the Mariana Islands. The scope of training matched the evolution of tactical plans, as individual and small unit training gave way to battalion exercises, and these, in turn, were followed by regimental and division maneuvers. The 2d Marine Division, encamped on the island of Hawaii, did its training in an area that closely resembled volcanic Saipan. After its conquest of northern Kwajalein, the 4th Marine Division arrived at the island of Maui to begin building its living quarters and ranges—tasks which coincided with training for FORAGER. Both construction and tactical exercises were hampered by the nature of the soil, a clay which varied in color and texture from red dust to red mud. The 27th Infantry Division, on the island of Oahu, emphasized tank-infantry teamwork and the proper employment of JASCO units during amphibious operations. The XXIV Corps Artillery was in the meantime integrating into its ranks the coast artillerymen needed to bring the battalions to authorized strength, conducting firing exercises, and learning amphibious techniques.

Amphibious training got underway in March, when the 2d Marine Division landed on the shores of Maalaea Bay, Maui. The 4th Marine Division, Corps Troops, and the 27th Infantry Division received their practical instruction during the following month. The climax to the indoctrination scheduled by General Watson for his 2d Marine Division was a "walk through" rehearsal held on dry land. An outline of Saipan was drawn to scale on the ground, the various phase lines and unit boundaries

[1] Unless otherwise noted, the material in this chapter is derived from: *TF 51 OpRpt; TF 56 OpRpt; TG 52.2 OpRpt; NTLF OpRpt;* 2d MarDiv OpRpt Phase I, FORAGER (incl a six-part narrative, four-part SAR, and SplCmts), dtd 11Sep44, hereafter *2d MarDiv OpRpt; 4th MarDiv OpRpt; 27th InfDiv Op Rpt;* CominCh, *The Marianas;* Crowl, *The Marianas;* Hoffman, *Saipan;* Morison, *New Guinea and the Marianas.* A complete file of CinCPac-CinCPOA and JICPOA translations is available from OAB, NHD.

were marked, thereby enabling the Marines to see for themselves how the plan would be executed. "Yet," the commanding general recalled, "only a few commanders and staff officers of the thousands of men who participated in this rehearsal knew the real name of the target."[2]

On 17 and 19 May, the two Marine divisions took part in the final rehearsals of Northern Troops and Landing Force. The first exercise, conducted at Maalaea Bay, saw the Marines land on the beaches and advance inland, following the general scheme of maneuver for the Saipan operation. The second rehearsal was held at Kahoolawe Island, site of a naval gunfire target range. Although the roar of naval guns added realism to the exercise, the assault troops did not go ashore. After the landing craft had turned back, shore fire control parties landed to call for naval salvos against the already shell-scarred island. The 27th Infantry Division completed its rehearsals between 18 and 24 May. The independent 1/2 and its reinforcing elements climaxed the training cycle with landings at Hanalei Bay.

The rehearsals were marred by a series of accidents en route to Maui that killed 2 Marines, injured 16, and caused 17 others to be reported as missing. In the early morning darkness of 14 May, heavy seas caused the cables securing three Landing Craft, Tank (LCTs) to part, and the craft plummeted from the decks of their parent LSTs. Only one of the boats lost overboard remained afloat. The LCTs mounted 4.2-inch mortars, weapons which would have been used to interdict the road between Garapan and Charan-Kanoa and protect the flank of the 2d Marine Division.[3]

Since there was not enough time to obtain replacements for the lost mortars, Admiral Turner decided to rely on the scheduled rocket barrage by LCI(G)s for neutralization of beach defenses. He ordered those LSTs and the LCT that carried the heavy mortars and their supply of ammunition to unload upon their return to Pearl Harbor. As the mortar shells were being put ashore, tragedy struck again.

On 21 May, one of the 4.2-inch rounds exploded while it was being unloaded, touching off a conflagration that enveloped six landing ships. Navy fire-fighting craft tried valiantly to smother the flames, but, though they prevented the further spread of the blaze, they could not save the six LSTs from destruction. The gutted ships had carried assault troops as well as weapons and equipment, so losses were severe. The explosion and fire inflicted 95 casualties on the 2d Marine Division and 112 on the 4th Marine Division. Replacements were rushed to the units involved in the tragedy, but the new

[2] LtGen Thomas E. Watson ltr to Dir Div PubInfo, dtd 9Jun49, quoted in Hoffman, *Saipan*, p. 31. No copy of this letter has been found.

[3] The arming and employment of these mortar craft was a project jointly developed by Admiral Hill and the CinCPac gunnery staff. Their intended mission was "cruising back and forth along a lighted buoy line close to the beach between Charan-Kanoa and Garapan and maintaining a constant barrage on the road connecting those two points throughout the first two or three nights after the landing." *Hill comments Saipan.*

men "were not trained to carry out the functions of those lost." [4] The destroyed ships, equipment, and supplies were replaced in time for the LST convoy to sail on 25 May, just one day behind schedule. The lost time was made up en route to the objective.

ONWARD TO SAIPAN

The movement of Northern Troops and Landing Force plus the Expeditionary Troops reserve from Hawaii to Saipan was an undertaking that required a total of 110 transports. Involved in the operation were 37 troop transports of various types, 11 cargo ships, 5 LSDs, 47 LSTs, and 10 converted destroyers.[5] Navy-manned Liberty ships, vessels that lacked adequate troop accommodations, were pressed into service as transports for a portion of the 27th Infantry Division. LSTs carrying assault troops, LVTs, and artillery from both Marine divisions set sail on 25 May. Two days later, transports bearing the remainder of the 4th Marine Division and Headquarters, Expeditionary Troops departed, to be followed on 30 May by elements of the 2d Marine Division. Because of the shortage of shipping, portions of XXIV Corps Artillery were assigned to the transports carrying the assault divisions. Garrison units and Expeditionary Troops reserve were the last units to steam westward.

The transports carrying the Marines sailed to Eniwetok Atoll where they joined the LST convoy. Here additional assault units were transferred from the troop ships to the already crowded landing ships for the final portion of the voyage. One observer, writing of the journey from Eniwetok to Saipan, has claimed that because of the overcrowding, "aggressiveness was perhaps increased," for "after six crowded days aboard an LST, many Marines were ready to fight anybody." [6] By 11 June, the last of the ships assigned to stage through Eniwetok had weighed anchor to begin the final approach to the objective. Meanwhile, the vessels carrying the 27th Infantry Division had completed their last-minute regrouping at Kwajalein Atoll.

While the vessels bearing General Holland Smith's 71,034 Marine and Army troops were advancing toward Saipan, the preparatory bombardment of the island got underway. The 16 carriers of Task Force 58 struck first, launching their planes on 11 June to begin a 3 1/2-day aerial campaign against Saipan, Tinian, Guam, Rota, and Pagan—the principal islands in the Marianas group. These attacks were originally to have started on the morning of the 12th, but Vice Admiral Marc A. Mitscher, the task force commander, obtained permission to strike one-half day earlier. Mitscher felt that the enemy had become accustomed to early morning raids, so he planned

[4] MajGen Louis R. Jones ltr to HistBr, HQMC, dtd 8Feb50, quoted in Hoffman, *Saipan*, p. 34. No copy of this letter has been found.

[5] The concentration of such an armada was a tribute to Navy planners, for the movement toward Saipan coincided with or immediately followed landings at Biak in the Schouten Islands, the sailing of the convoy that would carry Southern Troops and Landing Force to Guam, and the invasion of France.

[6] Hoffman, *Saipan*, p. 34*n*.

to attack in the afternoon. A fighter sweep conducted by 225 planes accounted for an estimated 150 Japanese aircraft on the first day, this insuring American control of the skies over the Marianas.

After the Grumman Hellcats departed, a member of the Saipan garrison noted in his diary that: "For two hours the enemy planes ran amuck and finally left leisurely amidst the unparalleledly inaccurate antiaircraft fire. All we could do was watch helplessly." [7]

On 12 and 13 June, bombers struck with impunity at the various islands and at shipping in the area. The only opposition was from antiaircraft guns like those on Tinian which "spread black smoke where the enemy planes weren't." One of Tinian's defenders glumly observed: "Now begins our cave life." [8]

Admiral Mitscher's fast battleships opened fire on 13 June, but their long-range bombardment proved comparatively ineffective. With the exception of the USS *North Carolina*, which a naval gunfire officer of Northern Troops and Landing Force called "one of the best-shooting ships I ever fired," [9] the new battleships tended to fire into areas or at obvious if unimportant targets, rather than at carefully camouflaged weapons positions. Neither crews of the ships nor aerial observers who adjusted the salvos had been trained in the systematic bombardment of shore emplacements. Although these battleships did not seriously damage the Japanese defenses, Admiral Spruance nonetheless believed that their contribution was valuable. The shelling by fast battleships, he later pointed out, "was never intended to take the place of the close-in fire of the [old battleships] to which it was a useful preliminary." [10]

Seven old battleships with 11 attendant cruisers and 23 destroyers relieved the fast battleships on 14 June to begin blasting Saipan and Tinian. The quality of the bombardment improved, but all did not go according to plan, for the neutralization of Afetna and Nafutan Points proved difficult to attain. Although aircraft assisted the surface units by attacking targets in the rugged interior, the preliminary bombardment was not a complete success. The size of the island, the lack of time for a truly methodical bombardment, the large number of point targets, Japanese camouflage, and the enemy's use of mobile weapons all hampered the American attempt to shatter the Saipan defenses.

On the morning of 14 June, underwater demolition teams swam toward Beaches Red, Green, Yellow, and Blue, as well as toward the Scarlet Beaches, an alternate landing area north of Tanapag Harbor. This daylight reconnaissance was a difficult mission. Lieutenant Commander Draper L. Kauffman, leader of one of the demolition teams, had told Admiral Turner that "You don't swim in to somebody's

[7] CinCPac-CinCPOA Item No. 10,238, Diary of Tokuzo Matsuya.

[8] CinCPac-CinCPOA Item No. 11,405, Diary of an Unidentified Japanese NCO.

[9] LtCol Joseph L. Stewart ltr to CMC, dtd 9Jan50, quoted in Hoffman, *Saipan*, p. 36. No copy of this letter has been found.

[10] Adm Raymond A. Spruance ltr to CMC, dtd 17Jan50, quoted in Hoffman, *Saipan*, p. 37. No copy of this letter has been found.

beaches in broad daylight," but swim they did—in spite of Kauffman's prediction of 50 percent casualties.[11] Despite a screen of naval gunfire, which had difficulty in silencing the weapons sited to cover the waters of the Blue and Yellow Beaches, the teams lost two men killed and seven wounded, approximately 13 percent of their total strength. The swimmers reported the absence of artificial obstacles, the condition of the reef, and the depth of water off the beaches. On D-Day, members of these reconnaissance units would board control vessels to help guide the assault waves along the prescribed boat lanes. (See Map 16.)

The heavy naval and air bombardment directed against the Marianas were only a part of the preparations decided upon for FORAGER. Wake and Marcus Islands had been bombed during May in order to protect the movement of Admiral Turner's warships and transports. Bombs thudded into enemy bases from the Marshalls to the Kuriles in an effort to maintain pressure on the Japanese. Finally, on 14 June, two carrier groups cut loose from Task Force 58 to attack Iwo Jima, Haha Jima, and Chichi Jima in the Volcano-Bonin Islands. These strikes were designed to prevent the enemy from making good his aerial losses by transferring planes from the home islands to the Marianas by way of the Bonins.

Like the attacking Americans, the Japanese defenders were completing their preparations for the Saipan landings. Fully alerted by the air and naval bombardment, the Saipan garrison realized that it soon would be called upon to fight to the death. Lieutenant General Hideyoshi Obata and Vice Admiral Chiuchi Nagumo awaited the arrival of the Marines so that they could execute their portion of the *A-GO* plan, which called for the destruction of the invaders on the beaches of Saipan.

THE DEFENSE OF SAIPAN[12]

Saipan had long figured in Japanese military plans. As early as 1934, the year before her withdrawal from the League of Nations, Japan had begun work on an airfield at the southern end of Saipan. By 1944, this installation, Aslito airfield, had become an important cog in the aerial defense mechanism devised to guard the Marianas. A seaplane base at Tanapag Harbor was completed in 1935, and during 1940-1941 money was appropriated for gun emplacements, storage bunkers, and other military structures.

On the eve of World War II, the *Fourth Fleet*, with headquarters at

[11] Cdr Francis D. Fane and Don Moore, *The Naked Warriors* (New York: Appleton-Century-Crofts, Inc., 1956), p. 88.

[12] Additional sources for this section include: CinCPac–CinCPOA Items Nos, 9,159, Organization of CenPac AreaFlt, n.d., 10,145, Thirty-first Army Stf, TransRpt, dtd 18Mar44, 10,638, O/B for Thirty-first Army, 1942–1944, n.d., 10,740, Location and Strength of Naval Land Units, c. Apr44; HqFEComd, MilHist Sec, Japanese Research Div, Monograph no. 45, *Imperial General Headquarters Army Section, mid-1941–Aug45;* HistSec, G–2, GHQ, FEComd, Japanese Studies in WW II Monograph no. 55, *Central Pacific Operations Record, Apr–Nov44;* Northern Marianas GruO A–4 (with maps), dtd 24May44 in 4th MarDiv RepTranslations made on Saipan, hereafter *4th MarDiv Translations.*

Truk, had responsibility for the defense of the Marianas. The work of building, improving, and maintaining the island fortifications was the task of the *5th Base Force* and its attached units, the *5th Communications Unit* and *5th Defense Force.* Logistical support of the Marianas garrison was turned over to the *Fourth Fleet Naval Stores Department* and the *4th Naval Air Depot,* both located at Saipan. Originally the Marianas forces were to strengthen the defenses of the area and ready themselves for a possible war, but once Japan had begun preparing to strike at Pearl Harbor, the *5th Base Force* received orders to lay plans for the capture of Guam.

War came, Guam surrendered, and the Marianas became a rear area as Japanese troops steadily advanced. Since Saipan served primarily as a staging area, a sizeable garrison force was not needed. In May 1943, when the Gilberts marked the eastern limits of the Japanese empire, only 919 troops and 220 civilians were stationed on Saipan. As American forces thrust westward, reinforcements were rushed into the Marianas area.

During February 1944, Kwajalein and Eniwetok Atolls, both important bases, were seized by American amphibious forces. Within the space of three weeks, Saipan became a frontline outpost rather than a peaceful staging area. That portion of the *5th Special Base Force* [13] located at Saipan, a contingent which now numbered 1,437 men, was too weak to hold the island against a determined assault.

After the collapse of the Marshalls defenses and the withdrawal of fleet units from Truk, the Japanese established the *Central Pacific Area Fleet* under the command of Vice Admiral Chiuchi Nagumo, who had led the Pearl Harbor raid, the successful foray into the Indian Ocean, and still later the ill-fated expedition against Midway. Nagumo's headquarters, charged with the defense of the Marianas, Bonins, and Palaus, was subordinate to Admiral Toyoda's *Combined Fleet,* now based at Tawi Tawi in the Philippines. The *Fourth Fleet,* relieved of overall responsibility for the Mandated Islands, retained control over Truk and the other eastern Carolines, as well as the isolated Marshalls outposts. (See Map I, Map Section.)

Nagumo's command, however, was an administrative organization unable to exert effective tactical control over the *Thirty-first Army*, the land force assigned to defend the various islands in the Marianas, Bonins, and Palaus. Initially, Nagumo was appointed supreme commander throughout this sector, but *Headquarters, Thirty-first Army* objected to being subordinated to a naval officer. By mid-March, Nagumo and Lieutenant General Hideyoshi Obata, the army commander, had sidestepped the issue, each one pledging himself to refrain from exercising complete authority over the other.

Instead of regarding the various island groups as an integrated theater under a unified command, the two officers, in keeping with an Army-Navy

[13] On 10 April 1942, the *5th Base Force* was reorganized and redesignated the *5th Special Base Force.* Chief, WarHistOff, DefAgency of Japan, ltr to Head, HistBr, G-3, HQMC, dtd 9Mar63.

agreement worked out by *Imperial General Headquarters*,[14] chose to treat each island as an individual outpost, to be commanded by the senior Army or Navy officer present. At Saipan, for example, Rear Admiral Sugimura in command of the *5th Special Base Force*[15] was originally given control over the defense of the island, but Obata reserved the right, in case of an American attack, of either commanding in person or designating a land commander of his own choice. Thus, the compromise left the general free to assume complete charge of the ground defense of any island in immediate danger of being stormed by Americans. Obata could assume overall responsibility for troop dispositions, coastal defense batteries, antiaircraft defenses, beach defenses, and communications. The employment of aircraft and the use of radar, however, would remain beyond his jurisdiction.[16]

This revision of the Central Pacific command structure reflected the increasing concern with which the Japanese high command regarded the defenses of Saipan and the other islands which lay in the path of the American offensive. Between February and May, two divisions, two independent brigades, two independent regiments, and three expeditionary units were rushed to the Marianas to form the *Marianas Sector Army Group* of Obata's *Thirty-first Army*. Naval strength in the islands was augmented by the arrival of the *55th* and *56th Guard Forces*[17] as well as antiaircraft and aviation units.

Prowling American submarines preyed upon the convoys that carried these reinforcements westward. One regiment of the *29th Division*, destined for Guam by way of Saipan, lost about half its men when a transport was torpedoed. Submarines also destroyed a vessel carrying some 1,000 reinforcements to the *54th Guard Force*, the unit which had garrisoned Guam since the capture of that island in December 1941. Five of the seven transports carrying elements of Lieutenant General Yoshitsugu Saito's *43d Division* to Saipan went down en route to the Marianas, but the ships that stayed afloat managed to rescue most of the survivors. Units in this convoy lost about one-fifth of their total complement, most of these casualties from a single regiment. Also destroyed were numerous weapons and a great deal of equipment. These successful undersea operations, strange to relate, resulted in the arrival at Saipan of some unscheduled reinforcements. About 1,500 troops, originally headed for Yap, were rescued when their transports were torpedoed and were added to the garrison of the Marianas bastion instead. Other survivors, members of units bound for the *Palau Sector Army Group*, also were put ashore at Saipan. In addition to these men, approximately 3,000 troops destined for garrisons on other islands of the Marianas and Carolines, were present on Saipan.[18]

Work on additional fortifications in the Marianas was handicapped by the

[14] *Ibid.*

[15] *Ibid.*

[16] CinCPac–CinCPOA Item No. 12,058, Thirty-first Army Stf Diary, 25Feb44–31Mar44.

[17] Japanese comments Saipan, *op cit.*

[18] *Ibid.*

deadly submarines which destroyed vital cargos as efficiently as they claimed Japanese lives. Obata's chief of staff acknowledged the double effect of the underwater attacks. "The special point of differentiation in the Saipan battle," he observed late in the campaign, "is that units sunk late in May [the troops intended for Yap and the Palaus] and the 8,000 men who landed on 7 June [members of the *43d Division*] eventually landed not up to full combat strength. . . Moreover, as they were still in the process of reorganization at the time of attack, our fighting strength on Saipan was in the process of flux." [19] Could these ill-equipped troops be put to work building obstacles and gun emplacements? The answer was an emphatic "No." As the chief of staff pointed out, "unless the units are supplied with cement, steel reinforcements for cement, barbed wire, lumber, etc., which cannot be obtained in these islands, no matter how many soldiers there are, they can do nothing in regard to fortification but sit around with their arms folded, and the situation is unbearable." [20]

The submarine campaign did not reach peak intensity in time to prevent the Japanese from building airfields throughout the Marianas. By June 1944, Guam boasted two operational fields and two others not yet completed, Tinian had three airfields with work underway on a fourth, and both Rota and Pagan were the sites of still other flight strips. At Saipan, the old Aslito airfield, now less important than the new Tinian bases, was capable of handling extensive aerial traffic. One emergency strip was built near Charan Kanoa, but another such field, begun at Marpi Point, was as yet unfinished. Work on land defenses, however, was not as far advanced as airfield construction.

The defenders of Saipan planned to defeat the invaders on the beaches, but General Obata also hoped to prepare "positions in depth, converting actually the island into an invulnerable fortress." [21] The coastal defenses sited to cover probable avenues of approach were completed. Five Navy coastal defense batteries on Saipan and one at outlying Maniagassa Island guarded the approaches lying between Agingan and Marpi Points. Two of these batteries, one armed with a 120mm and the other with a 150mm gun, could join twin-mounted 150mm pieces near Tanapag in engaging targets off the northwest coast. A 40mm battery of three guns protected Marpi Point, while Magicienne Bay was blanketed by the fires of four batteries, two of them mounting 200mm weapons. A lone battery of two 150mm guns guarded Nafutan Point. Army and Navy dual-purpose antiaircraft weapons reinforced the fires of these batteries, as did the Army artillery units located in southern Saipan.[22]

[19] NTLF G-2 Rpt, p. 65, in *NTLF OpRpt*.

[20] CofS, Thirty-first Ar, Rpt of Defenses of Various islands, dtd 31May44, in NTLF Translations of Captured Documents (FMFPac File, HistBr, HQMC).

[21] Japanese Monograph No. 55, *op. cit.*

[22] CinCPac–CinCPOA Items Nos. 12,250, Army and Navy AA, Dual Purpose, and Coastal DefBtrys on Saipan, n.d., 12,251, Order of Change of Location of Army AA Btrys on Saipan, dtd 10May44, and 12,252, Disposi-

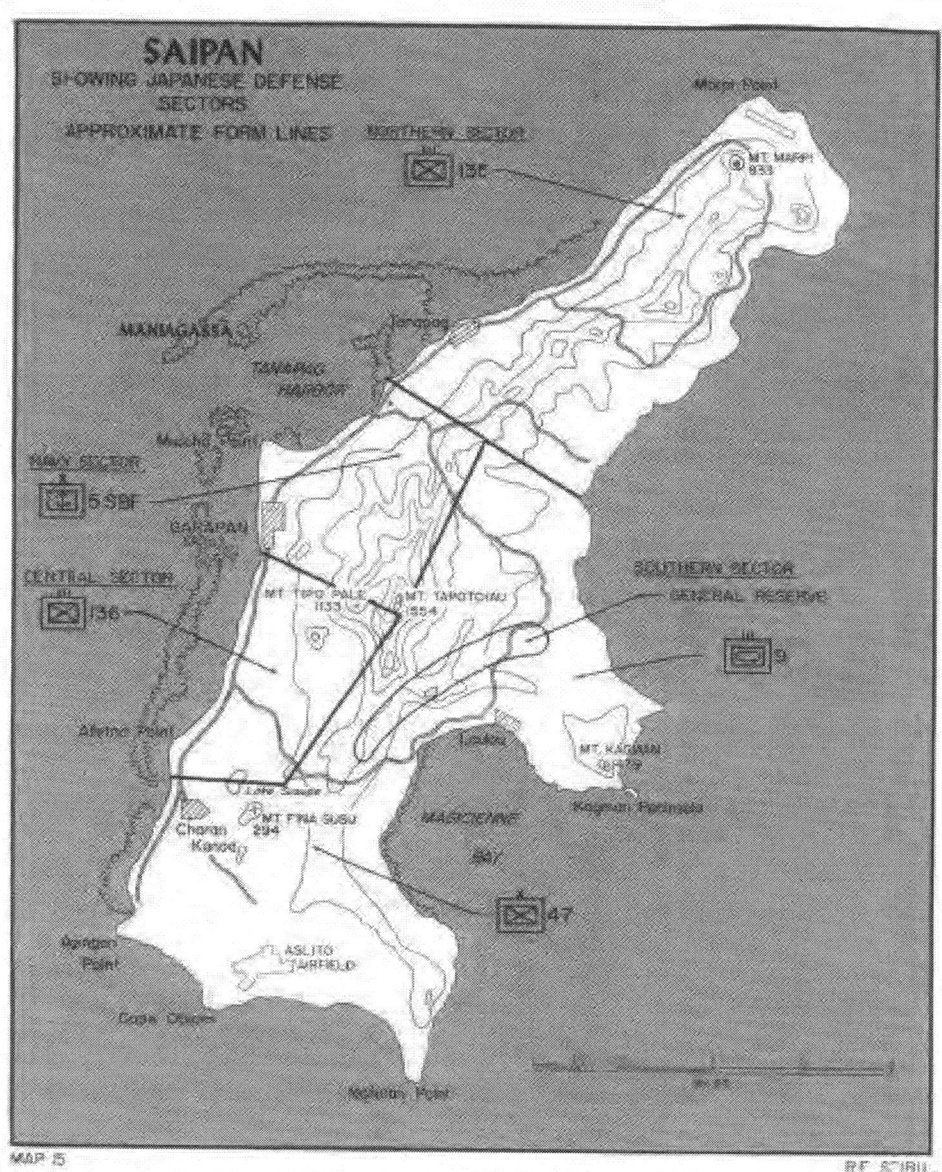

MAP 5

R.F. STIBIL

Those beaches judged best suited for amphibious landings were guarded by powerful forces backed by comparatively feeble local reserves. A short distance inland, the enemy had prepared a second line designed to contain penetrations of the coastal perimeter until a counterattack could be organized. A tank regiment shouldered the main burden of eradicating any American salient, but Obata also held out four rifle companies and two shipping companies as a general reserve to join in counterthrusts. When the Japanese commander turned his attention to the rugged interior of Saipan, he discovered himself to be short of critical building materials, vital time, and necessary engineer units. The invulnerable fortress depicted by Obata was not fully realized, but he nevertheless selected certain redoubts, most of them in forbidding terrain. If the Americans smashed the first two lines, caves, gorges, and dense thickets would have to serve as pillboxes, antitank barriers, and barbed wire.

The Japanese plan of deployment divided Saipan into four sectors, three under Army command and one nominally entrusted to the Navy. Since 25,469 soldiers and only 6,160 sailors [23] were serving on the island, the division of responsibility seems equitable, but many of the naval units specialized in supply or administration, so Army troops were stationed in all areas. The northern sector, which lay beyond a line drawn across the island just south of Tanapag, was protected by two battalions of the *135th Infantry Regiment* plus reinforcing elements. South of this zone, bounded on the east by a line drawn down the axis of the island and on the south by another line that stretched inland from just south of Garapan to include Mount Tapotchau, was the Navy sector, manned by a reinforced battalion from the *136th Infantry Regiment* and the *5th Special Base Force*. The naval unit included the recently arrived *55th Guard Force* as well as the *Yokosuka 1st Special Naval Landing Force*, which had served at Saipan since the autumn of 1943. The central sector, defended by elements of the *136th Infantry Regiment*, included that portion of Saipan that lay west of the spine of the island and north of a line drawn below Afetna Point. The remainder of the island was organized as the southern sector. Here Obata concentrated the bulk of his artillery and antiaircraft units, the *47th Independent Mixed Brigade*, the *9th Tank Regiment*, and the remainder of his *43d Division*, which included a general reserve, certain shipping companies, and stragglers from several miscellaneous units. (See Map 15.)

Although Saipan and her neighboring islands were heavily reinforced, Japanese planners felt that the Palaus rather than the Marianas would be Nimitz' next objective. According to Admiral Toyoda, commander of the *Combined Fleet*, "while the possibility of your offensive against the Marianas was not ignored or belittled, we thought the probability would be that your attack would be directed against Palau, and that was the reason for our adop-

---

tions of Navy Dual Purpose and Coastal Def Btrys on Saipan and Tinian, dtd 15May44.

[23] Crowl, *The Marianas*, p. 454. These revised figures will be used in preference to the estimates made by NTLF during the Saipan battle.

tion of the *A–GO* operation plan, which was to our advantage because of the shorter distance involved and would eliminate the need of tankers to some extent. . . ." [24]

General MacArthur's sudden descent upon Biak in the Schouten Islands off New Guinea, an operation that began on 27 May, diverted attention from both the Palau chain and the Marianas. Since Biak possessed airfields from which planes could attack American ships moving northward into the Palaus, the Japanese prepared the *KON* plan, a scheme for reinforcing the threatened island. The first attempts to aid the embattled garrison ended in frustration, so Toyoda decided to commit the modern battleships *Yamato* and *Musashi*, the most powerful surface units of the Japanese fleet. As this strengthened *KON* task force was assembling, American carriers hit the Marianas, so the enemy admiral left the reinforcement of Biak to destroyers, barges, and other small craft and ordered his forces to execute *A–GO*.

On 13 June, Admiral Jisaburo Ozawa led his *Mobile Fleet*, the *A–GO* striking force, from Tawi Tawi toward the Marianas. The *Yamato* and *Musashi*, with their attendant warships, steamed northward to a refueling rendezvous in the Philippine Sea, where they would join Ozawa's armada. Nimitz' blow at the Marianas caught the enemy somewhat off balance, for the ships dispatched toward Biak and the planes massed in the Palaus and eastern Carolines would have to be redeployed if they were to take part in the scheduled annihilation of the American expedition. The shifting of the *A–GO* battlefield from the Palaus northward also forced Ozawa to steam a greater distance, pausing en route to refuel at sea. Yet, an American attack on the Marianas was not unexpected. *A–GO* could succeed, provided the Saipan garrison held firm and the 500 land-based planes promised to augment Ozawa's carrier squadrons actually arrived in the Marianas.

When American battleships arrived off Saipan, General Obata was absent from his headquarters on a tour of inspection of the Palaus. When he realized that Saipan was in peril, Obata tried to return, but he got only as far as Guam. Tactical command passed to General Saito of the *43d Division*. The savage pounding by naval guns and carrier planes battered the defenders but did not destroy their will to resist. One Japanese admitted that the naval bombardment was "too terrible for words," but he nevertheless was "pleased to think" that he would "die in true Samurai style." [25] A naval officer found momentary respite from his worries when he and a few of his men paused amid the ruins to bolster their spirits with five bottles of beer.[26]

On 14 June, in the midst of the holocaust, Admiral Nagumo issued a warning that "the enemy is at this moment en route to attack us." He went on to predict that American amphibious forces would land no later than July.

[24] *USSBS Interrogation* Nav No. 75, Adm Soemu Toyoda, IJN, II, p. 316.

[25] CinCPac–CinCPOA Item No. 10,051, Extracts from the Diary of an Unidentified Soldier.

[26] CinCPac–CinCPOA Translations and Interrogations, No. 29, Item B–1938, Diary of a Naval Officer, Jun–Jul44.

After pointing out that the Marianas were the Japanese first line of defense, he directed each man to "mobilize his full powers to annihilate the enemy on the beach, to destroy his plan, and to hold our country's ramparts."[27] Along the western beaches of Saipan, members of frontline units were better informed than the admiral, for they could see the buoys which were being set out to aid in controlling the next day's assault.

The Saipan garrison had suffered from the preliminary bombardment, but the defenders were willing to fight. If humanly possible, they would defeat the Marines on the beaches. In the meantime, Ozawa's ships were beginning their voyage toward the Marianas. The portion of *A-GO* that called for aerial surface, and submarine attacks on the advancing American convoy had already gone awry. Possibly, the attackers could be wiped out before a beachhead was established. If not, merely by holding for a comparatively brief time, Saito's men might nevertheless set the stage for a decisive sea battle.

[27] ComCenPacFlt memo, dtd 14Jun44, in *NTLF Translations*.

Saipan: The First Day[1]

The final reports from underwater demolition teams were encouraging, for Kauffman's men had found the reef free of mines and the boat lanes clear of obstacles. As dawn approached, the Americans noted that flags, probably planted after the underwater reconnaissance, dotted the area between the reef and the invasion beaches. These markers, intended to assist Japanese gunners in shattering the assault, were probably helpful to the troops manning the beach defenses, but the artillery batteries, firing from the island interior, were so thoroughly registered and boasted such accurate data that the pennants were unnecessary.[2] Whatever their tactical value, the flags served as a portent of the fierce battle that would begin on the morning of 15 June.

FORMING FOR THE ASSAULT

The transport groups carrying those members of the 2d and 4th Marine Divisions who had not been crammed into the LSTs took station off Saipan at 0520. Two transport divisions steamed toward Tanapag Harbor to prepare for the demonstration to be conducted by the 2d and 24th Marines along with the orphaned battalion, 1/29. The other vessels, however, waited some 18,000 yards off Charan Kanoa. At 0542, Admiral Turner flashed the signal to land the landing force at 0830, but he later postponed H-Hour by 10 minutes.

The preparatory bombardment continued in all its fury as the LSTs approached Saipan and began disgorging their LVTs. Smoke billowed upward from the verdant island, but a short distance seaward, the morning sun, its rays occasionally blocked by scattered clouds, illuminated a gentle sea. Neither wind, waves, nor unforeseen currents impeded the launching of the tractors or the lowering of landing craft.

[1] Unless otherwise noted, the material for this chapter is derived from: *TF 51 OpRpt; TF 56 OpRpt; NTLF OpRpt; 2d MarDiv OpRpt; 4th MarDiv OpRpt;* 2d Mar SAR, Saipan, hereafter *2d Mar SAR;* 6th Mar SAR (Saipan), dtd 18Jul44, hereafter *6th Mar SAR;* 8th Mar SAR, dtd 20Jul44, hereafter *8th Mar SAR;* 23d Mar Final AR, Saipan, dtd 6Sep44, hereafter *23d Mar AR;* 24th Mar Final Rpt on Saipan Op, dtd 28Aug44, hereafter *24th Mar Rpt;* 24th Mar Small URpts, dtd 5May45; 25th Mar Final Rpt, Saipan Op, dtd 18Aug44, hereafter *25th Mar Rpt;* 1/8 Rpt on Ops, Saipan, dtd 17Jul44, hereafter *1/8 OpRpt;* 2/23 Final Rpt (Saipan), n.d., hereafter *2/23 Rpt;* 3/23 Rpt of Saipan Op, dtd 10Jul44, hereafter *3/23 OpRpt;* 2/24 Narrative of Battle of Saipan, 15Jun–9Jul44, n.d., hereafter *2/24 Narrative;* 1/25 Rpt on Saipan, dtd 19Aug44, hereafter *1/25 Rpt;* 3/25 Cbt Narrative of Saipan Op, n.d., hereafter *3/25 Narrative;* 3/25 Saipan Saga, n.d., hereafter *3/25 Saga;* Crowl, *Marianas Campaign;* Hoffman, *Saipan;* Morison, *New Guinea and the Marianas.*

[2] LtCol Wendell H. Best ltr to CMC, dtd 8Jan50, quoted in Hoffman, *Saipan*, p. 45. No copy of this letter has been found.

Nearest the beaches that morning were the two battleships, two cruisers, and six destroyers charged with the final battering of the defenses which the Marines would have to penetrate. Beyond these warships, some 5,500 yards from shore, the LSTs carrying the assault elements of both divisions paused to set free their amphibian tractors. Control craft marked by identifying flags promptly took charge of the LVTs and began guiding them into formation. Farthest out to sea were the landing ships that carried field artillery and antiaircraft units and the LSDs that had ferried to Saipan the tank battalions of both divisions.

As the landing craft swarmed toward the line of departure, their movement was screened by salvos from certain of the fire-support units. Other warships lashed out at those areas from which the enemy might fire into the flanks of the landing force. Agingan Point and Afetna Point shuddered under the impact of 14-inch shells, while to the north, the *Maryland* hurled 16-inch projectiles into Mutcho Point and Maniagassa Island. The naval bombardment halted as scheduled at 0700 for a 30-minute aerial attack. When the planes departed, Admiral Hill, the designated commander of the landing phase, assumed control of the fire support ships blasting the invasion beaches. The naval guns then resumed firing, raising a pall of dust and smoke that made aerial observation of the southwestern corner of Saipan almost impossible.[3]

At the line of departure, 4,000 yards from the smoke-shrouded beaches, 96 LVTs, 68 armored amphibian tractors, and a dozen control vessels were forming the first wave. These craft were posted to the rear of a line of 24 LCI gunboats. The remaining waves formed seaward of the line of departure to await the signal to advance toward the dangerous shores. Beyond the lines of tractors, the boats carrying reserve units maneuvered into position for their journey to the transfer area just outside the reef, where they would be met by tractors returning from the beaches. The LSTs assigned to the artillery units prepared to launch their DUKWs and LVTs, while the tank-carrying LCMs got ready to wallow forth from the LSDs. The control boats organizing these final waves rode herd on their charges to insure that the beachhead, once gained, could be rapidly reinforced.

At 0812, the first wave was allowed to slip the leash and lunge, motors roaring, toward shore. Ahead of these LVTs were the LCI(G)s which would pass through the line of supporting warships to take up the hammering of the beaches. Within the wave itself, armored amphibians stood ready to thunder across the reef and then begin their own flailing of the beaches. Overhead were the aircraft selected to make the final strike against the shoreline.

To the left of Afetna Point, looking inland from the line of departure, Gen-

[3] "Control of the fire support ships reverted to Adm Turner at 0910, following the termination of the 'landing phase.' Thereafter, the control of fire support remained with Adm Turner except during periods of darkness when Adm Turner retired to the eastward of Saipan with ships not actually being unloaded." *Hill comments Saipan.*

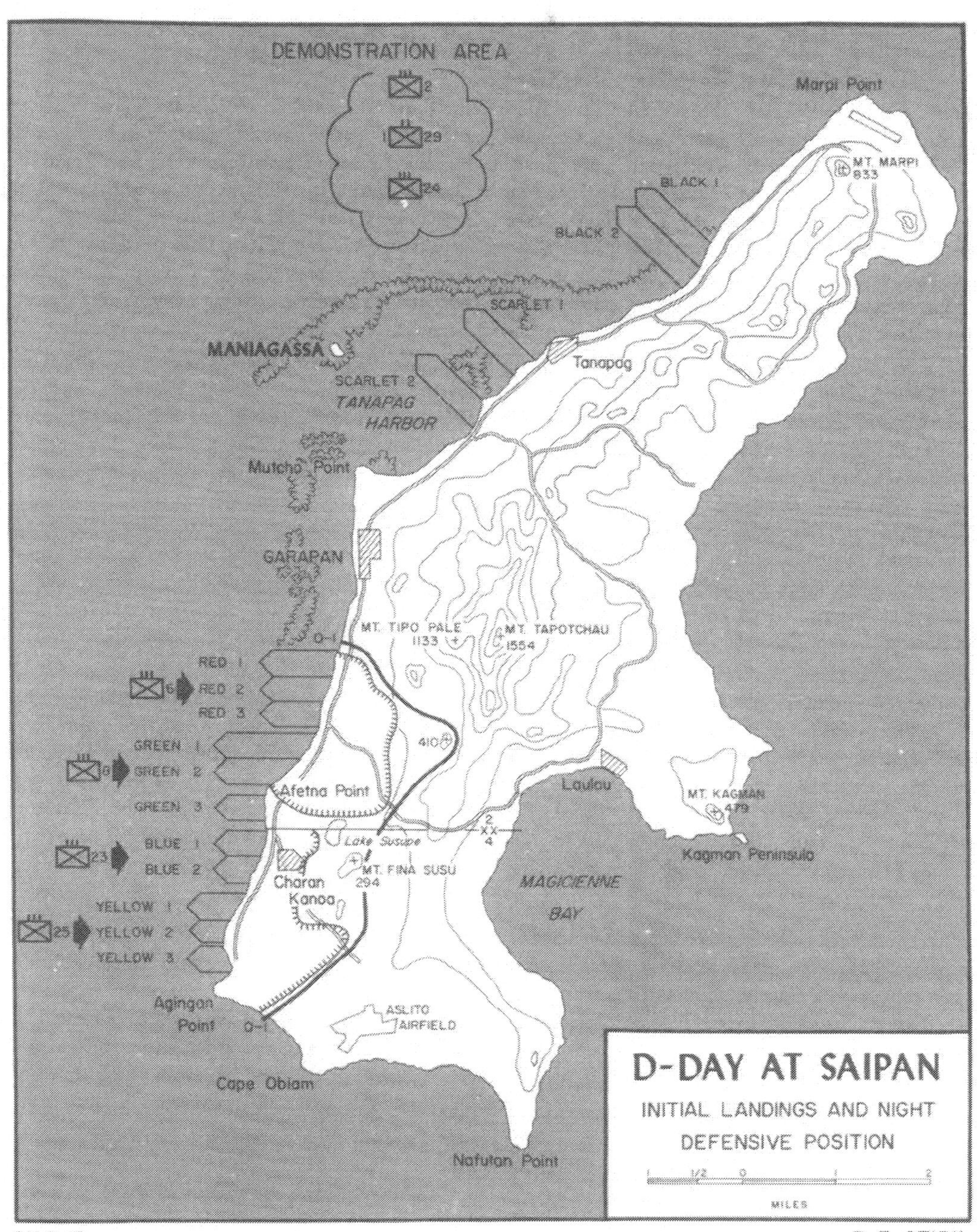

MAP 16

R. F. STIBIL

eral Watson's 2d Marine Division, two regiments abreast, surged toward the Red and Green Beaches. Farthest left was the 6th Marines, commanded by Colonel James P. Riseley. The assault battalions of the regiment were to storm two 600-yard segments of the coast labeled Red 2 and Red 3. On Riseley's flank, beyond a 150-yard gap, was Colonel Clarence R. Wallace's 8th Marines, also landing on a 1,200-yard, two-battalion front. Included in the 8th Marines zone, divided into Beaches Green 1 and 2, was the northern half of Afetna Point. To the right of General Watson's troops lay 800 yards of comparatively untroubled ocean, but off Charan Kanoa the seas were churned white by the LVTs carrying General Schmidt's 4th Marine Division. Next to the gap, within which two fire-support ships were rifling high explosives into the island, was the 23d Marines, under the command of Colonel Louis R. Jones. Separated by a lane of 100 yards from Jones' two assault battalions were the two battalions that were leading Colonel Merton J. Batchelder's 25th Marines toward its objective. The 23d Marines was to seize Beaches Blue 1 and 2, while the 25th Marines crossed Yellow 1 and 2. The frontage assigned each battalion was 600 yards. The right limit of Yellow 3, southernmost of the beaches, lay a short distance north of Agingan Point. (See Map 16.)

These two divisions were Admiral Turner's right hand, his knockout punch. As he delivered this blow, he feinted with his left hand, the units that had been sent toward Tanapag Harbor.

THE TANAPAG DEMONSTRATION

Since 14 June, two old battleships, a cruiser, and four destroyers had been shelling the coastline from Garapan to Marpi Point. While the assault waves were forming off Charan Kanoa on the morning of the 15th, the transports lying off the entrance to Tanapag Harbor began lowering their landing craft. Except for the intelligence section of the 2d Marines, no troops embarked in these boats, which milled about approximately 5,000 yards from shore and then withdrew. By 0930, the craft were being hoisted on board the transports.

The maneuvering of the landing craft drew no response from Japanese guns, nor did observers notice any reinforcements being rushed into the threatened sector. A prisoner captured later in the campaign, an officer of the *43d Division* intelligence section, stated that the Japanese did not believe that the Marines would land at Tanapag Harbor, for on D minus 1 the heaviest concentrations of naval gunfire, as well as the bulk of the propaganda leaflets, had fallen in the vicinity of Charan Kanoa. The enemy, though, was not absolutely certain that he had correctly diagnosed Admiral Turner's intentions, so the *135th Infantry Regiment* was not moved from the northern sector.[4] Admiral Turner's demonstration had immobilized a portion of the Saipan garrison, but it had not forced the Japanese to weaken the

[4] NTLF Spl Interrogation of Maj Kiyoshi Yoshida, IJA, dtd 11Jul44, app I to NTLF G–2 Rpt, pt II, in *NTLF OpRpt*.

concentration of troops poised to defend the southwestern beaches.

THE LANDINGS

Although the demonstration drew no fire, the enemy reacted violently to the real landing. A few shells burst near the line of departure as the LVTs were starting toward shore, but this enemy effort seemed feeble in comparison to the American bombardment which was then reaching its deafening climax. Warships hammered the beaches until the tractors were within 300 yards of shore, and concentrated on Afetna Point until the troops were even closer to the objective. Carrier planes joined in with rockets, 100-pound bombs, and machine gun fire when the first wave was 800 yards from its goal. The pilots, who continued their attacks until the Marines were ashore, carefully maintained a 100-yard safety zone between the point of impact of their weapons and the advancing LVTs.

Bombs, shells, and rockets splintered trees, gouged holes in Saipan's volcanic soil, and veiled the beaches in smoke and dust. The scene was impressive enough, but one newspaper correspondent nonetheless scrawled in his notebook: "I fear all this smoke and noise does not mean many Japs killed." [5] The newspaperman was correct. From the midst of the seeming inferno, the Japanese were preparing to fight back.

As soon as the tractors thundered across the reef, they were greeted by the fires of automatic cannon, antiboat guns, artillery pieces, and mortars. To the men of the 2d Marine Division it seemed that the shells were bursting "in an almost rhythmical patter, every 25 yards, every 15 seconds . . . ." [6] Japanese artillery units had planned to lavish 15 percent of their ammunition on the approaching landing craft and an equal amount on the beaches.[7] Some of these projectiles were bound to find their mark. Here and there an LVT became a casualty. Such a victim "suddenly stood on end and then sank quivering under a smother of smoke. Bloody Marines twisted on its cramped deck, and in the glass-hatched driver's cabin another Marine slumped among the stained levers." [8] In spite of their losses, the assault waves pressed forward, and by 0843 the first of the troops were ashore.

The 2d Marine Division, bound for the beaches on the left, landed somewhat out of position. Since control craft could not cross the reef, the LVTs were on their own during the final approach. Drivers found it difficult to maintain direction in the face of deadly fire, and a strong northerly current, undetected by the previous day's reconnaissance, further complicated their task. Commander Kauffman's underwater scouts had landed during different tidal conditions, so they did not encounter the treacherous current. Thus, the drift of sea, the inability of control vessels to surmount the reef, and the Japanese fusillade combined to

[5] Robert Sherrod, *On to Westward: War in the Central Pacific* (New York: Duell, Sloan, and Pearce, 1945), p. 47.

[6] Johnston, *Follow Me!*, p. 179.

[7] CinCPac–CinCPOA Item No. 9,604, Saipan ArtyPlan, n.d.

[8] Johnston, *Follow Me!*, p. 179.

force the division to land too far to the left.

The 6th Marines was scheduled to cross Red 2 and 3, but 2/6, commanded by Lieutenant Colonel Raymond L. Murray, and Lieutenant Colonel John W. Easley's 3/6 came ashore some 400 yards north of their goals, arriving on Red 1 and 2 respectively. In the zone of the 8th Marines, the situation was more serious. Lieutenant Colonel Henry P. Crowe's 2/8 and 3/8, under Lieutenant Colonel John C. Miller, Jr., landed on Green 1, some 600 yards from the regimental right boundary. Since the enemy had dropped a curtain of fire over the beaches, this accidental massing of troops contributed to the severe losses suffered during the day.

The 4th Marine Division landed as planned, with 3/23, commanded by Lieutenant Colonel John J. Cosgrove, and 2/23, under Lieutenant Colonel Edward J. Dillon, seizing footholds on Blue 1 and 2, while Lieutenant Colonel Lewis C. Hudson's 2/25 and Lieutenant Colonel Hollis U. Mustain's 1/25 landed on Yellow 1 and 2. Once ashore, the attached LVTs and armored amphibians were to have fanned out to overrun Agingan Point, Charan Kanoa, and the ridge line some 2,000 yards inland of the coast. Enemy fire, however, prevented the coordinated thrust upon which General Schmidt had counted. Portions of the division advanced as far as the ridges, but other units were forced to abandon their tractors at the beaches.[9] All along the western beaches, the attack was losing its momentum. The next few hours could prove critical.

THE FIGHT FOR THE RED BEACHES.

During the planning of the Saipan operation, General Watson had expressed doubts concerning the soundness of the Northern Troops and Landing Force scheme of maneuver. The Commanding General, 2d Marine Division, did not believe that the LVTs could scale the embankments, thread their way through the rocks, or penetrate the swamps that in many places barred the exits from the beaches. Instead of having the tractors advance to the 0–1 Line, he wanted the LVT(A)s to move a short distance inland and keep the defenders pinned down while the first wave of LVTs cleared the beaches and discharged their troops. Succeeding waves would halt on the beaches, unload, and return to the transfer area. Watson was convinced that the tractors should not attempt to advance beyond the railroad line running northward from Charan Kanoa. General Holland Smith accepted these suggestions and permitted the 2d Marine Division to attack on foot from the railroad to 0–1. General Schmidt, however, chose to rely on his LVTs to execute the original scheme of maneuver in his division zone.

---

[9] An LVT battalion commander attributed the lack of success of this use of tractors as combat vehicles ashore not so much to Japanese resistance but rather "to our inexperience in this type of assault, compounded by inadequate preparations, particularly in ensuring the coordinated movement of troop-carrying LVTs with the supporting LVT(A)s." Col Victor A. Croizat ltr to Head, HistBr, G–3, HQMC, dtd 5Mar63.

That General Watson obtained a modification of the plan was fortunate, for intense enemy fire and forbidding terrain halted the tractors near the beaches. On Red 1 and 2, the initial thrust of the 6th Marines stalled about 100 yards inland. The captured strip of sand was littered with the hulks of disabled tractors. Here the wounded lay amid the bursting shells to await evacuation, while their comrades plunged into the thicket along the coastal highway.

For the most part, the Marines were fighting an unseen enemy. A Japanese tank, apparently abandoned, lay quiet until the assault waves had passed by and then opened fire on Lieutenant Colonel William K. Jones' 1/6, the regimental reserve, as that unit was coming ashore. Rounds from a rocket launcher and rifle grenades permanently silenced the tank and killed its occupants. From the smoke-obscured ground to the front of 3/6, a machine gun poured grazing fire into the battalion lines. Equally impersonal, and perhaps more deadly, were the mortar and artillery rounds called down upon the advancing Marines by observers posted along the Japanese-held ridges that formed the 0–1 Line.

Occasionally, small groups of Japanese from the *136th Infantry Regiment* suddenly emerged from the smoke, but the enemy preferred mortar, artillery, and machine gun fires to headlong charges. A few minutes after 1000, as Colonel Riseley was establishing his regimental command post on Red 2, between 15 and 25 Japanese suddenly materialized and began attacking southward along the beach. The bold thrust accomplished nothing, for the enemy soldiers were promptly cut down.

Light armor from Colonel Takashi Goto's *9th Tank Regiment* made two feeble counterattacks against the 6th Marines. At noon, two tanks rumbled forth from their camouflaged positions to the front of 2/8 and started southward along the coastal road. Evidently the tank commanders were bewildered by the smoke, for they halted their vehicles within Marine lines. The hatch of the lead tank popped open, and a Japanese thrust out his head to look for some familiar landmark. Before the enemy could orient himself, Marine rocket launcher teams and grenadiers opened fire, promptly destroying both tanks. An hour later, three tanks attempted to thrust along the boundary between the 1st and 2d Battalions. Two of the vehicles were stopped short of the Marine positions, but the third penetrated to within 75 yards of Colonel Riseley's command post before it was destroyed.

The first few hours had been costly for Riseley's 6th Marines. By 1300, an estimated 35 percent of the regiment had been wounded or killed. Lieutenant Colonel Easley, though wounded, retained command over 3/6 for a time. Lieutenant Colonel Murray, whose injuries were more serious, turned 2/6 over to Major Howard J. Rice. Rice, in turn, was put out of the fight when, for the second time within five hours, a mortar round struck the battalion command post. Lieutenant Colonel William A. Kengla, who was accompanying the unit as an observer, took over until Major LeRoy P. Hunt, Jr., could come ashore.

In spite of the losses among troops and

leaders alike, the attack plunged onward. By 1105, the shallow initial beachhead had been expanded to a maximum depth of 400 yards. Twenty minutes later, Lieutenant Colonel Jones' 1/6 was ordered to pass through 3/6, which had been severely scourged by machine gun fire, and attack to the 0–1 Line, where it would revert to reserve by exchanging places with the units it had just relieved. This planned maneuver could not be carried out. The 1st Battalion could not gain the ridge line, and as the 6th Marines moved forward, the regimental frontage increased until all three battalions were needed on line.

During the day's fighting, a gap opened between the 6th and 8th Marines. Colonel Riseley's troops, manning a dangerously thin line and weary from their efforts, could extend their right flank no farther. Colonel Wallace's 8th Marines, which had undergone a similar ordeal, was in much the same condition.

THE GREEN BEACHES AND AFETNA POINT

The key terrain feature in the zone of the 8th Marines was Afetna Point which straddled the boundary between the 2d and 4th Marine Divisions. Since the company charged with capturing Afetna Point would have to attack toward the flank of General Schmidt's division, about half the unit was issued shotguns. These short-range weapons would not be as dangerous as M1s to friendly troops, and their wide patterns of dispersion would make up for their comparative inaccuracy. The attackers, Marines of Company G, also carried their regularly assigned weapons for use after the point had been secured.

While coming ashore, Wallace's command had suffered "miraculously few LVT casualties" [10] in spite of the ponderous barrage falling on and near the beaches. Both assault battalions, Crowe's 2/8 and Miller's 3/8, landed on the same beach, Green 1, and their component units became intermingled. In the judgment of the regimental commander, "If it had not been for the splendid discipline of the men and junior officers, there would have been utter confusion." [11] The various commanders, however, could not be certain of the exact location and composition of their organizations.

After a brief pause to orient themselves, the companies began fanning out for the attack. On the right, Company G of Crowe's battalion, its flank resting upon the Charan Kanoa airstrip, pushed southward along Green 2 toward Afetna Point. The advance was bitterly opposed. Japanese riflemen fired across the narrow runway into the exposed flank of the company until they were killed or driven off by Marine mortars and machine guns. On the opposite flank were emplaced nine antiboat guns. Fortunately for Company G, the Japanese gunners doggedly followed their orders to destroy the incoming landing craft, so the Marines were able to attack these emplacements from the rear. By darkness, when the company dug in for the night, all but two of the gun positions had been overrun, and all of Green

[10] *8th Mar SAR*, p. 1.

[11] *Ibid.*

2, including the northern half of Afetna Point, was in American hands. In his report of the Saipan operation, Colonel Wallace expressed his belief that because of the confused landing, the capture of the point was delayed by 24 hours.

While one company was battling to join forces with the 4th Marine Division and secure use of the boat channel that led to Green 3, the rest of 2/8 was advancing toward the marsh extending northward from Lake Susupe. Elements of the battalion crossed the swamp, only to discover they were isolated, and had to fall back to establish a line along the firm ground to the west. On the left, 3/8 pushed directly inland from Green 1.

The regimental reserve, 1/8, commanded by Lieutenant Colonel Lawrence C. Hays, Jr., was ordered ashore at 0950. One of Hays' companies was sent toward the airstrip to protect the left flank of the unit attacking Afetna Point. The two remaining rifle companies were committed along the boundary between the 2d and 3d battalions.

The next landing team to reach the Green Beaches was Lieutenant Colonel Guy E. Tannyhill's 1/29, the division reserve. Lieutenant Colonel Tannyhill's Marines, who had taken part in the feint off Tanapag, came ashore early in the afternoon and were attached to the 8th Marines. Company B was ordered to seal a gap in the lines of 2/8, but the reinforcing unit became lost, and Company A was sent forward in its place. This second attempt was thwarted by Japanese forward observers who promptly called for artillery concentrations which halted the Marines short of the front lines. While the men of Company A were seeking cover from the deadly shells, Company B found its way into position to close the opening.

The 8th Marines had battled its way as far inland as the swamps. On the left, the opening between Wallace's regiment and the 6th Marines was covered by fire. The actual lines of the 8th Marines began in the vicinity of the enemy radio station near the regimental left boundary, continued along the western edge of the swamp, and then curved sharply toward Afetna Point. In carving out this beachhead, the regiment had suffered about the same percentage of casualties as had the 6th Marines. Because of the intermingling of the assault battalions, Colonel Wallace could not at the time make an accurate estimate of his losses. The problem of reorganizing 2/8 and 3/8 was complicated by the grim resistance and the loss of both battalion commanders. Lieutenant Colonels Crowe and Miller had been wounded seriously enough to require evacuation from the island. Command of 2/8 passed to Major William C. Chamberlain, while Major Stanley E. Larson took the reins of the 3d Battalion.

CHARAN KANOA AND BEYOND

South of Afetna Point and Charan Kanoa pier lay the beaches assigned to Colonel Jones' 23d Marines. At Blue 1, eight LVTs, escorted by three armored amphibians and carrying members of Lieutenant Colonel Cosgrove's 3/23, bolted forward along the only road leading beyond Charan Kanoa. The column exchanged shots with Japanese

FIRE TEAM *member dashes across fire-swept open ground past a dud naval shell as Marines advance inland from Saipan beaches. (USMC 83010)*

JAPANESE MEDIUM TANKS *knocked out during the night counterattack on 17 June at Saipan. (USN 80-G-287376)*

snipers who were firing from the ditches over which the highway passed, but it encountered no serious resistance in reaching Mount Fina Susu astride the 0–1 Line. The troops dismounted and established a perimeter atop the hill, a position exposed to direct fire from Japanese cannon and machine guns as well as to mortar barrages. The LVT(A)s, which mounted flat-trajectory weapons that might have aided the unit mortars in silencing enemy machine guns, halted at the base of the hill. No friendly units were within supporting distance on either flank, but the Marines managed to foil periodic attempts to infiltrate behind them. After dark, the defenders of Fina Susu were ordered to abandon their perimeter and withdraw to the battalion lines.

A similar breakthrough occurred at Blue 2, where five LVT(A)s and a trio of troop-carrying tractors followed the Aslito road all the way to 0–1. Again, the remainder of the battalion, in this case Lieutenant Colonel Dillon's 2/23, was stalled a short distance inland. The advanced outpost had to be recalled that evening.

The 23d Marines was unable to make a coordinated drive to the 0–1 Line. In the north, the Lake Susupe swamps stalled forward progress, and to the south a steep incline, rising between four and five feet from the level beaches but undetected by aerial cameras, halted the tractors. Because of the gap between divisions, the regimental reserve came ashore early in the day to fill out the line as the beachhead was enlarged. At 1055, 1/23, commanded by Lieutenant Colonel Ralph Haas, landed and moved into an assembly area 300 yards inland of Blue 1. The beachhead, however, did not expand as rapidly as anticipated, so the battalion spent the morning standing by to protect the left flank or reinforce the front. After sundown, Haas' troops were ordered to relieve Cosgrove's 3d Battalion.

Although few men actually gained the 0–1 Line, the 23d Marines nevertheless managed to gain a firm hold on the Blue Beaches, in spite of the violent fire and formidable natural obstacles which it encountered. Japanese mortar crews and cannoneers created havoc among the amphibian tractors which were attempting to find routes through either the swamp or the embankment. Yet, the Marines cleared the beaches to battle their way toward the ridges beyond. The ruins of Charan Kanoa were overrun and cleared of snipers. A consolidated beachhead some 800 yards in depth was wrested from a determined enemy. The 23d Marines was ashore to stay.

ACTION ON THE RIGHT FLANK

Agingan Point, south of the beaches upon which Colonel Batchelder's 25th Marines landed, was a thorn in the regimental flank throughout the morning of D-Day. On Yellow 1, the beach farthest from the point, Lieutenant Colonel Hudson's 2/25 landed amid a barrage of high explosives. Approximately half of the LVTs reached the railroad embankment, which at this point ran diagonally inland between 500 and 700 yards from the coastline. LVT(A)s from the Army's 708th Amphibian Tank Battalion spearheaded the drive, pushing steadily forward in spite of small arms fire from the eastern side of the rail line. These Japanese riflemen fell back, but artillery pieces and dual-purpose antiaircraft guns kept pumping shells into the advanced position. A bypassed pair of enemy mortars now joined in the bombardment. Since no friendly troops were nearby, Navy planes were called in to destroy the weapons.

To the south, the assault waves of Lieutenant Colonel Mustain's 1/25 were stopped a dozen yards past the beach. Enfilade fire from Agingan Point inflicted many casualties and prevented the survivors from moving forward. LVTs of the Army's 773d Amphibian Tractor Battalion barely had room to land the succeeding waves. Since bursting shells were churning the narrow strip of sand, the tractor drivers retreated as quickly as they could, sometimes departing before communications gear and crew-served weapons and their ammunition could be completely unloaded.

Focal point of enemy resistance was Agingan Point, a maze of weapons positions, and the patch of woods adjacent to that promontory. About 800 yards to Mustain's front, four or more artillery pieces slammed shells directly into the crowded beachhead. Gradually, however, the Marines worked their way forward, finally reaching 0-1 late in the afternoon.

At 0930, the Japanese made their first attempt to hurl 1/25 into the sea. While troops advanced across the ridge that marked the 0–1 Line, another enemy force attacked from Agingan Point in an effort to roll up the narrow beachhead. The battalion commander called for air strikes and naval gunfire concentrations which ended the threat for the time being. The defenders, however, persisted in their efforts. Early in the afternoon, tanks from the 4th Tank Battalion joined Mustain's infantrymen in wiping out two Japanese companies, thus crushing the strongest counterattack of the day against the division flank.

Immediately upon landing, Lieutenant Colonel Justice M. Chambers' 3/25, the regimental reserve, sent reinforcements to Mustain. In the confusion of landing, portions of two rifle companies, instead of one complete company, were directed toward Agingan Point. The remainder of the reserve moved forward, mopping up in the wake of the advancing assault battalions. About 700 yards inland, Chambers' men took cover along the railroad embankment. From the comparative safety of this position additional reinforcements were despatched to Agingan Point, where 1/25 had by now seized the initiative from the elements of the *47th Independent Mixed Brigade* that had been posted there.

Progress on the southern flank was slow, for a powerful enemy contingent occupied the point. Like the Eniwetok Island garrison, these soldiers had dug and carefully camouflaged numerous spider holes. The defenders waited until a fire team had passed them, then emerged from concealment to take aim at the backs of the Marines. One of the companies detached from Chambers' battalion reported killing 150 Japanese during the afternoon.

In spite of the battering it had received from artillery located in the island's interior, the 25th Marines made the deepest penetration, over 2,000 yards, of the day's fighting. Its battalions had reached the 0–1 Line throughout the regimental zone, but an enemy pocket, completely isolated from the main body, continued to cling to the tip of Agingan Point. Both divisions had gained firm holds on the western beaches.

SUPPORTING WEAPONS AND LOGISTICAL PROBLEMS [12]

After the preliminary bombardment had ended, ships and aircraft continued to support both divisions. Planes remained on station throughout the day. Once the liaison parties ashore had established radio contact with the agencies responsible for coordinating and controlling their missions, the pilots began attacking mortar and artillery positions as well as reported troop concentrations.

Warships played an equally important role in supporting the Marines. From the end of the preparatory shelling until the establishment of contact with the battalions they were to support, the fire-support units blasted targets of opportunity. Subsequent requests from shore fire control parties were checked against calls for air strikes to avoid duplication of effort and the possible destruction of low-flying planes. Perhaps the most striking demonstration of the effectiveness of naval gunfire in support of the day's operations ashore was the work of the battleship *Tennessee* and three destroyers in helping to halt the first counterattack against Mustain's troops.

The 2d Tank Battalion, commanded by Major Charles W. McCoy, and the 4th Tank Battalion, under Major Richard K. Schmidt, also assisted the riflemen in their drive eastward. Armor from McCoy's battalion crawled from the LCMs, plunged into the water at the reef edge, and passed through the curtain of shellfire that barred the way to Green 1. Since the enemy still held Afetna Point, the boat channel leading to Green 3 could not be used as planned. The last tank lumbered ashore at 1530, 2½ hours after the first of the vehicles had nosed into the surf. One company of 14 Sherman medium tanks helped shatter the positions blocking the approaches to Afetna Point. A total of eight tanks were damaged during the day, but seven of these were later repaired.

Heavy swells, which mounted during the afternoon, helped complicate the landing of the 4th Tank Battalion. Company A started toward Blue 2, but en route the electrical systems of two tanks were short-circuited by seawater. Another was damaged after landing. Four of the 14 Shermans of Company B survived shells and spray to claw their way onto the sands of Blue 1. Six tanks of the company were misdirected to Green 2, but only one actually reached its destination, the rest drowned out in deep water; the sole survivor was promptly commandeered by the 2d Tank Battalion. Company C, which landed on Yellow 2 without

[12] Additional sources for this section include: 10th Mar SAR (incl Bn SARs), dtd 22Jul44, hereafter *10th Mar SAR;* 14th Mar Final Rpt, Saipan Op, dtd 31Aug44, hereafter *14th Mar Rpt;* 20th Mar Final Rpt, n.d., hereafter *20th Mar Rpt;* 1/13 Observer's Rpt, Saipan, dtd 13Jul44; 2/18 Narrative Account of Saipan Op, dtd 21Jul44, hereafter *2/18 Narrative;* 4th TkBn CbtRpt (incl CoRpts), dtd 20Aug44.

losing a single tank, supported the advance of Dillon's 2/23. Company D landed 10 of its 18 flame-throwing light tanks, but these machines were held in an assembly area. As far as the 4th Division tankers were concerned, the crucial action of the day was the smashing of the afternoon counterthrust against 1/25.

Two 75mm pack howitzer battalions landed on D-Day to support General Watson's division. Lieutenant Colonel Presley M. Rixey's 1/10 went into position to the rear of the 6th Marines, while the 2d Battalion, commanded by Lieutenant Colonel George R. E. Shell, crossed the airstrip in order to aid the 8th Marines. In crossing the runway, Shell's men were spotted by the enemy, but the ensuing counterbattery fire did not destroy any of their howitzers. Colonel Raphael Griffin established his regimental command post before dark, but none of the 105mm battalions were landed.

South of the 2d Marine Division beachhead, all five battalions of Colonel Louis G. DeHaven's 14th Marines landed on the Blue and Yellow beaches. The 2d Battalion had the greatest difficulty in getting ashore, for its elements were scattered along three different beaches. During reorganization on Blue 2, casualties and losses of equipment to both the sea and hostile fire forced the battalion commander, Lieutenant Colonel George B. Wilson, Jr., to merge his three 75mm batteries into two units. The other pack howitzer battalion, Lieutenant Colonel Harry J. Zimmer's 1/14, was forced to disassemble its weapons and land from LVTs, as the DUKWs that were scheduled to carry the unit failed to return as planned after landing a 105mm battalion. The only firing position available to 1/14 on Yellow 1 was a scant 100 yards from the water. Firing from Yellow 2 was 3/14, a 105mm battalion commanded by Lieutenant Colonel Robert E. MacFarlane, the first element of the 14th Marines to go into action on Saipan. Immediately after landing on Blue 2, Lieutenant Colonel Carl A. Youngdale's 4th Battalion lost four 105s to Japanese mortar fire, but the artillerymen managed to repair the damaged weapons. From its positions on Blue 2, 5/14, commanded by Lieutenant Colonel Douglas E. Reeve, temporarily silenced a Japanese gun that was pounding the beachhead from a range of 1,500 yards.

Supplying the landing force did not prove as difficult at Saipan as it had at Tarawa. By having certain of the LVTs dump boxes of rations and medical supplies, cases of ammunition, and cans of water onto the beaches as the later waves were landing, supply officers were able to sustain the assault troops. Unlike the cargo handlers at Betio, the Saipan shore parties soon had sufficient room to carry out their tasks. Early in the afternoon, supplies began flowing from the transports, across the beaches, and to the advancing battalions. Japanese fire and a lack of vehicles, however, did handicap the D-Day supply effort.

Enemy artillery and mortar concentrations also endangered the lives of the wounded Marines who were waiting on the beaches for the tractors that would carry them out to sea. About 60 percent of the wounded were taken directly to the transports. Although no accurate accounting was made until 17

June, as many as 2,000 men may have been killed or wounded on D-Day.

THE SITUATION ASHORE: THE EVENING OF D-DAY

By darkness on D-Day, the two Marine divisions had succeeded in establishing themselves on the western coast of Saipan. Approximately half of the planned beachhead had been won, but the enemy still held the ridges that dominated captured segments of the coastal plain. The 2d Marine Division manned a line that stretched from the coast about one mile south of Garapan to the middle of Afetna Point. The maximum depth of the division beachhead was about 1,300 yards. Before dark, Colonel Walter J. Stuart had landed one battalion and part of another from his 2d Marines. These reserves provided added strength in the event of a counterattack. Also ashore was General Watson, who now commanded operations from a captured munitions dump inland of the coastal road on the boundary between Red 1 and Red 2. The cached explosives were removed during the night and following morning. (See Map 16.)

In the 4th Marine Division zone, those elements of the 23d Marines that had reached the 0–1 Line fell back some 800 yards during the night. After this adjustment, the front moved from the coastline 800 yards inland along the division boundary, turned south past Charan Kanoa, and then bulged eastward to 0–1. In the right half of General Schmidt's zone of action, a band of Japanese entrenched on Agingan Point prevented the Marines from occupying all the territory west of the critical ridge line. Colonel Franklin A. Hart's 24th Marines was ashore, with elements of its 1st Battalion committed between 2/23 and 2/25, while the rest of the regiment occupied assembly areas. General Schmidt had moved into a command post on Yellow 2.[13]

THE JAPANESE STRIKE BACK

As soon as American carrier planes had begun to hammer the Marianas in earnest, Admiral Toyada signaled the execution of *A-GO*. On 13 June, as it was starting northward from Tawi Tawi, Ozawa's task force encountered the submarine USS *Redfin*, which reported its strength, course, and speed. Another submarine, the USS *Flying Fish*, sighted Ozawa's ships on 15 June, as they were emerging from San Bernardino Strait between Samar and Luzon. The Japanese were by this time shaping an eastward course. On this day, the submarine USS *Seahorse* observed the approach of the warships diverted from Biak, but the enemy jammed her radio, and she was unable to report the sighting until 16 June.

Admiral Spruance was now aware that enemy carriers were closing on the Marianas. Japanese land-based planes also were active, as was proved by an unsuccessful attack upon a group of

---

[13] General Schmidt recalled that several of his staff officers went ashore with him after dark and "after getting dug in, it was suddenly discovered that we were in a supply dump of bangalore torpedoes. We decided to get out quick. An armored vehicle was sent us and we arrived shortly in the temporary CP." Gen Harry Schmidt cmts on draft MS, dtd 4Jun63.

American escort carriers on the night of 15 June. After evaluation of the latest intelligence, Spruance decided on the following morning to postpone the Guam landings, tentatively set for 18 June, until the enemy carrier force had either retreated or been destroyed.

While Ozawa was steaming nearer, the Japanese on Saipan were preparing to carry out their portion of *A-GO*. As one member of the *9th Tank Regiment* confided to his diary, "Our plan would seem to be to annihilate the enemy by morning." [14] First would come probing attacks to locate weaknesses in the Marine lines, then the massive counterstroke designed to overwhelm the beachhead.

The heaviest blows delivered against General Watson's division were aimed at the 6th Marines. Large numbers of Japanese, their formations dispersed, eased down from the hills without feeling the lash of Marine artillery. The two howitzer battalions, all that the division then had ashore, were firing urgent missions elsewhere along the front and could not cover the avenue by which the enemy was approaching. The *California* received word of the movement and opened fire in time to help crush the attack. Before midnight, the Japanese formed a column behind their tanks in an effort to overwhelm the outposts of 2/6 and penetrate the battalion main line of resistance. Star shells blossomed overhead to illuminate the onrushing horde. Riflemen and machine gunners broke the attack, and the *California* secondary batteries caught the survivors as they were reeling back. Although this first blow had been parried, the Japanese continued to jab at the perimeter.

At 0300, regimental headquarters received word that an attack had slashed through the lines of 3/6, but the company sent to block this penetration found the front intact. A similar report received some three hours later also proved false. The enemy, however, maintained his pressure until a platoon of medium tanks arrived to rout what remained of the battalion which the *136th Infantry Regiment* had hurled against the beachhead. In eight hours of intense fighting, the 6th Marines had killed 700 Japanese soldiers.

The 8th Marines was harassed throughout the night by attacks that originated in the swamps to its front. These blows, weak and uncoordinated, were repulsed with the help of fires from 2/10. The enemy did not employ more than a platoon in any of these ill-fated thrusts.

Throughout the sector held by General Schmidt's 4th Marine Division, the Japanese made persistent efforts to shatter the American perimeter. Approximately 200 of the enemy advanced from the shores of Lake Susupe, entered the gap between the divisions, and attempted to overwhelm 3/23. The battalion aided by Marine and Army shore party troops, held firm.

The 25th Marines stopped one frontal attack at 0330, but an hour later the Japanese, advancing behind a screen of civilians, almost breached the lines of the 1st Battalion. As soon as the Marines discovered riflemen lurking behind the refugees, they called 1/14 for artillery support. This unit, out of

[14] JICPOA Item No. 10,238, Diary of Tokuzo Matsuya.

ammunition, passed the request to 3/14, which smothered the attack under a blanket of 105mm shells. The only withdrawal in the 25th Marines sector occurred when a Japanese shell set fire to a 75mm self-propelled gun. Since the flames not only attracted Japanese artillery but also touched off the ammunition carried by the burning vehicle, the Marines in the immediate area had to fall back about 200 yards.

The Japanese had been unable to destroy the Saipan beachhead, but the battle was just beginning. The *Thirty-first Army* chief of staff admitted on the morning of 16 June that "the counterattack which has been carried out since the afternoon of the 15th has failed because of the enemy tanks and fire power." Yet, he remained undaunted. "We are reorganizing," his report continued, "and will attack again."[16] While the battle raged ashore, an enemy fleet was bearing down on the Marianas. If all went as planned, Admiral Ozawa and General Saito might yet trap the American forces.

[16] NTLF G-2, Tgs Sent and Received by Thirty-first Army Hq on Saipan, dtd 25Jul44, p. 4.

214-881 O-67—19

The Conquest of Southern Saipan[1]

THE CAPTURE OF SOUTHERN SAIPAN

On the morning of 16 June, Admiral Spruance visited Admiral Turner's flagship, the *Rocky Mount*, to inform his principal subordinates how he intended to meet the threat posed by the approaching enemy fleet. Spruance wanted the vulnerable transports and other amphibious shipping to stand clear of Saipan until the Japanese carriers could be destroyed. General unloading over the western beaches was to stop at dusk on 17 June, after which transports that were not vital to the operation and all the LSTs would steam eastward from the island. If the cargo carried in any of the ships that had been withdrawn was later needed by the landing force, the necessary vessels, carefully screened by warships, could be sent back to Saipan.[2]

Spruance left the aerial support of operations ashore to planes based on the escort carriers. All of Task Force 58, the faster escort carriers included, was to concentrate on defeating Ozawa's approaching battle fleet. Certain cruisers and destroyers were freed from their mission of protecting Admiral Turner's amphibious force so they could reinforce Admiral Spruance's striking force. Rear Admiral Jesse B. Oldendorf was to station his old battleships, along with their screen of cruisers and destroyers, about 25 miles west of Saipan to shatter a possible night attack by Japanese surface units.[3] In order to detect the kind of surprise blow against which Oldendorf's giants were guarding, Navy patrol bombers flew westward from Eniwetok and, on the night of 17 June, began operating from

[1] Unless otherwise noted, the material in this chapter is derived from: *TF 51 OpRpt; TF 56 OpRpt; NTLF OpRpt; 2d MarDiv OpRpt; 4th MarDiv OpRpt; 27th InfDiv Op Rpt; 2d Mar SAR; 6th Mar SAR; 8th Mar SAR; 10th Mar SAR; 14 Mar Rpt; 23d Mar AR; 24th Mar Rpt; 25th Mar Rpt;* 105th Inf OpRpt, dtd 20Sep44 (WW II RecsDiv, FRC, Alexandria, Va.), hereafter *105th Inf OpRpt;* 106th Inf OpRpt, n.d., hereafter *106th Inf OpRpt; 165th Inf OpRpt; 1/8 OpRpt; 2/23 Rpt; 2/23 OpRpt; 2/24 Narrative; 1/25 Rpt; 3/25 Narrative; 3/25 Saga;* 1/29 SAR, dtd 1Sep44, hereafter *1/29 SAR;* Edmund G. Love, *The 27th Infantry Division in World War II* (Washington: Infantry Journal Press, 1949), hereafter Love, *27th InfDiv History;* Crowl, *Marianas Campaign;* Hoffman, *Saipan.*

[2] Admiral Hill, who remained in the unloading area off Saipan during this period, recalled that "each evening, after consultation with VAC, [he] informed Adm Turner what ships he desired at the anchorage at daylight the following day. In the main, this plan operated very well, and no serious shortages developed in supply to the three divisions ashore." *Hill comments Saipan.*

[3] Units of Oldendorf's Covering Group could be recalled for specific fire support assignments for troops ashore and were so used in several instances during this deployment. *Ibid.*

off the west coast of Saipan. Spruance also ordered the transports carrying the Guam expedition to get clear of the Marianas and take up station east of their objective. The recapture of Guam could wait until the *Imperial Japanese Navy* had played its hand.

"Do you think the Japs will turn tail and run?" asked General Holland Smith as the meeting was coming to a close.

"No," Admiral Spruance answered, "not now. They are out for big game. If they had wanted an easy victory, they would have disposed of the relatively small force covering MacArthur's operation at Biak. But the attack on the Marianas is too great a challenge for the Japanese Navy to ignore." [4]

While Spruance's thoughts turned to the enemy ships advancing eastward toward him, Generals Watson and Schmidt plotted the conquest of the southern part of Saipan. The overall scheme of maneuver called for the two divisions to pivot on Red 1 to form a line stretching from the west coast, across the island, to a point just south of the Kagman Peninsula. While the 2d Marine Division held off any attacks from the vicinity of Mounts Tipo Pale and Tapotchau, the 4th Marine Division was to smash through to the shores of Magicienne Bay.

THE CAPTURE OF AFETNA POINT

On the morning of 16 June, Lieutenant Colonel Easley, wounded the day before, turned command of 3/6 over to Major John E. Rentsch and was evacuated from the island. Major Hunt had by this time assumed command of 2/6 from Lieutenant Colonel Kengla, the observer who had temporarily replaced the wounded Major Rice. Strengthened by the arrival of its self-propelled 75mm guns and 37mm anti-tank weapons, Colonel Riseley's 6th Marines, the pivot for the entire landing force, spent the day mopping up the area it already had overrun. Fighting flared whenever Marines encountered Japanese die-hards, only to end abruptly once the enemy soldiers had been killed.

The 8th Marines zone also was quiet in comparison to the frenzy of D-Day. The 2d Battalion, however, saw sustained action while driving the enemy from Afetna Point and pushing toward Lake Susupe. At the point, the going was comparatively easy, for many of the defenders had either fled inland or been killed during the fruitless night counterattacks. By 0950, 2/8 had established contact with the 23d Marines at Charan Kanoa pier.

The company that had cleaned out Afetna Point then reverted to battalion reserve. Japanese artillery began relentlessly stalking the unit, even when it occupied positions screened from observers on the 0–1 ridges. Some days later the culprit was found, an enemy soldier who had been calling down concentrations from his post in one of the smokestacks that towered over the ruined Charan Kanoa sugar refinery.[5]

[4] Quoted in Gen Holland M. Smith and Percy Finch, *Coral and Brass* (New York: Charles Scribner's Sons, 1949), p. 165. hereafter Smith and Finch, *Coral and Brass*.

[5] The commanding officer of the 23d Marines recalled that Japanese troops, who infiltrated from the north, repeatedly occupied this re-

While this one company was securing Afetna Point, and later dodging shell bursts, the remainder of the battalion advanced to the western edge of Lake Susupe.

D plus 1 also saw the further strengthening of General Watson's 2d Marine Division. The remainder of Lieutenant Colonel Richard C. Nutting's 2/2 came ashore to serve for the time being with the 6th Marines. All of Lieutenant Colonel Arnold F. Johnston's 3/2 had landed on D-Day. Because of the volume of hostile fire that was erupting along the northern beaches, 1/2, commanded by Lieutenant Colonel Wood B. Kyle, was diverted to the zone of the 4th Marine Division. Northern Troops and Landing Force intended that the battalion serve with the 4th Marine Division, but Kyle learned only of the change of beaches when he reported to the control vessel. As a result, when his men landed, Kyle marched them north and rejoined the 2d Marine Division. Once the move had been made, NTLF decided that a return to 4th Division territory and control was undesirable and 1/2 remained with its parent regiment.

Originally scheduled to be supplied by parachute after the contemplated landing at Magicienne Bay, Kyle's battalion had placed its 81mm mortars and .30 caliber water-cooled machine guns on board an escort carrier. The torpedo planes that were sent to deliver the weapons after the battalion had landed flew so low over the Charan Kanoa airstrip that the parachutes did not open completely. As a result, almost all the equipment was damaged. With his 2d Battalion attached to the 6th Marines and 1/2 presumably under control of the 4th Marine Division, Colonel Stuart had been assigned to command a composite force made up of 3/2 and 2/6. Since 2/6 had fought desperately to repel the previous night's counterattack, Stuart ordered the tired unit into reserve, relieving it with his other battalion.

The second day of the Saipan operation also saw the landing of two 105mm howitzer battalions of the 10th Marines. Late in the afternoon, the DUKWs carrying Lieutenant Colonel Kenneth A. Jorgensen's 4/10 and Major William L. Crouch's 3/10 crossed Green 3. Jorgensen's battalion went into position near the radio station, while Crouch's unit prepared to fire from an area 200 yards inland from the southern limit of Green 2. The 2d 155mm Howitzer Battalion, detached from VAC Artillery, did not come ashore because adequate firing positions were not available. To the south, the arrival on the Blue Beaches of General Harper and the advance parties of all four XXIV Corps Artillery battalions gave promise of increasingly effective fire support as the battle progressed.

THE 4TH MARINE DIVISION BATTLES FORWARD

General Schmidt had decided that a strong effort in the center of his zone of action offered the best chance for success. Before launching his attack,

finery, which controlled the boat channel off Green 3. He noted that the danger was finally eliminated by the mopup action of two companies of Army troops assigned to the shore party on the Yellow Beaches. MajGen Louis R. Jones ltr to ACofS, G-3, HQMC, dtd 13Feb 63, hereafter *Jones ltr.*

scheduled for 1230 on 16 June, the 4th Marine Division commanding general parceled out elements of Colonel Hart's 24th Marines in order to strengthen his position. The 3d Battalion, led by Lieutenant Colonel Alexander A. Vandegrift, Jr., was attached to Colonel Batchelder's 25th Marines in order to shore up the right-hand portion of the division front, relieving the weary 1/25, while Lieutenant Colonel Richard W. Rothwell's 2/24 moved into positions from which to protect the left flank. The remainder of the 24th Marines took over the center of the beachhead. As 1/24 was moving forward, mortar fragments claimed the life of Lieutenant Colonel Maynard C. Schultz, the battalion commander, who was replaced by his executive officer, Major Robert N. Fricke.

The artillery battalions which were to support the attack also came under enemy fire. Lieutenant Colonel Reeve of 5/14 reported that by 1730 on 16 June, all but two of his 105mm howitzers had been knocked out. "When I say 'knocked out'," he continued, "I mean just that—trails blown off, recoil mechanism damaged, *etc.* By 1000, with the help of division ordnance and by completely replacing one or two weapons, we were back in business—full strength—12 guns." [6] Early in the morning 4/14 also came under accurate counterbattery fire. After the Marine cannoneers had blasted a 30-man patrol, hostile gunners retaliated by silencing one of the battalion's howitzers, killing or wounding every member of the crew.

Although Agingan Point was secured early in the day, the attack of the 4th Marine Division was not a complete success, for darkness found the enemy clinging stubbornly to a portion of the 0–1 Line. The longest gains were made on the right by the 25th Marines. While Vandegrift's attached unit moved forward, Mustain, commander of 1/25, released control of those elements of 3/25 that had been entrusted to the 1st Battalion on the previous day. Once his 3d Battalion had been restored as a team, Chambers sent tanks and infantrymen against pockets of resistance to his rear. The Marines silenced five machine guns and two howitzers, killing in the process some 60 Japanese.

When this task had been finished, Chambers lent assistance to 2/25, which was trying to destroy a quartet of antiaircraft guns located on the reverse slope of the 0–1 ridge. In spite of help from tanks and two of Chambers' rifle companies, Hudson's Marines could not dislodge the enemy, for the Japanese were able to place grazing fire along the crest. Still, the 25th Marines, with Vandergrift's attached battalion, was able to claw its way to within a half-mile of Aslito airfield. In the center and on the left, the 24th and 23d Marines fought a similar tank-infantry battle against equally resolute Japanese of the *47th Independent Mixed Brigade.* The division front line by nightfall formed a crescent around the southern shore of Lake Susupe, bulged eastward almost to 0–1, crossed the critical ridge near the center of the zone of action, and continued to a point almost 1,000

[6] LtCol Douglas E. Reeve, ltr to Maj Carl W. Hoffman, dtd 6Jan49, quoted in Hoffman, *Saipan*, p. 82. No copy of this letter has been found.

yards east of Agingan Point. (See Map 17.)

ADDITIONAL REINFORCEMENTS [7]

Ozawa's appearance east of the Philippines caused Admiral Spruance to order the American transports to safer waters, a withdrawal that would begin at darkness on 17 June. General Holland Smith was thus presented the choice of either landing his Expeditionary Troops reserve at Saipan or allowing it to disappear over the eastern horizon. Since the fierce battle on D-Day had served notice that the conquest of Saipan would be a difficult task, he released General Ralph Smith's 27th Infantry Division, less one regiment and its supporting artillery battalion, to Northern Troops and Landing Force and then ordered one of the Army regimental combat teams to land at once.

During the night, the 165th Infantry went ashore, came under General Schmidt's control, and got ready to pass through 3/24 and extend the 4th Marine Division right flank during the next day's attack. The 105th Infantry would land on Holland Smith's order, while the 106th Infantry, formerly scheduled to join Southern Troops and Landing Force at Guam, remained afloat as Expeditionary Troops reserve. Three of the 27th Division field artillery battalions, the 105th, 106th, and 249th, were ordered to disembark and serve under the direction of XXIV Corps Artillery, and by the middle of the following morning, all of them were ready for action. While these reinforcements were crossing the darkened beaches, an advance party from Northern Troops and Landing Force headquarters arrived to select a site for Holland Smith's command post.

THE TANK BATTLE [8]

During daylight on 16 June, the 2d Marine Division had not engaged in the savage kind of fighting endured by the 4th Marine Division. Once darkness arrived, their roles were reversed, for General Saito chose to hurl the *9th Tank Regiment, 136th Infantry Regiment,* and *1st Yokosuka Special Naval Landing Force* at the northern half of the beachhead. Because of the gains which the Marines had made during the past two days, the Japanese general could not hope to crush General Watson's division at a single stroke. Instead of simply issuing orders to drive the Americans into the sea, Saito directed his troops first to recapture the site of the Saipan radio station, some 400 yards behind the lines held by the 6th Marines. Once this initial objective had been gained, the Japanese would promptly launch further blows that would bring the Americans to their knees.

---

[7] An additional source for this section is: 27th InfDiv G-3 Periodic Rpt, 16–17Jun44.

[8] Additional sources for this section include: CinCPac–CinCPOA Items Nos. 9304, 9th TkRegt O/B, dtd 15May44, 9983–9985, Thirty-first ArHq outgoing msg file, msg no. 1039, and 10531, Excerpts from a Notebook of FOs; LtCol William K. Jones memo to Dir DivPubInfo, n.d., subj: "Campaign for the Marianas, comments on"; Maj James A. Donovan, "Saipan Tank Battle," *Marine Corps Gazette*, v. 32, no. 10 (Oct48).

Colonel Goto's *9th Tank Regiment,* which boasted new medium tanks mounting 47mm guns as well as older light tanks, was to spearhead the effort, attacking westward directly toward the radio station. Two of Goto's companies and part of a third had been sent to Guam, but 3½ companies were on hand at Saipan. Although one of these units had been almost wiped out during the earlier fighting, Goto was able to muster about 44 tanks, most of them mediums.

On the heels of the tank attack, Colonel Yukimatsu Ogawa's *136th Infantry Regiment,* which already had suffered serious losses, was to attack toward Charan Kanoa. From the north, Lieutenant Commander Tatsue Karashima's *1st Yokosuka Special Naval Landing Force* would advance from Garapan along the coastal road. Although Saito directed the naval unit to cooperate with his Army troops in the eventual capture of Charan Kanoa, he apparently was unable to impose his will on Admiral Nagumo. What was to have been a serious effort to penetrate the lines of the 2d Marines and push southward along the highway did not materialize. Colonel Stuart's regiment, subjected to scattered mortar fire, beat off "minor counterattacks" [9] but encountered no real peril from the direction of Garapan. To the south, however, Japanese Army troops delivered a blow which, in the opinion of Lieutenant Colonel Jones of 1/6, "could have been fatal to the division's fighting efficiency." [10]

Before darkness, American aerial observers had spotted several enemy tanks in the area inland of the 2d Marine Division beachhead, so the troops were alert to the possibility of an armored attack. At 0330 on the morning of 17 June, the Marines of 1/6 heard the roaring of tank motors. Star shells illuminated the darkened valley from which the noise seemed to be coming, a company of Sherman medium tanks was alerted, and supporting weapons began delivering their planned fires. Within 15 minutes, the hostile tanks, with Ogawa's infantrymen clinging to them, began rumbling into the battalion sector.

"The battle," wrote Major James A. Donovan, Jr., executive officer of 1/6, "evolved itself into a madhouse of noise, tracers, and flashing lights. As tanks were hit and set afire, they silhouetted other tanks coming out of the flickering shadows to the front or already on top of the squads." [11] Marine 2.36-inch rocket launchers, grenade launchers, 37mm antitank guns, medium tanks, and self-propelled 75mm guns shattered the enemy armor, while rifle and machine gun fire joined mortar and artillery rounds in cutting down the accompanying foot soldiers.

Between 0300 and 0415, when the battle was most violent, 1/10 fired 800 75mm rounds in support of 1/6. The battalion fired another 140 shells between 0430 and 0620, as the action waned. Additional support came from a 4/10 battery of 105mm howitzers.

Although directed primarily at 1/6, the attack spilled over into the sector manned by 2/2, which was still

[9] 2d MarDiv D-3 Rpt, 16–17Jun44.

[10] Jones memo, *op. cit.*

[11] Donovan, "Saipan Tank Battle," *op. cit.*, p. 26.

attached to the 6th Marines. Here three of Goto's tanks were disabled. By 0700, the hideous din had ended all along the front, but the quiet of the battlefield was broken by the bark of M1 rifles as Marines hunted down survivors of the night's bitter clash. Atop a hill in front of Jones' battalion, a Japanese tank, smashed by naval gunfire as it attempted to escape, lay wreathed in black smoke. At least 24 of the 31 armored vehicles whose charred hulks now littered the area were destroyed while attempting to pierce the lines of 1/6.[12] "I don't think we have to fear Jap tanks any more on Saipan," remarked General Watson. "We've got their number." [13] The Marines had handled their antitank weapons so effectively that only a handful of Goto's vehicles survived the massacre. These few tanks, however, would strike again before the battle ended.

The *136th Infantry Regiment* also suffered intensely at the hands of 1/6 and 2/2. Neither battalion estimated the number of Japanese killed on that hectic morning. Judging from reports made to division on the following evening, Colonel Ogawa must have lost about 300 men. The Japanese had suffered a bitter reverse. Commented the commanding officer of the *135th Infantry Regiment* in northern Saipan: "Despite the heavy blow we dealt the enemy, he is reinforcing his forces in the vicinity of Charan Kanoa. . . ." [14] Such was the epitaph to General Saito's counterattack.

In all but destroying the *9th Tank Regiment* and a 500-man detachment of infantry, 1/6 had suffered 78 casualties, more than one-third of a full-strength rifle company. The company from 2/2 that helped Jones' Marines shatter the attack lost 19 men killed and wounded. The battalions of the 10th Marines, whose positions had been carefully plotted during the day by Japanese observers, suffered many casualties, including the wounding of the commander of 2/10, Lieutenant Colonel Shell. The two battalions also lost a great deal of equipment to counterbattery fire. By dawn on 17 June, four of the 4th Battalion 105s were temporarily out of action, and only three of the 2d Battalion 75s were capable of firing.

In spite of these losses, which brought NTLF total casualties to approximately 2,500, the efficiency of the command was considered excellent. Now the two Marine divisions, aided by Colonel Gerard W. Kelley's 165th Infantry, would renew their efforts to break out from the coastal plain. While the 2d Marine Division sent the 2d and 6th Marines north toward Garapan and Tipo Pale and the 8th Marines eastward to 0–1, the 4th Marine Division and its attached Army regiment was to continue toward Aslito field.

---

[12] Because of the darkness and confusion, the troops involved could not accurately estimate the number of tanks they had destroyed. The tally made after the battle may have included some vehicles that were knocked out prior to the night attack.

[13] Quoted in Robert Sherrod, *On to Westward, War in the Central Pacific* (New York: Duell, Sloan and Pearce, 1945), p. 68.

[14] NTLF G–2 Rpt, p. 13, in *NTLF OpRpt.*

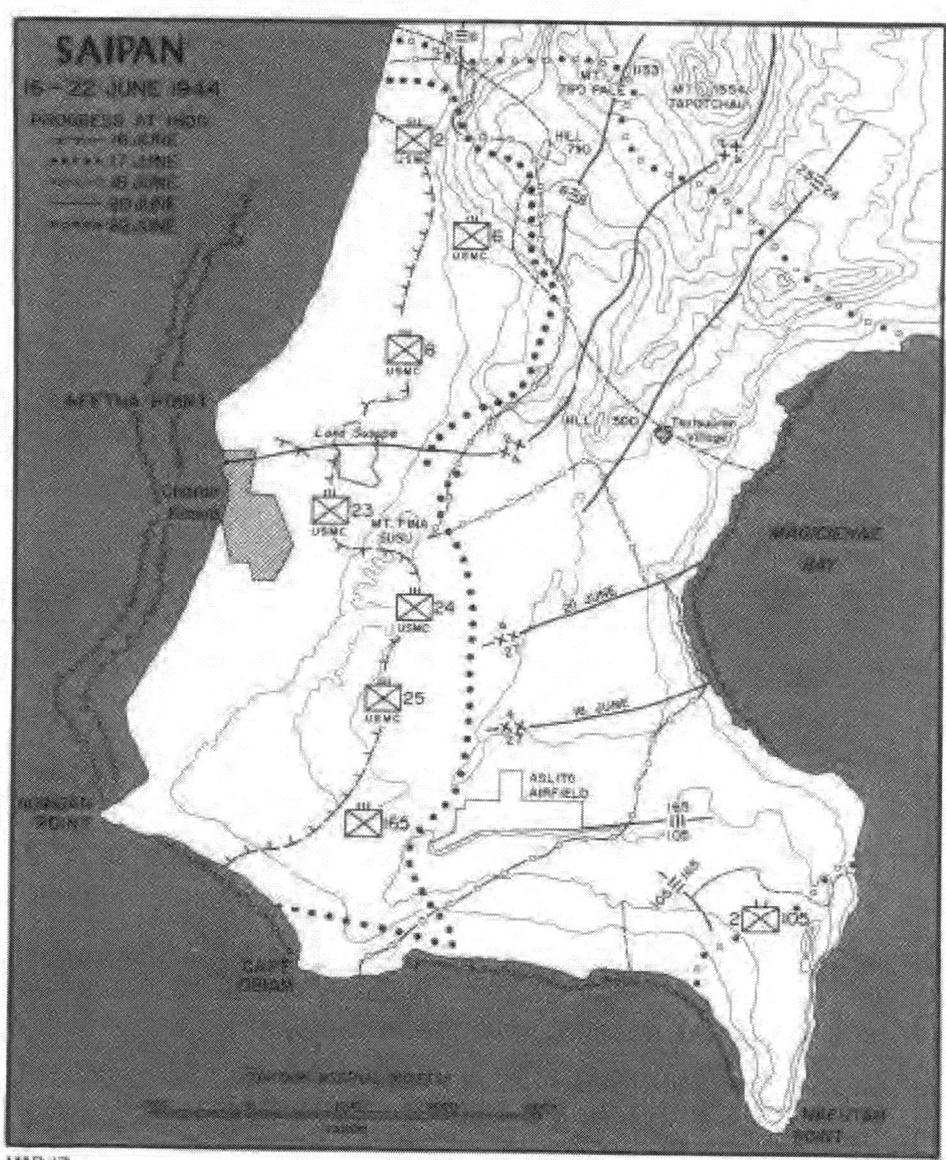

MAP 17

R F STIBIL

17 JUNE: THE ATTACK OF THE 2d MARINE DIVISION

The attack planned for 0730 on 17 June called for the 2d Marines to extend their beachhead almost halfway to Garapan, while the 6th Marines fought its way to a line drawn south and slightly east from the base of 1,133-foot Mount Tipo Pale. The 8th Marines had the mission of advancing across the Lake Susupe marshes to seize the O-1 ridges to its front. A planned 90-minute preparation by aircraft, warships, and artillery batteries was within 15 minutes of completion when General Holland Smith directed General Watson to postpone the attack until 0930. Convinced that he did not have time to inform all his infantry units of the delay, Watson allowed the three regiments to surge forward.

Colonel Stuart's 2d Marines, composed at the time of 3/2 and 2/6, advanced 400 yards within three hours. Resistance continued very light, and by 1800 the unit was digging in at its objective, about 1,000 yards south of Garapan. One company of 2/6, the regimental reserve, was inserted on the right of 3/2 to insure contact with the 6th Marines. General Watson then released to Stuart's control a company from 1/2, the division reserve, and this unit was attached to 2/6.

The men of the 6th Marines had barely finished ferreting out the snipers left behind as the Japanese counterattack receded, when they began an attack of their own. The regiment fought its way upward to the foothills of Mount Tipo Pale. Since the Japanese within the 6th Marines zone of action had been mauled during the night's fighting, the survivors could offer only slight resistance, but difficulty in maintaining contact with the 8th Marines slowed Colonel Riseley's command. By the time the objective had been captured, 1/6, 3/6, and 2/2 were on line, with the regimental scout-sniper platoon the only available reserve. To add strength to the position, Colonel Riseley received one company from 1/2.

Colonel Wallace's 8th Marines moved rapidly forward in the left of its zone of action, as the 1st and 3d Battalions seized their assigned portions of the disputed ridgeline. Lieutenant Colonel Tannyhill's 1/29, however, soon found itself mired in the bog that extended northward from Lake Susupe. Japanese snipers lurking in the swamp killed or wounded many of the floundering Marines, while enemy troops firing from a hill on the O-1 Line and an adjacent grove of palm trees inflicted their share of the 80 casualties suffered by the battalion. Among those wounded and evacuated was Lieutenant Colonel Tannyhill, who was replaced during the afternoon by Lieutenant Colonel Rathvon McC. Tompkins.

No sooner had Tompkins arrived on the scene than four medium tanks came thundering along the one good road leading through the zone. He commandeered the Shermans, and their 75mm guns kept the defenders crouching in the trenches until Marine infantrymen could overwhelm the enemy and gain the crest. The tanks then rumbled up the hill to fire directly into a cave where a number of Japanese were holding out, killing the occupants. Although the hill itself was firmly in American hands, the nearby coconut grove defied reduction. Whenever the battalion 81mm mortars took the stand of trees under fire, the Japanese responded with a torrent of bullets. To guard against a possible counterattack, four self-propelled 75mm guns were rushed onto the hill. No further advance was attempted that day.

On 17 June, the 2d Marine Division had almost doubled the area under its control. From a point on the coast 1,000 yards south of Garapan, the front curved inland past the approaches to Mount Tipo Pale, embraced the hard-won hill in the 8th Marines zone, and swung sharply toward Lake Susupe. The three regiments were in contact with one another, but a gap existed between divisions. To refuse the dangling flank, 2/8 dug in for the night facing generally to the south. (See Map 17.)

THE APPROACH TO ASLITO AIRFIELD

Two battalions of Colonel Kelley's 165th Infantry landed before dawn of 17 June, passed through the lines of 3/24, and attacked toward Aslito field. Vandegrift's Marine battalion, although it had become division reserve, remained in position until midmorning, when Kelley's remaining battalion, 3/165, was safely ashore. Attacking with its 1st and 2d Battalions abreast, the Army regiment thrust forward against comparatively weak defenses. As the day wore on, Japanese opposition along the southern coast increased in ferocity. Near Cape Obiam, 1/165 inched its way to the crest of the ridge that barred access to the airfield, only to be driven back down the slope by a determined counterattack. The battalion then paused for the night at the base of the ridge. Since this unit could advance no farther, Colonel Kelley halted 2/165 along the high ground just short of the Aslito runways. The 2d Battalion occupied a position that afforded excellent fields of fire and insured contact with 1/165 on the right. The regimental losses for the day's action totaled 15 killed and 57 wounded?

Just to the north, Colonel Batchelder of the 25th Marines sent a column of battalions knifing forward with 2/25 in the lead. The drive netted some 1,500 yards, as the Marines secured the area due north of the airfield and occupied the ridge beyond. Although Marine patrols discovered that Aslito field had been abandoned, Colonel Kelley was unwilling to occupy it with elements of his 165th Infantry. He believed that any further advance would have involved exchanging a good defensive position for a poor one, and in the process assuming a frontage too wide for his pair of battalions. Thus, the successful 2/25 found itself about 600 yards forward of the unit on its right. Colonel Batchelder now ordered his 3d Battalion to form a line facing southward toward the vacant runways and battered buildings. A narrow gap, however, still existed between the Marine and Army regiments.

The 24th Marines, in the left-center of the 4th Marine Division zone of action, had been plagued throughout the early morning by incoming fire from mortars, artillery, and machine guns, as well as by air bursts from

[16] 27th InfDiv G-1 PeriodicRpt, 17Jun44.

40mm antiaircraft guns. A 15-minute artillery preparation did not silence all these Japanese weapons, many of which continued to inflict casualties during the day. Because the division front was growing wider as the Marines moved inland, the 24th Marines had to shift its axis of advance slightly to the north, with 1/24 making a wide turning movement to come abreast of the 2d Battalion. A deluge of shells from antiaircraft guns, probably located near Nafutan Point, delayed completion of the planned maneuver by 1/24. By 1000, the long-range fire had subsided, so that Major Fricke's men, supported by mortars and by LVT(A)s, could overwhelm light opposition to capture their objective.

The attack of 2/24 also was slow in gaining momentum. On the battalion left, the 23d Marines found itself confronted by a defiant and well-entrenched group of Japanese from the *47th Independent Mixed Brigade*. These resolute defenders not only prevented the Marine units from maintaining contact along the regimental boundary but also turned a 3-inch dual-purpose gun against 2/24. The weapon was soon silenced, but almost an hour passed before Rothwell's battalion could extend to its left, make contact, and begin moving forward in the face of mortar concentrations and intermittent 40mm fire, probably from the same antiaircraft guns that were harassing the 1st Batallion.

At 1500, a barrage of 4.5-inch rockets, fired in support of the adjacent 23d Marines, struck the battalion, causing 20 casualties.[16] In spite of this accident, Rothwell's Marines continued their attack in the face of increasing resistance. The battalion was poised to make a final lunge toward its objective, when a savage concentration of enemy fire erupted from the face of a cliff only 300 yards to the front. Caught in the open and unable to gouge foxholes in the hard coral, the Marines twice fell back, yielding some 600 yards in order to obtain a suitable defensive position. The 2d Battalion had suffered 53 casualties in advancing slightly more than 2,000 yards.

The 23d Marines was in the meantime being fought to a standstill southeast of Lake Susupe, in front of a hill that jutted from the objective ridge. All went well until the 1st Battalion attempted to cross the valley beyond Mount Fina Susu. Lieutenant Colonel Haas' Marines were stopped in their tracks, even though the 2d Battalion, attacking on the right, was able to push steadily forward. The two units soon lost contact, so Colonel Jones was forced to use 3/23, his regimental reserve, to restore the line. Since the 2d Marine Division had been stymied in the southern part of its zone, the two divisions were unable to form a continuous line. (See Map 17.)

General Schmidt's Marines and soldiers had gained the central portion of their objective, but on either flank the line receded somewhat. Although Aslito airfield had not been occupied, Colonel Kelley's soldiers seemed capable

[16] A provisional rocket detachment had been attached to the headquarters battalions of both Marine divisions. The missiles were fired from multiple launchers mounted on trucks.

of overrunning that installation come morning. Also, the arrival of Lieutenant Colonel Marvin H. Floom's 2d 155mm Howitzer Battalion, detached from the 10th Marines, indicated that the 4th Marine Division would receive additional fire support when the attack was renewed.

THE AMERICAN BUILD-UP CONTINUES

While the fighting raged a few thousand yards inland, command posts were being occupied along the western coast. Marine observation planes were preparing to operate from the Charan Kanoa flight strip, and additional troops were pouring ashore. At 1530 on 17 June, General Holland Smith entered the ruins of Charan Kanoa to direct the efforts of Northern Troops and Landing Force. Meanwhile, General Harper had chosen a site from which to direct his corps artillery, but his 155mm battalions would not land until the next day. General Ralph Smith also arrived on the island to assume command of his 27th Infantry Division. The 165th Infantry, however, was to remain attached to General Schmidt's 4th Marine Division for the time being.

Colonel Leonard A. Bishop's 105th Infantry landed during the day. Although all three rifle battalions came ashore ready to enter the fight, most of the regimental communications gear and transport as well as some elements of its headquarters troops were left behind on their transport.[17] Since that vessel promptly set sail as Admiral Spruance had directed, a week passed before the missing men and equipment landed at Saipan. Acting upon orders from Holland Smith, Bishop posted 2/105 to the rear of the 4th Marine Division to guard against an enemy breakthrough.

Another Army unit which had difficulty in landing was the 864th Antiaircraft Artillery Group. The reef blocked the progress of its landing craft, so the group was unable to move into position until the next day. The 106th Infantry remained afloat as Expeditionary Troops reserve.

Although the transports began weighing anchor as darkness approached, the impending departure of these ships and the redeployment of Turner's combat vessels caused no great concern among the troops ashore, except for the equipment-short 105th Infantry.[18] Over 33,000 tons of cargo had been unloaded to sustain Northern Troops and Landing Force until the convoy returned. The situation was far different from that faced less than two years before by General Vandegrift's 1st Marine Division, when its amphibious shipping suddenly vanished from the dangerous waters off Guadalcanal.

Although the combat efficiency of Northern Troops and Landing Force was regarded as "very satisfactory, impaired to a certain degree by a considerable number of casualties," Colonel

[17] When it became evident that he would not be able to land much of his regimental headquarters and unit supplies, Colonel Bishop moved ashore ahead of schedule with some key personnel and communications gear. Until its ship returned, the 105th used some repaired Japanese trucks to meet its need for motor transport. Col Leonard A. Bishop, USA, ltr to Head, HistBr, G-3, HQMC, dtd 28Feb63.

[18] *Ibid.*

Robert E. Hogaboom, corps G-3, detected in the events of 17 June "the first signs of weakening enemy resistance." [19] The 2d Marine Division estimated 2,650 enemy dead in its zone, while the 25th Marines claimed that it alone had killed 1,550 Japanese.[20] The defenders of southern Saipan had suffered numerous casualties, though probably fewer than the Americans believed. Whatever the actual total of enemy dead, the coastal perimeter defenses had been shattered.

During the night of 17–18 June, a few weak counterattacks were crushed by Northern Troops and Landing Force. Weariness, casualties, and severed communications prevented the *136th Infantry Regiment* and *47th Independent Mixed Brigade* from making another serious effort to break through to the western beaches. Instead of a frontal attack, the Saipan garrison attempted a counterlanding behind the Marines—a move which had been anticipated.[21] Of about 35 landing craft that took part in the ill-fated venture, 13 were sunk by fire from patrolling warships or from the 105mm howitzers of 4/10. The Japanese who survived the shelling turned back to scurry ashore near Tanapag.

Enemy aircraft reappeared during the evening to flail away at American shipping. Truk-based planes launched torpedoes at units of the Southern Attack Force, nervously awaiting W-Day at Guam, and damaged one LCT so badly that it had to be scuttled. Fighters, dive bombers, and torpedo craft from Yap damaged an LST that was retiring with the transports. Later the same airmen struck an escort carrier group, inflicting grave damage on the *Fanshaw Bay*.

18 JUNE—"THE BEGINNING OF OUR SHOWDOWN FIGHT."

As the third day of the battle for Saipan drew to a close, the Japanese premier, Hideki Tojo, radioed those in command of the beleaguered garrison that: "Because the fate of the Japanese empire depends on the result of your operation, inspire the spirit of officers and men and to the very end continue to destroy the enemy gallantly and persistently; thus alleviate the anxiety of our Emperor." [22] Although this broadcast was intended to be inspirational, it all but conceded the eventual loss of Saipan. The *Thirty-first Army* headquarters, however, framed a reply that indicated "Have received your honorable Imperial words," read the message, "by becoming bulwark of the Pacific with 10,000 deaths, we hope to requite Imperial favor." [23]

The *Thirty-first Army* acting chief of staff,[24] at a command post in the mountains east of Garapan, knew few details concerning the actual progress

[19] NTLF G-3 PeriodicRpt, 16–17Jun44.

[20] NTLF G-2 PeriodicRpt, 16–17Jun44.

[21] RAdm Herbert B. Knowles ltr to ACofS, G-3, HQMC, dtd 30Jan63.

[22] NTLF G-2, Thirty-first Ar Incoming Msg File, no. 115, pt. 1, hereafter *Thirty-first Ar Incoming Msg*.

[23] NTLF G-2, Thirty-first Ar Outgoing Msg File, no. 1046, hereafter *Thirty-first Ar Outgoing Msg*.

[24] Major General Tamura, the army chief of staff, was trapped with General Obata on Guam.

of the battle. After piecing together the few reports that reached him, he concluded that the situation facing the defenders on the morning of 18 June was bleak indeed. He had no information on troop dispositions in the south except the assurance that three reserve companies were in the immediate vicinity of *43d Division* headquarters. Rumor had it that General Saito was dead. Nothing, it seemed, could stop the American advance across Aslito airfield, and a powerful attack toward Garapan appeared in the making.[25] If the Japanese high command had on the morning of the 18th any flickering hopes of a victory ashore, such sentiments would not last the day.

Two American attacks were planned for 18 June. At 1000, both Marine divisions would strike, to be followed two hours later by the 27th Infantry Division. General Ralph Smith was granted the additional time to move his 105th Infantry into position along the coast on the right of the 4th Marine Division.

The close of the previous day's fighting had found the 8th Marines in possession of a hill that overlooked a staunchly defended coconut grove. Now, while the rest of General Watson's division dispatched patrols and improved positions, the 8th Marines renewed its effort to seize this objective. Supported by a 15-minute preparation by 2/10, a barrage that had to be carefully adjusted to avoid hitting nearby elements of the 4th Marine Division, 2/8 stormed through the stand of splintered trees. At the same time, 1/29 pushed forward, keeping abreast of Chamberlin's battalion. Because of these gains, contact was temporarily restored between divisions, but by dusk a gap had been reopened.

At 1000, the 4th Marine Division also attacked, plunging forward with three regiments abreast. On the left, the 23d Marines underwent a hasty reorganization prior to crossing the line of departure. The attached 3/24 passed through 1/23, while Colonel Jones attached the rifle companies and mortar platoon of 3/23 to the 2d Battalion. Finally, the 1st Battalion and the headquarters elements of the 3d were merged, charged with supporting the advance from the slopes of Fina Susu, and entrusted to Lieutenant Colonel Cosgrove. The assault battalions soon were stalled by machine gun and mortar fire, but the prompt commitment of Cosgrove's composite unit sent the attack rolling forward once again. Although patrols reached the regimental objective, Japanese machine gunners prevented the Marines from occupying the ridge. In order to obtain ground suited to the defense, Jones had to withdraw his regiment to positions 400 yards east of Lake Susupe. This withdrawal severed the link so recently re-established with the adjacent 8th Marines.

During the afternoon, as the 23d Marines were advancing steadily, a 75mm half-track, supporting the 2d Battalion, fired into a cave in which the Japanese were manufacturing picric acid. A cloud of sickening fumes enveloped the infantrymen crouching near the opening; two men were overcome with attacks of violent nausea, and the gas alarm was sounded. Since

[25] *Thirty-first Ar Outgoing Msg* No. 1047.

the Marines made a habit of discarding gas masks as soon as they landed, many an uneasy moment passed before the cloud evaporated. Riflemen resolved never again to part with their gas masks, and an anonymous lieutenant dashed into the division command post to ask a startled General Schmidt if he had a mask to spare.[26]

While the men of the 23d Marines were recovering from the confusion caused by the providentially false gas alarm, the 24th and 25th Marines were fast closing on the eastern shore of Saipan. Both regiments advanced swiftly, even though the 24th Marines had to deal with a desperate counterattack. At 1615, after the regimental supporting armor had retired to take on ammunition and gasoline, two Japanese tanks rumbled close to American lines and raked the Marines with fire from cannon and machine guns. Although the vehicles were driven off by artillery and bazookas, the sudden foray resulted in 15 American casualties. In spite of the hectic exchange, both regiments reached Magicienne Bay, thus isolating Nafutan Point from the rest of the island.

In order to reach the coast, the 24th Marines bypassed a fortified cliff that extended southward into its zone of action. The mission of securing this stronghold was assigned to 1/25, the division reserve. Because of the steep slope and the numerous caves, a coordinated sweep was impossible. The fight for the cliff was a series of separate actions in which four or five Marines, hugging the jagged rocks to avoid enemy fire, climbed close to the mouth of a cave and attempted to kill the defenders with a burst from a flamethrower or with demolitions charges. If the cave proved so deep or tortuous that the enemy troops could escape the effects of fire or of concussion, the attackers blasted shut the entrance and left the entombed Japanese to suffocate.

The 165th Infantry began the day amid some confusion, for Colonel Kelley was not certain whether his command was to operate as a part of the 4th Marine Division or of the 27th Infantry Division. He directed his S–3 to call General Schmidt's D–3 section, but the ensuing conversation did not clarify the status of the Army regiment. After discussing the matter with General Ralph Smith, Kelley concluded that he was again part of the Army division and would receive a formal order to that effect from General Schmidt. Although no such order arrived, the 165th Infantry attacked at 0730 to secure the ridge in the right-hand portion of its zone and 30 minutes later began advancing on the airfield. By 1000, both objectives were firmly in American hands. The regiment then paused to regroup and in doing so came under fire from dual-purpose guns located on Nafutan Point. These weapons soon were silenced by American artillery, so that the soldiers could advance to the ridge that overlooked Magicienne Bay.

The 105th Infantry, attacking along the southern coast, encountered difficulty from the outset. Both 3/105 and 1/105, which were to have relieved Kelley's 1st and 3d Battalions by noon, were about 45 minutes late in carrying

[26] Gen Harry Schmidt ltr to CMC, dtd 8Jan 50, quoted in Hoffman, *Saipan*, p. 104. No copy of this letter has been found.

out their assignments. When the advance finally got underway, the troops entered a maze of dense undergrowth broken by coral peaks, some of them 90 feet high. Even though the Japanese chose not to oppose the passage, Colonel Bishop's men gained no more than 200 yards.

As darkness came on 18 June, Northern Troops and Landing Force held approximately three-fourths of southern Saipan. The enemy still clung to the approaches to Garapan, the foothills of Mount Tipo Pale, and a salient extending from Hill 500 toward Lake Susupe, as well as the jungle-clad hills southeast of Aslito field. Since the 4th Marine Division had captured a small segment of the eastern coast, the Japanese troops who had retreated toward Nafutan Point were isolated from their companions in the north. Elements of XXIV Corps Artillery, assisted by observation planes from the Charan Kanoa strip, had begun firing. General Holland Smith, concerned that his stockpile of supplies might become dangerously low, requested that the transports return on the following day, but Admiral Turner would release only those vessels carrying critical items. Two hospital ships made rendezvous on the 18th with the transports in order to pick up the seriously wounded. All in all, the situation looked as promising to American eyes as it seemed grave to the Japanese. (See Map 17.)

The *Thirty-first Army* now informed Tokyo that:

> The Army is consolidating its battle lines and has decided to prepare for a showdown fight. It is concentrating the 43d Division in the area east of Tapotchau. The remaining units (two battalions of the 135th Infantry, one composite battalion, and one naval unit) are concentrating in the area east of Garapan.[27]

The new defensive line would extend inland from a point just south of Garapan, past the cliffs guarding the approaches to Mount Tapotchau, to the shores of Magicienne Bay. Perhaps the only consolation to the enemy was the knowledge that Saito was alive to direct the impending battle.

Among the missions assigned the troops manning the new line was that of preventing the Americans from using Aslito field, a task that would have to be accomplished by infiltration since most of the Japanese artillery had been destroyed. In addition, the defenders were to hold the Marpi Point airstrip and portions of the northern beaches so that supplies and reinforcements could be landed.[28] Along with these instructions, Tokyo broadcast further words of inspiration. "Although the front-line officers and troops are fighting splendidly," read a message from the Emperor, "if Saipan is lost, air raids on Tokyo will take place often; therefore you will hold Saipan." [29] On the same evening that this exhortation arrived, the *Thirty-first Army* intelligence section began burning all but a few of its code books to prevent their capture.[30]

19 JUNE: THE FIGHTING ASHORE

On the morning of 19 June, after passing the quietest night since D-Day, the Marine and Army divisions continued their attack. In the north, the 2d

[27] *Thirty-first Ar Outgoing Msg* No. 1050.
[28] *Thirty-first Ar Incoming Msg* No. 150; *Thirty-first Ar Outgoing Msg* No. 1054.
[29] *Thirty-first Ar Incoming Msg* No. 152.
[30] *Thirty-first Ar Outgoing Msg* No. 1057.

214-881 O-67—20

HANDGRENADES *are tossed by riflemen at Japanese positions as the battle lines move across Saipan from the invasion beaches. (USMC 83366)*

SKIRMISH LINE *of 27th Division infantrymen moves out to mop up the enemy on Nafutan Point. (USA SC210608)*

Marine Division reorganized and sent patrols ahead of the battle position. The division reserve, 1/2, passed to regimental control. One company from 2/6, attached to the 2d Marines, was returned to its parent regiment. The patrols dispatched that day resulted in the destruction of three enemy tanks and the occupation of Hill 790, in the 6th Marines zone, a formidable height which the Japanese obligingly yielded. The 8th Marines probed the defenses to their front while simultaneously looking for routes over which to supply future operations. (See Map 17.)

As the 4th Marine Division was preparing to launch its attack, Japanese infantrymen and their armored support were seen massing in the vicinity of Tsutsuuran. Artillery fire promptly dispersed the enemy force, and the Marines lunged forward. While the 24th and 25th Marines consolidated their positions, the 23d Marines, with 3/24 attached, advanced almost to Hill 500. During the attack, Vandegrift's 3/24 was pinched out of the line, and Dillon's 2/23 along with Cosgrove's composite unit assumed responsibility for the entire regimental zone. The day's gains, however, could not be held because of the danger that the enemy might infiltrate along the division boundary, so the 23d Marines withdrew about 400 yards. Among the casualties was Lieutenant Colonel Cosgrove, wounded by a sniper's bullet; he was replaced by Major Paul S. Treitel.

Along the southern coast, the 27th Infantry Division was having difficulty in keeping its lines intact. The 165th Infantry pushed the remaining distance to Magicienne Bay, thus keeping contact with the 4th Marine Division on the left. The attack of the 105th Infantry, however, bogged down in front of a sheer cliff. A gap opened between the two regiments, but the 165th Infantry patrolled the area. In addition, 1/105 and 3/105 could not keep abreast, and 1/165 had to be employed to restore the line.

During the early morning of 20 June, the Japanese struck back with local counterattacks. A force of 75 stormed the positions of 2/24 and was repulsed, but not before 11 Marines had been wounded. In the 6th Marines sector, some 15 Japanese infiltrators did little damage.

While jabbing at American lines, the defenders were falling back to the line selected on the previous day. General Saito, however, had suddenly become concerned lest the flank of this new position be turned. He directed the *118th Infantry Regiment*, reinforced by tanks, to guard against a landing in the vicinity of Laulau on Magicienne Bay.[31] (See Map 18.)

During the course of the day, *Thirty-first Army* headquarters received additional reports from the units fighting in southern Saipan. According to the army chief of staff's tally, 3½ of *43d Division's* 7 infantry battalions and two-thirds of its artillery had been destroyed. The *47th Independent Mixed Brigade* no longer had any artillery and was unable to estimate the number of infantry units still capable of offering organized resistance. Of three or more rifle battalions that had been organized from the miscellaneous units stranded

[31] CinCPac-CinCPOA Item No. 10531, excerpts from a Notebook of FOs.

at Saipan when the invasion took place, approximately one battalion remained intact. Three engineer companies had been shattered completely, and only one antiaircraft battery remained in action. Although a few artillery pieces survived, all the Army artillery battalions were disorganized. Between 15 and 20 percent of the total Army strength was dead, wounded, or prisoners of war.[32]

The land operations that took place on 19 June were important. Steady pressure had been maintained on the Japanese by Marine infantry units. Corps and division artillery blanketed with fire enemy troop concentrations and weapons positions, while Army antiaircraft guns went into position to protect Aslito field. The retreating Japanese, it seemed, were showing the effects of the constant hammering. Northern Troops and Landing Force intelligence officers could take heart from the surrender of five thirsty Japanese, who claimed that the garrison was dependent on rainfall for its water supply.[33] Yet, the most significant events of 19 June took place at sea.

THE BATTLE OF THE PHILIPPINE SEA [34]

As his warships neared the Marianas, Admiral Ozawa realized that his attempt to cripple Spruance's fleet depended upon the cooperation of land-based Japanese aircraft and the determination of Saipan's garrison. General Saito's men, though they were fighting desperately, had been driven from the ridges that dominated the western beaches. The invaders had landed enough men and supplies to enable them to dispense temporarily with their transports. Yet, the Japanese soldiers had been partially successful. While the vulnerable transports had withdrawn to the east, Saipan was far from secured, so Spruance's ships remained, in a very real sense, tied to the beachhead.

The promised attempt by the Japanese to fly land-based bombers into the Marianas was less than a partial success. American raids on the Volcano-Bonin Islands destroyed some of the enemy planes, and others were stopped by adverse weather. A few aircraft succeeded in getting through, and these took part in a series of raids launched on 18 June.

At 0540 on that day, the Japanese attacked the old battleships, inflicting no damage, but later in the day three American oilers were damaged, one seriously. The last effort of the day, directed against the escort carriers, cost the Japanese five fighters. Unfortunately, 19 of the intercepting American planes crashed while attempting to land on the carriers after dark.[35] A Japanese naval aviator, who had helped dispatch 120–130 aircraft, most of them manned by student pilots, from Japan to the Marianas, later estimated that only 40 percent of the pilots

[32] *Thirty-first Ar Outgoing Msg* No. 1060.

[33] TF 56 G–3 Periodic Rpt, 19–20Jun44.

[34] An additional source for this section is: CinCPac–CinCPOA Ops in POA, Jun44, Anx A, pt. VII.

[35] *TF 52.2 OpRpt*, p. 124.

and virtually none of the planes survived the aerial battles of 18 June.[36]

As the Japanese planes were limping back to their bases, Admiral Spruance maneuvered to prevent the enemy from getting past his ships to attack Saipan. Recent submarine sightings and interceptions of enemy radio messages by high frequency direction finders seemed to indicate that Ozawa would divide his forces, but this was not the case. The Japanese admiral was intent on destroying the American carriers.

On the morning of 19 June, the enemy launched the first of four powerful raids. When the gigantic air battle ended, 330 of the 430 planes launched by the enemy carriers had been destroyed. American attacks against airfields on Guam accounted for another 50 Japanese planes. In contrast, Mitscher's carriers lost 30 planes, 13 of them as a result of the sweeps over Guam.

Ship losses on 19 June also indicated that an American victory was in the making. Japanese bombers had slightly damaged the battleship *South Dakota* and scored near misses on two carriers and a cruiser, while an enemy plane had crashed the *Indiana*, but two of Ozawa's nine carriers were sunk by submarines. The Japanese ships now altered course to avoid the lash of Mitscher's planes long enough to refuel.

As a result of the enemy's change of course, American airmen were unable to deliver an attack of their own until late in the afternoon of 20 June. Admiral Spruance then took a calculated risk in ordering the strike, for it was certain the planes could not get back to the carriers until after dark. The flyers sunk one carrier, slightly damaged two others, and downed 65 Japanese planes. American losses numbered 100 planes, but only about 20 of these were destroyed by the enemy. The rest crashed on or near the carriers, while attempting night landings. Although the waiting ships turned on their lights to aid the pilots, many planes were so low on gasoline that the first approach, good or bad, had to be the last one. Many of the pilots and crewmen who crashed were rescued from the sea that night and on the next day.

Ozawa had been crushed. Admiral Turner now was free to concentrate on supporting the troops ashore. On 20 June, as the Japanese were reeling under the blows of Mitscher's airmen, the transports unloaded 11,536 tons of supplies. In the following several days, increasing numbers of ships returned from the deployment area and the volume of supplies unloaded rose swiftly. With the defeat of the Japanese fleet, the eventual doom of the enemy garrison was assured. The defenders could no longer win the battle for Saipan, but they would fight as valiantly as though their triumph was certain.

THE CAPTURE OF HILL 500

On 20 June, while Navy planes were seeking out Ozawa's fleet, Marine and Army troops continued their pressure on the retreating enemy at Saipan.

[36] *USSBS Interrogation* Nav No. 91, Captain Akira Sasaki, IJN, dtd 23Nov45, II, p. 396.

In the northern part of the American beachhead, the 2d and 6th Marines continued patrolling, while the 8th Marines and the 4th Marine Division attempted to complete the turning movement that would enable the invaders to begin advancing north along the island spine. Meanwhile, in the south, the 27th Infantry Division persisted in its efforts to destroy the Japanese troops entrenched at Nafutan Point.

Patrols from the 2d Marines penetrated almost to the southern outskirts of Garapan, while those sent out by the 6th Marines discovered that the enemy had withdrawn at least 500 yards. Neither regiment, however, could advance until the 8th Marines had pushed forward. On 20 June, Colonel Wallace's unit completed its portion of the turning movement, with 2/8 advancing to the left of Hill 500 and gaining its part of the objective line against light opposition. The change in direction shortened the regimental front so that Colonel Wallace could move 3/8 and 1/29 into reserve, leaving 1/8 and 2/8 to man the line.

General Ralph Smith's troops had taken over responsibility for mopping up the shores of Magicienne Bay. Able to concentrate on its drive north, the 4th Marine Division made impressive gains during the day. The attack, originally scheduled for 0900, was delayed for 90 minutes to allow the 25th Marines, less its 1st Battalion, to replace the weary 23d. Critical terrain in the division zone was Hill 500, which was to be taken by Colonel Batchelder's regiment.

Because of the narrow frontage assigned him, Colonel Batchelder decided to attack in a column of battalions, Lieutenant Colonel Chambers' 3d Battalion leading the way. While the 1st Provisional Rocket Detachment, the regimental 37mm guns, and the battalion 81mm mortars joined the 1st and 3d Battalions, 14th Marines, in blasting the hill, Chambers' men, concealed by a smoke screen, advanced across the level ground to Laulau road, some 500 yards from the objective. Here the battalion reorganized, and, as the preparatory barrage increased in severity, the Marines began moving through wisps of smoke toward the crest. Although the summit was gained about noon, the sealing or burning out of bypassed caves took up most of the afternoon. When the action ended, 44 Japanese bodies littered the hill, while an unknown number lay entombed beneath its surface. The Marines lost 9 killed and 40 wounded, comparatively few casualties in contrast to the number the enemy might have exacted had he chosen to defend the hill more vigorously.

While Chambers' men were seizing Hill 500, the 1st and 2d Battalions, 24th Marines were racing forward a distance of 2,700 yards. Although these units encountered fire from rifles, machine guns, grenade launchers, and occasionally from mortars, the Japanese had withdrawn the bulk of their forces to man the new defensive line. Assisted by medium tanks, armored LVTs, and light tanks mounting flamethrowers, Lieutenant Colonel Rothwell's 2/24 executed what the battalion commander termed "the best coordinated tank and infantry attack of the operation." [37] In doing so, the battalion suffered 32 casualties.

During the 4th Marine Division swing toward the north, Lieutenant Colonel Mustain's 1/25 was battling to drive the enemy from the cliff bypassed on 18 June by the 24th Marines. On the 19th, Mustain's battalion attacked directly toward the sheer western face of the objective, gaining little ground and suffering 26 casualties. Now, on the 20th, he struck from the south. Twice American tanks thundered forward in an unsuccessful effort to draw enemy fire. The third time the armor advanced, Marine infantrymen also moved forward, and the Japanese greeted Mustain's troops with a deluge of fire. In spite of this violent opposition, the attackers moved steadily northward. Flamethrower operators and demolitions teams cleaned out those caves that could be approached on foot, while tanks fired into the openings cut into the face of the cliff.

In an accident reminiscent of the blockhouse explosion on Namur, a Japanese ammunition dump hidden in a cane field at the base of the cliff exploded, temporarily disorganizing Company A. The blast may have been caused by a Japanese shell, but it could have occurred when the flames set by American tracers reached the stockpile of explosives. Mustain's losses for the entire day totaled 31 killed or wounded.

South of the cliff, General Ralph Smith's soldiers continued their attack toward Nafutan Point. Two battalions of the 165th Infantry, attacking from the north and northwest, gained about 1,000 yards, but 3/105 had a difficult time in overcoming the cave defenses dug into the cliffs and ridges within its zone. On 20 June, the last of General Ralph Smith's regiments, Colonel Russell G. Ayres' 106th Infantry, came ashore. General Holland Smith felt that he might need the regiment at Saipan, even though it had been earmarked as reserve for the Guam landings. Admiral Turner agreed with the Marine general, but he specified that as much as possible of the unit equipment be left on board ship to speed the eventual movement to Guam. Once ashore, the 106th Infantry became Northern Troops and Landing Force reserve, thus releasing 2/105 to division control.

Other heartening changes in the tactical situation took place on 20 June. The first American plane to use Aslito field [38] touched down that evening. Also, the 155mm guns of XXIV Corps Artillery began firing at targets on Tinian. With all of southern Saipan except Nafutan Point under American control, General Holland Smith could devote his full attention to shattering General Saito's mid-island defense line.

[37] *2/24 Narrative*, p. 5.

[38] Upon its capture by the 165th Infantry, Aslito field had been renamed in honor of Colonel Gardiner J. Conroy, regimental commander killed at Makin. Later the installation was officially christened Iseley Field in memory of Commander Robert H. Isely, a naval aviator killed during a preinvasion strike. For unexplained reasons, the original spelling of Isely was not retained by the XXI Bomber Command. See Craven and Cate, *Guadalcanal to Saipan*, p. 515n.

PREPARATIONS FOR THE DRIVE TO THE NORTH

Prior to launching their blows against the newly formed Japanese line, the 2d and 4th Marine Divisions spent 21 June, D plus 6, in reorganizing, moving up supplies, and probing for enemy strongpoints which would have to be reduced when the attack began. Japanese troops who had taken refuge in the swamps surrounding Lake Susupe had been a source of trouble throughout the operation. A patrol attempted on 19 June to flush out these stragglers, killing seven of them and capturing a pair of machine guns before being forced to withdraw. A larger patrol returned the following day, but it too was not powerful enough to complete the task. On 21 June, 1/106 received orders to secure the area, and the Army unit went into action the following day. Although the soldiers conscientiously searched the marsh, they were unable to kill all the Japanese hiding there. The area remained a haven for enemy infiltrators for some time to come.

Elsewhere in the 2d Marine Division zone, patrols sought information on enemy defenses. A strong position was located south of Garapan, but neither the 6th nor 8th Marines was able to learn anything concerning General Saito's recent preparations. The 8th Marines took advantage of the lull to relieve 2/8 with 1/29, and the 2d 155mm Howitzer Battalion reverted to control of the 10th Marines.

On the right, the 4th Marine Division also paused before attacking along the east coast. Since 31 Japanese had been slain on Hill 500 during the previous night, 3/25 once again probed the caves that scarred its slopes. Lieutenant Colonel Chambers talked two enemy soldiers captured on the hill into trying to convince their comrades to surrender. Two wounded men yielded to their arguments, but four others emerged from their caves rifles ready and had to be killed.

At Nafutan Point, the 27th Infantry Division on 21 June continued its slow, cave-by-cave advance. In the midst of the day's action, an order arrived at Ralph Smith's headquarters directing the division, less one battalion and a light tank platoon, to pass into Northern Troops and Landing Force reserve and assemble northwest of Aslito field. A separate sub-paragraph assigned the reinforced battalion the mission of mopping up Nafutan Point and protecting the airstrip.[39] This assignment was made in the third paragraph of the order, the place, according to both Army and Marine Corps usage, where the commander stated the mission of his subordinate units.

Almost five hours after receiving the order, Ralph Smith telephoned Holland Smith to urge that a regiment rather than a battalion be assigned the job of reducing Nafutan Point. The Marine general approved the employment of the more powerful force, provided that one battalion was available for operations elsewhere on the island. At 2000, the Commanding General, 27th Infantry Division, ordered the 105th Infantry to "hold present front line facing Nafutan Point, with two battalions on line and one battalion in regi-

[39] NTLF OpO 9–44, dtd 21Jun44.

mental reserve."[40] The regiment was to relieve by 0630 on 22 June those elements of the 165th Infantry manning the front lines, reorganize, and resume the attack by 1100 on the same day. The reserve battalion of the 105th Infantry could not be employed without General Ralph Smith's approval.

At 0830 on 22 June, a modification of the previous NTLF order reached 27th Infantry Division Headquarters. The major change was the selection of a regiment, obviously the 105th Infantry, instead of a battalion to "continue the mission . . . of clearing up remaining resistance and patrolling [the] area."[41] Although the revised order from Northern Troops and Landing Force varied only slightly from the instructions issued by Ralph Smith, the fact that two commanders issued different orders to the same unit later served as partial justification for the relief of the Army general.

Holland Smith's original order had in its third paragraph detailed a reinforced battalion to eliminate the Japanese resistance on Nafutan Point. The Marine general considered this proof enough that the unit involved was under Northern Troops and Landing Force control. The substitution of a regiment for a battalion did not alter the command situation. Apparently his Army subordinate assumed otherwise, for Ralph Smith issued his field order for 22 June as though the Nafutan force were responsible directly to the 27th Infantry Division. Technically at least, he had contravened an order of his Marine superior. Also, Ralph Smith had specified that the 105th Infantry hold its present positions until late the following morning, even though the change to the NTLF order, which arrived after the division had assigned the regiment its mission, directed that the attack be continued. Both generals looked forward to taking the offensive, but by going on the defensive for even a few hours, Ralph Smith, his Marine corps commander later maintained, had countermanded a lawful order.[42]

At dusk of 21 June, while the two generals were in the midst of issuing the series of orders which would become so controversial, the frontline troops steeled themselves for the usual night infiltration. Scarcely had the sun gone down, when infiltrators managed to touch off a 2d Marine Division ammunition dump on Green 1. Explosions continued to spew shell fragments over the beach throughout the night, but the Marines along the front lines passed a quiet night. Clashes between patrols and minor attempts at infiltration occurred, but there was no major counterattack.

In the 4th Marine Division sector, four more Japanese were killed at Hill

[40] 27th InfDiv FO No. 45A, dtd 21Jun44.

[41] 27th InfDiv G-3 Jnl, 22Jun44, msg no. 14.

[42] *Cf.* Testimony of MajGen Ralph C. Smith, dtd 31Jul44, p. 6, Exhibit AAA to Proceedings of a Board of Officers Appointed by Letter Orders Serial AG 333/3, 4Jul44, HQ, USAF CPA, hereafter *Army Inquiry;* CG, Expeditionary Trps memo to CTF 51, dtd 24Jun44, Subj: Authority to Relieve Army Officers from Command, Exhibit D to *Army Inquiry;* CG, NTLF memo to CTF 51, dtd 27Jun44, Subj: Summary of Events Leading to Relief from Command of MajGen Ralph C. Smith, USA, Exhibit E to *Army Inquiry.*

500, and enemy bombs crashed harmlessly to earth in the vicinity of General Schmidt's command post. The same 12-plane flight that attacked the beachhead also tried to destroy the transports but was thwarted by a smoke screen. On the following morning, the fight for central Saipan would begin, as the Marines advanced toward some of the most formidable terrain on the entire island—the jumble of peaks that extended from the vicinity of Mount Tapotchau onto Kagman Peninsula. (See Maps 17 and 18.)

The Fight for Central Saipan[1]

On the evening of 21 June, the day before the attack northward was scheduled to begin, Northern Troops and Landing Force reported its combat efficiency as "very satisfactory,"[2] in spite of the 6,165 casualties incurred since 15 June. During the fight for southern Saipan, the 2d Marine Division had suffered 2,514 killed, wounded, and missing, while the losses of the 4th Marine Division totaled 3,628. The 27th Infantry Division, which had not taken part in the costly assault landings, lost 320 officers and men in overrunning Aslito field and seizing the approaches to Nafutan Point. General Harper's XXIV Corps Artillery and the provisional antiaircraft group had yet to lose a man. Force troops had suffered two casualties, both men wounded in action.

THE ATTACK OF 22 JUNE

Numerous as these casualties had been, General Holland Smith believed his two Marine divisions were capable of advancing a maximum distance of 4,000 yards by dusk on 22 June. The 2d Marine Division was to move forward a few hundred yards along the western coast, to seize Mount Tipo Pale in the center of its zone, and on the right to capture Mount Tapotchau, some 3,000 yards forward of the line of departure. While General Watson's troops wheeled past Mount Tapotchau, General Schmidt's 4th Marine Division would keep pace by securing the series of ridges along the division boundary, driving the enemy from Hill 600, and capturing the two terrain features which lay southeast of Mt. Tapotchau that later bore the ominous names of Death Valley and Purple Heart Ridge. If the divisions became extended over too wide an area, Holland Smith planned to commit the 27th Infantry Division, less the regiment which was in action at Nafutan Point. Uncertain where the Army troops might be needed, the corps commander directed Ralph Smith to select routes over which his men might march to the assistance of either frontline division. A total of 18 artillery battalions was to support the main attack.

At 0600 on 22 June, after a 10-minute artillery preparation, the Northern Troops and Landing Force offensive got underway. In the 2d Marine Division zone, the 2d Marines stood fast along the coast, while the 6th and 8th

[1] Unless other noted, the material in this chapter is derived from: *TF 51 OpRpt; TF 56 OpRpt; 2d MarDiv OpRpt; 27th InfDiv OpRpt; 4th MarDiv OpRpt; 2d Mar SAR; 6th Mar SAR; 8th Mar SAR; 10th Mar SAR; 14th Mar Rpt; 23d Mar Rpt; 24th Mar Rpt; 25th Mar Rpt; 105th Inf OpRpt; 106th Inf OpRpt; 165th Inf OpRpt; 1/8 OpRpt; 2/23 Rpt; 3/23 OpRpt; 2/24 Narrative; 1/25 Rpt; 3/25 Rpt; 3/25 Narrative; 3/25 Saga; 1/29 SAR;* Love, *27th InfDiv History;* Crowl, *Marianas Campaign;* Hoffman, *Saipan.*

[2] NTLF G-3 Periodic Rpt, 20–21Jun44.

Marines plunged into a tangle of brush-covered ridges and deep gullies. Attacking in the center of the division zone, the 6th Marines had to maintain contact with the stationary 2d Marines on the left as well as with the advancing 8th Marines. To solve this problem, Colonel Riseley let the progress of the 8th Marines, which had a greater distance to travel, determine his pace. No resistance was encountered until early afternoon, when the 6th Marines began advancing up the slopes of Mount Tipo Pale.

One rifle company sidestepped a ravine strongpoint near the base of the hill and moved unopposed to the summit. The remainder of 3/6 followed the same route to the top of the 1,100-foot peak, but a sheer drop, not shown on the maps, and accurate enemy fire prevented the battalion from moving down the northern slope. While 3/6 made its ascent, the strongpoint below was proving more powerful than anticipated.

The 6th Marines' scout-sniper platoon was the first unit to attack the ravine which 3/6 had bypassed. These few Marines soon discovered that the Japanese had tunneled into several steep bluffs separated by ravines which extended like fingers from the massive hill. The earlier action had disclosed only one of several mutually supporting positions. A rifle company from 2/2, still attached to Riseley's command, took over from the scouts the task of reducing the strongpoint. After destroying a few Japanese emplacements, the unit found itself caught in a deadly crossfire and had to withdraw. The enemy would cling to these formidable positions for two additional days before retreating to the north. The presence of this band of determined Japanese caused Riseley to bend his lines back along the fringe of the strongpoint, so that 2/2 faced more to the east than to the north.

In the 4th Division zone meanwhile, General Schmidt, prior to launching his attack, selected an intermediate objective line drawn near the base of Hill 600. Here the regiments could pause to reorganize before advancing the final 2,000 yards that separated them from the day's objective chosen by General Holland Smith. The rugged terrain as well as the distance to be covered compelled General Schmidt to employ this additional means of controlling the advance.

The 4th Marine Division moved forward with Colonel Batchelder's 25th Marines on the left, Colonel Hart's 24th Marines along the east coast, and Colonel Jones' 23d Marines in reserve. In front of Batchelder's troops lay the most jumbled terrain in the division zone, a series of four ridges that had for control purposes been labeled as 0–A, 0–B, 0–C, and 0–D. The last of these coincided with the intermediate objective. Fortunately, the regimental frontage was narrow enough to permit Batchelder to attack in a column of battalions, a formation that gave him a great degree of flexibility. Should he have difficulty in keeping contact with adjacent units, he would have enough reserve strength to extend his lines.

Lieutenant Colonel Chambers' 3/25 led the column, occupying 0–A by 0630. While the unit was reorganizing, the enemy counterattacked, triggering a violent fight that cost the Japanese 90 dead. Three successive commanders of Company K, the Marine unit hardest hit, were either killed or wounded, but the American attack quickly rolled forward. 0–B, only lightly defended, was captured, and by 1400, 3/25 had overcome increasing resistance to seize 0–C. During the advance, Colonel Batchelder had committed Lieutenant Colonel Hudson's 2d Battalion to seal a gap on the regimental right flank.

As Chambers' men approached 0–D, each of the two assault companies kept physical contact with elements of the flanking battalion, but not with each other, thus opening a hole in the center of the line. The battalion commander inserted his reserve into the gap, but he soon had to call for additional help, a company from Mustain's 1/25, to extend his line still farther to the right. The attack on 0–D was halted short of its goal by fire from caves dug into the ridge itself and from a patch of woods just south of the objective. The 3d Battalion had gained almost 2,000 yards during the day.

Late in the afternoon, an ammunition dump exploded near Chambers' observation post. The battalion commander was stunned by the blast, and Major James Taul, the executive officer, took over until the following day when Chambers resumed his duties. The major launched another attack toward 0–D, but his men were unable to dislodge the Japanese from the woods at the base of the objective.

While Colonel Batchelder's regiment was fighting for the succession of ridges within its zone, Colonel Hart's 24th Marines were advancing along the shore of Magicienne Bay. Gullies leading toward the beach and outcroppings of rock slowed the unit, but Hart's men nevertheless made steady progress. Although the frequent detours caused by the broken terrain opened numerous gaps within the regiment, General Schmidt was more concerned about the difficulty that Hart's Marines were having in keeping contact with Batchelder's troops. At midday, he ordered the 23d Marines, the division reserve, into line between the two regiments.

At 1500, after marching 2,500 yards from its assembly area, Colonel Jones' regiment attacked in a column of battalions. Lieutenant Colonel Haas' 1/23 was in the lead; the 2d Battalion, under Lieutenant Colonel Dillon, followed, while 3/23, commanded by Major Treitel, served as regimental reserve. The formidable terrain rather than the ineffectual enemy resistance slowed the advance, so that by dusk, 1/23 had halted some 200 yards south of the day's intermediate objective.

As darkness drew near, the 4th Marine Division completed adjusting its lines to thwart Japanese attempts at infiltration. On the right, 2/24 was inserted between the 1st and 3d Battalions, but this move did not restore the regimental line. Along the division boundary, the shift of one company from 3/25 caused Taul's battalion to lose contact with Lieutenant Colonel Tompkins' 1/29, on the right of the 2d Division. A company from 1/25 went into position to prevent the enemy from exploiting the break.

On 22 June, General Holland Smith decided to commit his corps reserve, the 27th Division. His operation order for that date fixed the next day's objective. This line included the village of Laulau on the east, the central stronghold of Mt. Tapotchau, and a point on the west coast about 1,000 yards south of Garapan. General Ralph Smith's soldiers were to pass through the lines of the 25th Marines and at 1000 on 23 June attack toward this line. When the objective had been taken, the division would continue its effort upon order from Northern Troops and Landing Force. Since the corps commander was releasing the 106th Infantry to division control, General Ralph Smith elected to attack with two regiments abreast, Colonel Ayres' 106th on the left and Colonel Kelley's 165th on the right.

Holland Smith, on the afternoon of 22 June, decided that a single battalion should be able to clean up Nafutan Point. Ralph Smith felt otherwise, expressing belief that the Japanese might pierce the thin American line to storm Aslito field. Nevertheless, he prepared to execute the decision of his superior commander. At 2100 on 22 June, he issued a field order to 2/105, which was at that time in corps reserve, directing that unit and its attached tanks "to continue operations to mop-up remaining enemy detachments in the Nafutan Point area." After the Nafutan pocket had been reduced, the battalion would, read the Army general's directive, revert to corps control as corps reserve.

At 2330 on 22 June, the division CP received a practically identical order from the corps commander. It included the subject of reversion to corps control. There was just one difference between the two directives: Holland Smith indicated that the attack would begin "at daylight," whereas Ralph Smith omitted those words. The corps commander subsequently objected that the Army general had issued an order to a unit not at the time under his tactical control. A relative fact was that the division commander had not been granted authority regarding use of the corps reserve.

This was Ralph Smith's second mistaken order to a unit not under his tactical control, the previous instance occurring on 21 June and also involving the 105th Infantry.

While preparations to resume the Nafutan Point mop up were underway, Colonels Ayres and Kelley were already selecting the routes which their regiments would follow to move into the front lines to the north. In the south, 2/105 extended its lines, while the remainder of the regiment reverted to corps reserve.[3]

On 22 June, the two Marine Divisions had advanced half the distance to the day's objective at a cost of 157 casualties.[4] The Americans, however, now faced General Saito's main line of resistance. Here, the enemy had concentrated some 15,000 men, two-thirds of them from the *43d Division* and the remainder either sailors or stragglers whose "fighting ability is reduced by

---

[3] NTLF OpO 10–44, dtd 22Jun44; NTLF G–3 Jnl, msg no. 743, dtd 1550, 23Jun44; 27th InfDiv FO No. 46, dtd 22Jun44; MajGen Ralph C. Smith, Notes on Ops of 27th Div at Saipan, Anx I to PreliminaryRpt on Ops of 27th Div at Saipan, dtd 11Jul44, Exhibit M to *Army Inquiry;* MajGen Ralph C. Smith memo to CG, NTLF, dtd 23Jun44, Subj: Hostile Forces on Nafutan Point, Exhibit VVV to *Army Inquiry.*

[4] NTLF G–1 Rpt, App I in *NTLF OpRpt.*

lack of weapons."[5] When the NTLF attempted to overcome these defenders, the number of Americans killed and wounded was bound to soar.

At Nafutan Point, most of the day was spent in adjusting the front line. As a result of the shifting of its components, 2/105 had to yield some of the ground it already had captured. Opposing the reinforced battalion were approximately 1,000 Japanese soldiers and civilians, a force about equal in numbers to the Army unit.

On the morning of 22 June, Army Air Forces fighters (P–47 Thunderbolts) of the 19th Fighter Squadron landed at Aslito field. The planes, which had been launched from escort carriers, were refitted with launching racks and armed with rockets by ground crews already at the airstrip. By midafternoon, eight of the P–47s had taken off on their first support missions of the Saipan campaign.[6]

By Saipan standards, the night of 22–23 June was comparatively quiet. Four Japanese who attempted to infiltrate along the division boundary were killed in a hand-to-hand struggle. The 14th Marines and 106th Infantry were shelled by enemy batteries located near Mount Tapotchau, and artillery pieces on Tinian damaged an LST off the Green Beaches before they were silenced by counterbattery fire.

Japanese aircraft also saw action. Late in the afternoon, a torpedo plane scored a hit on the *Maryland*, forcing that battleship to steam to Pearl Harbor for repairs. A night aerial attack on the Charan Kanoa anchorage did no damage to American shipping.

23 JUNE: INCREASING RESISTANCE

The corps attack of 23 June was a continuation of the previous day's effort. Once again, the 2d Marines served as pivot for the 2d Marine Division. In the adjacent 6th Marines zone, Lieutenant Colonel William K. Jones' 1/6 also held its ground to enable 3/6, commanded by Major Rentsch, to come abreast. The 3d Battalion advanced about 400 yards, but the pockets of resistance on Tipo Pale could not be eliminated. During the day, 2/2 was pinched out as the frontage became more narrow. This unit was returned to Colonel Stuart's 2d Marines in exchange for Major Hunt's 2/6, which was reunited with its parent regiment.

Colonel Wallace's 8th Marines benefited from an aerial search by observation planes of VMO–2 for routes leading to Mount Tapotchau. The reconnaissance disclosed a suitable supply road, but the observer also discovered that the only feasible avenue by which to approach the summit, a ridge near the division boundary, was dominated by a towering cliff not yet in American hands.

Resistance in the 8th Marines zone proved light at first, but the attack had to be halted at 1130 because the adjacent 106th Infantry had not yet crossed its line of departure. Until the Army regiment began moving forward, Tompkins' 1/29 would be unable to advance. At 1345, General Watson ordered the

[5] *Thirty-first Ar Outgoing Msg* No. 1081.

[6] AAF Hist Studies No. 38, OpHist of the Seventh AF, 6Nov43–31Jul44, p. 55 (MS at USAF Archives, Maxwell AFB, Ala.).

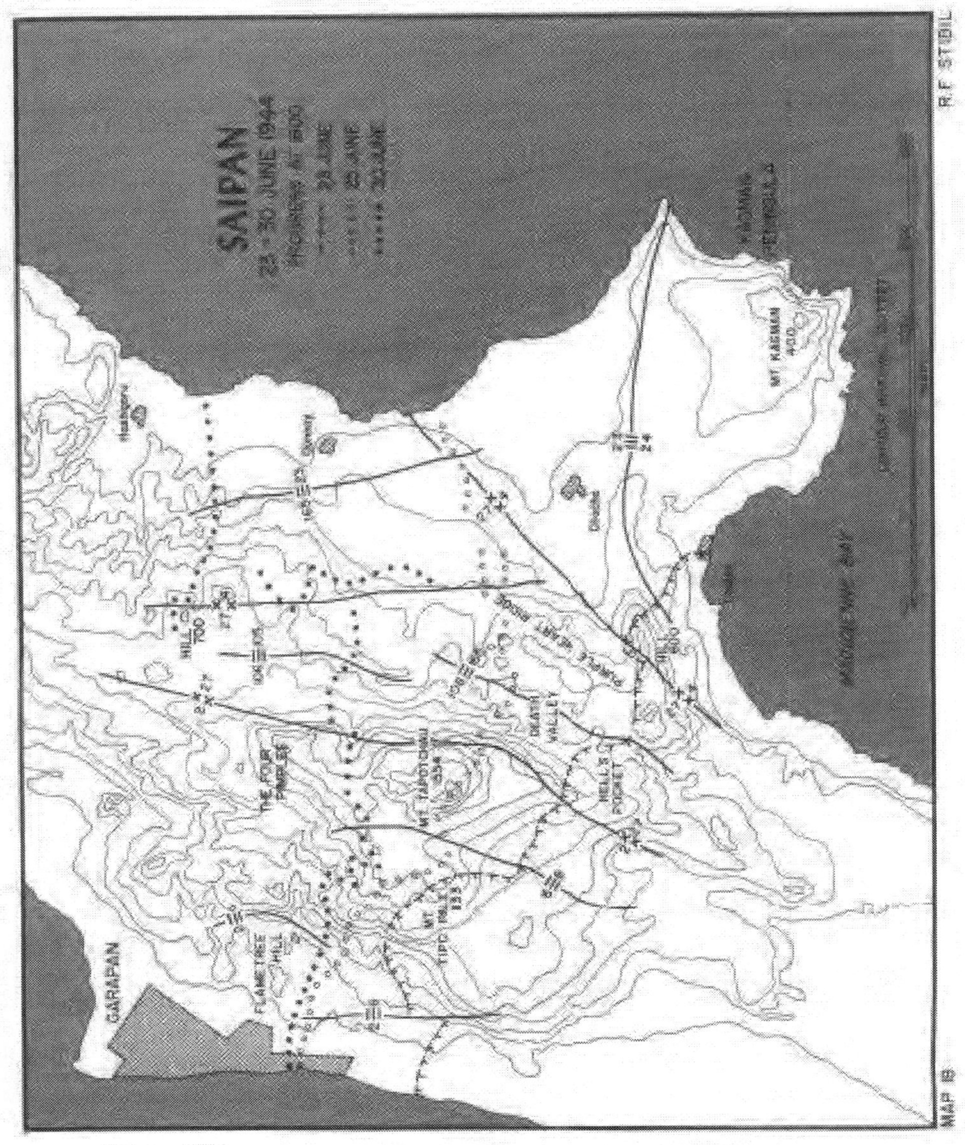
SAIPAN
23 - 30 JUNE 1944
GARAPAN
MAP 19
R.F. STIBIL

8th Marines to continue its attack. Colonel Wallace shifted Chamberlin's 2/8 to protect the exposed flank, and the Marines surged forward. Although 1/29 easily secured the cliff that barred the way to Mount Tapotchau, Lieutenant Colonel Hays' 1/8 collided with a force of 30 Japanese supported by six machine guns. These defenders, entrenched in a ravine at the left of the regimental zone, succeeded in halting the battalion advance.

As dusk approached, Chamberlin visited Major Almerin C. O'Hara at the 2/106 command post in an attempt to establish contact with the Army unit. General Ralph Smith soon arrived on the scene and permitted Chamberlin to borrow O'Hara's Company F in order to post it on the right flank of the 8th Marines. The officers involved reasoned that the Army battalion could more easily maintain contact with one of its own companies than with a Marine unit, but such was not the case. Although the additional company gave added protection to Wallace's flank, O'Hara could not extend far enough to the left to seal the opening, and for the next few days F/106 fought as a part of the 8th Marines.

The NTLF operation order for 23 June called for the 27th Infantry Division to assume responsibility for the center of the corps front by relieving the 25th Marines. The Marine regiment would then pass into Northern Troops and Landing Force reserve. The two Army regiments, the 106th and 165th Infantry, selected for the relief marched from their assembly areas at 0530, 4½ hours before the offensive was to begin. Within an hour, however, elements of the 165th Infantry had cut across the road over which the other regiment was marching, and the approach had to be halted until the tangled units could be sorted out.

In spite of the confusion, 2/165 and 1/165, Colonel Kelley's assault battalions, were in position by 1000 in the right of what had been the 25th Marines zone. The colonel recalled that one of the Marine officers judged the combination of terrain and enemy within the zone to be "about the worst he had run into yet."[7] To the front was a series of ridges and gullies that were dotted with camouflaged weapons positions. With the zone of the 106th Infantry on the left and parallel to the division line of advance was a steep slope, overshadowed by Mount Tapotchau and honeycombed with caves.[8] The 165th Infantry launched its attack against this formidable defensive network at about 1015, but the adjacent Army regiment was not yet in position. Colonel Ayres' unit, arriving one company at a time, did not move forward until 1055.

Throughout the zone of the 27th Infantry Division, the enemy made a determined fight. Colonel Kelley, like Colonel Wallace of the 8th Marines, suspended his advance to enable the 106th Infantry to come abreast, but Colonel Ayres' soldiers were stopped at the regimental left, Ayres could not strongpoint dubbed Hell's Pocket. Although 2/106 was ordered into line on the regimental left, Ayers could not maintain physical contact with the

[7] Testimony of Col Gerard W. Kelley, Exhibit PPP, p. 1, *Army Inquiry*.

[8] *Ibid.*, pp. 1–2.

company dispatched to the 8th Marines. The 165th Infantry, however, enjoyed some success, nearing the southern extremity of Purple Heart Ridge before being fought to a standstill.

The 4th Marine Division attacked with the 23d and 24th Marines on line. Lieutenant Colonel Dillon's 2/23 seized Hill 600. "This was very difficult terrain," reported the battalion commander, "and it was hard enough scaling the hill, let alone fighting up it." [9] From the summit, Marine observers could view the whole of Kagman Peninsula, the area to be seized during the next phase of the battle. While Dillon's men were destroying the defenders of Hill 600 with grenades and flamethrowers, Colonel Hart's 24th Marines pushed all the way to Laulau. Because the adjacent 27th Infantry Division had been stalled, the 4th Marine Division line was echeloned to the left rear, extending from Laulau past Hill 600 to the right flank of the 165th Infantry.

The NTLF operations map was little changed from the previous day. Although the 2d Marine Division had made gains on either flank of Tipo Pale and the 4th Marine Division had advanced about 1,000 yards along the coast, the Army division had accomplished very little. At Nafutan Point, the situation was practically unchanged. At day's end, one platoon manned a temporary perimeter atop Mount Nafutan, but otherwise the battle position was the same as before. (See Map 18.)

[9] *3/23 OpRpt*, p. 3.

"THE COMMANDING GENERAL IS HIGHLY DISPLEASED"

During the afternoon of 23 June, General Watson had two rifle companies formed from among his divisional shore party units. As more of the cargo handlers became available, additional units would be formed to serve as part of the division reserve. Since the 2d Marine Division soon would be advancing upon Garapan, the 2d Marines removed the minefield sown to block the coastal road.

Northern Troops and Landing Force, like General Watson's headquarters, turned its attention to maintaining a strong reserve. The 25th Marines, relieved by the 27th Infantry Division, withdrew to Hill 500 to await further orders.

The headquarters area of the 10th Marines and its fire direction center were heavily shelled during the night. The regimental executive officer, Lieutenant Colonel Ralph E. Forsyth, was killed and several key officers and noncommissioned officers were wounded. Communication facilities were badly damaged and 1/10 had to take over the direction of artillery support for the 2d Division. It was the 27th Infantry Division, however, that saw the fiercest action between darkness on 23 June and dawn of the 24th. Five of six Japanese tanks that attempted to knife along the boundary between the Army regiments were destroyed; a later enemy attack proved more damaging. Five tanks accompanied by infantry struck the lines of the 106th Infantry, and again all but one of the vehicles were destroyed. The survivor, however, burst through the American defenses to set fire to a stockpile of ammunition. The resultant explosions forced the 3d Battalion to fall back until the flames had died away. An attack up the western slopes of Hill 600 was beaten off by the 23d Marines, assisted by 1/165. Japanese aerial attacks during the afternoon and evening accomplished nothing at the cost of two enemy planes, but an early morning raid on the vessels off Charan Kanoa resulted in 18 American casualties at no loss to the marauders.

Holland Smith was angered by the failure of the 27th Infantry Division to advance. During the afternoon of the 23d, as the attack was grinding to a halt, he had discussed the situation with Major General Sanderford Jarman, the Army officer in command of the Saipan garrison.[10] The NTLF commander asked Jarman to visit Ralph Smith at the 27th Division command post to see what could be done to get the unit moving. Later Jarman recalled:

> I talked to General [Ralph] Smith and explained the situation as I saw it and that I felt from reports from the corps commander that his division was not carrying its full share. He immediately replied that such was true; that he was in no way satisfied with what his regimental commanders had done during the day and that he had been with them and pointed out to them the situation. He further indicated to me that he was going to be present tomorrow, 24 June, with this division when it made its jump-off and he would personally see to it that the division went forward. . . . He appreciated the situation and thanked me for coming to see him and stated that if he didn't take the division forward tomorrow he should be relieved.[11]

Both Holland and Ralph Smith agreed that the Army division would have to press its attack more vigorously.

On the following morning, the Marine general sent a dispatch that left his Army subordinate with no doubts concerning his attitude toward the recent performance of the 27th Infantry Division:

> Commanding General is highly displeased with the failure of the 27th Division . . . to launch its attack as ordered at King Hour [1000] and the lack of offensive action displayed by the division in its failure to advance and seize the objective 0–5 when opposed only by small-arms and mortar fire. . . .

The NTLF commander then pointed out that, because the 27th Infantry Division had failed to advance, the two Marine divisions were forced to suspend offensive operations to prevent "dangerous exposure of their interior flanks." Finally, he directed that "immediate steps" be taken to get the stalled Army unit moving forward.[12]

In stating that the 27th Infantry Division had been stopped by fire from small arms and mortars, Holland Smith underestimated the opposition which the soldiers had encountered. Tanks and mountain howitzers supported those portions of the *118th* and *136th Infantry Regiments* entrenched in front of the Army division. The strength of the *136th Infantry Regiment*, which also occupied positions in the 2d Marine Division zone, was less than 1,000 men. The other regiment, strung out from Mount Tapotchau to

[10] MajGen Sanderford Jarman statement, dtd 23Jun44, p. 1, Exhibit J to *Army Inquiry*.

[11] *Ibid.*

[12] CTF 56 disp to CG, 27th InfDiv, dtd 0836, 24Jun44, Exhibit G to *Army Inquiry*.

Magicienne Bay, was far under its peak strength of 2,600.[13]

Although their ranks had been seriously depleted, the defenders were posted on terrain ideal for their purpose. Any unit attempting to push rapidly along the floor of Death Valley would be exposed to fire from the slopes leading from Mount Tapotchau on the left and from Purple Heart Ridge on the right. On 23 June, Colonel Ayres, whose 106th Infantry zone included a portion of Tapotchau's slopes as well as part of the valley itself, had refrained from bypassing Hell's Pocket to advance across the level ground beyond. When asked what would have happened had he attacked vigorously up the valley, Ayers responded: "My candid opinion is that the regiment would have disappeared." [14]

Death Valley, then, was ringed with strong defenses, and the task facing Ralph Smith's division was more difficult than the corps commander realized at the time. Yet, the 27th Infantry Division commanding general, who had toured his front lines on 23 June, accepted Holland Smith's criticism, as reported by General Jarman, and admitted his own displeasure with the actions of some of his subordinates. Ralph Smith apparently believed that the fortifications to his front were strong but not impregnable. The next day's attack, he had vowed, would be both promptly and vigorously executed.

On the ground overlooking Death Valley, the Japanese were equally determined to stop the renewed attack. General Saito's line was threatened in three places—around Tipo Pale, at the mouth of Death Valley, and along the shores of Magicienne Bay. In spite of this pressure from the front and the increasing bombardment of rear areas, General Saito was confident that his men would make the best possible use of the rugged terrain of Saipan. "The 43d Division units, with the firm decision to hold out until the last," the Saipan headquarters reported, "expect to smash the enemy." [15]

THE BATTLE RENEWED

The next objective of Northern Troops and Landing Force was a line extending from the southern part of Garapan due east to the opposite coast of the island. Between the present front lines and this distant objective lay Mount Tapotchau, Death Valley, Purple Heart Ridge, and the whole of Kagman Peninsula. The 2d Marine Division was to enter Garapan, advance some 2,000 yards beyond Mount Tipo Pale, and overrun Tapotchau. The 27th Infantry Division, would continue advancing on the 2d Marine Division right, while the 4th Marine Division would pivot to the east, capture Kagman Peninsula, and then pass into corps reserve. Thus, when the objective line was reached, the battle would enter a new phase, with two-thirds of Saipan in American hands and two divisions moving forward to secure the remainder.

On 24 June, the principal compo-

[13] CinCPac-CinCPOA Item No. 10,531, Excerpts from a Notebook of FOs.

[14] Testimony of Col Russell G. Ayres, Exhibit CCC to *Army Inquiry*.

[15] *Thirty-first Ar Outgoing Msg* No. 1092, dtd 23Jun44.

nents of Northern Troops and Landing Force were to begin their drive toward the objective. Colonel Batchelder's 25th Marines would remain in the area of Hill 500, sending out patrols to eliminate the Japanese lurking around Lake Susupe. At Nafutan Point, 2/105 was to continue its operations against the isolated Japanese pocket.

In the 2d Marine Division zone, General Watson ordered Colonel Stuart's 2d Marines to take the offensive. After a swift initial advance, the regiment encountered more vicious fighting as it neared the town. Lieutenant Colonel Kyle's 1st Battalion, on the right of Major Harold K. Throneson's 3/2, soon collided with a Japanese outpost located on a ridge southeast of Garapan. Fire from mortars and 105mm howitzers enabled the Marines to gain the crest, but the enemy promptly counterattacked. Fortunately for the Americans, the north face of the ridge was so steep that it was almost impossible to scale. "Difficulties notwithstanding," one observer has written, "the Japanese made the effort; but, with two hands required to scale the slope and another to throw grenades or wave sabers, they were one hand short from the outset." [16]

After beating back the counterattack, the Marines began digging foxholes overlooking Radio Road in the southern part of Garapan. The 3d Battalion was in the midst of its preparations for the night when seven enemy tanks, unprotected by infantry, charged from the ruined town. Medium tanks and self-propelled 75s destroyed six of the vehicles and sent the seventh fleeing for cover. The arrival of the two-company provisional battalion that had been formed from among the shore party gave added depth to the regimental defenses.

In the center of the 2d Marine Division zone of action, 1/6 advanced 900 yards over comparatively open terrain, but 3/6, on the regimental right, was slowed by cliffs and ravines. The pocket north of Tipo Pale defied efforts to destroy it, but the southern and eastern slopes of the mountain were by now secured. Because the broken ground on the right had caused such uneven progress, Major Rentsch's 3/6 ended the day holding a 1,500-yard frontage. Since a rifle company from 2/2 and another from 2/6 had joined its three rifle companies on line, the 3d Battalion was able to establish contact throughout its zone.

On the division right, where the 8th Marines were battling toward Mount Tapotchau, Lieutenant Colonel Hays' 1/8 again attacked the pocket of resistance that had stopped the previous day's advance. While infantrymen attempted to keep the defenders pinned down, engineers armed with demolitions, rocket launchers, and flamethrowers crawled across the jagged coral to seal or burn enemy-infested caves. By late afternoon, the strongpoint had been reduced, freeing the battalion to continue its advance.

While Hays' men were eliminating the strongpoint within their zone, Major Larsen's 3/8 and, on the far right, Tompkins' 1/29 were approaching Tapotchau as rapidly as the terrain and the need to protect their flanks would permit. Tompkins' unit was pushing along an uneven plateau, a coral for-

[16] Hoffman, *Saipan*, p. 141.

mation that resembled a stairway leading toward the mountain. Trees and vines choked a part of the battalion zone, and, as happened so often during the Saipan campaign, the unit became overextended. Colonel Wallace then ordered 2/8 to protect Tompkins' flank and also posted his 37mm antitank guns along the ridge separating the 2d Marine and 27th Infantry Divisions.

As the 2d Marine Division surged forward, General Ralph Smith launched an attack which, he believed, would redeem the failure of the day before. The results, however, were disappointing. On the right, Colonel Kelley detoured 3/165 through the area already overrun by the 23d Marines so that the battalion could take up a position on the eastern slopes of Purple Heart Ridge. Although the Army regiment gained little ground, it was now in position to exert pressure from two directions against the formidable ridge.

In spite of pressure from the division commanding general, the 106th Infantry again failed to penetrate beyond Hell's Pocket. The action was much sharper than before. Prior to its relief during the afternoon, 3/106 suffered 14 killed and 109 wounded, more than twice the number of casualties it had endured on the previous day. In summing up the action of 24 June, Colonel Ayres stated that his regiment had been "thrown back onto the original line of departure." [17]

Having for two days bloodied his fists against the gates of Death Valley, Ralph Smith now decided upon a new approach. By making an attack along Purple Heart Ridge, he hoped to knife past the valley and establish firm contact with the Marine divisions, leaving one of Ayres' battalions to contain the bypassed Japanese.[18] Holland Smith however, simultaneously issued orders to continue the attack up the valley.[19] Before the two men had resolved this conflict, Ralph Smith had been relieved. General Jarman, the new commanding general, would decide to try the scheme of maneuver proposed by his predecessor.

While the main body of the 27th Infantry Division was hammering at the defenses to its front, the battalion at Nafutan Point was making little headway. On 25 June, 2/105 was to continue its attack under the command of Colonel Geoffrey M. O'Connell, chief of staff of the island garrison force. Responsibility for reducing the stronghold now rested with the Saipan Garrison Force.[20]

On 24 June, General Schmidt's 4th Marine Division began pivoting toward Kagman Peninsula. The 23d Marines, on the left, moved around an enemy outpost near Hill 66 to advance onto the peninsula itself. This turning movement, carried out against moderate resistance, exposed the left flank of the unit, which was separated by almost 1,000 yards from the positions held by the adjacent 165th Infantry. The 24th Marines, turning on a shorter radius, kept pace, so that by dusk the

[17] Ayres Testimony, *op. cit.*, p. 6.

[18] 27th InfDiv FO No. 48, dtd 1800, 24Jun44.

[19] NTLF OpO 12–44, dtd 1800, 24Jun44.

[20] CG, NTLF ltr Concerning Conduct of Ops by 2/105 in the Nafutan Point Area, Exhibit H to *Army Inquiry;* Col Geoffrey M. O'Connell memo to Gen Richardson, dtd 2Jul44, Subj: Ops of 2/105, Saipan, on Nafutan Point, p. 1, Exhibit WW to *Army Inquiry*.

division front lines formed an arc that encompassed almost one-third of Kagman Peninsula.

To the weary Japanese, the oncoming Marines seemed invincible. General Saito's chief of staff reported that "300–400 troops along with four or five tanks have broken through Chacha in the area of the eastern foot of Tapotchau [near the base of Kagman Peninsula]." He went on to confess that, though the *43d Division* was doing its best, the forces in the area were "reduced to the condition where we cannot carry out this plan [holding the cross-island line] with our present fighting strength." The enemy officer then repeated a call for reinforcements which he had made on the previous day.[21]

THE RELIEF OF GENERAL RALPH SMITH [22]

In his conversation with General Jarman, Holland Smith had predicted that summary relief of an Army officer, if such an incident should take place, was bound to stir up a controversy. On 24 June, however, the corps commander decided, come what may, to embark on "one of the most disagreeable tasks I have ever been forced to perform." [23] The Marine general, in a conference with Admirals Turner and Spruance, stated the problem, and Spruance, the overall commander, directed Holland Smith to replace Ralph Smith with Jarman.[24] "No other action," the Fifth Fleet commander later observed, "seemed adequate to accomplish the purpose." [25]

In requesting authority to relieve Ralph Smith from command of the 27th Infantry Division, Holland Smith stated that such action was necessary to give the corps commander "sufficient authority to cause Army units within landing forces to conduct operations in accordance with his own tactical orders." As examples of his subordinate's failure to follow orders, the Marine general cited the two instances, on 21 and 22 June, when Ralph Smith had issued instructions to units under corps control and the fact that the attack of 23 June had been delayed because of the late arrival of components of the 27th Infantry Division.[26] The basic reasons he stressed, however, for this drastic action were the "defective performance" of the Army division and its need of "a leader would make it toe the mark." [27]

No sooner had the relief been accomplished than the expected storm of

[21] *Thirty-first Ar Outgoing Msg* No. 1096.

[22] An Army-oriented account of the relief and the controversy that followed is contained in Crowl, *Marianas Campaign*, pp. 191–201.

[23] Smith and Finch, *Coral and Brass*, p. 173.

[24] *Moore comments Saipan.* At this same meeting, the three officers decided to maintain the 1st Provisional Marine Brigade as floating reserve for possible use at Saipan and to send the 3d Marine Division to Eniwetok, where it would await the Guam operation.

[25] Comdr Fifth Flt ltr to CinCPOA, dtd 29Jun44, Subj: Summary of Events Leading up to the Relief from Command of Major General Ralph C. Smith, p. 1, Exhibit C to *Army Inquiry*.

[26] CG, ExpeditionaryTrps ltr to CTF 51, dtd 24Jul44, Subj: Authority to Relieve Army Officers from Command, Exhibit D to *Army Inquiry*.

[27] Smith and Finch, *Coral and Brass*, p. 173; *Moore comments Saipan.*

controversy began to break. Although not included in the chain of command for the Marianas operation, Lieutenant General Robert C. Richardson, Jr., ranking Army officer in the Central Pacific, apparently was angered that the change had been made without his knowledge. When Ralph Smith reached Hawaii, Richardson appointed him, as a gesture of confidence, commanding general of the 98th Infantry Division. The Army lieutenant general also convened a board of officers to inquire into the circumstances surrounding what some of his fellow officers considered "the slur on their service implied by the relief of Ralph Smith." [28]

The board, headed by Lieutenant General Simon B. Buckner, Jr., could examine only those reports contained in Army files and hear testimony only from Army officers. Yet, in spite of its *ex parte* nature, the Buckner board concluded that Holland Smith had the authority to relieve his Army subordinate and that he issued in a proper fashion the orders effecting that relief. The investigators, however, objected that the Marine general, unaware of the resistance that the 27th Infantry Division had encountered, had taken action that was "not fully justified by the facts." [29] The report of the board proceedings also contained recommendations that Ralph Smith's future assignments not be adversely affected by the Saipan incident, that the senior Army commander be fully informed of theater and JCS policies concerning command relationships, and finally that "when it is necessary to combine elements of two or more services into one major unit, the most careful consideration be given to the personality and qualifications of the senior commanders concerned." [30] The Army board thus implied that Holland Smith, though his action was legally correct, had been more vigorous than circumstances warranted. The Marine general, the board members appear to have believed, was singularly lacking in tact.

Upon studying the Buckner report, certain of General Marshall's advisers, though they did not approve of Holland Smith's action, admitted that the 27th Infantry Division had not been performing as well as it should have been, principally because certain of Ralph Smith's subordinates lacked vigor. These officers nonetheless believed that the relief of the division commanding general was not necessary.[31] Once again, the Marine general's judgment and tact were questioned rather than his right to effect a change of command.

During the hearings, General Richardson visited Saipan, ostensibly to inspect Army forces, and engaged in a heated argument with Holland Smith. The spirit of harmony that had so far characterized the Central Pacific campaign was fast evaporating. In order to remove the occasion of the friction, the War Department had Ralph Smith recalled from Hawaii and eventually assigned to the European Theater of Operations. After Saipan had been declared secured, Holland Smith assumed command of Fleet Marine Force,

---

[28] Crowl, *Marianas Campaign*, p. 192.

[29] Rpt of *Army Inquiry*, p. 10.

[30] *Ibid.*

[31] Crowl, *Marianas Campaign*, p. 195.

Pacific, a post in which he would have no control over Army divisions. No effort was made, however, to alter the command structure for the imminent Guam operation. There the untested 77th Infantry Division would fight effectively when included with Marine units in a corps commanded by Marine Major General Roy S. Geiger.

The Saipan controversy, by no means typical of interservice relations in the Pacific, seemed destined to be fought to its conclusion in an atmosphere of comparative secrecy. Unfortunately, somewhat distorted accounts of Ralph Smith's relief slipped past the censors to touch off a journalistic battle that flared intermittently until 1948. The volcano of adverse publicity that erupted after the Saipan campaign, specifically the article in *Time* magazine that claimed the soldiers "froze in their foxholes"[32] at the entrance to Death Valley, had a crushing effect on the morale of the 27th Infantry Division. The story itself caused a flood of anger, but the arrival of letters from friends and relatives in the United States, who accepted the article as completely accurate, was a cruel blow to the men of the division.[33]

In November 1944, after the entire Marianas operation had ended, General Marshall suggested to Admiral King that both Nimitz and Richardson, as senior representatives of their services in the Central Pacific, should thoroughly investigate the incident in order to prevent future discord. King refused, for he believed that Richardson's previous inquiry had prolonged the strife instead of ending it. In the opinion of the Chief of Naval Operations, the record of the Buckner board contained intemperate outbursts against Holland Smith, and statements that did not pertain to the issue under investigation. The admiral was convinced that any new inquiry would degenerate into a clash of personalities, and Marshall apparently adopted a similar point of view, for no further official action was taken.[34]

The Smith against Smith controversy was caused by failure of the 27th Infantry Division to penetrate the defenses of Death Valley. Holland Smith had told the division commanding general that operations in the area had to be speeded up. Ralph Smith, who was thoroughly familiar with the tactical situation, informed Jarman of his own annoyance with the slow progress of his unit. He told the island commander that he intended to press the attack, but he postponed making the changes in command which, according to Jarman, he intimated might be necessary. The NTLF commander, after stating that the objective had to be taken, saw that no significant progress had been made on 24 June and promptly replaced the officer responsible for the conduct of the Army division. The Army Smith offered his subordinates another chance, but the Marine Smith did immediately what he felt was necessary, without regard for the controversy he knew would follow.

[32] *Time*, v. 44, no. 12 (18Sep44), p. 26.

[33] Love, *The 27th InfDiv History*, pp. 522–523.

[34] CofS, USA memo for Adm King, dtd 22 Nov44; FAdm King memo for Gen Marshall, dtd 23Nov44, Subj: Article in *Time* magazine (Smith-Smith File, HistBr, HQMC); Crowl, *Marianas Campaign*, p. 196.

25 JUNE: MOUNT TAPOTCHAU AND KAGMAN PENINSULA

During the evening of 24 June, as the relief of Ralph Smith was taking place, Japanese planes attacked ships clustered off Saipan. Neither antiaircraft guns nor Army night fighters (P–61 Black Widows) in the first of 105 sorties these planes would fly during the campaign, were able to down any of the attackers. The Japanese bombs, however, did no damage.[35]

Ashore, the enemy made repeated attempts to infiltrate the lines of General Watson's 2d Marine Division. Colonel Stuart's 2d Marines, which saw sporadic action through the night, killed 82 Japanese at a cost of 10 casualties. Enemy artillery and mortar fire forced Major Rentsch, in command of 3/6, to find a new location for his command post and disrupted the battalion communications. Elsewhere the night was quiet, although marred from time to time by the flash of rifle fire or the bursting of a grenade as Japanese attempted to slip through the corps lines.

Although little ground was captured on 25 June by elements of the 2d Marine Division, General Watson's troops dealt the Japanese a jarring blow. While the 2d Marines stood fast on the outskirts of Garapan and the 6th Marines hammered at the Tipo Pale pocket of resistance, the 8th Marines captured Tapotchau, the finest observation post in central Saipan. During the attack against this key height, some 200 members of the shore party finished their tasks at the beaches and formed replacement units for the 6th and 8th Marines.

Some portion of Tapotchau's bulk lay in the zone of each of the four battalions of Colonel Wallace's 8th Marines. The western slopes were to the front of the 1st Battalion, commanded by Lieutenant Colonel Hays, and Major Larsen's 3d Battalion. Lieutenant Colonel Tompkins faced the arduous task of leading his 1/29, still attached to the 8th Marines, directly toward the summit, while 2/8, under Major Chamberlin, had responsibility for securing the eastern slopes. Two routes led toward the mountain crest. Tompkins was to attack through the densely wooded valley and up the steep southern face. Chamberlin would follow the ridge in his sector until abreast of Tapotchau and then veer to the left, advancing up the eastern slope.

By 0930, after two hours of fighting, 1/29 became bogged down in the woods, halted by impassable terrain and enemy fire. On the right, however, 2/8 pushed rapidly to the cliff that marked the eastern terminus of the crescent-shaped mountain. Chamberlin ordered one platoon to scale the cliff, and this unit encountered no opposition. A smaller patrol advanced almost to the crest without being challenged. Since Hays and Larsen were maintaining unceasing pressure on the Japanese defending the western slopes, Tompkins decided to move to his right, pass through Chamberlin's lines, and approach the summit from the east rather than from the south.

Taking with him a detachment from the division reconnaissance company, the commander of 1/29 moved through the adjacent zone of action, scaled the

---

[35] AAF Hist Studies no. 38, *op. cit.*, p. 61.

cliff, and gained the summit. He then left the scouts to hold the objective and returned to his battalion. During the afternoon, Tompkins withdrew two of his three rifle companies from contact with the enemy. Companies A and C formed in single file for the march to the crest of the mountain, while Company B remained in the valley.

While Tompkins was shifting his main body, the Japanese awoke to the danger and launched a series of counterattacks against the perimeter manned by the reconnaissance detachment. During the afternoon, the enemy suffered 40 casualties and the Marines 3. The Japanese also turned upon the platoon which Chamberlin had posted atop the cliff. A violent concentration of mortar fire forced the 2/8 unit to fall back from its exposed position, but this withdrawal did not affect Tompkins' plans.

The sun was about to set by the time 1/29 was ready to climb the mountain. The battalion 81mm mortars blanketed the northern slopes with smoke, while Chamberlin's mortars and 105s of 3/10 blasted possible enemy positions. Tompkins' riflemen ascended the cliff, crossed the pinnacle at the eastern end of the hill mass, passed through a saddle, and moved up the gradual slope leading to the summit. Not a man was lost during the climb.

Atop the mountain, the Marines found that their entrenching tools could scarcely dent the rocky surface. The fortunate few whose positions were located on patches of soft earth scratched out foxholes, but the rest used loose stones to build low parapets over which to fire. Shortly before midnight, the Japanese attacked from the northwest. The hastily prepared defenses proved adequate, as evidenced by the 18 Japanese dead discovered the following morning.[36]

In the center of the island, the 27th Infantry Division, now commanded by General Jarman, attempted to bypass and isolate Death Valley. The 2d Battalion of Colonel Kelley's 165th Infantry captured the southern third of Purple Heart Ridge, then yielded its conquests in order to obtain a better defensive position. Meanwhile, 3/165, poised to attack to the west from positions along the division right boundary, had been halted by a pocket of resistance. Since the 3d Battalion could make no headway, Colonel Kelley ordered the unit to swing through the area already cleared by the 4th Marine Division in order to establish contact with the 23d Marines. By nightfall, the Army battalion was digging in astride the division boundary just west of the Chacha-Donnay road. (See Map 18.)

Colonel Ayres' 106th Infantry, less the 2d Battalion which was maintaining pressure on the southern entrance to Death Valley, had the mission of circling to the right of Purple Heart Ridge and making contact with the 2d Marine Division in the vicinity of Mount Tapotchau. Had this maneuver succeeded, the powerful strongpoint would have been isolated and an integrated corps front restored. Unfortunately, the enveloping force did not

[36] In addition to the reports of the units involved, the account of the capture of Mount Tapotchau is based on Capt James R. Stockman, "The Taking of Mount Tapotchau," *Marine Corps Gazette*, v. 30, no. 7 (Jul46).

reach its attack position until midafternoon. The 1st Battalion attempted to advance to the northwest along the road leading from Chacha across the valley, but the drive was stopped by enemy fire. The remainder of Ayres' maneuver element, 3/106, started toward Chacha, was delayed by road traffic, and elected to return to its original assembly area south of Death Valley. The 2d Battalion, in the meantime, pushed directly into the valley, enjoyed brief success, but finally was driven back to its line of departure.

Although Jarman's plan had failed, the 4th Marine Division managed to overrun Kagman Peninsula. General Schmidt's attack was delayed 45 minutes, for the supporting tanks had difficulty in negotiating the trails leading to the front lines. At 0815, the 23d and 24th Marines crossed the line of departure to begin their surge toward Mount Kagman and the coast. Although Colonel Hart's 24th Marines swept forward against extremely weak opposition, Colonel Jones' 23d Marines encountered a number of stragglers and was taken under fire by a field piece located in the 27th Infantry Division zone. After coordinating with the Army unit, the 14th Marines opened fire and succeeded in temporarily silencing the weapon.[37] By late afternoon, the entire peninsula was in American hands, but the task of mopping up had just begun.

The day's fighting in central Saipan resulted in important gains. Although the attackers had been unable to seal off Death Valley, Tapotchau had fallen and organized resistance on Kagman Peninsula had been shattered. At Nafutan Point, however, the Japanese made good use of broken terrain and heavy underbrush to stall 2/105, but not until after the battalion had pierced the main defenses. During the afternoon, the 40mm and 90mm antiaircraft guns assigned, on the previous day, to support Colonel O'Connell's troops, registered to fire air bursts in preparation for the attack of 26 June.[38]

The night of 25 June saw the foiling of a Japanese attempt to send reinforcements from Tinian. An infantry company, moving on 11 barges toward the Saipan coast east of Chacha, was detected by the destroyer USS *Bancroft* and the destroyer escort USS *Elden*. One of the barges was reportedly sunk, while the others were frightened back to Tinian.

Except for that incident and the fight atop Mount Tapotchau, the night was quiet. The defenders had been seriously weakened by 11 days of sustained fighting. Even had the Japanese troops been rested and more numerous, the lack of communications probably would have prevented a coordinated counterattack.

During 25 June, the Japanese *Thirty-first Army* Headquarters could account for a total of about 950 combat troops remaining in the *135th*, *136th*, and *118th Infantry Regiments*. The *47th Independent Mixed Brigade* was believed reduced to 100 men and the *7th Independent Engineer Regiment* to approximately 70 effectives. The *3d Independent Mountain Artillery Regi-*

[37] Later, when this piece opened fire again, a patrol of the 23d Marines destroyed gun and crew. *Jones ltr.*

[38] O'Connell memo, *op. cit.*, pp. 1, 3.

ment had no field pieces, and the *9th Tank Regiment* only three tanks.[39] These estimates took into account only those Army units in communication with Saito's headquarters. Many other detachments, isolated from the army command post, were fighting savagely. Yet, to General Saito the destruction of the Japanese garrison force seemed inevitable. "Please apologize deeply to the Emperor," he asked of Tokyo headquarters, "that we cannot do better than we are doing." [40]

26 JUNE: THE ADVANCE BEYOND TIPO PALE

The action on 26 June centered around Tipo Pale, where the 6th Marines had been stalled since the afternoon of the 22d. Instead of attacking along the stubbornly defended draw, Colonel Riseley's regiment bypassed the pocket, left one company to mop up, and continued advancing to the north. Lieutenant Colonel Jones' 1st Battalion moved into position to support by fire the capture of the next objective, a ridge that extended west from Mount Tapotchau. While crossing an open field, 2/6 came under deadly fire from the ridge and was forced to break off the action.

East of Tipo Pale, Tompkins' 1/29 strengthened its hold on the summit of Tapotchau. The company left behind in the valley succeeded in joining the rest of the battalion, but a patrol sent to the northernmost pinnacle of the jagged mountain was beaten back by the Japanese. Elsewhere in the 8th Marines zone, progress was slow. Along the western approaches to Mount Tapotchau, the 1st and 3d Battalions battled through dense woods to drive the enemy from a seemingly endless succession of ravines and knolls. For most of the day, these Marines clawed their way forward, dodging grenades and often diving for cover to protect themselves from the plunging fire of machine guns. East of the mountain, 2/8 extended its lines to the rear along the rim of Death Valley, but Chamberlin's battalion, with its adopted Army company, could not make physical contact with the 106th Infantry.

At the entrance to Death Valley, the battle was beginning again. After gaining ground on the previous day, 2/106 had fallen back under cover of darkness to its original position. General Jarman decided to shift slightly the axis of his main attack, but the most difficult tasks again were assigned to the 106th Infantry. While the 1st Battalion tried to reduce Hell's Pocket, the other two battalions, instead of circling completely around the valley, were to attack along the western slope of Purple Heart Ridge, then extend to the left in order to close the gap in the corps front. Meanwhile, 2/165 was to mop up the eastern slope of the ridge.

The attack of the 106th Infantry got off to a confused start, and for this reason General Jarman decided to entrust the regiment to Colonel Albert K. Stebbins, his chief of staff.[41] By the end of the day the 2d and 3d Battalions occupied all but the northern tip of the ridge. The defenders of Hell's Pocket,

[39] *Thirty-first Ar Outgoing Msg* No. 1102.

[40] *Ibid.*, No. 1097.

[41] Jarman Statement, *op. cit.*, pp. 1–2.

MARINES DIG IN *night defensive positions on 24 June as the attack moves northward on Saipan. (USN 80-G-234720)*

COMPANY CP *in the smoking ruins of Garapan stays in close touch with its attacking platoons. (USMC 85174)*

however, hurled 1/106 back on its line of departure.

The 4th Marine Division, charged with mopping up Kagman Peninsula, had been pinched out of the corps line to revert to Northern Troops and Landing Force reserve. Although harassed by artillery fire from enemy batteries in the vicinity of Death Valley, the Marines wiped out the Japanese forces that had survived the fighting of 24 and 25 June. As General Schmidt's troops were assembling at the close of day, the division, less the 25th Marines at Hill 500, was ordered to re-enter the lines. In place of the 25th Marines, General Schmidt was given Colonel Kelley's 165th Infantry, now composed of 1/165, 3/165, and 1/105.

The 26th of June also marked the beginning of the systematic hammering of Tinian by ships and planes as well as by artillery. Since 20 June, 155mm guns, first a battery and then an entire battalion, had been shelling the adjacent island. Now aircraft and cruisers joined in the bombardment. Tinian was divided into two sectors. Each day, the planes would alternate with the ships in blasting both portions of the island. XXIV Corps Artillery was to fire upon any suitable targets not destroyed by the other arms. The naval shelling, however, proved unsatisfactory, for the guns of the cruisers were ill-suited to area bombardment.

SEVEN LIVES FOR ONE'S COUNTRY [42]

Operations at Nafutan Point were speeded on 26 June, for O'Connell's men already had broken the enemy's main defensive line. Advancing against light opposition and supported by antiaircraft weapons, tanks, and naval gunfire, the soldiers secured Mount Nafutan. Late in the afternoon, the Japanese, their backs to the sea, began resisting more vigorously. Since the attacking companies had limited fields of fire, they withdrew before digging positions for the night. The American line was porous, with a gap on the left flank, and no more than a line of outposts on the right.

The enemy's slow response to the pressure applied by the Army battalion did not indicate that these disorganized Japanese were beaten. Captain Sasaki, commander of the *317th Infantry Battalion* of the ill-fated *47th Independent Mixed Brigade* gathered together some 500 soldiers and sailors, survivors from the various units that had helped defend southern Saipan, and issued orders to break out at midnight from the Nafutan Point trap. The men, "after causing confusion at the airfield," were to assemble at Hill 500, formerly the site of brigade headquarters but now the bivouac area of the 25th Marines. "Casualties will remain in their present positions and defend Nafutan Mount," Sasaki continued. "Those who cannot participate in combat must commit suicide. Password for the night of 26 June [is] *Shichi Sei Hokoku,* (Seven Lives for One's Country)." [43]

The enemy passed undetected through O'Connell's line of outposts. The first indication of a *banzai* attack

[42] An additional source for this section is: O'Connell memo, *op. cit.*, pp. 2–3.

[43] Quoted in NTLF G–2 Rpt, p. 34, in *NTLF OpRpt.*

came when a group of Japanese attacked the command post of 2/105. The marauders were driven off after killing 4 Americans and wounding 20 others at the cost of 27 Japanese dead. At 0230, the main force stormed across Aslito field, destroying one P–47 and damaging two others.[44] Three hours later, the Japanese reached Hill 500, where the 25th Marines greeted them with a deadly barrage of grenades and bullets. Fragments of Sasaki's group struck positions manned by the 14th Marines and 104th Field Artillery, but both units held firm.

On 27 June the 25th Marines mopped up the Japanese who had survived the night's action, while 2/105 overran the remainder of Nafutan Point. The soldiers discovered some 550 bodies within their zone. Some of the dead had been killed during the earlier fighting; others had committed suicide in obedience to Sasaki's instructions. Thus, in a burst of violence, ended the wearisome battle for Nafutan Point.

27 JUNE: THE ADVANCE CONTINUES

Considering the effect it had upon the Japanese in central Saipan, the Nafutan Point action might as well have been fought on another planet. If General Saito was aware that 1,000 members of his Saipan garrison had perished within the space of a few days, such knowledge could not have altered his plans. The general already had selected his final line of resistance, a line that stretched diagonally across the island from Tanapag village past Tarahoho to the opposite coast. Here the battle would be fought to its conclusion.[45] (See Map 19.)

On 27 June, the 2d Marine Division, composed of the 2d, 6th, and 8th Marines plus 1/29, readjusted its lines. Along the coast, the 2d Marines waited for orders to seize the town of Garapan. North of Tipo Pale, the 6th Marines repulsed an early morning counterattack, moved forward, but again was stopped short of the ridge that had previously stalled its advance. On the right, 1/29 secured the remainder of Mount Tapotchau, while 2/8 sent patrols into the area east of the mountain. During the morning, Lieutenant Colonel Hudson's 2/25 passed to control of the 2d Marine Division. General Watson attached the battalion to Colonel Wallace's 8th Marines. Hudson's men then relieved Chamberlin's troops of responsibility for guarding the division right flank.

In the 27th Infantry Division zone, the 106th Infantry made important gains. Two rifle companies of the 1st Battalion circled around Hell's Pocket to gain the crest of the ridge that formed the division left boundary. Meanwhile, at the northern end of Death Valley, the 2d and 3d Battalions succeeded in forming a line across the valley floor. On the eastern slopes of Purple Heart Ridge, 2/165 pushed forward to dig in to the right of 2/106.

Although the advance of the Army division had been encouraging, the most spectacular gains of the day were those made by the 4th Marine Division. On the east coast, the 23d Marines by-

---

[44] AAF Hist Studies No. 38, *op. cit.*, p. 59; *USAF Comments.*

[45] *Thirty-first Ar Outgoing Msg* No. 1120.

passed a minefield and advanced against intermittent fire to overrun the villages of Donnay and Hashigoru, capture a supply dump, and gain its portion of the corps attack objective. The attached 165th Infantry, made up of 1/165, 3/165, and 1/105, fared almost as well. By dusk, General Schmidt's lines ran west from the coast and then curved toward the division left boundary, along which 1/165 had encountered stubborn resistance. To maintain contact between that battalion and the units at Death Valley and Purple Heart Ridge, 2/24 was shifted to the 4th Marine Division left flank.

By the coming of darkness on 27 June, the gaps which had marred the corps front were well on their way to being closed. Although Japanese planes bombed both the Charan Kanoa roadstead and Aslito field, there was little infiltration during the night. A truck loaded with 12 enemy soldiers and civilians drove toward the lines held by the 23d Marines, but an antitank gun destroyed the vehicle and killed its occupants. On Purple Heart Ridge, 2/165 was shelled and its commander wounded. Sporadic mortar fire fell in the lines of the 2d Marines near Garapan, but, all in all, the night was quiet.

28 JUNE: MAINTAINING PRESSURE ON THE ENEMY

The Japanese, under steady pressure all along the front, were now preparing defenses to make the area north of Donnay and around Tarahoho secure. While these positions were being completed, those elements of the *118th* and *136th Infantry Regiments* that were opposing the 27th Infantry Division were to fight to the death. Checking the rapid advance of the 4th Marine Division was the task assigned the *9th Expeditionary Unit* and a 100-man detachment from the *9th Tank Regiment.*[48]

The tempo of action in the 2d Marine Division zone remained fairly slow during 28 June. While the 2d Marines conducted limited patrols, aircraft, supporting warships, and artillery pounded suspected strongpoints which might be encountered when the regiment resumed its advance. One preparatory air strike resulted in 27 Marine casualties, when a pilot mistook a puff of smoke for the bursting of the white phosphorous shell that was to mark his target and accidentally fired his rockets into a position manned by 1/2.

The 6th Marines made scant progress, for the 2d Battalion could not drive the Japanese from the ridge to its front. The longest gain made by Colonel Riseley's regiment was about 200 yards. To the rear, however, the bypassed Tipo Pale pocket was at last completely destroyed.

Colonel Wallace's 8th Marines, with 2/25 again withdrawn to corps control, found itself up against a formidable barrier, four small hills, one lying within the zone of each battalion. Because of their size in comparison to Tapotchau, the hills were dubbed the Four Pimples. To make identification easier, each of them was given the nickname of the commander of the battalion that was to capture it. Thus, Major William C. Chamberlin of 2/8 was responsible for Bill's Pimple, Lieutenant

[48] *Ibid.*, No. 1123.

Colonel Rathvon McC. Tompkins of 1/29 for Tommy's Pimple, Major Stanley E. Larsen of 3/8 for Stan's Pimple, and Lieutenant Colonel Lawrence C. Hays, Jr., of 1/8 for Larry's Pimple. (See Map 18.)

As the 8th Marines approached the four hills, enemy resistance increased, so that darkness found the regiment short of its objective. Chamberlin's 2d Battalion faced an especially difficult problem in logistics. Because of the rugged terrain, eight stretcher bearers were needed to evacuate one wounded Marine. Thus, a single bullet or grenade could immobilize most of a rifle squad. The battalion, however, did not passively accept enemy fire, for 100 Japanese perished during the day.

Beyond the ridge to the right, Army units again attempted to come abreast of the Marine divisions. Major General George W. Griner, dispatched from Hawaii by General Richardson, assumed command of the 27th Infantry Division on the morning of 28 June, and General Jarman returned to his assigned duties with the garrison force. Griner's first day of command on Saipan saw the 106th Infantry push a short distance forward in the north, at the same time crushing organized resistance in the bypassed Hell's Pocket. The regimental gains were made costly by accurate mortar fire and by a daring enemy foray in which two tanks killed or wounded 73 members of the 1st and 2d Battalions.

Because of the accumulated losses, Griner shifted his units. With only 100 riflemen present for duty,[47] 3/106 was replaced by the 1st Battalion of the 106th. Company F, which had been under Marine control, now returned to 2/106. On the right, 3/105, idle since its relief at Nafutan Point, entered the battle. With the new battalion came the regimental headquarters, and, as a result, 2/165, which was trying to destroy the knot of resistance at the northern tip of Purple Heart Ridge, was detached from Stebbins' command and attached to the 105th Infantry.

The 4th Marine Division, which had made such impressive gains on the 27th, paused to adjust its lines. While the 23d Marines sent patrols 500 yards to its front, the 165th Infantry occupied Hill 700 at the corner of the division's zone of action. Neither regiment encountered serious opposition, but Colonel Kelley was wounded by mortar fragments and replaced in command of the Army unit by Lieutenant Colonel Joseph T. Hart. Along the left boundary, the attached Army regiment, assisted by 1/24 and 3/24, was unable to make physical contact with General Griner's division. At dusk on 28 June, the 4th Marine Division lines formed an inverted L, with the 23d Marines and part of 3/165 facing north, while the rest of 3/165, 1/105, and the two battalions of the 24th Marines faced west.

The darkness of 28–29 June was pierced by the flash of rifles, bursting of grenades, and explosion of aerial bombs. Once again enemy planes raided both the anchorage and the airfield. In a typical night action, the 6th Marines killed 10 members of a

[47] 106th Inf Jnl, msg no. 609, dtd 1010, 28 Jun44 (WW II RecsDiv, FRC, Alexandria, Va.).

Japanese patrol. The 23d Marines, however, encountered an unusual situation when a 10-truck enemy convoy, lights ablaze, came rumbling toward the front lines. The Japanese realized where they were heading and beat a hasty retreat before the Marines could open fire.

SUCCESS IN DEATH VALLEY

On 29 and 30 June, the corps line remained almost stationary on its flanks, even though the fighting still blazed in its center. "With the operation two weeks old, *everyone* on the island felt the weight of fatigue settling down," a historian of the campaign has written. The Japanese after a succession of bloody reverses, were badly worn, and the American divisions resembled "a runner waiting for his second wind." [48]

Although tired, the Marines and soldiers were determined to finish the grim job at hand. Near Garapan, this determination resulted in a cleverly delivered blow against a formidable Japanese redoubt. About 500 yards in front of the 2d Marines lines, an enemy platoon had entrenched itself on Flametree Hill. During the day, the defenders remained in caves masked by the orange-red foliage that covered the hill. If the regiment should attempt to advance through Garapan, the Japanese could emerge from cover and rake the attackers with devastating fire. Either the enemy had to be lured onto the exposed slopes and scourged with long-range fire, or the hill itself would have to be captured, probably at a large cost to the attackers.

On the morning of 29 June, Marine artillery blasted Flametree Hill, and machine guns raked the tree-covered slope, while mortars placed a smoke screen in front of the objective. When the barrage stopped, the defenders dashed from their caves to repel the expected assault. Since rifle fire could be heard from beyond the smoke, the Japanese opened fire. Suddenly the American mortars began lobbing high explosives onto the hill, the machine guns resumed firing, and artillery shells equipped with time fuzes started bursting over the trenches. When the deluge of bullets and shell fragments ended, the weapons on Flametree Hill were silent.

Another accident befell the 2d Marines on 30 June. A Navy torpedo plane, damaged by enemy fire, crashed into the positions of 1/2, injuring seven infantrymen. The pilot escaped by taking to his parachute at an extremely low altitude.

During the last two days of June, the 6th Marines patrolled the area to its front. Colonel Riseley's men made no spectacular gains, but the 3d Battalion managed at last to seize the ridge from which the enemy had blocked the advance. Major Rentsch's troops gained a foothold on 29 June and, on the following day, secured the remainder of the objective. The capture of this ridge, which lay just north of Tapotchau, placed the regiment "on commanding ground in the most favorable position for continuation of the attack since D-Day." [49]

[48] Hoffman, *Saipan*, p. 180.

[49] *2nd MarDiv OpRpt*, Sec VI, p. 19.

The 8th Marines devoted these two days to finding a route over which tanks could move forward to support the attack against the Four Pimples. On 30 June, while moving toward Stan's Pimple, the 3d Battalion captured a road which could be improved adequately by bulldozers. On the far right, Chamberlin's 2/8 overcame light resistance and seized Bill's Pimple late in the afternoon of the 30th. The other hills, though blasted by shells and rockets, remained in enemy hands. Prospects for the 8th Marines, however, seemed excellent, for by the evening of 30 June, Army and Marine tanks had reached the front lines, supplies were arriving to sustain the regiment, and the gap along the division boundary was being patrolled by elements of the 106th Infantry.

The 2d Marine Division, which had suffered 4,488 casualties since D-Day, was employing all three of its regiments on line when the fight for central Saipan came to an end. Since replacement drafts had not yet arrived, support units had been organized to serve as the division reserve. A total of five such companies were available to General Watson on the evening of 30 June.

Success at last crowned the efforts of the 27th Infantry Division, for on 29 and 30 June the soldiers burst through Death Valley and drew alongside the 8th Marines. The 106th Infantry joined the 105th in overrunning the valley, a company from 1/106 wiped out the stragglers trapped in Hell's Pocket, and 2/165 eliminated the die-hards entrenched on Purple Heart Ridge. Looking back upon the one-week battle, General Schmidt, who later succeeded General Holland Smith as corps commander, observed that: "No one had any tougher job to do."[30] In clearing Death Valley and Purple Heart Ridge, the Army unit sustained most of the 1,836 casualties inflicted upon it since its landing.

Although no further advance was attempted, the 4th Marine Division continued to send patrols beyond its positions. Marine units made only occasional contacts with small groups of Japanese. The 165th Infantry, which yielded some of its frontage to the 23d Marines, exchanged long-range fire with the enemy.

On the 29th, the 1st and 3d Battalions of the 24th Marines protected the division left flank, while 2/24 mopped up Japanese infiltrators. Lieutenant Colonel Vandegrift, who had been wounded two days earlier, was evacuated. Command of 3/24 then passed to Lieutenant Colonel Otto Lessing, formerly the executive officer of the 20th Marines.

By dusk on 30 June, the 27th Infantry Division had advanced far enough to relieve 1/24 of responsibility for the southern segment of the left flank. The 4th Marine Division, however, continued to man an L-shaped line, though it encompassed less territory. The 25th Marines remained at Hill 500 in corps reserve. To date the division had suffered 4,454 casualties.

Central Saipan was now under American domination. The front stretched from Garapan past the Four Pimples to the 4th Marine Division left boundary.

---

[30] Gen Harry Schmidt ltr to MajGen Albert C. Smith, USA, dtd 10Jan55, quoted in Crowl, *Marianas Campaign*, p. 230.

Here the lines veered sharply northward to Hill 700, and then extended along a generally straight line from that hill to the eastern coast. Behind the lines, the hectic pace of the first few days had slowed. Of all the supplies carried for the assault troops, all but 1,662 tons had been unloaded by 28 June. (See Map 18.)

In spite of the long routes of evacuation and the difficult terrain, casualties were being moved speedily to the hospitals established on the island. Evacuating the wounded from the combat zone was a more difficult problem after the departure on 23 June of the last of the hospital ships. Transports and cargo vessels, some of them poorly suited to the task, were pressed into service. Since the corps casualty rate declined toward the end of June, these ships, supplemented by planes flying from Aslito field, proved adequate.[51]

By the evening of 30 June, the Japanese had begun withdrawing to their final defensive line. During the next phase of the Saipan operation, General Holland Smith planned to thrust all the way to Tanapag. Near Flores Point, the 2d Marine Division would be pinched out, leaving the 27th Infantry Division and 4th Marine Division face to face with Saito's recently prepared defenses. (See Map 19.)

---

[51] CominCh, *The Marianas*, pp. 5:19–5:20.

Northern Saipan: End of the Campaign[1]

THE PICTURE ON 1 JULY

The scene of Saito's last stand had been sketched out on 27 June by *Thirty-first Army* Headquarters; "The defense force . . . is at present setting up with a line between Tanapag—Hill 221—Tarahoho as the final line of resistance."[2] Withdrawal to the line was ordered by Saito on 2 July.

In contrast, on the same date, the 2d Marine Division moved forward more rapidly than ever since the landings. Holland Smith's objective line, fixed on 1 July, ran from Garapan up the west coast to Tanapag, then eastward across northern Saipan. Three American divisions—the 2d Marine Division on the left, the 4th Marine Division on the right, and the 27th Infantry Division between—were intent upon concluding the battle. Before executing the last moves, they turned to a straightening of the corps line.

On 1 July the 2d Division did not attempt to advance its left flank regiment, the 2d Marines, from favorable high ground outside Garapan, but awaited the advance of the 6th and 8th Marines on the right. The 27th Division, held up for five hours by opposition from several previously unknown enemy strongpoints, advanced 400 to 500 yards. The 4th Division held fast and supported the Army units by fire. Marine patrols found no Japanese up to 1,800 yards forward of the 4th Division line.

To the west, however, the 2d Division, like the 27th, encountered the enemy. In the division center, Marines of 3/6, moving toward the coast above Garapan, reached a wooded ravine defended by three Japanese field pieces, supported by rifles and machine guns. After briefly probing the strong point, Colonel Riseley bypassed it. He left Company B to destroy the resistance, a mission accomplished the next day.

On the division right, the 8th Marines picked up speed across relatively

[1] Unless otherwise noted, the material in this chapter is derived from: *TF 51 OpRpt; TG 52.2 OpRpt; TF 56 OpRpt; NTLF OpRpt; 2d MarDiv OpRpt; 4th MarDiv OpRpt; 27th InfDiv OpRpt; 2d Mar SAR; 6th Mar SAR; 8th Mar SAR; 10th Mar SAR; 23d Mar Rpt; 24th Mar Rpt; 25th Mar Rpt; 105th Inf Op Rpt; 106th Inf OpRpt; 165th Inf OpRpt;* Saburo Hayashi and Alvin D. Coox, *Kōgun: The Japanese Army in the Pacific War* (Quantico: The Marine Corps Association, 1959), hereafter Hayashi and Coox, *Kōgun;* Capt James R. Stockman and Capt Philips D. Carleton, *Campaign for the Marianas* (Washington: HistDiv, HQMC, 1946), hereafter Stockman and Carleton, *Campaign for the Marianas;* Crowl, *Marianas Campaign;* Hoffman, *Saipan;* Isely and Crowl, *Marines and Amphibious War;* Johnston, *Follow Me!;* Love, *27th Inf Div History;* Morison, *New Guinea and the Marianas;* Proehl, *4th MarDiv History;* Sherrod, *Marine Air History;* Smith and Finch, *Coral and Brass.*

[2] *Thirty-first Ar Outgoing Msg* No. 1120.

even terrain, where, better than around Tapotchau, the tanks could serve the infantry. At 0730 on 1 July, the 1st Battalion, 29th Marines, in a well executed tank-infantry thrust, overran Tommy's Pimple with no casualties. The battalion then advanced, in conjunction with 2/8, toward the Tanapag Harbor area.

To the regimental left, on 1 July, the 1st Battalion, 8th Marines, was joined by 2/2, relieving 3/8, and the two battalions reported good progress. The day's action included seizure of the last two Pimples, Larry's and Stan's.

By sunset of 1 July, then, the corps line had been straightened considerably. There was no longer any reason to delay the thrust toward Tanapag. The corps commander issued the appropriate order.

VICTORY AWAITS NORTHWARD

At 2245 on 1 July, the 2d Marines received attack orders from division ordering an advance into Garapan. The next morning at 1030, Colonel Stuart began to move out, the 1st Battalion on the right, the 3d on the left.

By 1200 the 3d Battalion, supported by Company C, 2d Tank Battalion, was 800 yards inside the town, finding grim evidence of what artillery, aircraft, and naval guns could do. Yet Japanese soldiers were still there—not many, but some—and hostile fire was encountered. Some American war correspondents reported that at Garapan the Marines experienced their first street fighting of World War II. According to division accounts, however, "actually there was little, if any, of this type of fighting compared to European standards. . . . The town had been leveled completely."[8] Garapan had been the second largest town of the Marianas, next only to Agana in Guam. Before the first World War it had been headquarters of the German administration, and a village centuries before that.

Twisted metal roof tops now littered the area, shielding Japanese snipers. A number of deftly-hidden pillboxes were scattered among the ruins. Assault engineers, covered by riflemen, slipped behind such obstacles to set explosives while flamethrowers seared the front. Assisted by the engineers, and supported by tanks and 75mm self-propelled guns of the Regimental Weapons Company, the 2d Marines beat down the scattered resistance before nightfall. On the beaches, suppressing fire from the LVT(A)s of the 2d Armored Amphibian Battalion silenced Japanese weapons located near the water.

Advancing to the coast above the town, 1/2 sliced through scattered enemy defenses. Southeast of Garapan, riflemen of Company A seized Flametree Hill, where, despite the blasting by Marine artillery on 29 June, the enemy had continued to hold out.

While the 2d Marines was moving into Garapan, the 6th Marines attacked the high ground overlooking the town and overcame moderate resistance. Company A of 1/6 joined men from Companies A and B of 1/2 in silencing the fire from caves on rockbound Sugar Loaf, a distinctive hill on the regimental boundary.

Inland, the 8th Marines continued toward Tanapag Harbor. Progress of 2/8 and 1/29, however, was stopped on

[8] *2d MarDiv OpRpt*, SplCmts, p. 3.

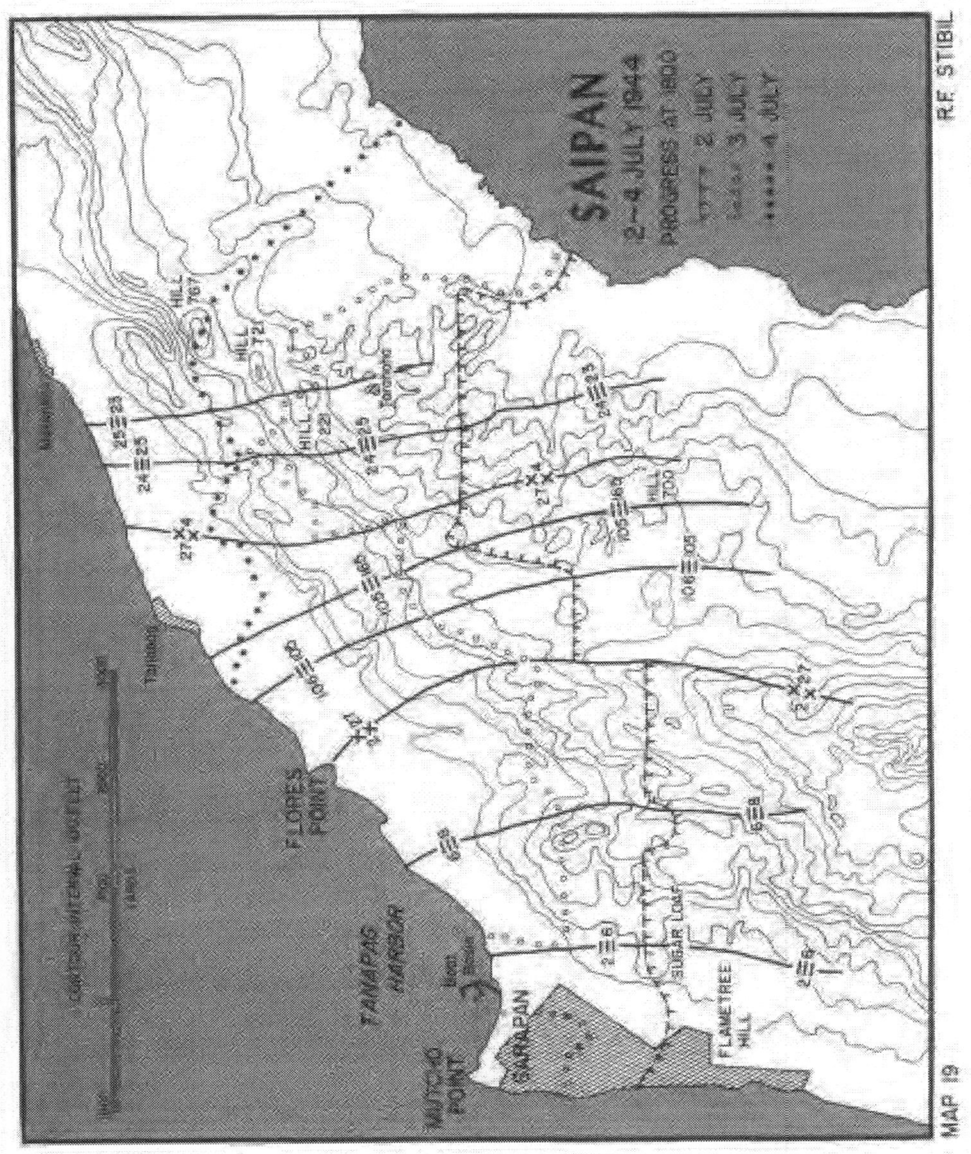
SAIPAN
2–4 JULY 1944
PROGRESS AT 1800
2 JULY
3 JULY
4 JULY
MAP 19
R.F. STIBIL
FLORES POINT
TANAPAG HARBOR
MUTCHO POINT
GARAPAN
SUGAR LOAF
FLAMETREE HILL
HILL 700
HILL 721
HILL 767
HILL 221
Tanapag

the afternoon of 2 July, when machine guns opened up from a coral-limestone hill to the right of the flat terrain. Fire enfiladed the entire front of 2/8 and much of that of 1/29. Lieutenant Colonel Tompkins, commanding 1/29, was wounded by a shell fragment, and was immediately replaced by Lieutenant Colonel Jack P. Juhan, the 8th Marines executive officer.

The strongpoint lay too close to Marine lines for artillery fire. Major Chamberlin, commanding 2/8, attempted an envelopment, swinging Company F to the east along the only available route. The Marines cut through entangling underbrush hoping for surprise, but the enemy anticipated the move and turned the attackers back with a withering fusillade.

In the early evening, tanks and flamethrowers were employed against the hill. They did some damage, but emplacements were so well dug-in, and the caves so well fortified and arranged, that nothing but slow yard-by-yard demolition would neutralize the position. It was decided, therefore, to bypass the strongpoint and resume the advance the next morning, leaving Company F to contain it. On the evening of 3 July, the 2d Provisional Company, one of the units formed from the shore parties, came up to relieve Company F of its task, and the latter rejoined its battalion, then 1,000 yards ahead.

Opposition to the progress of the 4th Division was markedly less than that met by the 2d Division. On 2 July, ending its brief pause, and with the 23d and 24th Marines in assault, the 4th Division advanced toward the northwest coast. A gain of some 1,600 yards was reported for the day against such light resistance that only one Marine was wounded. At 1345, the division dug in until units of 3/165 on the left could catch up.

The 27th Infantry Division, rejoined the day before by the 165th Regiment, spent 2 July mopping up rough terrain in its zone of advance. Five enemy tanks emplaced as pillboxes were encountered and knocked out by the 106th Infantry. Heavy machine gun fire delayed 3/105, creating a risky gap between 3/165 on its right and the 106th Infantry on its left. General Griner ordered 1/105 to wheel around the 3d Battalion combat area and march north to close the gap, a mission it accomplished before dark. The division reported gains for 1,400 to 1,800 yards on 2 July and made contact on the right with the 4th Marine Division. (See Map 19.)

A NEW LOOK AT THE MAP

Satisfied with the overall situation, Holland Smith felt it was time to execute certain changes. Desiring to rest the 2d Marine Division for the Tinian campaign, he altered direction of the corps attack late on 2 July, bending it left, more to the northwest. Under this plan, the 2d Division would be pinched out above Garapan, while the 27th Infantry Division inclined west toward the sea near Tanapag, forming a barrier against Japanese retreat northward. The 4th Marine Division, after reaching the west coast above Tanapag, would veer east, to compress the enemy in the remaining northern area. On 3 July, the 25th Marines was released from corps reserve, enabling

General Schmidt to attack on a three-regiment front.

Marines of the 2d Division spent a busy Fourth of July, prior to leaving the lines. The 3d Battalion of the 2d Marines took Mutcho Point by 0900, eliminating a small Japanese garrison. The only headache of the operation was an enemy heavy antiaircraft gun farther up the shore, which delivered air bursts uncomfortably close to the attacking troops.

By 1000 on the same day, the 6th Marines was on the beach at Tanapag Harbor, and at 1320, the 8th Marines reached the same vicinity. During the afternoon, Marines of 1/2 employing LVTs cleaned out the few enemy trapped in the boat basin. About the harbor the hulks of damaged ships sheltered some Japanese snipers. Up and down the coast there were still a number of undestroyed concrete pillboxes.

With the action of 4 July, most of the 2d Marine Division passed into NTLF reserve. The 4th Marine Division and the 27th Infantry Division were assigned to conclude the campaign. During the afternoon of 4 July, the Army division shifted its frontline units and prepared for the drive toward Marpi Point. The 105th Regiment was shortly to be joined by its 2d Battalion, which started marching north on the 4th after its release from duty at Nafutan Point. Soldiers of the 106th took over the shell-wrecked seaplane base at Flores Point on 4 July.

The NTLF commander addressed an Independence Day "well done" to all troops. The day was appropriately noted in 4th Division reports by the capture of "Fourth of July Hill," a heavily wooded knob on the eastern side of Hill 721. The higher hill was, in fact, the more significant rise, since the Japanese there observed much surrounding terrain. Efforts by 3/23 to capture Hill 721 on 3 July had been violently opposed. The infantrymen, deprived of tank support by mines, were stopped short. Colonel Jones, commanding the 23d Marines, therefore ordered the battalion pulled back some 300 yards, to permit night bombardment by howitzers of the 14th Marines. Next morning, 1/23 passed through 3/23 and swept to the top of Hill 721 against surprisingly light small arms and machine gun fire. Most of the battered enemy had departed for a more healthful area. (See Map 19.)

A neighboring hill, 767, was taken by a strong combat patrol from 1/23 without meeting enemy fire. This hill marked the deepest thrust of the American advance on 4 July. Around it, the 25th Marines tied in with the 23d, while the 24th Marines—with Hill 221[4] pocketed the day before—drew up on the left. Marines were by now practically neighbors of General Saito, for Hill 767 was next door to Paradise Valley (labeled "Valley of Hell" by the Japanese), site of the last Japanese headquarters on Saipan. (See Map 20.)

WRAPPING UP THE CAMPAIGN

General Smith fixed noon of 5 July as jump-off hour for the final push on

[4] The hill was nicknamed Radar Hill by the Marines because of Japanese radar installations there.

northern Saipan, involving the 4th Marine Division on the right and the 27th Infantry Division on the left. Army troops were by then on the Tanapag plain. Prior to the attack, the 106th Infantry went into reserve. The last advance was assigned to the other regiments, the 105th on the left and the 165th on the right. These soldiers near the eastern shore were due for some of the toughest combat experienced on the island.

In the middle interior, the 4th Marine Division advanced so rapidly to Karaberra Pass on 5 July that the corps commander resolved upon a change of missions. He felt concerned that the 27th Infantry Division, which was moving against stiffer resistance, would get too far behind. At 0900 on 6 July, therefore, he ordered the 27th Division to alter its direction of advance from northeast to north, and he moved the left flank of the 4th Division to the northwest. (See Map 20.)

When Army troops reached the coast near the village of Makunsha the 27th Division would be pinched out. The 4th Division was then to pick up the advance to Marpi Point, northern tip of the island. The new zone of the 27th Division extended up the coast from Tanapag to just above Makunsha and partially inland. It included a canyon, shortly to be dubbed Harakiri Gulch, and Paradise Valley. Everything northeast of the Army sector was assigned to the 4th Division. On 6 July, the 2d Marines was attached to that division and charged with destroying any Japanese that slipped away from the Army vanguard. On the same day, 1/29 passed to control of General Jarman's Garrison Force.

On the afternoon of 6 July, the 25th Marines, advancing with 13 tanks, got as far northeast as Mt. Petosukara, which was taken after digging out Japanese from cliffs en route. The day's action included surrender of a group of more than 700 civilians shortly before dark. The 24th Marines, to the left, gained up to 1,800 yards without difficulty, but the 23d Marines, probing the fringes of Paradise Valley, was delayed by fire from caves and underbrush, much of it at their backs. Contact with the 24th Marines was lost, but connection was made with elements of the 27th Division. Next day, the 2d Marines was put into line between the 23d and 24th Marines.

With the 106th Infantry going into reserve on 4 July, the 105th pursued the advance up the west coast while the 165th moved through the adjoining interior. The 2d Battalion, 105th passed through the ruins of Tanapag unopposed on 5 July, but beyond there its advance was blocked by machine gun fire. Shortly after moving out on 6 July, the battalion was stopped by a hail of small arms fire coming from the immediate front. The source, at first undetectable, proved to be a shallow ditch just 150 yards ahead. It seemed a suitable target for 60mm mortars, but ammunition was lacking. A rifle squad rushed what appeared to be the most active machine gun position, but the squad leader was wounded and the bold effort repulsed. Three roving Army tanks then turned up and joined the fight, with the result that some 150 Japanese soldiers jammed along the ditch were killed. The action freed the advance of 2/105.

On 6 July, 3/105, operating farther

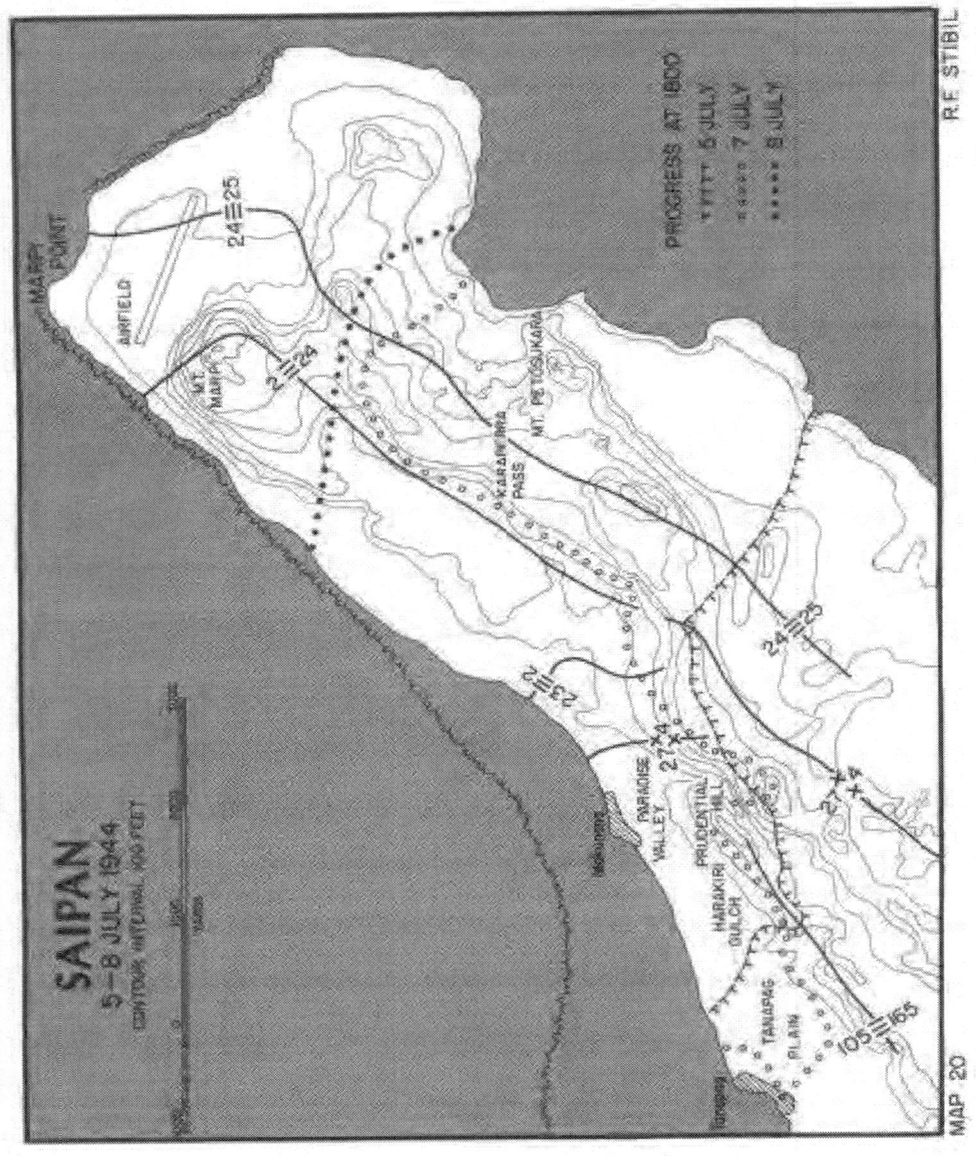
SAIPAN
5–8 JULY 1944
CONTOUR INTERVAL 100 FEET
PROGRESS AT 1800
5 JULY
7 JULY
8 JULY
MARPI POINT
AIRFIELD
MT. MARPI
PASS
MT. PETOSUKARA
PARADISE VALLEY
PRUDENTIAL HILL
HARAKIRI GULCH
TANAPAG PLAIN
MAP 20
R E STIBIL

inland, approached the edge of Harakiri Gulch and was stopped by fire. This canyon, stretching 400 yards long, east to west, and 50 yards wide, lay astride the regimental boundary. On 5 July, a company of the 165th Infantry had collided with a nest of enemy defenders in the same area and had been driven back by a veritable wall of enemy fire. On the next day, when 1/165 attempted the canyon, its men were surprised by a series of explosions inside some straw shacks, caused when about 60 Japanese committed suicide with grenades. The event did not indicate capitulation, however. Resistance by other Japanese kept the gulch impenetrable to 1/165, as well as to 3/105 which attacked it on the west. A relieving battalion, 3/106, aided by a platoon of tanks, was at last able to secure the floor of the canyon by 0900 on 7 July.

The enemy's defense of Harakiri Gulch, grim though it was, becomes obscured by the size and fury of what befell the 1st and 2d Battalions of the 105th Infantry on 7 July. A desperate scheme had been evolved by the weary and ill Saito—then "a pitiful sight," an enemy staff officer recalled.[4] Oppressed by reports of ground lost—of Saipan, last hope of Imperial victory, slipping away—he meditated upon cheerless alternatives. Saito's sixth and last command post, taken up on 3 July, was the smallest. A cave sheltered by jungle, it cut low into a hill of Paradise Valley. More like a refuge than a headquarters, the cave lay some 1,000 yards inland from Makunsha. It was miles and days from the village schoolhouse in Charan Kanoa, Saito's first command post of the Saipan campaign.

SAITO'S LAST DECISIONS

By now the valley of Saito's despair was raked daily by Marine artillery and naval gunfire. The general himself had been wounded by a shell fragment. After consulting staff officers and Vice Admiral Nagumo—likewise at a cave headquarters—the Japanese commander plotted a last grand *banzai* attack to start before the dawn of 7 July. It was one alternative to waiting for destruction; no tactical accomplishment was apparently expected. "Whether we attack or whether we stay where we are, there is only death" said the general's melancholy summary of affairs.[5] Saito, the advocate of mobile defense, was at last immobilized. In repayment to the Empire for the loss of Saipan, he exhorted each Japanese soldier to exact seven lives for one.

Saito would lead the advance, he proclaimed, but actually he had other plans for himself. The order prepared, the general adjourned to a farewell meal of *sake* and canned crabmeat. Next morning, leaving the attack to hands less old and tired, Saito committed *harakiri*.[6]

The desperate assault was expected by General Smith, among others. On 6 July, the corps commander, accompanied by General Watson, visited the

[4] *NTLF OpRpt*, Encl D, Pt II, App G, p. 2.

[5] From the text of Saito's last order, quoted in *NTLF OpRpt*, Encl D, Pt I, pp. 57–58.

[6] Japanese prisoners said the command of the attack devolved upon Colonel Eisuke Suzuki of the *135th Infantry Regiment*, who was reported killed.

27th Division CP and cautioned General Griner to be on the alert for a Japanese counterattack. The Tanapag plain furnished the most likely avenue of approach if the enemy chose to attack. Holland Smith also indicated that when the Army division lines had advanced about a mile to the north, he intended to pass Watson's 2d Division through the 27th and continue the attack with the 2d and 4th Divisions on line.[7]

Just before sunrise, about 0400 on 7 July, like some ancient barbaric horde, the Japanese soldiers started down the Tanapag plain from around Makunsha. The attack route followed mainly along the coastal railroad. The men were led by officers, and they were equipped with machine guns, mortars, and tanks. Yet it was like military order gone awry, replaced by individual passion, a fearful charge of flesh and fire. Savage and primitive, the assault reverted to warfare of centuries before. Some of the enemy were armed only with rocks or a knife mounted on a pole. (See Map VI, Map Section.)

Whatever it was that drove the Japanese, or inspired them, they came on and on, straight into the muzzles of opposing guns. "They just kept coming and coming," recalled Major Edward McCarthy, commanding 2/105. "I didn't think they'd ever stop. It was like a cattle stampede."[8] The exact number of the attackers will probably never be known, but it was believed to approximate 3,000, including remnants of every unit on the island, even walking wounded.[9]

First to receive the impact of the bloody attack were the isolated positions of the 1st and 2d Battalions, 105th Infantry, which had dug in for the night 1,200 yards south of Makunsha. At 0530 Colonel Bishop, commanding the Army regiment, telephoned to division that terrific mortar fire was falling on the two battalions.

The soldiers fought for their lives as tremendous masses of the enemy flooded into a 300-yard gap between the battalions, discovered by enemy patrols the night before. Both American battalions had pressed their attack until 1630 on 6 July, too late to consolidate their lines before nightfall.[10]

By 0635, after a night of fierce combat, the Japanese had overrun the 1st and 2d Battalion. Lieutenant Colonel William J. O'Brien, commanding 1/105, symbolized the high courage of the resistance. After emptying pistols held in each hand, and though seriously wounded, he turned a machine gun against the enemy until, like so many other officers and men, he was cut down in the hopeless struggle. Shortly

[7] MajGen George W. Griner, Jr., USA, ltr to ACofS, G-3, dtd 12Jan63.

[8] Quoted in Love, *27th InfDiv History*, p. 443.

[9] From its positions on the heights inland, the 23d Marines could see "considerable numbers of Japanese moving southward down the plain. These were taken under fire by machine guns and mortars. *Jones ltr.*

[10] Colonel Bishop indicated that the day's plan of attack called for units advancing on the right of the 105th to seize objectives which would have pinched out or at least narrowed the zone of his regiment. When this attack failed to reach its objective, he wrote, "We found ourselves off balance and with a gap between our 1st and 2nd Bns." Col Leonard A. Bishop ltr to Head, HistBr, G-3, HQMC, dtd 28Feb63.

before he was killed, he had radioed the 105th Infantry command post: "There are only 100 men left from the 1st and 2d Battalions of the 105th. For God's sake get us some ammunition and water and medical supplies right away." [11] But four jeeploads of ammunition could not get through.

An offshoot wave of the enemy attack broke against the positions of 3/105 at Harakiri Gulch, but here the Americans, holding high ground, beat off the Japanese. The 1st and 2d Battalions, what was left of them, were pushed back across the plain to Tanapag, where house-to-house fighting ensued.

About 500 yards southwest of Tanapag on that fateful morning was the 3d Battalion, 10th Marines. Nearby was 4/10. The two battalions, now attached to the 4th Division, had moved into position the day before, to provide supporting fires for the 23d Marines. About 0515, just at daybreak, enemy were identified at 400 yards, moving upon the most forward battery, H, of 3/10. It then seemed hardly minutes before nearly 500 Japanese, employing machine guns, rifles, grenades, and tanks assaulted the entire battalion position.

Only Battery H, on the left of the railroad tracks, was able to fire its 105mm howitzers. The guns of other batteries were silent, forced to hold their fire by the fact that Americans were positioned to their front. Artillerymen of Battery H cut their fuses to four-tenths of a second; shells exploded less than 50 yards forward of the muzzles. At such a range the artillerymen swung one howitzer around to destroy a Japanese medium tank approaching from the rear.

A number of the Marine cannoneers were shot in position, crippling the firing effectiveness of the battery. Finally, an enemy breakthrough at 0700 in a wooded ravine to the left forced the survivors of Battery H to withdraw about 50 yards across a road. There the unit set up a defense in an abandoned Japanese machinery dump, where the Marines held out with carbines, an automatic rifle, a pistol, and eight captured Japanese rifles until relieved around 1500 by Army troops.

Personnel of Headquarters and Service Battery, set up behind Battery H, were run over at the crest of the drive. The battalion commander, Major William L. Crouch, was killed in the vain defense. Battery I repelled a light brush with the enemy at 0455 before the full weight of the assault was felt. Thereafter, however, the supply of artillery and small arms ammunition dwindled rapidly, and, after removing the firing locks from their howitzers, the artillerymen fell back to the positions of Battery G. There the two batteries stood ground, joined at mid-afternoon by elements of the 106th Infantry.

Southeast of 3/10, Marines of the 4th Battalion defended their own firing positions, killing 85 Japanese who were on the fringe of the tide. Several men of the battalion also helped bring some small arms ammunition to 3/10 and evacuate wounded from that area. A group of 12 men and 1 officer of the battalion joined counterattacking Army troops. Of the 4th Battalion casual-

[11] Quoted in Love, *27th InfDiv History*, p. 447.

ties on 7 July—three enlisted men killed and seven wounded—most were the Marines who were helping out forward.

Following the assault of the Marine battery positions, the Japanese swept on, approaching the regimental command post of the 105th, about 800 yards south of Tanapag. Here, however, they could not get through the defenses. The enemy spearhead was beginning to show the blunting accomplished by the desperate fighting of the units that had been overrun. At 1130, the depleted and tiring enemy was considered pretty well stopped, but fighting dragged on through the afternoon. By then, however, the impetus had entirely vanished from the attack, and some of the Japanese were turning grenades upon themselves.

The end, at last, was due both to Army and Marine efforts. After sending every available tank to support the beleaguered battalions of the 105th, General Griner had issued orders at 0700 to the 1st and 2d Battalions of the 106th, the division reserve, moving them into position at Flores Point to attack north along the railroad. The 2d Battalion was already at a Flores Point assembly area. In support of the attack by the 106th, a provisional company of tanks was attached. The two Army battalions were to relieve the 105th and help regain the Marine battery positions. They would be supported by the three 105mm howitzer battalions of 27th Division artillery.

The counterattack got under way at 1000, the movement of units hampered by communication difficulties. The 2d Battalion of the 106th had been particularly directed to reinforce the Marines, and its Company F helped retake the Battery I position by 1135.

Entire reoccupation of the Marine positions was accomplished during the afternoon, and a line was then formed from the beach to the left of the 4th Division. Upon request by the 27th Division, a Marine battalion, 3/6, was attached and it helped solidify the new line. By 1800, most of the ground lost by the *banzai* attack was again in friendly hands.

Saito's farewell order had cost the two 105th Infantry battalions 406 killed and 512 wounded. The 3d Battalion, 10th Marines, lost 45 killed and 82 wounded, and in turn killed more than 300 Japanese. Survivors would never forget "the raid," as they termed it.

A staggering total of Japanese were killed. In the area of the *banzai* attack, 4,311 Japanese corpses were found. Some of these dead were undoubtedly the victims of artillery or naval gunfire prior to the attack, but the vast majority were killed in the climatic, fanatical charge of the Saipan garrison.

POSTSCRIPT TO "THE RAID"

The casualty-ridden 27th Division (less the 165th Infantry) was relieved by Holland Smith at 0630 of 8 July, and put into corps reserve. To accomplish the mop up of the devastated west coast, the corps commander recommitted the 2d Marine Division, attaching to it the 165th. He also brought 3/6, 3/10, and 4/10 back to the 2d Division.

Some resistance was met by the 6th Marines on 8 July, from about 100

Japanese well entrenched east of Tanapag. After a brief exchange, Company F was left behind to eliminate the pocket. Otherwise, the Marines found the enemy along the coast to be poorly armed and disorganized. Inland hills, however, still contained hidden defenses where few could kill many, and Japanese holdouts here slowed the progress of the 8th Marines.

The 1st Battalion of the 165th, after a costly action at Harakiri Gulch, moved through the canyon area and reached the west coast on 9 July. Paradise Valley was bypassed by 2/165, which left the 3d Battalion to destroy the Japanese still there. The 2d Battalion went on to Makunsha, by then a center of enemy stragglers.

In those last days, the spirit of the *banzai* attack flamed again occasionally. Japanese would charge from a hiding place, reckless of the consequences. Some, of course, were simply trying to escape. At the beaches a number of Japanese swam hopelessly out to coral outcroppings, where they either were killed or destroyed themselves.

The 4th Marine Division, which on 6 July took over the entire front beyond Makunsha, found the advance toward Marpi Point eased somewhat as a result of the *banzai* attack—there were fewer Japanese. The 2d Marines, attached on 8 July, went into line between the 23d Marines to the left and the 24th and 25th to the right. Thus, with four Marine regiments abreast on a 6,300-yard front, the division attacked toward the northwest on 8 July. (See Map 20.)

The 2d and 24th Marines secured their beach area at 1530. The advance of the 23d Marines was marked by destruction of stubborn resistance on a cliff overlooking Karaberra Pass. As Marines struggled against the enemy there on 7 July, they nicknamed the high ground "Prudential Hill" because it resembled an American insurance company trademark. But, unlike the peaceful scene of Gibraltar, there were mines hidden at the base of the hill. The area was masked for fire by 4th Division artillery, and in order to provide support, truck-mounted rocket launchers were lowered over the cliff with their rate of descent controlled by tanks chained to the trucks. Once they reached the base of the cliff, the launchers fired into its face to beat down Japanese resistance.[12] Offshore, rocket gunboats joined in the deluge of fire turned on the caves that held the enemy holdouts.

Reduction of "Prudential Hill" insured that Marines moving across the coastal plain would not be fired upon from the rear. By 1410 of 8 July, the 23d Marines reached the shore above Makunsha. The next morning the regiment was assembled in division reserve and assigned to mopping up along the coast. A detachment of the 2d Armored Amphibian Battalion helped demolish lingering cave resistance.

With the total good progress on 8 July, General Schmidt prepared to unleash a thrust to Marpi Point. At 1330 on that day, he directed the 25th Marines to seize commanding ground 700 yards forward of positions then held, in order to gain observation of the Marpi Point area. The move was made against practically no resistance.

[12] *Jones ltr.*

214-881 O-67—23

TRUCK-MOUNTED ROCKET LAUNCHERS *fire at Japanese strengpoints in northern Saipan. (USMC 104069)*

FLAME TANK *spews a stream of liquid fire toward an enemy cave during the mop-up action on Saipan. (USMC 85829)*

THE ISLAND SECURED

The next morning, the 25th, the 24th, and the 2d Marines, from right to left, attacked with the mission of securing the last objective line on the island. Some scattered counterattacks were beaten off by the 2d Marines at Mt. Marpi, but much of the hill was bypassed. The airfield beyond was found utterly wrecked by bombardment.

By 1615 all three Marine regiments were at the coast, having advanced a total of 2,500 yards that day. The fire-scarred earth of Saipan lay behind. Vice Admiral Turner declared the island secured, putting the time at 1615 of 9 July. The next day an official flag-raising took place at Holland Smith's headquarters in Charan Kanoa.

For the Marines at Marpi Point, a tragic sight took the edge off a happy occasion. Hundreds of Japanese civilians, fearful of the Americans, committed suicide by jumping from the seaside cliffs. Some took their children with them. Efforts to stop them fell upon ears deafened by Japanese propaganda. Fortunately, many civilians had previously surrendered amicably, entrusting their fate to Marine and Army civil affairs officers, and were grateful for the care and safety found in the internment camps.

Many of the more than 42,000 Japanese, Korean, and Formosan civilians on the island had been evacuated before the battle. Starting in March 1944, following seizure of the Marshalls, some 5,000 persons on Saipan, mostly Japanese women and children, were sent home. Of the Japanese still on the island, 9,091 were in camp by 15 July. Half of the number were children. The interned Korean civilians, at the same date, came to 1,158, including children.

A 1937 census of the native islanders of Saipan showed 3,143 Chamorros and 1,037 Kanakas.[18] When the United States and Japan fought over their home in 1944, the natives were mostly sympathetic to the Americans and glad to come under their protection. Nearly 3,000 of the islanders, mainly the Chamorros, were in an internment camp by the end of the battle.

The total number of POWs held on Saipan as of 9 July stood at 736, counting 438 Koreans. But the post-campaign mopping up raised the total to 1,734 by 27 July, including 838 Koreans.

After Saipan was secured, a miniature amphibious operation took place on 13 July. A small island in Tanapag Harbor—Maniagassa Island—was seized by the 3d Battalion, 6th Marines, which landed from LVTs after a preparation by artillery and naval gunfire. In taking over the island, the Marines received hostile fire from only one pillbox, where a light machine gun was manned. Of the small enemy garrison, which numbered 31, 15 were taken prisoner, including 2 Army laborers and 10 Koreans who could not speak Japanese. The rest of the garrison was killed. One Marine was wounded.

The taking of Maniagassa ended Marine action on Saipan. Leaving the hunt for Japanese stragglers to the Army Garrison Force, the 2d and 4th Marine Divisions prepared for their

---

[18] Tadao Yanaihara, *Pacific Islands Under Japanese Mandate* (London: Oxford University Press, 1940), Table I, p. 30.

next mission, due very soon—the capture of neighboring Tinian.

SAIPAN IN RETROSPECT

Letters from home were already telling Marines how the American press evaluated the campaign just over. "One of the bloodiest battles in U.S. military annals," said *Time* magazine.[14] The Marines knew well that Saipan was costly, but most of them understood something more. There were compensations, if there could be any to the bereaved, for the loss of more than 3,000 American lives. NTLF accounting on 5 August put American casualties at 3,225 killed, 13,061 wounded, and 326 missing;[15] known enemy dead were recorded as 23,811.

Holland Smith felt that Saipan was "the decisive battle of the Pacific offensive." Capture of the island, he said, "breached Japan's inner defense line, destroyed the main bastions, and opened the way to the home islands."[16] During the battle, Saito wrote that "the fate of the Empire will be decided in this one action."[17] It would have comforted many American homes to hear in 1944 what Vice Admiral Shigeyoshi Miwa said after the final Japanese surrender: "Our war was lost with the loss of Saipan."[18]

The B–29s could now bomb the Japanese homeland from Saipan. Indeed, the Army Air Forces lost no time building on the island it had coveted. The first B–29 airdrome there was begun on 24 June 1944, and on 24 November, a hundred B–29s departed Saipan for a strike at Tokyo, the first since Lieutenant Colonel Doolittle's daring raid in 1942.

The advance of United States naval power, permitted by the capture of Saipan, also worried the Japanese. As early as 26 June 1944, Emperor Hirohito expressed to Foreign Minister Shigemitsu a desire for diplomatic settlement of the war, and the actual fall of Saipan caused the resignation of Premier Tojo and his entire cabinet on 18 July. Yet, except for the Emperor, there was still no one in Japanese officialdom, including the new Premier Koiso, who dared to suggest peace. The military, as they continued the hopeless struggle, drew some tactical lessons from Saipan. The chief lesson was to organize defenses in depth, a change which would be noticed on Peleliu, Iwo Jima, and Okinawa.

For American planners there were various lessons derived from the campaign. One of the most vital concerned the proper employment of supporting aviation. Complaints had arisen on Saipan that Navy and Army planes did not arrive in time, because of faulty control procedures. Re-

[14] *Time*, v. 44, no. 2 (10Jul44), p. 33.

[15] A final medical summary, dated 9 September 1944, in *TF 56 OpRpt*, listed American casualties as 3,100 killed, 13,099 wounded, and 326 missing. Hoffman, *Saipan*, pp. 268–269, contains a unit breakdown of Army and Marine casualties in Saipan. Final official Marine casualty totals are contained in Appendix H.

[16] Smith and Finch, *Coral and Brass*, p. 181.

[17] NTLF G–2, Tgs Sent and Received by Thirty-First ArHq on Saipan, dtd 25Jul44, p. 16.

[18] *USSBS Interrogation* Nav No. 72, VAdm Shigeyoshi Miwa, IJN, II, p. 298. Miwa was successively Director, Naval Submarine Department, and Chief, Sixth (Submarine) Fleet.

quested missions were often cancelled because infantry had advanced past the target before the planes appeared. One result of the criticism was that close air support techniques, many of them pioneered and developed by Marines, received better attention after Saipan. On Luzon, where Army ground troops were supported by both Marine and Army Air Forces squadrons, close air support would really come into its own.

In the Saipan campaign, Marine aviation was represented by two observation squadrons, VMO–2 and VMO–4, which performed artillery spotting for the Marine divisions. On 17 June, for the first time, the OYs—the monoplanes called "grasshoppers"—took off from carriers. Landing at Yellow Beach or Charan Kanoa, they moved to Aslito airfield after its capture on 18 June. The little observation planes served valuably, not only in artillery spotting missions but also in gathering intelligence. Another Marine aviation unit, Air Warning Squadron 5, also operated with ground troops at Saipan, one detachment serving with corps troops, and two others with the Marine divisions. Not a single enemy aircraft slipped by the alert radar units of the squadron.

Naval gunfire seemed to impress the Japanese most at Saipan. The fire came from assorted vessels—LCI gunboats to battleships—and from guns 20mm to 16 inches in caliber. The gunfire ships supported troops on call, laid down preparatory fire, and illuminated the battlefield with star shells.

More than 8,500 tons of ammunition were expended on troop support missions. The ships could maneuver better than land-based artillery, but the flat trajectory of naval guns proved somewhat limiting, particularly against a reverse slope target. In addition to requested naval gunfire, certain destroyers, which were designated "sniper ships," cruised near the coast, picking out targets of opportunity.

Saito was so vexed by the incessant shelling from the sea that he wrote: "If there just were no naval gunfire, we feel we could fight it out with the enemy in a decisive battle." [19] The statement seemed to support Navy claims for their guns and marksmanship. It is probable that the Japanese switch to defense in depth, after Saipan, was due partly to fear of naval gunfire. Certainly it was a lesson of the campaign that naval gunfire could be enormously effective. Every previous operation had shown the necessity for more of it.

Saito's lament regarding naval gunfire could well also have been: "If there just were no artillery. . . ." As Holland Smith recalled, "never before in the Pacific had Marines gone into action with so much armament, ranging from 75's to 155's." [20] Marine and Army artillery shook the island.

General Smith felt, however, that the available wealth of artillery was not sufficiently appreciated by infantry commanders, at least at the beginning of the campaign. On 1 July, he specifically ordered that "massed artillery fires will be employed to support infantry attacks whenever practicable. Infantry will closely follow artillery

[19] CinCPac–CinCPOA Item No. 9983–85, Disps Sent and Received by Thirty-first ArHq.

[20] Smith and Finch, *Coral and Brass*, p. 191.

concentrations and attack ruthlessly when the artillery lifts." [21]

In every war the foot soldier has been skeptical of the cannoneer's marksmanship. There were instances on Saipan of friendly artillery fire hitting the lines and causing casualties. But such incidents did not detract from the praise accorded both Marine and Army artillery on Saipan. The destruction of Japanese water points was, in itself, quite decisive. The enemy's shortage of water—and food also—became truly desperate. Rain, cursed for the mud it caused, was blessed for the thirst it quenched.

The directing of artillery fire by forward observers and air spotters was sometimes hampered on Saipan by the rapidity of infantry movement. Under the hard-driving Holland Smith "the Japanese never got a minute's peace," as he said himself.[22] "The Saipan campaign followed a definite pattern of continuous attack," said a 4th Division summary.[23]

Some regimental commanders objected, however, that corps insistence on unrelenting pressure upon the enemy, often to a late hour, was not always a good thing. Extreme pushing of the attack could bring a unit to untenable ground. The policy of jumping off right after dawn sometimes prevented sufficient reconnaissance of caves and density of underbrush, features which seldom showed on a map. Inadequate reconnaissance could also result in a waste of fire on areas containing few, if any, of the enemy.

It was felt by some Marine officers that the factors of time and space were not always sufficiently considered by NTLF headquarters. "Progress through heavy canefields, through dense underbrush, and over extremely rough terrain, such as was encountered, cannot be made at 'book' speed," said one regimental commander.[24] Yet the incessant urgency which marked command policy on Saipan quite likely shortened the campaign and saved lives. "I was determined to take Saipan and take it quickly," said Holland Smith afterwards.[25]

Logistics progress kept up, breathlessly, with General Smith's impelling campaign. Unusual conditions encountered during the battle accounted partly for some faulty supply estimates. The 81mm and 60mm mortars, which were sparingly used on the small land areas of the atolls, were much in demand on Saipan for close infantry support. The unit of fire tables which sufficed for previous Central Pacific battles did not here provide for enough mortar ammunition. Extreme shortages resulted. In particular, the early commitment of the 27th Division taxed initial supplies of ammunition.

It was, in fact, the early debarkation of the Army division that led to a classic example of wholesale beach congestion. The imminence of a naval battle, added to the hard combat ashore, hastened the landing of the corps reserve, but evidently no plans had been formulated for landing in that partic-

[21] NTLF OpO 19–44, dtd 1Jul44.

[22] Smith and Finch, *Coral and Brass*, p. 167.

[23] *4th MarDiv OpRpt.*

[24] *Ibid.*

[25] Smith and Finch, *Coral and Brass*, p. 185.

OBSERVATION PLANE *flys over northern Saipan near Marpi Point, scene of the final battles on the island. (USN 80–G–238386)*

SUPPORTING CRUISER *fires at targets on Tinian as LVTs carrying assault troops head toward the White Beaches. (USMC 88102)*

ular area, directly behind the 4th Marine Division. Beach parties were consequently overwhelmed by supplies piling up and getting mixed up. There was not enough time to sort and separate, and some Marine equipment got into Army dumps. Soldiers received utility clothes marked USMC, and much of the 27th Division artillery ammunition turned up in Marine dumps.[26]

A certain opportunism marked the unloading, which did not help the beach parties any. There was a tendency, once a beach was in friendly hands, to shove all supplies over that beach, rather than risk the artillery and mortar fire which harassed unloading elsewhere. The plan relative to general unloading did permit supplies to be put off on any beach, but organic equipment was to be landed only on properly assigned beaches. "In practice, however, this was not done," said a 4th Division report,[27] and misappropriation resulted. After the Saipan experience, Admiral Hill felt that matters would improve if a permanent corps shore party was organized. It would be solely responsible for the movement of supplies from the beach to the dumps and for issue therefrom to the divisions.

On top of the other headaches was a special circumstance which delayed unloading. The Battle of the Philippine Sea was in the making, and the danger of air or surface attack by the approaching Japanese fleet required caution. Admiral Turner ordered all transports and landing ships except Admiral Hill's flagship, the *Cambria*, to retire for the night of D-Day, 15–16 June, and not to return until daylight. The next two nights there were only a few ships with high priority cargo permitted to stay and continue unloading. Then, until the naval battle was over, most of the transports stayed at sea both day and night, interrupting the flow of supplies.

Once the ships were unloaded, the battle of corps dimensions absorbed equipment at unprecedented rates. Estimates of resupply requirements proved much too low. Each signal unit loaded 20 days of equipment, but the campaign showed that on an objective like Saipan the supply would not last for 20 days' of combat. A shortage of radio batteries was not overcome. Such errors were not forgotten, however, and for the battles yet to be fought the logistic lessons of Saipan were useful.

The campaign also imposed tactical demands new to the Pacific war. It was a battle of movement on a sizable land mass, but movement was complicated by the Japanese system of caves. The enemy had defended caves before —on Tulagi, Gavutu, and Tanambogo —but never so extensively. On Saipan the caves were both natural and manmade, and often artfully hidden by vegetation. To cope with them, the Marines perfected various methods of approach. Where terrain permitted, a

---

[26] Admiral Hill recalled that when the complaints began to come in from the two divisions, he talked to the NTLF chief of staff, General Erskine, who sent out orders that there would be common dumps for all except organic equipment and supplies. The admiral pointed out that as soon as the supplies of the Army division "began to flow, there was no real problem." *Hill comments Saipan.*

[27] *4th MarDiv OpRpt.*

flame-throwing tank [28] would advance under cover of fire from medium tanks or half tracks. In terrain where armor could not be moved up, the infantymen would cover for the engineers who placed demolition charges. Sometimes a cave proved so inaccessible that engineers had to lob satchel charges from cliffs above it.

In other approaches, Marines fired automatic weapons or hurled grenades directly into the cave entrance. It was always dismaying to find that a cave which had been seared or blasted could bristle with live Japanese the next day. The enemy's clever use of caves was prophetic of Peleliu, Iwo Jima, and Okinawa and showed detailed planning. A number were well-stocked with supplies. Some had steel doors which were opened periodically to loose bursts of machine-gun fire.

Where a cave defense was not available, the enemy built emplacements of concrete or coconut logs, covered with earth and vegetation. A coconut grove often contained some Japanese strongpoint. Reserve slope defenses were popular, and the wooded valleys favored the enemy's talent for digging in. The canefields were a favorite hiding place for Japanese snipers, until the growth was flattened by a bulldozer. But sniping from trees, a common practice on other Pacific islands, seldom occurred on Saipan. Marines believed that perhaps the enemy feared artillery air bursts in the wooded areas.

From the beginning of the Saipan campaign the Japanese did not organize a true main line of resistance. Instead, they defended strong points which were not connected. For the most part, they made piecemeal counterattacks, attempted by relatively small groups of platoon or company size.

Infiltration was a beloved tactic. Nearly every night a handful of Japanese ventured out, bearing demolitions, grenades, and mines. But such enemy behavior was familiar to the Marines, who reported, in fact, that "no new tactics were observed" on Saipan.[29]

Night or day, except at rash moments, the Japanese cautiously respected their opponents. Prisoners expressed wonder at the accurate and tremendous firepower of Marine units. That included not only what was delivered by artillery but also by other weapons, not the least of which was the infantryman's rifle. On Saipan the M1 continued as an excellent weapon, more durable than the carbine, and, although much heavier, it was preferred by most Marines. A carbine bullet would not always stop an enemy soldier, and the weapon rusted too easily.

Next to his rifle, the infantryman cherished the tank, which, like a lumbering elephant, could either strike terror into a foe or be a gentle servant to a friend. On the open field, hospital corpsmen, moving behind a tank, could get to the wounded and safely bring them off. In attack, the Marine tank-

[28] The flame-throwing tank, recommended after Tarawa, appeared first on Saipan. Actually, it was the M3A1 light tank, mounting a flamethrower. Although the 318th Fighter Group pioneered in the use of napalm fire bombs during operations in the Marianas, no napalm was yet on hand for flamethrowers, only fuel oil, and the range was still too short.

[29] *4th MarDiv OpRpt.*

infantry team felt itself unbeatable, and the Saipan experience added confidence. The medium tank would precede the riflemen who, in return, protected the tank from Japanese antitank grenades. Each half of the team needed the other.

Such interdependence, which marked the tank-infantry team, was illustrated in a thousand ways at Saipan, where Marines and soldiers fought a hard campaign side by side. The controversy arising from the relief of General Ralph Smith, which was to have repercussions beyond the war years, should not obscure the fact that on the battlefield itself there was neither place nor time for interservice rivalry. The merits of the relief, however much they were argued at headquarters throughout the chain of command back to Washington, were largely academic to the men locked in combat with the enemy. What they looked for was mutual support and cooperation—and they got it. To an infantry unit desperate for artillery support, it mattered little if the shells that crashed down ahead were fired by Marine or Army batteries—only that they exploded when and where they were needed.

The same analogy applied to every phase of combat on Saipan, where the measure of value was how well each man stood his share of the common burden, not what his uniform color was when he stood clear of the mud and dust.

In truth, there could be no other answer to success in combat than interservice cooperation. The longer Army and Marine units fought together as partners with the Navy in the amphibious assaults in the Central Pacific, the surer would be the grounds for mutual understanding and respect. Admiral Nimitz, a man who was in an unrivalled position to assess the effect of the Smith against Smith controversy on future operations, noted that he was "particularly pleased that . . . the Army and Marine Corps continued to work together in harmony—and in effectiveness." [30]

[30] FAdm Chester W. Nimitz ltr to ACofS, G-3, HQMC, dtd 8Jan63.

PART V

Assault on Tinian

The Inevitable Campaign[1]

For Marines who had made the 3,200-mile voyage from Hawaii to Saipan, the trip to the next objective was a short one. Just three miles of water separate Tinian from Saipan. In the Pacific war, such proximity of the objective was unusual, but there were also other details of the Tinian assault which made it unique. Here was one of those military enterprises that observers like to term classic. Admiral Spruance called Tinian "probably the most brilliantly conceived and executed amphibious operation of World War II."[2] General Holland Smith saw gratifying results of the amphibious doctrine he helped develop before the war. Tinian, he wrote afterwards, was "the perfect amphibious operation in the Pacific war."[3] Marines in the battle for Tinian profited by the flexible application of amphibious warfare techniques so laboriously evolved during the practice landings of the 1930s.

WHY TINIAN?

Capture of the island was a military necessity. It was, of course, unthinkable that Japanese troops remain on Tinian, next door to Saipan. But there also existed a more positive reason for wanting Tinian–its usefulness for land-based aircraft. The island is the least mountainous of the Marianas, the one which was most suited for new American long-range bombers. It was from Tinian that the B–29s rose to bomb Hiroshima and Nagasaki in August 1945.

The Japanese knew the military

[1] Unless otherwise noted, the material in this chapter is derived from: *TF 51 OpRpt; TF 56 OpRpt;* TF 52 Rpt of Tinian Op, dtd 24Aug44, hereafter *TF 52 OpRpt; NTLF Op Rpt;* NTLF OPlan 30–44 (FORAGER, Phase III), dtd 13Jul44, hereafter *NTLF OPlan 30–44;* VAC ReconBn OpRpts, Saipan-Tinian, dtd 5Aug44, hereafter *VAC ReconBn OpRpts;* 4th MarDiv Representative Translations made on Tinian, hereafter *4th MarDiv Translations (Tinian);* MCS, Quantico, Va., "Study of the Theater of Operations: Saipan-Tinian Area," dtd 15Sep44; LtCol Richard K. Schmidt, "The Tinian Operation: A Study in Planning for an Amphibious Operation," MCS, Quantico, Va., 1948–1949; Lt John C. Chapin, *The Fourth Marine Division in World War II* (Washington: HistDiv, HQMC, Aug45), hereafter Chapin, *4th MarDiv in WW II;* Maj Carl W. Hoffman, *The Seizure of Tinian* (Washington: HistDiv, HQMC, 1951), hereafter Hoffman, *Tinian;* Crowl, *The Marianas;* Isely and Crowl, *Marines and Amphibious War;* Johnston, *Follow Me!;* Morison, *New Guinea and the Marianas;* Proehl, *4th MarDiv History;* Sherrod, *Marine Air History;* Smith and Finch, *Coral and Brass;* Stockman and Carleton, *Campaign for the Marianas.* Unless otherwise noted, all documents cited are located in the Marianas Area OpFile and Marianas CmtFile, HistBr, HQMC.

[2] Adm Raymond A. Spruance ltr to CMC, dtd 27Nov50.

[3] Smith and Finch, *Coral and Brass*, p. 201.

value of Tinian. They had used the island for staging planes and as a refueling stop for aircraft en route to and from the Empire. On Ushi Point they had constructed an airfield which was even better than Aslito on Saipan. The two excellent strips on this field were labeled by American intelligence as Airfield No. 1 and Airfield No. 3. The older north strip, the site of the main airdrome, was 4,750 feet long. Two villages adjoined the activity, housing the personnel. On Gurguan Point was another airstrip which extended 5,060 feet (Airfield No. 2). Northeast of Tinian Town lay Airfield No. 4, still under construction. Already 70 percent surfaced, it could be used for emergency landings. These airfields drew the bulk of Japanese defensive weapons on Tinian. The enemy had sited a number of heavy and medium antiaircraft and light machine guns in the vicinity of each field, particularly the prized Ushi Point strips. (See Map 21.)

American photographic reconnaissance of Tinian, begun on a carrier strike of 22–23 February 1944, focused on the airfields, though not to the neglect of the rest of the island. Perhaps no other Pacific island, not previously an American possession, became so familiar to the assault forces because of thorough and accurate mapping prior to the landings. Documents captured on Saipan were also informative, because the Japanese, as well as the Americans, had linked the two islands in their military plans.

In the whole field of intelligence, the Tinian operation benefited from early planning, general though it was. Detailed planning for Tinian had to yield precedence to that for Saipan and Guam, but once the end of the Saipan campaign was in sight, NTLF headquarters began daily conferences regarding the assault on the nearby target.

DESCRIPTION OF TINIAN

The island the commanders talked about was scenically attractive, observed from either a ship or a plane. In fact, it was said that naval and air gunners were sorry to devastate the idyllic landscape of Tinian. It consisted mainly of small farms, square or rectangular, which, viewed from the air, appeared like squares of a checkerboard. Each holding was marked off by bordering ditches, used for irrigation, or by rows of trees or brush, planted for use as windbreaks.

Tinian measures about 50 square miles. It extends 12-1/4 miles from Ushi Point to Lalo Point but never is more than 5 miles wide. In the wettest months (July to October) of the summer monsoon, the island is drenched by nearly a foot of rainfall per month. Ninety percent of the area is tillable. In 1944, the population of 18,000 consisted almost entirely of Japanese, for all but a handful of the native Chamorros had been moved off to lesser islands of the Marianas. Most of the Japanese had been brought to Tinian by a commercial organization to produce sugar, the chief island product. Tinian produced 50 percent more sugar cane than Saipan. Tinian Town was the center of the industry and had two sugar mills which received the raw product, mostly freighted over a small winding railroad. A good net-

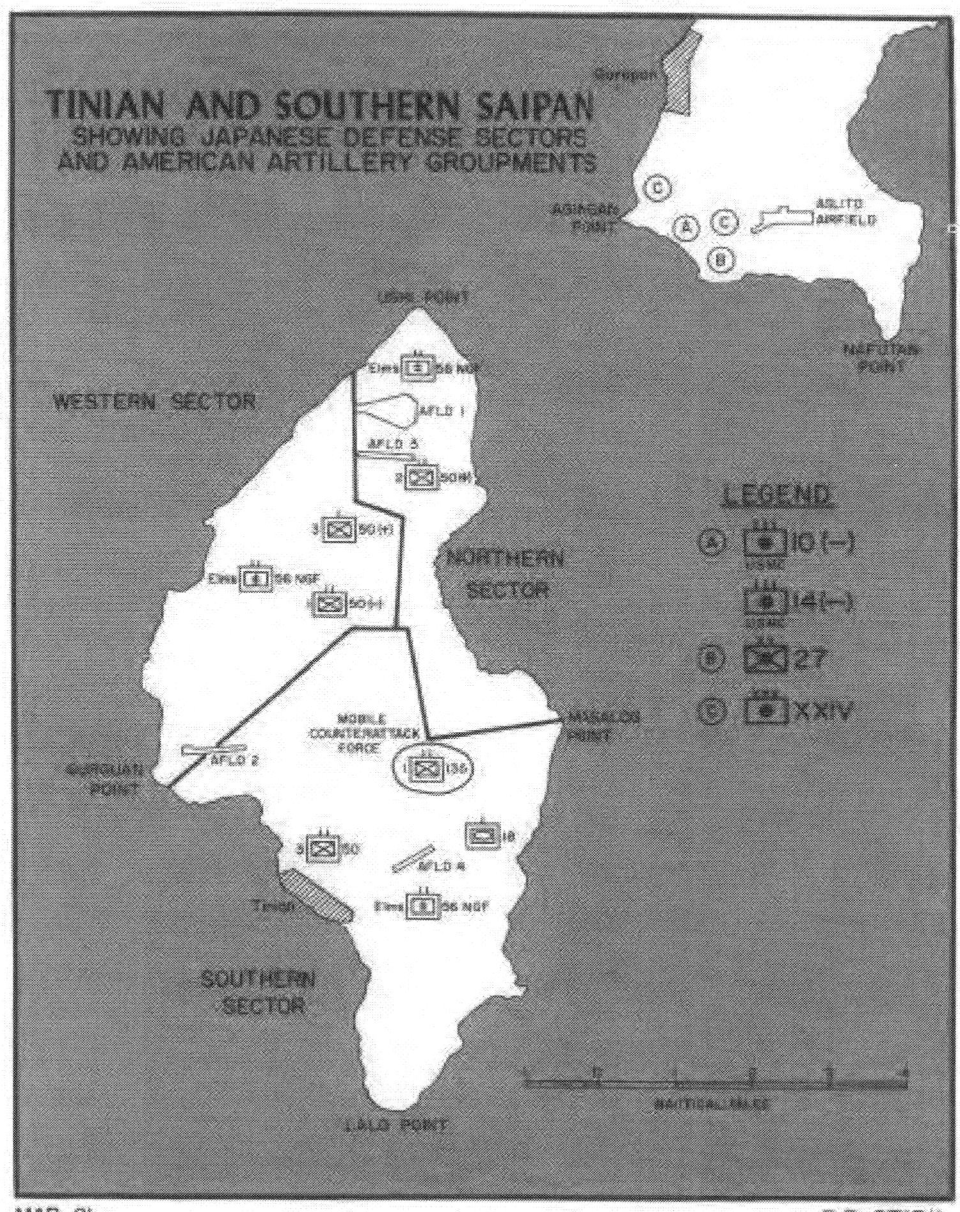

MAP 21 R.F. STIBIL

work of roads also served the transportation needs of the island.

Basically, Tinian was a pleasant and prosperous island. The sole forbidding aspect, except for the Japanese military installations that summer of 1944, were the coral cliffs which rise from the coastline and are a part of the limestone plateau underlying Tinian. A few hills jut up from the plateau, but the principal one, Mt. Lasso, in the center of the island, is only 564 feet high, just a third the size of Mt. Tapotchau on Saipan; Mt. Maga, in the north, measures 390 feet, and an unnamed elevation in the south is 580 feet high. The cliffs which encircle the plateau vary in height, from 6 to 100 feet. Breaks along the cliff line are few and narrow, putting beach space at a premium.

It was, in fact, the question of landing beaches which particularly dominated the planning for Tinian, even more than it usually did for other island campaigns. The Japanese knew they could not escape an assault of Tinian—but where would the landings be made, when, and in what force? Concerning these matters, the enemy had to be kept in the dark until the invasion actually began.

Along the entire coastline of Tinian, only three areas have beaches worthy of the name. One is the vicinity of Sunharon Harbor near Tinian Town, where the several sandy stretches are the widest and most suitable beaches for invasion. On the opposite side of the island, along Asiga Bay, the cliff line breaks off, resulting in a beach approximately 125 yards wide. On the northwest coast, below Ushi Point, are two stingy strips of beach 1,000 yards apart. Intelligence reported one to be about 60 yards wide and the other about 160. Japanese civilians had found the white, sandy beaches pleasant and the water there good enough for swimming and, in fact, had called them the White Beaches. This happened also to be the code name assigned to the two beaches by invasion planners, while the Asiga Bay beach was designated Yellow Beach. (See Maps 22 and 23.)

Colonel Keishi Ogata,[4] commander of the *50th Infantry Regiment* and responsible for the defense of Tinian, believed the Americans would land either near Tinian Town or at Asiga Bay. The colonel's "Defense Force Battle Plan," issued from his command post in a Mt. Lasso cave on 28 June, showed only such expectations.[5] He did not, of course, ignore the northwest beaches, but he anticipated only a small landing party there, at the most. To meet such a remote contingency, the colonel directed that some troops be positioned inland of the beaches. A "Plan for the Guidance of Battle," issued to those troops on 7 July, was captured by Marines the day after the landing on Tinian. In that plan, Colonel Ogata ordered his men to be ready to counterattack on the larger White Beach (White 2). But he scorned the smaller beach (White 1) as being unworthy of consideration.

An enemy strongpoint, American intelligence reported, was located

[4] Chief, War HistOff, Def Agency of Japan ltr to Head, HistBr, G–3, HQMC, dtd 9Nov63.

[5] DefFor BatPlan, dtd 28Jun44, in *4th Mar Div Translations (Tinian)*, hereafter *DefFor Plan*.

about 500 yards northeast of the White Beaches. It included the usual trenches, dugouts, and light machine gun or rifle positions in a wooded area. Among the heavier weapons emplaced here were a 37mm antitank gun, a 47mm antitank gun, and two 7.7mm machine guns.

JAPANESE TROOPS AND EQUIPMENT

Colonel Ogata had only about 8,900 men to dispose judiciously before the Americans came. The mainstay of the Tinian garrison was the well-trained *50th Infantry Regiment,* with a strength of about 3,800 men.[6] The regiment consisted of headquarters, three infantry battalions (each with 880 men, organized into a headquarters detachment, three rifle companies, and a battalion gun platoon with two 70mm guns), one 75mm mountain artillery battalion (three four-gun batteries, one to each infantry battalion), supply, signal, and medical companies, one antitank platoon (six 37mm guns), and a fortification detachment.

Other Army elements included the *1st Battalion* of the *135th Infantry Regiment,* the tank company of the *18th Infantry Regiment,* a detachment of the *29th Field Hospital,* and a motor transport platoon. The infantry battalion had been engaged in amphibious exercises off Tinian when, at the approach of Task Force 58 on 11 June, it was detached from its parent regiment on Saipan and put into the defense system for Tinian.[7] As a result, just about half of the strength available to Colonel Ogata was made up of Army personnel.

The naval complement on Tinian consisted chiefly of the *56th Naval Guard Force* (*Keibitai*), numbering about 1,400 men that had been partially trained as infantry. Most of the sailors of the *Keibitai* were assigned to the coastal defense and antiaircraft guns, but some of them comprised a *Coastal Security Force* which operated patrol boats and laid beach mines. The *233d Construction Battalion* came to about 600 men, while other miscellaneous construction personnel totaled around 800. The antiaircraft units of the *56th Keibitai* were later identified as the *82d* and *83d Air Defense Groups,* each numbering between 200 and 250 men, the former unit being equipped with 24 25mm antiaircraft guns and the latter with 6 dual-purpose 75mm guns. Other naval units included a detachment of the *5th Base Force* and the ground elements of seven aviation squadrons.[8]

[6] The *50th Infantry Regiment* had been transferred from Manchuria in March 1944. It had been scheduled to leave Tinian on 15 June to defend Rota, but the arrival of Task Force 58 in the Marianas on 11 June caused a change of plans. On 7 July, *Imperial General Headquarters* switched the responsibility for the defense of Tinian from the *Northern Marianas Army Group* on Saipan to the *Southern Marianas Army Group* on Guam. Consequently, the *50th Infantry Regiment* came under the *29th Division,* whose headquarters was on Guam.

[7] Japanese records in Tokyo indicate that other elements of the *135th Infantry* moved from Saipan to Tinian on 29 May, 31 May, and 1 June and that the strength of the regiment on Tinian may have been greater than the 900 men listed in NTLF G-2 estimates. Japanese War HistOff Cmts Tinian, *op. cit.*

[8] *Ibid.*

214-881 O-67—24

In charge of all naval personnel on the island was Captain Goichi Oya, though the senior naval officer present was Vice Admiral Kakuji Kakuda, commanding the *First Air Fleet,* whose headquarters was on Tinian. Kakuda, however, was more interested in transferring his command elsewhere, and he left the Tinian naval duties to Captain Oya, who, a week before the invasion, moved his command post from Tinian Town to high ground near the town. Captain Oya was supposed to report to Colonel Ogata, but he was inclined to act independently.

Colonel Ogata had marked off the island into three defense sectors. The southern, the largest of the three, comprised the entire area below Mt. Lasso and included Tinian Town. The northern sector covered the Ushi Point air strips and Asiga Bay. To defend each of those sectors, Ogata assigned a battalion of the *50th Infantry Regiment* and a platoon of engineers. In the western sector, however, where the northwest beaches lay, he left only the *3d Company* of the *1st Battalion* and an antitank squad. The rest of the *1st Battalion* was put into reserve just south of Mt. Lasso. Though in the western sector, these troops were positioned much closer to Asiga Bay than to the northwest beaches. (See Map 21.)

A "Mobile Counterattack Force"—the *1st Battalion* of the *135th Infantry Regiment,* actually another reserve—was located in the southern sector, centrally stationed to move either toward Asiga Bay or the Tinian Town area. The force was called mobile because it would "advance rapidly to the place of landings, depending on the situation, and attack."[9] Mobility, in fact, was a keynote of the Japanese plan. Each sector commander would be prepared not only "to destroy the enemy at the beach" but also "to shift two-thirds of the force elsewhere."[10]

Incorporated within the defense were the twelve 75mm mountain guns of the artillery battalion of the *50th Infantry Regiment* which, when reinforced by the 70mm guns of the infantry battalions, made up a Mobile Artillery Force. The artillery battalion would rapidly deploy to support jointly a counterattack with the tank company of the *18th Infantry Regiment,* whose 12 light tanks were the enemy's only armor on Tinian. This unit, positioned in the southern sector, also possessed two of the rare Japanese amphibian trucks.

Naval personnel on the island were variously employed by Colonel Ogata. They guarded the airfields, particularly at Ushi Point, and protected the harbor installations of Asiga Bay and Sunharon Harbor. Naval gunners manned most of fixed artillery on the island and its antiaircraft weapons. The former included ten 140mm coast defense guns—three of them on Ushi Point, three on Faibus San Hilo Point, and four commanding Asiga Bay.

Three of the ten 120mm dual purpose guns on Tinian shielded the Ushi Point air strips, while three more served the airfield at Gurguan Point. Behind Tinian Harbor stood four more 120mm dual purpose guns, in addition to three 6-inch naval guns of British 1905 make. These 6-inch guns were so artfully con-

[9] *DefFor Plan.*

[10] *Ibid.*

cealed in a cave that until they opened up on the day of the landing their presence was unknown. The prized Ushi Point airfield was solicitously guarded by antiaircraft weapons, including 6 13mm antiaircraft and antitank guns, 15 25mm twin mounts, 4 20mm automatic cannons, and 6 75mm guns.

Miscellaneous types rounded out the Japanese arsenal of weapons on Tinian. In 23 pillboxes which ringed Asiga Bay there were machine guns of unknown caliber. They never took a Marine's life, however, for they—like a number of the other guns pinpointed by reconnaissance—were destroyed by bombardment prior to the invasion.

PREPARATORY BOMBARDMENT

Tinian received a more thorough going over than most other island objectives of the Pacific war, chiefly because the usual naval and air bombardment was augmented for weeks by the fires of artillery. On 20 June, hardly a week after the landings on Saipan, Battery B of the 531st Field Artillery turned its 155mm guns, the "Long Toms," upon Tinian. Other units were added thereafter until, by the middle of July, a total of 13 battalions of both Marine and Army artillery were drawn up on southern Saipan, under the command of the Army Brigadier General Arthur M. Harper, General Holland Smith's valued artillery officer. (See Map 21.)

The Corps Artillery thus emplaced in position and firing on Tinian included the 10th Marines (less the 1st and 2d Battalions, attached to the 14th Marines); the 3d and 4th Battalions, 14th Marines (Headquarters and the 1st and 2d Battalions stayed with the 4th Division); and the 4th 105mm Howitzer Battalion, VAC, hitherto serving with the 14th Marines. These five battalions of 105mm howitzers were attached to XXIV Corps Artillery on 15 July and were designated Groupment A, under control of Headquarters, 10th Marines. The artillery of the 27th Infantry Division (less the 106th Field Artillery Battalion) was likewise attached on 15 July and comprised Groupment B. Five battalions of XXIV Corps Artillery formed Groupment C, and they set up the long-range 155mm guns and 155mm howitzers. The other Army and Marine battalions were equipped with 105mm howitzers. The Marines' four 75mm pack howitzer battalions were not used but were reserved more suitably for the invasion, where they would furnish close support for the assault divisions to which they were attached.

There was quite enough steel and powder to support the operation. The artillerymen used up 24,536 rounds prior to the landings. A total of 1,509 preinvasion fire missions included counterbattery, harassing, and area bombardment. Corps Artillery kept a valuable file of intelligence data on Tinian which was used by both aviators and naval gunners. A Corps Artillery intelligence section worked very closely with the Force G–2 at the NTLF command post on Saipan. Light spotter aircraft were assigned to the artillery units to observe fire results and to collect target intelligence data for either immediate or future use.

The sea bombardment of Tinian began before artillery was ashore on

Saipan. On 13 June, fire support ships of Task Force 52, which could be spared from the pounding of Saipan, were employed against Tinian. The chief object then was to forestall interference with the Saipan operation by Tinian guns or aircraft. Destroyers started a relentless patrol of Saipan Channel, turning their 5-inch guns upon shore batteries and harassing Ushi Point airfield.

Destroyer activity was rapidly extended to other waters. Star shells were placed over Tinian Harbor to prevent movement from the area of Tinian Town. That whole vicinity, especially the airfield, received harassing fire. Here, because of a shortage of destroyers, much of the responsibility fell to destroyer escorts (DEs), destroyer transports (APDs), or destroyer minesweepers (DMSs), whose crews enjoyed the change of routine. On 25 June, two DEs, the *Elden* and the *Bancroft*, spotted a few Japanese barges attempting to leave Sunharon Harbor and blocked their escape by shelling and destroying them. Destroyer escorts also roved to the northwest to harass Gurguan Point airfield by gunfire.

Starting 26 June, the cruisers *Indianapolis*, *Birmingham*, and *Montpelier* undertook a daily systematic bombardment of point targets, which lasted a week and paid special attention to the area of Tinian Town. Intensive bombardment by cruisers was resumed the last few days before the Tinian landings when the *Louisville* and the *New Orleans* delivered main and secondary battery fires. Both of these ships, like the *Indianapolis*, were heavy cruisers. The *Louisville* served as flagship for Rear Admiral Jesse B. Oldendorf, who commanded the fire support ships for Tinian.

During the numerous naval gunfire missions at Tinian, a variety of shells were utilized. On 18 and 19 July, two destroyers attempted to burn the wooded areas on Mt. Lasso with white phosphorus projectiles. Results were disappointing, however, evidently because of dampness due to rain. But since such fire proved terrifying to the enemy on Saipan, destroyers continued to employ it against caves on Tinian. LCI gunboats also shelled the cliffside caves of Tinian with their 40mm guns.

Nowhere on the island were the Japanese left at peace. Starting on 17 July, destroyers from the Saipan Channel patrol delivered surprise night fire at irregular times on the beaches of Asiga Bay, where the enemy was working feverishly to install defenses. The airfields of Tinian were incessantly harassed to deny their use by the enemy. On a single day, 24 June, the battleship *Colorado* shelled every airfield on the island.

The routine of bombarding Tinian, stepped up on 16 July, was climaxed on the 23d, the day before invasion, when 3 battleships, 5 cruisers, and 16 destroyers were involved. Only the beaches due to receive the landings were slighted; for the sake of deception, they were subjected to merely casual fire from the *Louisville* and the *Colorado*. The old battleship surpassed all its previous efforts by destroying, on the same day, the three 140mm coast defense guns of Faibus San Hilo Point, with 60 well-placed 16-inch shells. On the day of the landings, however, the *Colorado* was herself to suffer tragically from the fire of a coastal battery which had not been destroyed.

From the beginning, the Japanese did not suffer the naval shelling without reacting violently, and their return fire caused damage and casualties to a few of the fire support ships. The enemy's defenses were, as usual, well dug in, and some were able to survive the heaviest shelling. Ships found a position difficult to destroy totally except by a direct hit. Because there was a lack of profitable or suitable targets for the largest gunfire support ships, the naval bombardment was suspended for a week between the securing of Saipan on 9 July and 16 July. The only exception was night heckling of the enemy by DEs in the area of Tinian Town.

Provisions for naval gunfire were tied into the overall bombardment plan for Tinian. Efficient interlocking of the three supporting arms was served by a daily conference at NTLF headquarters on Saipan attended by representatives of artillery, air, and naval gunfire. Responsibility for daily assignments was left mainly to the fire direction center of XXIV Corps Artillery because of its collected intelligence data and excellent communication setup. Here the targets were allocated as appropriate to each of the supporting arms. If there was a unique aspect to direction of preliminary fires for Tinian, it was that artillery was a decisive factor.

The big land guns were, of course, aimed mostly at northern Tinian. The 155mm guns could stretch to the southern part of the island, but they seldom attempted it, leaving that half to aircraft instead. Sometimes all three supporting arms had a go at a target. The area of Tinian Town, perhaps the most punished of all, was such an example, though naval gunfire did the most damage there. Use of the island road network was virtually denied to the enemy by the shells and bombs which came from everywhere, isolating some sections and destroying others.

It was neither naval gunfire nor artillery, however, which started the preparatory bombardment of Tinian. On 11 June, Vice Admiral Mitscher's fast carrier task force sailed into the Marianas. Its immediate object was to support the Saipan operation. To minimize interference from the airfields, antiaircraft guns, and shore batteries on Tinian, these installations were bombed and strafed. The Battle of the Philippine Sea began on 17 June, and TF 58 steamed there to join the battle. Five days later, however, surface elements returned to the Tinian assignment. They then were joined by CVE-based aircraft of Task Force 52 and P–47 fighters based on Aslito. All landbased aircraft (except the Marine observation squadrons), as well as the carrier-based planes engaged in the Tinian operation, operated under Commander Lloyd B. Osborne, Commander Support Aircraft, who was embarked in the *Cambria*, flagship of Admiral Hill.[11]

[11] Commander Osborne acted as Commander Support Aircraft off Saipan during the periods when Admiral Turner was on night retirement and Admiral Hill was SOPA. Osborne assumed control of air operations over Tinian when Admiral Hill relieved Turner as CTF 52. Adm Harry W. Hill interview and cmts on draft MS, dtd 20Nov63.

Marine observation planes of VMO–2 and VMO–4 also took part in the pre-invasion activity over Tinian. For several days prior to the landings, pilots operating from Aslito flew over the island, learning about it and searching for targets of opportunity. The little unarmed and unarmored OYs, veterans of Saipan, would again serve infantry and artillery missions on Tinian. Until then, they performed some spotting for the artillerymen shelling Tinian from Saipan.

After the capture of Saipan, aerial reduction of the enemy's defensive positions was undertaken by rocketry, glide bombing, and strafing. Targets included railroad junctions, pillboxes, roads, covered artillery positions, and canefields. The beaches, in the main, were ignored to keep the enemy puzzled, but the airfields were unreservedly raked. Enemy air strength had already been decisively cut down. Of the 107 planes estimated as being based at Tinian airfields prior to 11 June, 70 were destroyed on the ground by carrier strikes long before the capture of Saipan.

Something new was added to air bombardment at Tinian. On 19 July, an enthusiastic Navy commander arrived on Saipan with an impressive Army Air Forces film showing what happened when napalm powder was mixed with aircraft fuel. He showed the pictures to Admiral Hill and General Schmidt and both were enthusiastic about the possibilities of the new fire bomb.[12] Enough napalm and detonators were at hand for a trial run by P–47s. Oil, gasoline, and napalm were mixed in jettisonable fuel tanks. The formula was later improved, but initial results proved good enough to ensure acceptance.[13]

The first use of the fuel-tank bombs was an attempt to burn off wooded areas which, due to dampness, had previously resisted white phosphorus and thermite. The new incendiary badly scorched the trees, but left the needles only partially burned. Because many of the trees were of hardy and indestructible ironwood, the experiment was inconclusive. Much better results were obtained when napalm was used on canefields. Smouldering piles and smoking ashes appeared where once grew flourishing cane stalks. On 23 July, two fire bombs burned out considerable underbrush in the White Beach area.

Pilots of the P–47s, which dropped most of the bombs, objected that such missions required extremely low flying, at a risk of attracting heavy ground fire. They also reported too much upward flash, which decreased the incendiary value, and too brief a burning time—60 to 90 seconds. The idea was a promising one, however, and, when the correct formula was evolved, the new fire bomb became one of the most formidable weapons in the American arsenal.

Thus while planes, ships, and artillery foreshadowed the land battle, organization and plans for the invasion were concluded. A few command changes took place, mostly because the

[12] Adm Harry W. Hill interview with Midn Stephen S. Toth, *ca.* 1963, encl to *Ibid.*

[13] Napalm was a thickener, consisting of a mixture of aluminum soaps, used in jellying gasoline.

recapture of Guam would be attempted at the same time.

COMMAND STRUCTURE AND ASSIGNMENTS

Admiral Hill took command of a reconstituted Northern Attack Force (TF 52) on 15 July, relieving Admiral Turner, who could now exercise more fully his responsibility in command of the Joint Expeditionary Force (TF 51). Hill had been Turner's able second-in-command at Saipan. General Holland Smith was relieved on 12 July and ordered to assume command of the newly-established Fleet Marine Force, Pacific, whose functions included administrative control of all Marines in the Pacific. Command of NTLF and of the V Amphibious Corps was assigned to General Schmidt, who was relieved as 4th Division commander by Major General Clifton B. Cates, one of the few officers who had commanded a platoon, a company, a battalion, a regiment, and now a division in battle. General Watson continued in command of the 2d Division.

General Smith, in retaining command of Expeditionary Troops (TF 56), also continued in overall command of ground forces in the Marianas operations. But neither he nor Admiral Turner was present at the Tinian landings. They sailed on 20 July, on board the *Rocky Mount*, to witness the invasion of Guam the next day. Though they returned on 25 July, they left the direction of the Tinian campaign to Admiral Hill and General Schmidt.

As landing force commander for Tinian, General Schmidt would be in tactical control of the troops. Admiral Hill, the attack force commander, was responsible to Admiral Turner for the capture of Tinian. Slated to command Tinian garrison troops was Marine Major General James L. Underhill. He would have the job of developing Tinian as an air base.

The 2d and 4th Marine Divisions would still compose the assault troops of NTLF, which would make the first corps-sized Marine shore-to-shore operation under combat conditions. The two divisions, however, had suffered grievously on Saipan, and only one replacement draft of 1,268 officers and enlisted men was due before the battle on Tinian. After designating the 4th Division to make the assault landing, General Schmidt reapportioned the men and equipment available. On 18 July, he took all the armored amphibian and amphibian tractor battalions, reorganized them under a Provisional Headquarters, Amphibian Tractors, V Amphibious Corps, and attached it to the 4th Division.[14]

The 2d Division lost its 1st Amphibian Truck Company, its 2d Tank Battalion, and its last two battalions (the 1st and 2d) of the 10th Marines to a heavy buildup of the 4th Division, whose final reinforced strength included also the 1st Joint Assault Signal Company, the 1st Provisional Rocket Detachment, the 2d Amphibian Truck

[14] The 2d Division gave up its 2d Armored Amphibian Battalion, 2d Amphibian Tractor Battalion, 5th Amphibian Tractor Battalion, and 715th Amphibian Tractor Battalion (Army). The Provisional LVT Group also embraced the 10th Amphibian Tractor Battalion, already with the 4th Division, and the 534th and 773d Amphibian Tractor and 708th Amphibian Tank Battalions (Army).

Company, and the 1341st Engineer Battalion (Army).

While such reorganization served to mass combat power in the assault division for seizing and holding the beachhead, it left the 2d Division markedly understrength for its own mission of landing in support. The only reinforcing units left to General Watson were the 2d Joint Assault Signal Company and the 2d Provisional Rocket Detachment.

The reserve 27th Infantry Division was to be prepared to embark on four hours' notice to land on Tinian. The Army division, however, had been much reduced by casualties and detachments and could muster only about half of its original strength. One of its regiments, the 105th Infantry, was required for garrison duty on Saipan, and its division artillery had been detached earlier to the XXIV Corps Artillery for the reduction of Tinian.

In the two-weeks' interval between the capture of Saipan and the invasion of Tinian, the battle-experienced Marines enjoyed a break, except for Japanese sniper activity. No rehearsal for the Tinian assault was held, nor was one regarded as necessary, but a certain amount of reorganization went on. On 11 July, the 2d Marines reverted to the 2d Division and moved to an assembly area near Garapan. The next day, the 165th Infantry was returned from the 2d Division to the 27th Division. On 13 July, the 4th Division began moving to a rest area on southeast Saipan, behind the beaches where the Marines had landed. The 23d Marines stayed in northern Saipan to destroy remaining pockets of Japanese resistance until relieved on 16 July by the 105th Infantry. At the same time the Army relieved the 6th Marines, which ended its mopping up activities and followed the 2d Marines into the assembly area. The orphan 1st Battalion, 29th Marines, which had been detached from the 2d Division to Island Command on 6 July, was due to stay on Saipan for garrison duty.

While the Saipan veterans relaxed, speculation as to the next objective was relatively absent. Troops knew the next island would be Tinian. They were left to guess where the landings would be made. That, in fact, was a question unresolved by the high command itself even after Saipan was secured.

THE ISSUE OF WHERE TO LAND

On 8 July, as time for the invasion of Tinian approached, Admiral Turner put the question to the man due to take over the Northern Attack Force. Admiral Hill was directed to prepare assault plans, suggesting where the landings be made. The debate over which beaches to use was months old. Admiral Turner's inclination to favor the beaches around Sunharon Harbor was opposed by the majority. Four possible landing spots had been noted in this area; those just off Tinian Town had been designated Red Beach and Green Beach, with the Orange and Blue Beaches nearby. The width of all these beaches totaled 2,100 yards. They were certainly the best on Tinian and superior to those on Saipan. Each was wide, with a gentle slope, and the offshore reef contained numerous channel openings. Although Tinian Harbor was a poor anchorage and would have to be developed shortly after the landing, it was the only feasible place for the landing of heavy equipment.

There were, however, two major obstacles to an invasion through the Tinian Town area. One, of course, was that the Japanese expected just such an eventuality and had concentrated so many troops and weapons that a landing there could be made only at a heavy cost to the assaulting forces. The other drawback, related to the first, was that an attack there—or at Asiga Bay—would deprive the Tinian operation of just about the only element of surprise yet left to it. "The enemy on Saipan," said Colonel Ogata on 25 June, "can be expected to be planning a landing on Tinian. The area of that landing is estimated to be either Tinian Harbor or Asiga Harbor." [15]

From the beginning, therefore, American planners had viewed with interest the two small beaches on the northwest coast which, to the Japanese, seemed entirely incapable of supporting a major landing. Perhaps surprise would be obtainable there. What also favored the beaches was their proximity to Saipan for resupply purposes, and the fact that artillery support of the landing would be possible from that island.

Admiral Turner, however, saw another side to the problem. True, the beaches were close to Saipan, but that also meant a long advance down the island, once the landing on Tinian had been accomplished. A shore-to-shore movement involved risking small craft to the vagaries of uncertain weather conditions, whereas at Tinian Town there was a protected harbor, favorable to small boats and unloading operations. Weather could also prevent the rapid displacement to Tinian of field artillery when the troops outran the range of such support from Saipan.

The limited size of the northwest beaches was the chief concern of Admiral Turner, and hardly less so for General Smith and others, though they were not so dubious about them. Everyone felt that the utmost knowledge would have to preface a decision. And, at best, such beaches could serve only as paths or routes, rather than as landing beaches in the usual sense of the word.

When reporting later to Fleet Admiral King, Admiral Hill stated, in simple terms, what the problem was: could two divisions of troops be landed and supplied across beaches the size of White 1 and White 2? Of uppermost concern was whether the amphibian tractors would be able to get ashore, move up to unload, and then turn around. Intelligence sources reported that on White 1 there were only about 60 yards usable for passage of amphibian vehicles and that on the wider White 2 only the middle 65 yards were free of coral boulders and ledges.

To check such findings and also to obtain better knowledge of the beaches,[16] a physical reconnaissance was necessary. It would have to include Yellow Beach on Asiga Bay, should White 1 and White 2 be found unusable. On 3 July, men of the VAC Amphibious Re-

[15] Tinian GarFor OpO A58, dtd 25Jun44, in *4th MarDiv Translations (Tinian)*.

[16] Part of the assorted intelligence on the beaches of Tinian had been obtained by interrogation of native Chamorro and Kanaka fishermen captured on Saipan.

connaissance Battalion were told to be ready for a mission on some Tinian beaches. On 9 July, General Holland Smith issued the operation order to Captain James L. Jones, specifying the beaches. The mission had the approval of both Admiral Turner and Admiral Hill. The latter ordered the participation of Naval Underwater Demolition Teams.

The mission of the Marines was to investigate the beaches, measure the cliffs, and note the area just beyond, including the exits. They were to report the trafficability of the beaches for LVTs and DUKWs in particular. The naval UDT-men were to do the hydrographic reconnaissance—measuring the height of the surf and the depth of the water, observing the nature of the waves, checking the reefs and beach approaches, and looking for underwater obstacles.

After a rehearsal on the night of 9–10 July, off the beaches of Magicienne Bay, Saipan, the Marines and the Navy teams boarded the transport destroyers *Gilmer* and *Stringham.* Company A of the Marine battalion, under the command of Captain Merwin H. Silverthorn, Jr., was assigned to investigate Yellow Beach, while Company B, commanded by Lieutenant Leo B. Shinn, would undertake a reconnaissance of the White Beaches.

The evening of 10 July was very dark when the men debarked into their small rubber boats about 2030. Moonrise occurred at 2232, but fortunately a cloudy sky obscured the moon until almost midnight. Thus the final 500-yard swim to the beaches could be made under the cover of darkness.

Reconnaissance of Yellow Beach, by 20 Marines and 8 UDT-men went off just right, but the reports were unfavorable. On each side of the 125-yard beach, swimmers observed forbidding cliffs which were 20 to 25 feet high. They found also that approaches to the beach contained floating mines, anchored a foot under water off the reef, and that many underwater boulders and potholes would endanger a landing. Craft might be affected also by the relatively high surf, which is generally whipped by prevailing winds.

On the beach itself, the enemy had strung double-apron barbed wire. After working his way through it, Second Lieutenant Donald F. Neff advanced about 30 yards inland to locate exit routes for vehicles—a bold mission, for a night shift of Japanese was busy constructing pillboxes and trenches nearby, and their voices could be plainly heard. Noises resembling gunfire, which had been puzzling while the Marines were moving to the beaches, could now be identified as blasting charges. The enemy seemed vaguely conscious of something strange. Three Japanese sentries were observed peering down at the beach from a cliff, and a few lights flashed seaward. Though all went well for the reconnaissance teams, such imminence of danger caused a suggestion after the reconnaissance that, instead of being unarmed, the swimmers should be equipped with a lightweight pistol or revolver which could be fired even when wet. Captain Silverthorn reported that the reef at Yellow Beach appeared suitable for the crossing of LVTs and DUKWs, but the sum total of the situation at Yellow Beach was plainly unfavorable to a landing.

Exploration of White 1 and White 2

got off to a bad start. Whereas the current off Yellow Beach was negligible at the time, it was so unexpectedly rough on the northwest coast that it pushed the rubber boats off course. The men who were scheduled for White 2 landed instead on White 1, which they reconnoitered. The men headed for White 1 were swept about 800 yards to the north, where there was no beach.

Reconnaissance of White 2 was delayed, therefore, until the next night, when Company A undertook the mission, sending 10 swimmers ashore. On the previous night, the operation had been handled by the *Gilmer*, but this time the *Stringham* took the Marines and the UDT-men toward White 2, leaving the pickup to the *Gilmer*. Radar, which the *Stringham* possessed, enabled it to guide the rubber boats and to send course corrections over an SCR-300 radio.

Findings at the White Beaches were relatively encouraging. They showed that the measurements indicated by air and photographic coverage were approximately correct. White 1 proved to be a sandy beach about 60 yards wide, and White 2 more than twice that size. At the larger beach, however, there were coral barriers which averaged 3½ feet high. They formed the beach entrance and restricted it, for vehicles, to about 70 yards, though infantry could scramble over the barriers. On the beach itself was found a man-made wall sloping up about 1½ to 2 feet but judged passable by vehicles.

Primarily, the physical reconnaissance verified that LVTs, DUKWs, and tanks could negotiate the reef and land. Moreover, it showed that LCMs and LCVPs could unload on the generally smooth reef which extends about 100 yards from the shore. It appeared, however, that White 1 would be able to receive just 8 LVTs—and then only if some unloaded opposite the cliffs. At White 2, the landing of a maximum of 16 LVTs seemed possible, if about half unloaded in front of the adjacent cliffs.

While the cliffs at the White Beaches were not more than 6 to 10 feet high—lower than those at Yellow Beach—and could be scaled by ladders or cargo nets, they were nevertheless rocky and sharp and were deeply undercut at the bottom by the action of the sea. There were a number of breaks in the cliff walls at both beaches, where it appeared that Marines could land single file without aid and move inland. In effect, while the avenues of approach for amphibious vehicles were severely limited, the landing area was fairly wide—on White 1 a probable 200 yards; on White 2, 400 yards.

Marines who debarked from LCVPs on the reef would be able to wade ashore without risk underfoot except from small holes and boulders. No dangerous depths were found, nor any line of mines or man-made underwater obstacles. The scouts, however, were not equipped to detect the buried mine—not easy to set into coral but quite practical in the gravel at the shore edge.

Bearing such detailed reports, UDT and VAC Reconnaissance Battalion officers went to see Admiral Turner on board the *Rocky Mount* early on 12 July. They felt that landings could be made on the White Beaches, and that successful exits were possible inland.

With the reports at hand, Admiral Turner opened a meeting on board the

Rocky Mount on 12 July. He started the conference by talking about the various beaches of Tinian from a naval standpoint. General Schmidt and Admiral Hill then were asked for their opinions, and they spoke unreservedly in favor of the White Beaches, expressing the following views:

(1) landings at the Tinian Town area would prove too costly;

(2) the artillery on Saipan could well support an invasion over the northern beaches;

(3) capture of Ushi Point airfield, a primary objective, was immediately desirable for supply and evacuation;

(4) tactical surprise might be possible on the White Beaches;

(5) a shore-to-shore movement would be feasible; and

(6) most supplies could be preloaded on Saipan and moved directly on wheels and tracks to inland dumps on Tinian.

Admiral Turner was agreeable to the idea. He remarked, later, that he had already become reconciled to the White Beaches and, like everyone else, was awaiting favorable reconnaissance reports. General Smith had been a pioneer advocate of the White Beaches, so now, with the consent of Admiral Spruance, the issue was settled.[17]

The next day, Admiral Turner released his operation plan, later promulgated as an order. Some 300 copies of the plan were circulated, starting in motion the troops and equipment which had hardly been idle since the capture of Saipan. On 20 July, Admiral Spruance confirmed J-Day as the 24th, but he authorized Admiral Hill to alter the date, if necessary, for the weather had to be right. This invasion was going to take place at the height of the summer monsoon period, a time of suddenly appearing typhoons and thunderstorms. The exacting logistical effort would require at least three calm days following the landings, a period to be forecast by Fifth Fleet weather reconnaissance.

LOGISTICS: PLANS AND PROBLEMS

Logistics, indeed, formed the heart of the operation plan, for upon that branch of military art depended, more than usually, a victory on the battlefield. The invasion of Tinian was going to test whether Navy and Marine Corps amphibious tactics were sufficiently flexible. In the attempt to land and supply two divisions over a space of less than 200 yards, there would be the risk of a pile-up at the beaches, which would be tragically compounded if the hope of surprising the enemy proved false.

A great variance from normal ship-to-shore procedure would be the fact that, except for the initial emergency supplies on LSTs, all supplies were to be landed in a shore-to-shore movement from Saipan. The usual beachhead dumps were going to be out of the question. No supplies could be landed initially except those which could roll across the beaches in LVTs or DUKWs from the LSTs to inland dumps. The

[17] Admiral Spruance, in commenting on this passage, wrote that it was the reports of the reconnaissance and UDT teams that "decided us on the change in plan. This operation, like that later on at Inchon in Korea, required trained and efficient troops like our Marines to be successful." Adm Raymond A. Spruance ltr to ACofS, G-3, HQMC, dtd 4Oct63.

supply plan also envisaged a shuttle of resupply by LCMs and LCTs carrying preloaded cargo trucks and trailers from Saipan and by several LSTs devoted to general reserve supplies. All equipment and supplies required were on Saipan except for petroleum products and certain types of food and ammunition, which were available on vessels in Tanapag Harbor.

To permit vehicular access over the coral ledges adjoining the beaches—and thus, in effect, widen them—a Seabee officer[18] on Saipan designed an ingenious portable ramp carried ashore by an LVT. Six of the 10 constructed by the 2d Amphibian Tractor Battalion were used at Tinian after being transported to the island by the LSD *Ashland* on the morning of the assault.

Even vehicles as heavy as the 35-ton medium tank could cross the ramp, which was supported by two 25-foot steel beams. These beams could be elevated 45 degrees by the LVT to reach the top of the 6-to-10 foot cliffs. As the LVT backed away, a series of 18 timbers fell into place on the beams, forming a deck for the ramp. The other end of the beams then dropped and secured in the ground at the base of the cliff, breaking free of the LVT. Such ramps were used to land vehicles until pontoon causeways were put into use. After 29 July, however, bad weather, caused by a "near-miss" typhoon, precluded unloading by anything but the agile and hardy DUKWs.

[18] This inventive officer was Captain Paul J. Halloran, CEC, USN, Construction Officer, NTLF. He submitted an interesting, detailed report which may be found as Enclosure B to *TF 52 OpRpt.*

At Tinian, the amphibian trucks were the prized supply vehicles. They were better suited to the roads than were LVTs, which clawed the earth. For the amphibian tractors, the engineers often constructed a parallel road.

The shipping and amphibious craft employed for moving troops and supplies were impressively numerous at Tinian, considering the size of the operation. Every available LST in the Saipan area, 37 of them—including a few from Eniwetok—was drafted to lift the troops of the 4th Division for the landing and the initial supplies for both Marine divisions. Ten LSTs were to be preloaded for the 4th Division, 10 for the 2d Division, and 8 for NTLF. In command of the Tractor Flotilla was Captain Armand J. Robertson.

Most of the ships were loaded at Tanapag Harbor at whose excellent docks six could be handled in a day. Troops of the Saipan Island Command acted as stevedores. Beginning on 15 July, they loaded the top decks of the LSTs with enough water, rations, hospital supplies, and ammunition to last the landing force three and a half days. The assault Marines would not land with packs at Tinian. In their pockets would be emergency rations, a spoon, a pair of socks, and a bottle of insect repellant. Ponchos were to be carried folded over cartridge belts.

Four of the LSTs each loaded one of the 75mm howitzer battalions on 22 July, off the Blue Beaches. Individual artillery pieces were stored in DUKWs on board to permit immediate movement to firing position by the two amphibian truck companies assigned to the 4th Division. Full use of the pack howitzers was especially desired, for the division did not have its 105s.

Another four of the LSTs each loaded 17 armored amphibians in their tank decks while at Tanapag anchorage on 23 July. Two of these four ships carried medical gear stowed on their top decks. In fact, all of the preloaded cargo on the LSTs was placed topside, and as much as possible remained in cargo nets. On each LST were two cranes to expedite loading and unloading over the sides into LVTs and DUKWs.

Because all the LSTs were needed to lift 4th Division troops and supplies, one regiment of the 2d Division—the 6th Marines—would remain on Saipan until 10 LSTs could unload their troops and return to Tanapag Harbor from Tinian. The 2d and 8th Marines were to be moved on seven transports for a J-Day feint off Tinian Harbor. No general cargo was loaded on the transports, but each carried organizational vehicles of the units embarked.

On 21 and 22 July, two LSDs, the *Ashland* and the *Belle Grove*, loaded at the Charan Kanoa anchorage most of the tanks assigned to the 4th Division. The LSDs each took 18 medium tanks, each tank carried in an LCM. Their other cargo included flamethrower fuel and ammunition received from the merchant ship *Rockland Victory* on 19–20 July and a supply of water.

General Schmidt rounded up 533 LVTs, including 68 armored amphibians and 10 LVTs which were equipped with the special portable ramp.[19]

[19] *TF 52 OpRpt*, Encl A, p. 6 indicates that 537 LVTs were used; the lower figure is contained in landing force reports.

DUKWs available came to 130. Landing craft employed for the Saipan-to-Tinian lift numbered 31 LCIs, 20 LCTs, 92 LCMs, and 100 LCVPs. Nine pontoon barges were loaded on 19 July with fuel in drums received from the merchant ships *Nathaniel Currier* and *Argonaut*. These barges would be towed to positions off the reef to service amphibious vehicles and landing craft. Captured Japanese gasoline and matching lubricants were stocked on the barges. Five additional barges were loaded on 24 July from the *Currier*.

After the initial landing, the shuttle system for resupply would begin to operate between the 7th Field Depot dumps on Tinian. Twenty LCTs, 10 LCMs, and 8 LSTs were allotted for such use. Also assigned to the resupply system were 88 2½-ton trucks and 25 trailers.

On 22 and 23 July, 32 of the LSTs sailed into the Saipan anchorage, where they embarked LVTs, DUKWs, and troops of the 4th Division from the Blue, Yellow, and Red Beaches of Saipan. Herein lay another variance from the usual ship-to-shore movement of an island campaign, for, except for the two regiments of the 2d Division embarked in transports, all other Marine units were moved to Tinian in LSTs or smaller craft.

The remaining five LSTs, still at Tanapag Harbor, embarked other troops and LVTs or DUKWs at the same time. Of the 20 LCTs available, 10 were designated for the 4th Division vehicles, which were loaded at the seaplane base in Tanapag Harbor on 23 July. Five additional LCTs were loaded the same day at the same place—three with medium tanks, four to each LCT, and two with bulldozers and cranes. The remaining five LCTs were loaded in the forenoon of J-Day at the seaplane base with 2d Division vehicles.

Of the 92 LCMs available, 36 bearing medium tanks were loaded onto the two LSDs. Ten were loaded with 4th Division armor on 23 July at the steel pier of Tanapag Harbor; these LCMs would make a direct passage. Forty-one of the remaining 46 loaded medium tanks and waited off the Blue Beaches for movement on 24 July or shortly after. Five of these LCMs moved to Tinian directly, and 36 were loaded on board the 2 LSDs when they returned from Tinian. The other five LCMs took on 2d Division vehicles at the seaplane base on 24 July for direct transfer to Tinian. The available 31 LCIs were used to carry troops and vehicles of the 2d Division to the 7 transports at Tanapag Harbor on 20–23 July, while the 100 LCVPs loaded 4th Division vehicles at the Green Beaches on 23 July for direct movement to Tinian.

While all such loading went on, the shore party of the 4th Division prepared for its modified task on Tinian. Usually, the shore party is responsible for first dumps off the beaches, but in this case not a pound of ammunition or other supplies could be landed on the sand. Still, there would be plenty to do. The shore party at Tinian was expected to keep supply traffic moving to the inland dumps, where some of its men would be working. It was also to provide equipment and personnel to expand and improve the beaches. Farther inland, responsibility for the trails and roads fell to Seabees of the 18th and 121st Naval Construction Battalions and to the assault engineers of the 1st Battalions of the 18th and 20th Marines.

Lieutenant Colonel Nelson K. Brown's 4th Division Shore Party for Tinian was composed of the pioneers of the 2d Battalion, 20th Marines on White Beach 2 and the Army 1341st Engineer Battalion on White 1. The Force Beachmaster was Commander Carl E. Anderson. The 2d Division did not operate a shore party, since there were already enough men on hand. But a platoon of 2/18 pitched in on White Beach 2, and the rest of the battalion worked at the division dumps.

On 26 July, an NTLF Shore Party Headquarters, commanded by Colonel Cyril W. Martyr of the 18th Marines, was superimposed upon the 4th Division Shore Party. The change indicated recent attention to consolidating shore party activities. The NTLF Shore Party Headquarters, with a strength of 6 officers and 8 enlisted men, was taken from the Headquarters of the V Amphibious Corps and of the 18th Marines.

A departure from Saipan supply practice took place at Tinian, where the unit distribution system was used. Small arms and mortar ammunition were not delivered to the regiments. Instead, those units drew from the division dumps and delivered by truck to the battalions. This practice on Tinian was in keeping with logistical procedures employed on a smaller island, and, as a result, regimental supply dumps did not have to be moved as often as they were on Saipan.

ATTACK PLANS

Because of the unusual shore-to-shore operation, involving constricted beach area, logistics monopolized much of the planning effort for Tinian. But mastery of the supply details could only enable that victory which arms must secure. In preparing for this battle, Marine commanders had a rare opportunity for reconnaissance. A number of them were taken on observation flights over the island or on cruises near its shores.

Under General Schmidt's attack plans, the 4th Division, upon pushing inland over the White Beaches, was to seize Objective 0–1, which included Mt. Maga. The main effort would be toward Mt. Lasso, reaching to a line which would include Faibus San Hilo Point, Mt. Lasso, and Asiga Point. (See Map 22.)

General Cates issued his operation order to the 4th Division on 17 July. He planned to use the 24th Marines in a column of battalions on White Beach 1 and the 25th Marines with two battalions abreast on White Beach 2. The 23d Marines would be held in division reserve and wait immediately offshore.

Because the beaches were so narrow, only Company D of the 2d Armored Amphibian Battalion was to be employed in the assault landing. One platoon would precede troop-carrying LVTs toward White 1, while the other two platoons led the attack against White 2. When the naval gunfire lifted, the armored amphibians would fire on the beaches and then turn to the flanks at a distance 300 yards from shore, where they would fire into adjacent areas. The first wave of Marines would continue forward to the beaches alone, except for the fire of the .30 caliber machine guns mounted on the LVTs.

The 2d Division was to satisfy, partially, Colonel Ogata's belief that the major landing was due in the Tinian Town area. The 2d and 8th Marines, as part of a naval force, would execute a feint off Tinian Town at the hour of the actual landing to divert attention from the northern part of the island. Included in the show by the Demonstration Group would be the battleship *Colorado*, the light cruiser *Cleveland*, and the destroyers *Remey* and *Norman Scott*, delivering a "pre-landing" bombardment. Following the demonstration off Tinian Town, the 2d Division Marines would return northward to land on the White Beaches in the rear of the 4th Division. General Watson planned to put the 2d Marines ashore on White 2 and the 8th Marines on White 1, while the 6th Marines would land over either beach. The command post of the 2d Division was set up on board the assault transport *Cavalier* at 0800 on 23 July, as the men made ready for both a fake landing and the real thing.

The 27th Division, less its 105th Infantry and division artillery, would be ready to embark in landing craft on four hours' notice to land on Tinian. Though Army infantry was never committed there, Army aircraft, artillerymen, amphibian vehicles, and engineers helped invaluably toward success of the Tinian operation.

THE MOUNTING THUNDER

Since 20 June, artillerymen on southern Saipan had been hammering Tinian. On 23 July, Corps Artillery fired 155 missions, and for J-Day, General Harper planned a mass bombardment by all 13 artillery battalions just before the landing—a crescendo of fire against every known installation on northern Tinian, every likely enemy assembly area, and every possible lane of approach by land to the White Beaches.

The Army Air Forces was likewise dedicated to seizing Tinian, which, of course, was to become particularly theirs. On the day before the landing, P–47s of the 318th Fighter Group flew 131 sorties against targets on the island,[20] joining carrier aircraft from the *Essex*, *Langley*, *Gambier Bay*, and *Kitkun Bay*, which made 249 sorties. The same day saw the arrival of a squadron of B–25s on Saipan, which were shortly to join the battle for Tinian.

In order to permit heavy air strikes on southern Tinian on 23 July, naval gunners withheld their own fire for three periods of up to an hour. Off northern Tinian, the *Colorado* and the *Louisville* also ceased fire at 1720 to allow a napalm bombing mission on the White Beach area, where two fire bombs burned out some underbrush. The naval gunfire of 23 July, started at sunrise, was partly destructive, partly deceptive. Yellow Beach and the beaches around Sunharon Harbor received fire intended chiefly to mislead the enemy. At Tinian Town, particularly, care was taken to confuse Colonel Ogata. Mine-sweeping and UDT reconnaissance of the reef off Tinian Town were conducted, both without findings or incident. In fact, minesweepers operating in Tinian waters prior to the landings reported no obstacles to shipping, though, later on, 17 mines, previously located by UDT reconnaissance, were swept from Asiga Bay.

Before 1845 on 23 July, when all but a few fire support ships left the area for night retirement, the *Tennessee* and the *California* had fired more than 1,200 14-inch and 5-inch shells into the vicinity of Tinian Town, already nearly demolished. A notable fact of the naval bombardment on 23 July was the comparative sparing of the Asiga Bay coastline. It would have been folly to invite Colonel Ogata's reserves to an area quite near the White Beaches, when it was better to keep them farther south.

After 1845, night harassing fire was assumed by the light cruiser *Birmingham* and five destroyers. The *Birmingham* and three of the destroyers covered road junctions between Faibus San Hilo Point and Gurguan Point on the western half of the island, besides shelling areas of enemy activity to the southwest. The destroyer *Norman Scott* was assigned to isolate road junctions on the east side of the island and the Yellow Beach vicinity.

To the rumbling of such gunfire, General Cates took the 4th Division command post on board *LST 42* at 1500

[20] Dr. Robert F. Futrell, USAF HistDiv, ltr to Head, HistBr, G–3, HQMC, dtd 29Nov63.

on 23 July.[21] Admiral Hill, in his attack order of 17 July, had fixed H-Hour at 0730. The sun would rise at 0557 on a day which would tell whether Tinian planners had gambled wisely when they picked such a landing area as the White Beaches. Another question was interjected shortly before dawn of J-Day, when a UDT mission on White Beach 2 was defeated by a squall which scattered the floats carrying explosives. The men had been sent from the *Gilmer* to blast boulders and destroy boat mines on the beach, the latter spotted by reconnaissance aircraft.

The squall was a phase of the rain which fell upon the assault-loaded LSTs the night of 23 July, when at 1800 they moved out to anchor. Nearby lay the line of departure, about 3,000 yards off the White Beaches, where Marines were to find every answer.

[21] The 4th Division chief of staff recalled that this unusual use of an LST as a command ship "worked very well . . . ," as, "we got in quite close to the beach and could see what was going on there." MajGen William W. Rogers ltr to ACofS, G-3, HQMC, dtd *ca.* 24Oct63.

J-Day and Night[1]

STRATAGEM AT TINIAN TOWN

While Marines of the 4th Division waited in their LSTs for the morning of 24 July, transports lifting Marines of the 2d Division sailed from the anchorage off Charan Kanoa at 0330 in darkness appropriate to their secret mission. Just before sunrise, the transports and their fire support ships —the battleship *Colorado*, the light cruiser *Cleveland*, and the destroyers *Remey*, *Norman Scott*, *Wadleigh*, and *Monssen*—moved into the waters opposite Tinian Town. The 2d and 8th Marines were on board the transports *Knox*, *Calvert*, *Fuller*, *Bell*, *Heywood*, *John Land*, and *Winged Arrow*, with Captain Clinton A. Mission on the *Knox* in command of the Demonstration Group. Two patrol craft, *PC 581* and *PC 582*, rounded out the task group.

The fire support ships were to deliver neutralizing and counterbattery fire on Tinian Town, and on the high ground north and south of the town, to divert the enemy further. The heavy cruiser *New Orleans* and the light cruiser *Montpelier* would meanwhile execute a similarly deceptive mission at Asiga Point, delivering 30 minutes of airburst fire over the vicinity of Yellow Beach.

Shortly after 0600, the Demonstration Group, lying about four miles off Tinian Town, began the planned deception. The commander of the *Calvert* logged the action at 0612: "Stopped ship. Commenced lowering landing craft. Simulated debarkation of landing team."[2] By 0630 all 22 boats of the *Calvert* were waterborne. Shortly before 0700, Navy planes swept over the vicinity of Tinian Town, bombing and strafing.

From each transport the Marines descended cargo nets into the landing craft and then climbed up again. No troops remained in the boats, but to the Japanese on shore it may well have

[1] Unless otherwise noted, the material in this chapter is derived from: *TF 51 OpRpt; TF 52 OpRpt; TF 56 OpRpt; NTLF OpRpt;* NTLF OPlan 30–44; NTLF Jnl, 23Jul–8Aug44, hereafter *NTLF Jnl;* 2d MarDiv Rpt of Ops (Pts 1 & 2), Phase III, FORAGER, dtd 11Sep44, hereafter *2d MarDiv Op Rpt Tinian;* 4th MarDiv Rpt of Ops (incl Rpts on Admin, Intel, Ops, Sup and Evac, Sig, SpecCmts and Recoms, 14th, 20th, 23d, 24th, and 25th Mar and 4th TkBn, dtd 25Sep 44, hereafter *4th MarDiv OpRpt Tinian;* 4th MarDiv D–4 Jnl, 21Jul–3Aug44; *4th MarDiv Translations (Tinian);* Capt John W. Thomason, III, "The Fourth Division at Tinian," *Marine Corps Gazette*, v. 29, no. 1 (Jan45), hereafter Thomason, "Tinian;" Chapin, *4th MarDiv in WW II;* Crowl, *Marianas Campaign;* Hoffman, *Tinian;* Isely and Crowl, *Marines and Amphibious War;* Johnston, *Follow Me!;* Morison, *New Guinea and the Marianas;* Proehl, *4th MarDiv History;* Smith and Finch, *Coral and Brass;* Stockman and Carleton, *Campaign for the Marianas.*

[2] USS *Calvert* (APA 32) AR, dtd 4Aug44.

appeared so. At 0730, the hour set for invasion of the White Beaches, coxswains guided their craft rapidly shoreward under cover of naval gunfire. Soon the *Calvert* reported seeing "splashes from large caliber shells 1500–2000 yards off starboard quarter,"[3] and Captain Mission confirmed that artillery and heavy mortar fire was being received in the boat lanes. Under orders from Admiral Hill not to jeopardize the men, he withdrew the boats to reform. A second run was then started, to make the deception realistic enough. Fire from Japanese shore batteries was again received, and some of the landing craft were sprayed with shell fragments. But no casualties resulted, and the boats moved to within 400 yards of the beaches—impressively close—before turning back. About 1000, the transports began recovering the landing craft, and an hour later all ships were under way to the transport area off the White Beaches.

Was the demonstration a success? Measured by the results intended, it was. The Japanese did believe, for a while, that they had foiled an attempted landing. Colonel Ogata sent a message to Tokyo, claiming that he had repelled more than 100 landing barges. The feint served to hold Japanese troops in the Tinian Town area, freezing the *3d Battalion, 50th Infantry* and elements of the *56th Naval Guard Force* while Marines moved inland over the White Beaches.

Not only was Colonel Ogata briefly deceived but so also were his soldiers. One Japanese infantryman of 1/135 wrote in his diary: "Up to 0900 artillery fire was fierce in the direction of Port Tinian, but it became quiet after the enemy warships left. Maybe the enemy is retreating."[4]

Two of the American warships suffered grievously from the violent Japanese response. Air photos of Tinian, good as they were, had failed to show the battery of three 6-inch naval guns behind Sunharon Harbor. At 0740, when the *Colorado* had moved to within 3,200 yards west of Tinian Town, she received the first of 22 direct hits in a period of 15 minutes. Casualties were many, totaling 43 killed and 176 wounded. Of the Marines on board, 10 were killed and 31 wounded.[5] The ship was badly damaged but was able to make it back to Saipan. The destroyer *Norman Scott*, while attempting to protect the *Colorado*, suffered 6 hits from the same guns and had 19 men killed and 47 wounded. Not until four days later was this Japanese battery destroyed by the battleship *Tennessee*.

INVASION—THE REALITY

Unlike those at Tinian Town, the fire support ships off the White Beaches (two battleships, one heavy cruiser, and four destroyers) were never in danger from guns on shore. The big ships here were given a special mission before H-Hour, after the underwater demolition team assigned to destroy the ominous mines on White Beach Two had lost its explosives in an

[3] *Ibid.*

[4] Diary of Takayoshi Yamazaki, in *4th Mar Div Translations (Tinian)*.

[5] USS *Colorado* AR, ser 0033 of 12Aug44. (OAB, NHD).

inopportune offshore squall. The *California*, *Tennessee*, and *Louisville* fired directly on the beach. Still, because of the smoke and dust there, it was difficult to determine whether the mines had been detonated, so another approach, at closer range, was tried. At 0625, the naval and artillery bombardment of the area was lifted for 10 minutes in order that orbiting call-strike aircraft might ensure destruction of the mines. This air strike, which involved 12 fighters and 2 torpedo bombers, was partially successful. Observers reported that 5 of the 14 known mines were detonated. At the time of the strike, some of the LVTs were already waterborne. They had begun emerging from the LSTs at 0600, at the same time that minecraft began sweeping the waters off the beaches.

In order to obscure Japanese observation of such prelanding activity, a battery of 155mm howitzers on Saipan began firing a concentration of smoke shells at 0600 on Mt. Lasso, the site of Colonel Ogata's command post. Corps artillery also struck the woods and bluffs just beyond the beaches to prevent any Japanese there from observing offshore activity. Operations off the White Beaches went like clockwork until shortly before 0700, when the control group commander informed Admiral Hill that initial assault waves were not forming as rapidly as planned. H-Hour, therefore, was delayed 10 minutes—to 0740.

Shortly before H-Hour, a wind change caused the smoke and dust over the target to shift offshore, where it covered the boat lanes. Adding to this hazard to the landing was a strong tidal current running northward at a right angle to the lanes. In order to guide the initial assault waves to the beaches, two P–47s were assigned to fly at low altitude in the direction the LVTs were to move.

At 0721, 24 LVTs took the first wave of Marines across the line of departure. In eight of the craft, Company E, 2d Battalion, 24th Marines was embarked, ready to land on White Beach 1. There, the attack was to be by a column of battalions. The other 16 LVTs lifted Company G, 2d Battalion, 25th Marines and Company I of the 3d Battalion, to land them abreast on White Beach 2. Only scattered rifle and machine gun fire was received as the troops approached the shore. Preceded by armored amphibian tractors and supported by LCI gunboats firing rockets and automatic cannon, the Marines of both RCTs hit the beach almost simultaneously. Gunfire ships and corps artillery supported the landing, but because of the long-range artillery fire on the beach area the usual strafing attack to cover the initial assault was omitted.

THE SITUATION AT THE BEACHES

At 0747, the eight tractors bearing Company E, 2/24, ground to a halt, and Marine riflemen got their first look at the cupful of sand that was White 1. The beach was just wide enough to accept four of the LVTs; the others had to debark their troops opposite the ledges adjacent to the beach. Surprise was not complete; a small beach defense detachment offered resistance. The handful of enemy troops gave some trying moments to the Marines, especially to those who had to climb over the jagged coral ledges from waist-deep water. Marines who crossed the beach were able to tread safely above a dozen horned mines which the Japanese, expecting no landing here, had permitted to deteriorate.

The beach defenders employed hand grenades, rifles, and machine guns against the Marines. During a brief but bitter fight, Company E destroyed the Japanese in their cave and crevice defenses and then pushed inland. The attackers had to move swiftly, not only to keep the beaches cleared for successive waves, but also to keep the enemy off-balance and prevent them from counterattacking.

On the heels of Company E, the rest of the 2d Battalion landed in a column of companies—A (attached for landing only), G, F, Headquarters, and Shore Party (1341st Engineers). Company A turned left behind Company E to await its parent battalion. By 0820, Major Frank E. Garretson had his entire 2d Battalion on Tinian. Lieutenant Colonel Otto Lessing's 1st Battalion got ashore by 0846 and immediately veered left.

The advance of the two battalions was opposed by intermittent mortar and artillery fire and by small arms fire. Coming from thick brush and caves, the source of the fire was hard to spot. Yet, after the first 200 yards, progress toward the 0–1 Line eased to what Major Garretson called a "cake walk."[6] At 0855, Lieutenant Colonel Alexander A. Vandegrift, Jr., received orders to land the reserve 3d Battalion, and upon reaching the shore at 0925, moved his unit to an assembly area about 300 yards inland. Marine commanders considered the opposition on White Beach 1 to be "light," and it was, when contrasted to the situation during the Saipan landing or to the moderate resistance encountered on White Beach 2.

While the Japanese hardly expected any sort of landing at White Beach 1, the same was not quite true at White Beach 2, for Colonel Ogata had cautioned against the possible appearance of a small landing party there. The result was a more vigilant force and improved defenses. The known antiboat mines on the beach had not deteriorated; a few had been exploded by aircraft, but the bulk of them had escaped destruction. The better Japanese defense here was built around two pillboxes situated to put crossfire on the beach. They had not been damaged by the bombardment. Because of the perils on shore, it was decided not to send LVTs over the beach until engineers could get at the mines. Initial waves were to avoid the beach; instead, the first troops would have to climb rocky ledges, which rose 3 to 10 feet above a pounding surf.

There had been some talk on Saipan that the Tinian beaches would be easy to take, but the battle-tried men of Colonel Merton J. Batchelder's 25th Marines expected no simple landings as they crowded into the LVTs. While Lieutenant Colonel Justice M. Chambers' 3/25 would go ashore on the left in a column of companies, a different procedure was planned for 2/25, commanded by Lieutenant Colonel Lewis C. Hudson, Jr. After the 2d Battal-

[6] Maj Frank E. Garretson ltr to Maj Carl W. Hoffman, dtd 17Aug50.

ion's Company G landed, the other two companies were to be put ashore abreast. It was believed that such a formation would permit the greatest speed in crossing the beach with the least loss of control.

Despite choppy water, some LVTs were able to edge near enough to the ledges so that two Marines, standing at the bow, could help their comrades catch a handhold on the jagged rocks along the top. The other assault companies of the regiment landed at scheduled intervals. Even the reserve 1st Battalion, 25th Marines (Lieutenant Colonel Hollis U. Mustain) was entirely ashore by 0930.

PROGRESS OF THE ATTACK [7]

In the area of White Beach 1, 2/24 gained 1,400 yards, reaching its objective line by 1600, unopposed except by occasional small arms fire. Elements of the battalion reached the western edge of Airfield No. 3 and cut the main road from Airfield No. 1 to other parts of the island. To the left, however, 1/24's advance was delayed at the shore, some 400 yards short of 0–1, because of Japanese resistance being offered from positions in caves and brush. Though armored amphibians were employed from the water to fire into the caves and flamethrower tanks burned out vegetation, the Japanese still would not be routed.

At 1630, the reserve 3d Battalion, 24th Marines went into line to close a gap that had opened between 1/24 and 2/24. Shortly before dark, the 1st Battalion, 8th Marines (Lieutenant Colonel Lawrence C. Hays, Jr.), after waiting in the *Calvert*, was landed. It thereupon became 4th Division reserve and took a position to the rear of 2/24.

While Colonel Franklin A. Hart's 24th Marines dealt with sporadic resistance at White Beach 1, Colonel Batchelder's 25th Marines encountered better-organized opposition. The enemy defenses included mortars and automatic weapons, located in pillboxes, shelters, caves, ravines, and field entrenchments. From Mt. Lasso some artillery pieces that had survived the preparation fires dropped shells into the beach area.

Fewer prepared defense positions were met as the 25th Marines progressed inland, but continual fire from small, well-hidden knots of Japanese held back the day's advance, keeping it to approximately 1,000 yards short of 0–1. The two pillboxes that commanded White Beach 2, and the rifle and machine gun pits which protected the fortifications, were bypassed by the initial assault waves, which were more concerned with getting a foothold inland. Other Marines reduced the two strongpoints and found 50 dead Japanese around antiboat and antitank guns.

The entire vicinity of White Beach 2 had been methodically seeded with mines, including the powerful antiboat types on the beach and deadly antipersonnel mines and booby traps inland. Experienced Marines avoided even the tempting cases of Japanese beer, but, despite all precautions, two LVTs which

[7] Additional sources for this section include: 8th Mar SAR, FORAGER, Phase III, dtd 19Aug44, hereafter *8th Mar SAR Tinian;* 1/8 Rpt of Ops, dtd 13Aug44, hereafter *1/8 Rpt Tinian;* 2d TkBn SAR, dtd 14Aug44, hereafter *2d TkBn SAR.*

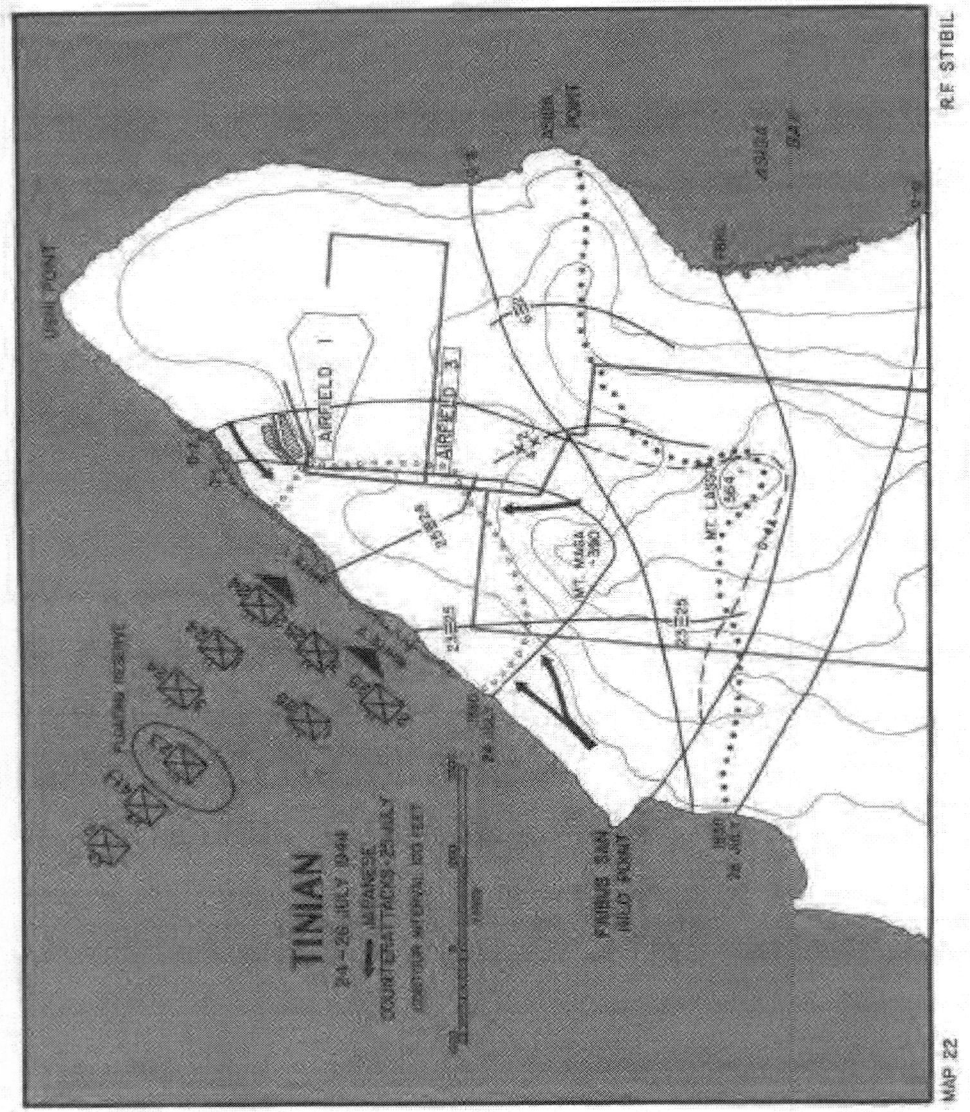
TINIAN
24–26 JULY 1944
JAPANESE
COUNTERATTACKS-25 JULY
CONTOUR INTERVAL 100 FEET
FLOATING RESERVE
USHI POINT
AIRFIELD 1
AIRFIELD 3
MT. MAGA
390
MT. LASSO
564
FAIBUS SAN
HILO POINT
24 JULY
26 JULY
ASIGA
POINT
ASIGA
BAY
MAP 22
R.F. STIBIL

ventured inland were blown up 30 yards from the shore, and a third detonated a mine while attempting to turn around on the beach. Removal of the mines required the diligent efforts of UDT-men, bomb disposal teams, and engineers. Not until 1337 could the infested White Beach 2 be reported clear of mines.

Resistance to the Marine landings on Tinian had been comparatively light—casualties on J-Day numbered 15 killed and 225 wounded, including casualties in the destroyed LVTs; the known enemy dead came to 438. Still, General Cates believed it wise to land the division reserve, the 23d Marines, the first day. His understrength regiments were occupying a beachhead which was, by the end of the day, some 3,000 yards wide, and at its maximum depth, approximately 1,500 yards. Moreover, an enemy counterattack was expected momentarily.

The Marine division commander was more interested in being ready for such a counterattack than in simply reaching the entire 0–1 Line. He therefore ordered his regiments to cease the attack about 1630 and begin digging in. Marines strung barbed wire along the entire division front and stacked ammunition near their weapons. Machine guns were emplaced to permit interlocking bands of fire, while 60mm and 81mm mortar target areas were assigned. Bazookamen were stationed at every likely tank approach, and 37mm gun crews got ready with canister and armor-piercing shells.

All troops of the 23d Marines, though not their vehicles, were ashore by 1630, landed over White Beach 2. At 1030, division had ordered the debarkation into LVTs. At 1300, Colonel Louis R. Jones received word from General Cates, written one hour before, specifying the mission. In effect, the 23d Marines was to pass through right elements of the 25th Marines along the coast and take up a frontline position in contact with 2/25. The cramped sector assigned to the 23d, however, permitted room for only the 2d Battalion in the line. The 1st Battalion dug in behind 2/23, creating valuable depth, for here seemed the "most probable counterattack zone." [8] To the 3d Battalion fell the role of division reserve.

A few vexations marked the landing of the 23d Marines. First, there was a series of communication difficulties which delayed getting the troops ashore. Then, the fact that other units, 1/25 and elements of 2/25, were still in the immediate vicinity crowded an area where artillery and tanks also kept landing. Finally, as the 2d Battalion, 23d was moving into position, it had to subdue violent resistance from lurking Japanese employing machine gun and rifle fire. Tanks, lumbering through brush and cane fields, helped to hunt down the enemy.

These tanks belonged to Company C, 4th Tank Battalion, and were among the many landed on J-Day upon an island so suitable for armor that the Marines eventually employed more tanks here than they had on any previous amphibious operation. For the Tinian campaign, Major Richard K. Schmidt's 4th Tank Battalion had received 13 new medium tanks from the 7th Field Depot. The older tanks of the battalion, how-

[8] BGen William W. Rogers ltr to CMC, dtd 20Dec50.

ever, were hard-used veterans of the Saipan campaign.

The tank-infantry teamwork on Saipan had been potent, so every effort was made to get the tanks ashore on Tinian at the first opportunity. By H-Hour the LSD *Ashland* had unloaded an initial cargo of tank-bearing LCMs and was off to Saipan to pick up armor of Major Charles W. McCoy's 2d Tank Battalion. At 0800, the LSD *Belle Grove* departed Tinian for the same purpose, and both ships were back before noon.

The landing of the 4th Tank Battalion, however, was somewhat hampered by conditions at White Beach 2, where the fissures in the reef proved more treacherous than was expected. One of the bulldozers, which were to prepare the beach, became irretrievably caught in a pothole shortly after emerging from an LCT. The threat of holes in the reef, when added to the danger of beach mines, led to a temporary re-routing of bulldozers and tanks over White 1 instead; there, at low tide, the LCTs and LCMs could safely unload their cargo on the fringing reef. From the smaller beach, some of the dozers and tanks crawled the 1,000 yards overland to White 2. Most of the division vehicles were likewise put ashore over White 1; in fact, except for some tanks, all vehicles were unloaded on the reef at White 1 on J-Day after a mishap to two vehicles—they were lost when the LCMs transporting them swamped at the reef edge off White 2.

In view of the crowding at the smaller landing area, one LCM debarked a tank for a trial run to White 2 at 1100. But the tank required 45 minutes to negotiate the 100 yards from reef to beach. Moreover, upon arrival, the crew learned that the beach was not yet quite cleared of mines; a tank crewman and a reconnaissance man had been wounded when a jeep ran over and detonated an undiscovered mine. By afternoon, however, the beach was confirmed as clear, and one entire tank company (Company A) was landed over White 2 without loss.

Except for Company A, the 4th Tank Battalion landed over White 1; there, the LCMs could move to within 15 yards of the beach at some places. Eight of the medium tanks belonging to Company B, which was attached to the 24th Marines, led infantry attacks on J-Day. Tanks of Company C, attached to the 23d Marines, were landed after direct movement from Saipan in LCTs. Company D, the light (flamethrowing) tank unit, also made a shore-to-shore journey in LCMs and one LCT.[9] Once ashore, light tanks were divided among the three medium tank companies. The initial unit ashore was the 1st Platoon, which landed at 1345 with two M5 tanks and four flamethrower tanks, and was attached to Company B. One of the M5s and two of the flame tanks were immediately dispatched to 1/24, in the area just north of White 1, where canister from the M5s helped to clean the enemy out of some heavy underbrush. The 2d Platoon was attached to Company C, and the 3d Platoon, landed at 1630, joined Company A. One platoon of four flame tanks of the 2d Tank Battalion reached Tinian in

[9] The flamethrowing tanks used at Tinian were M3A1 light tanks, which had their turret-mounted 37mm guns and ammunition racks removed and Ronson (Canadian) flamethrowers installed to replace them.

two increments at 1700 of J-Day and early the next morning. No other 2d Battalion armor came ashore on J-Day, but all organic tanks of the 4th Marine Division were on Tinian before dark. The Headquarters and Service Company of the 4th Tank Battalion, embarked in an LST, landed during the late afternoon.

J-DAY LOGISTICS

In addition to the tanks, four battalions of artillery and the 75mm half-tracks belonging to the weapons company of each assault regiment were put ashore on J-Day. The artillery pieces were successfully carried in DUKWs from the LSTs directly to firing positions. The 1st Battalion, 14th Marines landed at 1315 on White 2. The unit went into position about 300 yards inland from the southern end of the beach, and by 1430 its 75mm pack howitzers were supporting the 25th Marines. The two battalions of the 10th Marines, which were attached to the 14th Marines, landed next—2/10 at 1630 to reinforce fires of 2/14, and 1/10 at 1635 to serve with 1/14. At 1600, Colonel Louis G. DeHaven set up the regimental command post behind an abandoned railroad embankment north of White Beach 2.

After landing the artillery, the DUKWs unloaded ammunition from ships off the White Beaches, and drivers became expert at taking the loaded DUKWs through surf that ranged from four to six feet in height. It was the amphibian trucks and LVTs which were used principally for the landing of priority combat supplies on J-Day. No heavy trucks and none but essential light vehicles, such as jeeps, were put ashore. As it turned out, the small White Beach 1 had to absorb most of the landing effort on J-Day, and it inevitably became somewhat congested. Still, LVTs and DUKWs got up to the front line with ammunition, barbed wire, rations, and water.

Two pontoon causeway piers had been assembled at Saipan. These floating docks were towed to Tinian the afternoon of J-Day, but it was 0600 on 25 July, after Seabees worked all night on the job, before the first one, installed at White Beach 1, went into use. The pier for White Beach 2 was not emplaced until three days later, because of heavy enemy artillery and mortar fire, which impeded the task and caused several casualties. Each pier carried two tractors, anchors, chains, and mooring wire. Once the causeways were secured in place, LSTs and smaller landing craft could drop their ramps on the pier ends and run loaded cargo trucks ashore.

Advance elements of the shore and beach parties landed with the assault battalions and were ashore with communications established by 0830. All other men of these units were landed by 1000, for the need of prompt beach development was well realized. Shore party equipment had been preloaded in two LCTs, and all of that was ashore by 1400, much of it routed over White Beach 1 and thence overland to the other beach.

The bulk of the shore party on White 2—Major John H. Partridge's 2d Battalion, 20th Marines—was kept busy inland at first, unloading LVTs and DUKWs for the 4th Division dumps. In general, these dumps were located

MEDIUM TANKS *move through a beach exit at WHITE 1 toward Ushi Point airfields. (USN 80-G-237343)*

2D DIVISION MARINES *advance through a canefield on Tinian flushing enemy snipers as an OY stands guard overhead. (USMC 87890)*

inland of White 2, whereas the 2d Division dumps were set up behind White 1. Just below White 1 late on J-Day, a special portable LVT ramp was set up; it helped some of the tracked vehicles climb onto the land. An LVT carrying another ramp struck a coral head on the edge of the reef and turned over. The next day eight ramps were launched, though two were swamped.

During J-Day, the 4th Division Shore Party commander, Lieutenant Colonel Nelson K. Brown, commanding officer of the 20th Marines, and the Group Beachmaster, Lieutenant Samuel C. Boardman, USN, supervised operations on shore, both personally and from a radio-equipped tender. The next day the shore party headquarters was established at a point between the two beaches. General Cates remained at his command post on *LST 42* the first day, finding communications excellent from a radio jeep on deck.

Landing operations were discontinued for the night after 1/8 and the 23d Marines were fully ashore. The only elements of the division support group to land on J-Day were the Headquarters and Service Company and Company D of the 4th Medical Battalion, which, preloaded in LVTs on Saipan, landed over White 1 about 1630. Before that time, battalion and regimental aid stations had been set up, of course, and they handled the early evacuation of casualties to transports, usually by jeep ambulance loaded in an LVT. The division engineers, except for a platoon attached to each landing team, did not get ashore until 25 July. Aside from the division command ship, the control vessels, and three LSTs retained for emergency unloading, all vessels and landing craft retired to Saipan for the night. Sufficient initial supplies of ammunition, water, rations, and medical necessities had been landed prior to darkness.

J-Day had been a memorable day in the history of Navy and Marine Corps amphibious accomplishments. More than 15,600 American troops and their primary combat materiel had been put ashore efficiently over beaches which the Japanese had regarded as impassable for a major landing—and, in fact, mostly over the very beach at which the enemy utterly scoffed.[10] On the next day, the 2d Marine Division would begin landing on its own vast scale. Before then, however, the Marines of the 4th Division, remembering the great *banzai* charge on Saipan, waited seriously but calmly for the Japanese counterattack, expected to come the first night.

THE JAPANESE COUNTERATTACK [11]

No one supposed that Colonel Ogata had remained long in ignorance of

---

[10] Some idea of the logistical miracle of J-Day on Tinian may be gained from the following partial list of items unloaded prior to 2148: 99 DUKWs loaded with 75mm howitzer ammunition; 48 DUKWs loaded with 75mm gun ammunition; 48 medium tanks; 15 light tanks; 6 jeeps; 6 radio jeeps; 20 1-ton trucks; 12 2½-ton trucks; 7 bulldozers. TransDiv 7 AR, dtd 1Aug44.

[11] Additional sources for this section include: NTLF G–2 Jnl, 27May–13Aug44, hereafter *NTLF G–2 Jnl;* Anx A to NTLF G–2 Periodic Rpt No. 46, dtd 30Jul44; 4th MarDiv D–2 Periodic Rpt, dtd 25Jul44; 2/23 Jnl, 7May–24Aug44, hereafter *2/23 Jnl;* 2/24 Narrative of Bat of Tinian Island, n.d., hereafter *2/24 Narrative Tinian.*

where the Marines were invading. At his command post atop Mt. Lasso, he dwelt as a virtual prisoner of American gunfire, behind thick smoke, which clouded observation but not discernment. At 1000 on J-Day, he issued orders for the reserve *1st Battalion, 135th Infantry Regiment*, the Mobile Counterattack Force, to move north to the Mt. Lasso area. Still, he could not quite divorce himself from the set belief that at Tinian Town lay the greatest threat. There, while Captain Oya, hearing of the landing, fumed because his fixed guns could not be turned northwest, word was received from Colonel Ogata for the *3d Battalion, 50th Regiment* to stay in position. Also kept static initially, as the Marines moved inland, was the *2d Battalion*, assigned to defend the Asiga Bay vicinity. In the colonel's judgment, that was the only alternate invasion area.

Elements of the *56th Naval Guard Force* had been stationed in various parts of the island or had operated patrol boats off shore. The naval troops in southern Tinian were kept at their coastal defense artillery or antiaircraft guns on J-Day. Near the White Beaches, there were also some components of the force, chiefly antiaircraft personnel trained additionally as infantry. Other Japanese units at hand for a counterattack were the tank company of the *18th Infantry Regiment* and a company of engineers (both attached to the *50th Infantry Regiment*). The engineers were trained to double as riflemen. Finally, Ushi Airfield harbored 600 to 1,000 naval troops, who were charged with maintenance and defense of that base.

Marines had already met the well-trained soldiers of the *50th Infantry* early on J-Day besides some naval and aviation personnel who bore no unit identification. The Americans had also felt the fire of the *2d Artillery Battery*; then, on the night of 24 July, the *1st Artillery Battery*, under cover of darkness, lugged its pieces from the Asiga Bay area.

That the *1st Artillery*, or some of the other Japanese units, moved at Colonel Ogata's bidding is improbable, for enemy communications were extremely poor after the American bombardment. Every Japanese commander remembered, however, what he had been told a month before—that when the Marines land, "destroy the enemy at the beach." [12] Lacking any contrary order, his duty appeared plain.

The most sizable enemy movement was that of the *1st Battalion, 135th Infantry Regiment*—the Mobile Counterattack Force—believed to number more than 900 men, assembled near the Marpo radio station, about two miles northwest of Tinian Town. So cleverly did Captain Izumi move his battalion more than four miles that only once did a hedge-hopping aerial observer see some marching men beneath the trees. Unobserved fire fell along the route, but the troops plodded on, moving by squads, chiefly along tree lines between cane fields, avoiding the open roads.

A day of periodic drizzle was followed by a night of pitch darkness. Close to midnight the Marines, who were waiting for the expected enemy attack, noticed that the incoming fire

[12] *DefFor Plan.*

changed from an occasional mortar shell to an increased number of rounds from heavy field guns. At 0200, men of the 1st Battalion, 24th Marines made out a compact group of the enemy some 100 yards away and opened fire. The battalion was then occupying the extreme left flank of the Marines' front, anchored on the coast. The officer commanding Company A expressed the belief later that the Japanese, who had come from the Ushi Airfield vicinity, were marching along the beach road, quite innocent of the fact that they were so near the invaders they sought.[13]

Startled to receive Marine fire sooner than they expected, some 600 Japanese naval troops hastily deployed to attack. Their small figures emerged from the darkness into the bright light of Marine flares. The enemy here was not Colonel Ogata's professional infantry; in fact, the white gloves of some of the naval officers gave a curious dress formality to the scene of carnage after more than three hours of bitter fighting left 476 Japanese dead.

At the beginning of the battle, the Japanese tried to rush the prepared Marine positions, charging into the canister of 37mm guns, machine gun fire, mortar shells, and rifle fire.[14] The enemy's weapons consisted mainly of rifles, hand grenades, and machine guns taken from aircraft. The Marines' Company A received the most pressure, being reduced to about 30 men with usable weapons, but the company well answered the enemy fire—the next morning showed that most of the Japanese dead lay forward of its lines.

In the declining phase of the battle, a platoon of Marine medium tanks moved up, while Marine artillery of 2/14 (Lieutenant Colonel George B. Wilson, Jr.) registered on the area behind the enemy, preventing retreat or reinforcement. A number of Japanese suicides by grenade signified collapse of this section of the enemy counterattack, and by 0700 the Marines were through mopping up. In that job, armored amphibians helped.

The counterattack on the left had been repulsed with no enemy breakthrough, but at the center the boundary between the 24th and 25th Marines proved insufficiently covered. The enemy's approach to the center, heralded by artillery fire, was observed shortly after midnight by a 15-man combat outpost of 2/24, stationed about 400 yards to the front. They reported Japanese in great numbers.

At 0230 the vanguard of this enemy force, including a few tanks, attacked near the boundary of the two Marine regiments, specifically on the left flank of 3/25, but was stopped by a fusillade of small arms, mortar, and 37mm fire. The battalion commander, Lieutenant Colonel Chambers, reported that "in the light of subsequent develop-

[13] Maj Irving Schecter interview with Hist Div, HQMC, dtd 2Jan51.

[14] In preparation for the counterattack, Lieutenant Colonel Lessing of 1/24 had emplaced attached half-tracks and 37mm guns directly in the front line, where, that night, "gun crews fought in the dual role of gunners and riflemen." Col Otto Lessing ltr to CMC, dtd 11Dec50.

ments" this thrust "appeared as a feint."[15]

A second attack followed, which involved elements of the *1st* and *2d Battalions, 50th Infantry,* and of the *1st Battalion 135th Infantry*—equipped with new rifles and demolition charges. About 200 of these well-trained foot soldiers broke through the lines of Company K on the extreme left of 3/25. After getting to a swamp which was not covered by machine guns, they paused and divided into two groups.

One group headed straight for the artillery positions of 2/14 near the beach. Here Battery D, receiving the impact of the charge, employed not only howitzers but also machine guns to stem it. As the surviving Japanese stole doggedly closer, despite the fire, gunners of Batteries E and F turned infantrymen, leveling enfilading fire from their .50-caliber machine guns into the area forward of Battery D. That fire was conclusive—it "literally tore the Japanese to pieces," said the battalion executive officer.[16]

At 0400 Colonel Hart, commanding the 24th Marines, asked that 1/8 dispatch a company to help protect division artillery. The riflemen of Company C found the situation quite improved; a platoon of tanks also arrived to mop up any surviving Japanese behind the lines. The morning light, replacing flares and star shells,[17] showed some 100 enemy dead in the area. The cost to the Marines had been two men of Battery D, killed while manning machine guns.

The other group of the Japanese breakthrough force fared no better. After turning west into the rear areas of the 25th Marines, the enemy was stopped by a support platoon of 3/25, employing machine guns. Some of the Japanese, caught in a wooded area near Company K, were destroyed by 60mm mortars shells lobbed into their midst.

The enemy's push at the center of the Marine line had cost them approximately 500 dead. Many of these were Japanese that got caught on the barbed wire forward of the line and were cut down by machine gun fire. Identification of the Japanese that fell while attacking the center showed that most of them were of the *1st Battalion, 135th Regiment.* Some were engineers, armed and fighting as infantry, and just as dangerous with the rifle, bayonet, or grenade.

The counterattack on the right, or south, took the form of a mechanized thrust. Up the coastal road, which was crossed near the end by the lines of the 23d Marines, moved five or six Japanese tanks,[18] each transporting some infantrymen and camouflaged with leaves and branches. Other Jap-

[15] "Combat Narrative of Tinian Operation," Encl B to 3/25 AR (Saipan-Tinian), dtd 19Aug44.

[16] Maj William McReynolds ltr to CMC, dtd 8Jan51.

[17] Naval supporting ships lit up the entire length of the Japanese counterattack—left, center, and right. The destroyers *Monssen*, *Eaton*, and *Conway* provided constant illumination.

[18] 4th MarDiv IntelRpt, Anx B to *4th Mar Div OpRpt Tinian*, gives six as the number of attacking Japanese tanks; the battalion commander of 2/23 has stated that only five tanks actually took part, an opinion supported by other witnesses. Colonel Edward J. Dillon interviews with HistDiv, HQMC, dtd 25Sep50 and 22Jan51.

anese soldiers followed on foot, marching over the hard white coral in the total darkness of the night. The tanks represented half the armor of the tank company attached to the *50th Regiment*—in fact, half of the enemy's entire armor on Tinian, all of which consisted of light tanks mounting 37mm guns and 7.7mm machine guns.

Marine listening posts reported the approach of the tanks; the stepping up of Japanese artillery fire and patrol activity had already indicated that some sort of attack was due here. At 0330 the enemy tanks were observed 400 yards forward of the Marine perimeter, specifically that section guarded by 2/23. The Japanese column then ploughed right into the weird daylight created by naval star shells, to receive at short range the fire of bazookas, 75mm half-tracks, and 37mm guns.[19] The scene could be described only by someone that had seen it:

> The three lead tanks broke through our wall of fire. One began to glow blood-red, turned crazily on its tracks, and careened into a ditch. A second, mortally wounded, turned its machine guns on its tormentors, firing into the ditches in a last desperate effort to fight its way free. One hundred yards more and it stopped dead in its tracks. The third tried frantically to turn and then retreat, but our men closed in, literally blasting it apart. . . . Bazookas knocked out the fourth tank with a direct hit which killed the driver. The rest of the crew piled out of the turret, screaming. The fifth tank, completely surrounded, attempted to flee. Bazookas made short work of it. Another hit set it afire, and its crew was cremated.[20]

Such was the fate of five of the tanks, which, being visible over a wide area, received fire even from the attached 37mm platoon of the regimental reserve, 1/25.[21] Despite the concerted Marine fire, however, a sixth tank at the far rear was believed to have fled undamaged.

The catastrophe which befell their armor did not break the fighting will of the surviving infantry, dedicated veterans of the *1st* and *2d Battalions, 50th Regiment*, and of the *1st Battalion, 135th Regiment.* Unwavering before the canister of 37mm guns and machine gun fire, they charged the lines of 2/23 and 2/25, the former unit receiving the hardest thrust of the assault. A few of the Japanese even got through to engage at savage combat the Marines of the regimental reserve, 1/23 (Lieutenant Colonel Ralph Haas), positioned to provide depth here. But total destruction was the fate of the enemy's infantry, no less than of their tanks. In the last hopeless moments of the assault, just at dawn, some of the wounded Japanese destroyed themselves by detonating a magnetic tank mine, which produced a terrific blast. Evidently, these men had been ordered to break through and de-

[19] Col Louis R. Jones had fortunately reinforced the position of the 2d Battalion with the 37mm gun platoon of the 3d Battalion, thus doubling the 37mm firepower of 2/23.

[20] Lt Jim G. Lucas, AsstDiv PubRelO, quoted in Proehl, *4th MarDiv History*, p. 101.

[21] The parent unit of a 37mm gun platoon was the regimental weapons company, which, besides three 37mm platoons (one ordinarily assigned to each battalion), had a platoon of 75mm half-tracks. Bazookas, which showed well what they could do in stopping a light tank attack, were carried by teams of a rifle company.

214-881 O-67—26

molish Marine tanks at the rear of the lines. Some of those tanks were of Company C, 4th Tank Battalion, which helped to mop up the forward area after the battle.

The Japanese effort here had cost them 267 casualties. The number of counted enemy dead resulting from the total counterattack came to 1,241, of which some 700 were irreplaceable infantrymen of organized units. Such a loss represented one-seventh of Colonel Ogata's entire defense force and signaled the virtual extinction of the Mobile Counterattack Force.[22] The percentage did not include those casualties which the enemy suffered during the bombardment of Tinian or from the landing and initial advance of the Marines on J-Day.

In retrospect, General Cates felt that by more than withstanding the organized counterattack, the Marines "broke the Jap's back in the battle for Tinian."[23] The victory certainly proved decisive, yet on the morning of 25 July no Marine believed the fight was over. As a matter of fact, the *50th Infantry Regiment* was still largely intact and composed of well-equipped troops. Its entire *3d Battalion* had not yet been committed.

[22] Only two POWs were taken from *1/135* in the counterattack. They said that their battalion was practically annihilated, that only a few stragglers could have remained.

[23] *4th MarDiv OpRpt Tinian*, Sec IV, p. 25, dtd 25Sep44.

Southward on Tinian[1]

THE 2D DIVISION GOES ASHORE

The landing of the 2d Division on 25 July was partly accomplished before the 4th Division resumed its advance that morning. The Japanese counterattack had depleted ammunition stocks and necessitated some reorganization of the Marine units that had been involved; the attack hour was delayed, therefore, from 0700 to 1000.

First to land was the 8th Marines, less its 1st Battalion already on Tinian. A double column of LCVPs carried the men from the transports to the reef off White Beach 1, where they waded the last 100 yards to the shore. The 2d Battalion had landed by 0922, and by 1100, Colonel Clarence R. Wallace's entire regiment was ashore. During the landings, Japanese mortar and artillery fire, directed from the enemy observation post on Mt. Lasso, plagued the troops and caused some boat damage. American naval gunfire and artillery eventually quieted the enemy guns.

With the remainder of the 8th Marines coming ashore, 1/8 reverted to its parent regiment at 0920. The battalion had begun the day under the control of the 24th Marines, which had ordered it to relieve 1/24 along the coast on the extreme left flank of the beachhead. As the 8th Marines landed, the regiment was attached to the 4th Division and given the northernmost sector of the front.

At noon, after the 8th Marines had cleared the beach, the 2d Marines began landing and by 1755 was bivouacked some 500 yards inland.[2] The 6th Marines completed the loading of personnel and vehicles in LSTs at Saipan on 25 July and moved to the transport area off Tinian, but, except for 2/6, the regiment stayed on board ship until the next day. At 1745 of the 25th, the 2d Battalion was ordered to land on White Beach 2 and, upon moving to an assembly area 700 yards inland, it was detached from the 6th Marines and designated division reserve. General

[1] Unless otherwise noted, the material in this chapter is derived from: *TF 51 OpRpt; TF 52 OpRpt; TF 56 OpRpt; NTLF OpRpt;* NTLF OpOs 31–44 through 37–44, dtd 24–30Jul44; NTLF G–2 Periodic Rpt 46, Anx A, dtd 31Jul44; NTLF G–3 Periodic Rpts 41–46, dtd 25–31Jul44; *2d MarDiv OpRpt Tinian;* 2d MarDiv D–2 Periodic Rpts 75–80, dtd 27–31Jul 44; 2d MarDiv D–3 Rpts 72–77, dtd 25–30Jul 44; *2d MarSAR; 6th Mar SAR; 8th Mar SAR Tinian;* 4th MarDiv D–2 Periodic Rpts 71–78, dtd 24–31Jul44; *4thMarDiv Translations (Tinian);* Chapin, *4th MarDiv in WW II;* Crowl, *The Marianas;* Hoffman, *Tinian;* Isely and Crowl, *Marines and Amphibious War;* Johnston, *Follow Me!;* Morison, *New Guinea and the Marianas;* Proehl, *4th MarDiv History;* Smith and Finch, *Coral and Brass;* Stockman and Carleton, *Campaign for the Marianas.*

[2] Pending the commitment of the 2d Division, it was, except for the 8th Marines, in the status of NTLF reserve.

Watson, commanding the 2d Division, left the *Cavalier,* and at 1600 of 25 July, set up his command post on land. Division armor moved ashore during the evening. The landing of the 2d Tank Battalion had not been rushed; the tanks were not immediately required, and the beach congestion would not permit rapid deployment.

Before nightfall on the 26th, the 2d Division shore party, 2/18, had unloaded from LSTs a two-day reserve of rations, water, hospital supplies, and three units of fire. The work then went on under floodlights as a round-the-clock schedule was begun. On Saipan, the resupply machinery started to function when LCTs, with preloaded trucks and trailers, flowed toward Tinian.

THE SECOND DAY OF THE BATTLE[3]

The objective of the 4th Division on 25 July was the 0–2 Line, which began at a point about 1,200 yards north of White Beach 1; it extended south through the middle of northern Tinian and formed a juncture with the Force Beachhead Line (FBHL), which lay like a relaxed rope below Mt. Lasso, crossing the island east to west. The 0–2 Line had first been mentioned in General Schmidt's operation order for 25 July, when he directed General Cates to seize the 0–1 Line and then "on division order seize the division 0–2 line and be prepared to seize FBHL on NTLF order."[4] The landing plan had designated but two objectives: the 0–1 Line and the FBHL. (See Map 22.)

The withdrawal of Japanese troops from some areas to a new line, which Colonel Ogata had fixed south of Mt. Lasso, considerably eased the Marines' task on 25 July. The advance of 1/8 up the coast, however, was hindered by coral rocks and thick undergrowth and was not made easier by certain survivors of the counterattack, who harassed the Marines with rifle and machine gun fire from holes and caves. At 1115, a pocket of 20 to 25 well-hidden Japanese briefly checked the advance at a spot where tanks could not operate and where the fire of armored amphibians was not effective.

With the front of the 8th Marines expanding, Colonel Wallace committed the 2d Battalion on the right of 1/8 and ordered it to attack to the east. Units of 2/8 were soon at Airfield No. 1 and found the prized area weakly defended; most of the Japanese had left to join the counterattack of the night before, never to return. The battalion reached the middle of the airfield, and at the end of the day, made contact with 1/24, some 400 yards to the south. Colonel Hart had taken the 1st Battalion out of reserve to cover a gap between 3/24 and the 8th Marines for the night's defense.

The 3d Battalion, 24th Marines, moving out to the east, had reached the 0–2 Line at 1025 with no opposition. The unit then turned south along the

[3] Additional sources for this section include: *1/8 Rpt Tinian;* 1/24 Rpt of Ops, dtd 25Aug44; 2/24 AR, dtd 5May45; *2/24 Narrative Tinian;* 3/24 AR, dtd 5May45; 3/24 Narrative of Tinian Is Op, dtd 5May45; 1/25 Rpt on FORAGER, Phase III, dtd 19Aug44; 3/25 CbtNarrative of Tinian Op, n.d.; *2d TkBn SAR.*

[4] NTLF OpO 31–44, dtd 24Jul44.

objective to support 2/24, which was receiving small arms fire while advancing toward Airfield No. 3. By mid-afternoon, the 2d Battalion was at the 0–2 Line, which crossed the airfield. The strip and the adjoining buildings were found to be abandoned.

At the end of the second day on Tinian, the 24th Marines was in contact with the 25th Marines on the right and the 8th on the left. That night, the Japanese attempted only petty infiltration, but a sharp clash occurred when Marines of a regimental combat outpost near a road junction ambushed an enemy patrol. Manning the outpost was a platoon of the division reconnaissance company, attached to the 24th Marines for the night.

By the evening of 25 July, the 23d Marines had advanced halfway to Faibus San Hilo Point. The 1st Battalion had relieved 2/23, which passed to division reserve, and then moved through cane fields and underbrush against the light opposition of Japanese stragglers from the counterattack. The 0–1 Line was reached at 1637, and a position in advance of it was secured before dark. The 3d Battalion, the regimental reserve which had followed 1/23 during the daylight hours, moved up for the night to relieve left elements of the 1st Battalion.

The hardest fighting on 25 July took place at Mt. Maga and involved the 25th Marines, advancing at the center of the division line.[5] Mt. Maga lay just inside the 0–1 Line and stood astride the path of the regiment. That side of the hill which rose before the advancing Marines was the most precipitous one; Colonel Batchelder saw that a frontal assault would be arduous, and probably costly in Marine lives. He settled, therefore, upon the tactic of a double envelopment, using the 1st Battalion on the left and the 3d Battalion on the right. The 2d Battalion would hold to the front of the hill, delivering suppressive fire upon it.

While the Japanese were retiring from other sections of their defenses, they still clung to Mt. Maga. Marines of 1/25 were able to get safely into position at the foot of the hill, but when they tried to climb the east side, they were opposed immediately by such a hail of rifle and machine gun fire that Lieutenant Colonel Mustain ordered withdrawal. A road to the peak was then discovered, and engineers searched it for mines. When the path had been cleared, tanks made a strike on top of the ridge, but after being unable to locate the well-concealed enemy firing upon them, the vehicles were ordered down from the ridge. A second attempt by 1/25 drew the same violent response as the first, but now the sources were spotted. The battalion commander then employed 81mm mortar fire on the top of the ridge, while tanks fired from the hill base into pillboxes and caves in the face of the cliff. These fires did the trick. At 1200, the infantrymen again started up the hill, encountering much less resistance. Once at the top, however, the Marines received considerable fire from Japanese positions to the front. As there were yet no friendly units either on the right or left, Colonel Batchelder ordered 1/25 to hold up the

[5] Battalion action reports referred to Mt. Maga as Hill 440, but a captured Japanese map put its height at 390 feet.

attack. At 1330, the enemy succeeded in setting up machine guns and mortars on the open right flank, forcing 1/25 to withdraw 200 yards under this fire. The ground was soon retaken, however, after Marine mortar and machine gun fire, helped by 75mm fire from the tanks, destroyed the Japanese positions. Two hours later, Colonel Hart ordered the 1st Battalion to continue the attack to the right front, encircle Mt. Maga, and join forces with the 3d Battalion before digging in for the night.

As 3/25 had started along its envelopment route, the movement was delayed by enemy fire from the hill, causing Lieutenant Colonel Chambers to order tanks and combat engineers forward of the leading Company L. The fire by the tanks, added to the work of flamethrowers, bazookas, and demolitions employed by the engineers, appreciably lessened the resistance; another delay ensued, however, while 3/25 waited for restoration of contact with the approaching 23d Marines. During the hold-up, the battalion commander requested naval and artillery gunfire upon the west slope of Mt. Maga. Under such cover, combat patrols destroyed three unmanned 47mm guns near the foot of the hill.

When the 23d Marines came abreast, Chambers ordered resumption of the attack, and by 1600, all companies of 3/25 reported being at the top, where they established contact with 1/25. The 0–1 Line in the center of the division perimeter was secured by 1715. After dark, a few bypassed Japanese attempted vainly to get through the Marine lines. The mop up of the Mt. Maga area was left to 2/25, which finished the task by noon the next day, when the battalion was put into regimental reserve.

Casualties in the assault of Mt. Maga had been light, but a tragic toll resulted elsewhere on 25 July, when at 0920, a Japanese 75mm shell exploded on the tent pole of the Fire Direction Center, 1st Battalion, 14th Marines. The battalion commander, Lieutenant Colonel Harry J. Zimmer, was killed, as were the intelligence officer, the operations officer, and seven assistants. Fourteen other members of the battalion headquarters were wounded. Major Clifford B. Drake, the executive officer, assumed command.

During the same morning, enemy artillery fire was laid upon the pier, under construction by Seabees, at White Beach 2, causing several casualties there. As the shells were believed to be coming from Mt. Lasso, the 14th Marines directed counterbattery fire at caves in the face of the hill. During the afternoon, however, the Japanese guns were again active for a few minutes, setting fire to one DUKW and causing more casualties among men at the beach. An air strike that morning had supposedly destroyed two guns at the base of Mt. Lasso, and fire support ships had been directed to search for and silence Japanese guns in the vicinity. It was evident, however, that some well-concealed weapons had escaped the best efforts to destroy them.

Japanese power in the Mt. Lasso area, both of guns and men, was hard to measure. At a point 1,000 yards northwest of the hill, Marine air spotters saw a force, reported of battalion size, moving south. The 14th Marines took the enemy under fire, reducing the force by an estimated 25 percent; the rest of the enemy scattered into the cane fields where hiding was easy. The evasive Japanese soldier and the well-hidden gun would continue to be obstinate threats on Tinian. At the end of the second day, however, the Marines' attack was proceeding beyond expectations.

PREPARING TO DRIVE SOUTH

General Schmidt's operation order for 26 July took note of a rapidly changing picture. Although the southern half of the 0–2 Line and the entire FBHL had not been reached, the Marine commander omitted both objectives from the order. Instead, he spoke of 0–3 and 0–4 for the first time. He drew the 0–3 Line from the shore 1,000 yards south of Faibus San Hilo Point to the coast at a nearly equal distance north of Asiga Point. The line almost converged with 0–4 on the west, but the two lines diverged increasingly toward the east, finally becoming nearly 5,000 yards apart. General Schmidt put the 0–3 Line across the width of the island because it appeared that the 2d Division would reach the east coast with relative ease. After that, the two Marine divisions would be in position for the sweep to the south. The FBHL stretched across the island between 0–3 and 0–4, but the beachhead line was now omitted as being incompatible with a change of tactics then being considered at General Schmidt's headquarters. On 26 July, the 4th Division was to move toward Mt. Lasso, encompassed by 0–4A. The 2d Division, leaving NTLF reserve, would take over the left sector of the front, advance east to the coast, and envelop Airfield No. 1 in the process. (See Map 22.)

Prior to the attack hour of 0800, General Watson regained control of those of his units which had been under 4th Division control. The 1st and 3d Battalions, 6th Marines had begun landing at 0630 over White Beach 2 and were moving inland to an assembly area to await attack orders. Over the same beach, during the morning, the 2d Tank Battalion completed landing, and its elements went up to positions from which the battalion could support the 2d Division attack.

General Watson's 1st and 2d Battalions, 2d Marines relieved 1/24 and 3/24 as the battle for Tinian went into the third day. The two battalions of Colonel Hart's regiment were put into division reserve, but 1/24 was designated at a later hour as NTLF reserve. The 2d Battalion, 24th Marines was attached to the 25th Marines and committed to the left flank of that regiment to maintain contact with the 2d Division. To the right of the 25th Marines was the 23d, ready to push further down the west coast. On the left of the front, the 8th Marines waited to bring Airfield No. 1 entirely into American hands for early use.

The pace of the advance on the morning of 26 July led General Schmidt to amend his operation order shortly before noon. Instead of requiring that the division commanders wait for NTLF orders before advancing to the 0–4 Line, he permitted them to continue south of the 0–3 Line at their own discretion.

The 8th Marines crossed Airfield No. 1 on 26 July, finding it abandoned, wet, and cluttered with Japanese planes wrecked on the ground by the American bombardment. The adjoining village, which housed airfield personnel, was likewise deserted. The Marines left the airstrip to the 121st Naval Construction Battalion, and after just a few hours of clean up and repair, the Seabees had the field usable for small observation planes. Two days later, on 28 July, the first P-47 landed and took off from airfield No. 1 with no difficulty.

In the rapid advance of 26 July, Colonel Wallace had his assault battalions, the 1st and 2d, followed by 3/8, on the east coast at the 0–3 Line at 1140. That afternoon, the 8th Marines became division reserve. The next day, 27 July, the regiment took up position as NTLF reserve, but the 2d Battalion continued in division reserve the 8th Marines, Colonel Stuart took the 2d Marines to the east coast by 1230 on 26 July, at which time he realigned his regiment to begin the attack southward.

On the right of the corps front, the 23d Marines was at the 0–3 Line by 1200, despite thick cane fields and densely wooded areas along the coast. Once at the 0–3, Colonel Jones pushed on to a point well below Faibus San Hilo. His 2d Battalion was relieved from division reserve and mopped up the rear areas as the attack progressed. The resistance encountered by the 23d Marines was not heavy; it consisted mostly of isolated machine guns or individual riflemen employing hand grenades.

For the 25th Marines, Mt. Lasso was the chief objective on 26 July. As the 1st and 3d Battalions moved out from the Mt. Maga area, they expected considerable resistance on the higher hill, whose steep approaches made it a better citadel than Mt. Maga. Moreover, Colonel Ogata's command post had been set up on Mt. Lasso, and the guns on the hill had been effectively employed since J-Day. To the Marines' surprise, however, they were able to occupy Mt. Lasso without opposition; the enemy had pulled out during the night.[6]

While 1/25 climbed Mt. Lasso, the 3d Battalion, on the right, gained the 0–4A Line which circled around the hill. Lieutenant Colonel Chambers requested permission to advance to the 0–4 Line, some 1,000 yards farther south, but the regimental commander felt that a contact problem would result. The 3d Battalion 0–4A sector lay in a depression commanded by enemy positions visible on the 0–4 ridge, so the unit was pulled back 450 yards to a more favorable location. As the 1st Battalion dug in for the night, the men put a ring of defense around the summit of Mt. Lasso.

With the advance to Mt. Lasso, the Marines on Tinian had begun to outdistance the support of artillery on Saipan. Consequently, on 26 July, the 3d Battalion, 14th Marines moved across to Tinian where it was assigned the mission of general support. The 105mm howitzers of the battalion were the first artillery heavier than 75mm to land on Tinian. The next day, 3/14 was followed ashore by the 3d

[6] A Japanese POW said that Colonel Ogata switched his command post from Mt. Lasso to a cave about two miles northeast of Tinian Town.

and 4th Battalions of the 10th Marines and the 4th 105mm Artillery Battalion, VAC.[7] Colonel Raphael Griffin of the 10th Marines set up his command post on Tinian, signifying the break up of Groupment A of the Corps Artillery which he had commanded on Saipan. As the colonel's regimental units landed on Tinian, they reverted at once to control of the 2d Division.

Movement of Corps Artillery 155mm howitzers from Saipan was begun on 27 July, and the next day the first of these guns began firing from Tinian positions. General Harper, commanding XXIV Corps Artillery, moved his headquarters to Tinian on the same day, leaving on Saipan only the long-range 155mm guns which could reach any part of Tinian. The increasing abundance of Marine and Army artillery on the island was reflected in the complaint of one Japanese POW: "You couldn't drop a stick," he said, "without bringing down artillery." [8]

As the Japanese withdrew under the pressure by Marine infantry, who were now supported by intensified artillery fire, any repetition of the initial enemy counterattack seemed most unlikely. Yet, on the night of 26–27 July, there were attempts to get through the lines of Lieutenant Colonel Richard C. Nutting's 2d Battalion, 2d Marines from both the front and rear. While enemy troops probed and poked along the entire battalion front, other Japanese, presumably some that had been bypassed, tried to break through the Marine rear areas, evidently hoping to get back to their units. A party of about 60 such Japanese, armed with a light machine gun and grenades, fell upon Company F from the rear and was destroyed. The enemy's activity cost him 137 dead, while the Marine battalion suffered 2 men killed and 2 wounded.

Despite the incident involving 2/2, the withdrawal of the enemy was becoming obvious. Marine patrols lost contact, so rapidly were the Japanese pulling back before the American advance. General Schmidt had his troops well forward, and the two Marine divisions were now spread across the width of the island.

After appraising the situation on 26 July, the NTLF commander decided to use elbowing tactics. In other words, he would not employ both divisions equally each day, but instead, would charge just one division with the main effort while the other made the secondary attack. On the following day, the roles would be switched; it would be like a man elbowing his way through a crowd.

By adopting such tactics, General Schmidt could put the bulk of the artillery support behind a single division. Each was to have a different attack hour; that is, the division chiefly involved that day would jump off at 0700 or 0730, while the other waited until 1000 to attack. The 0–4 Line lay much farther from the 0–3 on the east than on the west, so General Schmidt picked the 2d Division to receive the strongest support the next day. Then, looking ahead to 28 July, he drew the 0–5 Line farther from the 0–4 on the west than on the east, permitting a shift of em-

[7] The 4th 105mm Artillery Battalion, VAC, was referred to in 14th Marines reports as 5/14, its original designation.

[8] *4th MarDiv Translations (Tinian).*

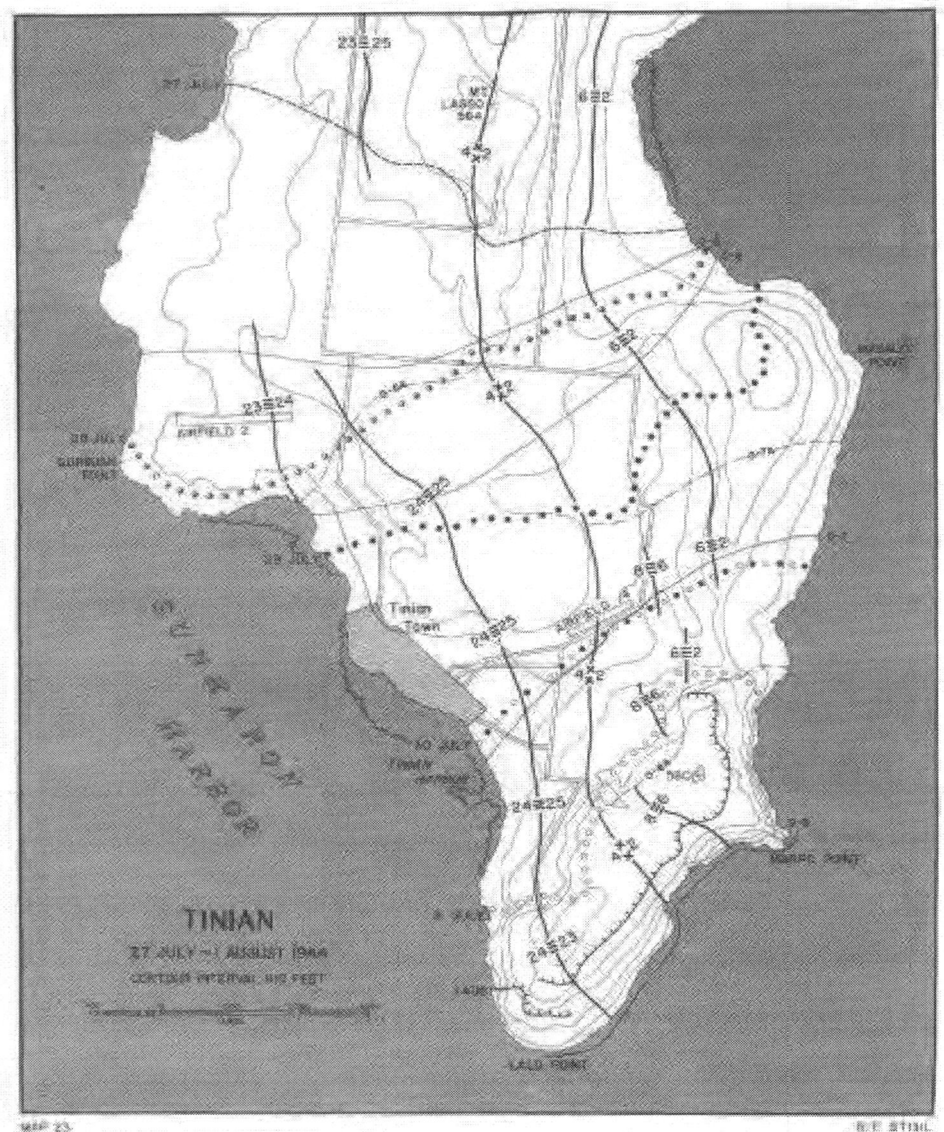
TINIAN
27 JULY–1 AUGUST 1944
Tinian Town
AIRFIELD 2
MAP 23
R.F. STIBIL

phasis to the 4th Division. Two days of elbowing tactics would be tried. After that, plans would have to be reviewed against the situation. (See Map 23.)

27–28 JULY—"MAGNIFICENT WORK" [9]

At 0730 on 27 July, General Watson moved out to the attack, employing the 2d Marines on the left, along the east coast, and the 6th Marines on the right. The advance lay mostly across rolling farm country, marked by cane fields and patches of woods. Only scattered rifle and machine gun fire was encountered, and by 1345 the two assault regiments were at the 0–4 Line. The 1st Battalion, 2d Marines, on the extreme left, had moved 4,000 yards along the coast of Asiga Bay. Marines then sent patrols forward about 500 yards; only five Japanese were found by a patrol from 2/2. The strong positions near Asiga Bay had been abandoned without a fight, thanks considerably to naval gunfire. Just the day before, the battleship *Tennessee* had demolished a blockhouse by main battery fire.

In the 4th Division zone on 27 July, the 23d Marines continued he attack at 0950, and meeting no enemy resistance, was at the 0–4 Line an hour later. Both the 1st and 3d Battalions then sent patrols up to 1,000 yards forward of the line, but none of them reported enemy activity, and the regiment consolidated positions on 0–4 for the night. To the left, along the division boundary, the 25th Marines had moved out at 1000, with 2/25 following the assault battalions at 500 yards. Opposition was negligible, and the 0–4 Line was gained by 1200. The progress of the 2d Division during the day closed the previously existing gap on the 4th Division left flank, so 2/24 was pinched out and moved into regimental reserve, still attached to the 25th Marines.

It was planned to bring the 24th Marines back into the lines on 28 July, and with a view to that, General Cates took 3/24 out of division reserve. The 1st Battalion was returned from NTLF reserve to its parent regiment. At 1800 of 28 July, 2/24 was detached from the 25th Marines and replaced 1/24 in the corps reserve. Lieutenant Colonel Richard Rothwell, the regularly assigned commander of 2/24, had been able to rejoin his battalion, relieving Major Garretson, on 27 July, after being in the hospital during the first three days of the campaign.

Scarcity of opposition to the 4th Division advance on 28 July—the 0–5 Line was reached by 1250—made it unnecessary to use any unit of the 24th Marines that morning. Not until 1300 was the regiment put into attack position between the 23d and 25th Marines, and it was then utilized because the island expands to its widest where the 0–5 Line was located. About two miles south of the line, however, a bay cuts into the coast, narrowing the island. Here was where General Cates desired to end the day's advance, at a shorter and more defensible line, and he received permission to go beyond 0–5 to a line he designated 0–6A.

After naval gunfire and artillery pre-

[9] General Holland Smith departed for Guam at 1800 on 28 July. Before leaving, he sent a message to General Schmidt: "Magnificent work. Keep the heat on." NTLF G–3 OpDisps 27–29Jul44.

pared the forward area, the 24th Marines moved out at 1325, initially in a column of battalions with 1/24 in assault. As the advance progressed, 3/24 was put into the expected gap that developed between the 24th and 25th Marines.

With resistance light, 0–6A was reached at 1730, when the 23d Marines, pinched out by the bay indentation, reverted to division reserve. The regiment had moved 7,300 yards on 28 July; the spectacular advance was "accomplished in blitz fashion," said the regimental report, "with troops riding on tanks and in half-tracks." [10] Progress of the regiment beyond the 0–5 Line had enveloped Airfield No. 2 near Gurguan Point. The field was secured at 1420 against ineffective fire from Japanese small arms and light automatic weapons.

Throughout the day, the armor of the 4th Tank Battalion had led the attack, demolishing cane stalks and other tropical vegetation to gain the infantry fields of fire. At such work the medium tanks were joined by flame tanks. The fast-moving Marine infantry set a merry pace for both armor and artillery. Units of the 14th Marines had to displace frequently to avoid getting out of range.

The 2d Division jumped off at 1024 on 28 July. The 1st Battalion, 2d Marines, which the day before had advanced 4,000 yards, now found itself restrained by division to a gain of only 350 yards, the distance from 0–4 to 0–5. With 2/8 attached for the day, the 2d Marines reached the 0–5 Line at 1130, just an hour after moving out. Patrols roved 500 yards forward of the line and encountered no Japanese, but the regiment was kept at 0–5 for the night. At 1730, 2/8 was returned to its parent regiment, when the 8th Marines, less its 3d Battalion retained in NTLF reserve, became division reserve. The 6th Marines, which had more ground to cover than did the 2d Marines, reached the 0–5 Line by early afternoon. Both regiments dug in at 0–5 and reported minimum enemy activity that night: two Japanese soldiers tried to infiltrate the perimeter of the 2d Marines, and two small enemy patrols were discovered in cane fields near the 6th Marines.

LOGISTICS VERSUS WEATHER

Progress on Tinian had been all that could be wished; more than half of the island was already in American hands. It seemed too much to expect that the weather would likewise stay favorable. In fact, Admiral Hill had originally been hopeful of no more than three days of relatively quiet sea.

On the afternoon of 28 July, the period of moderate wind and rain suddenly ended. The Marianas felt the edge of a typhoon born in the Philippine Sea, and the storm caused such heavy swells off the White Beaches that unloading had to be suspended at 1800. The next day, the whipping winds prevented unloading except by LVTs and DUKWs; then, at last, by the amphibian trucks alone.

An LST ventured to dock at the pier on White 2; it debarked 24 loaded trucks with their drivers and took on a number of casualties. While retract-

[10] 23d Mar SAR, Anx H, p. 23, to *4th MarDiv OpRpt Tinian.*

ing, however, the ship was seized by a squall and broached, then ran hard aground on the reef. The casualties were transferred to another ship, but all efforts to refloat *LST 340* proved unsuccessful. The same squall washed a control craft, *LCC 25473*, onto the reef north of White Beach 1, where it was salvaged the next day.

The causeways at each of the beaches held until the night of 29 July. Then the pier at White 1 broached when the anchor chains parted, and the pier at White 2 split. The causeway at the smaller beach was restored by the evening of 31 July, but it was then broached for a second time by the heavy surf.[11]

The entire burden of unloading could not be put upon the DUKWs, efficient as they were, and besides, Admiral Hill did not propose to do that. He had foreseen and prepared for a change in the weather. Plans included the readying of approximately 30 tons of varied supplies for delivery by parachute drop. Moreover, the admiral called forward a previously alerted Army Air Forces squadron of cargo aircraft at Eniwetok to supplement the planes available on Saipan for transporting supplies to Tinian.

On 29 July, the 9th Troop Carrier Squadron at Eniwetok sent its C–47s (Douglas Skytrains) to support the emergency air-supply plan for Tinian.[12] Except for a few other priority items, only rations were actually delivered by air; on 31 July, approximately 33,000 (99,000 meals) were flown to Tinian. On the way back, the planes carried wounded men to Saipan. The 30 tons of parachute drop material, while valuable emergency resources, were never needed on Tinian.

By 28 July, the day when the good weather ended, supply requirements on Tinian consisted only of rations, ammunition, and fuel. A fourth indispensable, water, was never a problem; Marines were well taken care of by their initial supply and by the output of engineer distillation units, which at the beginning of the campaign, used a small lake near the White Beaches. As to rations, a reserve supply of approximately two days was kept undiminished, thanks partly to the airlift. The demands for ammunition rose with the flow of artillery ashore, but here again no shortages were suffered. Two ammunition ships, the *Rockland Victory*, which arrived on 26 July, and the *Sea Witch*, which anchored on the 27th, remained off shore until the island was secured, and DUKWs shuttled back and forth to keep the guns firing.

The only near supply shortage occurred in the matter of fuel. Here, the rapid advance of the Marines stepped up the estimated requirement of 400 drums a day. Beginning 27 July, a daily supply of 600 to 800 drums of fuel was provided via pontoon barge, from which the oil would be delivered to the dumps by amphibian tractors. A satis-

---

[11] The wrecking of these piers prevented the landing of the 4th Battalion, 14th Marines, which was kept on board the *Cambria* until 1 August. The unit then landed over the beaches at Tinian Town.

[12] The two-engine C–47, known as the R4D by the Navy and Marine Corps, was the aerial workhorse of World War II, useful for transporting either soldiers or cargo.

factory reserve had not been built up on shore before the weather reverse made further unloading into the tractors too risky. Only the DUKWs could then be relied upon; so, in addition to their other chores, the tough amphibian trucks undertook the transporting of fuel. Their service, coupled with the fact that much gasoline was captured from the Japanese, averted a major fuel shortage on Tinian.[13] Delivery of fuel by air, though contemplated, did not become necessary.

PROGRESS ON 29 JULY

The logistics of the Tinian campaign were spared the complication of a pressing enemy. Until the withdrawing Japanese made a stand beside their comrades in southern Tinian, the path was devoid of collective opposition. General Schmidt, who moved his headquarters to Tinian on 28 July, desired to put no restraints upon his fast-moving Marines. Let the advance be as rapid as practical—such was the essence of the orders for 29 July. The elbowing technique was abandoned; both divisions would again move out at the same time, 0700, and their commanders, after seizing the 0–6 Line, could advance to the 0–7 Line as they saw fit. The usual preparatory fires were not to be delivered on the morning of 29 July. It seemed idle to draw upon the depleted supply of artillery shells left on Saipan, or waste naval gunfire on areas largely deserted by the enemy.

General Watson did not expect to gain the 0–7 Line on the 29th because of the distance involved; 0–7 lay nearly 5,000 yards forward of the 2d Division line of departure. Instead, he fixed an intermediate 0–7A Line, 3,000 yards away. The 2d Marines and the 6th Marines both reached the 0–6 Line about 0800 with no difficulty; after that, however, fire was received periodically along the entire division front. Local resistance developed near the east coast when the 1st Battalion, 2d Marines, on the regimental left, approached a 340-foot hill on Masalog Point and was met by machine gun and mortar fire. In the center, 2/2 made good progress, and the same was true for 2/8, which had been again attached to the 2d Marines on 29 July.[14] Such relatively easy advances put those two units a few hundred yards past the 1st Battalion, prompting Colonel Stuart to bring up two companies of his reserve 3d Battalion to attack the Masalog Point elevation from the right. The companies moved through the area cleared by 2/2 and 2/8. By 1715, much of the high ground had been taken by the 2d Marines, and the entire capture of it was left to 3/2 for the next morning. The regiment dug in between 0–6 and 0–7A. The day's advance had been mostly across thick cane fields; Colonel Stuart reported a number of casualties from heat exhaustion.

Resistance to the advance of the 6th Marines on 29 July was erratic, as enemy groups kept up a constant fire from machine guns and mortars but fell back whenever units of the assault

[13] The 8th Marines, for example, captured 1,600 gallons of Japanese gasoline, 90 octane.

[14] 3/8 continued in NTLF reserve and 1/8 in division reserve.

battalions, 1/6 and 3/6, deployed to attack. By 1500, the regiment was on line just short of 0–7A but on the commanding ground of the area, so no further advance was attempted that day. During the night, a patrol of 20 Japanese tried to break into the lines of the 6th Marines; otherwise, there was no enemy activity.

To the right of the 2d Division, the advance of the 25th Marines lay across dense cane fields which impeded progress, especially when crossed diagonally. As the Marines pushed through, in the heat of the day, units had difficulty keeping contact. Scattered nests of Japanese, well-hidden in the fields, harassed the advance with rifle fire and occasional machine gun fire. Still, the 3d Battalion reached the 0–6 Line at 1030, and the 1st Battalion was there shortly after.

The 25th Marines chief encounter with the enemy on 29 July occurred after the 3d Battalion had gained 0–6 and been ordered to continue the attack. While moving along an unimproved road, Marines of the battalion came upon a number of well-dug-in Japanese, and a heavy firefight ensued, resulting in several Marine casualties before the resistance was overcome. The tanks supporting 3/25 were involved in the fight, and one light tank was knocked out by a mine. The crew was evacuated with one casualty.

Near the west coast, the 24th Marines reported no opposition before passing the 0–6 Line at 0900 on 29 July. Then the 1st Battalion, on the regimental right, encountered an enemy strongpoint, consisting of a series of mutually-supporting bunkers. They were believed to be defensive works meant to oppose a landing at Tinian Town. It took a tank-infantry assault to destroy the resistance offered by rifle fire and crossfire from automatic weapons. Company B, 4th Tank Battalion reported that the area "had to be overrun twice by tanks." [15] By 1300, the Marines of 1/24 were again able to move freely. The 3d Battalion had preserved contact with the 25th Marines, and when the 24th Marines halted for the day, at 1525, a company of the regimental reserve, 2/24, moved into a gap which had developed between 1/24 and 3/24.[16] The 23d Marines (less 2/23) continued in division reserve; to keep up with the assault regiments, it had displaced twice during the day.

As the Marines of the 4th Division dug in for the night on 29 July, some of them could see Tinian Town from their foxholes and gun emplacements. The town and Airfield No. 4 lay inside the 0–7 Line. East of Tinian Town, a valley stretched across the island. Cheerful prospects of the campaign ending were dampened, however, by a night of heavy rain which soaked everything from the ground up. Added to this unpleasantness was incessant enemy artillery and mortar fire, which kept Marine gunners replying throughout the wet and dark hours. In front of 3/25, the rustling sounds of enemy movement were heard and silenced. The next morning, 41 Japanese dead were found, victims chiefly of Marine mortar fire.

---

[15] 4th TkBn Rpt, Encl B, p. 3 to *4th MarDiv OpRpt Tinian*.

[16] 2/24 had reverted to regimental control at 0600 of 29 July, being relieved as NTLF reserve by 2/23.

30 JULY—TINIAN TOWN AND BEYOND

Inasmuch as 0–7 still lay ahead, General Schmidt had fixed no additional objective line for 30 July; he ordered simply that the divisions continue operations to complete the mission assigned. He set H-Hour at 0700 but moved it later to 0745, following a request from the 25th Marines. Colonel Batchelder had asked a delay to 1000 to permit the men time to clean and service weapons.

Preparatory artillery fire, omitted the day before, was resumed on the morning of 30 July. All battalions of the 14th Marines took part, opening 10 minutes of fire at 0735 and shelling areas just ahead of the Marine front lines. Then, at 0800, the gunners delivered five minutes of fire, lifting it to areas 400 yards farther out.

Two destroyers were assigned a preparation fire mission on the slopes just south of Sunharon Harbor from 0745 to 0845, and a cruiser was assigned to 1/24 to deliver preparatory fires in support of the attack on Tinian Town. By 1100 of 30 July, however, as Admiral Hill reported, the Marines "had advanced so rapidly that only four square miles of the island remained for safe firing by ships not supporting battalions (i.e., not with shore spotter)." [17]

For the 4th Division—specifically, the 24th Marines—Tinian Town was a significant objective on 30 July. For the Japanese, the coming of the Marines by land to the area where they had first been expected by sea must have been a regretful event; against it they could summon only a shadow of the resistance once available. Indeed, the only tangible opposition to the 24th Marines as the regiment approached Tinian Town came not from the area itself but from caves along the coast north of the town, where Japanese machine gunners and riflemen were holed up. The 1st Battalion, 24th Marines had progressed about 600 yards from its line of departure when the left flank was stopped by enfilade fire. The resistance was overcome with the help of tanks and of armored amphibians offshore. Flame tanks seared the caves, and following that, combat engineers employed demolitions. It was the approach that had become a Marine classic, and it was extremely effective. In one cave the Marines destroyed a 75mm gun sighted toward Tinian Town.

At 1000, Colonel Hart committed his reserve 2d Battalion in a column of companies, between 1/24 and 3/24, in the vicinity of the cliff line. When 1/24 resumed the advance at 1100, the 2d Battalion was assigned to follow the attack. The 3d Battalion, ordered to preserve contact on the left with the 25th Marines, advanced rapidly; in fact it got so well forward that it stretched the battalion lines, causing a temporary gap inside the unit.

At 1420, the 24th Marines reached Tinian Town, to find it virtually leveled by the American bombardment and almost entirely deserted—the population had left, and only one Japanese soldier was discovered. By 1700, Marine infantrymen had thoroughly combed the ruins and gone on to occupy the 0–7 Line south of the town. The

[17] *TF 52 OpRpt*, Pt VI, p. 78.

only enemy fire received came, it was believed, from tanks in the distance and caused no harm. Enemy emplacements in the town had been evacuated as useless, for the guns were trained to repel an attack from the Tinian Town beaches. Nearby, there were also some deserted emplacements and abandoned caves. In the streets of the town, the Japanese had left barriers, such as log barricades or timber cribs filled with stones, but none of these obstacles was sufficient to stop a medium tank.

Where the peril lay for the Marines was in the mines which the Japanese had planted. From the Tinian Town area, the engineers removed a new type of antipersonnel mine. It consisted of a wooden box containing 10 to 12 pounds of dynamite. A pressure type of igniter required an estimated pressure of 200 pounds, while a companion pull type seemed explosive with just 8 pounds of pull. The enemy had also conceived a device to make the beach mines more dangerous. Some of the horned mines—78 were removed from the Red Beach alone—were joined by rods about 20 feet in length fastened to the horns. Pressure applied by a vehicle to any part of the rods could theoretically detonate two or three of the mines simultaneously.

While the 24th Marines found Tinian Town deserted, the 25th Marines beheld the same forlorn scene at Airfield No. 4, though enemy mortar fire from beyond it was received as the Marines pushed across the strip, gaining the 0–7 Line at 1430. Prior to reaching the airfield, the Marines had met little else but scattered enemy sniper fire. The area of advance was dotted with concrete dugouts and emplacements but few Japanese.

The strip surface of Airfield No. 4 consisted of rolled coral, pocked with holes from artillery hits but repairable. Marines reported finding one small Zero-type plane. In a supply room, the Japanese had left some flying suits, helmets, and goggles. The field was still under construction; prisoners said that until the Marines came, it was being rushed to completion upon orders from Tokyo, to bring help by air. Such rumors gave enemy morale a needed lift.

The seizure of the Tinian Town airfield marked the last battle action of the 25th Marines on the island. The regiment, less its 3d Battalion assigned to division reserve, was put into NTLF reserve and continued there for the rest of the campaign. The 23d Marines relieved elements of the 25th Marines on the 0–7 Line at 1600 of 30 July; 3/23 reverted from NTLF reserve to its parent regiment.[18] The 1st Battalion, 25th Marines was relieved by 1/8 at 1800, shifting the division boundary.

On 30 July, the 2d Division had encountered fitful opposition, sometimes amazingly strong, which the Japanese offered as their hold on the island slipped away. Shortly before the attack hour, a 1/2 patrol, pinned down only 500 yards from the front lines by enemy machine gun fire, was rescued by a platoon of tanks. The offending strongpoint was destroyed by Marine artillery, removing it from the path of the battalion, which then moved rapidly

[18] 3/23 had relieved 2/23 as NTLF reserve at 1000 of 30 July.

214-881 O-67—27

south along the coast below Masalog Point.

A task for 3/2 that had been left unfinished on 29 July was the capture of the Masalog Point high ground, but most of the Japanese there had withdrawn under cover of darkness, and the Marines gained the objective early on 30 July. The battalion then hastened to catch up with 1/2 on its left. The advance of 3/2 was briefly delayed by a Japanese 70mm howitzer, which was destroyed by a combat patrol. At 1345, the 2d Marines reached the 0–7 Line. The 2d Battalion was then detached and designated as NTLF reserve, relieving 3/8. The attached 2/8 had been pinched out before noon, so both battalions of the 8th Marines were back with their parent regiment, whose 1st Battalion, however, was still in division reserve.

At 1700, after the 2d Marines dug in for the night along the 0–7 Line, the 3d Battalion began to receive enemy machine gun and mortar fire from caves in the cliffs to its rear. The positions were reduced by flamethrowers and demolitions prior to darkness. The night was quiet except for some movement to the front of 3/2, apparently from small groups of the enemy digging in caves. The Japanese attempted no fire upon the Marine positions.

The advance of the 6th Marines was mostly uneventful on 30 July. By 1245, the 1st Battalion, on the right, had reached the 0–7 Line. The 3d Battalion, however, received fire from a Japanese field piece which caused a brief delay. A combat patrol sent out to locate it was unsuccessful, but the weapon did become silent; it was probably the 70mm howitzer which Marines of 3/2 had knocked out. At 1604, 3/6 reported its position on the 0–7 Line. The 2d Battalion had been detached at 1345 when General Watson ordered it into division reserve. The 3d Battalion, 8th Marines relieved 1/6, which then went into regimental reserve. By 1830, the 8th Marines, with its 2d Battalion as regimental reserve, was in position and wired in for the night, relieving extreme left elements of the 4th Division and the rightmost elements of the 6th Marines.

THE FIGHT STILL AHEAD

The commitment of additional 2d Division troops took care of a widening in the division front, as General Schmidt prepared the concluding moves of the campaign. His operation order issued at 1730 on 30 July was more detailed than usual, and it had a single purpose—"to annihilate the opposing Japanese," now cornered in southern Tinian.[19] The two divisions, jumping off at 0830, were to seize 0–8, the southeast coastline between Lalo Point and Marpo Point.

Preparatory naval gunfire of exceptional intensity was scheduled to begin at 0600; assigned to deliver the goods were the battleships *California* and *Tennessee*, the heavy cruiser *Louisville*, and the light cruisers *Birmingham* and *Montpelier*. At 0710, the ships would cease firing for a period of 40 minutes, to permit a bombardment by 126 air-

[19] NTLF OpO 37–44, dtd 30Jul44.

craft.[20] Corps and division artillery were to step up their fires of the previous night. Once the Marines moved to the attack, all three supporting arms would be on call.

There were good reasons, indeed, for General Schmidt's cautious preparations. In the first place, the Japanese would be making their last desperate stand on Tinian, and experience indicated that it would be a very tough one. Marines speculated on whether the enemy would wait concealed, to exact a high price for the last one-fifth of the island, or stage a counterattack in a final *banzai*, the most likely tactic. A third but lesser possibility was a mass suicide by the enemy, using ammunition dumps and hoping to take some of the Marines with them.

According to a 4th Division intelligence report of 7 August,[21] based upon interrogation of Japanese prisoners, Colonel Ogata issued his last order on 29 July, directing Army and Navy units to assemble in the wooded ridges of southern Tinian, to make their last stand. It was to that area below Tinian Town that the Japanese commander moved his CP the same day. Captain Oya, supposedly, issued his own orders to the naval troops; they were to defend the high ground of southeast Tinian. A rumor among Oya's men was that their commander had received a radio message from Tokyo on 29 July, advising that the Imperial Fleet was en route.

The Battle of the Philippine Sea, a month before, had rendered such help most illusory, for with its air arm destroyed, the Japanese fleet was crippled. Nevertheless, there were enough enemy troops left on Tinian to keep the Marines from undue optimism. A Japanese warrant officer of the *56th Naval Guard Force*, captured on 29 July, said there were about 500 troops left in that force. He believed the *50th Infantry Regiment* still had 1,700 to 1,800 men. Marines had met elements of the *50th*, but as the 4th Division D–2 emphasized, there was "no concrete evidence" that the regiment "has been committed in force." [22]

Another reason for General Schmidt's modified optimism on 30 July was the geography of Tinian at its southern end. Not only would the Marines probably experience their hardest fighting of the campaign, but most certainly they were approaching the most difficult terrain on the island. The gentle landscape around Tinian Town ended suddenly about a mile to the south. There, the ground rose to a high plateau, thick with brush and rock, measuring about 5,000 yards long and 2,000 yards wide, with altitudes over 500 feet. Approach was blocked by cliffs and jungle growth. Along the east coast, the cliffs rose vertically and were next to impossible to scale. In the center, a road leading to the plateau had to wind a tortuous way; a prisoner said it had been mined. Only on the

[20] Assigned for the air strike were 80 P–47s of the 318th Fighter Group, 16 B–25s of the 48th Bombardment Squadron (Medium), and 30 torpedo bombers from the escort carrier *Kitkun Bay*.

[21] SpecIntlRpt No. 4, D–2 Sec, 4th MarDiv, dtd 7Aug44, in *4th MarDiv Translations (Tinian)*.

[22] 4th MarDiv D–2 Periodic Rpt No. 77, dtd 30Jul44.

west were the cliffs relatively easy to negotiate.

Such was the picture as Marines dug in for the seventh night on Tinian. The land itself, not the Imperial Fleet, would try to save the Japanese. The toughness of that ground matched the enemy's will.

The Island Secured[1]

ORGANIZED RESISTANCE DECLINES[2]

At 0200 on 31 July, a Japanese force of company size led by three tanks stole through the darkness upon the lines of the 24th Marines on the division right. A heavy outburst of fire stopped the enemy thrust, knocking out one of the tanks and scattering the attackers. Japanese mortar fire fell along the entire corps front that night but was eclipsed at daylight by the thunder of American naval guns; they expended approximately 615 tons of shells in the preplanned bombardment. Aircraft dropped another 69 tons of explosives. For the cornered Japanese, the effect of such preparation fire was, according to prisoners, "almost unbearable."[3]

The cliff facing the 2d Division left and center was almost impossible to climb. A twisting road with hairpin turns led up to the plateau from the division right. General Watson's plans, therefore, were influenced by the terrain over which his troops had to fight. He set up an 0–8A Line which followed the base of the cliff except on the right. There he included not only the cliff but also 500 yards of plateau. On the division left, then, the 2d Marines would halt at the base of the cliff and remain in position to prevent Japanese escape along the east coast. In the center, the 6th Marines would not attempt the hopeless cliff but would turn west at the base and follow the 8th Marines up the road.

The 2d Marines moved out at 0830 and was opposed by sniper fire while advancing to the cliff, which was reached at noon. Large numbers of Japanese and Korean civilians who surrendered held up the advance much more than did enemy troops. In the center of the division line, the 6th

[1] Unless otherwise noted, the material in this chapter is derived from: *TF 51 OpRpt; TF 52 OpRpt; TF 56 OpRpt; NTLF OpRpt;* NTLF OpOs 38–44 and 39–44, dtd 31Jul and 1Aug44; NTLF G–2 Periodic Rpts 47–54, dtd 1–8Aug44; NTLF G–3 Periodic Rpts 47–57, dtd 31Jul–3Aug44; *2d MarDiv OpRpt Tinian;* 2d MarDiv OpOs 52–55, dtd 31Jul–10Aug44; 2d MarDiv D–2 Periodic Rpts 81–88, dtd 2–8Aug44; 2d MarDiv D–3 Rpts 78–85, dtd 31Jul–7Aug44; *2d Mar SAR; 6th Mar SAR; 8th Mar SAR Tinian; 4th MarDiv OpRpt Tinian;* 4th MarDiv D–2 Periodic Rpts 79–84, dtd 1–6Aug44, *4th MarDiv Translations (Tinian);* Thomason, "Tinian"; Chapin *4th MarDiv in WW II;* Crowl, *Marianas Campaign;* Hoffman, *Tinian;* Isely and Crowl, *Marines and Amphibious War;* Johnston, *Follow Me!;* Morison, *New Guinea and the Marianas;* Proehl, *4th MarDiv History;* Smith and Finch, *Coral and Brass;* Stockman and Carleton, *Campaign for the Marianas.*

[2] Additional sources for this section include: *1/8 Rpt Tinian;* 1/24 Rpt of Ops, dtd 25Aug44;2/24 AR, dtd 5May45; 3/24 AR, dtd 5May 45; 3/24 Narrative of Tinian Island Op, dtd 5May45; 2d TkBn SAR.

[3] *4th MarDiv OpRpt Tinian,* Anx C, p. 11.

Marines moved forward against scattered rifle and machine gun fire coming from positions on the cliff face and light mortar fire dropping from the plateau above. After his advance elements reached the 0–8A Line at 1330, Colonel Riseley received permission to pull back about 400 yards to better defensive positions. Later that afternoon, General Watson committed the 3d Battalion, 2d Marines in relief of 3/6 and moved that battalion behind the 8th Marines as division reserve.

It was the 8th Marines that bore the major responsibility on 31 July and that posed the greatest threat to the enemy. It was expected that the Japanese would concentrate defensive fires along the route into their positions, yet unless Colonel Wallace could establish a foothold on the plateau the division plan would fail.

The first yards of the day's advance were relatively easy. Supported by tanks, the 1st and 3d Battalions moved out across a flat land where cane fields, brush, rocks, and a railroad track embankment gave concealment to some Japanese riflemen and machine gunners. In one instance, 15 Japanese left their hiding place to make a *banzai* charge upon a Marine tank; they caused no damage but lost their lives. Following 1/8 and 3/8, the reserve 2d Battalion advanced, mopping up behind the attack.

At noon, the 3d Battalion reached the foot of the plateau. The 1st Battalion had more yards to cover, but by 1500 it was also at the cliff base, in contact with 3/8 on the left and the 4th Division on the right. In front of the 1st Battalion there was a road, the only practical route for tanks. The commander of the 3d Battalion, Lieutenant Colonel Gavin C. Humphrey, wanted to move his supporting tanks up this tortuous path but was denied permission because 1/8 had not yet secured the path.

The cliffs which faced the 8th Marines had the same innocent appearance as the enemy's fortified hills of Saipan, which the Marines remembered so well. Vegetation masked the deep caves and fissures where Japanese riflemen and machine gunners were waiting. Their vigil ended as Marines of 3/8 started scrambling up the rocky incline. The sudden outburst of Japanese fire prompted Humphrey to hold up the infantry assault and look to measures for reducing the opposition. Exploratory fire from medium tanks failed to find the enemy positions, and the flamethrower tanks were able to burn off only part of the vegetation. The fire of the half-tracks was equally ineffectual. Permission to withdraw the battalion 400 yards and to call down artillery fire upon the cliff had to be withheld by the regimental commander because it would involve danger to 1/8 on the right. The 3d Battalion was then forced to dig in for the night.

While the Marines of 3/8 had struck vainly against the cliff in their area, the 1st Battalion turned to its mission of opening the road. Engineers removed mines; tanks moved up, withstanding the fire of 37mm and 47mm antitank guns, and destroyed Japanese bunkers in the cliff; the infantrymen climbed step by step, opposed by rifle and machine gun fire and by hand grenades rolled downhill into their path. The thick vegetation alongside the road served both the enemy and the Marines. While it concealed the Japanese, it often obscured their view of the advancing men.

Movement was inescapably slow; by late afternoon it began to seem that the Marines would never get to the top that day. At 1650, however, Lieutenant Colonel Lawrence C. Hays, Jr., commanding the 1st Battalion, received the cheering report that a platoon of Company A was at the top. Several minutes later, a platoon of Company C dispatched the same good word.

Encouraged by such reports, Colonel Wallace ordered Hays to press the attack and get the entire battalion onto the plateau. The regimental commander, moreover, requested General Watson's permission to commit his reserve 2d Battalion, for the purpose of exploiting the success of 1/8 and gaining a surer foothold on the plateau before dark.

With the division commander's approval, the 2d Battalion, commanded by Lieutenant Colonel Lane C. Kendall, began moving up the road shortly after 1700. By then, all three rifle companies of 1/8 were represented at the top. The 2d Battalion received heavy mortar fire while moving into position on the left of the 1st Battalion. Part of Kendall's mission was to attempt physical contact with 3/8 at the base of the cliff. The 1st Battalion had lost contact with the 4th Division, to the regimental right; in fact, a gap of 600 yards developed as 1/8 shifted to the east while 4th Division elements moved westward.

BATTLE FOR THE PLATEAU

For the Marines on the plateau the situation was tense. The ground they intended to take was still commanded by hidden enemy positions, and the least motion invited a furious outburst of fire. Company E, leading 2/8 up the road, had just come upon the high ground when, at 1830, the Japanese openly attacked along the boundary between that company and Company A. The momentum of the assault forced a part of the Marine line back a few yards before it could be repulsed. Most of the 75–100 attackers were destroyed.

Company G of the 2d Battalion reached the base of the cliff at sunset, 1845, and went on to the top without delay. There it tied in immediately with Company E and disposed its line down the cliff to seek contact with 3/8. Still a gap of 350 yards existed between the 2d and 3d Battalions. Colonel Wallace was determined to remain on the plateau and elected to cover the gap with machine gun fire rather than commit the regimental reserve, Company F, with which he wanted to strengthen the forces at the top.

Two platoons of the reserve company and two 37mm guns established secondary positions at 2100 behind Company A, for if the enemy followed his usual tactics, he would direct another counterattack at the same spot. When, two hours later, the expected assault came, it was decisively broken up by the Marine guns. Yet the enemy persistently approached; Marine combat patrols fought groups of Japanese less than 20 yards from the front line. The night was foreboding; a major enemy counterattack was surely yet to come. The enemy kept probing.

While the Marines on the plateau waited, the situation on the road became equally suspenseful. Over the white coral surface, visible in the dark, cargo jeeps lugged ammunition, barbed wire, and supplies, while jeep ambulances evacuated casualties. Half-tracks and tanks labored up the winding road, adding to a traffic which was intolerable to the enemy.

An attempt to cut the supply route took form about 0100. After locating the open flank on the right of 1/8, by their favorite practice of forcing return fire, a well-organized group of at least 100 Japanese, armed with rifles and grenade launchers, infiltrated through the gap between 1/8 and the 4th Division and moved to the rear of 2/8. A part of the force appeared on the road, burned two jeep ambulances, and started to block the supply route. In the same hour, nearer the top of the road, a platoon of Japanese captured a few parked vehicles belonging to the 2d Battalion.

The command post of 2/8 was still at the base of the cliff, its headquarters personnel tied in with the left of Company G. On the plateau, the battalion executive officer, Major William C. Chamberlin, was organizing the defenses. When the Japanese attempted to cut the road, he took two platoons of Company F and elements of Company A, issued the simple oral order—"Let's go!"—and led the Marines in removing the threat. Major Chamberlin then positioned two platoons of Company F left of the road and a support platoon of Company G on the right, halfway down the cliff, as a preventive measure. Most of the infiltrating Japanese had been killed by the Marine counterattack, but an isolated group of 20 were discovered the next day on the hill—suicides by grenade. The vehicles captured by the Japanese were retaken intact.

The imperiling of Marine rear positions and the virtual certainty of a much larger enemy attack upon the front lines hastened commitment of the 2d Division reserve. Upon request by Colonel Wallace, 3/6 was attached to the 8th Marines at 0320, and the battalion at once started moving toward the cliff. Beginning then also, artillery fire by the 10th Marines was employed to prevent the enemy from bringing up reserves. Both the 2d Marines and the 6th Marines, whose sectors had been quiet, were alerted to the danger of a massive breakthrough in the wide gap which existed between the 8th Marines and the 4th Division. The two battalions of the 8th Marines were practically alone on the plateau, a fact probably understood by the Japanese, who struck before that situation could change.

At 0515, a well-organized force totaling more than 600 soldiers and sailors, equipped with nearly every weapon except tanks, charged the Marine positions, especially those of 2/8. Here the enemy tried to disable the two 37mm guns that strengthened the Company E position, but were unable to stop the fearful canister fire. Japanese 13mm machine guns tore holes in the upper shield of one of the Marine guns. Eight of the 10 37mm crew members were casualties of the assault, but other Marines kept the guns firing. "Without these weapons," said the regimental commander, "the position would have been overrun." [4]

The 1st Battalion received a lesser attack, numerically considered; about 150 Japanese charged the left flank, which adjoined 2/8, and were driven off "without great difficulty" by Marine fire.[5] The same *banzai* fervor which marked the larger attack excited these Japanese. Neither here, however, nor elsewhere along the front, was the enemy able to penetrate, though some of their number were killed just five yards from Marine positions.

In less than one hour of fighting, which cost the 8th Marines 74 casualties, the enemy suffered a loss of 200 killed—about half of the number fell in an area only 70 yards square. Despite the terrific repulse, however, the Japanese preserved organizational integrity and staged a very orderly withdrawal to the woods and cliffs to the southeast. The enemy rear guard was destroyed by Marine tanks.

The situation on the plateau appeared favorable to a steady Marine advance, but still unrectified was the gap which existed between the 2d and 4th Divisions. Progress of the 23d Marines on 31 July had been good until the left flank of the 1st Battalion, exposed by the gap, received machine gun and mortar fire from the cliff line. The attached tank platoon, advancing in front of the 1st Battalion, then suddenly ran into close range, high velocity antitank fire from beyond the tree line of the cliff. The left flank tank received six hits in rapid succession, one of them penetrating the turret. The tank commander backed off about 15 yards to a defiladed spot from which he fired two smoke shells to bracket the area in the cliff to the left front, where he believed the enemy gun was located. Rockets, naval guns, and tanks then plastered the suspected ground.

In the quiet that followed, the tanks went forward once more, this time with another tank on the left flank, the disabled vehicle following about 10 yards behind. When the replacement tank came to the spot where the other had been struck, it too was hit six times, three of the shells tearing through the armor plate. Then, however, the enemy gun was located; it was 30 yards to the left, beyond the tree line which paralleled the Marines' advance. The two battered tanks took their revenge. One of them fired a blinding smoke shell in front of the gun, while the other tank maneuvered behind the gun, knocked it out, and killed about 20 Japanese running out of the emplacement. The enemy position had been roofed over and enclosed on three sides with concrete. From an aperture a 47mm antitank gun was able to cover a fire lane about 10 yards wide. Into that lane the two Marine tanks had unhappily moved.

After the encounter, 3/25, then in division reserve, was assigned to the 23d Marines to form a perimeter defense around the tanks and service vehicles of 1/23 on the low ground. The 1st Battalion had secured the high ground in its zone by 1745, but mines along the only road to the top prevented moving the machines up until the next day. The regimental reserve battalion, 3/23, which had followed the advance, moved onto the high ground to the rear of the

[4] *8th Mar SAR Tinian*, p. 10.

[5] *Ibid.*, p. 8.

MACHINE GUN *on a half-track lays searching fire along a tree line on Tinian to cover advancing 4th Division Marines. (USMC 88220)*

PACK HOWITZER, *firmly lashed in position, fires on enemy caves in cliffs along the Tinian coast. (USMC 94660)*

1st Battalion. There the two battalions established a perimeter defense with flanks bent back and anchored on the cliff line. One company of 3/23 stayed on the low ground, however, for the protection of the left flank and to contain those Japanese that still remained in the gap between the Marine divisions. Patrols from the units of 3/23 on the high ground roved the gap but were unable to locate any elements of the 2d Division. The 23d Marines settled down to a night marked only by sniper fire and infiltration attempts.

To the right, next to the two battalions on the high ground was one company of 2/23, which had worked its way to the top before dark by moving through the zone of 1/23. Progress of the 2d Battalion on 31 July had been good until the afternoon. Then the attached tank platoon leading 2/23 reached a well-seeded minefield planted across the valley road which led to the high ground. Engineers started to clear lanes through the field for each tank of the platoon; two engineers walked in front of each tank, removing the mines as they were discovered. Suddenly, Japanese riflemen and machine gunners opened up from a trench 20 yards away and across the route of advance. The engineers, as well as the accompanying Marine infantrymen, were pinned down.

Tanks took the trench under fire; one of them started toward the end of the trench, to fire down the length of it. Traversing some ground which was judged the least dangerous, the tank got to just five yards from the objective when it hit a mine. The explosion shattered the tank's suspension system and injured the tank commander, the driver, and the assistant driver. Emboldened by this success, a number of Japanese darted from concealment to attack the Marines openly and were either killed or pinned down by point-blank fire. One of the tanks rescued the crew of the disabled vehicle and after pulling back saw the Japanese trying to set up machine guns in the wreck. To prevent this, the Marine tanks blew it apart.

Darkness approached before the strong enemy position could be reduced, so Lieutenant Colonel Edward J. Dillon, commanding 2/23, left one infantry company behind to contain the pocket. Other Marines of the battalion moved around to the left and went on to the base of the high ground, digging in there for the night. The tanks remained with the containing company until it had set up its defenses. During the night the Marines at the foot of the cliff received some enemy fire from caves near the base.

For the 24th Marines on 31 July opposition developed early along the west coast south of Tinian Town. The 1st Battalion received light artillery fire shortly after moving out; then at 1000 the Marines of Company C, advancing along the beach, were stopped short by rifle and machine gun fire coming from an isolated enemy position defended by 70 naval troops. It took an hour of hard fighting to subdue the Japanese. The mopping up of the beach area behind Company C was left to Company E of 2/24, which followed the advance at 600 yards.[6] In overcoming the resistance of the Japanese naval troops,

[6] 2/24, though then in division reserve, was under the control of the 24th Marines.

and of other enemy groups hiding in caves or jungle brush, the 1st Battalion was aided by armored amphibians offshore. Flame tanks seared enemy caves and also burned off some of the vegetation which entangled the individual Marines or hid Japanese positions.

The movement of 3/24 on 31 July was interrupted at noontime by a loss of contact with the 23d Marines on the left. After an hour or so the gap was closed, and 3/24 resumed the advance. At 1600, however, as the battalion neared the cliff, the enemy opened up with machine gun and mortar fire from the ridge line to the left front. Tanks and half-tracks were called upon to overcome the resistance, but the terrain forbade their movement except along the road to the high ground, a path which the enemy had thoroughly and meticulously mined. Engineers began the tedious and delicate work of removing the threat—they cleared 45 mines from an area 30 yards long. In view of the late hour, the battalion halted for the day. Neither here nor in the zone of 1/24 was the Marines' position especially good. Division reported that the troops dug in for the night "on the least unfavorable ground." [7]

1 AUGUST: THE NINTH DAY

Victory on Tinian was obviously near, but the situation of the Marines was momentarily difficult. General Watson ordered that the attack by the 2d Division on 1 August, scheduled for 0700, be delayed an hour, for not until daylight would there be more than two battalions, 1/8 and 2/8, at the top of the cliff, and both of those units had suffered a number of casualties from the Japanese counterattack. The 2d Battalion, the hardest hit, was put into regimental reserve when 3/6 reached the plateau at daybreak. The 3d Battalion, 8th Marines, which also climbed the hill that morning, took up a position to the left of 3/6.

By 0800 then, General Watson had three battalions ready to attack across the plateau—3/8, 3/6, and 1/8. The 2d Battalion, 6th Marines, which followed 3/6 up the cliff, would be committed when necessary. To the division left, the 2d Marines was kept at its mission of preventing any Japanese escape up the east coast. On the plateau, it was Colonel Wallace of the 8th Marines who would command the advance to the 0–8A Line.

The first battalion to reach the objective was 1/8; it was on the line just 15 minutes after moving out. The other two battalions were at the objective well before noon. Resistance was negligible and came mostly from isolated groups of Japanese. It had been planned that when 0–8A was reached the 6th Marines would be more fully committed with the regiment scheduled to assume responsibility for the left half of the division zone. Colonel Riseley was to take over the two left battalions, 3/8 and 3/6. The right half of the division zone would still be under Colonel Wallace, but with 2/8 in reserve, he would have only 1/8 in the assault. Such a weighting of the division line indicated the slant of the concluding push, due southeast, chiefly against Marpo Point.

To the 6th Marines, which had not suffered the violent enemy counterat-

---

[7] *4th MarDiv OpRpt Tinian,* Sec. IV, p. 31.

tack of the night before, went the major effort on 1 August. Neither 1/8 nor the 6th Marines, however, met any organized resistance while advancing to the cliff above the shore; one company of the 8th Marines reached the objective by 1455. The most significant fact revealed by the easy advance was that a tedious mop up would ensue after the campaign itself was over. Innumerable caves sheltered the remnants of the enemy force; these Japanese lacked command and organization, but they still possessed some weapons and ammunition.

Other caves had been the refuge of frightened civilians, but as Japanese resistance collapsed they began to emerge from hiding. In fact, the progress of the 6th Marines was interrupted on 1 August by the flock of civilians who approached waving white cloths. The large-scale surrender was partly in response to leaflets and voice broadcasts by Marine language personnel, who sought to avoid a repetition of the mass suicides which occurred on Saipan. Division intelligence had estimated that from 5,000 to 10,000 civilians were in hiding on the southeast part of the island. Some had been living in caves since J-Day.

Many of the civilians that surrendered were thirsty and hungry, but few lacked clothing. Some of them came forth lugging suitcases full of clothes, which they had taken upon leaving Tinian Town. A few wore their Sunday best, to greet the Americans they no longer feared. A number of the civilians needed medical attention, but remarkably few of the Tinian population had been wounded by the American bombardment of the island.

The outflow of confused humanity—they were all either Japanese or Korean—reached such a number in the path of the 6th Marines that at 1510 Colonel Riseley received orders to halt for the day, even though the regiment was short of the cliff above Marpo Point. No Japanese troops were observed, but the colonel took the precaution of committing his 2d Battalion on the regimental right, to tighten the lines. Moreover, when he received 1/6 back from division reserve at 1730, he put Company A on watch near the cliff where many Japanese soldiers were known to be hiding.

The processing of civilians that surrendered on 1 August was not a problem for the 6th Marines, because at 0600 that day the control of civilian internment was assumed by the Island Commander, Major General James L. Underhill, who took over a NTLF internment camp established south of the old 0–4 Line on 31 July. Few civilians or prisoners of war had been taken by the Marines until late in the campaign. By the evening of 2 August, however, NTLF G–1 reported that 3,973 civilians had been received, while 48 prisoners were in custody. By 4 August, the number of civilians had reached 8,491 and the prisoners totaled 90.[8]

Early in the campaign, the 2d Division had established a stockade, to care for both civilians and prisoners, near the Ushi Point Airfield. The 4th Division tried regimental stockades which were moved forward with the regimental CP. For Marines who had fought only in jungles and on barren atolls,

[8] NTLF G–1 Periodic Rpts No. 48, dtd 3Aug44, and No. 50, dtd 5Aug44.

the handling of civilians and their property was, even after Saipan, still a new experience. Not until Tinian did the 4th Division use civil affairs teams on a regimental level.

As it happened, the 4th Division met fewer civilians on 1 August than the 2d Division did, and those were mostly Korean field laborers. On the west side, the enemy soldier proved the more obstructive element. Until 1045 the 2d Battalion, 23d Marines was occupied at reducing the strong point encountered the day before. Tanks and vehicles then started up the hillside road which engineers had cleared of mines. As the battalion climbed, one rifle company was posted to prevent ambush of following troops and vehicles.

The other two battalions of the 23d Marines, already on the plateau, were harassed by considerable enemy machine gun and rifle fire as they adjusted frontage before moving out at 1000. About 50 Japanese ventured near the lines of 1/23 and were destroyed by Marine machine gunners hitting them from two sides. Enemy opposition the rest of the day consisted of rifle and machine gun fire from cane fields and tree lines. At 1715 the two battalions reached the furthest possible line of advance, a cliff overlooking the sea on the east coast. Patrols then reconnoitered routes to the low ground in front of the cliff and along the coast; they reported a honeycomb of caves and deep recesses, hiding Japanese. The two battalions encountered no immediate trouble, however; nor did 2/23, which had halted to the rear on commanding ground.

The 24th Marines reported that 1 August "was almost a prototype of the day before." [9] So it was, in the respect that more rocks and more of the same dense undergrowth kept the advance to a plodding pace. The sort of enemy resistance was much the same also—isolated groups, usually hidden by caves or vegetation and ever ready to fire or throw some hand grenades at the Marines.

The regiment had moved out at 0800, with 3/24 and 1/24 from left to right. Artillery of the 14th Marines fired a 5-minute preparation 600 yards forward of the front lines, and two others subsequent to the attack hour, the last one 1,200 yards forward of the line of departure. After that, however, the restricted area of combat made risky not only any artillery support but also call strikes by aircraft, or assistance from gunfire support ships.

At 1500 the 2d Battalion, released from division reserve, was committed to the attack. The Marines were then reaching curious terrain "of a palisade-like nature." [10] It consisted of three levels, descending from the cliff top to the sea. The regimental lines were consequently readjusted, all three battalions being deployed abreast. The 3d occupied the high ground on the regimental left, 2/24 moved onto the center level, and 1/24 stayed on the low ground, its flank on the coastline.

After such an adaptation the advance was resumed, turning from a southward to an easterly direction around the tip of the island. The advance was not rapid here; the Marines received intermittent machine gun and

[9] 24th Mar SAR, Anx I, p. 13, to *4th MarDiv OpRpt Tinian.*

[10] *Ibid.*

rifle fire before reaching the 0–8 Line on Lalo Point at 1800. Delay was once occasioned when several dozen civilians had to be removed from a cave. In some areas, progress was only by small fire groups, working their way through the obstructive vegetation between coral boulders. The Japanese took advantage of such difficulty. After the daylight turned to pitch darkness, some Marine casualties resulted from mines actually thrown down upon the men from the cliff.

FIGHTING OFFICIALLY ENDS

The hostile fire received by the Marines on 1 August did not suggest that organized resistance was over, but General Schmidt recognized the essential facts: Colonel Ogata's well-planned defense of Tinian had irrevocably collapsed; most of the Marines were either on or near the concluding objective line. At 1855 on 1 August, the American commander declared the island secured.

A statement like that, however, was a sort of partial truth on any Pacific territory captured from the Japanese. On Tinian, even more than elsewhere, the residue of the enemy force was troublesome. Some of the Japanese preferred self-destruction to surrender, but the proportion of soldiers and civilians that committed suicide on Tinian was smaller than on Saipan.[11] The Japanese soldier that chose to live was a die-hard type, able to hide out for months.

Most of the Marine casualties after 1 August were caused by those Japanese, who faithful to their military code, decided to forego security and die in combat. The 4th Division D–2 correctly predicted that the enemy would "sally forth from the caves in group *banzai* charges."[12] Just before the dawn of 2 August nearly 200 Japanese, armed with rifles, machine guns, and grenades, attacked the command post of the 3d Battalion, 6th Marines. The pistols, carbines, and two automatic rifles available to the Marines seemed insufficient against the do-or-die spirit of the Japanese, and the outcome was uncertain until the Headquarters Company commander obtained a medium tank, along with a rifle platoon, from nearby Company F of 2/6. Two hours of combat left 119 Japanese dead. The Marines lost their battalion commander, Lieutenant Colonel John W. Easley, and suffered other casualties. Major John E. Rentsch, the executive officer, assumed command of the unit.

The 2d Battalion, 6th Marines reported a similar assault upon their command post. Developing shortly before 3/6 was hit, the incident appeared to have been connected with the same enemy outbreak. The brunt of the Japanese attack here was borne by the Mortar Platoon of the Headquarters Company. Three Marine tanks, which had bivouacked for the night at the CP, proved handy. The Japanese pulled back, leaving 30 of their number dead, but they withdrew toward the com-

[11] Some of the Japanese civilians who killed themselves were members of the Civilian Militia, or *Zaigogunjin*, which had been quite inoperative. The men had received a bit of military training but no weapons, and their prime interest was fleeing to safety with their families.

[12] 4th MarDiv D–2 Periodic Rpt No. 79, dtd 1Aug44.

mand post of 3/6, either by design or mistake.

The next morning at 0530, 3/6 and 3/8 (still attached to the 6th Marines) had to withstand a second enemy attack, staged by a composite group of 150 Japanese soldiers and sailors.[13] Here also the enemy achieved only the wish to die in battle, rather than surrender; 124 Japanese lay dead after the attack. On succeeding days, the hopeless efforts were repeated. On 3 August, the 4th Division killed 47 of the enemy. On 4 August, when Battery I of the 14th Marines was attacked by 15 Japanese in a cane field, 12 of the enemy were killed.

The protracted chore of mopping up on Tinian went to the 8th Marines, which on 6 August became Ground Forces Tinian, under the command of Brigadier General Merritt A. Edson, assistant commander of the 2d Marine Division. He thereupon assumed tactical responsibility for the island. At the same time, he released one rifle company to the control of the Island Commander to assist the Civil Affairs Officer in the handling of prisoners and civilians on Tinian. The Japanese troops were removed to Hawaii.

On 25 October 1944, the 8th Marines went back to Saipan, but its 1st Battalion remained on Tinian until 1 January 1945. In the period from the securing of the island until the end of the year, more than 500 Japanese were killed during exchanges of fire, but such encounters cost the 8th Marines 38 killed and 125 wounded.

These losses appear high when compared to the Marine casualty figures for the campaign itself. The 2d Division reported 105 killed and 653 wounded;[14] the 4th Division, 212 killed and 897 wounded.[15] Marines missing in action came to 27 for the two divisions. NTLF records, which included Army casualties, show a total of 328 killed and 1,571 wounded.[16] For the Japanese, the price of the vain defense of Tinian had been extremely high; nearly 5,000 men were killed.[17]

While patrols of the 8th Marines hunted enemy survivors, units of the 2d Division departed for Saipan; by 7 August, 2/6 and 3/6 had left via LSTs. On Saipan the division was to relieve the 27th Infantry Division and remain on that island until the Okinawa campaign the next spring. The 4th Division went back to Camp Maui, the last units boarding ship on 14 August. In Hawaii, the division would prepare for its next battle: Iwo Jima.

At 1200 on 10 August, upon orders from Admiral Spruance, General Schmidt passed the command of all forces on Tinian to the Island Com-

[13] Some prisoners said that Colonel Ogata was involved and became a casualty. Marines did find his last command post, which afforded a view of Tinian Town, the harbor, and the airfield. American artillery and naval gunfire had demolished the headquarters; a prisoner said that Colonel Ogata had left the area on 31 July.

[14] 2d MarDiv G–1 Periodic Rpt No. 83, dtd 5Aug44.

[15] *4th MarDiv OpRpt Tinian*, Anx A, p. 1.

[16] NTLF G–1 Periodic Rpt No. 50, dtd 5Aug44, Anx B; *NTLF OpRpt*, Encl H (LVT Rpt), pp. 5–6. A unit breakdown of NTLF casualties is contained in Hoffman, *Tinian*, p. 150. Final official Marine casualty totals are contained in Appendix H.

[17] NTLF G–2 Periodic Rpt No. 48, dtd 2Aug44, gives the figure of 4,858 enemy killed prior to 1800 of 1 August.

mander, who thereafter handled the embarkation of men and equipment. NTLF was dissolved on 12 August. On the same day, TF 52 closed its books; Admiral Hill had eminently fulfilled his responsibility for the capture of Tinian.

LOGISTICS AT TINIAN TOWN

The garrison troops had begun landing the day after the island was secured, unloading by LCTs at Tinian Town.[18] The units went ashore over Green Beach, where once the land mines had been cleared and four wrecked Japanese small craft removed, three LCTs could be received simultaneously. One additional LCT could be handled alongside South Pier, which, being hardly damaged, was easily put into commission.

On 1 August, upon orders from the NTLF Shore Party Commander, the 4th Division Shore Party, less the 1341st Engineer Battalion, had left the weather-battered White Beach 2, to take over all supplies landed over the Tinian Town beaches.[19] The shore party immediately opened up and operated South Pier and that section of Green Beach which extended south of it. The depth of the water at South Pier did not permit landing LSTs either alongside or at the end, so pontoon causeways, brought from White Beach 1 and Saipan, were installed at the end. On 4 August, the first LST docked there.

North Pier had been severely damaged by bombardment, and like the other pier, rigged with booby traps; complete repair took until 3 August. The 2d Battalion, 20th Marines then began operating that pier, turning over South Pier to the Island Commander.[20] By 5 August, the stretch of beach between the two piers (the wider section of Green Beach) had been cleared of land mines and surfaced with coral. Here it was possible to land 15 LCMs simultaneously. The waters of Sunharon Harbor presented no problem except for the sunken Japanese hulks which had to be blasted by UDT men. The Japanese had not mined the waters.

American merchant ships soon crowded the harbor, bringing equipment to reconstruct Tinian. On 3 August, the Stars and Stripes had been officially raised over the island, marking its commission as a naval base. Primarily, however, Tinian would serve the Army Air Forces. They had wanted Saipan and Guam also, but Tinian was the most suitable of the three islands because of its relatively level ground. The Navy, on the other hand, had less interest in Tinian, whose Sunharon Harbor was unable to berth many ships.

Tinian would be "developed as an air base for . . . heavy, medium, and light aircraft," said General Underhill when

[18] The Tinian garrison force eventually numbered 5,235 men—2,527 Marines, 2,693 Navy, and 15 Army.

[19] White Beach 1 was closed shortly after. Yellow Beach was never used, even after the mines had been cleared away, because of the consistently heavy surf and the nature of the reef.

[20] Both piers were under the Island Commander after 10 August, when the 4th Division Shore Party was disbanded. Until then, however, the shore party helped embark Marines and their equipment.

214-881 O-67—28

he was designated Island Commander.[21] Navy and Marine aircraft did use the base, but the island became particularly a home for the Army Air Forces giant Superforts. Two wings of the Twentieth Air Force operated from Tinian, flying the B-29s nonstop to Japan itself. The island was developed into the largest B-29 combat base in the Pacific.

Immediate responsibility for construction and defense fell upon General Underhill, who reported to the Commander Forward Area Central Pacific, Task Force 57 (Defense Forces and Land-Based Air, Vice Admiral John H. Hoover). To prevent Japanese interference with construction, the Island Commander relied upon two Marine antiaircraft artillery battalions, the 17th and 18th, assigned there. They formed the Antiaircraft Defense Command. The 17th Battalion set up 90-mm guns and two platoons of automatic weapons for defense of the port area. Not until November, however, did any Japanese planes fly near the island, and never were any bombs dropped.

THE CAMPAIGN REVIEWED

The absence of enemy air or sea interference, following the Battle of the Philippine Sea, had been one of several ways in which Tinian differed from other Pacific island campaigns. The major differences arose, as we have seen, from the nearness of Saipan. Such proximity of the staging base to the objective permitted a shore-to-shore operation, the first large-scale one in the Central Pacific, and that, in turn, allowed the landing and supplying of two Marine divisions over the extremely narrow beaches.

The plan was not only a bold one, it was the only possible plan if the Marines were to be spared a bloody assault of the well-defended beaches at Tinian Town. Success of the attempt hung upon two unpredictable elements—the will of man and the fancy of the wind. If Colonel Ogata, by a flash of insight, had decided to wait for the Americans at the northwest beaches instead of near Tinian Town, or if the weather had suddenly changed, the logistically complex landings could have ended in disaster. To save Marine lives, the gamble was taken; fortunately, all went well. The tactical surprise unbalanced Colonel Ogata's defense plan beyond repair, leading to General Holland Smith's opinion that "our singular success at Tinian lay in the boldness of the landing." [22]

It was ironic that the Japanese were caught by surprise here on an island where they were absolutely certain of an American invasion. The loss of Saipan made that inevitable. Long before then, however, the Japanese had seen American planes flying over Tinian from the captured fields of Saipan or ships nearby—reconnaissance planes getting photographs and bringing Marine commanders for a view of their next battlefield, or P–47s and carrier planes bombing defense installations. Most of all, it was the preparatory bombardment which had destroyed any Japanese illusion that Tinian would not be invaded.

The nearness of Saipan made possi-

[21] Directive issued by CG, TG 10.12 and Prospective Island Comdr, dtd 15May44.

[22] Smith and Finch, *Coral and Brass*, p. 203.

ble the unusual bombardment of a Pacific island objective by land-based aircraft and artillery positioned on adjacent soil. The fact that artillery support would be available from Saipan had influenced the choosing of the northwest beaches, and, next to the landings, the preassault bombardment by artillery, ships, and planes was decisive. For never did a single island of the Pacific war receive a more prolonged and continuous pounding before the Marines landed. Afterwards, when the artillery was moved to Tinian, the Marines enjoyed the wealth of such support, especially at the last when the restricted area of combat made naval gunfire and air strikes impractical.

The task of naval gunfire was somewhat lightened at Tinian because land-based artillery joined the preparatory bombardment. Still, according to Japanese prisoners, there was plenty of hell from the sea. Naval gunfire had been improved by the Saipan experience and was even more effective than before. Call-fire procedure was carried out better than at Saipan. The TF 56 naval gunfire officer noted that ships and shore fire control parties "worked in far greater mutual understanding than on any prior operation."[23] He recommended the addition of another battalion spotter, to ensure best results.

The fire support ships did not have the complications with artillery which were occasionally reported by the pilots of P–47s. Field guns were sometimes firing into the same area assigned to planes. To avert such a difficulty in the future, Admiral Hill suggested a Combat Liaison Team, to be composed of air, naval gunfire, and artillery officers, each with his portable radio set. The team would move forward as a unit and decide just which weapon should be used on the target in question. Spotting and the checking of results was simplified at Tinian by the absence of Japanese ships or planes and by the next-door nature of the targets. The busy Marine observation planes, which did much of that work, were controlled entirely by artillery units.

The mission of preparatory air bombardment was vigorously executed by the P–47s on Saipan and by Navy planes from the escort carriers. After J-Day, the Army Air Forces and the Navy complied with requests for air support by a system of alternating, each furnishing four call strikes a day, assigned by Commander Support Aircraft in the *Cambria*.

The P–47s also undertook a new kind of mission at Tinian: the dropping of the napalm bomb, initially used there and then later on other Pacific islands in a more improved form. During the entire Tinian operation, 147 jettisonable tanks were dropped from 21 July to 1 August. Fourteen of them were duds, but 8 of the 14 were subsequently set afire by strafing runs. Owing to a shortage of napalm powder on Saipan, only 91 of the fire bombs contained the napalm mixture; the rest consisted of an oil-gasoline mixture.[24]

For the airmen, as for artillerymen and naval gunners, the relatively level terrain of Tinian made targets easier to

[23] TF 56 OpRpt, NGF Sec, p. 138.

[24] Figures for the employment of napalm bombs are derived from 318th FAGru IntelRpt, dtd 1Aug44, cited in Dr. Robert F. Futrell, USAF HistDiv, ltr to Head, HistBr, G–3, HQMC, dtd 29Nov63.

hit than was true on Saipan. Indeed, the nature of the ground appears hardly second to the nearness of Saipan as an influence upon the campaign. Except at the southern end, the landscape of Tinian is fairly gentle, offering little opposition to the advance of troops or vehicles. The Marines employed more tanks here than ever before on a Pacific island. Many of the enemy were killed in the open by medium tanks leading infantry attacks. Traversing the cane fields did impose a problem, however. Rows of high stalks obscured the already restricted vision of a tank platoon leader, who normally had to poke his head out the turret to observe his vehicles. The difficulty prompted one tank officer to suggest a new type periscope or a protected turret. Tank communication, however, was better on Tinian than before. The efficient SCR–500 series of push-button type radio had recently become standard for Marine tank battalions and was first employed at Tinian.[25]

The flat stretches of Tinian were favorable to wire communication; the Japanese had prepared the entire island for sending messages by telephone, only to have the system wrecked by the American bombardment. Moreover, until the last days of the Tinian campaign, the Marines advanced so rapidly that their communications men were hard-pressed to string wire fast enough.

In such a short campaign, however, contact by radio was often sufficient, the infantry again finding the SCR–300 a reliable set. This Army Signal Corps radio had become the standard field radio used by Army and Marine infantry in World War II. Tank commanders on Tinian also had an SCR–300 for communicating in infantry command nets. It was a portable radio set, adapted for carrying on a soldier's back. The SCR–536, a small hand-carried radio was also used at Tinian by platoon leaders and company commanders. The range of these field sets, however, did not exceed a mile or so; the water-proofing was inadequate for the almost daily rain; and transmission was often blanketed by other stations on the net.

In getting supplies across country to the fast-moving Marines, the level nature of the island was helpful. Moreover, the Japanese had constructed a good network of roads. Yet, logistically, the Tinian operation was constantly challenged: first by the beaches and then by the weather. Problems had begun at the planning stage. General Schmidt gathered enough LVTs to form a provisional LVT group, but he saw the necessity for a permanent LVT group organization for corps-size landings. The labors of the shore party, herocially performed, emphasized likewise the need for a permanent corps shore party organization, large enough for a major amphibious assault.

When the weather turned, it was the DUKW that saved the day. The tough amphibian truck again demonstrated its usefulness under conditions risky for other craft. Colonel Martyr, who commanded the NTLF Shore Party, said that without the DUKW "supply in this operation would have

[25] The Marine Corps Table of Organization for a tank battalion, T/O F–80, dated 4 April 1944, authorized the Army Signal Corps radios 508 and 528, the short-range, frequency-modulated sets expressly created for armored divisions and well liked by Army tankmen.

been practically impossible." [26] He recommended that henceforth amphibian trucks should be supplied not only to the artillery but also to the shore party—and in greater numbers. Admiral Hill advised that DUKWs, manned by Navy crews, replace the LCVPs then carried on deck by transports and attack cargo ships (AKAs).

Two of the four amphibian truck companies at Tinian were Army units: the 477th Amphibian Truck Company and the 27th Division Provisional Amphibian Truck Company. Much credit belongs to both Army and Marine drivers of the DUKWs, who worked long hours through a taxing surf.

The same weather reverse which forced reliance upon the DUKWs invoked the employment of C–47s for transporting rations and supplies from Saipan and evacuating wounded men. For the Tinian campaign, cargo delivery by air had been planned only as an emergency method, and no more that 60 tons of air cargo was actually delivered. It was enough, however, to show that cargo delivery by air was very practicable and open to future development.

The battle for Tinian had required logistic ingenuity from the very beginning of plans, but there was less demand for tactical adaptation once the troops were ashore. Because of the narrow beaches, General Schmidt had concentrated power behind a single assault division on J-Day, thus combining mass with economy of force. In the elbowing technique, he applied the same principles. Mostly, however, there was little necessity for tactical inventiveness at Tinian. The Japanese, disorganized by the preassault bombardment and the surprise landings, fell back upon their usual *banzai* attacks and cave warfare, tactics which the Marines were prepared by experience to meet.

After the Americans landed on Saipan, Colonel Ogata had prepared an elaborate battle plan, issued new rifles and other field equipment to a well-trained garrison, and hastened the construction of defenses. He was short of tanks, having only 12, but he possessed a large stock of other weapons; even on the last day of the battle, Marines encountered well-armed Japanese. A poor command relationship existed between Army and Navy officers, but whether it was consequential is hard to tell.

In preparing the defense of Tinian, Colonel Ogata worked in a sort of glass headquarters. Documents captured by Marines on Saipan revealed his strength and order of battle; photo reconnaissance, the best yet obtained of a Pacific objective, located every major Japanese installation. In the preassault bombardment a number of the defense positions were destroyed, one exception being the well-camouflaged guns which damaged the *Colorado* and the *Norman Scott* on J-Day. Many of the enemy artillery positions illustrated the Japanese art of camouflage. Guns were well-hidden in caves and wooded terrain, so that Marines were able to locate them only by observation of gunflashes and by sound ranging.

Among the objects visible to photo reconnaissance were Japanese planes idle on the fields of Tinian. The headquarters of the *First Air Fleet* and two

[26] *NTLF OpRpt*, Encl J (2), p. 9.

naval air groups had been identified as located on the island, but the pilots left in May or June for missions elsewhere, and survivors were unable to get back. Photo reconnaissance was unrestrained, therefore, except by Japanese antiaircraft fire.

Air observation was unfortunately limited to what lay above the ground. At Tinian, the Japanese seeded the earth with a larger number of mines than Marines had encountered elsewhere. The certainty of invasion allowed the enemy time for planting many antipersonnel, antitank, and antiboat mines, besides setting booby traps. The usual home-made types of mine appeared, but the only true novelty at Tinian was the interconnection of horned mines.

Some destruction resulted from the enemy's antitank mine; a Marine tank commander was killed by one which a Japanese lodged upon the hatch. The enemy sometimes buried a 500-pound bomb beneath the anticipated flight path of low-flying American aircraft. Then when a plane appeared they would electrically detonate the bomb from a remote vantage point. One Army flier was killed by such a device. Most of the time, the vigilance of Marines and the tireless efficiency of their engineers minimized casualties and damages. The antiboat mines resulted in far less damage than the Japanese expected they would.

No obstacle the enemy imposed, whether a mine underfoot or a hidden gun, equalled the well-trained Japanese soldier himself. On Tinian, he exhibited the usual professional skill in attack and a calm order in withdrawal which contrasted to the emotion of the *banzai* charge. The enemy withdrawals before the pressing Marine advance, which marked the battle on Tinian, illustrated, as did camouflage, the Japanese art of furtive action. Troops moved usually at night, in small groups and with few losses due to detection. The heavy Japanese casualties resulted from impatience to defeat the invader not by a well-concealed defense, at which they were masters, but by a hopeless open attack against superior firepower and Marine infantrymen who were second to none at close combat.

The opening fires of the American bombardment foretold the capture of Tinian. With a numerically inferior garrison, isolated from reinforcement, the Japanese commander fought a losing battle. Yet if he had made a more subtle judgment on where the Americans intended to land, the campaign would probably have been longer and the ratio of casualties different. As it was, Tinian became a model victory for Navy and Marine Corps amphibious tactics.

PART VI

The Return to Guam

Preparing for Guam

PLANS MADE—AND REVISED[1]

The battle to recapture Guam took place at the same time as the Tinian campaign, but it was the former that drew more attention from the American people. A possession of the United States since the island was taken from Spain in 1898, Guam had fallen to the Japanese just three days after Pearl Harbor.[2]

To regain the island was not only a point of honor; Guam was now definitely wanted for an advance naval base in the Central Pacific and for staging B–29 bomber raids upon Japan itself. Recapture of Guam had been scheduled as phase II of the FORAGER operation; it was slated to follow phase I immediately after the situation at Saipan permitted.[3] In May 1944, when the preparations for Guam were taking shape, the tentative landing date was expectantly set as 18 June, three days after the scheduled D-Day at Saipan. As it turned out, however, the landings on Guam did not come until 21 July.

There were several reasons for postponing W-Day, as D-Day of Guam was called. The first was the prospect of a major naval engagement, which evolved as the Battle of the Philippine Sea. Intelligence that a Japanese fleet was headed for the Marianas had been confirmed by 15 June, and Admiral Spruance cancelled W-Day until further notice to prevent endangering the transports and LSTs intended for Guam. These vessels were then ordered to retire 150 to 300 miles eastward of Saipan.

On 20 June the Battle of the Philippine Sea was over; Japanese ships and planes were no longer a substantial threat to American forces in the Marianas. By then, however, there were other facts to consider before W-Day could be reset. Japanese resistance on

[1] Unless otherwise noted, the material in this section is derived from: FifthFlt FinalRpt on Ops to Capture the Mariana Islands, dtd 30Aug44, hereafter *FifthFlt FinalRpt; TF 51 OpRpt; TF 56 OpRpt;* TF 53 Rpt on PhibOps for the Capture of Guam, dtd 10Aug44, hereafter *TF 53 OpRpt;* IIIAC SAR, dtd 3Sep44, hereafter *IIIAC SAR;* Maj Orlan R. Lodge, *The Recapture of Guam* (Washington: HistBr, G–3 Div, HQMC, 1954), hereafter Lodge, *Recapture of Guam;* Crowl, *Marianas Campaign.* Throughout this and succeeding chapters dealing with the Guam campaign, Lodge, *Recapture of Guam,* and Crowl, *Marianas Campaign,* have served as overall guidelines. For this reason, they will be cited only in direct reference hereafter. Unless otherwise noted, all documents cited are located in the Marianas Area OpFile and Marianas CmtFile, HistBr, HQMC.

[2] The Japanese seizure of Guam in December 1941 is related in Volume I of this series, pp. 75–78.

[3] For the background and planning of FORAGER, see Chapter 1 of Part IV, "Saipan: The Decisive Battle."

Saipan had required the commitment of the entire 27th Infantry Division, the Expeditionary Troops Reserve. The only available force was the 77th Infantry Division, which was then ashore in Hawaii.

The Marines assigned to recapture Guam had been deprived of their reserve; yet the dimensions of the approaching battle appeared to increase. Japanese prisoners and documents captured on Saipan confirmed what aerial photographs of Guam were indicating, that enemy strength on the island had been increased. Anticipating a campaign even more difficult than Saipan, Admiral Spruance saw the necessity for having an adequate reserve immediately available. At the same time, he realized that additional troops might yet be needed on Saipan, so the task force slated for Guam was retained as a floating reserve for the Saipan operation during the period 16–30 June.

Admiral Nimitz was willing to release the 77th Division to General Holland Smith as Expeditionary Troops Reserve, and on 21 June, he sent word that one regimental combat team would leave Hawaii on 1 July, with the other two following as the second echelon after transports arrived from Saipan. In a hastily assembled transport division of five ships, the 305th RCT and an advance division headquarters sailed from Honolulu on 1 July. On 6 July, General Holland Smith assigned the 77th Division to General Geiger's control. Admiral Nimitz then sent for the 26th Marines to serve as Expeditionary Troops Reserve for Guam, and the regiment departed San Diego on 22 July.

It was Nimitz' wish that Guam be attacked as soon as the 305th reached the area. Further postponement of the landings would give the Japanese more time to prepare. Besides that, the weather normally changed for the worse in the Marianas during late June or early July. The rainfall increased, and to the west of the islands, typhoons began to form, creating sea conditions unfavorable for launching and supporting an amphibious operation.

Just as anxious as Nimitz to avoid prolonged delay, Spruance reviewed the situation with the top commanders assigned to the Guam operation. At a meeting off Saipan on 30 June, they concurred in his judgment that "the Guam landings should not be attempted until the entire 77th Division was available as a reserve." [4] To that decision, Nimitz agreed. On 3 July, Spruance designated the 25th as tentative W-Day. On 8 July, after learning that the entire Army division would be at Eniwetok by the 18th, four days before it was expected, he advanced the date to 21 July, and there it stood.

The Fifth Fleet commander had postponed W-Day "with reluctance," [5] knowing that for the Marines due to land on Guam, it meant more days of waiting on board crowded ships under the tropical sun. Except for the replacement of the 27th Division by the 77th, the command and troop organization for the Guam campaign had not been changed, and troop movements until the middle of June had gone ahead as planned. The task

[4] *FifthFlt FinalRpt*, p. 5.

[5] *Ibid.*

force charged with the recapture of Guam sailed from Kwajalein for the Marianas on 12 June, to act as reserve at the Saipan landings before executing its primary mission.

COMMAND AND TASK ORGANIZATION [6]

The top commands for Guam were the same as those for Saipan and Tinian. Under Admiral Spruance, commanding Central Pacific Task Forces, Admiral Turner directed the amphibious forces for the Marianas, and General Holland Smith commanded the landing forces. Guam was to involve Admiral Turner's and General Smith's subordinate commands, Southern Attack Force (TF 53) and Southern Troops and Landing Force (STLF). At Guam, unlike Saipan, the hard-hitting senior Marine general would not take direct command of operations ashore, but would leave it to Major General Roy S. Geiger, whose III Amphibious Corps had been designated the landing force for the Guam campaign.

In direct command of the Southern Attack Force, activated on 24 May 1944, was Admiral Conolly, who had taken Roi and Namur in the Marshalls a few months before. Admiral Nimitz had assigned to TF 53 a number of ships from the South Pacific Force, which until 10 May, had been engaged in General MacArthur's Hollandia operation. As the attack plan for Guam envisaged simultaneous landings at two points, Admiral Conolly divided his task force into a Northern Attack Group, which he himself would command, and a Southern Attack Group, to be led by Rear Admiral Lawrence F. Reifsnider. To facilitate joint planning, Conolly and key members of his staff flew to Guadalcanal, arriving on 15 April, and set up temporary headquarters near the CP of the landing force.[7]

Admiral Conolly's task force was the naval echelon immediately superior to the Southern Troops and Landing Force. That organization traced back to the I Marine Amphibious Corps (IMAC) activated in November 1942. On 10 November 1943, after the successful start of the Bougainville operation, the then corps commander, Lieutenant General Alexander A. Vandegrift, left the Pacific to become 18th Commandant of the Marine Corps. He was relieved by Major General Geiger, who had led the 1st Marine Aircraft Wing at Guadalcanal.[8] On 15 April 1944, IMAC became the III Amphibious Corps (IIIAC), still under

[6] Unless otherwise noted, the material in this section is derived from: *TF 51 OpRpt; TF 56 OpRpt;* TF 56 OPlan 2–44, dtd 11Apr44; *TF 53 OpRpt; IIIAC SAR;* IIIAC OPlan 1–44, dtd 11May44, hereafter *IIIAC OPlan 1–44;* III Corps Arty SAR, dtd 2Sep44, hereafter *Corps Arty SAR.*

[7] This move by CTF 53 "permitted detailed planning between the two staffs of very great benefit to both, as well as cementing personal relations, neither feasible by dispatch communication." LtGen Merwin H. Silverthorn ltr to CMC, dtd 9Jun65.

[8] It was the introductory command of a large ground force for the veteran Marine airman, who had commanded a bombing squadron in World War I. General Geiger had been so much a part of Marine aviation since its early years that friends liked to say he had been "weaned on aviation gasoline." *Newsweek,* 31Jul44, p. 25.

Geiger and with headquarters on Guadalcanal.

The III Amphibious Corps consisted largely of the 3d Marine Division, the 1st Provisional Marine Brigade, and Corps Artillery. The division had returned to Guadalcanal in January 1944 after its first campaign, the battle for Bougainville, and had camped at Coconut Grove, Tetere. Few command changes took place. Major General Allen H. Turnage retained command for the Guam campaign; Brigadier General Alfred H. Noble became assistant division commander, relieving Brigadier General Oscar R. Cauldwell; and Colonel Ray A. Robinson relieved Colonel Robert Blake as chief of staff, the latter assuming command of the 21st Marines. On 21 April, Colonel Blake was relieved by Lieutenant Colonel Arthur H. Butler who was promoted to colonel shortly thereafter and then led the regiment on Guam. The other regiments that comprised the division were the 3d, the 9th, the 12th (artillery), and the 19th (engineer).

The 1st Provisional Marine Brigade was just a few months old, having been organized at Pearl Harbor on 22 March 1944, but the Marines that composed it were battle-tried men. The historic 4th Marines, with its traditions of Dominican and China service, and lastly of Corregidor,[9] had been reactivated on Guadalcanal on 1 February 1944, absorbing the famed Marine raiders, veterans of fighting in the Solomons.[10] Lieutenant Colonel Alan Shapley, who had commanded the 1st Raider Regiment, was assigned to command the 4th Marines; he led the regiment in the seizure of Emirau Island in March.

The other major unit in the brigade was the 22d Marines, which had fought at Eniwetok before coming to Guadalcanal in April 1944. Colonel John T. Walker, who had commanded the 22d in the Marshalls, became temporary commander of the brigade on 10 April 1944, when Brigadier General Thomas E. Watson, its first commander, was assigned to lead the 2d Marine Division. On 16 April, Brigadier General Lemuel C. Shepherd, Jr. assumed command of the brigade, but Colonel Walker remained as chief of staff, leaving the 22d Marines under Colonel Merlin F. Schneider. The new commander of the brigade had served with the old 4th Marine Regiment, having been its adjutant in Shanghai for a period during the 1920s. Now, with the new 4th Marines part of his command, General Shepherd arrived at Guadalcanal on 22 April from duty as ADC of the 1st Marine Division during the Cape Gloucester campaign on New Britain.

Planning for the Guam operation began immediately, but as General Shepherd later noted:

[9] For the story of the 4th Marines at Corregidor see Volume I of this series, pp. 155–202, and Kenneth W. Condit and Edwin T. Turnbladh, *Hold High the Torch, A History of the 4th Marines* (Washington: HistBr, G–3 Div, HQMC, 1960), pp. 195–240, hereafter Condit and Turnbladh, *Hold High the Torch.*

[10] In the reactivation of the 4th Marines, the Headquarters and Service Company of the 1st Raider Regiment became the same type of unit for the new regiment. The 1st, 3d, and 4th Raider Battalions became the 1st, 2d, and 3d Battalions, 4th Marines. The Regimental Weapons Company was formed from the 2d Raider Battalion.

> . . . the limited staff provided the 1st Brigade and lack of an adequate Headquarters organization, placed a heavy load on the Brigade Commander and his Chief of Staff. Since each of the two Regiments composing the Brigade had operated independently in previous campaigns the task of molding these infantry units and their supporting elements into a unified command presented many problems to the new commander and his embryo staff in the limited time available before embarkation for the Guam operation. With customary Marine sagacity, however, plans were completed and units readied for embarkation on schedule.[11]

The artillery component of IIIAC had been activated originally in IMAC on 13 April 1944 and then consisted of the 1st 155mm Howitzer Battalion and the 2d 155mm Gun Battalion, in addition to the 3d, 4th, 9th, 11th, 12th, and 14th Defense Battalions. Two days later, when IMAC was redesignated, the artillery organization became III Corps Artillery and the 2d 155mm Gun Battalion was redesignated the 7th.[12] For the Guam operation, it was decided to employ the two 155mm artillery battalions and the 9th and 14th Defense Battalions. Elements of the 9th were attached to the brigade, and units of the 14th would serve with the division. On 16 July, the 2d 155mm Howitzer Battalion of the V Amphibious Corps was added. It replaced the 305th Field Artillery Battalion (105mm) and the 306th Field Artillery Battalion (155mm howitzer), which were reattached to their parent 77th Division for the landing.

[11] Gen Lemuel C. Shepherd, Jr., cmts on draft MS, dtd 22Jun65.

[12] The designation "III Corps Artillery" appears in the organization's documents and will be used here.

Named to command the III Corps Artillery was Brigadier General Pedro A. del Valle. He had led an artillery regiment, the 11th Marines, in the battle for Guadalcanal. At Guam he would have control over all artillery and antiaircraft units in the STLF. Under his command also would be a Marine observation squadron (VMO-1), equipped with eight high-wing monoplanes.

Once Guam was again under the American flag, Marine Major General Henry L. Larsen's garrison force would begin its mission. The prospective island commander and part of his staff arrived at IIIAC headquarters on Guadalcanal on 29 May. The time proved somewhat early, considering the postponement of W-Day, but it helped to unify the total plans for Guam.

GUAM, 1898–1941 [13]

The delay of the Guam landings was not without some benefits. For one thing, it permitted American military intelligence to gain a better knowledge of the island and of Japanese defenses there. The easy capture of Guam by the enemy in 1941 followed years of neglect by the United States. In 1898 the American Navy had wanted

[13] Unless otherwise noted, the material in this section is derived from: TF 53 OPlan No. A162–44, Anx B, dtd 17May44, hereafter *TF 53 OPlan No. A162–44;* Capt Lucius W. Johnson (MC), USN, "Guam—Before December 1941," *USNI Proceedings,* v. 68, no. 7 (Jul42); Laura Thompson, *Guam and Its People* (Princton, N. J.: Princeton University Press, 1947), hereafter Thompson, *Guam and Its People.*

Guam chiefly as a coaling station for vessels going to the Philippines. The other islands of the Marianas, including Saipan and Tinian, were left to Spain, which sold them the next year to Germany.[14] In 1919, by the Treaty of Versailles, Japan received those islands as mandates, a fact that put Guam in the midst of the Japanese Marianas.

At this time, it seemed unlikely to many Americans that they would ever be at war with Japan. In 1923, when one of the worst earthquakes of history devastated Japan, Americans gave generously to relieve the suffering. A year before, the United States had joined with Japan, Great Britain, France, and Italy in a treaty that restricted naval armament and fortifications in the Pacific. As one result, the United States agreed to remove the six 7-inch coastal guns that had been emplaced on Guam. The last gun was removed by 1930.

Japan withdrew from the arrangement in 1936, but by then the treaty had quashed some ambitious planning by American naval officers to fortify Guam. The idea of turning the island into a major base had not been supported, however, either by the Secretary of the Navy or by the Congress, which was averse to large military appropriations.[15] As late as 1938, it refused to fortify Guam.

The collapse of efforts to transform Guam from a naval station into a major naval base did not, however, put an end to preparing plans on paper. In 1921, the Commandant of the Marine Corps approved a plan of operation in the event of war with Japan. From 1936 on, officers attending the Marine Corps Schools at Quantico bent over a "Guam Problem," dealing with capture and defense of the island.[16]

As for Guam, it remained a minor naval station, useful mainly as a communications center. It had become a relay point for the trans-Pacific cable, and the Navy set up a powerful radio station at Agana, the capital.[17] Few ships docked at the large but poorly improved Apra Harbor. In 1936, Pan American clippers began to stop at the island, bringing more contact with the passing world. No military airfield existed, although plans were underway to build one in late 1941.

The Governor of Guam was a naval officer, usually a captain, who served also as Commandant of the United

[14] The Spanish occupation of Guam had begun in 1668 when a few Jesuits and soldiers founded a mission on the island.

[15] In a letter to the Navy General Board on 10 June 1920, Secretary Josephus Daniels cited Rear Admiral Alfred T. Mahan, then the oracle of naval thought. It was Mahan's view that unless Guam were impregnably fortified, a costly project, military resources should not be squandered on the island simply for Japan to take if war came. Quoted in Stockman and Carleton, *Campaign for the Marianas.* "The Guam Operation," p. 1.

[16] Actually, the solution to the Guam problem taught at the Marine Corps Schools was rather out of date by 1944, so rapid had been the development of Marine Corps amphibious craft and techniques. A number of the officers that later served at Guam were grateful, however, for what they had learned about the island. Col William F. Coleman ltr to CMC, dtd 5Sep52.

[17] In the spelling of Agana and other Spanish proper names the anglicized usage of American reports has been followed.

States Naval Station. He controlled the small Marine garrison with its barracks at the village of Sumay, overlooking Apra Harbor. The Marines guarded installations such as Piti Navy Yard and the governor's palace. Ten Marine aviators and their seaplanes had been sent to Guam in 1921, but they were withdrawn in 1931 and no others came until 13 years later.[18]

The American, however, is a Robinson Crusoe on whatever island he finds himself. On Guam he fostered the health of the natives, developed compulsory education, and improved the road and water supply systems. The naval administration also took some interest in the economic welfare of the island, encouraging small industries such as those manufacturing soap and ice, but avoiding interference with the farmers' preferred old-fashioned methods. Little was exported from the island except copra, the dried meat of the ripe coconut. The largest market for native products was the naval colony itself. In addition, the Navy employed many Guamanians in the schools, the hospital, and other government departments. A heritage of that service was a devoted loyalty to the United States, which was not forgotten when war came and the Japanese occupied the island. The enemy made no attempt to use the conquered people as a military force but did press them into labor digging trenches, constructing fortifications, and carrying supplies.

The Guamanians are a racial mixture of the original islanders—the intelligent and gentle Chamorros, a Polynesian people from Asia—and Spanish or Filipino colonists. In 1940, the governor reported the native population as 21,502. It was concentrated near the harbor of Apra; about half of the number dwelt at Agana, and another 3,800 lived in villages close to the capital. Piti, the port of entry for Agana, located about five miles southwest of it, contained 1,175 inhabitants. Asan, a village between Agana and Piti, had 656. The municipality second in size to Agana was Sumay, with a population of 1,997, on the northeast shore of the Orote Peninsula. Here, in addition to the Marine barracks and rifle range, were the headquarters of the Pacific Cable Company and of Pan American Airways. The rest of the peninsula consisted of rolling terrain, marked by tropical vegetation, with some mangrove swamps and a few coconut groves. (See Map 24).

Six other villages in the southern half of the island accounted for 5,000 of the population: Agat, Umatac, Merizo, Inarajan, Yona, and Sinajana.[19] Most of the other natives lived on farms, some near rural centers like Talofofo in the south, or Dededo, Barrigada, Machanao, and Yigo in the north. Such centers included simply a chapel, a school, and a store. On the farms, most of which were located in southern Guam, the natives raised livestock, corn (the chief food staple), vegetables, rice, and fruit. Villagers, too, would sometimes have a plot of land that they tilled. The farmers took their products to market on carts

[18] Sherrod, *Marine Air History*, p. 27.

[19] On the southwest coast at Umatac was a marker claiming that in 1521 the explorer Magellan stopped there while on his famous voyage around the world. He is credited with discovery of the Marianas.

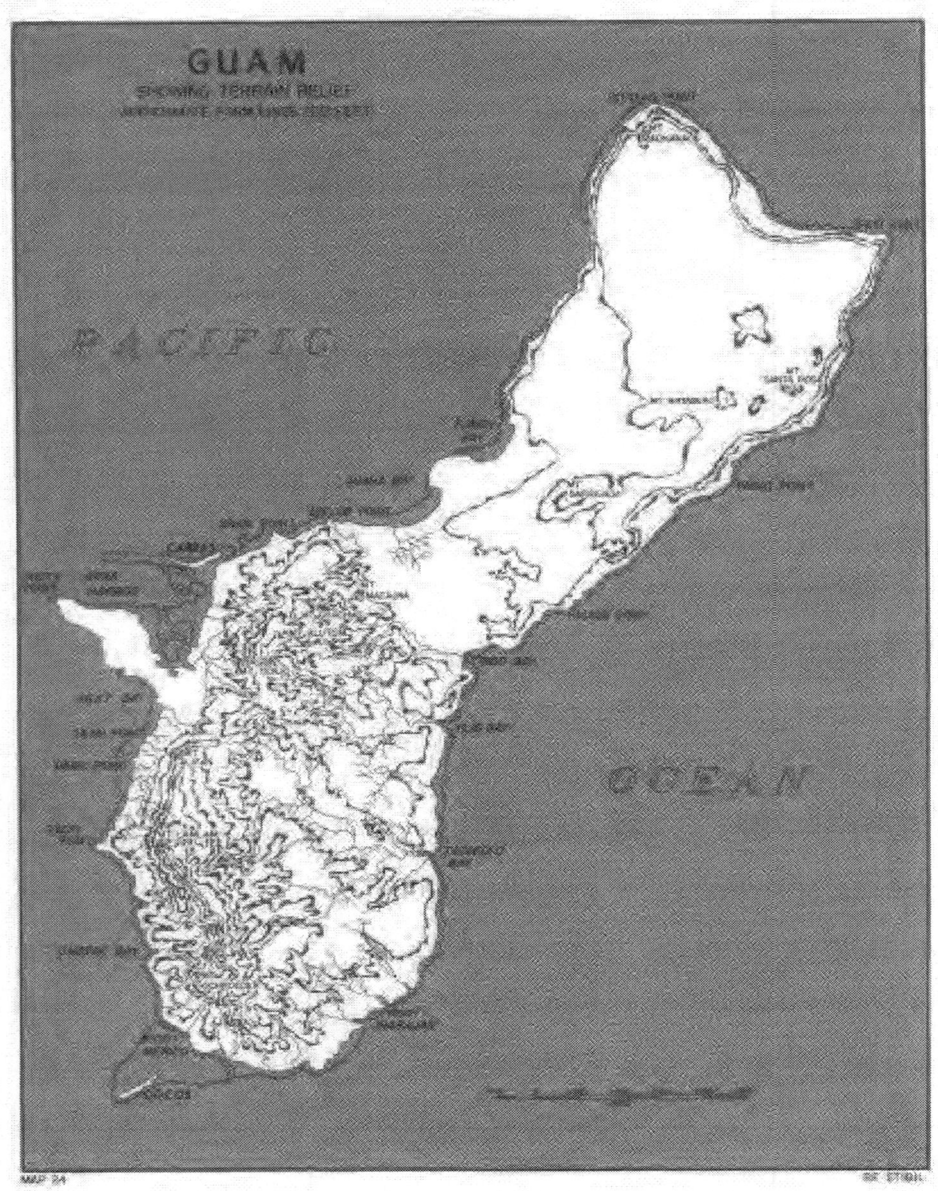

MAP 24

drawn by the carabao, a water buffalo.

The people of Guam were under the supreme authority of the governor, but not unhappily. They did not receive American citizenship, but they had the status of American nationals and their leaders served on the governor's staff of advisors. When the Japanese seized the island in 1941, they tried at first to preserve the contentment of the natives.[20] They offended the Guamanians, however, by changing the name of their homeland to "Omiyajima" (Great Shrine Island) and that of their capital to "Akashi" (Red or Bright Stone). The schools were ordered to teach Japanese instead of English. In 1944, as the Japanese rushed work on island defenses, they closed the schools and required labor even by children. Much of the island food supply was taken over by the expanded garrison. Native health and welfare was neglected because the Japanese became engrossed in preparing for the American invasion.

The enemy had a sizable territory to get ready. Guam is the largest island north of the equator between Hawaii and the Philippines. With an area of 225 square miles, it is three times the size of Saipan and measures 30 miles long by 4 to 8½ miles wide. The island is encircled by a fringing coral reef, ranging in width from 25 to 700 yards. For the most part, the coastline was familiar only to the native fishermen, although the United States Navy had prepared some good hydrographic charts.

Guam consists actually of two topographic entities, the north and the south, joined by a neck of land between Agana and Pago Bays. A small river starts in the Agana lowland and empties into the bay. North of that central strip, the island is largely a *cascajo* (coral limestone) plateau, covered with hardwood trees and dense tropical vegetation, but partly useful for agriculture. The southern half of Guam is the truly agricultural section, where streams flow through fertile valleys. Cattle, deer, and horses graze upon the hills. A sword grass, called *neti*, is common to the whole island.

A warm and damp air hangs over the land, but the temperature seldom rises much above the average of 87 degrees. Like the rest of the Marianas, Guam was often called "the white man's tropics." From July to December, however, the island is soaked by rain nearly every day. The road system became rough to travel when it rained. A mere 100 miles of hard-surfaced roads were joined by unsurfaced roads and jungle trails which turned to mud when wet.[21] The main road on the island ran from Agat along the west coast to

[20] Prior to the Japanese invasion, there had been only 39 Japanese adults on the island and 211 Japanese-Guamanian children. According to the same count, in 1940, only 13 Spaniards still dwelt on the island, though Spain had possessed Guam for nearly 200 years.

[21] The commanding officer of the 9th Marines noted that the hard-surfaced roads were actually very narrow with just enough room for two cars to pass. "They were surfaced with a combination of native cascajo and crushed stone which was rolled down. Apparently little maintenance had been done during the Japanese occupation and in addition ships' gunfire and air bombing had seriously damaged them in hundreds of places." LtGen Edward A. Craig ltr to ACofS, G-3, dtd 22Jun65, hereafter *Craig 22Jun65 ltr*.

Agana, thence northeast to Finegayan, east of Tumon Bay. There it split into two parallel branches, both ending near Mt. Machanao. Despite the high precipitation, problems of water supply had occurred until the Americans constructed some reservoirs. The water system was then centered in the Almagosa reservoirs around Agat.

A number of elevations, high and low, appear on the landscape, but there are no real mountains. Cliffs rim the shoreline of the northern plateau, from Fadian Point to Tumon Bay. Above the tableland itself Mt. Santa Rosa (840 feet) rises in the northeast. At the center of the plateau is Mt. Mataguac (620 feet), and near the northern tip lies Mt. Machanao (576 feet). Marking the southern edge of the plateau, Mt. Barrigada rises to 640 feet, and from its slopes a 200-foot bluff reaches west to upper Agana Bay. These hills are not so high as those of southern Guam, but they are comparably rocky at the top and covered on the sides with shrubs and weeds.

A long mountain range lies along the west side of southern Guam from Adelup Point south to Port Ajayan at the tip of the island. Parts of the mountain range, such as Chonito Cliff near Adelup Point, rise very close to the shore. Inland of Apra Harbor is a hill mass, with a maximum height of 1,082 feet. Here are Mt. Chachao, Mt. Alutom, and Mt. Tenjo. The highest hill on Guam—Mt. Lamlam (1,334 feet)—rises from the ridge line below the Chachao-Alutom-Tenjo massif. Conspicuous in the ridge, which starts opposite Agat Bay, are the heights of Mts. Alifan and Taene. Near Agana Bay and the central lowland, which links northern and southern Guam, lies Mt. Macajna.

Several prominent points of land that figured in the fighting jut from the west coast of the island—Facpi, Bangi, Gaan, Asan, and Adelup. On the northern end are Ritidian Point and Pati Point. The largest island off the coast is Cabras, a slender finger of coral limestone about a mile long; the island partly shelters Apra Harbor. Others, like Alutom, Anae, Neye, and Yona, are hardly more than islets. Rivers are numerous on Guam but most of them are small. The Talofofo and the Ylig are difficult to ford on foot, but others are easy to cross except when they are flooded.

Such geographic forms were known to Navy and Marine officers that had been stationed on Guam, and from those men was gleaned much of the intelligence necessary for planning the operation. Other sources were natives that were serving in the United States Navy at the time the Japanese seized Guam. Despite the fact that Guam had been an American possession for almost a half-century, the sum total of knowledge held by American authorities concerning the island was relatively small. In February 1944, the Office of Naval Intelligence issued a useful "Strategic Study of Guam;" the data was compiled by Lieutenant Colonel Floyd A. Stephenson, who had served with the Marine garrison on the island, and who returned in July 1944 as Commanding Officer, IIIAC Headquarters and Service Battalion.

The Marine Corps Schools had prepared some materials in connection with the "Guam Problem," and its map of the island was of particular use. It formed the basis of the maps drawn by the Joint Intelligence Center, Pacific Ocean Areas and furnished to the III Amphibious Corps. The cartographers at the Marine Corps Schools had worked from ground surveys made by Army engineers, but the Corps C–2 complained that the contours on the maps they received "did not portray anything like a true picture of the terrain except in isolated instances." [22] The road net, they said, was "generally correct" but did not show recent changes or roads built by the Japanese.

To correct such omissions and errors, aerial photography was called for. The first photo mission, flown on 25 April 1944, suffered the handicap of cloud cover, but subsequent flights in May and June produced somewhat better results, and photographic reconnaissance was kept up until after the landings. A naval officer, Commander Richard F. Armknecht, who had been a public works officer on Guam, prepared an excellent relief map, based largely upon his own thorough knowledge of the terrain. Admiral Conolly was so impressed with the map that he ordered several more made to give the fire support ships for study. By these means, information about the island was expanded, but knowledge of it was still deficient, especially regarding the areas of vegetation and the topography.[23]

Intelligence of the coastline was obtained by a submarine and by UDT men. The USS *Greenling* got some good oblique photographs of the beaches and also took depth soundings and checked tides and currents. The underwater demolition teams started their reconnaissance and clearing of the assault beaches on 14 July. The men destroyed 940 obstacles, 640 off Asan and 300 off Agat; most of these were palm log cribs or wire cages filled with coral. Barbed wire was sparsely emplaced, however, and no underwater mines were found. On the reef that men of the 3d Division would cross, the UDT men put up a sign: "Welcome, Marines!" [24]

THE JAPANESE ON GUAM, 1941–1944 [25]

The size of the Japanese force and the state of its recent defenses indicated that the enemy did not plan a

[22] *IIIAC SAR*, Encl C, p. 1.

[23] The lack of information regarding the reefs off the landing beaches and the belief that they might drop off sharply on the seaward side prompted the 3d Division to provide for a certain number of LVT(A)s, "after landing and providing support fire for the infantry . . . to return to the edge of the reef at the time of the landing of the tanks in LCMs. The LVT(A)s were then to provide 'anchors' for the LCMs and to guide the tanks onto the beach." BGen Louis Metzger memo to ACofS, G–3, HQMC, dtd 19Jul65, hereafter *Metzger memo*.

[24] Morison, *New Guinea and the Marianas*, p. 380.

[25] Unless otherwise noted, the material in this section is derived from: *TF 56 OpRpt; IIIAC SAR;* LtCol Hideyuki Takeda, IJA, "Outline of Japanese Defense Plan and Battle of Guam," encl to LtCol William F. Coleman ltr to CMC, dtd 4Oct46, hereafter *Takeda ltr I;* Mr. Hideyuki Takeda ltr to Dir of MarCorps Hist, dtd 20Feb52, hereafter *Takeda ltr II;* Japanese GSDF Staff School, "How the Guam Operation Was Conducted," translation of a series of articles published in the staff school journal, *Kambu Gakko Kiji* (Oct–Dec62), hereafter *GSDF Study*.

cordial welcome for the Marines.[26] After the Japanese seized Guam in 1941 they undertook no better preparation to defend it than the Americans had done. The enemy left only 150 sailors on the island—the *54th Keibitai*, a naval guard unit. Guam and other isolated Pacific islands were regarded merely as key points of the patrol network, not requiring Army troops for defense.

In late 1943, Japan became fearful of an American push through the Central Pacific and put new emphasis upon defense of the Marianas. As a result, the *13th Division*, which had been fighting in China since 1937, was slated for duty in the Marianas. In October 1943, an advance detachment of about 300 men sailed for Guam, but military developments in south China prevented the sending of the rest of the division. Instead, the *29th Division* was substituted. It took up the great responsibility indicated by advice from Tokyo: "The Mariana Islands are Japan's final defensive line. Loss of these islands signifies Japan's surrender." [27]

The *29th Division* had been a reserve for the *Kwantung Army*. In February 1944, while undergoing anti-Soviet combat training in Manchuria, the unit received its marching order for the Marianas. Horses were left behind,[28] the troops were supplied with summer uniforms, and on 24 February the division embarked in three ships at Pusan, Korea. On board the *Sakito Maru* was the *18th Infantry Regiment;* on another ship, the *50th Infantry Regiment;* and on the third, the *38th Infantry Regiment* and division headquarters.

Disaster befell the *Sakito Maru* when it was just 48 hours' sailing distance from Saipan. The American submarine *Trout* sank the ship by a torpedo attack at 1140 on 29 February. The regimental commander and about 2,200 of the 3,500 men on board ship were drowned. In addition, eight tanks and most of the regimental artillery and heavy equipment were lost.[29] The two destroyers of the convoy picked up survivors and took them to Saipan, where the *18th* was reorganized. The *1st Battalion* stayed on Saipan, and the tank company went to Tinian. The regimental headquarters, newly commanded by Colonel Hiko-Shiro Ohashi, and two battalions were sent to Guam, arriving there on 4 June. The regiment brought along two companies of the *9th Tank Regiment*. The two infantry battalions each had three rifle companies, a trench mortar company (seven 90mms), and a pioneer unit. Minus the battalion on Saipan, the regiment numbered but 1,300 men after the reorganization.

The *50th Infantry Regiment* went to Tinian. Division headquarters, with Lieutenant General Takeshi Takashina

[26] The C-2 of the IIIAC compiled a final Japanese order of battle summary, which proved remarkably accurate. The tabulation is the basis for Appendix VII, "Japanese Order of Battle on Guam," in Lodge, *Recapture of Guam*, pp. 196–197.

[27] *GSDF Study*, p. 68.

[28] Before departing for the Marianas, the *29th Division* was streamlined into an RCT type of organization. The engineer, cavalry, and transport regiments were dropped, and a tank unit was added. Each of the infantry regiments was assigned an artillery battalion and an engineer company.

[29] Losses according to *GSDF Study*, p. 68.

commanding, and the *38th Infantry Regiment* (Colonel Tsunetaro Suenaga) proceeded to Guam, arriving there on 4 March. This regiment numbered 2,894 men and included signal, intendance (finance and quartermaster), medical, transport, and engineer units. Its three infantry battalions each contained three rifle companies, an infantry gun company, and a machine gun company. Attached to each infantry battalion was one battery of 75mm guns from the regimental artillery battalion.

The second largest Army component sent to Guam was the *6th Expeditionary Force*, which sailed from Pusan and reached Guam on 20 March. This unit totaled about 4,700 men drawn from the *1st* and *11th Divisions* of the *Kwantung Army;* it comprised six infantry battalions, two artillery battalions, and two engineer companies. On Guam, the force was reorganized into the 2,800–man *48th Independent Mixed Brigade* (*IMB*), under the command of Major General Kiyoshi Shigematsu, who had brought the force to the Marianas, and the 1,900–man *10th Independent Mixed Regiment* (*IMR*), commanded by Colonel Ichiro Kataoka.

The infantry battalions of the *48th IMB* and the *10th IMR* included three rifle companies, a machine gun company, and an infantry gun company (two 47mm antitank guns and either two or four 70mm howitzers). The infantry battalions of the 38th Regiment had the same organization, except that the gun company had four 37mm antitank guns and two howitzers.

The total number of Army troops, including miscellaneous units, came to about 11,500 men.[30] The overall command of both Army and Navy units on Guam went to General Takashina, whose headquarters strength was estimated at 1,370. Upon arrival on the island, he had been given the Southern Marianas Group, which included Guam and Rota, and, after the fall of Saipan, also Tinian. Defense of the entire Marianas was the responsibility of General Obata, commander of the *Thirty-first Army,* into which the *29th Division* had been incorporated, but the general left the immediate defense of Guam to the division commander.

In February 1944, the Japanese naval units on Guam had comprised about 450 men. From then on, however, the *54th Keibitai* was steadily reinforced by additional coast defense and antiaircraft units, so that by July the organization totaled some 2,300 men commanded by Captain Yutaka Sugimoto, once island commander. Two naval construction battalions had 1,800 men relatively untrained for fighting. With nearly 1,000 miscellaneous personnel, the figure for naval ground units reached about 5,000. Naval air units probably held some 2,000 men.[31] Most accounts agree that the entire Japanese troop strength on

[30] Major Lodge gives this total in *Recapture of Guam,* p. 197. It is lower and more precise than the IIIAC C–2 figure of about 13,000 because the latter estimate included several units which, as it turned out, were not present during the battle for Guam. See *IIIAC SAR,* Encl C, Intelligence.

[31] This figure, cited in Lodge, *Recapture of Guam,* App. VII, p. 197 embraces facts revealed later than the IIIAC C–2 report which put enemy air unit strength at 600.

Guam totaled a minimum of 18,500 men.[32]

On 23 June, the *1st Battalion* of the *10th IMR*, with one artillery battery and an engineer platoon attached, moved to Rota for garrison duty. Shortly thereafter, the battalion was joined by a task force composed of the *3d Battalion, 18th Regiment*, supporting engineers, and amphibious transport units; the object, a counterlanding on Saipan. The condition of the sea made such a mission impossible, however, so *3/18* returned to Guam on 29 June. The *1st Battalion, 10th IMR* remained on Rota, but since it could possibly be transferred in barges to Guam, both American and Japanese listings included it as part of defensive strength of the larger island.

Documents showing enemy strength figures and unit dispositions fell into American hands with the capture of *Thirty-first Army* headquarters on Saipan. Such information helped IIIAC intelligence officers prepare a reliable sketch map indicating the main Japanese defensive dispositions as of late June. General Takashina had set them up on the premise that the landing of a division-level unit was possible on beaches in the Tumon Bay-Agana Bay-Piti coastal section and the beach of Agat Bay. The Japanese were expecting four or five American divisions, a force adequate for landing operations at two fronts. (See Map 25.)

The enemy's immediate concern was the defense of Apra Harbor and of the island airfields. Construction of a military airfield near Sumay on the Orote Peninsula (occupying the golf course of the former Marine Barracks) had not been started until November 1943. The Japanese based about 30 fighter planes here. In early 1944, construction was begun on two other airfields, one at Tiyan near Agana and the other in the vicinity of Dededo. The Tiyan (Agana) airfield became operational by summer. This was intended for use by medium attack planes; the Japanese had six of those on the island.

Assigned to the Agana sector, which covered that part of the west coast from Piti to Tumon Bay, were the four infantry battalions of the *48th Brigade*. The *319th Independent Infantry Battalion* was positioned inland, east of Agana, in reserve. The *320th Battalion* manned defenses near the coastline between Adelup Point and Asan Point. The *321st Battalion* was located around Agana Bay, and the *322d Battalion* at Tumon Bay. The Agana sector received most of the Army artillery: the brigade artillery unit and the two artillery batteries of the *10th IMR*, all under the control of the *48th IMB*.[33] In the Agana sector also were naval land combat troops holding the capital city, most of the *29th Division* service troops, and General Takashina's command post at Fonte. The *3d Battalion, 38th Infantry*, initially stationed in reserve behind the *48th Brigade* positions, was returned to Colonel Suenaga's control in July and moved south to rejoin its regiment.

The rest of the *38th Infantry* had

[32] The G-2 of TF 56 put the figure at 18,657, "excluding aviation." *TF 56 OpRpt*, Encl D, App H, p. 3.

[33] The brigade artillery unit was formerly the *3d Battalion, 11th Mountain Artillery Regiment, 11th Division.*

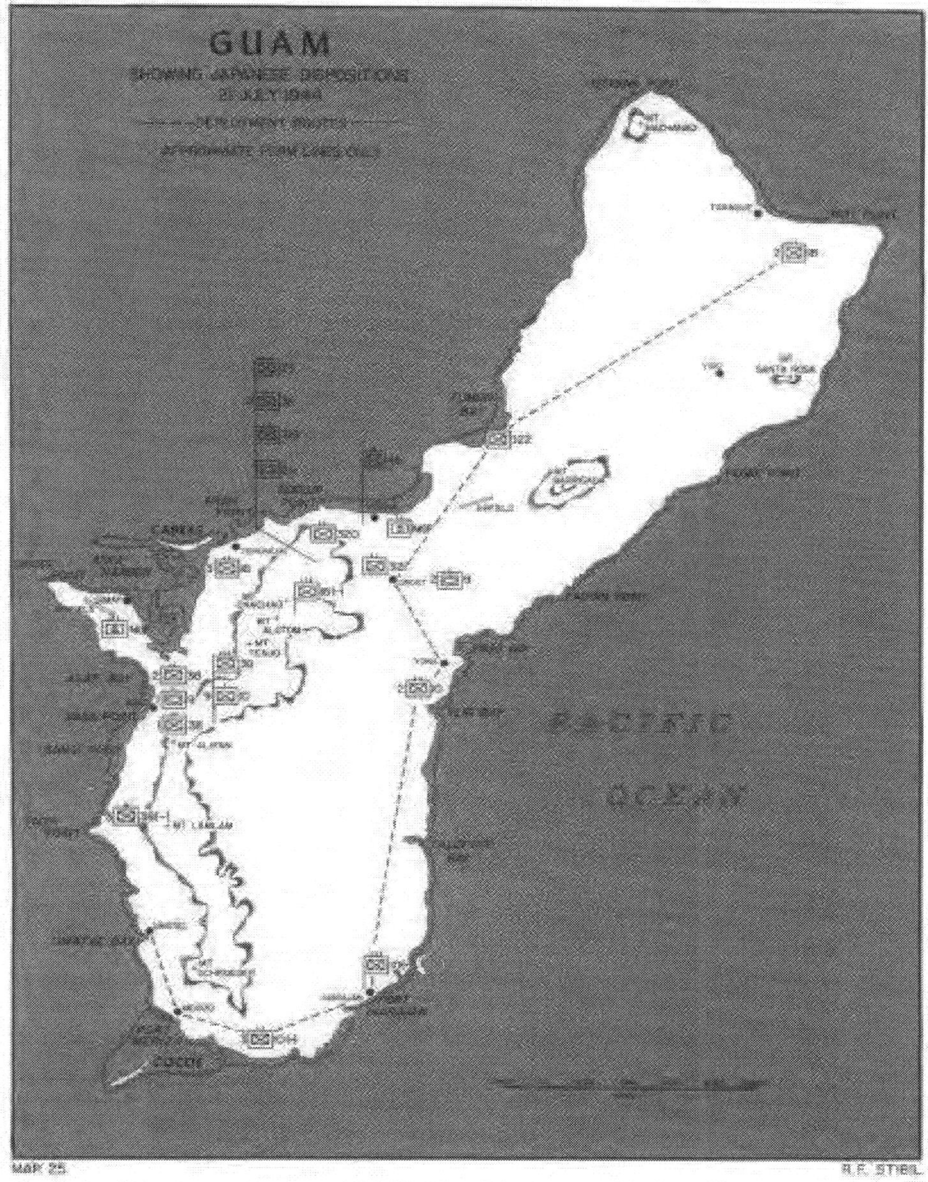

MAP 25 R.F. STIBIL

been put into the Agat sector, which stretched along the coast between Agat Bay and Facpi Point, with *1/38* covering the Agat Beach area. Colonel Suenaga's command post was located on Mt. Alifan. The Agat sector included the Orote Peninsula, where most of the naval infantry, the *60th Antiaircraft Defense Unit*, and coast defense elements of the *54th Keibitai* were stationed; *2/38* occupied the base of the peninsula. Completing the troop organization for the peninsula was the *755th Air Group*, which had reorganized its 700 men for ground combat.

Until General Takashina was fairly sure where the unpredictable Americans would land, he kept some troops in other parts of the island. Stationed in southeast Guam until July was the *10th IMR* (less *1/10* on Rota). In extreme northern Guam was the *2d Battalion, 18th Regiment*. The *3d Battalion* of the regiment, after failing to get to Saipan, took up defense positions between Piti and Asan Point in the Agana sector. General Shigematsu, commanding the *48th IMB*, had the responsibility for island defense outside the Agat sector, which was under Colonel Suenaga.

In early July, the Japanese virtually abandoned other defense positions and began to localize near the expected landing beaches on the west coast. The *10th IMR* went to Yona, thence to positions in the Fonte-Ordot area—ominously near the Asan beaches. *The 9th Company* of the regiment was ordered into a reserve position near Mt. Alifan to lend support to the *38th Regiment*. Most of the *2d Battalion, 18th Regiment* was brought south to back up the *320th Independent Infantry Battalion;* the *5th Company* of *2/18* was put to work constructing defenses in the hills between Adelup Point and Asan Point. These troop movements, made mostly at night, were handicapped by the American bombardment.

The enemy's armor was shifted around as the Japanese got ready. The tank units were positioned in reserve, prepared to strike the beachhead with the infantry. One was the *24th Tank Company*, assigned as the division tank unit, with nine light tanks (eight of its tanks had been lost in the sinking of the *18th Regiment* transport). That company was put at Ordot, inland of Fonte. The *2d Company, 9th Tank Regiment*, with 12 to 14 tanks, mostly mediums, was turned over to the *48th IMB*. The *1st Company, 9th Tank Regiment*, with 12 to 15 light tanks, was assigned to the *38th Regiment* and took up a position to the rear of the Agat beaches.

The Japanese fortification of Guam was, like the buildup of manpower on the island, a hasty development. Before the *29th Division* was stationed here, the enemy had only a few batteries on the island, and these were not dug in. The principal armament consisted of 75mm field guns, the largest caliber artillery was 150mm. Two cave-type dugouts for the communications center at Agana were under construction, and a concrete naval communications station was being built at Fonte.

In the fever of preparations after 1943, the Japanese armed the ground from Tumon Bay to Facpi Point, providing concrete pillboxes, elaborate trench systems, and machine gun emplacements. Mortars, artillery, and coast defense guns were positioned along the coast. The number of antiaircraft weapons was increased; the *52d Field Antiaircraft Battalion* was assigned to the Orote airfield and the *45th Independent Antiaircraft Company* to Tiyan. In the defense of the Orote Peninsula, the 75mm antiaircraft guns of the *52d* could serve as dual-purpose weapons, augmenting the artillery.

An unfortunate result of the postponing of W-Day was the extra time it afforded the Japanese to prepare. They overworked the naval construction battalions and native labor to bulwark the island, mostly in the vicinity of the beaches and the airfields. Some inland defenses were constructed, however, and supply dumps were scattered through the island.

American photo reconnaissance between 6 June and 4 July showed an increase of 141 machine gun or light antiaircraft positions, 51 artillery emplacements, and 36 medium antiaircraft positions. Better photographs may have accounted for the discovery of some of the additional finds; still, the buildup was remarkable considering the short period involved. The number of coast defense guns, heavy antiaircraft guns, and pillboxes had increased appreciably also. The distribution of weapons to Army forces on Guam was indicated from a captured document dated 1 June 1944:

14—105mm howitzers
10— 75mm guns (new type)
8— 75mm guns
40— 75mm pack howitzers (mountain)
9— 70mm howitzers (infantry)
8— 75mm antiaircraft guns (mobile)
6— 20mm antiaircraft machine cannon
24— 81mm mortars
9— 57mm antitank guns
30— 47mm antitank guns
47— 37mm antitank guns
231— 7.7mm machine guns
349— 7.7mm light machine guns
540— 50mm grenade dischargers [34]

Of grim significance in the enemy's defensive organization was their intention to deny land access to Orote Peninsula. A system of trenchworks and foxholes was constructed in depth across the neck of the peninsula and supported with large numbers of pillboxes, machine gun nests, and artillery positions. Rocks and tropical vegetation provided concealment and small hills lent commanding ground.

THE PREPARATORY BOMBARDMENT [35]

If the postponing of W-Day permitted the Japanese to put up more defenses, it also gave American warships and planes time to knock more of them down. Beginning on 8 July, the enemy was subjected to a continuous 13-day naval and air bombardment. It was the wholesale renewal of the first naval gunfire on 16 June, when two battleships, a cruiser, and a number of destroyers from Task Force 53 shelled the Orote Peninsula for some two hours, exciting Japanese fears of imminent invasion. Planes from Task

[34] *TF 56 OpRpt*, Encl D, App H, p. 7.

[35] Unless otherwise noted, the material in this section is derived from: *TF 51 OpRpt; TF 53 OpRpt; TF 53 OPlan No. A162–44; IIIAC SAR; Takeda ltr II;* Morison, *New Guinea and the Marianas.*

Force 58 had started bombing Guam on 11 June, hitting the enemy airfields particularly; by 20 June, the Japanese planes based there had been destroyed and the runways torn up.[36] On 27 June, Admiral Mitscher's airmen bombed Japanese ships in Apra Harbor. Then, on 4 July, destroyers of TF 58 celebrated the day by exploding 5-inch shells, like giant firecrackers, upon the terrain in the vicinity of Agana Bay, Asan Point, and Agat Bay.

Such gunfire, however, was a mere foretaste of what was to come from the sea and air. On 8 July, Admiral Conolly began the systematic bombardment, day after day, which was to assume a scale and length of time never before seen in World War II.[37] Destroyers and planes struck at the island, and on 12 July they were joined by battleships and cruisers. Two days later, Admiral Conolly, arriving in his command ship, the *Appalachian,* took personal charge of the bombardment. From then on it reached, the Japanese said, "near the limit bearable by humans."[38] The incessant fire not only hindered troop and vehicle movements and daytime work around positions, but it also dazed men's senses.[39]

Admiral Conolly, justifying his nickname "Close-in," took his flagship to 3,500 yards from the shore and went to his task with dedication. "He made a regular siege of it," wrote a naval historian.[40] On the *Appalachian,* a board of Marine and Navy air and gunnery specialists kept a daily check of what had been done and what was yet to be done. General Geiger, who was on board with Conolly, said that "the extended period of bombardment, plus a system of keeping target damage reports, accounted for practically every known Japanese gun that could seriously endanger our landing."[41] It was the belief of Admiral Conolly's staff that "not one fixed gun was left in commission on the west coast that was of greater size than a machine gun."[42]

Every exposed naval battery was believed destroyed; more than 50 percent of the installations built in the seashore area of the landing beaches were reported demolished.[43] A number of guns emplaced in caves, with limited fields of fire, were reduced in efficiency when naval shells wrecked the cave entrances. Some permanent constructions, however, which were thickly walled with concrete and *cascajo,* and at least partially dug into the earth, resisted even a direct hit.

Certain installations and weapons escaped. The Japanese reported later that their antiaircraft artillery on

[36] "As a result, there was no Japanese plane in the sky over Guam" on W-Day. *Takeda ltr I,* p. 3.

[37] From 8 July until W-Day the expenditure of naval ammunition against shore targets amounted to 836 rounds of 16-inch, 5,422 of 14-inch, 3,862 of 8-inch, 2,430 of 6-inch, and 16,214 of 5-inch shells. *TF 56 OpRpt,* Encl G, p. 71.

[38] *GSDF Study,* p. 117.

[39] There were "scattered outbreaks of serious loss of spirit." *Takeda ltr II,* p. 9.

[40] Morison, *New Guinea and the Marianas,* p. 378.

[41] *IIIAC SAR,* Encl G, p. 3.

[42] *TF 53 OpRpt,* Encl B, p. 11.

[43] These estimates of destruction were confirmed later by the Japanese. *Takeda ltr II,* pp. 7–9, contains a credible summary of what was and was not accomplished by the American preparatory bombardment.

Guam "sustained damage from naval gunfire only once," [44] and only once did water pipes receive a direct hit. Communications installations were constructed in dead spaces immune to bombardment, and practically no lines were cut by naval gunfire. Moreover, no damage was done to power installations because generators were housed in caves. The interior of the island was, of course, less the province of ships' guns than of roving aircraft; the Japanese claimed that naval gunfire had very little effect beyond four kilometers (roughly, two miles) from the shoreline. It thus did little damage to enemy construction in the valleys or the jungle.

While air bombardment and strafing was able to reach where naval gunfire could not, the Japanese mastery of the art of concealment still hampered destruction. On 28 June, Admiral Mitscher's aircraft began periodic strikes against Guam; then on 6 July, TF 58 and two carrier divisions of Admiral Conolly's TF 53 started the full-scale preparatory air bombardment. Targets included supply dumps, troop concentrations, bridges, artillery positions, and boats in military use. Most such craft were sunk by strafing, the rest by naval gunfire. Harbor installations were spared for use after the battle. In the period of the preparatory bombardment, the island was divided into two zones—naval gunfire and air alternated zones morning and afternoon. Aircraft were particularly useful at hindering Japanese troop movements; they were less effective against enemy gun emplacements.

[44] *Takeda ltr II*, p. 8.

On 12 July, before leaving Eniwetok for Guam, Admiral Conolly met with Admiral Mitscher, and they set up a schedule of intensified strikes, which were to take place from 18 July through W-Day. Mitscher greatly increased the number of aircraft available to Conolly for the final all-out attacks. Until 18 July, TF 58 had made strikes on Guam independently of Commander, Support Aircraft, Guam. For the period 18–20 July, the combined tonnage reached the figure 1,131, including bombs, depth charges, and rockets. The explosives were not delivered, however, without some losses to American aircraft. Sixteen naval planes were brought down by the Japanese antiaircraft fire before W-Day.

JAPANESE FORTUNE TELLING [45]

It was the focus of the intensified bombardment starting 18 July which tipped off American intentions. From the action of the ships at sea, rather than from any leaves in a teacup, the Japanese were able to foretell more specifically where the invaders would come ashore.[46] When UDT men

[45] Unless otherwise noted, the material in this section is derived from: CinCPac–CinCPOA Items No. 10452–10791, Translations of Captured Japanese Documents, dtd 29Sep44, hereafter *CinCPac–CinCPOA Item*, with appropriate number; *GSDF Study; Takeda ltr I; Takeda ltr II.*

[46] Some American naval officers felt that the bombardment on 16 June was a tip-off to the Japanese. The enemy knew, however, that the Orote Peninsula, as well as Apra Harbor, would be early objectives of an invasion, wherever it came. From the beginning, the Japanese had suspected that the landings would be on the west coast.

cleared obstacles from the chosen beaches, all doubt was removed.

In 1941, the Japanese had landed their main force at Tumon Bay, so at first they had supposed the Americans would attempt the same; the beach was ideal for an amphibious assault (at least two miles of sand), the reef was not impassable, and inland the ground rose gently. This judgment regarding the Tumon beaches did not give much weight to the factors that decided American planners against them—their distance from Apra Harbor and the highly defensible terrain that blocked the way to the harbor.

The enemy had not, however, really expected a repetition of their other landings elsewhere on the island, where neither the surf nor the ground was appropriate for a large-scale invasion. It was not until the middle of June, when the Americans began shelling the beaches below Tumon Bay, that the Japanese gave serious attention to fortifying the west coast south of the bay. Before then, they had viewed as dismaying to an invader the wide reef protecting the beaches here—"a reef varying in width from 200 to 500 yards offshore." [47] Moreover, on the commanding ground just inland, the defenders would have excellent observation for mortar and cannon fire.

As late as 16 July, General Shigematsu regarded the Agana sector as the probable area of invasion, with the Agat sector as a second target area if a two-front attack were staged. A landing force at Agat Bay could seize the Orote airfield. The white sandy beach along most of Agat Bay was comparable to that of Tumon Bay. Before the American bombardment shattered the picture, the beach was fringed with palm trees. The northern coastline of Agat Bay, along the Orote Peninsula, is different, however; there a fringe of cliffs ranges from 100 to 200 feet high.

The Japanese did not rule out a possible small American landing at Pago Bay on the east coast for the purpose of getting behind their lines, but General Takashina's defense efforts were almost wholly devoted to the west coast. *Imperial General Headquarters* doctrine insisted upon the destruction of the assault forces at the beaches, though Lieutenant Colonel Hideyuki Takeda, the perceptive operations officer of the *29th Division*, favored a deployment in depth at Guam.[48] If the Japanese beach defense units should fail to destroy the American landing force at the beaches, General Takashina had instructed the *10th IMR* and the two battalions of the *18th Regiment* to counterattack in force.

AMERICAN TACTICAL PLANS [49]

American tactical spadework for the assault on STEVEDORE, the code

[47] *TF 53 OpRpt*, Encl H, p. 5.

[48] *Imperial General Headquarters* soon modified its policy. In September 1944, Marines met a preplanned Japanese defense in depth for the first time at Peleliu.

[49] Unless otherwise noted, the material in this section is derived from: *FifthFlt FinalRpt; TF 51 OpRpt; TF 56 OpRpt; TF 56 OPlan A162-44; IIIAC SAR; IIIAC OPlan 1-44; Corps Arty SAR;* 1st ProvMarBrig Op and SAR, dtd 19Aug44, hereafter *1st ProvMarBrig SAR;* 1st ProvMarBrig OPlan 1-44, dtd 26May44, and Mod No. 5, dtd 12Jul44,

name assigned to Guam, had been started at Pearl Harbor as early as March 1944. General Geiger's staff prepared the tentative operation plan, which was approved by General Holland Smith on 3 April and shortly after by Admirals Turner and Spruance. The working out of details went forward on Guadalcanal, where, with the establishment of the 1st Provisional Marine Brigade command post on 2 May, every major unit of the corps was present.

On 17 May, General Geiger circulated the corps operation plan. As originally evolved at Pearl Harbor, it provided for a 3d Marine Division landing on beaches between Adelup Point and Asan Point, while to the south the 1st Provisional Marine Brigade was to go ashore between Agat village and Bangi Point. Subsequent junction of the beachheads was planned.

Early capture of the Orote Peninsula-Apra Harbor area was imperative to secure the use of the harbor and the Orote airfield. Here was, as General Holland Smith said, "the focal point of attack."[50] Upon General Shepherd's brigade fell the hard assignment of seizing the Orote Peninsula, a rock-bound fortress. In order to free the brigade for such a mission, the 305th Infantry of the 77th Division was attached on 10 July to follow the brigade ashore, while the rest of the Army division remained as corps reserve.[51]

hereafter *1st ProvMarBrig OPlan 1–44;* 77th InfDiv G–3 Jnl, 6Jun–10Aug44, hereafter *77th InfDiv Jnl.*

[50] Smith and Finch, *Coral and Brass*, p. 214.

[51] The attachment of the 305th RCT involved a modification of the brigade tactical plan, which had been issued on 26 May.

Major General Andrew D. Bruce, commanding the 77th, wanted to use his other two regiments for a secondary landing on northwest Guam about W-Day plus four to attack the Japanese rear, but it was felt that the Army troops should be kept in reserve, available for support at the beachheads.

The 305th Regiment was to relieve the brigade on the Force Beachhead Line (FBHL), which extended from Adelup Point along the Mt. Alutom-Mt. Tenjo-Mt. Alifan ridge line to Facpi Point. The brigade could then reorganize for the attack on the Orote Peninsula. Once that area was secured, the brigade would again take over the defense of the FBHL, while Army troops joined with the 3d Marine Division in seizing the rest of Guam.

The two assault points were five miles apart, creating the situation of two almost separate military operations by the same landing force. Owing to this distance, each of Admiral Conolly's two attack groups, northern and southern, would land and support its own assault troops.

In the north, the three regiments of the 3d Marine Division would land abreast on a 2,500-yard front—the 3d Marines on Beaches Red 1 and 2, the 21st Marines in the center on Beach Green, and the 9th Marines on Beach Blue. At one end of the front jutted Adelup Point and at the other, Asan Point; both had cave-like holes appropriate for enemy machine gun positions. Beyond the beaches lay dry rice paddies, yielding to the Fonte Ridge which overlooked the landing area. On 15 July, General Shigematsu moved his battle command post to this high ground. (See Map VII, Map Section.)

The 1st Provisional Marine Brigade would go ashore with two regiments abreast—the 22d Marines on Yellow Beaches 1 and 2; the 4th Marines on White Beaches 1 and 2, to the south. These beaches stretched more than a mile between Agat village and Bangi Point, with Gaan Point at the middle. The cliffs of the Orote Peninsula 2,000 yards to the north flanked the landing area. Neye Island, just off the peninsula, and Yona Island, near the White Beaches, rose from the water like enigmatic bystanders, probably carrying hidden weapons.

Two 155mm battalions of the III Corps Artillery were to land behind the brigade, whose artillery group included the 75mm pack howitzer battalion of each regiment and two other units to be attached on landing—the Army 305th Field Artillery Battalion and Battery C, 1st 155mm Howitzer Battalion. Artillery support for the 3d Division would be provided by the 12th Marines, comprised of two 75mm pack howitzer battalions and two 105mm howitzer battalions. The fires of the 12th Marines were to be reinforced by the 7th 155mm Gun Battalion firing from the southern beachhead, while the brigade artillery group would be backed up by the 1st 155mm Howitzer Battalion.

LOGISTICS [52]

The 105mm howitzers would be taken ashore in amphibian trucks of the IIIAC Motor Transport Battalion, which had been converted to a DUKW organization for Guam. DUKWs would also carry radio jeeps, 37mm antitank guns, and infantry ammunition; after that, they would be used for resupply. Of the 100 amphibian trucks in the battalion, the 40 of Company C were assigned to the brigade, while the remaining 60 would support the 3d Division.

Other supplies would be moved by amphibian tractors from the reef edge across the beaches to dumps inland. The 180 LVTs of the 3d Amphibian Tractor Battalion would serve the Marine division; the 4th Amphibian Tractor Battalion, with 178 LVTs, was attached to the brigade.[53] After the securing of the beaches, LSTs would anchor at the reef edge for unloading.

At the northern beaches the reef was dry at low water, and trucks would be able to run out from the shore to the edge. At the southern beaches the water over the reef was always too deep for trucks to operate; LVTs and DUKWs would have to bear the cargo, risking the usual coral heads and potholes.

Neither reef was covered at any time with water deep enough for shallow draft craft to pass over. In fact, nowhere along the entire coastline of Guam was the reef covered at high tide by more than two feet of water. Cranes could be operated on the northern reef, but only those cranes that

[52] Unless otherwise noted, the material in this section is derived from: *TF 51 OpRpt; TF 56 OpRpt; TF 53 OpRpt; IIIAC SAR;* 3d MarDiv SAR, dtd 19Aug44, hereafter *3d MarDiv SAR; 1st ProvMarBrig SAR; 77th InfDiv Jnl.*

[53] Both battalions had been reinforced: Company A of the 10th Amphibian Tractor Battalion was attached to the 3d, and Company A of the 11th to the 4th.

were mounted on pontoon barges would be usable on the southern reef. Forty-four 9 x 21-foot barges and twelve 6 x 54-foot pontoon causeways were to be carried to Guam on the sides of LSTs to save deck space for troop cargo; brackets for that purpose were installed on 17 of the landing ships.

Task Force 53 mounted out in the Solomons, where ships drew upon the storage dumps at the Naval Base, Tulagi, and the floating storage in Purvis Bay. The transports anchored close to Cape Esperance and Tetere, Guadalcanal, to be near the Marine camps to facilitate training and combat loading. Kwajalein and Roi Islands in the Marshalls served as the staging area, but owing to the postponement of W-Day it was necessary to restage at Eniwetok. The restaging involved topping off with fuel, water, provisions, and ammunition.

Adequate shipping had been provided to lift the units originally assigned to IIIAC, but additional units to be embarked required some reductions of cargo, particularly vehicles. On 4 May, for instance, Admiral Conolly was directed to take on board the entire first garrison echelon, comprising 84 officers and 498 enlisted men, an addition that somewhat complicated the allotment of space between assault and garrison troops.

In general, the logistic planning for operations on the large island of Guam had been so efficiently accomplished that no serious difficulty arose. The shipping available for FORAGER was never really enough, but miracles of adjustment were performed. Square pegs were practically fitted into round holes, and the distances between Guadalcanal and the chief sources of supply at Pearl Harbor, Espiritu Santo, and Noumea were telescoped by fast ships.

Outstanding and new in the logistic preparation for Guam was the IIIAC Service Group, an organization used again later at Okinawa. Staffs of the Corps Engineer and Corps Quartermaster formed the nucleus of the group, which shortly after W-Day, would include personnel of the engineer, construction, medical supply, and transport services. The Corps Engineer, Lieutenant Colonel Francis M. McAlister, was assigned to command the group; he would supervise the corps shore party operations once the Japanese port facilities had been seized. Until the garrison commander took over, the Service Group would operate the port to be established in Apra Harbor and also the airfields to be built. In a word, no time was going to be lost in transforming Guam into an advance base.

For landing the mountain of supplies, the harbor offered Piti Navy Yard and the seaplane ramp at Sumay as the best unloading points, at least at the start. The corps shore party planned to operate Piti with the 2d Battalion, 19th Marines, and Sumay with the two pioneer companies of the brigade. Two naval construction battalions, the 25th and the 53d, had been attached to IIIAC; initially the Seabees, along with corps engineers, would develop the road net in the beachhead area. After the battle was over, the 5th Naval Construction Brigade, comprised of three regiments, would begin its work under the Island Command.

Essential to the ambitious plans for developing a base, however, was the recapture of the island. The American ground forces to be engaged totaled 54,891 men:

3d Marine Division	20,328
1st Provisional Marine Brigade	9,886
77th Infantry Division	17,958
III Amphibious Corps Troops	6,719 [54]

A provisional replacement company (11 officers and 383 enlisted men) embarked with the assault troops. The unit would help with unloading until its men were needed to replace combat losses. A provisional smoke screen unit, formed to augment a Seabee battalion, was also to be available for frontline combat.[55] For the handling of casualties, the landing force had a corps medical battalion, which embarked with equipment and supplies to operate a 1,500-bed field hospital. In addition, there were two medical companies with the brigade and the division medical battalion. The 77th Division would bring an Army field hospital.

As at Saipan, the APAs would bear the initial casualty load from the beach assault. After treatment by frontline medical personnel, wounded men would be taken either by stretcher bearers or ambulance jeep to the beaches, where they would be received by beach medical parties and placed in an LVT or DUKW for movement to transports and LSTs equipped and staffed to handle the casualties.

[54] *TF 56 OpRpt*, Encl F.

[55] No replacements arrived while the Guam operation was in progress, but some 2,600 were en route directly to the 3d Marine Division from Administrative Command, FMFPac, when the operation ended. In later Pacific campaigns, Marine divisions took along an entire replacement battalion, and used its men for beach and shore party duties until they were required in combat.

TRAINING AND SAILING [56]

Most of the Marines that would fight on Guam were veterans of recent combat and experienced in an amphibious operation, but training on Guadalcanal was none the less intensive. Emphasis lay upon development of efficient tank-infantry teams. From 12 to 22 May, training included six days of ship-to-shore practice (three for each attack group), two days of air support exercises in conjunction with regimental landings, and two days of combined naval air and gunfire support exercises. On the 22d, the Northern Attack Group sortied from Guadalcanal and Tulagi, cruised for the night, and then made its approach to the rehearsal beach at Cape Esperance. All assault troops and equipment of the 3d Division were landed, supported by air and naval gunfire bombardment. Only token unloading of heavy equipment, such as tanks and bulldozers, was made. The Southern Attack Group conducted a similar rehearsal in the same area during 25-27 May. The practice was particularly designed to test communications and control on the water and on the shore.

Training on Guadalcanal was some-

[56] Unless otherwise noted, the material in this section is derived from: *TF 53 OpRpt; IIIAC SAR; Corps Arty SAR; 3d MarDiv SAR; 1st ProvMarBrig SAR; 77th InfDiv Jnl;* Cdr H. E. Smith (CEC), USN, "I Saw the Morning Break," *USNI Proceedings*, v. 72, no. 3 (Mar46), hereafter Smith, "I Saw the Morning Break"; Lt Robert A. Aurthur and Lt Kenneth Cohlmia, *The Third Marine Division* (Washington: Infantry Journal Press, 1948), hereafter Aurthur and Cohlmia, *The Third Marine Division;* Maj Frank O. Hough, *The Island War* (Philadelphia: J. B. Lippincott Co., 1947).

what handicapped because the island has no fringing reef, such as would be encountered at Guam. In the ship-to-shore phase, troops had to practice transferring from boats to tractors at an arbitrary point simulating the edge of the reef. Reality was lent to the rehearsals, however, by the use of live bombs and ammunition in the naval air and gunfire support exercises.

The Army troops due for Guam went straight from Hawaii to their staging area at Eniwetok, so they did not take part in the IIIAC training on Guadalcanal. The 77th Infantry Division had not yet experienced combat, but the men had been schooled in amphibious warfare, desert and mountain warfare, village fighting, and infiltration tactics at Stateside camps and then had spent some time at the Jungle Training Center on Oahu.[57] The 305th Infantry Regiment joined Task Force 53 at Eniwetok on 10 July, and the remainder of the 77th Division reached there a week later.

Marines of the 3d Division, their dress rehearsals over, embarked on transports and LSTs from docks at Tetere. Other ships loaded brigade troops at Kukum. On 1 June, the tractor groups left for the staging area at Kwajalein. The faster transport and support groups of TF 53, which included the *Appalachian* with IIIAC Headquarters on board, followed on 4 June. The ships stayed in the Marshalls long enough to take on fuel, water, and provisions and to transfer assault troops from transports to landing ships. By 12 June, Admiral Conolly's entire task force had left in convoy formation, bound for the Saipan area. For 10 days, from 16 June, Marines waited on board ships near Saipan, retiring every night and returning every morning, to be ready in the event they were needed on shore. On 25 June, Admiral Spruance sent ships of the Northern Transport Group, which was carrying the 3d Marine Division, to a restaging area at Eniwetok, but he detained the brigade for five more days before returning it to the Marshalls.

Among the Marines sidetracked at Eniwetok were men of Marine Aircraft Group 21. On 4 June, the forward echelon of MAG-21, then attached to the 4th Marine Base Defense Aircraft Wing, had sailed from Efate in the New Hebrides for Guadalcanal, expecting to go on to Guam. The pilots of Marine Corsairs were prepared to fly close support missions on Guam once Orote airfield was secured and made ready. To their dismay, the men were kept on board ship at Eniwetok from 19 June to 23 July.

While the ships lingered at Eniwetok, Marines were debarked, a few at a time, for exercises on sandy islets of the lagoon, but that was hardly a respite from the average of 50 days that troops had to spend on board the hot and overcrowded ships before getting off at Guam. Marines tried to shield themselves from the burning sun by rigging tents and tarpaulins on the weather decks of LSTs. As was common on every troop ship in the Pacific, men would leave the stuffy holds to seek a cool sleeping spot topside. In the ships

[57] LtGen Andrew D. Bruce, USA ltr to ACofS, G-3, HQMC, dtd 18Jun65. The Army's 77th, which was suddenly pitched into a hard Pacific campaign and fought like veterans, consisted mostly of draftees from the New York metropolitan area.

214-881 O-67—30

due for Guam, there were several platoons of war dogs, who shared the discomfort of the voyage but were not bothered by the dwindling supply of cigarettes. A variation of shipboard monotony occurred on 17 June when a formation of Japanese torpedo bombers approached the Northern Tractor Group; the attackers were turned away by the fire of LSTs and LCTs, which shot down three of the enemy planes. One of the prized LCI(G)s was hit; the gunboat was taken under tow, but finally had to be sunk by destroyer gunfire.[58]

General Geiger reported that "contrary to popular opinion, this prolonged voyage had no ill effect upon the troops."[59] Nevertheless, everyone breathed a sigh of relief when finally, beginning on 11 July, elements of Task Force 53 again sailed for Guam. The bulk of the troops, including RCT 305, departed in transports on 18 July. The ships which had been sent from Saipan to Pearl Harbor to pick up RCTs 306 and 307, arrived at Eniwetok just before the main force got underway for Guam. They continued on their long voyage to the objective on the 19th. On 20 July, the *Indianapolis*, bringing Admiral Spruance, joined the great task force, and, on the same day, Admiral Turner and General Holland Smith departed Saipan in the *Rocky Mount* to observe the Guam landings. The Japanese, viewing the armada from the crest of Mt. Tenjo, counted 274 vessels.

By the afternoon of 20 July, every ship that would be connected with the amphibious assault was either at or approaching its designated position off Guam. Prospects for success on W-Day appeared to be good, except for a flurry of concern lest an impending typhoon move near the area—and that worry was dismissed by Admiral Conolly's hurricane specialist. The weather prediction for W-Day was optimistic: a friendly sky, a light wind, a calm sea.

Admiral Conolly confirmed H-Hour as 0830. In a dispatch to the task force, he felt able to say, that because of the excellent weather, the long preparatory bombardment, and the efficient beach clearance, "conditions are most favorable for a successful landing."[60] Events of the next day would show whether he was right.

[58] The LCI(G)s had been used in their new role as gunboats at Saipan, but they were planned for wider use at Guam, again preceding the first assault wave to the shore. At Guam the reefs were near enough to the shore, so that the gunboats could fire their rockets successfully. These little vessels, with a five-foot draft, were armed with five 20mm cannon and three 40s, and their forward decks were packed tight with rocket frames. Originally, the LCIs had been used to carry assault infantry from shore to shore. The LCI(G) now carried a crew of 70 and 6 officers, about thrice its original T/O.

[59] *IIIAC SAR*, p. 2.

[60] *TF 53 OpRpt*, p. 11.

W-Day

THUNDER AT SUNRISE [1]

"My aim," Admiral Conolly had remarked, "is to get the troops ashore standing up." [2] In the preparation fires at Guam, he had left no shell unused if it would remove some peril to the landings. The same zeal to accomplish maximum results went into the bombardment on the morning of W-Day.

Some improvements upon fire support had been suggested by the experience at Saipan on D-Day. It was felt that the beach preparation there could have been enhanced by continuous deep fires along the high ground 1,500 yards in the rear of and overlooking the beaches, started well before H-Hour and kept up until the assault troops were reorganized ashore and had pushed out toward their objectives. That moment was anticipated to occur at H plus 90, or 1000. Such a procedure appeared especially worth trying at Guam, where the most serious opposition would probably come not from fixed defense guns at the beach—most of them were believed to be out of commission—but from mobile artillery inland which had not fired and had not been located. In addition, simultaneous naval gunfire and air bombardment was going to be attempted to increase overall volume and the shock effect upon the enemy. Finally, a greater use of rocket-equipped gunboats had been planned for Guam. Nine thousand 4.5-inch rockets were scheduled to be fired between 0530 and 1000.

The morning twilight of 21 July, beginning at 0445, erased a tropical sky "bespangled with stars." [3] At 0530, a half hour before sunrise, all fire support ships were on their assigned stations, and at 0535 four battleships off Orote Peninsula and Cabras Island opened fire with 12 14-inch guns each. Inside Agat Bay, the old *Pennsylvania* thundered at the cliff line of the peninsula. Other battleships, cruisers, and destroyers up and down the west coast immediately joined with slow and deliberate fire on the landing beaches, their flanks, and the areas just inland. Admiral Conolly in the *Appalachian* directed the bombardment of the Asan beaches, while Admiral Reifsnider in the *George*

[1] Unless otherwise noted, the material in this section is derived from: *TF 51 OpRpt; TF 56 OpRpt; TF 53 OpRpt; TF 53 OPlan A162–44; IIIAC SAR;* Smith, "I Saw the Morning Break."

[2] A hope expressed to General Geiger and recalled by the Corps C–2. Col William F. Coleman ltr to CMC, dtd 5Sep52.

[3] Smith, "I Saw the Morning Break," p. 409. War seemed out of place on a beautiful tropical night over the Pacific, and the sky would inspire poetic description by witnesses such as Commander Smith. His account of the events of W–Day morning is personal and vivid.

Clymer handled the shelling of the Agat beaches.[4]

By 0615, 12 fighters, 9 bombers, and 5 torpedo planes from the carrier *Wasp* were on station as a roving combat air patrol, an experiment at Guam. In the first air strike of W-Day, the nine bombers hit at buildings, machine gun nests, and antiaircraft emplacements on Cabras Island. The Commander, Support Aircraft, in the *Appalachian*, planned that such roving patrols be kept on station through most of W-Day, to seek out hidden guns and mortars in defiladed positions inaccessible to naval gunfire.

A spectacular sweep of the 14 miles of coastline from Agana to Bangi Point was executed between 0715 and 0815 by carrier planes flying parallel to the beaches. Assigned to the mission were 85 fighters, 62 bombers, and 53 torpedo planes. An unusual feature was that naval gunfire accompanied the attack. Under what was called Plan Victor, the firing calculations of the ships had to be adjusted so that the trajectory of their projectiles would bring them no higher than 1,200 feet. Pilots pulled out of their runs before reaching as low as 1,500 feet.

SHIP TO SHORE [5]

To the familiar sounds of the prelanding preparation,[6] Marines moved closer to the island. The 1st Provisional Brigade arrived in the transport area 12,000 yards east of Agat at approximately 0600. The ships carrying the assault troops of the 3d Marine Division stopped about the same time at an equivalent distance from the Asan beaches.

The brigade and the division each used 16 LSTs, and these moved into the launching area about 0700. There the landing ships opened their bow doors to disgorge LVT(A)s and LVTs carrying assault troops. On board the transports, Marines of the reserve battalions waited to debark into LCVPs. Once loaded with troops, the boats would proceed to the reef transfer line and stand by until the first waves of tractors returned from the beach. About the same time that the reserves began landing, the tanks that had moved to the target by LSD would start rumbling across the reef. Each LSD (two served the division and one the brigade) carried 20 medium tanks, loaded in a LCT and 14 LCMs.

Close on the tracks of the tanks, the direct support artillery would begin to

[4] One witness of the preassault bombardment, an officer of the 9th Marines, wrote later: "I was particularly impressed to see Japanese soldiers still alive right on the landing beaches after almost 24 hours of incessant bombardment by naval gunfire." LtCol Calvin W. Kunz ltr to HistBr, G-3, HQMC, dtd 27Feb52.

[5] Unless otherwise noted, the material in this section is derived from: *TF 51 OpRpt; TF 53 OpRpt; IIIAC SAR; 3d MarDiv SAR; 1st ProvMarBrig SAR;* 1stLt Millard Kaufman, "Attack on Guam," *Marine Corps Gazette*, v. 29, no. 4 (Apr45), hereafter Kaufman, "Attack on Guam"; Smith, "I Saw the Morning Break"; Aurthur and Cohlmia, *The Third Marine Division.*

[6] Marine veterans of the Pacific campaigns sometimes felt that "when you've seen one naval prelanding bombardment, you've seen them all." This quip is recalled in Hough, *op. cit.*, p. vii. At Guam, however, the innovations would seem to make such a comment less applicable.

land, either in DUKWs that carried 105mm howitzers direct from ship-to-shore or in LVTs that picked up 75mm pack howitzers at the reef edge. Detached from the LSTs that had carried them to the target, pontoon barges, some mounted with cranes, would move to the reef to facilitate the transfer of supplies and equipment. As soon as the situation ashore permitted, the LSTs themselves would nose up to the coral shelf and begin landing vehicles and supplies directly on the reef.

The ships off Guam on 21 July included the 12 transports of the 3d Marine Division and the 8 of the 1st Provisional Marine Brigade, besides the destroyers which screened the transports. Five assault cargo ships (AKAs) shared the task of supplying thousands of Marines on Guam. The 77th Infantry Division had 12 transports, 5 cargo ships, and 3 LSTs, which rounded out a weight of shipping which lay upon some of the deepest water of the Pacific Ocean.[7]

Waiting to lead the assault were 18 gunboats, the LCI(G)s—evenly divided between the Marine division and the brigade. These vessels had each been fitted with 42 rocket launchers, in addition to their 20mm and 40mm guns, for the Guam operation. The craft would form the vanguard at the landings, shelling the beaches and then swinging to the flanks when about 200 yards from the reef.

Following the gunboats would be the 1st Armored Amphibian Battalion (Major Louis Metzger), its turreted LVT(A)s firing their 37mm guns at targets on the beach. Running behind such interference, 360 LVTs were to land the assault troops almost on the heels of the first wave of LVT(A)s. Such was the usual pattern, and at Guam on the morning of 21 July, no hitches developed. "The ship to shore movement," Admiral Conolly proudly reported, "was executed with perfect precision and exactly on schedule."[8]

A few minutes before 0800, the gunboats crossed the line of departure and headed toward the beaches, followed seconds later by the wave of armored amphibians. Behind were six waves of LVTs, formed up and ready for the attack. H-Hour was just 30 minutes away. So far a silent enemy appeared dazed by the constant air and naval gunfire bombardment, and while there were no illusions about what could happen later, a minimum of resistance was expected to the landings.

On the northern front, as the LVTs took the assault troops shoreward, the smoke and dust of the bombardment obscured the beaches where the men were to land. The 2,500 yards of enemy-held coastline which lay between Asan Point and Adelup Point had been parceled out among the three infantry regiments of the 3d Marine Division, which were to land abreast in a column of battalions, each regiment keeping one battalion as a reserve afloat.[9] (See Map VII, Map Section.)

On the left, the 3d Marines, commanded by Colonel W. Carvel Hall, would go ashore over Red Beaches 1 and

[7] About 200 miles southwest of Guam, the ocean floor is five miles below the surface.

[8] *TF 53 OpRpt*, p. 11.

[9] The division, lacking a floating reserve, would have to rely upon the corps reserve (77th Infantry Division, less the 305th Regiment).

2. The immediate task of the regiment was to secure Adelup Point, Chonito Cliff, and the high ground southeast of the cliff, thus protecting the left flank of the division. The 21st Marines (Colonel Arthur H. Butler), landing on Green Beach, would seize the cliff line to its front and hold there until the division was ready to move inland. Upon securing the objective, Colonel Butler would assign one battalion as division reserve. On the right, the 9th Marines (Colonel Edward A. Craig) was to cross Blue Beach and take the low ridges just beyond. Colonel Craig's 3d Battalion, which was landing in assault, would become regimental reserve once the other two battalions were ashore, and it would be prepared, if so ordered, to make an amphibious landing on Cabras Island. The Piti Navy Yard, down the coast from Blue Beach, appeared also as a probable objective for the 9th Marines.

To the south, the assault troops of the 1st Provisional Marine Brigade moved toward narrower beaches than those that faced the 3d Marine Division. The lesser width was compensated for by more favorable ground immediately inland; the hills were lower and the terrain more open. This promise of an easier initial advance had played a large part in influencing the choice of beaches to be hit by the two major assault units of IIIAC. Despite its smaller size, the brigade was "a two-regiment division, if I ever saw one," said Admiral Conolly later, in tribute to its accomplishments.[10] Actually, the brigade was substantially a division once its reserve, the Army 305th RCT was called into action.

On the morning of W-Day, Colonel Schneider's 22d Marines was to land on Yellow Beaches 1 and 2, occupy Agat, and then turn north to seal off the Orote Peninsula. The 4th Marines under Lieutenant Colonel Shapley would go ashore over White Beaches 1 and 2, establish a beachhead, and protect the right flank of the brigade. A major and perhaps costly mission lay ahead of the brigade once the 305th was ashore—the seizure of the Orote Peninsula.

Japanese opposition to the oncoming waves of Marines was late in appearing. The enemy's coastal defense guns had either been destroyed by the bombardment or left unmanned. At 0800, the division air observer saw no activity inland of the beaches. Twelve minutes later, when the first LVTs in the assault waves were well under way, he reported "no enemy fire from the beach observed."[11] At 0810, the brigade air observer reported "no firing on our boats of the leading wave."[12] The American gunboats were then firing tremendous salvoes. At the southern beaches a number of the rockets fell short, but the division air observer reported at 0820 that "the rockets are landing and giving them hell."[13]

The armored amphibians of the lead-

[10] Quoted in "Combat Leadership," The John A. Lejeune Forum, compiled by Capt Robert B. Asprey, *Marine Corps Gazette*, v. 46, no. 11 (Nov62), p. 26.

[11] *TF 53 OpRpt*, Encl A, p. 27. These air observers were infantry officers assigned by the IIIAC. From carrier planes they reported to headquarters ships by voice radio (the SCR 694).

[12] 1st ProvMarBrig Jnl (App 2 to *1st ProvMarBrig SAR*), 21Jul44, hereafter *1st ProvMarBrig Jnl*.

[13] *3d MarDiv SAR*, p. 1.

ing assault wave, moving forward at 150 yards per minute, were then about 1,200 yards from the beaches—the scheduled time for air observers to drop their white parachute flares as a signal to the gunfire ships. Major caliber guns were then to raise their fire inland, while the rate of 5-inch gunfire would be stepped up until the armored amphibians started across the reef.

The white flares were also a signal for a special air strike by 32 Navy fighters. They were each to drop a depth bomb and then strafe the beaches until the Marines were almost on land. Following that strike, 12 other planes were to strafe just inland from the beaches until the troops set foot on the shore. Adding to the last violent preparation by naval shelling and air bombardment, the armored amphibians would fire their guns when crossing the reef, while, stationed on the flanks of the beaches, the gunboats employed their 20mm and 40mm weapons to disrupt any enemy movement sighted.

As the LVTs carrying the assault troops headed for the beaches, there was no sign of enemy activity. Admiral Conolly turned naval gunfire upon Gaan Point and Bangi Point, both of which were believed to contain well-hidden defenses, and upon Yona Island, where the brigade observer had noticed some firing. The gun there was later found to be a 75mm field piece. Except from such scattered positions, however, the Japanese did not return fire.

It was not until the Marines were within the last few yards of the beaches that the situation changed. The cumbersome amphibian tractors had negotiated the reef successfully, but they fared badly thereafter from enemy fire and mines, as the beach defenses suddenly came to life. Off the northern shore, the armored amphibians and the following wave of LVTs were nearly at the beach when they received fire from Japanese small arms and antiboat guns ranging from 37mm to 75mm in size. Several tractors were hit; at least one was disabled by .30 caliber armor-piercing bullets.[14] Admiral Conolly's hope of getting the troops ashore standing up took an ironic twist when Marines had to leave a crippled tractor and wade in to the beach. From the high ground just inland, the Japanese turned mortar and artillery fire upon other approaching LVTs; a number of the vehicles were damaged by shell fragments.

Off the southern beaches, 24 of the tractors serving the brigade were put out of commission either by enemy fire, by damage to the treads caused by jagged coral, or by mechanical trouble.[15] When the first wave of the 22d Marines was about 100 yards from the beaches, intense enemy fire was received.[16]

[14] A mountain gun located on Adelup Point hit several armored amphibians. The weapon was "silenced by a destroyer that closed to the reef edge in a beautiful bit of seamanship. The destroyer's action saved the lives of a lot of Marines." One antiboat gun located at the junction of Asan Point and the beach hit two armored amphibians. That weapon was silenced by LVT(A) 37mm guns. LtCol Louis Metzger ltr to CMC, dtd 29Oct52.

[15] The leading waves of the brigade had 10 LVT and LVT(A) casualties. The division reported nine LVTs and LVT(A)s destroyed by enemy fire during the landing.

[16] One officer of 3/22, which went ashore at noon on W-Day, recalled later that "the gun

"Looks like 75mm," the brigade air observer radioed. "Can you locate source of fire?" came the query in reply.[17] The source proved to be a concrete blockhouse on Gaan Point. Built deep into a small coral hill, the installation had evaded photographic detection. Here was a 75mm gun lodged below four feet of rock. A shelter for a companion 37mm gun was also walled with concrete.[18] A few of the LVTs bound for the Yellow Beaches were damaged by the enemy fire, and some of the Marines they carried were hit.

Crossfire from Gaan Point and from Yona Island raked White Beach 2, a 300-yard strip of sand where the 1st Battalion, 4th Marines was landing. Scattered resistance came from pillboxes between Agat and Bangi Point; other fire developed from well-concealed guns at Bangi Point and artillery on the south side of the Orote Peninsula. Some resistance to the landings was offered even by guns at Facpi Point, down the coast.

Despite such spirited attempts, however, the Japanese plan of stopping the American return to Guam at the beaches had been set back, thanks in large part to Admiral Conolly's efforts. His planes and guns had not destroyed as many enemy installations as he believed; still, as the Japanese explained later, it was "the interruptive operation of the severe bombardments" that upset their plan.[19]

Nowhere were the Marines prevented from landing on schedule. They were not delayed either by damage to tractors or by opposition from those enemy riflemen and machine gunners who had not yet deserted the shell-ridden beaches. Marines had a foretaste, however, of the hard fighting due on Guam; the 3d Marines received ominously heavy fire from the vicinity of Adelup Point.

At 0833, the division air observer, flying over the Asan beaches, reported: "Troops ashore on all beaches." [20] The brigade was on the island by 0832. Now, said the division report, "the capture of Guam was in the hands of the foot soldier." [21]

THE NORTHERN BEACHES [22]

Once the Marines were ashore, and at least until the end of W-Day, the battle for Guam shaped up as two

was firing on line of fire as landing craft passed without traversing." Maj Samuel A. Todd ltr to CMC, dtd 30Oct52.

[17] *1st ProvMarBrig Jnl*, 21Jul44.

[18] The guns "were in a double cave, one above the other. . . . The mouth of the caves could not be seen from the sea, and trees and shrubbery prevented them from showing in aerial pictures." Col Edwin C. Ferguson interview by HistBr, G–3, HQMC, dtd 28Nov52.

[19] *Takeda ltr I*, p. 3.

[20] *3d MarDiv SAR*, Encl A, p. 1.

[21] *Ibid.*

[22] Unless otherwise noted, the material in this section is derived from: *TF 53 OpRpt; IIIAC SAR;* SARs of 3d MarDiv regiments and organic and attached units enclosed with *3d MarDiv SAR*, hereafter cited separately as necessary, *e.g., 9th Mar SAR, 3d TkBn SAR;* 3d Mar Jnl, 21Jul–12Aug44, hereafter *3d Mar Jnl;* 1/3 Jnl, 21Jul–16Aug44, hereafter *1/3 Jnl;* 2/3 Jnl, 21Jul–24Aug44; hereafter *2/3 Jnl;* 3/3 Jnl, 21–31Jul44, hereafter *3/3 Jnl;* 9th Mar URpts, 21Jul–19Sep44, hereafter *9th Mar URpts;* 2/21 Jnl on Guam, hereafter *2/21 Jnl;* 3/21 Jnl, 21Jul–1Nov44, hereafter *3/21 Jnl;* Aurthur and Cohlmia, *The Third Marine Division.*

separate military operations on beaches miles apart. On the left of the 3d Division beaches, the 3d Marines had the hardest going on the morning of 21 July. The whole division was landing between what the Marines called "a pair of devil's horns"—Adelup Point and Asan Point.[23] The latter, on the right, had been dulled by the naval and air bombardment, but was still infested with enemy troops.[24] The devil's left horn, the reports understate, "still had some life in it." [25] To be more specific, the Japanese had weathered the terrific preassault gunfire and explosives, emerged from their caves and wooded folds on the reverse slopes of the high ground, and returned to their prepared gun and mortar positions on Chonito Cliff, which overshadowed the Red Beaches, and on the ridges to the south and southeast. (See Map VII, Map Section.)

Expecting grim resistance to the advance of the 3d Marines, Colonel Hall drew his first objective line across the enemy's well-defended high ground immediately inland. He was landing the 1st Battalion over Red Beach 2 and the 3d to the left over Red Beach 1. The 2d Battalion was to land in reserve and move to an assembly area behind Red 1. The regimental commander planned to put the reserve either at the center of the objective line once it was gained, or else to pass it through the left company of the 3d Battalion, to seize Adelup Point.

The immediate situation at the Red Beaches was not favorable. Minutes after the leading waves of the 3d Marines were ashore, the Japanese opened up in earnest, turning artillery, mortars, and machine guns upon the beaches and the reef, lobbing well-directed mortar shells squarely among the LVTs. Some of the Marines were casualties before getting on land; others were hit when they were barely on the beaches by an enemy enjoying perfect observation. At 0912, the commander of 3/3, Lieutenant Colonel Ralph L. Houser, reported "mortar fire and snipers very heavy," resulting in "many casualties." [26]

The optimistic hope of a dash to the initial objective, Chonito Cliff, before the enemy revived from the preassault bombardment dissolved into grim acceptance of the struggle ahead. The danger posed by the Japanese in their caves on Chonito Cliff led to some exaggerated news reports of its size. The cliff itself was only the seaward edge of the steep ridge which overlooked the whole length of the Red Beaches; it lay northeast of Red Beach 1. While Chonito Cliff's rugged terrain was a boon to its defenders, it was curiously obstructive to the Japanese on adjoining Adelup Point. Projecting to the edge of the water, Chonito Cliff walled off Red Beach 1 and restricted the enemy guns on Adelup Point to attacking the approaching LVTs rather than the Marines on the beach. That fire was fi-

[23] *3rd MarDiv SAR*, Encl A, p. 1.

[24] Asan Point and ridge running inland from it came alive with enemy fire as the 9th Marines advanced inland. Many Japanese held their fire as the assault troops passed and then opened up on support and CP echelons. *Craig 22Jun65 ltr.*

[25] *3d MarDiv SAR*, Encl A, p. 1.

[26] *3/3 Jnl*, 21Jul44. The spare litters were soon used up. Others were constructed from poles and ponchos.

3D DIVISION ASSAULT TROOPS *take cover along the Asan beaches as messengers crouch low to avoid enemy fire. (USMC 88167)*

MARINES *watch tensely as a flamethrower blasts an enemy dugout in the advance inland on Guam. (USMC 88072)*

nally silenced by a destroyer which moved up to "rock throwing" range,[27] but the Marines were not yet through with Adelup Point.

In the approximate 400 yards between Adelup Point and Chonito Cliff lay a deep dry stream bed where the beach road which followed the west coast went over a concrete bridge after cutting through Chonito Cliff. "The bridge and the ridge tip between the beach and the road formed an enemy strong point," recalled a Marine officer of 3/3. "The cut and the bridge afforded excellent protection from bombardment and bombing." [28] The Japanese had dug an ingenious tunnel system, permitting them to fire upon both the road and the beach. South of the cliff was a draw leading inland.

Company I, landing on the right of Red Beach 1, tried to get through the draw but was stopped by enemy fire. Company K crossed the stream bed and started up Chonito Cliff but without success. The support platoon of Company K then attempted to force a way through the cut but was badly hurt by machine gun fire and grenades. The enemy rolled some of the grenades down the cliff.

To break up the impasse, Lieutenant Colonel Houser employed flamethrowers and called upon tanks of Company C, 3d Tank Battalion, which took position along the beach road and fired squarely into the caves.[29] The battalion commander then committed his reserve, Company L, which "breeched the cut and pushed on to the flat land north of Chonito Cliff. This move required the entire company to move down the beach road with the sea on the left and the steep cliff face on the right." [30]

By noon, the danger of Chonito Cliff had been removed, and here, at least, the 3d Marines had reached its initial objective.[31] The situation permitted Colonel Hall to confer with battalion commanders on top of the cliff at 1300. That afternoon, Marines of 3/3, supported by tanks and armored amphibians, overcame some Japanese resistance on Adelup Point; a few of the enemy guns there had escaped the sea bombardment. Meanwhile, Lieu-

[27] BGen W. Carvel Hall ltr to CMC, dtd 4Dec52, hereafter *Hall ltr.*

[28] LtCol Royal R. Bastian, Jr., ltr to CMC, dtd 23Aug52, hereafter *Bastian ltr.*

[29] Company C had landed from LCMs on Red Beach 1 at H-Hour plus 29 minutes, and just a half hour later its tanks were the first to go into action on Guam. The rest of the division's 40 medium tanks were ashore by 1000. General Craig commented in regard to this feat: "The tanks did a wonderful and dangerous job in getting ashore. Transferring those big 45-ton tanks from Navy landing craft to a sheer reef edge in choppy seas and then driving them through rough coral spotted with deep potholes to the beach is an accomplishment which I believe deserves special note. The method devised of holding the Navy landing craft against the face of the reef by using LVTs and cables is also worthy of note. The tanks would probably have never made it if someone had not worked out this method." *Craig 22Jun65 ltr.*

[30] *Bastian ltr.*

[31] Later, while enlarging the beach road, engineers and Seabees altered the appearance of Chonito Cliff so that, as one Marine officer recalled, "the area was not recognizable when I returned several weeks after the landing." *Ibid.* General Craig noted that the same situation held true regarding Asan Point, where Army engineers set up a quarry and rock crushing machine and tore down most of the ridge leading from the point for road construction material. *Craig 22Jun65 ltr.*

tenant Colonel Houser moved the battalion command post from the beach to a bend of the road.[32] The subsequent movement of 3/3, however, was handicapped not only by enemy fire from the front but also, particularly, by the Japanese defenses on Bundschu Ridge, which lay in the path of 1/3, commanded by Major Henry Aplington, II.

Bundschu Ridge was one of those inherently worthless pieces of land which were emotionally remembered by the men who fought there in World War II. On board ship, before the landing, it had been named for Captain Geary R. Bundschu, commander of Company A, who had been assigned to take the ridge. It was also referred to in the reports as "Our Ridge." Similar to Chonito Cliff, but farther inland and beyond some rice paddies, the ridge stood near the boundary of the two Red Beaches, a rock pile 400 feet high and 200 yards square, thatched with jungle vegetation. It was so situated that even a handful of well-hidden men, using mortars and machine guns, could repel a much larger force moving up from below.

Captain Bundschu's company had already suffered from enemy fire while on the water and on the beach. Now, with but a few minutes for reorganization, he started the attack, moving across the rice paddies toward the ridge, with two platoons in assault and one in support. By 0920 the lead platoons were pinned down in a gully to the west of the ridge by Japanese mortar and machine gun fire, so the support platoon was committed to the left, or east side. Captain Bundschu was then able to get up to within 100 yards of the ridge top. At the same time, 1045, he called for more corpsmen and stretcher bearers. Company B was somewhat better off. Advancing on the right, it was delayed more by jungle and rock than by enemy fire; still the company lost five men killed in the day's action. Company C, the reserve, was not committed to the fighting on W-Day, but Major Aplington did receive permission to use two platoons for a combat outpost on the right flank.

The plight of Company A led the regimental commander to drop his original plan of massing 81mm mortar fire on Adelup Point, where enemy resistance had proved relatively minor.[33] Instead, at 1045, he reassigned control of the 1st Battalion 81s to Major Aplington. The platoon was pinned down, however, shortly after moving up to Bundschu Ridge. Its gunnery sergeant and four men were hit, and late that day the unit was still unable to move. Colonel Hall committed the reserve 2d Battalion, under Lieutenant Colonel Hector de Zayas, to the center of the regimental front and ordered renewal of the attack at 1500 along the entire line.

For Captain Bundschu, the situation had been frustrating and saddening, as the hidden enemy exacted a toll of Marines for every step taken. About 1400 he asked Major Aplington for permission to disengage, a request which had

[32] A branch of the beach road wound from near Adelup Point into the Fonte hill mass, where General Shigematsu's battle command post was located.

[33] The battalion mortar platoons were to land with their parent units and then combine into a mortar groupment near the boundary of Red Beaches 1 and 2. 3d Mar OPlan 3-44, dtd 27May44.

to be denied because the company was so involved. It was Colonel Hall's view that a second attack on the ridge should be attempted, but he "did not specify a frontal assault." [34]

Apprehensive about the results, Captain Bundschu reorganized what was left of Company A and prepared to undertake again the last 100 yards of the ridge. At nearly the end of a day oppressive with tropical heat, the Marines tried again, knowing the odds. They once more encountered the machine gun fire that had stopped the initial assault; now, however, with the effective support of 40mm guns of Battery I, 14th Defense Battalion, "a thin line of Company A men reached the crest." [35] Other Marines, shot en route up the steep slope, fell backwards to the ground far below. At the top of the ridge, enemy fire of savage intensity prevented a reorganization for defense of the ground gained; the foothold became untenable. The second attack on the ridge had cost the life of Captain Bundschu and further depleted Company A. At nightfall, the enemy still held Bundschu Ridge, and the Marines were reminded of this fact by the Japanese fire which kept up through the unhappy night.

While 1/3 was stalled at the initial regimental objective line, the 2d Battalion was past it, yet still short of the first division objective, which Colonel Hall had fixed as the goal of a renewed attack at 1500. The arc of steep hills which circled the Asan beachhead was everywhere well-defended by the enemy who had started moving up reserves from the Fonte area to combat the invasion.[36] This movement was impeded but not prevented by the fire of 75mm and 105mm howitzers of the 3d Division artillery regiment. The first battery of the 12th Marines had landed and registered by 1215. By 1640, all the division artillery was ashore. Close support artillery, however, was not available to the 3d Marines on W-Day; the range was too short, and the fire could not be seen by forward observers.[37]

W-Day had ended with the 3d Marines still out of contact with the 21st Marines on its right. Colonel Butler's regiment had landed on Beach Green in a column of battalions, in order 3d, 2d, and 1st. Nowhere were the results of the naval gunfire preparation more evident than here on Beach Green; it was "extremely effective." [38]

The Japanese had abandoned their

[34] *Hall ltr.*

[35] BGen James Snedeker ltr to CMC, dtd 28Sep52. Colonel Snedeker, executive officer of the 3d Marines, assumed fire direction of the 40mm guns. He recalled "sitting on a sand dune with a portable radio. From this position I could see the 40mm guns and the enemy, but neither could see the other. Enemy machine gun fire picked up the sand all about my exposed position." The 14th Defense Battalion was armed chiefly with antiaircraft weapons; however, with the absence of enemy planes over Guam, the weapons were handy for other uses.

[36] Even on W-Day, the 3d Marines were already opposed by an enemy force of at least three companies "with a large number of automatic weapons." 3d MarDiv D-2 Periodic Rpt No. 70, dtd 23Jul44.

[37] Once the Marines were farther inland, artillery was employed more often. In fact, the 3d Marines reported that "for close support, there is no substitute for artillery." *3d Mar SAR*, p. 7.

[38] *21st Mar SAR*, p. 1.

organized defenses in the beach area; no enemy dead were found there. The scene of wreckage included a demolished coconut grove along the beach; trunks of the trees lay across the road. The assault waves of the 3d Battalion encountered no resistance in landing but received mortar fire from the Japanese positions on the high ground just inland. Such fire on the beach area became more intense by the minute and resulted in a number of casualties. When the regimental headquarters landed in the 11th wave, it had to set up temporarily in a ditch near the beach to obtain cover.

At Guadalcanal, officers had been briefed on the "almost impossible" cliffs which the 3d Marines and the 21st Marines would face shortly after landing.[39] Colonel Butler had mapped out a tactical plan based on aerial photos which identified two defiles, or narrow passages—one at each end of the regimental zone—which permitted access to the cliff tops via the steeply rising ground inland of Beach Green. The defiles were related to the two forks of the Asan River, which joined to emerge into the rice paddies.

According to Colonel Butler's plan, the 2d Battalion, landing behind the 3d Battalion, would pass through the left of 3/21 when the latter had reached its first objective, a moderate height beyond the village of Asan. The 2d Battalion would then move up the defile on the left toward the steep cliffs, while the 3d Battalion undertook the other passage. The two units would not try for contact until they had gained the plateau, where they would extend to form a new line. Behind the advance to the cliffs, the regimental reserve, 1/21, would mop up and then revert to division reserve.

Starting up the Asan River valley, the 3d Battalion suffered casualties from enemy mortar fire. At one point, the advance was held up by a Japanese machine gun platoon which was so positioned that it could also fire southwest into the zone of 1/9 on the right of 3/21. Here Lieutenant Colonel Carey A. Randall, commanding 1/9, joined with Lieutenant Colonel Wendell H. Duplantis of 3/21 in removing enemy threats. He laid down preparatory fires for an attack on the machine gun position by 3/21, while naval gunfire, directed by 3/21, neutralized a mortar position on the objective of 1/9. "Approximately 14 machine guns, heavy and light, 6 mortars, and the entire supply of ammunition were seized in this section." [40] Two Japanese were captured in the machine gun position; they were "believed to be the first prisoners seized in the campaign." [41]

By midmorning, the 3d Battalion had reached the high ground behind Asan, and at 1250, the 2d Battalion passed through the lines of Company K. For 2/21, the ordeal of the cliff area, which was to drag out for days, began in earnest. Some Marines would remember it in total as the battle for Banzai Ridge.[42] Actually, the battle involved a

[39] Aurthur and Cohlmia, *The Third Marine Division*, p. 147.

[40] Col Wendell H. Duplantis ltr to CMC, dtd 30Oct52, hereafter *Duplantis ltr.*

[41] *Ibid.*

[42] See 1stLt Anthony A. Frances, "The Battle for Banzai Ridge," *Marine Corps Gazette*, v. 29, no. 6 (Jun45), hereafter Frances, "The Battle for Banzai Ridge."

series of cliffs, "where every ridge gained by the 21st Marines disclosed another pocket of the enemy behind it." [43]

After travelling almost a mile from the beach, the 2d Battalion, moving up through the defile, approached a steep 100-foot cliff which cut diagonally across the main axis of attack. The Japanese expected no one to be hardy or bold enough to attempt a frontal attack here, but the terrain required it; there was no room to maneuver troops. Upon Company F fell the burden of the assault. Company E was echeloned to the right rear, while Company G took its position below the cliff as the reserve.

The rifle platoons of Company F started up the rocky cliff face, climbing via three indentations which permitted some concealment. "Slowly the men pulled themselves up the cliff, clinging to scrub growth, resting in crevices, sweating" under the tropical sun—it was a story often to be repeated on Guam. "Scouts on the left drew the first enemy fire. The platoons kept climbing. The platoon on the right was nearly decimated." [44] Company E started two squads and a patrol up the cliff and also suffered casualties. Results of the shipboard confinement seemed to show here; a few of the men were unable to finish the arduous climb.[45] The Marines who did get to the top received machine gun fire there from a ridge less than 50 yards away, but they held on while the battalion commander, Lieutenant Colonel Eustace R. Smoak, set up the defense for the night; he put Company G on the left, Company F in the center, and Company E on the right flank. The battalion dug in on the objective under artillery and mortar fire from the ridge beyond.

The 3d Battalion, moving upon the high ground to the right, was able to tie in with 2/21 by outposts only; the jungle vegetation made contact difficult. The 1st Battalion, after mopping up to the rear and encountering few of the enemy there, reverted to division reserve. To the regimental left, a deep jungle-thick ravine separated the 21st and 3d Marines, leaving a gap of 150 yards, despite the efforts of patrols to make contact. Yet it was "a well neutralized gap," the division reported. "Enemy mortar fire kept the gap open; our own kept out the enemy." [46]

To the right, contact was well established between the 21st and the 9th Marines. Of the division infantry regiments, the 9th Marines had met the least resistance from the terrain, although as much from enemy troops. It was able to make the most actual progress on W-Day. The regiment landed in a column of battalions, with 3/9 in the assault, followed by 2/9 in support and 1/9 in reserve. The mis-

[43] *3d MarDiv SAR*, Encl A, p. 3. The Commandant of the Marine Corps, Lieutenant General Alexander A. Vandegrift, a veteran of jungle and mountain warfare in the Carribbean and in the South Pacific, later inspected the terrain here and in the zone of the 3d Marines; he called it "some of the most rugged country I have ever seen." *Ibid.*, p. 4.

[44] Frances, "The Battle For Banzai Ridge," p. 13.

[45] A shortage of water added hardship to a hot day. At 1730, however, 3/21 reported that "one canteen of water arrived at CP for each man." *3/21 Jnl*, 21Jul44.

[46] *3d MarDiv SAR*, Encl A, p. 3.

sion of the 3d Battalion (Lieutenant Colonel Walter Asmuth, Jr.) was to seize the high ground immediately inland, including Asan Point. The other two battalions would then pass through when so ordered, while 3/9 became regimental reserve. The 1st and 2d Battalions, the latter on the right, were scheduled to seize the next objective, a line 1,000 yards from the beach and just short of the Tatgua River.

The 9th Marines landed under Japanese mortar and artillery fire directed at LVTs in the water, on the reef, and on the beach; a considerable number of casualties resulted.[47] Once past the beach, the troops encountered negligible small arms fire while crossing the dry rice paddies. Further along, however, the southeasterly course of Company I on the right was slowed by fire from caves on Asan Point and along the ridge which extended from Asan Point to the mouth of the Nidual River, but no line of enemy resistance was set up. Lieutenant Colonel Asmuth used the reserve Company L to assist in taking and clearing the ridge, while tanks provided overhead fire support. Company K, on the battalion left, fared very well; after a steady advance across the rice paddies, it took the ridge to its front "with astonishing rapidity." [48]

[47] Here, as along the entire division landing area, the enemy fire benefited from perfect observation. The commander of the 9th Marines remarked later that "until the FBHL in the Mt. Alutom-Mt. Tenjo area was taken by us, direct observation of practically all our rear areas was possible by the enemy." LtGen Edward A. Craig ltr to CMC, dtd 30Sep52, hereafter *Craig ltr.*

[48] Col Walter Asmuth, Jr., ltr to CMC, dtd 11Sep52.

Following the seizure of the rice paddy area near the mouth of the Asan River, the 12th Marines (Colonel John B. Wilson) began setting up its firing batteries to support the infantry assault.

At 1350, the 3d Battalion, 9th Marines reached its objective, and 1/9 and 2/9 waited orders to pass through. At 1415, just eight minutes after receiving the word from the division commander, Colonel Craig attacked, advancing to within 400 yards of the Tatgua River by 1600. There the troops dug in for the night.

The progress of 1/9 and 2/9 had not been devoid of enemy resistance. Though Asan Point had been previously well covered by 3/9, there were still small groups of Japanese in concealed firing positions.[49] When the 2d Battalion crossed the bridge over the Nidual River, enemy machine guns on the point opened up, and the Marines had to fight to the rear a short distance in order to reduce the opposition.[50]

Colonel Craig had set up his advance command post immediately to the rear of the 3d Battalion, and it was fire from Asan Point apparently which wounded the regimental executive officer, Lieutenant Colonel Jaime Sabater, on W-Day.[51] A Marine antitank gun at the command post knocked out a con-

[49] Three weeks later, the Marines were still finding Japanese in the honeycomb of caves on Asan Point. *Craig ltr.*

[50] Besides the machine gun positions on Asan Point, there was a battery of three 8-inch naval guns in concrete emplacements. The battery covered the beaches and seaward to the west of Asan Point; Marines found it abandoned.

[51] The 9th Marines then had no executive officer until 30 July when Lieutenant Colonel Ralph M. King joined the regiment.

cealed Japanese antitank gun in the vicinity. Like the machine guns that covered the Nidual River bridge, the enemy weapon, manned by eight men, was so well camouflaged that it escaped detection by 3/9.

At 1830, the 9th Marines tied in with the 21st Marines. The progress of the 9th Marines on W-Day—the regiment had secured a beachhead 1,500 yards in depth—was dearly won, for casualties had been high.[52] Included in the figure of 231 were 20 officers killed or wounded. Lieutenant Colonel Asmuth of 3/9 was among the wounded; he was relieved on 22 July by Major Donald B. Hubbard. The commanders of Company I and Company K were both killed in action.

SUNSET OVER THE ASAN BEACHHEAD [53]

The first day on Guam had cost the 3d Marine Division 105 men killed, 536 wounded, and 56 missing in action. A number of these casualties had resulted from the mortar, artillery, and sniper fire which fell upon the beaches—handicapping but never stopping the movement of supplies.[54]

To get the immediate necessities ashore, every available man was employed; bakers of the 3d Service Battalion, who did not have to bake bread until later, turned to as boat riders and handled cargo. It was the 19th Marines, commanded by Lieutenant Colonel Robert E. Fojt, which formed the backbone of the division shore party. Company B, 5th Field Depot, of the Supply Service, FMFPac, had been attached to the division and at 1030 the unit landed on Red Beach 2 to operate the supply dumps. The 5th Field Depot, which was part of the Island Command, had been assigned a string of prospective dump sites on Guam, totaling more than 600 acres, but most of the areas "proved to be suitable for rice cultivation and not much else." [55]

As General Geiger reported, the ship-to-shore movement was "skillfully executed." [56] There were instances where some things could have been done differently and better, but they were relatively few in proportion to the size of the division landing.[57] Men trans-

[52] This was the maximum depth of the division beachhead at the end of W-Day. In width it measured 4,000 yards. Such figures are illusory, however, because of the numerous gaps in the line and the fact that the enemy held strong positions overlooking the beachhead.

[53] Unless otherwise noted, the material in this section is derived from: *TF 53 OpRpt; IIIAC SAR; 3d MarDiv SAR;* 5th FldDep SAR, dtd 25Aug44, hereafter *5th FldDep SAR;* Capt Edwin H. Klein, "The Handling of Supplies at Guam," *Marine Corps Gazette*, v. 29, no. 2 (Feb45), hereafter Klein, "The Handling of Supplies at Guam."

[54] Such enemy fire kept up all day and "had troop leaders been less aggressive in moving their units off the beach, casualties would have been much heavier." Lodge, *Recapture of Guam*, p. 47.

[55] Klein, "The Handling of Supplies at Guam," p. 26. Company B was attached to the 3d Service Battalion for the landing.

[56] *IIIAC SAR*, Encl B, p. 2.

[57] Teams of the 3d Joint Assault Signal Company (JASCO) landed less than 20 minutes after H-Hour, much sooner than necessary. They had their ship-to-shore and lateral beach communications set up by 1100, but there was no traffic until more than two hours later. Major John H. Ellis, the company commander, recommended after the campaign that "the

214-881 O-67—31

ferred tons of cargo from landing craft to LVTs and DUKWs, using large cranes mounted on pontoon barges anchored just off the reef. The amphibian tractors and trucks then took the cargo from the reef to the shore.[58]

The reef here extended at distances varying from 100 to 350 yards from the beaches. At high tide it was covered by 30 inches of water and at low tide by 6 inches. The edge dropped off abruptly; the reef detachment often worked in waist-high water. When fuel drums were deposited from landing craft at the reef edge they were floated in by wading Marines. Unloading was continued for some hours after dark—an unusual procedure on the day of a landing, for it required partial lighting on the ships—but the absence of enemy aircraft allowed such a risk.

By sunset of W-Day, the 3d Marine Division was well started on the battle to recapture Guam. At 1715, General Turnage assumed command ashore.

THE SOUTHERN BEACHES [59]

The 1st Provisional Marine Brigade on W-Day encountered more favorable terrain than the division. The enemy, however, supplied the resistance which the earth itself did not. In spite of the preassault bombardment, there were Japanese waiting for the Marines—deafened and shocked, but waiting grimly. The beach defenses, some of them intact although scarred by gunfire, included concrete pillboxes and a trench system with machine gun emplacements and tank traps. Casualties were numerous at Yellow Beach 2, where Marines received savage fire from the concrete blockhouse on Gaan Point—a cornerstone of the beach defense—and small arms, mortar, and machine gun fire from other well-concealed positions overlooking the beach. (See Map VII, Map Section.)

Brigade assault troops set foot on Guam at 0832. At the extreme left, the 1st Battalion, 22d Marines landed on Yellow Beach 1, while 2/22 went ashore on Yellow Beach 2 and the 3d Battalion, boated in LCVPs, marked time at the line of departure, in ready reserve. When ordered, the Marines of 3/22 would transfer at the edge of the reef to LVTs returning empty from

teams should not be landed until they are operationally useful. Four JASCO teams," he said, "took boat spaces of 80 infantrymen and then waited on the beach for more than four hours before their services were required." *3d JASCO SAR*, p. 1.

[58] Marines of the 2d Separate Engineer Battalion comprised the reef transfer battalion. For the immense job at the reef, they were helped by men of the 3d Service Battalion.

[59] Unless otherwise noted, the material in this section is derived from: *IIIAC SAR; 1st ProvMarBrig SAR; 5th FldDep SAR;* 1st ProvMarBrig URpts Nos. 1–12, 21Jul–1Aug44, hereafter *1st ProvMarBrig URpts;* 22d Mar Jnl, 21Jul–16Aug44, hereafter *22d Mar Jnl;* 1/4 WarD, 30May–9Sep44, hereafter *1/4 WarD;* 3/4 WarD, 21Jul–9Aug44, hereafter *3/4 WarD;* 1/22 Jnl, 21Jul–9Aug44, hereafter *1/22 Jnl;* 6th TkBn SAR, dtd 30Mar45, hereafter *6th TkBn SAR;* Condit and Turnbladh, *Hold High the Torch;* Charles O. West, *et. al.*, eds., *Second to None! The Story of the 305th Infantry in World War II* (Washington: Infantry Journal Press, 1949), hereafter West, *Second to None;* LtCol Max Myers, ed., *Ours to Hold it High: The History of the 77th Infantry Division in World War II* (Washington: Infantry Journal Press, 1947), hereafter Myers, *Ours to Hold it High.*

the beach.[60] Led by their share of the 37 armored amphibians assigned to the brigade, assault troops of the 4th Marines landed on White Beaches 1 and 2—the 2d Battalion on the left and the 1st on the right, with 3/4 in reserve.[61] At 0846 Lieutenant Colonel Shapley reported "battalions landed and received mortar fire on beaches." [62] The brigade had begun its battle for Guam.

The 22d Marines suffered a considerable loss of men and equipment while landing, but once the troops were some 200 yards inland, out of range of the Japanese guns aimed at the beaches, progress was easier—at least briefly so. The 2d Battalion (Lieutenant Colonel Donn C. Hart) had advanced to high ground about 1,000 yards inland before noon, when it began to receive artillery fire, a foretaste of the resistance beyond the beaches. Such fire increased as Lieutenant Colonel Hart reorganized on the high ground and prepared to move out at 1250 to seize his portion of the brigade objective, a line which included the crest of Mt. Alifan and the village of Agat. Progress that afternoon was measured by inches. When a Japanese dual-purpose gun stopped Company E, the battalion commander requested an air strike. But the strafing hit the front lines, and casualties resulted when several bombs fell in the vicinity of Company F. The accident prevented resumption of the attack before the battalion received orders to dig in for the night.

The 1st Battalion (Lieutenant Colonel Walfried H. Fromhold) had wheeled left toward Agat after landing. The villagers had long since deserted the town, but the rubble left by the naval and air bombardment was still inhabited—by Japanese snipers. The Marines expected to encounter organized resistance from the surrounding area, if not from the town itself. While Company A moved rapidly across the rice paddies, Company B, to the left, advanced up the beach. Both units reported little opposition, but Lieutenant Colonel Fromhold took the precaution of committing the battalion reserve, Company C, on the seaward flank.[63]

In the ruins of Agat, the Marines received some sniper fire, but at 1020 Lieutenant Colonel Fromhold reported: "We have Agat." [64] By 1130 the 1st Battalion was at Harmon Road, which led from the middle of Agat to the Maanot Pass on the northern shoulder of Mt. Alifan, and the regimental commander ordered the capture of the rest of the town.

Company C, on the extreme left of

[60] "The LVT waves had been ordered to proceed inland a distance of 1,000 yards from the beach before stopping to unload, but that was found to be impracticable, except in isolated places, due to obstacles and mines inland of the beach." *1st ProvMarBrig SAR*, p. 4.

[61] The 3d Battalion began landing on White Beach 2 at 0930. Due to failure of radio communications, the reserve battalion of the 22d Marines did not receive orders to land until 1236.

[62] *1st ProvMarBrig Jnl*, 21Jul44.

[63] Regimental headquarters then attached Company I as the 1/22 reserve, after landing the company on Yellow Beach 2 at 1010. The 3d Battalion headquarters and Company K landed at 1255 and moved to an assembly area. At 1615, Company L was attached to 2/22. At 1630, the 3d Battalion, less Companies I and L, moved to set up defenses for Yellow Beach 1.

[64] *22d Mar Jnl*, 21Jul44.

the brigade, had some rough going that afternoon. While attempting to flank an insignificant mound east of Agat, the Marines received machine gun fire from the beach 50 yards away, which forced their withdrawal to a series of trenches near the foot of the hill. Here the men were pinned down for an hour. When a reserve platoon of Company I was sent forward, the Marines renewed the attack, only to be turned back again by the intolerable fire of automatic weapons concealed in a maze of underbrush.

In graphic language, a Marine officer described the situation: [65]

> . . . the Marines didn't know where the emplacements were, and many of them died trying to find out. The men wondered and waited, and dug in for the night.
>
> Then occurred one of those inexplicable things known to every Marine who has fought Japs, and understood by none. Down a trail leading to the center of the trench marched 12 Japs. They carried the machine guns—three heavies and a light—which had held up the American advance all afternoon. The Japs were riddled by Marine bullets. 'Those Nips were so heavy with slugs we couldn't lift them,' said one of the men.

The fighting had depleted Company C. At 1705, the commander reported he had only 100 effectives, including the reserve platoon, and would "need help for tonight." [66] A second reserve platoon was moved up. The battalion commander ordered Company C to fall back 50 yards to a better position for the night and to tie in with Company B. At 2000, all companies of the battalion were dug in, believing they could hold their positions until morning.[67]

The 1st Battalion, 22d Marines had lost a number of men on W-Day. The handling of casualties had been complicated for hours after the landing because a shell from a Japanese 75mm field gun hit an aid station party, destroying medical supplies and injuring every member except one. Not until afternoon did the battalion have a doctor, but it was still short of corpsmen, stretchers, and bearers. Evacuation was hampered until an amphibian tractor was obtained.

It was the shortage of amphibian tractors, due to losses, that was chiefly responsible for the supply headaches that plagued the brigade on W-Day. Commanders called for more ammunition at the frontlines. When the situation did not improve, General Shepherd sent word to the commander of the Southern Transport Group and to the control vessel:

> Supplies not coming ashore with sufficient rapidity. Believe delay at transfer line at edge of reef. Expedite movement, with preference to all types ammunition.[68]

Getting supplies transferred at the reef was never a picnic; with insufficient LVTs, the difficulties were compounded and the tasks made even harder. Another handicap was the deposit of silt at the inner edge of the reef, which caused some of the amphibian tractors and DUKWs to bog down.

[65] Kaufman, "Attack on Guam," p. 3.
[66] *1/22 Jnl*, 21Jul44.

[67] General Shepherd had ordered that the brigade attack cease not later than 1700 and that particular attention be paid to defenses in depth and maintenance of a local reserve against possible counterattacks. 1st ProvMarBrig OpO No. 9, dtd 21Jul44.
[68] *1st ProvMarBrig Jnl*, 21Jul44.

Rubberboat causeways and ship life rafts partially helped to relieve the congestion on the reef, and every available man was put to work here.[69]

Among the brighter aspects of W-Day were the optimum conditions for use of armor. With the advance inland, 1/22 came to "good tank country" before noon and reported it "would like to use the tanks here." [70] The 22d Marines armor support had reached the reef at 0840 and run into mortar fire, mines, and shell holes while moving onto the beaches; two tanks submerged before getting ashore.

Due to the condition of the reef, the tank company of the 22d Marines had been ordered to land on the 4th Marines beaches and then travel along the waterline to join its regiment. The detour took time but it was not without benefit, for en route the tanks destroyed the troublesome Japanese emplacement at Gaan Point, knocking out one of the guns at a range of 50 yards.[71] Machine gun and mortar positions along the beach were also fired upon. The tanks reported to 2/22, according to orders, but the lack of opposition and the unsuitable terrain there suggested support of 1/22 instead, and armor led the afternoon attack by Company A. Toward evening, tanks were sent to reinforce the hard-hit Company C.

Before dark of W-Day, the Marines of 2/22 could see the 4th Marines to their right, across a deep gully. Lieutenant Colonel Shapley's regiment had moved rapidly inland after meeting negligible enemy resistance at the beaches. Up to an hour after the landing, casualties were still "very light." [72]

The immediate ground encountered by the 4th Marines was more flat than that the 22d Marines had met; in fact, the elevations were so low that the maps did not show them. The Japanese knew of them, however. One such rise—it was 10 to 20 feet high—lay in the path of the 2d Battalion, 4th Marines, at a distance of less than 100 yards from the beach. The Japanese were dug in on the reverse slope, and the pocket of resistance briefly delayed the advance of 2/4. By 0947, however, Lieutenant Colonel Shapley reported that the 2d Battalion (Major John S. Messer) was 700 yards inland.

The 1st Battalion (Major Bernard W. Green) had landed with Company A and Company B in the assault. When 30 yards from the beach, Company B, on the left, had two Marines killed and three wounded by machine gun fire before the pillbox from which it came was located and its five defenders killed. Company A reported less opposition, but a platoon leader was killed by enemy fire while crossing an open rice paddy.

When Companies A and B were some 700 yards inland, in contact with 2/4, the reserve Company C was landed and turned right to attack Hill 40 and

---

[69] Except for Company B, which went ashore on Red Beach 2, the assault echelon of the 5th Field Depot was landed over the Yellow and White Beaches and attached to the brigade. The pioneer companies of the 4th and 22d Marines passed to control of the brigade shore party.

[70] *1/22 Jnl*, 21Jul44.

[71] A misfortune of the trip was that two tanks got stuck in shell holes on the beach. In retrieving the vehicles, the Marines suffered several casualties from mortar and artillery fire.

[72] *1st ProvMarBrig Jnl*, 21Jul44.

Bangi Point. The latter had been heavily worked over by naval gunfire and was readily occupied, but Hill 40 was still bristling with live Japanese and machine guns whose fire halted the attack by Company C. When Company A, to the left, also caught some of the fire, Major Green called up two tanks, which supported a second and successful assault of the hill. At 1130 two companies of the reserve 3d Battalion (Major Hamilton M. Hoyler) started forward to relieve the 1st Battalion and free it to push on toward Mt. Alifan. Company K took over Hill 40 and Bangi Point, relieving Company C which reverted to regimental reserve. Company I moved up on the left flank in the battalion zone and relieved Company A. One platoon of the reserve Company L was assigned to seize Alutom Island off Bangi Point, which it found undefended. The rest of the company moved into a small river valley 300 yards upstream and straight east of Alutom Island.

Before noon, the two assault battalions of the 4th Marines had reached the initial regimental objective line, over 1,000 yards inland. At 1345, on brigade order, Lieutenant Colonel Shapley resumed the attack to seize the brigade objective, including the peak of Mt. Alifan. Scattered resistance was encountered as the Marines crossed open fields, but by 1700 they reached the rough and wooded ground at the foot of the mountain.

Digging in for the night, the men prepared for an expected counterattack. Company B set up a roadblock on Harmon Road; five tanks of the 4th Marines Tank Company were parked in a hollow just off the road, not far from the 2/4 CP. The regimental line stretched from heights above the Ayuja River around the lower slopes of Mt. Alifan to the beach at Bangi Point. It was a long line, measuring about 1,600 yards, and strongpoints had to be wisely located to cover the gaps with fire. Lieutenant Colonel Shapley also bolstered the line by tying in his Reconnaissance Platoon and an engineer detachment on the Company A left. Company C, kept in reserve near the regimental command post, would be ready for action if needed.

The brigade command post, located about 200 yards southeast of Gaan Point, had opened at 1350, when General Shepherd assumed command ashore. When he reported the brigade situation at the end of W-Day, the southern beachhead measured about 4,500 yards long and 2,000 yards deep:

> Own casualties about 350. Enemy unknown. Critical shortages fuel and ammunition all types. Think we can handle it. Will continue as planned tomorrow.[73]

The next day the brigade commander would have the reserve 305th Infantry Regiment at hand. Its 2d Battalion had landed on the afternoon of W-Day. With no LVTs to use—no Marine amphibian tractors were available and the Army had none—the soldiers had to wade ashore from their LCVPs, which could not cross the reef or negotiate the shallow waters beyond it. It was a blessing that the Japanese were too involved with the Marines to endanger the Army landing with fire, but the curse of sharp coral and deep potholes plagued the watery approach by foot. After reorganizing on White Beach 1,

[73] *1st ProvMarBrig Jnl*, 21Jul44.

the battalion moved to an assembly area about 400 yards inland from Gaan Point.

At 1430, General Shepherd ordered the rest of the 305th to land; owing to communication problems, the regimental commander (Colonel Vincent J. Tanzola) did not receive the message for an hour. He had only enough craft to move one battalion, and he turned to the 1st (Lieutenant Colonel James E. Landrum), but naval officers had received no landing instructions and refused to dispatch the boats to the reef. As a result, the men of 1/305 waited in their LCVPs until 1730 when the brigade confirmed the movement. With darkness fast approaching, Colonel Tanzola suggested suspension of the battalion landing. General Shepherd, however, desired that the reserve get ashore that night, so the 1st Battalion continued on to the beach. Again the troops had to wade ashore, but now the water had become chest-high from the incoming tide, and, though weighted with their gear, some soldiers attempted swimming. By 2130, 1/305 was digging in on land. The 3d Battalion followed; it was 0200 before the leading waves got to the reef, and 0600 before the last men got to shore. An hour later, the battalion was still wet and tired but reorganized.[74] The landing of the 305th had been a confused and dragged-out affair, but it revealed a stamina that was to be indicated again in battle.

One battery of the 305th Field Artillery Battalion was landed at dark on White Beach 1 and attached to the Brigade Artillery Group.[75] General Shepherd stressed the early landing of artillery, and he wanted the Corps 1st 155mm Howitzer Battalion to get ashore before the second day on Guam. At 1835, however, he could report that only three 155mm howitzers had been landed.[76]

The brigade's two pack howitzer battalions were in position with batteries registered before dark. The weapons of these battalions had been loaded on DUKWs, which delivered the goods despite jagged coral heads and potholes. Actually, there was only one point where it was practical for the amphibian trucks to move to the beach from the reef edge, so their traffic was restricted. As soon as they had delivered their loads of howitzers and ammunition, the DUKWs were pressed into service as cargo carriers, joining the LVTs at the transfer line.

Unlike the operations at the northern

[74] The commander of 3/305, was able to borrow five LVTs from the Marines, and the later waves of the battalion were taken across the reef by four of the vehicles while the fifth LVT was used as a control vehicle. Col Edward A. Chalgren, Jr., USA, ltr to Head, HistBr, G–3, HQMC, dtd 23Jan53.

[75] The battalion had been loaded on five different ships, complicating the task of getting ashore. The unit report describes some of the woe of landing on a tropical Pacific island: the battalion "was brought to the edge of the reef in LCMs, and then an attempt was made to drive across the reef. In most cases vehicles stalled and had to be towed. Equipment was soaked in salt water, and two howitzers were out of action for several days." 305th FldArtyBn OpRpt, dtd 14Aug44 (WW II RecsDiv, FRC, Alexandria, Va.)

[76] The 1st 155mm Howitzer Battalion of the Corps Artillery was to reinforce fires of the brigade. Battery C would be attached on landing to the Brigade Artillery Group.

beachhead, where the water was more shallow and it was possible to set up some cranes on coral heads at the sharp edge of the reef dropoff, all cranes off the Agat beaches had to be barge mounted. Most cargo transfer took place in deep water, utilizing the barges or makeshift raft platforms as floating dumps. The shortage of LVTs, as a result of W-Day casualties, was the crowning logistic difficulty and kept the supply situation tight on shore. Recognizing this, Admiral Conolly ordered unloading to continue through the night to insure that the brigade had adequate supplies for its mission. Regardless of the logistic situation, General Shepherd felt that his men could handle whatever the enemy should attempt that night or the next day.

Consolidating a Foothold

FIRST NIGHT AT AGAT[1]

The day had not gone well for the enemy's *38th Regiment.* Most of the men in the two *1st Battalion* companies that had tried to hold the Agat beach defenses were dead by noon, including the commander of *1/38* who was killed as he led his headquarters and reserve elements in a "*Banzai*" counterattack against 4th Marines assault troops. The guns of the two artillery batteries that had fired in direct support of the beach defenses had been demolished by naval gunfire and air bombardment. Only a few members of the gun crews survived the destructive fire.

On the northern flank of the beachhead, the 22d Marines had wiped out forward elements of *2/38* that tried to hold Agat. Most of the units of the enemy battalion were still intact, however, when darkness fell. Since the battalion commander had lost contact with regimental headquarters at about 1200, he had little knowledge of how the battle was going except on his own front, where it was going badly.

To the south of the Marine positions, the *8th Company* of *3/38* was committed early on W-Day to reinforce the *1st Battalion* platoons that had tried to hold Hill 40 and Bangi Point. The remainder of the *3d Battalion*, spread out through a defensive sector stretching to Facpi Point and beyond, was assembled by its commander by mid-afternoon, ready to move against the American beachhead. Marine intelligence officers considered the situation was ripe for a Japanese counterattack —and a counterattack was coming.[2]

From his command post on the slopes of Mt. Alifan, Colonel Tsunetaro Suenaga had seen the Americans overwhelm his defenses along the island shore. The resulting swift inland advance of Marine infantry and tanks threatened to make a mockery of the attempt by the *38th Regiment* to hold the Agat sector unless the Japanese commander regained the initiative. Suenaga, who felt that his only chance to retrieve the situation lay in an all-out counterattack, gave orders for his

[1] Unless otherwise noted the material in this section is derived from: *1st ProvMarBrig WarD; 1st ProvMarBrig URpts; 22d Mar Jnl; 1/4 WarD; 3/4 WarD; 6th TkBn SAR;* 77th InfDiv and 1st ProvMarBrig NGF LnOs Rpts to CGFMFPac, variously dtd 14–24Aug 44; *GSDF Study;* Myers, *Ours to Hold it High.*

[2] Contemporary intelligence studies by the brigade and IIIAC located *3/38* in reserve in the Agana area on W-Day. This error was repeated in text and maps in both Lodge, *Recapture of Guam*, and Crowl, *Marianas Campaign.* The best Japanese account of the battle, the *GSDF Study*, correctly places the battalion in the Agat sector and sheds fresh light on conflicting evidence of 1944, which was the basis of the original order of battle information.

battalions to prepare for a three-pronged assault against the center and both flanks of the 1st Brigade position. By word of mouth and runner, all *1st Battalion* survivors of the day's battles were ordered to assemble at regimental headquarters.

At about 1730, Colonel Suenaga telephoned the *29th Division* CP to inform General Takashina of his counterattack plans. At first, the general refused permission for the attack because the regiment had been "badly mauled during the day,"[3] but finally, in view of the overall battle situation, he reluctantly authorized the assault.[4] Takashina cautioned the colonel, however, to make plans for reassembling his men following the counterattack in order to continue the defense of Mt. Alifan. Doubt about the outcome of the attack was obviously shared by Suenaga, who, soon after this call, burned the colors of the *38th Regiment* to prevent their capture.

The pending Japanese counterattack was fully anticipated by General Shepherd's veteran troops. All along the Marine front lines as darkness deepened, company and battalion mortars registered their fire along possible avenues of approach. Taking position offshore, gunfire support ships checked into the control nets shared with the liaison officers and spotter teams. The six pack howitzer batteries of the Brigade Artillery Group made preparations for their part in the night's proceedings.

The early hours of the evening were tense but quiet. Occasional brief flare-ups of firing marked the discovery of enemy infiltrators. Finally, just before midnight, a flurry of mortar shells burst on the positions of Company K of 3/4, on the right flank of the brigade line. Japanese infantrymen, bathed in the eerie light of illumination flares, surged forward toward the dug-in Marines. The fighting was close and bitter, so close that six Marines were bayoneted in their foxholes before combined defensive fires drove the enemy back.[5]

This counterattack was but the first of many that hit all along the beachhead defenses during the rest of the night. Illumination was constant over the battlefield once the Japanese had committed themselves; naval gunfire liaison officers kept a parade of 5-inch star shells exploding overhead. Where the light shed by the naval flares seemed dim to frontline commanders, 60mm mortars were called on to throw up additional illumination shells. The attacking enemy troops were nakedly exposed to Marine rifles and machine

[3] *GSDF Study*, p. 151.

[4] A Marine officer, well acquainted with the Japanese accounts of this action and the personalities involved, commented: "In my judgement, permission to launch this piecemeal counterattack was given because the 38th Japanese Regiment was isolated and on the extreme right flank of the American landing (Japanese left flank). As this regiment was isolated and therefore not available to the overall attack which was planned for later, the Division Commander gave his permission, with hopes of turning the American flank or at least delaying the inland movement of the 1st Marine Brigade." *Metzger memo.*

[5] Bevan G. Cass, ed., *History of the Sixth Marine Division* (Washington: Infantry Journal Press, 1948), p. 15, hereafter Cass, *6th MarDiv History*.

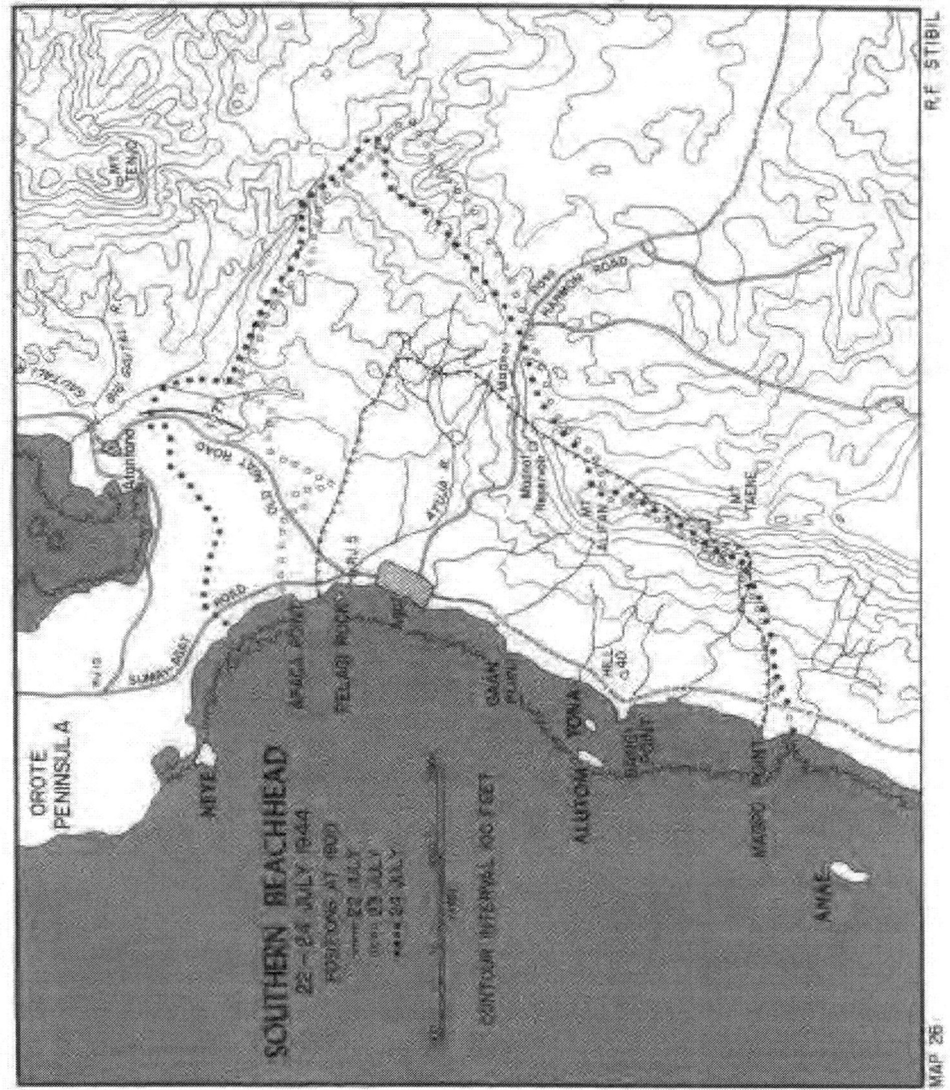
SOUTHERN BEACHHEAD
22 – 24 JULY 1944
CONTOUR INTERVAL 100 FEET
OROTE PENINSULA
AGAT
ALUTOM
YONA
BANGI POINT
MAGPO POINT
ANAE
MT. TENJO
MT. ALIFAN
MT. TAENE
RAINBOW ROAD
OLD AGAT ROAD
R.F. STIBIL
MAP 26

guns and the lethal bombardment directed by forward observers for heavier supporting weapons. The carnage was great, but the men of the *3d Battalion, 38th Infantry* kept trying to break through the American lines.[6]

Hill 40, 300 yards inland from Bangi Point, became the focal point of the *3/38* attack. The platoon of Company K holding the small rise was hard pressed and driven out of its positions twice, but rallied each time, counterattacked, and recovered its ground. Similar dogged efforts by Major Hoyler's men kept the remainder of the 3/4 defenses intact, but when small arms ammunition ran low in the forward holes, the Marines reserved their fire for sure targets. The defensive fusillade, however, had accomplished its purpose; there were few Japanese left alive in front of Company K.

In the confusion of the fighting, small groups of the enemy, armed with demolition charges, made their way through the lines headed for the landing beaches. Some of these Japanese stumbled into the night defensive perimeters of the 1st and 2d Battalions of the 305th Infantry; those that did were killed by the alert soldiers. Other Japanese made life miserable for the Marine artillerymen that were firing in support of the frontline troops. As one battery executive officer recalled:

> By 0130, we were up to our necks in fire missions and infiltrating Japanese. Every so often, I had to call a section out for a short time so it could take care of the intruders with carbines and then I would send it back into action again.[7]

Explosive-laden parties of enemy soldiers got as far as the beach road, where they disabled two weapons carriers and three LVTs before they were gunned down. A platoon from the Ammunition Company, 5th Field Depot intercepted and killed 14 Japanese headed for the brigade ammunition dump.[8] The 4th Marines Tank Company also had a rough night with infiltrators; 23 were killed in and around the service park.

Not all the Japanese that found their way into the rear areas of the brigade came through the thinly spread positions on the south flank of the beachhead. A few filtered through the 22d Marines lines on the north, and others were offshoots of the force that attacked the 4th Marines units dug in on the lower slopes of Mt. Alifan. Here, where Colonel Suenaga was in the forefront of the assault troops, the Japanese made an inspired effort to break through to the beach, but in vain. In the course of the fighting the enemy commander himself was killed.

Japanese probing attacks began hitting all along the lines of 1/4 shortly before midnight, but the fire fights that developed were just preliminaries to

---

[6] A Japanese prisoner taken on Guam aptly described the attacking troops' dilemma in a situation that occurred frequently on the island, noting: "We had been thinking that the Japanese might win through a night counterattack, but when the star shells came over one after the other we could only use our men as human bullets and there were many useless casualties and no chance of success." Quoted in ComInCh, *The Marianas*, Chap 3, p. 13.

[7] Capt Benjamin S. Read ltr to Capt Orlan R. Lodge, dtd 3Jan52, quoted in Lodge, *Recapture of Guam*, p. 55.

[8] *5th FldDep SAR*, p. 7.

the main event. At about 0230, the rumble of tanks was heard above the din of battle by the Marines guarding Harmon Road. A hurry-up call for Marine tanks was sent to the 2/4 CP, where a platoon of the regimental tank company was on alert for just such an eventuality. First two and later the remaining three mediums of the platoon moved up to the area where Company B held blocking positions on the road. At 0300, Marine infantry and tank machine guns opened up on a column of Japanese light tanks as they approached the American lines. When tracers located the targets, tank gunners and a bazooka team close by the roadside opened fire at pointblank range. The first two enemy tanks were hit by rockets before the bazooka gunner was cut down by the return fire. The 75s of the Shermans also hit both lead tanks and two others besides. Helped by the light of burning tanks and the flares which sputtered overhead, the men of Company B beat back the Japanese infantry that had accompanied the abortive tank thrust.

To the right of the Harmon Road positions, Company A had a hard night-long struggle to hold its ground against the Japanese troops that repeatedly charged down the heavily wooded slopes of the mountain. But the Marines did hold, despite casualties that reduced one rifle platoon to a strength of four able-bodied men.[9] By dawn, the worst part of the night's battle to hold the center of the brigade line was over. As the sun came up, a Japanese tank was spotted well up in the moutain pass near the Maanot Reservoir. A Marine Sherman, one of those that had helped repulse the night's attacks, fired four armor-piercing shells at a range later figured at 1,840 yards, and scored two hits, setting the tank afire.

Marine tanks, sharpshooting or otherwise, were not needed on the northern flank of the perimeter during the night's fighting. Although there was a constant drumfire from enemy infiltration attempts all along the 22d Marines lines, there was no all-out effort by the Japanese, since the commanding officer of *2/38* had received no orders to join in the counterattack of his regiment. Only his *6th Company*, which was positioned near Maanot Pass, got caught up in the *38th Infantry* attempt to break through the Marine lines. As a consequence, Company G on the right flank of the lines of 2/22 had a busy night of fighting, killing 30 enemy troops between 0100 and 0500. Bands of infiltrators that did get into the rear areas harassed the 22d Marines CP until daylight, when Colonel Schneider's headquarters troops mopped up the area.

Dawn brought a general cleanup of the surviving Japanese infiltrators throughout the brigade perimeter. Local attacks supported by tanks quickly restored the lines wherever they had contracted for better night defense during the height of the fighting. The brigade lost at least 50 men killed and twice that number wounded during the

[9] Company A operated with two platoons for the rest of the campaign since replacements were not available. Maj Orville V. Bergren ltr to CMC, dtd 6Jun47, hereafter *Bergren ltr*.

counterattack,[10] but counted over 600 enemy dead within, on, and in front of the perimeter.

After one day and a night of battle, the *38th Regiment* ceased to exist as an effective fighting force. Only its *2d Battalion* was still intact, and it now started to pull back from contact with the 22d Marines and retire toward Orote Peninsula. The dazed and scattered survivors of the counterattack, about 300 men in all, gradually assembled in the woods northeast of Mt. Alifan. There, the senior regimental officer still alive, the artillery battalion commander, contacted the *29th Division* headquarters. He soon received orders to march his group north to Ordot, the assembly point for Japanese reserves in the bitter struggle for control of the high ground that commanded the Asan-Adelup beaches.

BUNDSCHU RIDGE AND CABRAS ISLAND [11]

There were few members of the enemy's *320th Independent Infantry Battalion* left alive by nightfall on W-Day. Two of its companies, once concentrated in the Chonito Cliff area and the other at Asan Point, had defended the heights that rimmed the 3d Marine Division beaches. The third rifle company, originally located along the shore east of Adelup Point, had been committed early in the day's fighting to contain the attacks of the 3d Marines. The commander of the *48th IMB*, General Shigematsu, had also committed his brigade reserve, the *319th Battalion*, to the battle for control of the high ground on the left flank of the American beachhead.

According to plan, as soon as the landing area was certain, General Shigematsu assumed command of most of the *29th Division* reserve strength and began its deployment to the rugged hills above the Asan beaches. Elements of the *2d Battalion, 18th Infantry* plugged holes in the defenses in the center of the Japanese position, where they tangled with the 21st Marines. The *9th Company, 38th Infantry* reinforced the troops holding the well-concealed emplacements and trenches atop Bundschu Ridge. From positions near Ordot, the *2d* and *3d Battalions* of the *10th Independent Mixed Regiment* were ordered to move out to reinforce *2/18*, hard pressed by Colonel Butler's Marines who had seized a lodgement on the cliffs behind Green Beach.

American carrier planes spotted the movement of the *10th IMR* battalions as soon as they began to move out—about 1100. Although the regiment was only 2½ miles from its initial objective, it took most of the long, hot afternoon to reach it. Towards dusk, the leading elements of the *10th* began filing their way up the steep, brush-filled valley between Fonte Plateau and Mt. Macajna. (See Map 27.)

Just about the time that the *10th IMR*

---

[10] The American casualty total is an estimate based on unit accounts of the fighting. The casualty figures available in contemporary personnel records generally cover a longer period than the time encompassed by the counterattack.

[11] Unless otherwise noted the material in this section is derived from: *3d MarDiv SAR; 3d Mar Jnl; 1/3 Jnl; 2/3 Jnl; 3/3 Jnl;* 9th Mar R–2 Jnl, 21Jul–31Aug44, hereafter *9th Mar Jnl;* 21st Mar URpts, 22Jul–3Nov44, hereafter *21st Mar URpts; 3/21 Jnl; Craig ltr; GSDF Study.*

was reaching the relatively open ground along the Mt. Tenjo Road, the *321st* and *322d Independent Infantry Battalions* began moving toward the fighting, too. Leaving one company and a rapid fire gun unit to man its defenses at Agana Bay, the *321st* started south at 2000.[12] An hour earlier, the *322d*, which had 2–3 miles farther to travel, had left Tumon Bay on a forced march for the battlefield.

Using the Fonte River valley as their gateway to the heights, Japanese reserves continued to arrive at their assembly area on Fonte Plateau throughout the night. General Shigematsu, operating from his battle command post in a quarry not far from the road, dispatched reinforcements into the fight as they became available. Repeatedly, as the night wore on, small groups of enemy infantrymen charged out of the brush, hurling grenades and firing their rifles as they attempted to drive the defending Marines off the high ground. Japanese mortar fire tore at the thin American lines throughout these attacks, and casualties were heavy, but the men of the 21st Marines held. The brunt of the assaults fell on 2/21 along its right flank, but Lieutenant Colonel Smoak drew back his right company to the edge of the cliff where it held and beat back all comers.

Helped immensely by the constant flare light overhead, American mortar, artillery, and naval gunfire observers rained a holocaust on the determined attackers. According to Japanese estimates, during this one night's fighting, *2/18* had two-thirds of its men killed or wounded, *2/10* suffered comparably heavy losses, and *3/10* lost "approximately 200 men." [13] The remaining attackers drew off at dawn to join forces with the troops that General Shigematsu had called up from Agana and Tumon Bay.

Neither of the battalions of the *48th IMB* was able to make its way up through the Fonte valley in time to have a significant effect on the night's fighting. The *321st* in fact was "thrown into utter disorder" [14] by Marine artillery fire as the battalion struggled upward in the darkness, and was scattered again by strafing carrier aircraft after first light. The *322d Battalion*, which followed, could do little more than move into holding positions in the Fonte area, where it arrived near dawn, and wait for a more auspicious occasion to launch a counterattack. The focus of Japanese efforts to dislodge the Marines now shifted from the center to the left of the 3d Division lines.

The Marines of 2/3 and 3/3 that had seized Adelup Point and Chonito Cliff had a precarious hold on their prize terrain. Early in the morning, the men of the *319th* and *320th Independent Infantry Battalions*, who had lost the positions on the 21st, tried to win them back by an all-out counterattack. The situation was serious enough for Colonel Hall to commit all his strength and, at 0605, to request reinforcement from

[12] Convinced that Agana Bay remained a logical landing point for the Americans, the Japanese were reluctant to completely denude its defenses. In addition to the infantry that remained, naval troops continued to hold reserve positions at Tiyan airfield and in the vicinity of the ruins of Agana.

[13] *GSDF Study*, p. 153.

[14] *Ibid.*, p. 154.

the division reserve. One company of 1/21 was ordered to report to the 3d Marines immediately, the shore party on the Red Beaches was alerted to back up the lines on the left, and priority of air support was given to the 3d Marines. Offshore the fire support ships that had illuminated and fired harassing fires in the Agana area all night were anxious to give all the help they could, but the enemy was too close to the Marine lines. The commander of the destroyer *McKee* could see Japanese troops attacking the men on Chonito Cliff, but could not obtain permission to fire from control parties ashore.[15]

While some Japanese units made frontal attacks on the Marine positions, others found their way along the dry stream bed that cut between Adelup Point and the looming cliffs. These attackers moved through the 3/3 command post area and began climbing the slopes in the rear of the Marine foxholes. Fire from Lieutenant Colonel Houser's headquarters troops and from supporting LVTs eventually stopped this thrust and eliminated the remaining Japanese that had penetrated the lines.[16] By 0830, the enemy had started to withdraw and the threat of the counterattack was ended. On the heels of the retreating Japanese, the Marines began to advance but the enemy was able to throw up an almost impenetrable barrier of artillery, mortar, and small arms fire.

The nature of the Japanese counterattacks, and of the terrain that gave them added impetus, provided the pattern for the American response. Originally, the 3d Division had scheduled a three-regiment attack for 0700 on the 22d. Now the 21st Marines held fast, since any advance would dangerously expose its left flank. The 3d Marines had to come abreast of the 21st to make a concerted advance possible. The key to that advance appeared to be possession of Bundschu Ridge. Until the 3d Marines could win its way to the top of this well-defended salient, there could be little progress on the left or center of the 3d Division lines.

The situation was different on the right, where Colonel Craig's regiment fought its way into the flats beyond Asan Point and eliminated most of the defending company of the *320th Battalion* in the process. Elements of *3/18* then attempted to slow the Marine advance during the rest of the day. After nightfall, as the enemy battalion commander prepared to launch a counterattack, he was ordered instead to move most of his men, supplies, and equipment into the hills east of the 9th Marines positions. The Japanese were concentrating their remaining strength on the high ground, and the *18th Infantry* was to hold the left flank of the main defensive positions. As a result of this withdrawal, only small delaying groups countered the advance of 1/9 and 2/9 when they jumped off at 0715 on 22 July.

Inside of two hours the assault companies of both battalions were consolidating their hold on the day's first objective, the high ground along the Tatgua River. Resistance was light and plans were laid for a further advance which would include seizure of the villages of Tepungan and Piti. At

[15] CO, USS *McKee* ActRpt of Guam Op, dtd 7Aug44 (OAB, NHD).

[16] *Bastian ltr.*

1300, the battalions moved out again, and by 1700, 2/9 had captured both villages and the shell- and bomb-pocked ruins of the Piti Navy Yard as well. Inland, the 1st Battalion had kept pace with difficulty, as it climbed across the brush-covered slopes and gulleys that blocked its path. It was obvious that the Japanese had been there in strength; recently abandoned defensive positions were plentiful. The fire of the few enemy soldiers that remained, however, kept the advancing Marines wary and quick to deploy and reply in kind.

While 1/9 and 2/9 were driving forward to secure the coastal flats and their bordering hills, Colonel Craig was readying 3/9 for the assault on Cabras Island. The regimental weapons company, a company of Shermans from the 3d Tank Battalion, and 18 LVT(A)s from the 1st Armored Amphibian Battalion were all alerted to support the infantry, which would make a shore-to-shore attack mounted in LVTs. The morning advance of the regiment had uncovered an area, near the mouth of the Tatgua River, that Craig had designated for the assembly of troops and amphibious vehicles.

Shortly after 1400, the armored amphibians crawled out across the reef and began shelling the beaches on the eastern end of Cabras. The tractor-borne assault platoons followed, avoiding the mined causeway and moving across the reef and water. At 1425, they clattered ashore on the elongated islet.[17] There was no defending fire, but there was a defense. Marines soon found that the ground was liberally strewn with mines spread out beneath a blanket of thick brush; as a result, the going was cautious and slow. At 1650, Major Hubbard reported that 3/9 had advanced 400 yards without making enemy contact, but that the combination of mines and brambles would keep his men from reaching the end of the island before dark. On order, the battalion halted and set up for night defense; two platoons of tanks reinforced the beach positions. With no opposition in sight, the early capture of Cabras on 23 July seemed assured.

Although the situation of the 9th Marines was a favorable one, the regiment was fully committed and holding far longer lines than either the 3d or the 21st. Impressed by the need for strengthening his positions on the left and center of the beachhead and for maintaining the impetus of the attack, General Turnage asked General Geiger to attach a regimental combat team of the corps reserve to the 3d Marine Division. The one reserve infantry battalion that was available to the division commander was "40% depleted" [18] as a result of two days' combat, as 2/21 had been pulled out of the cliff positions it had defended so ably and replaced by 1/21 late on the 22d. Colonel Butler, wanting to give maximum effect to any 21st Marines attack on W plus 2, had requested the switch of battalions in division reserve and the last units were

[17] "Due to the steep banks in the landing area, it was necessary to make and issue scaling ladders and these were used." Although none of the assault troops were hit, one LVT was blown up by a mine, while it was returning to the main island; there were four casualties. *Craig 22Jun65 ltr.*

[18] *3d MarDiv Jnl*, entry of 1455, 22Jul44.

exchanged in place on the shell-battered heights at 2115.

When Lieutenant Colonel Williams' battalion moved into the lines, it had the responsibility of extending the zone of action of the 21st Marines 200 yards to the left of the regimental boundary, which had remained unchanged since the landing. This shift, which appeared to offer a better control point for contact, had been directed by General Turnage in order to improve the opportunity for the 3d and 21st Marines to link their frontline positions. The gap had opened and stayed open, not as a result of a lack of will but of a way, to close it. Patrols attempting to find a lateral route which joined the flanks of 1/3 and 2/21 (later 1/21) could find none that did not include a time-consuming return to the lower slopes back of the beaches. No amount of maps, terrain models, or aerial photographs, nor advance intelligence from former island residents, could do full justice to the nightmare of twisting ravines, jumbled rocks, and steep cliffs that hid beneath the dense vegetation.

With such terrain on its flanks and upper reaches, Bundschu Ridge was a natural fortress for the relative handful of Japanese troops that defended it. Throughout the fighting on 22 July, Major Aplington tried repeatedly to work some of his men up onto the high ground that appeared to lead to the enemy positions. Using Company C on the right and coordinating his attack with Company E of 2/3 on the left, the 1st Battalion commander maintained constant pressure on the Japanese, but could make no permanent headway. Despite some temporary success, Marine units that fought their way to the high ground could not hold what they had won in the face of punishing enemy mortar and machine gun fire. About the only encouraging event in the day's action came near nightfall, when the remnants of Company A were finally able to pull out of exposed positions on the nose of Bundschu Ridge, after Japanese fire, which had pinned them down, slackened and then ceased.

General Turnage planned an all-out attempt to erase the Bundschu salient on 23 July and to make sure that there was a firm and permanent juncture between the 3d and 21st Marines. In a way, the Japanese helped him by sacrificing more of their men in another fruitless attempt to break through the left flank positions of the 21st Marines on the night of 22–23 July. The counterattack that developed against 1/21 was not the one, however, that was planned. The operations officer of the *29th Division*, Lieutenant Colonel Hideyuki Takeda, had issued careful instructions to the commander of the *321st Independent Infantry Battalion* to work his assault units up close to the Marine lines, to throw grenades at the unsuspecting Americans, and then to withdraw in the resulting confusion.[19] In the heat of combat, the enemy assault platoon commanders ignored their orders and charged the Marines. The results were devastating. Japanese casualties were heavy, and only about 50 men of what had been a 488-man

[19] Curiously, *21st Mar URpts* state that this counterattack opened with "an intense light mortar barrage" and make no mention of the hail of grenades ordered by Lieutenant Colonel Takeda.

battalion remained when the last attackers pulled back at about 0300.

The losses suffered by the Japanese in this attack, and the steady attrition of two days of battle, were rapidly thinning the ranks of the *29th Division* and *44th IMB*. Although there were thousands of service and support troops of varying quality left alive to fight, the number of veteran infantrymen was fast shrinking. The valley between Fonte Plateau and Mt. Macajna, site of the division field hospital and that of the naval guard force too, was crowded with wounded men. Aggravating the medical situation was the fact that the Fonte River, which coursed the valley, was so fouled by blood and bodies that it could not be used as a source of drinking water. Thirsty Japanese troops holding the arc of Asan defenses received short water rations from the small supply that could be carried in from Sinajana.

The enemy situation was deteriorating and no one knew it better than General Takashina. His aggressive defensive tactics had cost him many casualties. Faced with what appeared to be almost certain defeat by a superior force, he had the choice of conserving his strength and prolonging the battle as long as possible or trying to obtain a decisive advantage by a massive, coordinated counterattack. By the 23d, the enemy division commander had made his decision, the key decision in the Japanese defense of Guam. He would stage a full-blown attack employing all the men and guns he could bring to bear on IIIAC positions, while he still had substantial strength in veteran troops. At 1300 on the 23d, he issued orders outlining areas of responsibility for combat and support units in preparing for the assault.

A lot of fighting, and a lot more casualties on both sides, occurred before the Japanese were ready to strike. On the morning of the 23d, the 3d Marines continued its attack to seize a firmer hold on the ridges which overlooked every part of the beachhead. To give Major Aplington more men, and thus a better chance to bridge the troublesome gap between the 3d and 21st Marines, Colonel Hall attached to 1/3 a provisional infantry company formed from his regimental weapons company. Referring to Bundschu Ridge, he reported:

> I am going to try to advance up that mess in front of me. What I really need is a battalion whereas I have only 160 men to use on that 500-yard slope. They might move to the top but they couldn't advance on. Company A is down to about 30–40 men with an air liaison officer in charge. Company E is down to half strength. They have no strength to push on.[30]

To give the new thrust as much impetus as possible, every available supporting weapon—naval guns and carrier air, field and antiaircraft artillery, half-tracks and tanks—bombarded the wooded slopes ahead of the 3d Marines before the regiment attacked at 0900. In the center, parallel drives by the 1st and 2d Battalions converged on the Bundschu strongpoint, but the Japanese position was strangely silent. During the night, the enemy had pulled back to fight again on some other ridge of the many that still lay ahead of the Marines. Defense of Bundschu had cost the *9th Company, 38th Infantry,*

[30] *3d Mar Jnl*, 23Jul44.

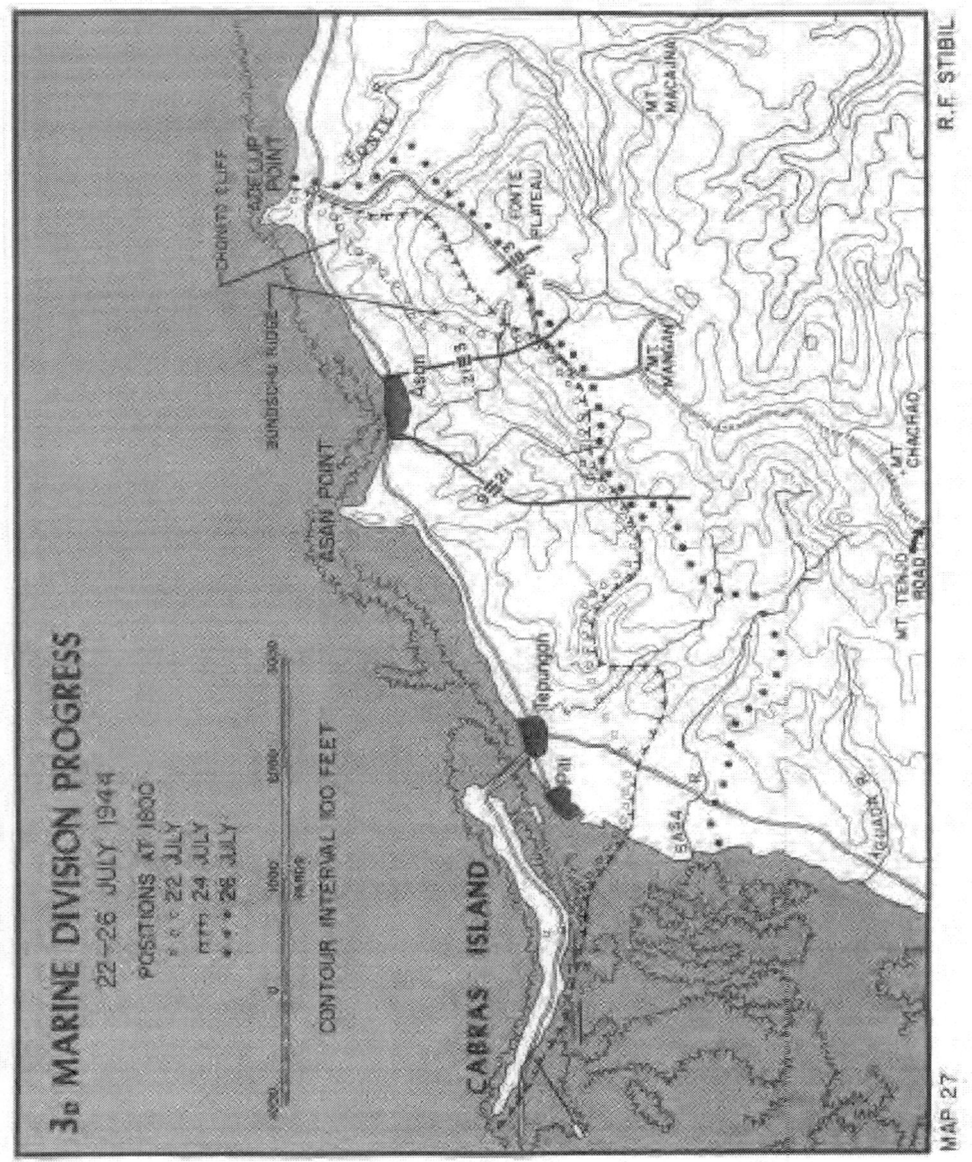
3d MARINE DIVISION PROGRESS
22–26 JULY 1944
POSITIONS AT 1800
22 JULY
24 JULY
26 JULY
CONTOUR INTERVAL 100 FEET
CABRAS ISLAND
Piti
Tepungan
ASAN POINT
Asan
BUNDSCHU RIDGE
CHONITO CLIFF
ADELUP POINT
FONTE PLATEAU
MT. MACAJNA
MT. ALUTOM
MT. CHACHAO
MT. TENJO ROAD
SASA R.
AGUADA R.
R.F. STIBIL

MAP 27

30 casualties, but the return exacted from the 3d Marines was far greater.

Assault platoons of 1/3 and 2/3 linked up atop the ridge at 1108, and the battalions spent the rest of the day cleaning out nests of enemy riflemen and machine gunners who held out in deftly hidden cliffside and ravine defenses within the Marine lines. The concealment offered the Japanese by the dense vegetation and the cover by numerous caves and bunkers made the task of consolidating the newly won positions a formidable one. The incredible complexity of the cut-up terrain in this relatively small area was clearly demonstrated by the failure of all attempts to make permanent contact on the frontline boundary between the 3d and 21st Marines. On the 23d, a 1/3 patrol in radio contact with both regiments moved out from the left flank of 1/21 and "attempting to rejoin its own lines in broad daylight, over a gap of a few hundred yards . . . was lost." The 3d Division comment on the plight of the patrol was sympathetic, noting that "the innumerable gulleys, valleys, and ridges might as well have been gorges and mountains." [21]

The continued existence of the gap plagued Marine commanders, but the Japanese did little to exploit its potential.[22] In fact, they, like the Marines, peppered the area with mortar fire at night to discourage infiltrators.

What the Japanese were really concerned about was readily apparent on the 23d, once 3/3 opened its attack. The enemy reaction was swift, violent, and sustained; a heavy fire fight ensued. Lieutenant Colonel Houser's battalion, by virtue of its hard-won positions at Adelup Point and Chonito Cliff, threatened to gain command of the Mt. Tenjo Road where it climbed to the heights. Once the Marines controlled this vital section of the road, tanks and half-tracks could make their way up to Fonte Plateau and bring their guns to bear on the enemy defenses that were holding back the units in the center of the 3d division line.

During the morning's fighting, Houser was hit in the shoulder and evacuated; his executive officer, Major Royal R. Bastian, took command of 3/3. At 1217, the major reported that his assault companies, I and K, had seized the forward slopes of the last ridge before the cliff dropped off sharply to the rear and the Fonte River valley. The Japanese used their positions on the reverse slope to launch counterattacks that sorely pressed the Marine assault troops. Major Bastian put every available rifleman into the front, paring down supporting weapons crews for reinforcements, and his lines held. By 1400, Colonel Hall was ordering all his units to consolidate their hold on the ground they had won and to tie in solidly for night defense.

The main thrust of the 3d Division attack on 23 July was on the left flank; the rest of the division kept pressure on the Japanese to its front. The battalion on the right of the 3d Marines, 1/21, had its hands full destroying a network of caves and emplacements that covered the sides of a depression just forward of its nighttime positions. The 3d Bat-

[21] *3d MarDiv SAR*, p. 4.

[22] The commander of 1/3, discounting the threat posed by the gap, did not believe "that the terrain made anything but minor infiltration possible." Maj Henry Aplington II, ltr to CMC, dtd 9Apr47, hereafter *Aplington ltr*.

talion, 21st Marines spent the day improving its positions, establishing outposts well forward of its lines, and tangling with small groups of Japanese, who themselves were scouting the American defenses. In general, the right half of Colonel Butler's zone of action was as quiet as it had been since the night of W-Day.

This absence of significant enemy activity carried over into the 9th Marines zone. Squad-sized Japanese units made sporadic harassing attacks both day and night, but there was little organized enemy opposition. The 3d Battalion finished its occupation of Cabras Island early in the morning and at division order, turned over control of the island to the 14th Defense Battalion at 0900. An hour later, Colonel Craig received word from division that his 2d Battalion would replace 2/21 in division reserve. The regimental commander ordered 3/9 to take over the lines held by Lieutenant Colonel Cushman's unit. The relief was effected at 1230, and Cushman moved 2/9 to the positions formerly occupied by 2/21.

Once it was released to Colonel Butler's control, Lieutenant Colonel Smoak's battalion moved to an assembly area near the 21st Marines left boundary. In the attack on the 24th, 2/21, which was all too familiar with the rugged terrain, would spark the drive to close the gap between regiments. The lone infantry battalion in reserve was all that General Turnage could spare from the front lines; he had learned earlier in the day that he could expect no immediate reinforcement from the IIIAC reserve. General Geiger had decided that the situation ashore did not warrant the landing of a 77th Division combat team in the Asan beachhead.

Enemy activity was markedly less after dark on the 23d than it had been on previous nights. Only 2/21 was seriously threatened, and the Japanese thrust at its lines was turned back by artillery and naval gunfire. Since most of the 3d Division front was held by strongpoints only, the support provided by the howitzers of the 12th Marines and the guns of destroyers and cruisers standing by offshore was vital. The constant harassing fire laid on enemy-held territory and the prompt interdiction of actual or suspected routes of approach to the American positions held the Japanese in check.

The fourth day of attacks to expand the 3d Division beachhead saw no spectacular gains, but Marine assault platoons were able to make steady progress. Yard by yard, they increased their hold on the high ground, and, on the left particularly, won positions that gave access to the Mt. Tenjo Road. Not unexpectedly, the hardest fighting took place in a densely wooded draw in the troublesome boundary area between the 3d and 21st Marines.

Lieutenant Colonel Smoak's battalion stirred up a hornet's nest when it attempted to center its drive to the heights on the draw. In it, Japanese troops were set up in mutually supporting cave positions whose machine guns drove the Marines to cover. Emboldened by this temporary success, the enemy made two counterattacks, which were readily beaten back. Assault units, moving upward on the flanks of the Japanese position, were able to bring fire to bear on the caves, but could not silence the enemy guns. A supporting air strike at 1205 enabled a platoon working its way up the bottom of the draw to advance 200 yards before heavy fire again forced a halt. This time the carrier planes, although they were mainly on target, dropped three bombs amongst the Marines, causing 17 casualties. Although this unfortunate accident marked the end of the day's advance, 2/21 had accomplished its mission. When Smoak adjusted his lines for night defense, he was able to tie in strongly with both 1/3 and 1/21. The gap was finally closed.

Highlighting the action on this same day, in the relatively quiet sector of the 9th Marines, was the first attempt to contact the brigade. In the morning, a 30-man patrol worked its way south along the Piti-Sumay Road, while a covey of six LVT(A)s guarded its Apra Harbor flank. Scattered rifle and machine gun fire coming from the high ground inland, coupled with fragments flying from a bombing and shelling of Orote Peninsula forced the patrol to turn back after it had gone 2,600 yards. It found evidence that the Japanese had once occupied the area in force and discovered "huge dumps of all classes of supply near the [Aguada River] power plant, enough to service a regiment, but no traces of the regiment." [23]

If the *18th Infantry* had disappeared from one shore of the harbor, there was ample evidence to show that there was no lack of Japanese on the other side. Soon after night fell on the 24th, the 9th Marines spotted enemy barges along the coast of Orote near Sumay. Star shells were fired by the call fire support ship assigned to the regiment, the destroyer *Franks*,[24] and the area of Japanese activity was hammered by newly emplaced 90mm guns of the 14th Defense Battalion on Cabras Island. At 2010, after receiving permission from the shore fire control party with 3/9, which was spotting for it, the *Franks* illuminated the suspected area with its searchlight in order to conserve star shells.[25] The light on the ship was shuttered when two 14th Defense searchlights on Cabras took over the sweeping search of cliff, beach, and water, looking for targets for the 90s. The night's events showed plainly that the Japanese on Orote Peninsula were stirring. The Marine observers who knew it best were those who were charged with its capture.

CLOSING OFF OROTE PENINSULA [26]

The heavy losses suffered by the enemy *38th Infantry* in its counterattack on the 1st Marine Brigade perimeter opened the way for a rapid advance on

[23] *3d MarDiv SAR*, p. 4.

[24] CO, USS *Franks* Rpt of Fire Support during Guam Occupation, dtd 16Aug44 (OAB, NHD).

[25] On 23 July, General Geiger had authorized the use of destroyer searchlights for night illumination "in view of limited star [shells] available." *3d MarDiv Jnl*, entry of 23Jul44.

[26] Unless otherwise noted, the material in this section is derived from: TG 53.2 OpRpt, dtd 11Sep44 (OAB, NHD), hereafter *TG 53.2 OpRpt;* 77th InfDiv OpRpt FORAGER, 21Jul–16Aug44, containing reports of all major component units (WW II RecsDiv, FRC, Alexandria, Va.) hereafter *77th InfDiv OpRpt; 77th InfDiv Jnl; 1st ProvMarBrig SAR; 1st ProvMarBrig Jnl; 1st ProvMarBrig URpts; 22d Mar Jnl; 1/22 Jnl; 2/22 Jnl; 1/4 WarD; 3/4 WarD; 6th TkBn SAR;* Cass, *6th MarDiv History.*

the 22d. Isolated from the rest of the Japanese garrison, the remaining troops were incapable of fighting a delaying action on all fronts. The enemy could muster strength enough to put up a stiff fight to block one route of advance—the road to Orote Peninsula. The task of opening that road fell to the 22d Marines; the rest of the brigade was charged with the mission of reaching and securing the Final Beachhead Line where it ran along the Alifan-Taene massif, crossed Maanot Pass, and reached the high ground leading to Mt. Tenjo. (See Map 26.)

General Shepherd's plan for the brigade operations on W plus 1 called for the 1st and 3d Battalions of Colonel Tanzola's 305th Infantry to pass through the left flank of the 4th Marines and attack to seize and hold Maanot Pass. The 2d Battalion of the 305th was to remain in brigade reserve. The 305th was given responsibility for maintaining contact with the 22d Marines, which was to move out echeloned to the left rear of the Army regiment, making its main effort on the left along the Agat-Sumay Road. The initial objective of the 4th Marines was the capture of Mt. Alifan and the seizure of the ridge leading toward Mt. Taene. Once the regiment secured this commanding ground, 3/4 was to drive south to take Magpo Point, extending the south flank of the beachhead 1,200–1,500 yards beyond Hill 40 and Bangi Point.

By 0740, it became apparent that 1/305 and 3/305 would need several hours to regroup and reorganize after the unavoidable delay and disorganization resulting from their nighttime landing. Consequently, General Shepherd ordered 2/305 to move forward and relieve 2/4 in position. The 4th and 22d Marines jumped off at 0900, and the 305th followed suit an hour later, passing through elements of both 2/305 and 2/4 and striking northeast through Maanot Pass. Colonel Tanzola's men found their first taste of combat an easy one to take. Except for scattered opposition by individuals and the sporadic fire of one mortar, the regiment met little resistance. The 3d Battalion, on the left, took its part of the day's objective by 1300, and the 1st Battalion, slowed by thick underbrush and more rugged terrain, came up on line at dusk. Most supporting units of the 305th RCT, including half-tracks, antitank guns, and tanks, came ashore during the day, and the 305th Field Artillery moved into firing positions and registered its 105mm howitzers.[27]

The terrain problems posed by the heavily wooded slopes that slowed the advance of 1/305 were multiplied in the zone of the 1st Battalion, 4th Marines. The day's objective included the top of Mt. Alifan and the direction of advance was up. The steep sides of the mountain were covered with dense, thorny undergrowth, and only a few trails wound their way through the sprawling tree roots and tangled vines. The mountain itself was a formidable obstacle, but the Japanese made it even more difficult. On the lower slopes, bunkers, reinforced with coconut logs, and some of the numerous caves contained Japanese defenders. These were methodi-

[27] Battery B of the 305th had landed late on W-Day but did not move into position and start firing until 0945 on the 22d. 305th FA Bn AAR, 21Jul–10Aug44 (WW II RecsDiv, FRC, Alexandria, Va.).

cally eliminated by the grenades and rifle fire of assault squads of Company C and Company G, attached to 1/4 after the 2d Battalion was relieved by the 305th Infantry.

At about noon, the climb for the mountain peak began, an ascent that grew steadily tougher as the Marines went higher. Fortunately, the Japanese did not contest the last stages of the advance when packs and all excess gear were discarded to lessen the burden on the sweating climbers. Finally, at 1530, a patrol reached the very top, where it could see the other side of the island. The peak proved to be indefensible, so night positions were dug in on the lower slopes, where 1/4 tied in with 1/305 on the left. On the right, where the lines of the battalion extended southwest along the ridge leading to Mt. Taene, the flank hung open.

In order to help block this gap, Company E of 2/4 was attached to 3/4 late in the afternoon of the 22d. Major Hoyler's companies had begun their attack at 1100 to extend the beachhead south. Resistance was light on all company fronts, and naval gunfire, artillery, and mortars helped discourage any Japanese attempt to hold in strength. Company K, advancing across the low, rolling ground along the shore, was supported by a platoon of Shermans, which knocked out enemy machine gun nests before they could do any damage. Once the battalion had reached and secured Magpo Point, extending its lines inland, the tanks set up close to the frontlines to bolster night defenses. There was no significant number of Japanese in front of 3/4, however, to stage a repeat of the wild counterattack on the first night ashore. The few survivors of *1/38* and *3/38* were already assembling behind Mt. Tenjo to move north to Ordot.

The pattern of light and scattered resistance, which marked the advance of the other regiment of the brigade, was repeated in the right portion of the 22d Marines zone of action. Moving out at 0900, 2/22 had little difficulty in eliminating the few Japanese it met; naval gunfire knocked out several pillboxes, which might have meant more serious opposition. The battalion was held up more by the extreme difficulty of getting supplies up to its assault platoons than it was by enemy activity. LVTs, which might have negotiated the broken, trackless ground, were in such short supply and so vital to the ship-to-shore movement that General Shepherd forbade their use inland except in emergency situations.

Along the shore, where 1/22 attacked astride the Agat-Sumay Road, the supply situation was not a problem but amphibian tractors were still needed. Here the call went out for LVT(A)s to act in lieu of tanks and half tracks. During the morning's action, mediums of the 22d Marines Tank Company helped clear the way through partially abandoned defenses outside Agat, where the enemy had held up the advance on W-Day. Armor had to stop at the Ayuja River, since the only bridge over it had been demolished and the banks were too steep for fording. When the request went back for engineers, LVT(A)s were asked for too, and a platoon was ordered up, to come in by sea if necessary, in order to join the advancing infantry. By late afternoon, Company C of 1/22 had taken Road Junction 5 (RJ 5) and won its way about 300 yards beyond, fighting through a nest of enemy pillboxes. Company A on the right flank had crossed Old Agat Road. At 1800, Lieutenant Colonel Fromhold ordered his men to dig in along a line about 50–100 yards back of their farthest advances in order to set up stronger defensive positions tied in with 2/22.

The second night ashore in the southern beachhead was a relatively quiet one. There were infiltration attempts at various points all along the perimeter and occasional fires from Japanese mortars and artillery emplaced on Orote Peninsula, but no serious threats to the perimeter. Should another large-scale counterattack come, however, it would be met by a markedly increased volume of supporting fires. Most of the men and guns of General del Valle's III Corps Artillery had landed during the day; the "Long Toms" of the 7th 155-mm Gun Battalion to support the 3d Division and the shorter range pieces of the 1st and 2d 155mm Howitzer Battalions to reinforce the fires of the 1st Brigade and the 77th Division.[28] The Light Antiaircraft Group of the 9th Defense Battalion had landed on the 22d also and sited its .50 caliber machine guns and 20mm and 40mm guns in positions where they could improve beach defenses. Lieutenant Colonel Archie E. O'Neill, commanding the 9th Defense Battalion, was placed in charge of all shore party, LVT, and LVT(A) units used to defend the beaches and inland beach area perimeter.[29]

On 23 July, General Geiger was prepared to send thousands of men and guns of the 77th Division ashore in keeping with the prelanding scheme of maneuver. The corps commander conferred with General Bruce early on the 22d and authorized the landing of all but one infantry regiment of the floating reserve. The 307th RCT, less its reinforcing artillery battalion, was to stay on board ship for the time being while the need for its commitment in the 3d Division beachhead was assessed. General Bruce issued warning orders for the landing to all units of his division at 1400 on the 22d and followed through with a request to Admiral Reifsnider that the 306th RCT be landed on White Beach 2 at the earliest practicable daylight hour on the 23d. The Army regiment, commanded by Colonel Aubrey D. Smith, was slated to relieve the 4th Marines in its positions along the southern flank of the beachhead.

At 0800 on the 23d, the 22d Marines and the 305th Infantry attacked to seize an objective line that ran across the neck of Orote Peninsula to Apra Harbor and then southeast to the ridge leading to Mt. Tenjo and south along commanding ground to Maanot Pass. The 305th, with the 1st and 3d Battalions in assault, encountered little opposition to its advance and secured its objective, part of the FBHL, without difficulty. By the day's end, Colonel Tanzola's regiment was digging strong

[28] *TF 53 OpRpt*, p. 12. The 2d 155mm Howitzer Battalion had been detached from VAC Artillery at Saipan on 14 July and reassigned to IIIAC to increase available firepower.

[29] 9th DefBn WarDs, Jul–Aug44, hereafter *9th DefBn WarDs*.

defensive positions along the high ground overlooking Orote Peninsula.

General Bruce had intended to relieve the 4th Marines with the 306th Infantry by nightfall on the 23d so that the Marine regiment could move north to take part in the brigade attack on Orote Peninsula. Since no LVTs or DUKWs could be spared from resupply runs, the soldiers of the 306th had to wade ashore, like those of the 305th before them. Admiral Reifsnider recommended that the men come in at half tide at noon, when the water over the reef would be about waist deep. This timing precluded the early relief of the 4th Marines. The first battalion to land, 3/306, began trudging through the water at about 1130. Three hours later, the Army unit, reinforced by a company of 1/306, began relieving 3/4 in place; a platoon of Marine 37mm guns and one of Sherman tanks remained in position as a temporary measure to strengthen night defenses. The remainder of Colonel Smith's combat team came ashore during the afternoon and went into bivouac behind the 4th Marines lines. Colonels Smith and Tanzola met with General Bruce in the 77th Division advance CP ashore at 1400 to receive orders for the next day's action, when the division would take over responsibility for most of the brigade-held perimeter.

Once it was relieved, the 3d Battalion, 4th Marines was ordered to move to positions near Agat and was attached to the 22d Marines as a nighttime reserve. One company of the 4th, F, had already been attached to the 22d as a reserve during the day, and a platoon of the 4th Marines tanks was also sent to back up the regiment driving towards Orote.

During the morning's advance, the 22d Marines had met only light resistance. The Japanese appeared to be falling back before the assault platoons of 1/22 and 2/22. Colonel Schneider's regiment keyed its movement on Company I, attached to 1/22, which had relieved Company C as the unit charged with fighting its way up the Agat-Sumay Road. The attack plan called for the companies on the right of Company I to swing north and west across the neck of the peninsula. By noon, tanks were again available to support the attack, since a tank dozer and tankmen armed with pick and shovels had built a causeway across the Ayuja.

Prior to the attempt to close off the neck of the peninsula, the attacking Marines paused while an intensive air, artillery, and naval gunfire preparation was laid on the difficult terrain that lay ahead. Much of the ground that lay between the Agat-Sumay and Old Agat Roads was covered with rice paddies interspersed with small hillocks and stretches of thick brush. It was terrain calculated to spread the attacking troops thin and to make contact and any concentration of unit strength difficult. The defending Japanese infantry, presumably from *2/38*, had organized the ground effectively, taking good advantage of natural obstacles. Enemy supporting artillery and heavy mortars on Orote Peninsula, well registered in the area of Marine advance, frequently timed their fires to coincide with American preparations, a practice that led to a rash of reports about American fires falling short into friendly lines.

Once the Marines jumped off, they found that the little hills ahead were infested with enemy riflemen and machine gunners. When squads of men advanced into the open paddies, small arms and light mortar fire pinned them down in the mud and water. Heavier guns positioned on Orote raked the lines with enfilade fire. Stretcher bearers and ammunition carriers attempting to reach the front lines were driven back by the hail of explosions, only to come on again with the needed aid. Supporting tanks could not manuever in the soft footing of the paddies, and when they tried to use the roads, one was knocked out by 37mm antitank fire and another was disabled by a mine. In a wearying afternoon marked by repeated but fruitless attempts to reach its objective, the 22d Marines suffered over a hundred casualties. As darkness approached, the units that had been pinned down were able to shake loose and pull back to better night defensive positions along the Old Agat Road, giving up about 400 yards of untenable ground in the process.

On the night of 23–24 July, there was still a considerable hole between the flank units of the 305th Infantry and the 22d Marines, but the Japanese took no advantage of the gap. Instead, at about 0200, counterattacks by small units, attempts at infiltration, and harassing fires from mortars and artillery were directed against the Marine positions along Old Agat Road at the boundary between 1/22 and 2/22. The Brigade Artillery Group was quick to respond to requests for supporting fire, and the fire support ships offshore joined in with increased illumination and heavy doses of 5-inch high explosive. The flurry of Japanese activity died away quickly beneath the smother of supporting fires.

General Shepherd and his operations officer, Lieutenant Colonel Thomas A. Culhane, Jr., worked out a plan of attack for the 24th that was designed to outflank the Japanese defensive complex encountered on 23 July. Issued at midnight, the brigade operation order called for two battalions of the 22d Marines to attack in column on a 200-yard front with the left flank resting on the coast. Once through the narrow corridor between the rice paddies and the sea, the trailing battalion would extend to the right, seize en route the troublesome hill defenses that had stopped the previous day's attack, and then drive for the shore of Apra Harbor on a two-battalion front. In an attack simultaneous with the main thrust up the Agat-Sumay Road, the remaining battalion of the 22d would advance on a 400-yard front on the right of the regimental zone, jumping off from Old Agat Road with an objective of seizing and holding the shore of Apra Harbor. The 4th Marines, when relieved by the 306th Infantry, would assemble in brigade reserve in the vicinity of RJ 5. One platoon of the 4th Marines tanks and a platoon of LVT(A)s would be attached to the 22d Marines to beef up the attack along the coast.

The time of the attack was set for 0900 following a lengthy softening-up of the target by air, naval gunfire, and artillery, with corps 155mm howitzers adding their heavier metal to the fires of the brigade 75s. The attack was delayed an hour to increase the effect of cruiser and destroyer bombardment along the southern coast of Orote Peninsula, where suspected and known Japanese positions could pour fire into the western flank of the attacking Marines. At 1000, Company C of 1/22 led off a column of companies driving forward from a line of departure at Apaca Point. The regimental tank company, reinforced by the platoon of the 4th Marines Shermans, moved out with the assault rifle squads.

The enemy reaction to the advance of 1/22 was immediate; artillery and mortar shells exploded among the leading units and automatic weapons fire whipped across the front. Taking advantage of natural cover and of the shelter provided by the tank armor, riflemen of Company C kept moving forward. When five enemy tanks suddenly appeared to block the advance, the Marine mediums made quick work of destroying them, and continued forward using their 75mm guns and machine guns to blast concrete and coconut log emplacements.[30] As the leading units reached the area beyond the rice paddies, fire from enemy guns concealed in the cliffs of Orote near Neye Island became so troublesome that two gunboats were dispatched to knock them out. In a close-in duel, both craft, *LCI(G)s 366* and *439*, were hit by enemy fire and suffered casualties of 5 killed and 26 wounded.[31] But their 20mm and 40mm cannonade beat down the fire of the Japanese guns, and a destroyer came up to add 5-inch insurance that they would remain silent.

At 1400, after the ship and shore gun battle had subsided, the rest of 1/22 started moving up on the right of Company C. The 3d Battalion (Lieutenant Colonel Clair W. Shisler), echeloned to the right rear of the 1st, now had maneuver room to attack and roll up the line of enemy positions that had held up the 22d Marines' attack across the rice paddies on W plus 2. Moving quickly, 3/22 took and demolished the strongpoint and then turned north toward the harbor. Lieutenant Colonel Shisler's companies encountered the same type of light-to-moderate small arms, artillery, and mortar fire that confronted 1/22 and the going over rugged terrain was slow. By dusk, the 1st Battalion was dug in on its objective, but the 3d Battalion had to set up its night defensive perimeter about 400 yards short of the harbor shore. This gap was well covered, however, as a result of the success of the attack by 2/22 on the right of the regiment.

Lieutenant Colonel Hart's battalion was getting ready to move out from Old Agat Road at 1000 when lead elements were hit by fire, which appeared to herald an enemy counterattack. At almost the same time, fragments from the heavy naval shelling in support of the Marine attack began hitting along the front lines. While the troops were waiting for this fire to be lifted and moved farther ahead, they spotted a column of about 100 Japanese moving across the front towards the flank

[30] The day's tank score was eight. At 1205, aerial observers discovered three tanks inland near Harmon Road at the center of the island. Commander, Support Aircraft ordered them attacked by fighter-bombers; all three were destroyed.

[31] LCI(G) Div 15 AR, dtd 17Aug44 (OAB, NHD).

of 1/22. Mortar and artillery fire was called down on the enemy, scattering the group, and the Marine battalion prepared for a counterattack, but none came. Once the confusion caused by the shelling and the abortive counterattack was straightened out, the attack was rescheduled. At 1300, 2/22 moved out with patrols to the front and overran a succession of small dumps and abandoned cave positions along the road; the latter were seared by flamethrowers to eliminate any stragglers. Only a few Japanese were encountered in the advance to the harbor and these were soon killed.

Once 2/22 had reached its objective, it was ordered to continue its advance east along the coast and to occupy the high ground at the road junction village of Atantano. In late afternoon, while it was moving into position through the dense underbrush which blanketed the area, the battalion was harassed by enemy fire. In view of its exposed position, 2/22 was reinforced for the night by Company F of 2/4, which marched into the Atantano perimeter at about 1850. The remainder of 2/4, attached to the 22d Marines as a nighttime reserve, was moved up after dark to the Old Agat Road, where it set up all-around defenses to plug the gap between 3/22 and 2/22.

All units of the 4th Marines were available to back up the 22d by the evening of 24 July as a result of the day's shifting of troops and reorganization of areas of responsibility within the southern beachhead. At 0800, the 77th Division assumed control of the entire perimeter east of Old Agat Road, and the 306th Infantry took command of the defenses formerly held by the 4th Marines. During the morning and early afternoon, elements of the 306th relieved companies of 1/4 in position. At 1400, while Lieutenant Colonel Shapley's Marines were shifting to a bivouac area north of Agat, General Bruce opened his CP ashore close to the area where the 307th Infantry was assembling after a rough passage to shore.

On the 23d, General Bruce had requested that two battalions of the 307th be landed and placed under his command so that he would have enough men to expand the perimeter to the originally planned FBHL. General Geiger felt that this expansion, which involved the movement of the southern flank over 3,000 yards south to Facpi Point, was no longer desirable or necessary. The move would also leave IIIAC with only one uncommitted infantry battalion in reserve. The corps commander did decide, however, that the situation now warranted the landing of the reserve, to remain under corps control. The 307th began crossing the reef at 1300 on the 24th. The luckless soldiers had to wade to the beach, like all 77th Division infantrymen before them. Their ship to shore movement was complicated by heavy ground swells raised by a storm at sea; two men were lost when they fell from nets while clambering down the sides of rolling transports into bobbing LCVPs.

The landing of the last major element of IIIAC on 24 July found both beachheads soundly held and adequately supplied. The price of that secure hold was high to both sides. The III Corps count of enemy dead consisted of the conservative figure of

623 bodies buried by the 3d Division and the 1st Brigade estimate of 1,776 Japanese killed. By enemy account of the four days' fighting, the casualty totals must have been significantly higher, particularly on the Asan front. In winning its hold on the heights, the 3d Marine Division had had 282 of its men killed, 1,626 wounded, and had counted 122 missing in action. For the same period, the 1st Provisional Marine Brigade casualty totals were 137 killed, 700 wounded, and 87 missing; the 77th Infantry Division had lost 12 men and had 20 wounded.

SUPPLY AND EVACUATION [32]

By nightfall of W plus 3, most of the logistical problems that had arisen during the first days of the assault phase had been solved. For Guam, the majority of such problems had been anticipated and countered by a proper mix of ships, service troops, and equipment. The veteran planners of TF 53 and IIIAC were well aware that the success of an amphibious operation depended as much upon rapid and effective unloading and distribution of supplies as it did upon the courage and aggressiveness of assault troops. Profiting from lessons learned in earlier campaigns, the task force vessels and shore parties were able to put an average of 5,000 tons of vehicles, supplies, and equipment ashore in both beachheads during each of the first four days.

The prime obstacle to unloading progress was the reef which denied landing craft access to shore. And the prime weapon in combatting the reef was the LVT. The III Corps logistics officer observed that without them "the unloading of assault shipping would have proceeded only under greatest difficulty." [33] Hampered only by the limitation that it could not operate effectively in rough or irregular terrain, the DUKW was almost equally useful. The amphibious vehicles were used everywhere on the reef and in the immediate beach and dump area, and, as most men of the 77th Division discovered, proved to be too valuable as cargo carriers to be used to transport troops after the assault waves landed. As a result of their almost continuous operation, many of the vehicles were deadlined by operational mishaps and mechanical failures. Herculean efforts by crewmen and mechanics kept the daily unserviceability rate to about 35 percent for amphibian tractors and 40 percent for amphibian trucks. Spare parts were at a premium, particularly for the newly acquired DUKWs of III Corps Motor Transport Battalion, and vehicles knocked out by enemy guns and others wrecked by surf and reef obstacles were cannibalized to keep cripples going.

The pontoon barges and cranes at reef edge were a vital part of the unloading process. In the shallower water over the coral shelf off Asan, versatile tractor-mounted cranes could maneuver in waist-deep water dragging, lifting, and carrying as the load to be landed required. Where the water was too deep off both Asan and

[32] Unless otherwise noted, the material in this section is derived from: *TF 53 OpRpt*; *IIIAC SAR*; IIIAC C-4 PeriodicRpts, 21Jul–10Aug44; 3d MarDiv D-4 Jnl, 21Jul–10Aug44; *1st ProvMarBrig SAR*.

[33] *IIIAC SAR*, Encl D, p. 5.

Agat, the barge-mounted cranes swung bulging cargo nets from boats to vehicles and lifted out the heavy drums of fuel and water that were often floated and pushed to the beaches by men of the reef transfer battalions.[34] Since few wheeled vehicles could make shore under their own power, tractors and LVTs were used to tow most trucks from the ramps of LSTs onto dry land.

By 24 July, nine LST unloading points had been opened on the reef off each beachhead and landing ships had been about half emptied. The transports and cargo vessels that had carried the assault units to the island were 90 percent unloaded, and those that had brought the 77th Division had landed 25 percent of their cargo. At 1700, Admiral Conolly reported that 15 of the big ships were cleared of landing force supplies, and preparations were made to return the first convoy of those emptied to Eniwetok on 25 July.

Many of the APAs that had served as casualty receiving stations during the first days of fighting were among those that were sent back. The hospital ship, *Solace*, which arrived according to plan on 24 July, took on board some of the most seriously wounded patients from the transports lying offshore. The transports *Rixey* and *Wharton*, both remaining in the area, loaded those walking wounded that would require no more than two weeks hospitalization. Once the major unit hospitals were fully established ashore, these men would be landed to recuperate on Guam and rejoin their units. Many of the 581 casualties that filled the *Solace* when she sailed on W plus 5 were men loaded directly from the beaches that had been hit in the heavy fighting on 25 and 26 July.

[34] On 24 July, the opening of a water point at springs near Asan brought an end to the need to land drummed water for the 3d Division. All ships were directed to dump the remainder of such water supplies that they carried.

Continuing the Offensive

ATTACK AND COUNTERATTACK [1]

General Geiger's original operation plan for the coordinated IIIAC advance on 25 July called for the 1st Provisional Marine Brigade to begin its assault on the Japanese defenses of Orote Peninsula. When it became obvious on the 24th that the brigade would not be able to get into position by nightfall to mount a two-regiment attack, General Shepherd sent a message to the corps commander stating that in view of the:

> . . . delay in the relief of the 4th Marines which was not completed until 1500 today, necessity for moving 4th Marines to assembly areas and relief of 22d Marines in line, reorganization and preparation for attack, strongly recommend assault Orote Peninsula be delayed until 26 July.[2]

General Geiger quickly concurred in Shepherd's recommendation and revised the order, setting forward the time of the Orote attack to 0700, 26 July. In the day gained, the brigade would attempt to seal the neck of the peninsula from sea to harbor.

After an uneventful night, marked only by harassing artillery and mortar fires falling on the lines of 1/22 and 3/22, Colonel Schneider's regiment prepared to attack at 0830 on the 25th. Moving out behind a 15-minute artillery preparation by brigade 75s, the two assault battalions immediately ran into heavy enemy small arms fire coming from covered emplacements in the low, irregular hills ahead. Again enfilade fire from Neye Island and the cliffs near it raked the front of the advancing Marines. An air strike was called down on Neye, naval gunfire and artillery added their firepower, and 40mm guns of the 9th Defense Battalion pounded the precipitous island shores from positions near Agat. Along the coast, half-tracks of the regimental weapons company moved to positions from which they could fire across the narrow stretch of water at the bend of the peninsula into caves and other likely gun positions which studded the cliffs.

The fury of supporting fires knocked out some but not all of the Japanese weapons. The attacking Marines, particularly those of Lieutenant Colonel Fromhold's 1st Battalion, which was advancing along the coast, were hard

---

[1] Unless otherwise noted, the material in this section is derived from: *3d MarDiv SAR; 3d MarDiv Jnl;* 3d MarDiv D-2 PeriodicRpt No. 72, dtd 26Jul44; *77th InfDiv OpRpt; 1st ProvMarBrig SAR; 1st ProvMarBrig Jnl; 1st ProvMarBrig URpts; 3d Mar Jnl; 1/3 Jnl; 2/3 Jnl; 9th Mar Jnl;* 12th Mar Jnl, 21Jul–15Oct44; *21st Mar URpts; 3/21 Jnl; 3d TkBn SAR; 22d Mar Jnl; 1/22 Jnl; 2/22 Jnl; 1/4 WarD; GSDF Study;* LtCol Robert E. Cushman, "The Fight at Fonte," *Marine Corps Gazette*, v. 31, no. 4 (Apr47), hereafter Cushman, "Fight at Fonte"; Frances, "Battle of Banzai Ridge."

[2] *1st ProvMarBrig Jnl*, entry of 24Jul44.

214-881 O-67-33

hit. At one point in the morning's bitter fighting, Fromhold committed his last reserve platoon, reinforced by 20 men from the 4th Marines, to back up Company C, which was driving up the Agat-Sumay Road. Just before noon, five Japanese light tanks, accompanied by infantry, were spotted ahead of Company C, and Shermans of the 22d Marines converged on the enemy armor. A short, sharp exchange left the Japanese tanks broken and aflame and scattered the enemy troops. Not long afterwards, bazookas and tanks of the 22d accounted for at least two more Japanese tanks that were attacking Marines on the right of the 1/22 zone.[3]

Although the 1st Battalion encountered the stiffest enemy resistance during the day's advance, 3/22 was also heavily engaged. As it swung into line and closed on the harbor shore, Lieutenant Colonel Shisler's unit met increasingly stronger Japanese fire. By early afternoon, all evidence indicated that the brigade had run up against the main defenses of Orote.

To bring fresh strength to bear in the attack ordered for 26 July, the 4th Marines began taking over the left of the brigade lines shortly after noon. Lieutenant Colonel Shapley was given oral orders to have his lead battalion, 1/4, mop up any Japanese resistance it encountered moving forward to relieve 1/22. General Shepherd moved his CP closer to the fighting and set up near RJ 5, not far from the bivouac area of the brigade reserve, 2/4. Well before dark, all brigade assault units were on the day's objective, firmly dug in, and ready to jump off the following morning. (See Map 28.)

Manning the left half of the newly won positions was the 1st Battalion, 4th Marines, with elements of three companies on line and a platoon of regimental tanks guarding the Agat-Sumay Road where it cut through the American defenses. The 3d Battalion, 4th Marines was in position behind 1/4 ready to move into the front line as the peninsula widened and allowed for more maneuver room. On the brigade right was 3/22, occupying a low rise that overlooked an extensive mangrove swamp along the shore of Apra Harbor. Backing the 3d Battalion was 1/22, which had moved after its relief to positions near the regimental boundary in the narrowed zone of the 22d Marines. To augment night defenses on the extreme right flank where the Piti-Sumay Road paralleled the harbor shore, Colonel Schneider attached Company E of 2/22 to the 3d Battalion.

During the day's fighting, the 2d Battalion of the 22d, operating from Atantano, patrolled extensively and mopped up enemy holdouts in the area between the Old Agat Road and the brigade front lines. Firm contact was established with the left flank of the 77th Division, which spent the 25th consolidating its hold on the FBHL and landing more of its supplies and equipment. Patrols from General Bruce's infantry battalions ranged the hills to the northeast, east, and south hunting down Japanese stragglers.

By the 25th, the 77th Division was also probing cautiously toward Mt. Tenjo, sending its patrols to scout approaches to the hill mass. The way

---

[3] Units accounts of the total bag of Japanese tanks vary, but all accord that at least seven were destroyed.

was rugged and the possibility of encountering enemy defenses in the high, broken ground seemed strong. Although the mountain peak was included within the 3d Marine Division FBHL in prelanding plans, the pattern of Japanese resistance indicated that it might fall easier to American troops attacking from the south rather than the north. No significant enemy opposition was developed by the Army patrols as they moved further toward the Asan beachhead. Their negative findings matched the experience of patrols from the 9th Marines and 2/22, which made contact along the harbor shore near Atantano about 1600.

The events of 25 July indicated that Japanese troops were scarce in the area bordering Apra Harbor, but there was ample evidence that the enemy was still plentiful and full of fight everywhere else on the heights confronting the 3d Division. Unknown to the Marines, the eve of the *29th Division* counterattack had arrived, and the bitter resistance met in the day's close combat by the 3d and 21st Marines had been furnished by units that were trying to hold jump-off positions for the night of 25–26 July.

General Takashina's orders to his troops were to concentrate in the general area from Fonte Plateau to "Mt. Mangan." The latter name was given by the Japanese to a 100-foot-high hill about 1,500 yards southwest of Fonte Plateau. Mt. Mangan marked the junction of the Mt. Tenjo Road with a trail that branched off to the head of the Fonte River valley. One principal enemy assembly area faced the positions of the 3d Marines, the other was in front of the lines of the 21st Marines. The *48th IMB,* with the *10th IMR* attached, was to launch its attack from a line stretching from the east side of Fonte Plateau to the east side of Mt. Mangan. The *18th Infantry* was to move out from a line of departure running west from Mt. Mangan along the Mt. Tenjo Road. The naval troops that had helped hold the approaches to Agana against the 3d Marines were to assemble in the hills east of the Fonte River and attack toward Adelup Point. Reinforcing the naval infantry, who were mainly former construction troops operating under the headquarters of the *54th Keibetai,* would be the two companies of tanks that had remained hidden near Ordot since W-Day. (See Map VIII, Map Section.)

Many of the veteran Japanese infantrymen scheduled to spearhead the counterattack were killed in the bloody fighting on the 25th. The bitterest contest was joined along Mt. Tenjo Road where it crossed Fonte Plateau. Here, the road fell mainly within the zone of action of Lieutenant Colonel Cushman's 2d Battalion, 9th Marines.

Cushman's outfit was attached to the 3d Marines at 0600 on the 25th and ordered to relieve 1/3 on the right of the regimental front line. By 0930, when the 3d Marines moved out in attack, the relief was completed, and 1/3 supported the advance of 2/9 by fire. Once their fire was masked, Major Aplington's badly depleted companies moved back to division reserve positions behind the 21st Marines and about 1,000 yards inland from Asan Point. Again General Turnage had only one understrength infantry battalion to back up the 3d Division front; the regiments, with all battalions committed, had no more than a company as reserve, the battalions frequently had only a platoon.

To both sides in the battle, the big difference in the fighting on 25 July was the presence of tanks on Fonte Plateau. Assault units of 3/3 and 2/3 blasted and burned their way through a barrier of enemy cave defenses and won control of the road to the heights within an hour after jump-off. Medium tanks of Company C, 3d Tank Battalion rumbled up the road soon after the attack began and joined the infantry in destroying Japanese positions that blocked passage upward. After engineers cleared the roadway of some bomb-mines, which temporarily stopped the Shermans, the advance resumed with infantry spotters equipped with hand-held radios (SCR 536s) pointing out targets to the buttoned-up tank gunners. In midafternoon, General Turnage authorized Colonel Hall to hold up the attack of 2/3 so that the enemy positions bypassed during the day's action could be mopped up prior to nightfall. At the same time, the 2d Battalion, 9th Marines, driving toward Fonte, was ordered to continue its attack.

No tanks reached the high ground where 2/9 was fighting until late afternoon; enemy fire and mines had slowed their arrival. Support for the infantry attacking the Fonte defenses was furnished by naval guns, artillery, and mortars, reinforced by a nearly constant fusilade from light and heavy machine guns. The return fire of the Japanese inflicted severe casualties on the assault troops, but failed to halt the Marine advance into the broken terrain of the plateau. The battalion battled its way across the Mt. Tenjo Road and drove a salient into the enemy defenses of Fonte. At 1700, the reserve company, G, was committed on the left flank to lessen a gap which had opened between 2/3 and 2/9 during the afternoon's advance.

As darkness approached, there was no letup in the ferocity of the enemy resistance and the close-in fighting continued to be costly to both sides. The situation prompted Lieutenant Colonel Cushman to pull back his forward elements on both flanks to secure better observation and fields of fire for night defense. While Companies E and G dug in close to the road, Company F in the center continued to hold a rocky prominence, about 150–200 yards to the front, that marked the limit of the day's advance. When four tanks finally arrived at 1825, it was too dark to use them effectively so they were placed in supporting positions behind the lines. At this time, as the battalion action report noted: "The enemy was within hand grenade range along the entire line to the front and retained strong positions in caves to the right Co's right rear." [4]

These caves, bypassed during the morning's advance, were left to the attention of a reserve rifle platoon. The resulting mop-up operation was only partially successful, and enemy troops continued to emerge from the caves for several days afterwards. Although these Japanese harassed the command post areas repeatedly, they were not in sufficient strength to have

[4] 2/9 SAR, dtd 15Aug44, p. 3, Encl M to *3d MarDiv SAR.*

a significant effect on the actions of 2/9.

In the zone of 2/21, which flanked that of 2/9, a similar pocket of enemy holdouts was left behind the lines when the 21st Marines attacked on 25 July. The Japanese, holed up in cave positions in the eastern draw of the Asan River, were wiped out by Company E during a morning's hard fighting; later over 250 enemy bodies were buried in this area, which had been the target of heavy American air strikes on the 24th. Company E, once it had completed the mop-up mission, moved back into the attack with the rest of 2/21. Every foot of ground that fell to Lieutenant Colonel Smoak's Marines was paid for in heavy casualties, and every man available was needed in the assault to maintain the impetus of the advance. When the 2d Battalion dug in just short of the Mt. Tenjo Road about 1730, all units were fully committed to hold a 1,000-yard front. There was no reserve.

Like 2/21, the 1st Battalion, 21st Marines ended its fifth day of battle with all companies manning frontline positions. The trace of the 1,200 yards of foxholes and machine gun emplacements held by 1/21 ran roughly parallel to the Mt. Tenjo Road where it looped south from Fonte to Mt. Mangan. Despite an all-out effort on the 25th, which saw assault elements reach and cross the all-important road at many points, the Marines were not able to hold most of their gains in the face of heavy and accurate enemy fire. In the morning, the attacking units were stopped and then driven back by the enfilade fire of Japanese artillery, well hidden in the brush and irregular terrain at the head of the Fonte River valley. In the afternoon, when tank support was available for the first time, some hill strongpoints were taken in the center of the line near a quarry which was a focal point of Japanese resistance.

The freshly arrived tanks, a welcome sight to the men of 1/21, reached the heights by means of a steep and twisting supply trail, which engineers had constructed through the draw that had been cleared by Company E of 2/21 that morning. Company B of the 3d Tank Battalion reported to Lieutenant Colonel Williams' CP at 1615, and he immediately set one platoon of Shermans to work hammering enemy defenses in the quarry area. A second platoon of the mediums spearheaded a limited objective attack on Mt. Mangan, which was recognized as the launching point for many of the Japanese night counterattacks that had plagued the 21st Marines. When the tanks swung behind the hill, a tremendous outpouring of fire from the reverse slope cut down most of the accompanying infantry. The tanks answered with cannon and machine guns, closing in on the Japanese positions to fire point blank on any targets that showed. When the Shermans returned to the Marine lines, the tank commanders were sure that they had hurt the enemy badly—and they had. Only about 40 men were left of the *321st Independent Infantry Battalion*, which had attempted to defend Mt. Mangan. Many of the luckless survivors of the *38th Regiment*, that had assembled at Mangan to take part in the counterattack, were also killed.

The enemy casualties inflicted by this

tank thrust into the heart of the Japanese defensive complex may well have altered the course of action later that night, for the Japanese were not strong enough to exploit limited penetrations in the 1/21 sector. The Marines holding positions opposite Mt. Mangan were too few in number to form a continuous defensive line. Instead, strongpoints were held—natural terrain features that lent strength to the fire of Marine small arms. Gaps between squad and platoon positions were covered by infantry supporting weapons, and artillery and naval guns were registered on possible enemy assembly areas and routes of approach.

Along the boundary between 1/21 and 3/21 a considerable interval developed during the day because the 1st Battalion was held up by enemy fire, and the 3d Battalion was able to move out to its objective within an hour after the regimental attack started at 0700. In contrast to the rest of the 21st Marines, 3/21 encountered no strong resistance on 25 July. All day long, however, sporadic fire from enemy mortars and machine guns peppered the battalion positions. Patrols scouring the hills in the immediate vicinity of the front line were also fired upon, but in general the Japanese hung back from close contact.

Despite the relative lack of opposition, Lieutenant Colonel Duplantis' situation was precarious, because he had only two companies to hold 800 yards of terrain that seemed to be nothing but ravines and ridges smothered in dense vegetation. Not only was there a gap between 1/21 and 3/21, there was also an 800-yard open stretch between the right flank of the 21st Marines and the left of the 9th Marines, which had pushed forward well beyond the 21st during the day. Just before dark, Colonel Butler, in an effort to ease the situation on his right flank, released his only reserve, Company L, to Lieutenant Colonel Duplantis. The 3/21 commander placed this company in positions that filled a weak spot in the center of his line and enabled the companies on either flank to tighten up their defenses. As a site for his command post, Duplantis chose the reverse slope of a 460-foot hill which stood squarely on the regimental boundary and in the path of any enemy attempt to exploit the yawning space between the 21st and 9th Marines.[5]

Colonel Craig's regiment made rapid progress on the 25th from the time the two assault battalions jumped off at 0700. By 0845, the regiment was on its day's objective, a line running generally along the course of the Sasa River. At 0915, the division ordered the attack to continue with the aim of seizing the high ground on the north bank of the Aguada River. The Marines encountered very few Japanese and moved out almost as fast as the rugged terrain would permit. In the hills on the extreme left flank, an outpost of 1/9 reported clashing with small groups of the enemy during the day, but the overall intelligence picture in the 9th Marines zone indicated that few Japanese were present. Under the circumstances, Colonel Craig thought that his regiment could have advanced easily and made contact with

[5] *Duplantis ltr.*

the brigade "at any time,"[6] but considered that such a move would have served no useful purpose. By limiting the advance on the division right to a front that could be held by two battalions, General Turnage was able to draw on the 9th Marines for reserves to use in the hard fighting on the left and center of the beachhead.

As night fell, the troublesome gap between 3/21 and 1/9 was partially blocked by small Marine outposts. Lieutenant Colonel Duplantis organized his CP defenses in a small depression on the left rear of Hill 460 and set out a blocking force armed with BARs and bazookas to guard a trail that skirted the hill on the right and led toward the beaches. A 25-man detachment of the 3d Division Reconnaissance Company held a strongpoint in the midst of the 500-yard open stretch between the hill and the shoulder of the ridge defended by the left flank company of the 9th Marines. The reconnaissance unit, composed of elements of two platoons, had been attached to the 9th since W-Day for the express purpose of maintaining contact between regiments, but on the night of 25–26 July it was simply too weak for the task at hand.

The start of the Japanese counterattack was heralded at 2330 by an artillery forward observer's report that enemy activity was developing in the gap between the 9th and 21st Marines. Very shortly after midnight, the 3d Marines called for artillery and naval gunfire support to silence enemy artillery, mortars, and machine guns that were hitting the left flank. Soon units all across the center and left of the beachhead perimeter were reporting Japanese probing attacks and patrol action close to the Marine lines. Darkness turned to half light as flares went up all along the front to help spot the Japanese. Cruisers and destroyers increased their rate of fire of 5-inch star shell and followed with 5- and 6-inch high explosive at the call of shore fire control parties. To aid local defenses, 60mm mortars of the rifle companies kept flares overhead wherever the front line was threatened.

The first serious assault on American positions was launched against the isolated Reconnaissance Company outpost. An enemy group, estimated at 50 men, attacked the Marine unit shortly after midnight, and during a brief, hot firefight killed four men and wounded five, over one-third of the American strength. Convinced that his position was untenable in the face of another attack by a superior enemy force that could hit from any direction, the Reconnaissance Company commander withdrew his men to the lines of Company B of 1/9, which held the left flank of the 9th Marines.[7]

Japanese troops that drove in the reconnaissance outpost were men of the *3d Battalion, 18th Infantry.* The enemy unit was assigned an objective of penetrating the Marine lines in the area held by 3/21. Surprisingly, instead of

[6] LtGen Edward A. Craig ltr to CMC, dtd 30Sep52, hereafter *Craig 30Sept52 ltr.*

[7] The Reconnaissance Company commander felt that the gap was so wide that the Japanese who hit his perimeter, and lost 35 men in the process, "could have bypassed our position at many points." PhibReconCo SAR, dtd 14Aug44, p. 2, Encl C to Encl Q, *3d MarDiv SAR.*

pouring full strength through the hole that had been left open, most elements of *3/18* continued to feel out the main defenses of Lieutenant Colonel Duplantis' battalion. Despite their superior observation of the American beachhead from positions on the Chachao-Alutom-Tenjo massif,[8] and the information supplied by their patrols, the Japanese did not really make use of their best opportunity for success.

The pressure of enemy units testing the Marine defenses along the rest of the division front increased as the long night wore on. Both the 1st and 2d Battalions of the 21st Marines were hit repeatedly and in gradually increasing strength. Apparently, the *2d Battalion, 18th Infantry* and the elements of the *48th Brigade* in the Mt. Mangan area were looking for a weak spot that would let them break through to the Asan River draws. The draws in turn would provide a path to the division rear areas. On the Fonte front, most of the Japanese troops pressing 2/9 were part of the *2d Battalion, 10th IMR;* almost all of the *3d Battalion* had been killed during the fighting on the 25th, as had the commander of *2/10*. The remainder of the Marine line on the left flank was harried by other elements of the *48th IMB* and by naval troops of the *54th Keibitai*. Throughout the Japanese ranks, the infantry was bolstered by service and support troops, by walking wounded that could still handle a weapon, and, in short, by everyone that could be mustered for the assault.

One important aspect of the Japanese counterattack plans went awry in the darkness—enemy tanks never reached the battle line. When night fell, the armored vehicles rumbled from their hiding places near Ordot and headed onto the trails leading to Agana. Inexplicably, the tanks got lost. Unable to find their way to the designated assembly area, the commanders of the *2d Company, 9th Tank Regiment* and the *29th Tank Company* led their units back to the Ordot area before dawn broke. Hidden again at daylight from the eyes of American artillery and air observers, the Japanese tanks bided their time for a more effective role in the fighting.

Undaunted by the absence of expected armored support, the *54th Keibitai* moved forward to attack in the early hours of 26 July. An intensive preparation fired by mortars and artillery crashed down on the positions occupied by 3/3 and 2/3. Led by the senior enemy naval officer, Captain Yutaka Sugimoto, the Japanese sailors launched the first of what proved to be a series of counterattacks. From Adelup Point and Chonito Cliff, Marine small arms fire crackled forth from well-dug-in foxholes and machine gun emplacements. Shells from company and battalion mortars exploded amidst those from the 105mm Howitzers of 3/12, and drove the onrushing enemy back. Captain Sugimoto was killed in the first outburst of defensive fire; later, his executive officer was felled by a shell burst. Despite repeated attempts to break through the

[8] The Commanding Officer, 9th Marines recalled that when this terrain was finally captured "three huge telescopes of 20 power were found. Looking through these scopes one could almost make out individual features of Marines below us. Practically every part of our lines and rear areas, as well as my own CP, could be seen through these glasses from this high ground." *Craig 30Sept52 ltr.*

Marine lines, the Japanese were unsuccessful and most of the attackers were dead by early morning. As day broke, the weary survivors, many of them wounded, fell back toward the low hills west of Agana.

One of the night's most bitter struggles was waged on Fonte Plateau, where Lieutenant Colonel Cushman's embattled battalion strove to hold its gains of the 25th. Local counterattacks flared all along the front of 2/9 and caused a constant drain of Marine casualties. At 2200, it was necessary to pull Company F back 50 yards from its salient in the center of the line in order to consolidate defenses. Because there was little letup in the pressure that the *10th Independent Mixed Regiment* applied, the expenditure of ammunition by all types of Marine weapons mounted alarmingly. Seven major counterattacks in succession ate away at the American line, but it held, often only by the slimmest of margins. The height of the battle was reached in the early morning hours when the Japanese seemed to come in unending waves and the din of weapons firing all at once mixed with the screams and yells of men caught up in the frenzy of close-quarter combat.

As "the first faint outline of dawn showed," and "ammunition ran dangerously low" [9] in the American positions, Marine tanks were able to play a significant role in the hard-fought battle. The platoon of Shermans that had spent the night behind the lines now moved to the front, where their cannon and machine guns helped break up the last desperate enemy thrust. Soon afterwards, another platoon of tanks escorting trucks loaded with ammunition passed through the lines of 3/3 and made its way up the road to Fonte Plateau. While the armor provided covering fire, riflemen and machine gunners grabbed belts and bandoliers of .30 caliber cartridges and mortar crews quickly stacked live rounds near their shell-littered firing sites. With tank support and adequate reserve ammunition, the Marines of 2/9 were solidly established and ready for renewed enemy attacks. Without the shield of darkness, the Japanese held off, however, for only about 100 men of the *10th IMR* had survived the night's fighting.

Not all of the Japanese that died on the night of 25–26 July were killed in front of the American lines. Some infiltrated through the widely spaced strongpoints of the 21st Marines and others found their way through the gap between 3/21 and 1/9. The positions manned by 1/21 and 2/21 lay generally along a low ridge that paralleled and ran north of the Mt. Tenjo Road. From this rise the ground sloped back several hundred yards to the edge of the cliffs. Over much of this area Marines waged a fierce, see-saw battle to contain enemy units that had broken through. In the thick of the fight was Company B of the 3d Tank Battalion, which was bivouacked behind the lines of the 1st Battalion, 21st Marines. Although the tanks were cut off for a time from Marine infantry support, they were able to fend for themselves with their machine guns and 75s. Apparently the Japanese infiltrators were more interested in other and easier targets, for only one tank, which had a track blown and its engine damaged,

[9] Cushman, "Fight at Fonte," p. 15.

was put out of action during the night's combat.

The coming of daylight brought a quick end to the limited Japanese penetration of the lines of 1/21. Fire from 60mm mortars sealed off the area where the enemy had broken through and ripped apart the groups of Japanese that tried to make a stand. Supported by tank fire, the Marines of Lieutenant Colonel Williams' battalion, reinforced by a company of engineers, counterattacked to restore their lines.[10] Infiltrators were hunted down relentlessly, and by 0800, the enemy had been cleared from the entire area between the edge of the cliffs and the original front line of 25 July. Along this segment of the embattled 3d Division front, the weary Marines could relax a bit and feel, as one did, that "the fireworks were over." [11]

Although the fighting on the heights had subsided by early morning, the conflict was far from settled in the division rear areas, particularly in the vicinity of the wooded draws that held the Nidual River and the west branch of the Asan. Most Japanese that found their way through the gap between 3/21 and 1/9, or infiltrated the fire-swept openings in the Marine front line ended up following these natural terrain corridors toward the sea. Directly in the path of the majority of these enemy troops, the elements of *3/18* that had skirted the right flank of 3/21, was Hill 460 and Lieutenant Colonel Duplantis' command post.

After feeling out the positions held by 3/21, the Japanese attacked in force about 0400 all along the battalion front and drove in a platoon outpost of Company K, which held the right of the line. The intensity and strength of the enemy assault mounted as dawn approached, and the Marines on the front line had all they could do to hold off the attackers. Consequently, Duplantis believed that he could not call on his rifle companies for help when the Japanese began attacking his command post. In fact, a reinforced rifle squad, the only reserve available to the commander of 3/21, was committed soon after the Japanese attacked to defend the area between the ridge positions of Company K and Hill 460. Like the Marines that held the trail block Duplantis had set out earlier that night, the outnumbered squad "went down fighting to a man," [12] overwhelmed by the enemy troops, who swept around both sides of the hill.

Perhaps nowhere else within the Marine perimeter was the situation so desperate as it was in the 3/21 CP as daylight approached. In most parts of the 3d Division beachhead, dawn gave the Marines a better chance to wipe out the infiltrators; on Hill 460, in contrast, the increasing light furnished the Jap-

[10] The action of these engineers from Company B, 3d Engineer Battalion was typical of those of many supporting units on 26 July. The Executive Officer of 1/21 stated that he "was particularly impressed with the number of automatic weapons they were able to produce (from organic vehicles). They advanced as a leading company in at least one of our attacks and performed many infantry duties with credit. This is another advantage of basic [infantry] training being given to all Marines." LtCol Ronald R. Van Stockum ltr to Head, HistBr, G–3, dtd 15Oct52.

[11] Frances, "The Battle of Banzai Ridge," p. 18.

[12] *Duplantis ltr.*

SHERMANS AND RIFLEMEN *of the 1st Brigade advance together on Orote Peninsula. (USMC 88152)*

FIRING LINE *of 3d Division riflemen engage the enemy in the hills above Asan. (USMC 88090)*

anese better targets. From positions on the crest of the hill, enemy machine guns raked the rear of Company K and small arms and knee mortar fire poured down on the CP, less than 75 yards away. The deadly hail that swept Duplantis' position took a heavy toll among the corpsmen and communicators, who made up a large part of the defending force. The headquarters group fought doggedly, keeping up a steady fire against the Japanese, who showed no disposition to charge the beleaguered Marines.

The task of eliminating the troublesome enemy strongpoint on Hill 460 fell to the 9th Marines. At 0655, a time when most of the division reserve and support forces were hotly engaged with infiltrators, General Turnage ordered Colonel Craig to shorten his front lines and pull back to the Sasa River and to send troops to recapture Hill 460. Craig, in turn, detailed his regimental reserve, Company L, to take the hill and assigned an officer familiar with the terrain as the temporary commander. Led by Major Harold C. Boehm, executive officer of 1/9, the men of Company L advanced toward the hill along the course of the Masso River. The approach march over difficult terrain was time-consuming, but the Japanese on 460 did not spot Boehm's command until the Marines were about 250 yards away and ready to attack. Aided by covering machine gun fire from Company B, 1/9, shortly before noon Company L launched an assault that carried the enemy position. Twenty-three Japanese were killed on the hill and the remainder were driven toward a firing line set up by Company K of 3/21. The few enemy that survived the attack fled down the Nidual River draw to annihilation at the hands of the Marine units then mopping up the rear areas.

The clash at Hill 460 was one of a series of hard-fought actions that took place behind the 3d Division front. Japanese infiltrators moving down the stream lines leading to the beaches continually harassed the perimeters of Marine units that stood in their paths. Throughout the night, gunners in artillery and mortar positions had to interrupt their supporting fires to beat off troops.[13] The neighboring command posts of the 12th Marines and of 3/12 were beset by snipers, who had infested the high ground above the camp areas. By midmorning, artillerymen acting as infantry, aided by two Shermans from the division tank park nearby, had destroyed this nest of enemy.

One of the most serious encounters behind the lines took place at the division hospital. At 0600, about 50–70 enemy troops opened fire on the hospital tents from the high ground on the west bank of the Nidual River. Doctors and corpsmen immediately began evacuating patients to the beach while other hospital personnel and 41 of the walking wounded formed a defensive line and returned the Japanese fire. As soon as word reached General Turnage that the hospital was being attacked, he ordered the division reserve commander, Lieutenant Colonel George O. Van Orden (Division Infantry

[13] Many of these enemy troops had infiltrated the Marine lines on the previous night and laid in hiding throughout daylight hours on the 25th with the intent of knocking out the artillery when the counterattack started. BGen John S. Letcher ltr to CMC, dtd 12Jun65.

Training Officer), to take two companies of pioneers and eliminate the threat.

Moving quickly, Van Orden's command reached the hospital area and joined the battle. After three hours of fighting, during which the enemy force was eventually surrounded, the pioneers killed 33 Japanese at a cost of three of their own men killed and one wounded. The 3d Medical Battalion had 20 of its men wounded, including two that later died of wounds, but only one patient was hit and he was one of the volunteer defenders. By noon, the hospital was back in operation, caring for the heavy influx of casualties from all parts of the Marine beachhead.

Even before the fighting was over at the hospital and at Hill 460, it was apparent that these two areas held the only sizeable enemy groups left within the perimeter. Any other Japanese still alive behind the lines were the subject of intensive searches by combat patrols of service and support troops. Along the front line, infantry units, often with tank support, scoured the woods and caves in their immediate areas to flush enemy stragglers. The mop-up and consolidation of defensive positions continued through the afternoon as Marine commanders made certain that their men were ready to face whatever the night might bring. While he was inspecting these defensive preparations, Lieutenant Colonel de Zayas, commanding 2/3, was shot and killed; the battalion executive officer, Major William A. Culpepper, immediately took command and continued the defensive buildup.

Conservative intelligence estimates indicated that the Japanese had lost close to 3,200 men, including 300 behind Marine lines, in the counterattack.[14] The comparable casualty total for the 3d Division and its attached units was approximately 600 men killed, wounded, and missing in action.[15] It appeared to General Turnage that the enemy was still capable of mounting another strong counterattack, and he directed that all units establish the strongest possible night defense.

Early on the 26th, General Turnage had requested reinforcements from the corps reserve, and during the afternoon, General Geiger dispatched one battalion of the reserve, 3/307, overland to the Piti area to be available immediately in case of need. On its arrival the Army unit was attached to the 9th Marines. As a further safeguard, Turnage directed the organization under Lieutenant Colonel Van Orden of a provisional division reserve composed of 1/3, a platoon of tanks, and elements of eight Marine and Seabee support battalions. Most of these units spent the night of 26–27 July manning defensive perimeters or standing by for employment as infantry reinforcements.

Actually, the Japanese commanders

[14] A careful Japanese accounting of the battle places their casualty total at about "3000 persons." *GSDF Study*, p. 189.

[15] The exact figures are buried somewhere in the casualty statistics for 25–27 July, since many units completed a head count too late on the 26th to be accurately reflected in that day's totals. The cumulative total casualties reported as of 1800 on the three days are: 25 July–315 KIA, 1,760 WIA, 132 MIA; 26 July–333 KIA, 1,869 WIA, 162 MIA; 27 July–481 KIA, 2,405 WIA, 166 MIA. *3d MarDiv Jnl.*

had no further massive counterattack in mind. To an extent not yet realized by American intelligence officers, the fruitless assault had broken the backbone of Japanese resistance on Guam. While there was no disposition to stop fighting on the part of the remnants of the *29th Division*, the ground within, in front of, and behind the Marine lines was littered with the bodies of the men and with the weapons, ammunition, and equipment that General Takashina sorely needed to prolong the battle. Undoubtedly the most damaging losses were those among senior combat unit leaders, whose inspirational example was essential to effective operations in the face of obvious and overwhelming American superiority in men and material. General Shigematsu, the *48th Brigade* commander, was killed on 26 July when tanks supporting the consolidation of Marine frontline positions blasted his CP on Mt. Mangan. The regimental commander of the *18th Infantry* was cut down in the forefront of his counterattacking troops, and both of the battalion commanders were killed after having led their men through the Marine lines. The body of one was discovered in the Asan River draw; the other was found in the Nidual River area.[16]

As the senior Japanese officer on Guam, General Obata had the unpleasant duty of reporting the failure of the counterattack to Tokyo. At about 0800 on the 26th, the *Thirty-first Army* commander radioed *Imperial General Headquarters*, stating:

> On the night of the 25th, the army, with its entire force, launched the general attack from Fonte and Mt. Mangan toward Adelup Point. Commanding officers and all officers and men boldly charged the enemy. The fighting continued until dawn but our force failed to achieve the desired objectives, losing more than 80 percent of the personnel, for which I sincerely apologize. I will defend Mt. Mangan to the last by assembling the remaining strength. I feel deeply sympathetic for the officers and men who fell in action and their bereaved families.[17]

The following day Tokyo acknowledged Obata's message and commended the general and his men for their sacrifice, emphasizing that the continued defense of Guam was "a matter of urgency for the defense of Japan." [18] After this, Generals Obata and Takashina and the few surviving members of their staffs concluded that their only practical course of action was to wage a campaign of attrition, whose sole purpose would be, in the words of Lieutenant Colonel Takeda, "to inflict losses on the American forces in the interior of the island." The *29th Division* operations officer explained the factors influencing this decision as:

> 1. The loss of commanders in the counterattack of 25 July, when up to 95% of the officers (commissioned officers) of the sector defense forces died.
>
> 2. The personnel of each counterattacking unit were greatly decreased, and companies were reduced to several men.
>
> 3. The large casualties caused a great drop in the morale of the survivors.
>
> 4. Over 90% of the weapons were destroyed and combat ability greatly decreased.

[16] Near the body of the commander of *2/18* was found a map which showed the Japanese plan for the counterattack. This information is included on Map VIII, Map Section.

[17] Quoted in *GSDF Study*, pp. 188–189.

[18] *Ibid.*, p. 193.

5. The rear echelons of the American forces on Agat front landed in successive waves and advanced. There was little strength remaining on that front and the strength for counterattacks became non-existent.

6. The Orote Peninsula defense force perished entirely.

7. There was no expectation of support from Japanese naval and air forces outside the island.[19]

Part of the Japanese estimate of the situation was based on a lack of knowledge of the exact situation on Orote. All communication with the Japanese command on the peninsula was lost by the evening of 25 July, but the last messages received indicated that the defenders were going to take part in the general counterattack.[20]

CAPTURE OF OROTE PENINSULA [21]

Commander Asaichi Tamai of the *263d Air Group* was the senior officer remaining in the Agat defense sector after W-Day. The death of Colonel Suenaga elevated Tamai, who had been charged with the defense of Orote Peninsula, to the command of all sector defense forces, including the *2d Battalion, 38th Infantry*. During the period 22–25 July, the Army unit fell back toward Orote, fighting a successful delaying action against the 22d Marines.

When the 1st Brigade closed off the neck of the peninsula on 25 July, it sealed the fate of some 2,500 Japanese soldiers and sailors who were determined to die fighting in its defense. Although more than half of Commander Tamai's troops were lightly armed and hastily trained aviation ground crewmen and engineers, he had a strong leavening of experienced ground defense units of the *54th Keibitai*. Even if many of the Japanese were not trained in infantry tactics, they were apparently experts in the use of pick and shovel and well able to man the fixed defenses, which they had helped build. Their handiwork, a formidable belt of field fortifications, stretched across the peninsula just beyond the marsh and swamp area and generally along the 0–3 Line, the initial brigade objective in its attack on 26 July. (See Map 28.)

Before the men of the 1st Brigade could test these hidden defenses—whose presence was suspected but not yet confirmed—they had to deal with the Japanese troops that took part in the general counterattack ordered by the *29th Division*. In contrast, the counterattack on the brigade defenses was made by about 500 men and the action was concentrated in a narrow sector near the regimental boundary. The left flank unit of the 22d Marines, Company L, bore the brunt of the Japanese thrust, helped by the withering fire of the right flank platoon of the 4th Marines from Company A.

The assembly area for the Japanese attack force, principally men of *2/38*, was the mangrove swamp in front of the 22d Marines. In apparent preparation for the assault, *sake* was passed about freely, and the Marines manning forward foxholes could plainly hear the

[19] *Takeda ltr II*.

[20] *Ibid*.

[21] Unless otherwise noted the material in this section is derived from: *IIIAC SAR; 1st ProvMarBrig SAR; 1st ProvMarBrig Jnl; 1st ProvMarBrig URpts; 1/4 WarD; 3/4 WarD; 1/22 Jnl; 2/22 Jnl; 6th TkBn SAR; Takeda ltr II*.

resulting riotous clamor as the Japanese worked themselves up to fever pitch. Finally, just before midnight, a tumultuous *banzai* charge erupted out of the swamp as a disorganized crowd of yelling, screaming men attacked the positions held by Company L. The resulting carnage was almost unbelievable, as artillery forward observers called down the fire of brigade, 77th Division, and III Corps artillery on the surging enemy troops. At one point, the shells of pack howitzers of the 22d Marines were brought to within 35 yards of the front lines in order to check the Japanese.[22] The few scattered groups that won their way though the barrier of exploding shells crisscrossed by machine gun fire were killed in frenzied hand-to-hand fighting with Marines of Company L. By 0200, the action died down, and all supporting weapons resumed normal fires for the night defense.

Daylight revealed a gruesome scene, for the mangled remains of over 400 enemy dead lay sprawled in front of the Marine lines in the impact area where over 26,000 artillery shells had fallen during the counterattack. Marine casualties in Company L were light, despite the close-quarter combat, and the flanking platoon of the 4th Marines did not lose a man, although it counted 256 Japanese bodies in the vicinity of its position. Any information that might reveal the exact cost of the counterattack to the Japanese, who evacuated their wounded during the night, was buried with the Orote garrison. There was no doubt, however, that *2/38* ceased to exist as an effective fighting force. Save for small groups of soldiers that continued to fight on, enemy naval troops now had the main responsibility for the defense of Orote.

The night's counterattack had no effect on General Shepherd's attack plan for 26 July. A thorough air, naval gunfire, and artillery preparation exploded on enemy-held areas, and at 0700, the 4th Marines moved out in a column of battalions, 1st in the lead, supported by the regimental tank company. On the right of the brigade front, the assault elements of 3/22 and 2/22 were heavily shelled as they were preparing to jump off. The Marines were convinced that their own supporting ships and artillery were off target, although subsequent investigation indicated that Japanese artillery was again taking advantage of American preparatory fires to strike some telling blows without detection. Regardless of its source, the effect of the fire was demoralizing to the 22d Marines, and it was 0815 before the assault units were reorganized and ready to move out.

The delay in the attack of the 22d Marines opened a gap between the regiments, which was bridged by Company L of 3/4. Another 3d Battalion company, I, followed in trace of the swiftly advancing tank-infantry spearheads of 1/4 to mop up any bypassed enemy. Major Green's 1st Battalion met only light resistance until it approached the 0–3 Line, where heavy brush on the left and the threat of enemy fire ripping across the more open ground on the right slowed the advance. Anxious to maintain the impetus of the attack and to make more

[22] Col Edwin C. Ferguson comments on draft of Lodge, *Recapture of Guam*, dtd 28Nov52.

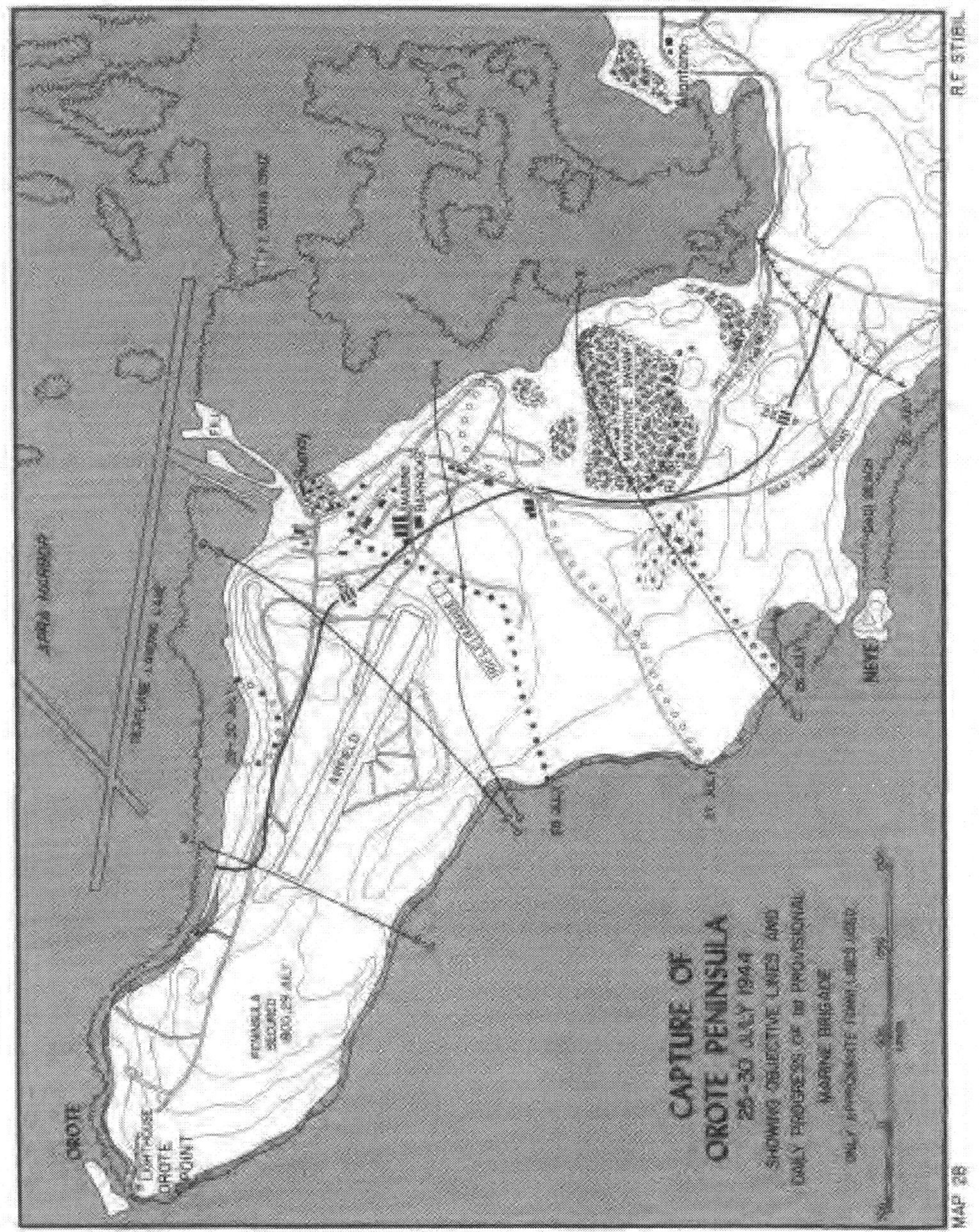

214-881 O-67—34

effective use of the comparatively fresh units of the 4th Marines, General Shepherd at 1145 ordered a change of regimental boundary that would enable Lieutenant Colonel Shapley to employ all his battalions in assault. All terrain east of the road from RJ 15 to Sumay went to the 22d Marines, while the 4th Marines took responsibility for the wider zone to the west.

Initial resistance to the 22d Marines, once its attack was launched, was slight, and 2/22 patrols, wading deep into the mangrove swamp, encountered only snipers. Along the Sumay Road, where there was room to maneuver and firm ground to support their weight, the regimental tanks moved out with 3/22. At 1220, the 3d Battalion reached RJ 15 and discovered that the Japanese had planted an extensive field of aerial bomb-mines across the 200-yard corridor between the swamp and a wide marsh lying west of the road junction. Unable to advance further, the Shermans set up a firing line along the high ground that overlooked the junction and the minefield beyond.

The mined area was covered by a nest of Japanese machine guns, which the assault infantry did not discover until a sudden outburst of automatic weapons fire pinned the lead platoon down in the midst of the mines. Spotting the Japanese strongpoint, a cluster of brush-covered bunkers northwest of the road junction, the tanks fired low over the heads of the ground-hugging infantry to hit gun ports and disrupt the enemy fire. When Japanese mortars opened up from defilade positions behind the bunkers, the tank company commander called down high-angle artillery fire to silence them, which also set afire an ammunition and a supply dump in the area. With the aid of the tanks, the Marines of 3/22 were able to pull back to relative safety, but too late in the day for any further attempt to force the minefield.

On the left of the brigade line, tanks also played a prominent part in the afternoon's advance. The 4th Marines, maneuvering to get three battalions on line, began to move into heavy vegetation as forward elements approached the 0–3 Line. The Shermans broke paths for accompanying infantry where the going was toughest and helped beat down the scattered opposition encountered. In midafternoon, heavy enemy machine gun and mortar fire hit 2/4 as it was moving into the center of the regimental front. Shortly thereafter, leading elements of the 1st Battalion were raked by intense fire from enemy positions in the dense undergrowth ahead. Japanese gunners had a clear shot at the Marines along well-prepared fire lanes cut through low-hanging branches and thick ground cover, often before the Americans were aware that they were exposed. It was readily apparent that an extensive and gun-studded belt of Japanese defenses had been encountered. At 1730, when brigade passed the order to dig in, both regiments consolidated their positions along 0–3 except on the right, where the 22d Marines set up in the swamp, refusing its flank and covering the resulting gap with artillery and mortar fire.

After a quiet night with no unusual enemy activity, the brigade attacked in the wake of an extensive air, naval gunfire, and artillery preparation. Neither this fire nor the night-long program of harassment and interdiction by American supporting weapons seemed to have much effect on the dug-in Japanese. The 4th Marines had as its attack objective an unimproved trail, about 700 yards forward of 0–3, that stretched completely across the regimental zone. Except along the Sumay Road, the intervening ground was covered with a thick tangle of thorny brush, which effectively concealed a host of mutually supporting enemy pillboxes, trenches, and bunkers well supplied with machine guns, mortars, and artillery pieces.

In the narrow corridor forward of RJ 15, 3/4 faced a low ridge beyond the marsh area, then a grass-choked grove of coconut palms, and beyond that another ridge, which concealed the ground sloping toward the old Marine Barracks rifle range and the airfield. On the 22d Marines side of the Sumay Road, the mangrove swamp effectively limited maneuver room beyond RJ 15 to an open area about 50 yards wide.

The terrain and the enemy dispositions gave special effect to the attack of Major Hoyler's battalion. With Companies I and L in the lead, and a platoon of tanks moving right along with the assault troops, 3/4 broke through the enemy defenses along the first low ridge to its front during a morning of heavy and costly fighting. The tank 75s played the major role in blasting apart the Japanese gun positions. During the afternoon, the tank-infantry teams made their way through the coconut palms at a stiff price to the unprotected riflemen. By the time 3/4 had seized and consolidated a secure position on its objective, Company L alone had suffered 70 casualties.

On the far left of Lieutenant Colonel Shapley's zone, the enemy resistance was lighter than that encountered by 3/4, and the 1st Battalion was on its part of the objective by 1100. Led by path-making tanks, the 2d Battalion reached the trail about a half hour later. Both units then set up defenses and waited for the 3d Battalion to fight its way up on the right. At about 1500, while he was inspecting dispositions in the 1/4 area, the regimental executive officer, Lieutenant Colonel Samuel D. Puller, was killed by a sniper.

Shortly after this, when tanks supporting Hoyler's battalion ran out of ammunition, Shermans from the platoons that had advanced with the left and center of the Marine line moved over to cover 3/4 while it was digging in. From firing positions ahead of the infantry, these tanks spotted a company of Japanese moving in the open along a road atop a ridge some 300 yards away. Cannon and machine gun fire tore apart the enemy column and scattered the luckless troops. At 1830, their job well done for the day, the tanks returned to their bivouac for maintenance and replenishment.

Armor also played a significant role in the day's action on the 22d Marines front. Supporting tank fire helped Company G, leading the regimental assault in the narrow zone between swamp and road, to thread its way through the minefield that had held up the advance on the 26th. Under cover of smoke shells fired by the Shermans, the regimental bomb disposal officer disarmed enough mines to clear a path through the field for the tanks to move up with the lead riflemen. Engineers then cleared the rest of the mines while 2/22 continued its advance, meeting the same type of determined opposition that had slowed 3/4 on its left. Fire from the barrier of Japanese positions, which confronted the brigade all across the peninsula, took a heavy toll of Marines, particularly unit leaders. Three of the four company commanders were hit during the day's fighting as was the new battalion commander, Major John F. Schoettel.[23]

At 1415, after 2/22 had won some maneuver room east of the road to Sumay, the 3d Battalion moved up on the right to join the battle. Since much of the zone assigned 3/22 was swamp, there was only room for Company L in assault. This company tried unsuccessfully to outflank the enemy defenses by moving along the coast, but was stopped by vicious automatic weapons fire. Despite the determined Japanese defense, the 22d Marines kept inching ahead, utilizing tanks to blast bunker firing ports so that accompanying infantry could move in with flamethrowers and demolitions.

At 1700, the brigade ordered the 22d to dig in on commanding ground for the night. In an effort to seize the most defensible terrain, Colonel Schneider stepped up his attack, calling for increased artillery support and for carrier air to bomb and strafe the Japanese. The response was prompt, sustained, and effective. The wing guns of the attacking aircraft sprayed enemy defenses close enough to the American lines for 2/22 to report it as "too close" for safety at 1802, and thankfully as causing "no casualties, but plenty close" at 1810 when the planes drew off.[24] Whatever the precipitating cause—bombing, strafing, artillery fire, or steady unrelenting tank-infantry pressure—about 1835 the enemy troops confronting the 22d Marines suddenly bolted from their defenses and ran. Taking swift advantage of the unusual Japanese action, a rout almost unprecedented in Pacific fighting, the Marines surged forward close on the heels of the fleeing enemy.

The approach of darkness stopped the attack as the 22d Marines reached high ground overlooking the Marine Barracks area. The precipitate advance opened a 500-yard gap between 2/22 and 3/4, which Company C of 1/22 closed. Two men of the company were killed and 18 wounded in a flurry of Japanese mortar fire that struck the Marine unit as it set up defenses in the flare-spotted darkness. The remainder of the 1st Battalion, which had moved from Atantano to reserve positions near RJ 15 during late afternoon, was alerted to back up Company C. There was no further significant enemy reaction, however, anywhere along the front that night.

The preparation for the brigade attack on the 28th included 45 minutes of air strikes, 30 minutes of naval gunfire, and a final 30 minutes of artillery fire. Perhaps as a result, when the 22d Marines attacked at 0830, the regiment swept forward against little opposition.

[23] Lieutenant Colonel Hart, who was reassigned as brigade liaison officer with IIIAC, was relieved by Major Schoettel at about 1430 on 27 **July.**

[24] *2/22 Jnl*, entry of 27Jun44.

At 1005, Colonel Schneider reported his troops had reached the 0–4 Line, and General Shepherd ordered the advance to continue, "echeloning units to the left rear as necessary to maintain contact with the 4th Marines."[25] Major Schoettel's battalion concentrated its drive on the capture of the Marine Barracks area, while Lieutenant Colonel Shisler's 3d Battalion entered the battered ruins of Sumay. Tanks supported the assault troops of both battalions, but found the litter and rubble in Sumay so minestrewn that support had to be confined to overhead fire until engineers could clear the streets. Before the armor halted, one Sherman and its crew were completely destroyed when it hit a 1,000-pound bomb-mine. In the face of desultory opposition, the 22d Marines was able to seize the barracks ruins, the whole of Sumay, and the cliffs along the harbor shore before dusk. For stronger night defense, 3/22 pulled back to high ground east of the town and dug in at 1750.

In contrast with their weak defense on the 22d Marines front, the Japanese facing the 4th Marines were ready and able to make the Americans pay dearly for every foot of ground they won. The enemy defenses were arrayed in depth, along a 300-yard stretch of ridgeline guarding the approaches to the rifle range and airfield. Beneath the thorn bushes and other varieties of densely-clustered jungle growth lay almost 250 emplacements and bunkers, many of them strong pillboxes constructed of coconut logs, cement, and earth. Minefields were cleverly hidden amidst the brush along all approaches to the enemy positions.

Both 2/4 and 3/4 had run up against the outskirts of this defensive complex in the previous day's fighting. The difficulties imposed by the terrain and the pattern of Japanese defending fires and minefields prevented the Marines from outflanking the enemy position, and left Lieutenant Colonel Shapley no choice but to order a frontal assault. The extensive preparatory fires for the attack on 28 July appeared to have made no impression on the Japanese: there was no letup in the volume of enemy fire. When the regiment advanced, a slugging match ensued in which Companies E and I spearheaded the determined assault. Throughout the morning and early afternoon, riflemen working closely with tanks gradually forced their way into the nest of enemy emplacements. At 1545, about 20–30 Japanese charged out of the remaining key strongpoint in a futile attempt to drive the Marines back; every attacker was quickly killed. Shortly thereafter, in an attack that General Shepherd had personally ordered during a visit to the front lines, two platoons of Marine mediums and a platoon of Army light tanks led a 4th Marines advance that smashed the last vestiges of the Japanese defenses and swept forward to positions just short of the rifle range. Tied in solidly with the 22d Marines at the Sumay Road by nightfall, the regiment was ready to carry out the brigade commander's order to seize the rest of the peninsula on the 29th.

Assigned missions for the attack on 29 July gave the 22d Marines responsibility for cleaning the Japanese out of

[25] *1st ProvMarBrig SAR*, p. 12.

the barracks area, the town of Sumay, and the cliff caves along the coast. The prime objective of the 4th Marines was Orote airfield. To make sure that the attack would succeed, Shepherd arranged for a preparation that included the fires of eight cruisers and destroyers, six battalions of artillery (including one from the 12th Marines), and the heaviest air strikes since W-Day. To increase direct fire support for the infantry, the Marine commander asked General Bruce for another platoon of Army tanks, which would work with the one already assigned to the brigade, and for a platoon of tank destroyers as well.

When the brigade attacked at 0800, following a thunderous and extended preparation, there were few Japanese left to contest the advance. By 1000, General Shepherd was reporting to General Geiger: "We have crossed our 0–5 Line and are now rapidly advancing up the airstrip meeting meager resistance." [26] An hour later, Shepherd ordered the 22d Marines to hold up its attack at the 0–6 Line and directed the 4th Marines to take over there and capture the rest of the peninsula.

In moving toward 0–6, the 3d Battalion, 4th Marines encountered and handily overcame resistance from a strongpoint located near the ruins of the airfield control tower. This proved to be the only significant opposition that developed during the day. The relief of the 22d Marines on 0–6 took place without incident at 1500; and Lieutenant Colonel Shapley held up his advance about 500 yards beyond this objective line. At 1600, while the infantry dug in, two platoons of riflemen mounted the regimental tanks and a reinforcing platoon of Army lights and made a combat reconnaissance to Orote Point. Only two Japanese were sighted and they were killed. When the tank-infantry patrol reported back, General Shepherd declared the peninsula secured.

Earlier on the 29th, at 1530, a ceremony took place at the ruins of the Marine Barracks that had special significance to all Americans on the island and on the waters offshore. To the accompaniment of "To the Colors" blown on a captured Japanese bugle, the American flag was officially raised on Guam for the first time since 10 December 1941.[27] Present to witness this historic event were Admiral Spruance and General Holland Smith, ashore on an inspection trip, and Generals Geiger, Larsen, and Shepherd as well as the brigade regimental commanders and those few officers and men that could be spared from the fighting still going on. Fitting honors for the occasion were rendered by a platoon of

[26] *1st ProvMarBrig Jnl*, entry of 28Jul44.

[27] This ceremony, which usually took place in the Central Pacific after an objective was secured, came as a surprise to the naval officers attending. One officer not present who was particularly disappointed at this early flagraising was Captain Charles J. Moore, Admiral Spruance's executive officer. His father, Lieutenant Charles B. T. Moore, USN, had raised the flag over Guam on 23 January 1899 on the occasion of the take-over of the island government by the Navy Department. It had been Admiral Spruance's intention to suggest to General Geiger that Captain Moore be accorded the privilege of raising the first official flag over Guam. RAdm Charles J. Moore cmts on draft MS, dtd 6July65.

the men that had repossessed the barracks for the Marine Corps. In a brief address, General Shepherd caught the spirit of the event, saying:

> On this hallowed ground, you officers and men of the First Marine Brigade have avenged the loss of our comrades who were overcome by a numerically superior enemy three days after Pearl Harbor. Under our flag this island again stands ready to fulfill its destiny as an American fortress in the Pacific.[28]

Last-gasp resistance by the scattered enemy survivors was confined to sniping and bitter-end defense of caves and dugout hideaways, principally in the cliffs that bordered Apra Harbor. Many Japanese committed suicide when American troops approached; others tried to escape the peninsula by swimming to the low-lying ruins of Ft. Santa Cruz in the middle of harbor.[29] The swimmers were shot, captured, or turned back by a watchdog platoon of LVT(A)s. On the opposite side of the peninsula, Neye Island, long a source of galling enemy fire, was scouted by an LVT-borne patrol of the 9th Defense Battalion and found deserted. Brigade intelligence officers conservatively estimated that at least 1,633 enemy troops had been killed on Orote by 30 July. The cost of those deaths to the brigade was 115 Marines killed, 721 wounded, and 38 missing in action.

The end of the battle for possession of the peninsula coincided with a realignment of the IIIAC battle line. While the brigade had been clearing the Japanese from Orote, the 3d Division had fought its way to complete control of the Fonte heights, and the 77th Division had patrolled all of southern Guam looking for enemy troops. While the two divisions prepared to drive north in line abreast and wipe out the remaining Japanese, the brigade was to take an active role in reserve, guarding the corps rear area, mopping up the peninsula, and hunting down enemy stragglers in the southern mountains.

Nothing signified the change of ownership of Orote Peninsula better than the landing on its airfield of a Navy TBF from the *Chenango* on 30 July. Touching down first to test the surface of a 2,000-foot-long strip cleared by six hours of feverish engineer activity, the plane circled and came down again at 1650.[30] Once the field proved ready, the escort carriers USS *Sangamon* and *Suwanee* each launched two VMO–1 observation planes to become the first elements of what was eventually to

[28] Quoted in Kaufman, "Attack on Guam," p. 63.

[29] On 21 June 1898, the first American flag was raised over Guam at Ft. Santa Cruz; this ceremony signified the bloodless capture of the island from the Spanish garrison.

[30] This test landing by a Navy plane spoiled the plans of several Marine officers to have the first American plane to land on Guam be one from VMO–1, whose craft were poised on the escort carriers offshore. Col Frederick P. Henderson ltr to CMC, dtd 21Nov52. Actually, the officer that called down the TBF to land was a Marine, Colonel Peter F. Schrider, commanding MAG–21, who was present on the strip with an advance detachment of his air group. Sherrod, *Marine Air History*, p. 253. As it happened, the first American plane to operate from Guam was an Army liaison aircraft assigned to the 77th Division Artillery. This plane took off from an improvised airstrip at 1310 on the 30th. 77th InfDiv Arty AAR, 21Jul–10Aug44 (WW II RecsDiv, FRC, Alexandria, Va.)

become a powerful American aerial task force based on Guam.[31]

FONTE SECURED [32]

When the 3d Marine Division reopened a full-scale attack to secure the Fonte heights on 27 July, there was little evidence of the Japanese decision to withdraw to the northern sector of Guam. The enemy seemed as determined as ever to hold his ground, and the day's fighting, focused on the left center of the division front, cost the Marines over a hundred casualties. Holding out, often to the last man, Japanese defenders made effective use of the broken terrain which was honeycombed with bunkers, caves, and trenches.

The twisted and broken remnants of a powerline, which cut across Fonte Plateau and ran in front of Mt. Mangan, became the initial objective of 2/3, 2/9, and 2/21, which bore the brunt of the assault. (See Map VIII, Map Section.) The battalions flanking the plateau fought their way forward to the line shortly before noon and then held up awaiting the advance of 2/9. Lieutenant Colonel Cushman's unit had been strafed and hit by bombs falling short during the aerial preparation for the morning's attack, and the resulting reorganization had held up the assault companies for 80 minutes. About 1300, just after it finally came up on line with 2/3 and 2/21, 2/9 was hit hard by a surging counterattack, which boiled up out of the thick brush that blanketed the plateau. Company G, on the left of the battalion front, met most of this thrust by 150–200 Japanese troops. The tanks working with the infantry played a large part in the repulse of the attack, which finally subsided after almost two hours of hot, close-quarter action.

Shortly after this fight died down, Cushman recommended that his battalion stop its advance and dig in strongly for the night. A formidable strongpoint, a large cave-rimmed depression, which appeared to be the key to the remaining Japanese defensive system, lay just ahead in the path of 2/9. When division authorized a halt for the night, Cushman sent out scouts to find the best way to attack the strongpoint, issued replenishment supplies, and built up reserve ammunition stocks for the next day's drive.

While the fighting on the flanks of 2/9 was not so frenzied as it was on the plateau itself, there was ample evidence here too that the Japanese had not lost their will to fight. Neither 2/3 nor 2/21 could advance much beyond the powerline without being exposed to enemy flanking attacks. Toward the center of the division line, tank-infantry teams of 1/21 were heavily engaged all day in cleaning out enemy troops holed up in caves and dugouts in the vicinity of some demolished radio towers. Some Japanese still manned defenses in the quarry area near the center of the battalion zone, even after three days of

[31] CTU 53.7 (ComCarDiv 22) Rpt of Marianas Ops, 12Jun–1Aug44, ser. 0047 of 3Aug44, p. 10 (OAB, NHD).

[32] Unless otherwise noted the material in this section is derived from *3d MarDiv SAR; 3d MarDiv Jul44 WarD; 3d MarDiv Jnl;* 3d MarDiv D-2 and D-3 PeriodicRpts, 26–31Jul44; *77th InfDiv OpRpt; 77th InfDiv Jnl; 9th Mar Jnl; 2/3 Jnl; GSDF Study; Craig 30Sep52 ltr;* Cushman, "Fight at Fonte."

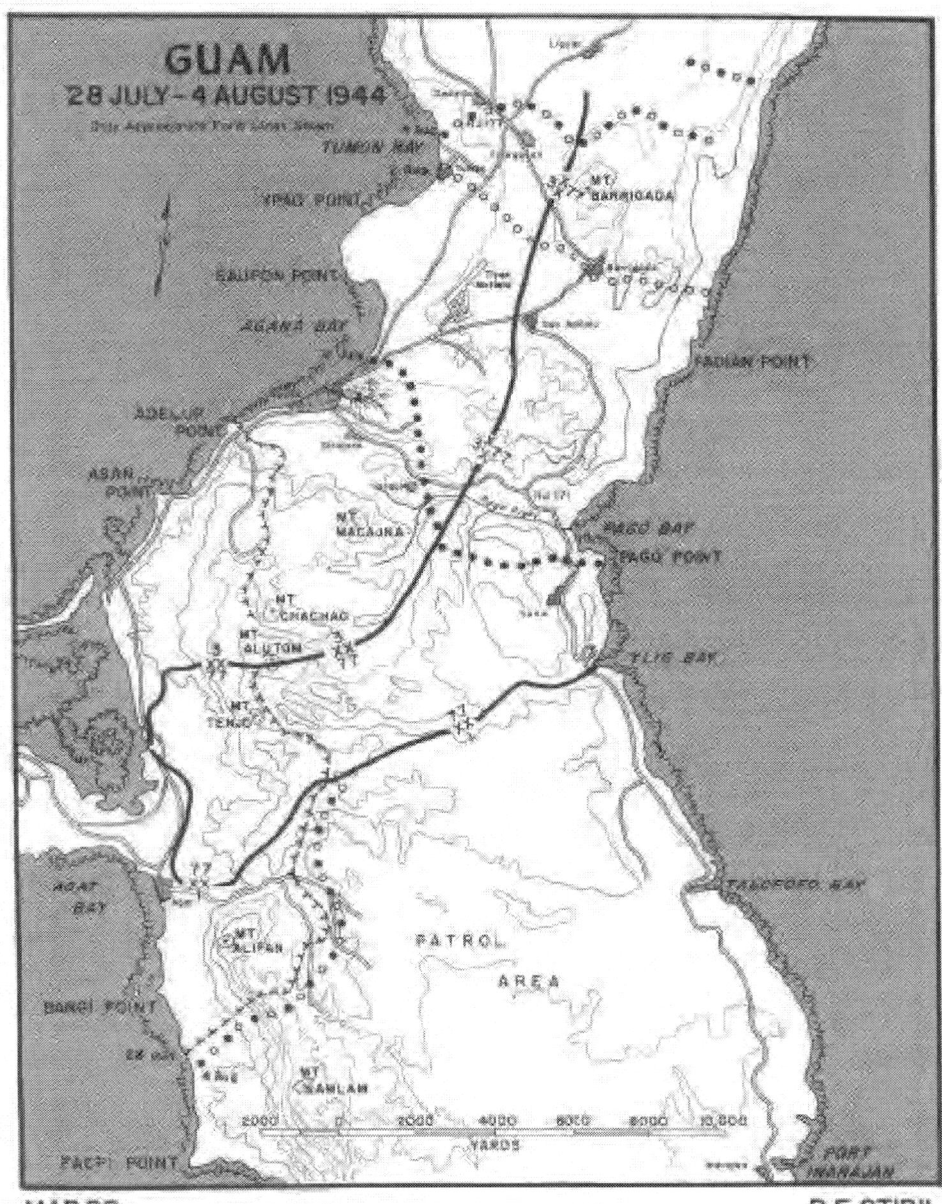

MAP 29 R.F. STIBIL

constant attacks with explosives, gunfire, and flame. Despite the spirited enemy resistance, both here and on the plateau, the heavy Japanese losses foretold the end. The 3d Division attack order for 28 July called for all three regiments to seize the FBHL in their zone.

The 9th Marines was to have the difficult task of driving south up the rugged slopes of Mts. Chachao and Alutom and along the ridge approaches to Mt. Tenjo. The crest of Tenjo was made an objective of the 77th Infantry Division, and the boundary between divisions was altered to show this change from the original landing plan. (See Map 29.) The axis of the Marine attack was plotted in the zone of 3/21, and, on the 27th, 3/9 moved into positions behind Lieutenant Colonel Duplantis' battalion, ready to pass through on the 28th. The 3d Battalion, 307th Infantry, attached to Colonel Craig's regiment, relieved 3/9 on the right of the Marine line so that Major Hubbard's men could spearhead the regimental assault on the peaks that loomed ahead.

The III Corps attack on the 28th was successful on all fronts, and the day ended with the Final Beachhead Line from Adelup to Magpo Point in American hands. At 0800, in a bloodless advance which culminated a week of patrol and mopping-up action in the hills between the two beachheads, a company of 1/305 seized the peak of Mt. Tenjo. The 2d Battalion of the 307th Infantry then moved up to occupy the mountain and extend its lines north toward the new division boundary. Patrols of the 9th Marines made contact with 2/307 on the heights during the afternoon.

The 3d Battalion, 9th Marines relieved 3/21 on position at 0800, and at 0910 began moving forward behind tightly controlled artillery and machine gun support. The 21st Marines battalion was attached to the 9th Marines to guard the open left flank behind Major Hubbard's assault companies. Inside of a half hour, 3/9 was on its initial objective and abreast of 1/9 and 3/307. An hour later, Colonel Craig ordered a general advance of the three battalions toward the Chachao-Alutom massif. Although Craig had not received the IIIAC map overlay showing the new division boundary, he could plainly see the Army infantrymen on Mt. Tenjo's slopes, so he contacted the commander of 2/307, while both officers were checking the flank positions of their units, and arranged a mutually satisfactory boundary.[33] When the corps overlay arrived, its boundary was found to coincide with that worked out by the two commanders.

The only serious resistance encountered by the 9th Marines was centered on a strongpoint located on Mt. Chachao. Manned by a company of Japanese troops, presumably remnants of the *18th Infantry*, this complex of machine gun nests and trenches guarded the trail along the ridge leading to Mt. Tenjo. Once 3/9 had developed this position, Major Hubbard called down artillery fire on the defenders to cover the infantry approach and conceal the movement of tanks to the rear to cut off the enemy escape route. When the artillery fire lifted, Companies I and K,

[33] *Craig 22Jun65 ltr.*

rushing the Japanese, drove steadily forward along the ridge, destroyed everything in their path, and charged the last emplacements with bayonets behind a shower of hand grenades. When the battle was over, 3/9 counted 135 Japanese dead in its zone. The victory enabled the 9th Marines to secure its objective from Apra Harbor to the 21st Marines boundary near Mt. Mangan.

In its drive to the FBHL, Colonel Butler's regiment overran the *29th Division* headquarters caves, located near the head of the Fonte River valley close to the wrecked radio towers, and wiped out the last defenders of Mt. Mangan as well. In both actions, tanks were in the forefront of the fighting and the Japanese tried desperately to knock them out with grenades and hand-carried antitank mines. Well covered by riflemen of 1/21 and 2/21 and their own machine gun fire, the tanks escaped unscathed from numerous fanatic attacks by individuals and small groups. By the time the 21st Marines were ready to dig in at dusk, all was quiet around the radio towers. The reverse slope defenses of Mt. Mangan were finally silenced.

The only other area of enemy opposition to the 3d Division advance on 28 July was located in the depression on Fonte Plateau. Here, Lieutenant Colonel Cushman's careful preparations paid off in a smoothly executed attack. Utilizing tank, machine gun, and bazooka firing positions that had been pinpointed by reconnaissance the previous afternoon, 2/9 cut loose with a deadly crossfire which blanketed every cave entrance in the pit. Under cover of this fire, a picked assault group with flamethrowers and demolitions worked its way down from the rim and methodically destroyed every enemy position without losing a man. Once this strongpoint was reduced, 2/9, working with 2/3 and 2/21, was able to clear the rest of the plateau area and secure its share of the FBHL. Cushman's battalion, in four hectic and wearying days of hard fighting for control of Fonte had lost 62 men killed and 179 wounded, but it had captured the anchor position of the enemy defenses.

As night fell across the island on the 28th, reports came in from all along the new Marine positions that scattered Japanese holdouts, who had purposely or unknowingly been bypassed during the day's advance, were trying to infiltrate the lines heading north. This attempted exodus from the Marine beachhead by a relative handful of enemy survivors reflected the orders that had been passed by the *29th Division* following the unsuccessful counterattack of 26 July.

The able-bodied fighting men were directed to disengage on the night of the 28th and withdraw through Ordot to prepared positions near Barrigada and Finegayan, there "to engage in delaying action in the jungle in northern Guam to hold the island as long as possible."[34] All sick and wounded combatants and Japanese civilians not attached to fighting units were started north on the night of the 27th, the division hospital and its patients to a position behind Mt. Santa Rosa and the civilians to "a safe area further north."[35] Accompanying this first

[34] *GSDF Study*, p. 193.

[35] *Ibid.*

echelon was the *Thirty-first Army* commander, General Obata, and three of his staff officers, who left the Fonte headquarters at midnight on the 27th to move to Ordot. At the same time, one of General Takashina's staff was also sent to Ordot to marshal all available motor transport and move rations and supplies to storage areas in the jungle north of Mt. Santa Rosa.

General Takashina and Lieutenant Colonel Takeda remained behind at the Fonte headquarters when the withdrawal began, and as a result were directly involved in the fighting on the 28th, when Marine tanks attacked the *29th Division* cave CP area. At about 1100, as it became increasingly apparent that the dwindling number of Japanese defenders could not stop the rampaging tanks, General Takashina decided to make a break while there was still a chance to escape north. Then, as Takedo recalled the events, the two Japanese officers:

> . . . stole out of the headquarters cave and ran straight between some enemy tanks and jumped from a cliff. The U. S. tanks, sighting the two persons, fired volleys of tracer bullets. Fortunately for the two, they managed to escape into a dead angle of the tank guns. About 1400 hours they reached a small stream at the northern foot of Mt. Macajna when the division commander was shot by machine gun fire from a U. S. tank, and died a heroic death, his heart having been penetrated by a bullet.[39]

With Takashina's death in battle, the tactical command of all Japanese forces remaining on Guam was assumed by General Obata. He had only a few senior officers remaining to rally the surviving defenders and organize cohesive units from the shattered remnants of the battalions that had fought to hold the heights above the Asan-Adelup beaches. All through the night of 28 July, Japanese troops trudged along the paths that led from Fonte to Ordot, finding their way at times by the light of American flares. At Ordot, two traffic control points guided men toward Barrigada, where three composite infantry companies were forming, or toward Finegayan, where a force of five composite companies was to man blocking positions. As he fully expected the Americans to conduct an aggressive pursuit on the 29th, General Obata ordered Lieutenant Colonel Takada to organize a delaying force that would hold back the Marines until the withdrawal could be effected.

Contrary to the Japanese commander's expectations, General Geiger had decided to rest his battle-weary assault troops before launching a full-scale attack to the north. The substance of his orders to the 3d and 77th Divisions on 29 July was to eliminate the last vestiges of Japanese resistance within the FBHL, to organize the line of defense, and to patrol in strength to the front. All during the day, small but sharp fights flared up wherever 3d Division Marines strove to wipe out the isolated pockets of enemy defenders that still held out within the beachhead perimeter. A very few Japanese surrendered, and most of these men were dazed, wounded, and unable to resist further. Almost all the enemy died fighting instead.

Although they made few contacts with retreating Japanese, Marine and Army patrols began to encounter in-

[39] *Ibid.*, pp. 195–196.

creasing numbers of Guamanians, who started to move toward the American lines as the enemy relaxed his watch. Intelligence provided by the natives confirmed patrol and aerial observer reports that the Japanese were headed for northern Guam. There was no strong defensive position within 2,000 yards of the FBHL, and there were ample signs of a hasty withdrawal. Patrols found a wealth of weapons, ammunition dumps, and caves crammed with supplies of all types in the area ringing the III Corps position. The discoveries emphasized the sorry plight of the ill-equipped and ill-fed men, who were struggling north through the jungle, punished by constant harassing and interdiction fires by Corps Artillery and the machine guns and bombs of carrier air.

The 2d Battalion, 9th Marines was relieved on line by 1/3 on the 29th and was placed in division reserve for a short and well-earned rest. As the 3d Marines was readied for a new phase in the battle, the regiment received a new commander, Colonel James A. Stuart, who had been the D–3. As a part of a division-wide shift in individual command and staff responsibilities, Colonel Hall was reassigned duties as the D–4.[37] The changes seemed to be in keeping with the aura of preparation and reorientation that was prevalent throughout the IIIAC positions. Everywhere the assault troops and the service and support units were refurbishing equipment and stockpiling ammunition and supples for the drive into the northern jungle.

Although the Japanese were no longer in close contact with the Americans, the patrols sent out on the 30th ran into scattered enemy fire as soon as they began to move up from the belt of lowland between Agana and Pago Bays and onto the northern plateau. All reconnaissance and other intelligence indicated that the Japanese were ready to defend the road that forked north of Agana, one branch leading to Finegayan and the other to Barrigada. (See Map 29.)

A BASE OF OPERATIONS[38]

Before General Geiger was ready to launch a drive north on an island-wide front, he needed assurance that his rear was secure from attack. Equally as well, he had to be certain that he possessed the supplies and support forces necessary to sustain an advance by two divisions through extremely difficult country against an opponent that was battered but by no means beaten.

To answer one requirement, knowl-

[37] As a result of the same order, Lieutenant Colonel Ellsworth N. Murray (D–4) replaced the D–2, Lieutenant Colonel Howard J. Turton, who became D–3. Colonel Robert G. Hunt (Division Inspector) was given the additional duties of Liaison Officer to IIIAC and Lieutenant Colonel Ralph M. King (Assistant D–3) was assigned as executive officer of the 9th Marines to replace Lieutenant Colonel Jaime Sabater, wounded on 21 July. The order also confirmed the appointment of Major Irving R. Kriendler as D–1 on 22 July, following the death of Lieutenant Colonel Chevy S. White in a shelling of the division message center.

[38] Unless otherwise noted, the material in this section is derived from: *IIIAC SAR;* IIIAC C–4 PeriodicRpts Nos. 1–44 to 23–44, dtd 21Jul–14Aug44, hereafter *IIIAC C-4 Rpts;* LtCol F. Clay Bridgewater, USA, "Reconnaissance on Guam," *The Cavalry Journal*, v. LIV, no. 3 (May–Jun45).

PONTOON BARGE *loaded with fuel drums at the reef transfer line off Guam. (USA SC210553)*

GUAMANIAN WOMEN *wash their clothes in a shell hole in the midst of a refugee tent city behind American lines. (USMC 92233)*

edge of the enemy situation in southern Guam, the 77th Division sent infantry patrols deep into the mountains and jungle in the vicinity of the FBHL. On the 27th, General Bruce ordered the 77th Reconnaissance Troop to investigate reports that the Japanese might still be present in strength, particularly in the center of the island near Mt. Lamlam. Five small patrols set out, two for objectives on the east coast, two to the southeast, and one down the southwest coast. Although the sickness of one member forced the patrol to Ylig and Pago Bays to turn back after it had covered 8,000 yards, the others stayed out three days checking all signs of the Japanese. Scattered opposition was encountered from snipers and small units by the patrols when they moved south along the mountainous spine of the island, but there was no evidence of enemy resistance in strength. As the mission of the patrols was reconnaissance not combat, the soldiers evaded most of the enemy troops they spotted, noting that the Japanese were all headed north. Other patrols sent out on the 29th and 30th travelled along the 77th Division proposed route of advance to Pago Bay. They gained valuable terrain intelligence to aid General Bruce in planning the difficult movement of his regiments east and then north through the jungle to come up on line with the 3d Marine Division.

Once General Geiger knew that no significant Japanese force was present in southern Guam, he assigned responsibility for its control and pacification to his smallest major tactical unit, the 1st Provisional Marine Brigade. Plans were formulated for General Shepherd's regiments to relieve elements of the 77th Division when the fighting on Orote ended.

In many ways, the assault phase of the Guam operation was partially over when IIIAC was ready to launch its northern offensive. Apra Harbor, the key objective of the dual landing operation, had been secured and was being converted into a major anchorage. Seabees and engineers had cleared beaches that had been battlegrounds and had rebuilt and replaced roads and bridges to handle heavy vehicular traffic. Extensive supply dumps, repair facilities, and other service installations had begun to take on the appearance of order and permanence.

On 26 July at 1300, General Geiger had opened his headquarters ashore near Agat, and on the following day, the Corps Service Group under Lieutenant Colonel Francis M. McAlister had begun operations by taking over control of the 5th Field Depot, 53d Naval Construction Battalion, and the Corps Medical Battalion. At the same time, the success of combat operations enabled the Corps Shore Party to begin unloading garrison force supplies over Dadi Beach near the foot of Orote Peninsula. On the eve of the second phase of the Guam operation, the Corps Service Group had grown in size and complexity to include many of the Seabee, engineer, pioneer, amphibian tractor, motor transport, and service units that had originally been part of the brigade and division shore parties. On hand and ready to issue in 5th Field Depot and 3d and 77th Division dumps were an average of 13 days' supply of rations, 15 days' gasoline and other petroleum products, and at least 3 units of fire for all weapons. Facilities were being prepared at Piti for the unloading of ships; the first vessel, a cargo type carrying 3,000-man resupply blocks of all classes, was slated to start discharging on the 31st.

The cost of the fight thus far had been heavy. From H-Hour until midnight on 30 July, the III Amphibious Corps had lost 989 men killed in action and had had 4,836 wounded; in addition, 302 men were missing and unaccounted for. On the Japanese side of the grim tally sheet, 6,205 dead had been counted. Several thousand more were estimated to have been killed, their bodies lying sealed in caves or hidden by folds of ground and thick brush in the battle area. Only 50 prisoners had been taken despite the desperate, hopeless nature of the Japanese situation. The rugged terrain to the north, coupled with the discipline and tenacity of the defenders, promised further heavy casualties to both sides before the battle for Guam would end.

Seizure of Northern Guam

TIYAN AND BARRIGADA [1]

From their hard-won positions on the Fonte heights and from the slopes of the mountain ridges that marked the trace of the FBHL, the assault troops of III Corps could easily see the broad expanse of jungle, which covered the northern plateau. Just forward of the 3d Marine Division positions, the ground fell away sharply to a brush-covered lowland studded with small hills where the Agana River flowed into a large swamp southeast of the capital. The terrain in front of the 77th Infantry Division left flank was badly cut up by ravines formed by stream tributaries of the Pago River. Most of the rest of General Bruce's zone of advance was also high ground, trackless and dotted with barriers of thick vegetation, which gradually grew denser on the approaches to Pago Bay. (See Map 29.)

The IIIAC scheme of maneuver for the 31 July attack called for the 77th Division to move northeast from its FBHL positions, pivoting on left flank units to come abreast of the 3d Division on a cross-island objective line, which ran just north of Agana, turned south through the Ordot area, and then headed east to the coast at Pago Point. General Bruce's plan directed an advance in regimental columns to effect the quickest possible passage of the 10 miles of hill country that separated Mt. Tenjo from the objective. Thorough reconnaissance had indicated that no significant enemy opposition would be encountered—and none was.

On schedule at 0630 on the 31st, the 77th Division moved out from the FBHL with the 307th Infantry in the lead. At first the soldiers were able to follow a fresh-cut road that led along the axis of advance, but the head of the column soon passed the sweating engineers and their bulldozers, which were engaged in a running battle with the rain-sodden ground. Striking out cross-country, the long, snaking line trudged over the rugged terrain in a march that seemed at times to involve more up and down movement than it did forward progress. One marcher later described his experiences graphically:

> The distance across the island is not far, as the crow flies, but unluckily we can't fly. The nearest I came to flying was while descending the slippery side of a mountain in a sitting position. . . . After advancing a few yards you find that the handle of the machine gun on your shoul-

[1] Unless otherwise noted, the material in this section is derived from: *IIIAC SAR; 3d MarDiv SAR; 3d MarDiv Jnl;* 3d MarDiv D-2 and D-3 Periodic Rpts, 31Jul-3Aug44; *77th InfDiv OpRpt; 77th InfDiv Jnl; 1st ProvMarBrig SAR; GSDF Study;* MajGen Andrew D. Bruce, USA, "Administration, Supply, and Evacuation of the 77th Infantry Division on Guam," *Military Review*, vol. 24, no. 10 (Dec44).

214-881 O-67—35

COLUMN OF SOLDIERS *of the 305th Infantry advances cross-island on 31 July at the start of the attack on northern Guam. (USA SC272338)*

JUNGLE TRAIL *is scouted by Marine tanks with covering infantry during the advance in northern Guam. (USMC 91166)*

> der, your pack and shovel, canteens, knife, and machete all stick out at right angles and are as tenacious in their grip on the surrounding underbrush as a dozen grappling hooks. Straining, sweating, and swearing avails you nothing so you decide on a full-bodied lunge—success crowns your efforts as all the entangling encumbrances decided to give up the struggle simultaneously. Just before you hit the ground a low swinging vine breaks your fall by looping itself under your chin, almost decapitating you and snapping your helmet fifteen yards to the rear. . . . You untangle your equipment, retrieve your helmet, and move on. The flies and mosquitos have discovered your route of march and have called up the reinforcements including the underfed and undernourished who regard us as walking blood banks. We continue to push on. . . .[2]

Despite the difficult terrain, the 307th kept up a good pace and reached the Pago River early in the afternoon. Patrols discovered an unguarded concentration camp on the banks of the river and released a group of about 2,000 happy Guamanians. As the natives started moving back toward Agat along the column of soldiers, the Americans shared their rations, cigarettes, and whatever else they could spare with the hungry men, women, and children.

The lead unit of Colonel Tanzola's regimental column, the 3d Battalion, 305th Infantry, moving to the right rear of the 307th, met the only opposition that was offered to the advance of the 77th Division. As scouts of 3/5 approached the village of Yona late in the afternoon of the 31st, a number of Japanese hidden amidst the buildings opened fire. Deploying quickly, the leading company attacked and overran the village, killing 5 enemy and scattering the remainder of a force estimated at 50 men. Moving on, the 3d Battalion reached Pago Point before nightfall and set up on a hill there in all-around defense. Companies of the other battalions of the 305th established blocking positions along the division southern boundary from that point to the FBHL, where the 4th Marines had relieved the regiment earlier in the day. With the advance of the 77th Division to the east coast, the 1st Brigade, temporarily reinforced by the 306th Infantry, assumed responsibility for pacifying the southern half of Guam.

Like the 77th Division, the 3d Marine Division encountered little opposition on 31 July. General Turnage attacked with three regiments abreast at 0630, and by nightfall, when the advancing units held up, they had seized 4,000–5,000 yards of important terrain, including 4,000 yards of the crucial Agana-Pago Bay Road. The capital of the island was once again in American hands, and in the town plaza, amidst the shattered remnants of houses and government buildings, an advance division supply dump was operating at a brisk pace.

The honor of liberating Agana fell to 3/3, which advanced up the coastal road. At 1030, scouts of its assault platoons entered the town cautiously, threading their way through the rubble and the mines, which were strewn everywhere. Within 15 minutes, the battalion had reached the central plaza, and by noon was through the ruins and had set up in the northern outskirts on the regi-

[2] Quoted in HistDiv, WD, *Guam, Operations of the 77th Division (21 July–10 August 1944)* (Washington, 1946), p. 65, hereafter HistDiv, WD, *77th Div on Guam*.

mental objective. The rest of the 3d Marines was soon up on line with the 3d Battalion. At the start of the day's advance, 1/3 had held positions along the northern lip of the Fonte River valley, where its lines were soon masked by the forward movement of 2/21. Temporarily in reserve, 1/3 then followed up the assault units of 2/3, which reached and secured its portion of the Agana-Pago Bay Road by noon.

The road was also the initial objective of the 21st Marines, but the lead elements of 2/21 and 3/21, with a longer distance and more rugged terrain to travel, did not reach their goal until 1350. Moving along the right boundary, 3/21 tangled with a small force of Japanese holding a pillbox near Ordot and wiped out the defenders, one of the few such clashes during the day's advance. In the same vicinity, the 3/21 command post group, moving to a new forward position, scattered a force of 15–20 Japanese it encountered behind the lines. The enemy were evidently unaware that the Marine advance had passed them by, a tribute to the nature of the terrain.

On the right of the division zone of action, the 9th Marines had the farthest distance to go on 31 July over ground as bad as that any unit encountered. Like 3/21 on its left flank, 1/9 met and overcame resistance from a small outpost of enemy troops left behind to impede movement through the Ordot supply area. Then, at 1415, two enemy light tanks appeared out of the brush, spraying the advancing Marines with machine gun fire, killing one man and wounding three others. An alert bazooka team took care of both tanks, and the advance resumed. At 1510, the 9th Marines had reached its objective, which was partially along the cross-island road. A platoon of the division tank battalion was able to join the regiment and add strength to the antitank defenses.

Earlier in the afternoon, General Geiger had authorized the 3d Division to continue moving forward as long as firm contact was maintained across the front. Although General Turnage alerted all units for a possible further advance, only the 3d Marines actually moved out. The 21st Marines committing 1/21 on the left, was engaged in shifting its frontline units to the right, in order to maintain contact with the 9th Marines, while the 9th was consolidating its link with the 307th Infantry along the division boundary.

Major Bastian's 3/3 had little difficulty in advancing from Agana once it jumped off at 1545. Before dusk, it had seized 1,200–1,500 yards of the roads northeast of the capital, one of which led to Finegayan and the other to Barrigada through the road junction village of San Antonio. On the right, 1/3 passed through the lines of the 2d Battalion and almost immediately entered an extensive swamp, which was the source of the Agana River. The going was slow and rough, and it was dark before the lead elements could dig in on dry land. The assault battalions set up perimeter defenses for the night about a mile and a half apart, with 2/3 in reserve positions blocking the gap.

As 3/3 was digging in, two small jeep-mounted patrols of Reconnaissance Company cleared the forward positions of the battalion and moved along the road to San Antonio. Their mission was to check the trail network leading to Tumon Bay and Tiyan airfield. Both patrols sighted small groups of Japanese, hiding out in houses along the routes followed, and exchanged fire with them before turning back to the outskirts of Agana. As a combined patrol, the group then drove along the shore road until it was stopped by a tangle of trees blown across the path. In the gathering darkness, the reconnaissance force turned back again with little more to report to General Turnage than that some Japanese were located forward of the Marine lines. At the division CP, plans were laid for new motorized patrols the following day, this time bolstered by half-tracks and tanks.[3]

After a quiet night with no enemy activity, the 3d Division attacked at 0700 on the 1st with a scheme of maneuver calculated to pinch the 21st Marines out of the front line of a narrowing division zone. General Turnage ordered the 3d Marines to hold its positions once the 1st Battalion moved out of the swamp area and came abreast of 3/3. In advancing, 1/3 extended toward its right, closing in front of 1/21. The 21st Marines moved out with the 2d and 3d Battalions abreast, but 2/21 soon halted and withdrew to reserve positions when it was covered by 3/21. At 0910, the 3d Battalion was ordered to pivot on its right and occupy the boundary between the 3d and 9th Marines until its position was masked by the advance of the 9th. By 1400, Colonel Craig's regiment had completed this maneuver, after moving with difficulty through heavy brush and irregular terrain to seize the remaining portion of the Agana-Pago Bay Road in its zone. The 21st Marines, less 3/21, which was attached to the 3d Marines, was ordered into division reserve during the day, replacing 2/9, which reverted to regimental control.

With the readjustment of lines completed, the division resumed its general advance at 1500. Forward progress was steady and enemy resistance negligible. By 1745, when General Turnage called a halt, the center of the division was just short of the airfield dispersal area and the right was within easy striking distance of San Antonio.

The greatest problem facing the 3d Division on 1 August was the hundreds of mines that the Japanese had planted on all roads leading north. The bomb-disposal teams of the 19th Marines were hard put to find, let alone remove or destroy, all the lethal explosives the enemy had buried. Inevitably, several vehicles were blown apart and their passengers killed or wounded during the day. When an armored reconnaissance patrol was attempted at 1745, its nine tanks and a half-track were turned back by a profusion of mines on the coastal road to Finegayan. The armored vehicles were able to drive along the entire western side of Tiyan airfield, where it ran along a low cliff, but they could find no usable roads that led down toward the coast. Near the far end of the airstrip, an undetermined number of Japanese opened fire on the tanks from concealed positions in the brush, but the patrol avoided a fire fight in order to return to friendly positions before dark. Like the motorized reconnaissance on the previous

[3] The original patrol makeup on 31 July had included two half-tracks, but both broke down en route to the objective.

afternoon, this patrol on 1 August developed little vital information on Japanese dispositions or strength in the 3d Division zone.

Undoubtedly, the most significant accomplishment of the day was the seizure of the Agana-Pago Bay Road along its entire length. This feat provided the solution to formidable logistical problems, which would otherwise have plagued the 77th Division. The Army assault regiments, the 307th and 305th Infantry, had jumped off at 0700 on the 1st and had crossed the Pago River soon after. Inexplicably, the Japanese had failed to destroy the main bridge over the river, and 3/305 seized it without incident at 0800. Within two hours, both regiments had secured the cross-island road in their zones, a stretch including RJ 171, where an intersecting road curved north through the jungle to San Antonio. The soldiers, keeping their direction by compass bearings, pushed on through the dense vegetation, taking advantage of trails wherever they occurred and blazing new paths where there were none. All assault units were short of rations and water, but were well supplied with small arms ammunition; the Japanese had provided scant opportunity to do much firing. By nightfall, the 305th Infantry was located in perimeter defenses one and a half miles northeast of RJ 171, and the 307th was generally on line with it and in contact with the 9th Marines near San Antonio. The 306th Infantry, less 2/306 in corps reserve, was set up near RJ 171, having marched there during the day after being relieved on the FBHL by the 22d Marines.

The 77th Division began to use the Agana-Pago Bay Road as its main supply route (MSR) almost as soon as it was captured. In planning the IIIAC drive to seize northern Guam, General Geiger had counted on the 77th Division to cut a new road from the Agat beachhead to the east coast road near Yona. Terrain difficulties, compounded by frequent rains, and the time factors involved forced abandonment of the road-building project late on the 31st. The only practical alternative to construction of a new MSR was for both divisions to use the same road, a solution that General Bruce has noted was unorthodox enough for "the books [to] say it can't be done, but on Guam it was done—it had to be." [4] At 1620 on the 1st, General Geiger issued an order assigning the 77th Division priority over all traffic on the west coast road between Agat and a turnaround north of Adelup Point and equal priority with the 3d Division on the road beyond Adelup as far as the division boundary.

Moving throughout the afternoon and on through the night (with headlights as far as Agana and blackout lights beyond), a steady procession of 77th Division trucks, jeeps, and trailers moved supplies and equipment across the 3d Division zone. Three battalions of artillery and the light tank company of the 706th Tank Battalion also travelled the route on the 1st. General Bruce ordered the medium tank companies that were attached to his RCTs, the division artillery headquarters, and the remaining 155mm howitzer battalion to make the move as early as possible on the morning of the 2d. The

[4] Bruce, *op. cit.*, p. 8.

general wanted as much support available as he could get, for intelligence sources all indicated that the Japanese were present in force in the Barrigada area, the next 77th Division objective.

In order to pinpoint the suspected enemy positions, the division commander ordered an armored reconnaissance made. Fourteen light tanks moved out along the road to San Antonio at 0630 on 2 August, a half-hour prior to the general division attack. About 800 yards beyond San Antonio on the road to Barrigada, the tanks were fired on by enemy troops concealed in the thick bordering jungle. After replying to this opposition with machine gun and cannon fire, the tanks returned to the American lines at 0730. They were soon sent out again, but this time got as far as Barrigada without meeting any resistance.

At the road fork in the village, the tanks at first turned left to move toward Mt. Barrigada on the road to Finegayan. Opposite the mountain the tanks encountered a trio of trucks, backed by enemy riflemen, blocking the way. The tank gunners made short work of both trucks and defenders, killing an estimated 35 Japanese. Returning to Barrigada, the armored column moved northeast along a road that appeared to swing around the other side of the mountain. Within 1,000 yards of the village, the track had dwindled to the size of a foot trail, and the lead tank got hung up on a stump. At this moment, Japanese troops began firing from all sides with rifles, machine guns, and 20mm guns. Some enemy soldiers tried to rush the tanks, but they were swept away by heavy fire from bow and turret guns. Once the stranded tank was able to work itself loose, the armored patrol withdrew without having suffered any losses.

As the day wore on, this morning tank action proved to be the sparring session before the main event. The 307th Infantry, after pausing along the road to San Antonio to distribute badly needed water and rations, moved out again at 1030 and ran head on into a bristling enemy defensive position covering approaches to the mountain and the village. The day's plan of attack called for 1/307 on the left to move through the jungle, cross the road to Finegayan north of Barrigada, and seize the western slopes of the mountain; 3/307 was to move through the village and attack the southern slopes. Both units met increasingly steady resistance from Japanese manning prepared positions in the jungle and amidst the scattered village houses. Companies that had been assigned wide attack zones were crowded together as withering defensive fire channelized the assault. The regimental reserve, 2/307, was committed to the fight, and both light and medium tanks were called up to support the advance. Tank fire-support, particularly the destruction wrought by the 75s of the Shermans, helped smash an opening in the defensive barrier; tank armor shielded wounded infantrymen being evacuated under Japanese fire. When the 307th dug in for the night, it held positions in Barrigada just beyond the road junction.

On the right of the division zone, the 305th Infantry ran up against the eastern extension of the enemy position at Barrigada. Hidden in the jungle, well camouflaged and dug in, the Japanese held their fire until the assault platoons of 3/305 and 1/305 were almost upon them and then shot with deadly accuracy. This tactic frustrated all attempts to outflank the enemy covering the open ground near Barrigada, and the battle resolved itself into a grinding tank-infantry action where gains of a few yards often took hours to win. Like the 307th, the 305th was finally able to fight its way past the Barrigada road junction and into the midst of the Japanese defenses when the approach of darkness forced a halt. The 77th Division, its combat experience thus far limited to minor patrol and defensive clashes, had had a rough introduction to the offensive in jungle warfare. In fighting often confused and frustrating, 29 men had been killed and 98 wounded, but the soldiers had proved their mettle.[5]

One unfortunate result of the day's action was that a gap developed between the 3d and 77th Divisions. In the wild tangle of trees and undergrowth along the boundary, the Marines and soldiers lost sight of each other after the morning attack began. The company of the 307th charged with maintaining contact spent most of its time out of touch with its own regiment as well as with the 9th Marines. General Turnage attached 2/21 to the 9th for the night to guard the open flank, and Lieutenant Colonel Smoak disposed his men along an unimproved trail that stretched from San Antonio to the positions that 3/9 had reached opposite Mt. Barrigada.

The 3d Marine Division did not encounter any significant opposition on 2 August for the third day in a row. As a result, the 9th Marines overran its objective, Tiyan airfield, by 0910. On order from division, Colonel Craig held his troops up at the north end of the field until the 3d Marines could come up on line.

During this lull, a Japanese tank caused quite a bit of excitement when it broke through the extended 9th Marines lines and raced through the airfield dispersal area toward the rear of the 3d Marines. As the tank roared by the CP of 1/3, one of the crew, scorning the main armament "opened the turret and began to shoot wildly with a pistol" [6] at the Marines, who were scurrying to take cover. When the tank careened into a ditch several hundred yards farther on, the crew abandoned it and escaped into the brush. Marine mediums came up later in the afternoon and destroyed the enemy vehicle.

About the time the Shermans were blasting the hulk of the enemy tank, the 3d Marines was striving to take as much ground as it could before dark. Colonel Stuart's regiment had been slowed all day by dense vegetation and mines along the few roads and trails in its zone. It was 1400 before the 3d came abreast of the 9th. At that time all division assault units continued the attack with the Japanese offering only sporadic and ineffectual resistance. As the 3d Marines wrestled its way through the jungle along the road to Finegayan, 3/21 covered the left flank of the regiment, reconnoitering the

[5] This fight, covered in full and interesting detail in HistDiv, WD, *77th Div on Guam*, pp. 75–102, furnishes an excellent study of small unit action.

[6] Maj Henry Aplington II ltr to CMC, dtd 9Apr47.

bulging cape formed by Saupon and Ypao Points. Where the division zone narrowed at Tumon Bay, 3/21 was pinched out of line and reverted to control of the 21st Marines as part of the reserve. So difficult was the problem of contact in the jungle that the 3d Marines continued advancing after dusk until it could reach a favorable open area to hold up for the night. There, Colonel Stuart and his executive officer, Colonel James Snedeker, personally helped tie in the positions of assault units by the light of a full moon.

An armored reconnaissance patrol cleared the front lines of the 3d Marines at 1815, its mission the same as that of the similar group sent out the previous evening—find the Japanese. After moving about 1,200 yards toward Finegayan, the patrol spotted several groups of the enemy, but did not engage them, turning back instead on order at 1845. This sighting confirmed previous intelligence that the Japanese were located in the vicinity of Finegayan, but there was still no strong evidence of their numbers or dispositions.

No one at III Corps or 3d Division headquarters doubted that the lull in the battle was temporary. The Japanese already were defending bitterly one road that led to Finegayan in the 77th Division zone, and the 3d Division was advancing astride the other main artery, which led to the junction near the village. Beyond Finegayan, the principal roads led to Ritidian Point and Mt. Santa Rosa. While there was no indication that the Japanese would defend the northernmost point of the island, aerial reconnaissance, captured documents, prisoner interrogations, and information supplied by Guamanians all pointed to Mt. Santa Rosa as the center of resistance.

Concerned though he was with the immediate struggle to break through the outpost defenses at Barrigada and Finegayan, General Geiger was also looking ahead to the capture of Mt. Santa Rosa. Once it had driven past the coastal indentation of Tumon Bay, IIIAC would be operating in a wider zone, one just as choked with jungle growth and as hard to traverse as any area yet encountered. Geiger planned to use the 77th Division to reduce enemy positions in the vicinity of Santa Rosa, leaving the capture of most of Guam north of the mountain to the 3d Division. Under the circumstances, the corps commander believed that he could make good use of the 1st Brigade in the final clean-up drive, which would narrow the zones of attack and enable Generals Bruce, Turnage, and Shepherd to employ their men to best advantage in the difficult terrain.

Oral instructions were issued on the morning of 2 August for the brigade to be prepared to move to the vicinity of Tiyan airfield in corps reserve. General Shepherd in turn issued an operation order at 1030 directing the 4th Marines (less two companies on distant patrol) to assemble at Maanot Pass ready to move north by 0800 on the 3d. The 22d Marines (less 1/22) was ordered to continue patrolling and to prepare to move on 5 August.[7] Corps planned to shift responsibility for the

[7] Adding an amphibious note to the reconnaissance along the west coast were several long patrols mounted by elements of the 1st Armored Amphibian Battalion. *Metzger memo.*

security of southern Guam from the brigade to a task force composed in the main of 1/22, the 9th Defense Battalion, and the 7th Antiaircraft Artillery (Automatic Weapons) Battalion, all under the defense battalion commander, Lieutenant Colonel O'Neil.

Following its relief of the 77th Division, the brigade had sent out deep, far-ranging patrols to continue the hunt for Japanese stragglers and to locate and encourage Guamanians to enter friendly lines. The patrols were made strong enough—all were at least reinforced platoons—to handle any potential opposition. Although the Marines found a considerable number of defensive positions wherever units of the *38th Infantry* and *10th IMR* had been stationed prior to W-Day, only a few enemy troops were discovered and these were swiftly eliminated. On 2 August, a 4th Marines patrol moving toward Talofofo Bay ran across a group of about 2,000 natives, who were directed to report in to the corps compound near Agat. Civil affairs officers there were already caring for approximately 5,000 Guamanians, most of whom had filtered into American lines since 31 July. In the 3d Division zone, an additional 530 civilians were being fed and housed in a temporary camp, and the number coming in increased sharply as the Japanese retreated to the north.

The problems involved in handling thousands of civilians were new to Marines in the Pacific, but they were anticipated. Whenever prelanding civil affairs plans went awry, there was a will to find and apply alternate solutions. Much improvisation was necessary, the corps C–1 recalled, because the supplies intended for the Guamanians, which "were loaded on a ship with a low unloading priority . . . reached the beach after fifteen thousand civilians were within our lines."[8] As a result of this situation, effective emergency measures were taken. As soon as the first natives were contacted:

> . . . every piece of canvas which could be spared by units of Corps, was turned over to the Civil Affairs Section and a camp was established south of Agat. 350 shelter tents were borrowed from the 3d Marine Division. The Army loaned tentage for a 100-bed hospital which the Corps Surgeon borrowed from the Navy. The Corps Medical Battalion made 250 beds available for civilians. A Marine officer was assigned to build the Agat camp. 36 military police from the Corps military police were assigned to guard the camp. Badly needed trucks were borrowed from the motor pool and from two to six trucks worked constantly at hauling captured enemy food supplies and materials salvaged from bombed buildings, including the Marine Barracks. All this was immediately put to use for civilian relief.[9]

The Agat Camp was soon crowded, but no one went hungry; everyone had at least a piece of canvas overhead, and adequate medical attention was assured. On 2 August, as it became increasingly apparent that there was no organized enemy activity in southern Guam, corps issued an order stating that all Guamanians living south of a line from Agat to Pago Bay would be encouraged to remain at their homes, resume their normal pursuits with emphasis on agriculture, and obtain food and medical attention as necessary

[8] *IIIAC SAR*, Encl E (Personnel), p. 3.
[9] *Ibid.*

from the Agat camp.[10] As soon as priority camp shelter construction was well started, the Corps Service Group began to employ some Guamanians as laborers. Plans were laid to organize a native police and patrol force. The rough terrain of Guam offered ample hiding places for individuals and small groups of the enemy. It was believed that native familiarity with the mountains and jungle would be of great value in hunting down any holdouts.

OBJECTIVE: FINEGAYAN-YIGO ROAD[11]

General Obata, after surveying the positions his men had prepared at Barrigada, determined that they were unsuitable for a sustained defensive effort. Although he considered that the jungle maze around the village would be an aid to ambush and outpost action, he also believed that the dense growth would hinder the establishment of effective firing positions and would work as well to bar counterattacks. The army commander's instructions to Major General Tamura, his chief of staff, were to fight a delaying action at Barrigada to gain time for the construction of final defensive positions in the Mt. Mataguac-Mt. Santa Rosa area. The hard fighting at Barrigada on 2 August showed how well the Japanese troops could carry out their orders to hold up the advance of the 77th Division in the eastern sector of the outpost defenses. On the 3d, the disposition of American forces, the terrain, and the roadnet combined to bring the 3d Marine Division into a head-on clash with the enemy deployed near Finegayan, guarding the western approaches to the final Japanese stronghold.

Ten days of hard-won experience had demonstrated that even the heaviest caliber guns had a difficult time making any impression on Japanese defenses dug into the rugged terrain of Guam. Where thick jungle cover added its mantle, the task of blasting out the enemy was doubly difficult. Impressed by the need to employ every available supporting weapon to maximum effectiveness, both Admiral Conolly and General Geiger took steps to muster a formidable array of ships, artillery, and aircraft to aid the advance to the north.

On 2 August, CTF 53 reorganized his gunfire support ships to cover operations along both coasts. Admiral Ainsworth, his flag in the light cruiser *Honolulu*, took station on the east side of the island with a battleship, another cruiser, and five destroyers. On the west, Rear Admiral C. Turner Joy in the heavy cruiser *Wichita* commanded a similar task unit, which was augmented by a third cruiser and four gunboats.[12] All the 155mm guns and howitzers of General del Valle's Corps Artillery were displaced forward by the morning of 3 August to positions where they could reinforce the fires of seven

[10] STLF GO No. 5, dtd 2Aug44, in 3d MarDiv D–1, 21Jul–10Aug44.

[11] Unless otherwise noted, the material in this section is derived from: *IIIAC SAR; 3d MarDiv SAR; 3d MarDiv Jnl;* 3d MarDiv D–2 and D–3 PeriodicRpts, 3–6Aug44; *77th InfDiv OpRpt; 77th InfDiv Jnl; 1st ProvMarBrig SAR; 3d TkBn SAR; GSDF Study.*

[12] The two battleships, *Pennsylvania* and *Colorado*, were detached on 3 August and returned to Eniwetok.

battalions of 75mm and 105mm and one of 155mm howitzers. Plans were laid to increase the aerial fire support available by supplementing carrier aircraft strikes with sorties by Seventh Air Force planes. The first deep support missions flown by Saipan-based B–25s and P–47s were directed against RJ 460 during the afternoon of the 3d. (See Map 30.)

There was heavy fighting in the 3d Marine Division zone on 3 August at RJ 177 where the roads from Agana and Barrigada crossed. Lieutenant Colonel Randall's 1/9 bore the brunt of the day's action as it advanced astride the road from Agana. At 0910, when the lead company (B) was about 500 yards from the junction, its men were driven to cover by a sudden burst of fire from Japanese dug in on both sides of the route. In a rough, close-quarter battle, two Marine tanks, an assault platoon of infantry, and plentiful supporting fire from all available weapons finished off the Japanese defenders at a cost of three men killed and seven wounded. Moving through the shambles of the enemy position, which was littered with 105 dead, 1/9 continued its advance on RJ 177. Continued opposition from Japanese troops hidden in the brush and ditches along the road was steady but light. By 1300, the battalion had driven past the junction. Shortly thereafter, as fresh assault troops relieved Company B, Lieutenant Colonel Randall received orders to dig in for the night.

On both flanks of 1/9, Marine units made good progress marked by clashes with small enemy delaying forces. The jungle and the constant problems it posed to movement and contact continued to be the most formidable obstacle. When it ended its advance along the coast, 3/3 was nearly 3,000 yards forward of the positions of the 9th Marines on the division boundary. There 3/9, now commanded by Major Jess P. Ferrill, Jr.,[13] held up when it reached the Finegayan-Barrigada road because the battalion had no contact with the Army units on its right. At 1615, in order to plug the gap between divisions, General Turnage attached 3/21 to the 9th Marines; the battalion moved to blocking positions along the boundary to the right rear of 3/9.

As the Marines were digging in near RJ 177 late in the afternoon, an armored reconnaissance in force was attempted. Organized earlier in the day under the 3d Tank Battalion commander, Lieutenant Colonel Hartnoll J. Withers, the patrol group consisted of the Shermans of Company A reinforced by battalion staff tanks, two half-tracks from Reconnaissance Company, four radio jeeps, and mounted in trucks, Company I, 3/21 and a mine-clearing detachment of the 19th Marines.

Originally, the motorized patrol was scheduled to clear American lines shortly after 1200, but it was held up by the fighting at RJ 177. When the patrol commander finally got the word to proceed at 1525, it was already too late to reach its original objective, Ritidian Point, and return during daylight. Lieutenant Colonel Withers was ordered instead to try to reach RJ 460

[13] On 1 August, Major Ferrill, who had commanded the Regimental Weapons Company, replaced Major Hubbard, who was wounded in the day's fighting.

before turning back and to complete his mission on the following day.

Shortly after 1600, the buttoned-up half-track leading the patrol point reached RJ 177 and veered right instead of left, heading east toward Liguan and Yigo. Approximately 400 yards past the junction, Japanese forces on both sides of the road opened up on the point vehicles. For nearly two hours the small Marine force was caught up in a fire fight and partially cut off from aid. The jungle terrain limited the maneuvering of American tanks and infantry and gave the advantage to well-emplaced enemy field guns and small arms. Eventually, covering fire from Shermans with the point was able to break loose the ambushed force. When the Marines pulled back to RJ 177, they left behind a destroyed half-track and a damaged truck and took with them 15 casualties.[14] Marine tank gunners reported that they had knocked out one Japanese tank, two 75mm guns, and several machine guns.

The wrong-way turn at RJ 177 furnished ample evidence that the Japanese would dispute strongly any attempt to use the road to Liguan. The ambush also effectively killed the idea of a reconnaissance of the roads to Ritidian Point for the time being, as the enemy could be defending them as well. The risk was too great. Lieutenant Colonel Wither's force was disbanded after it reentered American lines and its elements returned to parent units to take part in the general attack on 4 August.

Plenty of action after dark on 3 August underscored the resurgence of enemy activity in the 3d Marine Division zone. At 2200, two Japanese medium tanks roared down the Liguan road, crashed through the defenses set up by 1/9, firing steadily all the while, wheeled to the right at RJ 177, and sped west up the road toward Dededo. As they clattered through the positions held by 1/3, the enemy tanks continued to fire their 57mm guns and machine guns at any target that showed. Despite all the return fire directed at them, the enemy armor escaped. This incident was the dramatic highlight of a series of clashes, which occurred all across the Marine front in the several hours before midnight. Then at 2300, American artillery fire, "placed perfectly"[15] in an enemy assembly area forward of the regimental boundary, was responsible for breaking up a counterattack. After this, Japanese activity died away for the rest of the night.

In contrast with the situation in the 3d Division zone on 3 August, where resistance was steady all day long, the advance of the 77th Division was marked by sporadic clashes with the Japanese. When the 307th and 305th Infantry Regiments moved out from their hard-won foxholes and emplacements at 0730, the enemy units that had fought so doggedly to hold Barrigada the previous afternoon had disappeared. In their stead, scattered through the jungle were lone snipers and small automatic weapons groups

[14] From his examination of the enemy defenses which his regiment later reduced, General Craig concluded that "if the patrol had not pulled back when it did it would have been annihilated." *Craig 22Jun65 ltr.*

[15] *3d MarDiv Jnl*, entry of 2326, 3Aug44.

which were a constant irritant but no real threat to a steady advance. By 0930, the Army regiments had secured Barrigada and with it an all-important well, which could supply the 77th Division with 30,000 gallons of fresh water daily. After a pause to reorganize and regroup, the advance continued at 1330 behind a rolling barrage fired by all four battalions of division artillery. The 307th, with tanks breaking trail, struggled through the jungle, meeting little enemy opposition on its way to the crest of Mt. Barrigada. By 1500, 3/307 had secured the summit, and shortly thereafter it began consolidating positions for night defense.

The difficulties imposed by dense vegetation and a sparse trail network kept down the pace and extent of the advance. In an effort to speed the progress of the 305th Infantry through the lush jungle, Colonel Tanzola narrowed his zone of attack and covered much of the area between Mt. Barrigada and the coast with patrols. Complicating the problems of contact and control, the Japanese fought what the regiment reported was a "good delaying action."[16] They staged a series of ambushes, which forced the Americans to deploy and maneuver against a foe that vanished as often as he stayed to fight.

The heavy opposition encountered at Barrigada on 2 August had caused the left flank units of the 77th Division to fall behind the Marines. Although some of the ground was regained on the 3d, when the 3d Division too was slowed by enemy resistance, at nightfall the corps line still slanted back from RJ 177 to Mt. Barrigada. Despite persistent efforts by the 307th Infantry, the combination of jungle and Japanese had defeated all efforts to make contact. In late afternoon, a tank-infantry patrol that tried to reach Marine lines using the road to Finegayan was stopped by a roadblock and then a barrier of mines, both well covered by enemy fire. One tank was disabled and had to be abandoned and destroyed when the outnumbered patrol withdrew.

This encounter with the Japanese on the Finegayan Road had an unfortunate sequel on 4 August. General Bruce, anxious to re-establish contact with the Marines as soon as possible after the attack opened that day, issued orders for another force of tanks and infantry to push through to the Marine lines. This patrol, spearheaded by Shermans, blasted its way through two roadblocks and opened fire on a third about 1045. This time, however, Company G of 2/9 held the position, not the Japanese; seven Marines were wounded before the company commander succeeded in stopping the fire poured out by the tank guns.[17] Even after this unhappy incident, which was caused by a misunderstanding regarding recognition signals, there was still no contact between Army and Marine front lines. On both sides of the division boundary, assault units had already moved well

[16] *77th InfDiv Jnl*, entry of 3Aug44.

[17] The Army force was told that Marine units would identify themselves with red smoke grenades; the Marines, who were unaware of the significance of this signal, recognized the Army patrol when it began firing and hence held their own fire. *Craig 22Jun65 ltr.*

beyond the Finegayan road into the jungle.

In the 77th Division zone of action, where there were no roads and few trails paralleling the axis of advance, the main struggle on 4 August was with the rugged terrain. Shermans broke trail for the assault platoons of the 305th Infantry, and tank dozers cut roads behind the plodding forward companies. On the northern slopes of Mt. Barrigada, the soldiers of the 307th, cutting their way through the mass of brush, vines, and trees, could make no use of the crushing power of the tanks. Progress was agonizingly slow, despite the absence of any strong Japanese opposition. At noon, General Bruce ordered both assault regiments to concentrate their men in one or two battalion columns in order to speed passage through the jungle. If any mopping up had to be done, reserve units would handle the task. As if to emphasize the need for this decision, General Geiger informed Bruce about an hour later that III Corps was going to have to hold up the advance of the 3d Division until the 77th could come abreast. By 1710, when the 307th reported that all of Mt. Barrigada was within its lines, the forward positions of the two divisions were more closely aligned. Soon afterwards, corps headquarters ordered a vigorous advance all along the front for 5 August.

A factor contributing to General Geiger's order that held up the advance of the 3d Division on the 4th was the stubborn resistance of the Japanese defending the road leading to Liguan and Yigo. Assault units of the 1st Battalion, 9th Marines hammered at the enemy strongpoints but made little headway in the face of interlocking fire from machine guns and cannon hidden in the brush. Again the lay of the land prevented the Marines from outflanking the Japanese or from bringing the full power of supporting weapons to bear. There was only enough maneuver room for about one infantry company to take part in the fight to seize the vital road.

Elsewhere in the 3d Division zone on 4 August, the 3d Marines was able to secure its portion of the day's objective with little trouble. The 2d Battalion stood fast in its positions near the coast, and the 1st Battalion moved along the road through Dededo to seize a fork where the branches continued north in two trails about a mile apart. (See Maps 29 and 30). Both units sent strong patrols forward of their lines to range ahead in the jungle as far as 1,000–1,200 yards. The reconnaissance uncovered a formidable array of abandoned enemy defenses facing toward Tumon Bay, but discovered few Japanese.

On the afternoon of the 4th, in a move calculated to take advantage of the widening division zone of attack, the 21st Marines (less 3/21) reentered the front lines. Elements of the 1st and 2d Battalions replaced the left flank and center companies of 1/9 by 1730 in the area between the Dededo and Liguan roads. While the 3d Division was thus redeploying its units for an attack on a three-regiment front, the 1st Brigade was completing its move to northern Guam. General Shepherd's CP opened near San Antonio at 1200, and the 22d Marines (less 1/22) completed its move into a bivouac area near Tiyan airfield by 1530.

On 5 August, the focal point of battle in the 3d Marine Division zone continued to be the Japanese positions along the Liguan road. Again the 9th Marines bore the brunt of the fighting in jungle so thick that at one point an American tank passed within 15 yards of a Japanese medium and failed to spot it. Throughout the day, small arms fire stemming from many mutually supporting dugouts and trenches whipsawed the Marine riflemen, and well-sited antitank guns slowed the advance of vital supporting armor. The steady attrition of three days' fighting had taken its toll of the enemy, however. By dusk, when a Marine half-track knocked out the last Japanese 75mm gun, 1/9 was firmly in possession of the ground that it had fought so hard to win. On the right of the 1st Battalion, 2/9, which had passed through 3/9 during the morning's attack, was also astride the Liguan road, having fought its way forward through the jungle against moderate resistance.

Neither the 3d nor the 21st Marines faced anything like the organized opposition encountered by the 9th on 5 August. In the center and on the left of the division zone, small groups of the enemy that attempted to halt the advance of the infantry were quickly overrun. When 2/21, moving along the road to RJ 460, was pinned down by automatic weapons fire, a platoon of tanks made short work of the enemy defenders, the crews of two machine guns. It became increasingly apparent during the day's advance that the Japanese did not intend to hold the western sector of the island in any appreciable strength. Reconnaissance by Marine and Army light planes spotting for artillery and naval guns and sightings by carrier planes and the B–25s and P–47s attacking from Saipan pinpointed the Mt. Santa Rosa area as the center of enemy activity.

Although the assault units of the 77th Division found few Japanese during their arduous trek through the jungle on 5 August, there was no doubt that the final enemy bastion lay ahead of the soldiers. The flood of natives that entered American lines, the few prisoners that had been taken, and the supporting evidence of captured documents reinforced the reports of aerial observers. Much of the division effort on the 5th was directed toward moving troops into position to make a concerted drive on the Japanese forces known to be holding Mt. Santa Rosa and its outworks.

Committing the 306th Infantry at 0700, General Bruce ordered it to pass around the right flank of the 307th and attack in the zone formerly assigned to that regiment. Pinched out of the front lines by the advance of the 306th, the 307th Infantry was to replenish supplies and ammunition in preparation for a move to the center of the division zone of action and a drive against Yigo and Mt. Santa Rosa when ordered by division. The 306th, its attack formation a column of battalions, completed much of its planned maneuver on 5 August despite problems posed by the jungle, a lack of useful trails, and the maddening fact that available maps proved to be unreliable guides to terrain. With General Bruce's permission, the regiment held up for the night about 2,000 yards short of the division boundary after 1/306 and 3/306 had both secured

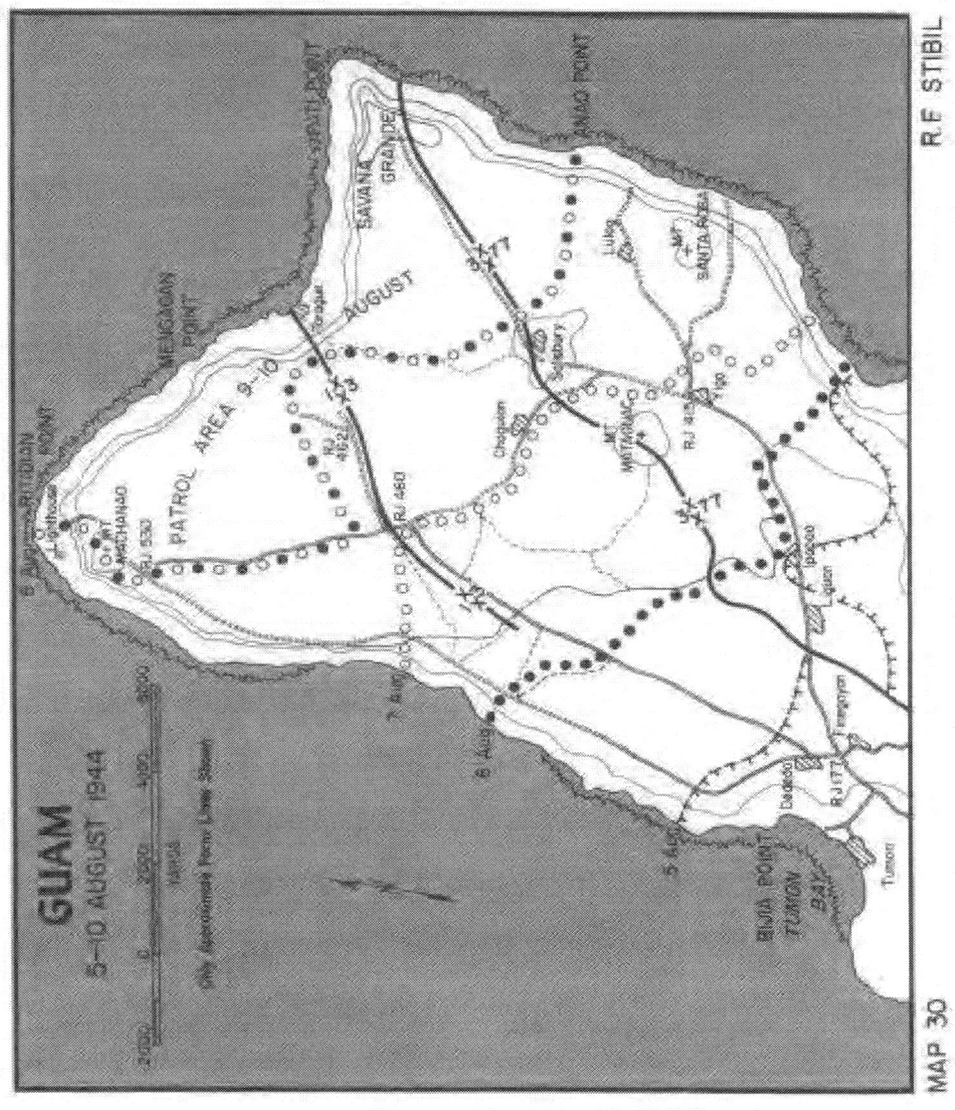
GUAM
5–10 AUGUST 1944
YARDS
PATROL AREA 9–10 AUGUST
RITIDIAN POINT
MENGAGAN POINT
PATI POINT
ANAO POINT
SAVANA GRANDE
SANTA ROSA
MT MATAGUAC
RJ 460
RJ 530
RJ 177
Dededo
Tumon
BIJIA POINT
TUMON BAY
5 Aug
6 Aug
7 Aug
8 Aug
MAP 30
R.F. STIBIL

portions of the road to Yigo near the village of Ipapao. (See Map 30.)

For the 305th Infantry, the pattern of attack on 5 August had a monotonous sameness with the actions of the previous day. Deeply enmeshed in the jungle, the two assault battalions of the regiment hacked their way forward behind trail-breaking tanks and half-tracks. Direction was maintained by compass bearings, and when 2/305, in the lead, reached what it thought was the day's objective about 1400, it had to cut a trail to the sea in order to verify its position. The 1st Battalion of the 305th set up about 1,000 yards to the rear of the 2d, and both units, lashed by a driving rain in the forest gloom, dug in as best they could for all-around defense. In the area occupied by 1/305, the coral subsurface was only six to nine inches below ground level; no satisfactory weapons emplacements or foxholes could be constructed.[18]

At 0200 on 6 August, two Japanese medium tanks, accompanied by a platoon of infantry, came clattering out of the darkness and attacked south down the trail that led into the 1/305 perimeter. A swift outpouring of small arms fire scattered the enemy riflemen, but the tanks ignored the bullets and broke through the American defenses, continuing down the trail and firing at targets on both sides. Much of the heavy return fire ricocheted off the armor and added to the lethal shower of lead and steel that lashed the surrounding brush. After one of the tanks collided with a Sherman parked on the trail, then backed off and crushed a jeep, both enemy vehicles turned and raced back the way they had come, firing steadily all the while. Behind them the Japanese tankers left 15 Americans dead and 46 wounded; many of the men were casualties because they had been unable to dig in and therefore lacked effective cover from both tank and antitank fire.

The grim saga of the Japanese tanks was not over when they broke out of the 1/305 perimeter. At 0630, scouts of 2/305 ran into them, too. In search of a better route of advance, the Army battalion was retracing its steps along the trail it had blazed on 5 August. Cannon and machine gun fire from the two tanks, which were protected by a small rise of ground, quickly swept the path clear of infantrymen. American mediums worked forward along the narrow and congested trail to join the fight, but the advantage was with the enemy armor in hull defilade. Tree bursts from the Japanese tank shells scattered deadly fragments about, pinning the American infantry to the ground. Eventually, an 81mm mortar crew was able to get its tube in action, find a clear path through the trees for its line of fire, and lob shells into the Japanese position. This silenced the enemy armor, and assault units that outflanked the tanks and came upon them from the rear found both abandoned. Three bodies were the only evidence of the defenders' strength. The cost to 2/305 of the sharp and unexpected clash was 4 dead and at least 14 wounded.

These two tank-infantry battles were the highlights of 77th Division action on 6 August. Enemy opposition was

[18] Col James E. Landrum, Jr., USA, ltr to CMC, dtd 22Oct52.

light and scattered otherwise, and all units spent the daylight hours getting into position for the attack on Mt. Santa Rosa. The basic scheme of maneuver planned by General Bruce's staff called for the 306th Infantry to make a wide sweep on the left of the division zone, advancing as rapidly as possible in column to reach the vicinity of Lulog village north of the mountain. The regiment would patrol to the division boundary to keep contact with the Marines. The 307th Infantry, with most of the 706th Tank Battalion attached, would make the main effort, attacking to seize Yigo and RJ 415, before turning eastward to take the mountain itself. The 305th Infantry (less 3/305 in corps reserve) would continue its attack toward the mountain with the objective of seizing the high ground south of it.

A map overlay outlining this operation plan and projecting a new division boundary beyond Liguan was distributed to all 77th Division units on 5 August. On the same date, the corps operation plan, incorporating the basic scheme proposed by the 77th Division, was also distributed. There was no overlay with the corps plan, but its language clearly stated that there was a change in the division boundary, and gave its map coordinates. Unfortunately, the new boundary and that shown on the 77th Division overlay did not coincide from the vicinity of Mt. Mataguac to the village of Salisbury. The result was that "the zone of action of the 306th Infantry, making its sweep around Mt. Santa Rosa on the division left was . . . partially within the 3d Marine Division's operational area." [19]

As part of its move into attack position, 1/306 closed to the division boundary northwest of Ipapao during 6 August and made contact with 2/9. The 3d Battalion of the 306th advanced to within 2,000 yards of Yigo, securing a large section of the road, which was to be the axis of the attack by the 307th Infantry. Behind the assault units of the 306th, the road was cleared of the enemy back to RJ 177. General Geiger authorized the 77th Division to use those portions within the zone of the 3d Division in order to move supplies and equipment from Barrigada dumps to the new forward area.

In late afternoon, while he was reconnoitering the site of a new CP near Ipapao, the 77th Division chief of staff, Colonel Douglas C. McNair, was killed by a sniper.[20] This incident grimly demonstrated the ease with which individuals and small groups of the enemy could avoid detection by mop-up forces.

The fact that many Japanese could hide out in areas supposedly secured by Army and Marine units was in part a result of enemy defensive tactics and American measures taken to combat them. Since *29th Division* forces were concentrated along key roads and trails, 3d and 77th Division attack plans were adapted to meet this situation. As a result, the approach of the 77th to Mt. Santa Rosa was made by means of a few strong battalion and

[19] Crowl, *Marianas Campaign*, p. 422.

[20] Colonel McNair's father, Lieutenant General Leslie J. McNair, had been killed 12 days before by a misdirected American bomb, while he was observing an infantry battalion attack in France.

regimental columns, which smashed their way north through the jungle whenever trails were not available. On 6 August at 0900, General Turnage issued orders for the 3d Division to advance in column along the roads and trails leading north, patrolling and mopping up for 200 yards on either side in dense vegetation, and to the edge of first growth in more open country. As was the case in the Army zone of action, contact between Marine assault units would be made at designated objectives, usually lateral trails or road junctions. The attack formations ordered by Generals Bruce and Turnage and approved by General Geiger were designed to keep maximum pressure on the Japanese and to deny the enemy any more time to build up his defenses.

Early on 6 August, before the 3d Division altered its attack formation for a more rapid advance, the assault regiments made local attacks to straighten the front lines and reach a predesignated line of departure. The 9th Marines killed the few Japanese that were still alive in the ruins of the defenses along the Liguan road. On the division boundary, a tank-infantry patrol of Company G, 2/9 moved out to destroy an enemy roadblock on a trail leading into the 77th Division zone. Part of the defending force was a Japanese tank, which scored three hits on a Marine medium before being knocked out by return fire. Enemy infantry fled the roadblock and was hunted down by the Marine force, which killed 15 men before it turned back after reaching a point about 1,000 yards inside the Army zone.

At 1045, the 3d Division moved out all across the front in what was essentially a series of parallel battalion columns. No longer deployed in skirmish line, the Marine units made great strides forward against minimal opposition. The 3d Marines, with 3/3 in assault, moved ahead 5,000 yards, along the road to Ritidian Point. When Major Bastian's lead units reached the day's objective, the 2d Battalion came up and extended to the left, while 3/3 moved to the right to contact the 21st Marines. After the new front line was occupied, 1/3 relieved the 3d Battalion in position so that 3/3 could shift to the right and pass through 2/21 on 7 August. This realignment was a preliminary maneuver to the entry of the 1st Brigade into the attack to the north.

The 21st Marines, like the 3d, shifted to a column of battalions when the new attack formation was ordered. With 2/21 in the van, the regiment moved 4,000 yards and reached its objective, a trail junction on the road to RJ 460 by 1300. Then the 1st Battalion moved up and extended to the right in rugged jungle terrain, while 2/21 contacted the 3d Marines to the left. The 9th Marines, with 1/9 preceding 2/9, followed a small trail that was the trace of the division boundary to the point where the boundary veered sharply northeast toward the coast between Pati Point and Anao Point. At this turning, 1/9 moved to the northwest along a trail that angled in the direction of the positions held by the 21st Marines. The battalion set up for the night without having made contact with 1/21. For the first time in three days, however, the right flank battalion of the 3d Division (2/9) was in visual contact with the left flank battalion

(1/306) of the 77th Division when the frontline units established their night defensive perimeters.

During the afternoon's advance, corps headquarters passed the word that its operation plan for the attack against Mt. Santa Rosa would be effective at 0730, 7 August. General Shepherd was notified that his brigade would pass through the positions held by 2/3 and 1/3 and assume responsibility for an attack zone that included the western part of the island and the northern end from Ritidian Point to the village of Tarague. (See Map 30.) Shepherd alerted the 4th Marines to make the relief of the 3d Marines battalions and to move out in assault the following morning.

In its narrowed zone of action in the center of the island, the 3d Marine Division was directed to continue its attack and to assist the 77th Division, which would be making the principal corps effort to destroy the remaining Japanese. Priority of fires of corps artillery and naval support ships was given the Army division. Targets assigned for morning strikes by Seventh Air Force planes were all picked with the aim of softening up the defenses of the key Santa Rosa heights. The heavy bombing and shelling of areas behind the enemy lines in northern Guam had been going on for days. As one Japanese survivor recalled the period, the bombardment was nerve-wracking and destructive, and often seemed all too thorough to the individual, since American aircraft:

> . . . seeking our units during daylight hours in the forest, bombed and strafed even a single soldier. During the night, the enemy naval units attempting to cut our communications were shelling our position from all points of the perimeter of the island, thus impeding our operation activities to a great extent.[21]

THE FINAL DRIVE[22]

Before the 77th Division launched its final drive on 7 August, assault units of the 306th and 307th Infantry advanced to occupy a line of departure closer to the attack objectives. Twenty P–47s from Saipan strafed and bombed Mt. Santa Rosa as the infantrymen were moving out. The 306th Infantry plunged into the jungle, following trails that would skirt Yigo on the west, while the 307th guided on the road to the village and reached the last control point on its approach march by 0900. As the leading company of 3/307 was nearing this area, 600 yards from the road junction at Yigo, its men were harassed by small arms fire. The Americans deployed and poured a heavy volume of return fire into the thick brush ahead. Within an hour all opposition had faded away.

Once all units were in position on the designated line of departure, General Bruce issued orders for the general attack to begin at noon. In preparation, 10 B–25s roared in over the mountain dropping 120 100-pound bombs on the south slopes and firing 75 rounds at Japanese positions from nose-mounted

[21] *Takeda ltr I.*

[22] Unless otherwise noted, the material in this section is derived from: *IIIAC SAR; 3d MarDiv SAR; 3d MarDiv Jnl; 77th InfDiv OpRpt; 77th InfDiv Jnl; 1st ProvMarBrig SAR; 1st ProvMarBrig Jnl; GSDF Study; Takeda ltr I; Takeda ltr II.*

75mm cannon.[23] For an hour before H-Hour, support ships pounded the heights and possible enemy assembly areas, and then in the final 20 minutes before jump-off, seven battalions of artillery fired a preparation on defenses in the vicinity of Yigo. As the fire lifted on schedule and the assault troops warily advanced, supporting tanks had not yet made their way through the barrier of troops, trucks, and jeeps on the narrow, crowded road. At 1215, the light tanks caught up with the leading elements of 3/307 about 400 yards from the village and passed through the infantry front lines.

Overrunning and crushing several enemy machine gun positions, the lights topped a small rise where the ground was sparsely covered with brush. A seeming hurricane of enemy fire struck the armor from hidden positions ahead, and a radio call for help went out to the mediums. When the heavier tanks came up, a raging duel of armor and antitank guns ensued. With their freedom of action hampered by the jungle, the tanks were channeled into the fire lanes of enemy guns. Two lights were knocked out, one medium was destroyed and another damaged, and 15 tank crewmen were casualties before the short, furious battle was over. Infantrymen that tried to outflank the Japanese strongpoint by moving through the jungle, which crowded the road, were driven to cover by deadly and accurate machine gun fire.

Suddenly, the fight ended almost as quickly as it began, when the enemy force, 100–200 men, was soundly beaten by elements of 3/306. Moving from his position on the left flank toward the sound of the firing, and keying his location to the distinctive chatter of the enemy machine guns, the battalion commander, Lieutenant Colonel Gordon T. Kimbrell, led a platoon of Company K through the jungle and rushed the Japanese position from the rear. Surprise was complete and the defenders were killed or routed. Other elements of 3/306 wiped out enemy infantry holding out closer to the village road junction. With the welcome aid of this flanking attack, which accounted for 105 Japanese, the 307th and its supporting tanks were able to sweep through the shell-pocked ruins of Yigo. As the 307th turned toward the mountain, 3/306 moved out up the road toward Salisbury. (See Map 30.)

The fighting near RJ 415 did not end until midafternoon, and when the two assault battalions of the 307th had moved into position to attack east, the day was already spent. On General Bruce's orders, the 307th dug in about a half mile beyond Yigo and made preparations to renew the attack at 0730 on the 8th.

On the right of the division zone of action, the 305th Infantry spent another hard day cutting its way through the trackless jungle toward the mountain. Enemy opposition to both assault battalion columns was light, but the rate of advance was maddeningly slowed by the difficult terrain. The troops ended the day close enough to their objective, however, to get caught in the fringe of an afternoon bombing attack; 2/305 suffered several casual-

[23] AAF in the Marianas Campaign, Operation Forager, Mar–Aug44, listing of 48th BombSqn(M) sorties. (USAF 105.1–3, USAF Archives, Maxwell AFB, Ala.)

ties from a misdirected bomb. The 3d Battalion of the 307th, strafed at about the same time, luckily escaped injury.

In an unexpected twist of fate, 3/305, in division reserve but under corps control, had one of the day's hardest fights. It was ordered to clean out the area near the new 77th Division CP, where Colonel McNair had been killed on the 6th. A platoon uncovered an enemy strongpoint deftly hidden in the jungle about 500 yards from the headquarters camp. A fierce fire fight broke out. Elements of two rifle companies and a platoon of mediums were called up to surround the Japanese, estimated at company strength. Six hours of desperate close-quarter fighting followed before the defenders were wiped out at a cost of 12 Americans killed and 21 wounded.[24]

The column of 1/306, which advanced on the division left on 7 August, made good progress after the noon jump-off time. When its leading platoons reached a trail junction near the division boundary about 1500, they ran into a strongpoint built around two machine guns and manned by 40–50 Japanese. The fight to eliminate this opposition took much of the rest of the afternoon, with the result that the 1st Battalion set up for the night just on the edge of the area that corps maps showed as part of the 3d Marine Division territory. If 1/306 had continued its advance, it would have encountered elements of the 9th Marines.

The 3d Division assault troops met little enemy resistance on 7 August. With bulldozers and tanks breaking trail where none existed and attack formations narrowed to battalion columns, all three regiments reached and secured their objectives by midafternoon. They held up along a trail from RJ 460 to the boundary near the village of Chaguian, generally 5,000 yards forward of their line of departure.

On the right, 1/9 and 2/9 moved out in attack with the 1st Battalion in the fore. A few isolated enemy stragglers were killed, and signs of fresh tank tracks were found by patrols that scouted toward the 21st Marines, advancing in the center of the division zone. The 21st, with 3/21 in the van, found no fresh evidence of the enemy as it struggled forward along the scant tract of meandering trails. On the far left, 3/3 led the 3d Marines attack along the road to RJ 460. At 0850, a few enemy artillery shells exploded among the advancing troops but with little effect. About two hours later, the Marines discovered the source of this fire when the tank-infantry point found a 75mm gun posted to hold a roadblock. After a brief flurry of fire, the defenders fled to the north and the advance continued at a good pace. When it dug in for the night, the regiment was in contact with the 4th Marines on the left and the 21st Marines on the right. The favorable reports by the 3d Marines of the day's action added to an already optimistic picture at the 3d Division CP. General Turnage ordered all assault units to continue their advance to the sea in the morning.

The situation in the 1st Brigade zone proved equally promising after the results of the advance on 7 August were evaluated. When the 4th Marines attacked along the roads to Ritidian

[24] 305th Inf AAR, 18Jun–9Aug44 (WW II RecsDiv, FRC, Alexandria, Va.)

ARMY TANKS *hit and aflame during the attack the Yigo village road junction near Mt. Santa Rosa. (USMC 92083)*

JUNGLE FOLIAGE *almost hides a Marine patrol from view as it nears Tarague on 9 August. (USMC 93894)*

Point, its progress was so rapid that General Shepherd alerted the 22d Marines to move forward behind the 4th, ready to join the assault as the zone widened to the north. At a trail junction about 2,000 yards short of RJ 460, Company L of 3/4, was fired upon by an enemy 75mm gun, which wounded the company commander and two men. A supporting platoon of mediums quickly demolished the gun and a mortar position nearby, and blew apart the roadblock they had covered. Inexplicably, the Japanese gunner had fired three ineffectual rounds of high explosive at the tanks, although over 100 armor-piercing projectiles lay nearby.[25] Aside from this brief encounter, little opposition developed. The 22d Marines reached a position behind 1/4 on the left of the brigade zone in late afternoon, ready to move into the attack on order.

The capstone to the good news of 7 August was furnished by planes of MAG–21.[26] During the day VMF–225 began flying routine combat air patrols from Orote, relieving Navy planes of this responsibility. At the same time, the Seventh Air Force command on Saipan was notified that Marine night fighters would take over all night air patrol duties. Although the Marine Corsairs and Hellcats were not slated to provide close support for ground troops, they could be called upon in that role if needed. With its own air defense garrison in operation, Guam was a long step forward in its development as a major Allied base for further moves against Japan.

Despite the cheering events of 7 August, none of the American commanders had any idea that the fight for the island was over. On the night of 7–8 August, it was the Japanese tanks, as it had been so often in the past few days, that added a fresh reminder of the enemy spirit. Harried by air attacks, artillery, and naval guns, the Japanese could not move any armor in daylight along major roads and trails, but at night, after the flock of Marine and Army observation planes had landed, the tanks could shift into attack position.

About 0300 on 8 August, the soldiers holding the northern sector of the 3/306 perimeter heard tanks rumbling down the road from Salisbury toward them. Three Japanese mediums with an undetermined force of accompanying infantry loomed out of the darkness, all guns blazing away. Alerted by the unmistakable clatter, the men of the 306th were ready and replied to the attack with every weapon they could muster. The enemy infantry was quickly driven off, one tank was knocked out by a rifle grenade and a second was stalled by heavy machine gun fire. The remaining medium abandoned the fight and towed the cripple away. Morning's light showed the Japanese losses to have been 18 men, including 3 officers, and the cost to 3/306 for holding its ground, 6 men dead and 18 wounded.

Unshaken by this attack, the 3d Battalion led the advance of the 306th on 8 August, heading cross-country by a narrow trail for Lulog. The few Jap-

[25] *6th TankBn SAR*, entry of 7Aug44.

[26] On 4 August, the night fighters of VMF(N)–534 had led the flight echelons of VMF–216, –217, and –225 from the CVE *Santee* into Orote airfield. Sherrod, *Marine Air History*, p. 253.

anese encountered appeared to be dazed and shocked by the downpour of bombs and shells that had preceded the attack. By 1040, the battalion reached the village, and patrols headed for the coast on General Bruce's order. In view of the slight opposition, the 77th Division commander had revised his plan for encircling Mt. Santa Rosa to include not only the movement of 1/306 along what was believed to be the division boundary to a blocking position at Salisbury, but also the advance of 2/306 through that village and on to the coast near Pati Point. As the 2d Battalion, following its orders, approached Salisbury, Marine artillery shells hit along the column and wounded several soldiers, an unfortunate incident attributable to the confusion of boundary overlays in the hands of the two divisions. The swift protest of the violation of his supposed zone of action by Colonel Aubrey D. Smith of the 306th brought an equally prompt and sure reply from the neighboring 9th Marines. This exchange led to the discovery of the cause of the boundary confusion and its resolution by Colonels Smith and Craig.

Before the 306th proceeded further with General Bruce's plan, the division commander saw a POW report—which later proved false—placing 3,000 Japanese in the area just north of Mt. Santa Rosa. A cautious view of this intelligence prompted Bruce to order 1/306 to close on the 3d Battalion at Lulog and 2/306 to stand fast in reserve 1,200 yards northeast of Yigo, ready to close any gap between the 306th and 307th Infantry.

A strong reason for believing that major Japanese forces were located in the zone of the 306th lay in the results of the 8 August attack by the other regiments of the 77th Division. No significant enemy opposition was developed by the 305th Infantry as it neared its objective; the 307th eliminated 35 bombardment-dazed Japanese on the lower slopes of Mt. Santa Rosa, but found no one manning defenses on the bare upper reaches. Patrols to the sea by the 305th and 307th uncovered few signs of the enemy.

Under the circumstances, Colonel Smith ordered 2/306 to fill a gap between the 307th on the mountain and the rest of his regiment dug in near Lulog. In the course of this move, designed to block possible Japanese escape routes through dense jungle, elements of the 306th and 307th mistook each other for the enemy troops and exchanged artillery and tank fire. Ten casualties were incurred before the mistake was discovered. After this unfortunate mishap, the night was quiet.

Less than 600 enemy dead had been counted in the two-day fight for Yigo and Mt. Santa Rosa. Since intelligence officers had estimated a tentative garrison strength of 1,500 soldiers, 1,000 sailors, and 2,500 laborers in the 77th Division zone, it seemed probable that many Japanese had slipped away into the jungle. Although all 7 of the enemy artillery pieces thought to be part of the defense had been accounted for, only 5 of the 13 tanks reported in the vicinity had been knocked out.

Some enemy elements, which might have been units fleeing the Mt. Santa Rosa action, cropped up in the zone of the 9th Marines on 8 August as 3/9, moving along the trail from Salisbury to the coast, met and overcame successive small pockets of resistance. By the time orders were passed to dig in, Major Bastian's men had reached a point about 800–1,000 yards beyond Salisbury. At that village, 2/9, which was following the path taken by the 3d Battalion, held up and established a strong blocking position. The 1st Battalion, in reserve, patrolled in the vicinity of Mt. Mataguac and killed 25 Japanese in scattered encounters. Colonel Craig, whose CP was located in the 1/9 patrol area close to the division boundary, notified the nearest Army unit that his men had sighted considerable enemy activity near a brush-covered hill just within the Army zone of action.[27] Available intelligence indicated that the enemy headquarters might well be located in this area.

In the center of the 3d Division zone, the 21st Marines passed into reserve at the start of the 8 August attack. Redrawn regimental boundaries pinched the 21st out of the front line, but gave it a large triangular patrol area, about 3,000 yards along each leg, to clear of Japanese. One patrol of the many threading their ways through the jungle discovered a truck, which contained the bodies of 30 Guamanians, who had been beheaded; in the same area, near Chaguian, 21 more bodies of natives, who had been as brutally murdered, were found the next day. Subsequent intensive investigation revealed that these victims had been impressed at the concentration camp near Yona to work on the defenses at Yigo. These gruesome discoveries spurred the Marines to a grim determination in their task of hunting down and eliminating the Japanese.

Although there were signs of recent Japanese activity throughout the jungled interior, particularly along the trails, relatively few enemy were found by the 3d Marines moving northeast on the left flank of the division. The 3d Battalion, which could follow a trail along the boundary, was able to make rapid progress. It reached RJ 460 and moved 1,500 yards further to the northeast before holding up for night defense. Patrols found their way to the cliffs overlooking the sea before returning to the perimeter for the night.

The 2d Battalion was not so fortunate as the 3d, for the trail it followed in the morning attack soon ended in a wall of jungle. A second trail which was supposed to intersect the first, a narrow pathway leading from Salisbury to the coastal village of Tarague, proved to be 1,300 yards away through the brush. Major Culpepper had no choice but to plunge ahead into the tangle, with relays of men cutting their way through the mass of vegetation, in order to reach his objective. All heavy weapons were left behind to come up with the bulldozers and tanks that followed the trace of the infantry column, building a wider trail, which could be used by trucks and jeeps. When 2/3 broke through to the Salisbury-Tarague trail, a patrol headed south to contact the 9th Marines. Not far from the new trail junction, an enemy blocking force was encountered and a fire fight broke out. When the last shots died away, 19 dead Japanese were found in the remnants of the enemy position, but it was too late to continue any further south. The patrol retraced its steps

[27] *Craig 22Jun65 ltr.*

and rejoined the battalion, which had moved north on the trail toward the coast. On Colonel Stuart's order, 2/3, still minus its supporting weapons, dug in along the trail at a point roughly two miles north of Salisbury. (See Map 31.)

Helped by good trails that paralleled its direction of attack, the 1st Brigade reached the northern tip of Guam on 8 August. On General Shepherd's order, the 22d Marines moved into the line on the left at the start of the morning's attack, relieving elements of 1/4. As the battalion columns advanced in approach march formation, there was little enemy resistance. Shepherd ordered 2/22 to send a patrol to Ritidian Point lighthouse, where air observers had reported Japanese activity. Company F drew the mission and advanced rapidly while carrier aircraft hit each successive road and trail junction to the front. By 1500, the company had reached Ritidian Point and had begun to work a patrol down a twisting cliff trail to the beach. A small force of Japanese tried to ambush the Marines but was easily eliminated. Following Company F, the remainder of 2/22 set up a defensive perimeter near Mt. Machanao. The 3d Battalion dug in on the road about halfway between RJ 530 and RJ 460.

The 4th Marines, experiencing little difficulty in seizing the day's objectives, set up night defenses in a series of perimeters, which stretched from the position held by 3/22 back down the road to RJ 460 and thence along the trail to Tarague as far as the defenses established by 3/3. Vigorous patrolling during the day had located few Japanese in the brigade zone of action. That night in a surprisingly honest broadcast, that might almost have been a IIIAC situation report, Radio Tokyo announced that American forces had seized 90 percent of Guam and were patrolling the remaining area still held by the Japanese.

Emphasizing the closeness of the end on Guam, IIIAC had placed restrictions on the use of supporting fires since 7 August. On that day, corps headquarters cancelled all deep support naval gunfire missions except those specifically requested by the brigade and divisions. Those headquarters could continue call fire on point and area targets, but had to coordinate closely and control each mission precisely. The last strike by Saipan-based P–47s was flown on the afternoon of the 7th. B–25s made their final bombing and strafing runs on Ritidian Point targets the next morning.[28] After the strikes in support of 2/22 approaching the northern tip of the island, carrier planes were placed on standby for possible supporting strikes but never called. For the last stages of the campaign, artillery was the primary supporting weapon, and battalions of brigade and division howitzers displaced forward on the 8th in order to reach firing positions that would cover the stretches of jungle that remained

[28] During 24 missions flown against targets on Guam between 3 and 8 August, Seventh Air Force squadrons lost one plane, a B–25, which crashed in the jungle on 5 August, killing the six-man crew and one observer. It is not known whether the ship was shot down or had an operational failure. AAF in the Marianas, *op. cit.;* 48th BombSqn(M) OrgHist, 1–31Aug44, pp. 1–2. (USAF Archives, Maxwell AFB, Ala.)

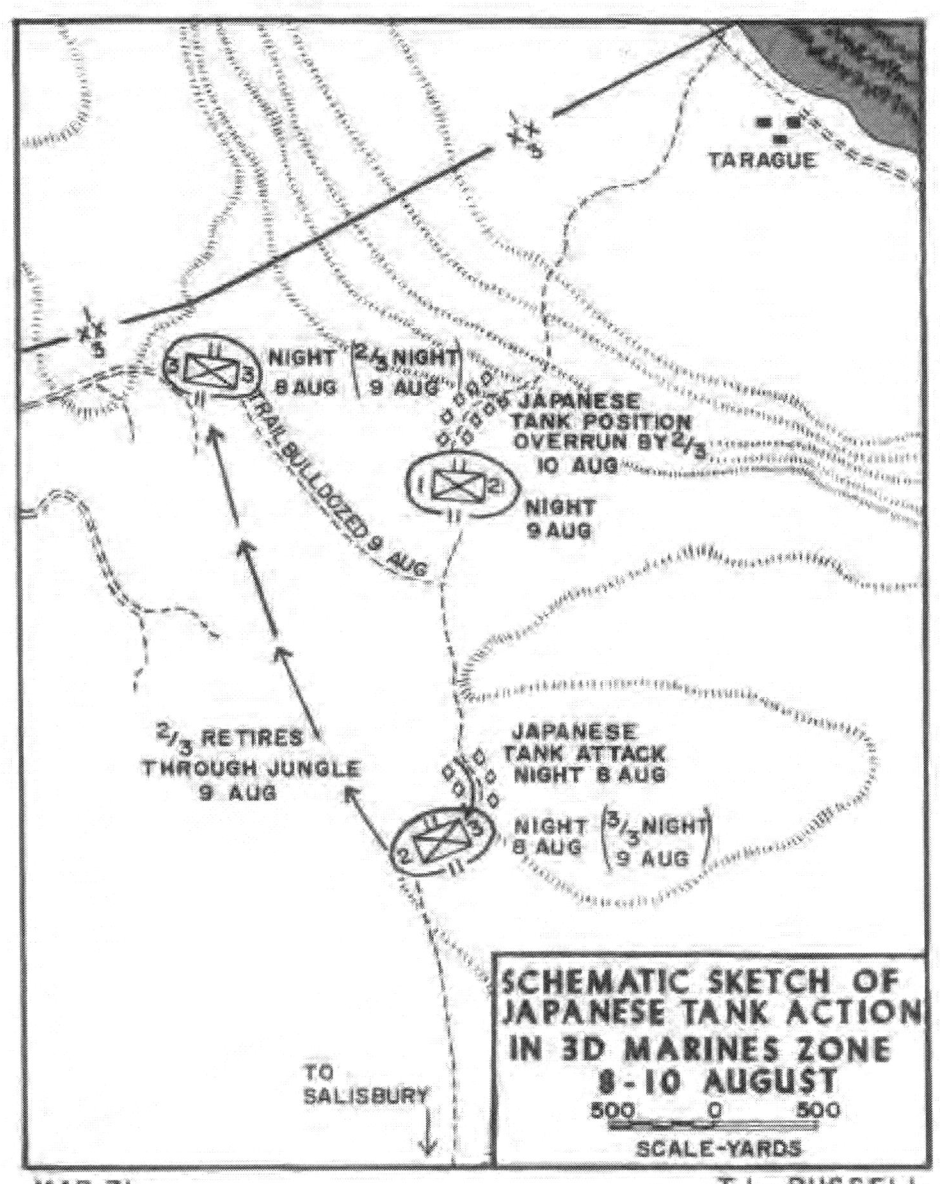
TARAGUE
XX 3
XX 5
NIGHT 8 AUG
(2/3 NIGHT 9 AUG)
JAPANESE TANK POSITION OVERRUN BY 2/3 10 AUG
TRAIL BULLDOZED 9 AUG
NIGHT 9 AUG
2/3 RETIRES THROUGH JUNGLE 9 AUG
JAPANESE TANK ATTACK NIGHT 8 AUG
NIGHT 8 AUG
(3/3 NIGHT 9 AUG)
TO SALISBURY
SCHEMATIC SKETCH OF JAPANESE TANK ACTION IN 3D MARINES ZONE 8-10 AUGUST
500 0 500
SCALE-YARDS
T. L. RUSSELL

MAP 31

in Japanese hands.

On the night of 8/9 August, the center of action was the position occupied by 2/3 on the Salisbury-Tarague trail. At 0130, enemy mortar fire crashed down in the perimeter, heralding a tank-infantry attack launched from the direction of Tarague. The Marines immediately took cover off the trail and opened fire with every weapon they had. The fury of defending fire succeeded in annihilating the Japanese riflemen. The tanks continued firing and edged forward without infantry support when bazooka rockets and antitank rifle grenades, both in poor condition from exposure to the frequent rains, proved ineffective against the Japanese armor.

At 0300, when three enemy mediums had advanced into the midst of his position, Major Culpepper ordered his company commanders to pull their men back into the jungle and to reassemble and reorganize in the woods behind his CP.[29] Miraculously, a head count taken 45 minutes later when the companies had found their ways through the dark jungle showed that there were no American casualties despite the prolonged firefight. Culpepper radioed Colonel Stuart of his actions, and as dawn broke, 2/3 struck out cross-country, cutting a trail toward the positions held by 3/3. (See Map 31.)

As his 2d Battalion fought its way through the jungle, Colonel Stuart, whose CP was located in the 3/3 perimeter, bent every effort toward getting heavy weapons onto the trail where the Japanese tanks were last reported. Bulldozers plowed their heavy blades through the thick growth and tanks followed, crushing or knocking down all but the biggest trees. By noon, a rugged track usable by tanks and antitank guns was cut through to the Salisbury-Tarague trail. Leaving a blocking force at this junction, 3/3 moved south with tank support toward the scene of the night's action. The enemy mediums had disappeared, however, and 3/3 set up where 2/3 had dug in the night before. At 1500, 1/21, which had been attached to the 3d Marines the night before, was ordered to move up to the trail and advance toward Tarague. When the battalion received the word to set up for the night, it was 1,500 yards from the coast. At the other end of the trail, 3/21, operating under regimental control, set up a blocking position at Salisbury. Completing the picture of a day of maneuvering to trap the enemy tanks, 2/3 reached the division boundary road after a hard trek through the jungle and established a night perimeter where 3/3 had been located on 8 August. (See Map 31.)

While the 3d Marines concentrated its efforts on destroying the Japanese armor, the 9th Marines advanced to Pati Point. With 2/9 leading, followed by 1/9, the regiment attacked along a trail on the division boundary and patrolled every intersecting path. The 3d Battalion in reserve sought the Japanese as aggressively as the assault units. About 1030, one of 3/9 patrols fought the day's major action when it discovered a trailblock built around a light tank and two trucks. A sharp, brief battle eliminated all opposition and accounted for 18 Japanese.

Colonel Craig was directed to hold up

[29] *2/3 Jnl*, entry of 8Aug44.

the advance of 2/9 in midmorning,[30] when a native reported that 2,000–3,000 Japanese were located in the vicinity of Savana Grande, a large, fairly open stretch of dwarf jungle growth, coconut trees, and tall grasses. Division artillery poured 2,280 rounds of 75mm and 105mm shells into the area, and the 7th 155mm Gun Battalion, the only Corps Artillery unit that could reach the target, added 1,000 of its heavier rounds in 2½ hours of concentrated firing. When the 9th Marines moved out in attack again, this time with the 1st and 2d Battalions abreast, it found few signs of the enemy. Those Japanese encountered were quickly killed. Patrols from both battalions reached the cliffs along the coast by 1800 and scouted Pati Point and Savana Grande without results. When Colonel Craig ordered all units to set up night defenses, the regiment had made contact with elements of the 306th Infantry patrolling to the south.

General Bruce was satisfied that his zone was free of organized Japanese opposition on 9 August. Feeling that there was no necessity for three regiments in the forward area, he ordered the 305th Infantry (less 3/305 in corps reserve) back to Barrigada. This action lessened the strain on supply agencies, which were forced to use the road through Finegayan in the 3d Division zone or the rugged bulldozer trail that marked the route of advance of the 305th. The 307th Infantry sent out patrols from its positions on Mt. Santa Rosa, but the bulk of the task of eliminating the remaining Japanese was given to the 306th. While there was little evidence of organized resistance, there were frequent clashes with small groups of Japanese throughout the jungle. Of particular interest were several ambushes that patrols from the 77th Reconnaissance Troop ran into when they approached Mt. Mataguac from the east and fire that 1/306 encountered moving in from the west. For the most part, the enemy troops there stayed hidden in caves and waited.

The northernmost part of Guam, that most lightly defended by the Japanese, was aggressively patrolled on 9 August. Brigade units met scattered opposition from small groups and snipers, but it was insignificant and easily overcome. The 2d Battalion, 22d Marines climbed down the cliffs at Ritidian Point and searched for the enemy along the beaches. The 4th Marines reached the north coast at Mengagan Point, sent patrols to contact the 22d along the cliffs and beaches, and scouted toward Tarague in the other direction. The patrol reports reaching General Shepherd from both assault and reserve units, and a personal visit to the regimental CPs in late afternoon, convinced him that his area of responsibility was secure. At 1800, he announced that all organized resistance had ceased in the brigade zone of action.

On the evening of 9 August, it appeared that the last strong pocket of Japanese lay in the 3d Marine Division zone. Plans were laid to make sure that it was eliminated early on the 10th. Once the troublesome Japanese tanks were accounted for, General Geiger was ready to declare the island secure, an announcement which would happily coincide with a visit of Ad-

[30] *Craig 22June65 ltr.*

miral Nimitz to inspect Guam and to discuss future operations.[31]

All night long howitzers of the 12th Marines fired on the probable tank bivouac area along the trail to Tarague and other paths leading into Marine defenses. Colonel Stuart gave 2/3 the job of tangling with the Japanese armor, this time with the support of the heavy weapons it had lacked on the night of 8–9 August. At 0730, Major Culpepper's assault platoons passed through the positions held by 1/21 and attacked toward the sea. About 400 yards up the trail, two enemy medium tanks opened fire on the point. A platoon of Shermans moving with the infantrymen returned the fire and destroyed the tanks. The advance continued past the pair of smoking hulks, which proved to be the last operational enemy tanks on Guam. Seven more enemy mediums, abandoned, were captured by 1030.[32] The Marines continued on to the coast against negligible opposition with 1/21 trailing 2/3. When the cliffs were reached in the afternoon, patrols from both units worked their way down to the beach, where they quickly disposed of about a platoon of the enemy.

At 1131, shortly after receiving reports that the Japanese tanks were accounted for, General Geiger announced that all organized resistance on Guam had ended. At 1423, a IIIAC operation order was issued, effective on receipt, establishing an enemy straggler line across the island from Fadian Point to the coast just north of Tumon Bay. All Japanese were to be contained north of that line and the brigade and the two divisions were directed to continue mopping up action in their respective zones against "numerous enemy remnants" estimated to be present in northern Guam.[33] The 77th Division was given a special mission of maintaining three motorized patrols ready on an hour's notice to reduce points of resistance south of the straggler line.

One final battle remained to be fought on Guam. On 10 August, when patrols of 1/306 checked the Japanese strongpoint near Mt. Mataguac first spotted by the 9th Marines, they encountered heavy fire from hidden cave positions. An earlier probing attack by Army troops, supported by mortars set up in Colonel Craig's CP area,[34] had been unsuccessful in penetrating the Japanese defenses. In a full-scale bat-

[31] Unit journals show that all corps commanding officers were well aware of the pending visit of CinCPac and his official party, as typified by General Geiger's message to General Bruce: "Admiral Nimitz and other officials arrive from Pearl Harbor 10 August. Push all Japs from Guam." *77th InfDiv Jnl*, entry of 9Aug44.

[32] The Commanding Officer, *24th Tank Company* reported that these tanks were "scuttled" because of lack of fuel or mechanical failure. 3d MarDiv POW Interrogation Rpt No. 396, Capt Hideo Sato, IJA, dtd 7Oct44.

[33] *3d MarDiv Jnl*, entry of 10Aug44.

[34] *Craig 22Jun65 ltr.* General Craig, vividly recalling the details of this action, noted that the Army battalion attacked soon after he notified them of the enemy activity on 8 August. The bulk of the fighting for the Japanese headquarters caves did not take place until 10–11 August, however, on the evidence of 1/306 records and contemporary Japanese accounts. Surprisingly enough, this hot action is very scantily covered in the records of higher headquarters. 306th InfRegt URpts, 23Jul–10Aug44 (WW II RecsDiv, FRC, Alexandria, Va.)

tle on the 10th, 1/306 lost 8 men killed and 17 wounded before it withdrew for the night.

Once his headquarters was discovered, General Obata knew that the end was at hand. Only three officers survived the day's fighting and a renewed attack in increased strength was certain. During the morning of the 10th, the *Thirty-first Army* commander composed his last messages to Japan. To the Emperor he sent personal thanks, not only for past special favors of the Imperial family since he had entered the military profession, but also his apologies for what he considered his personal responsibility in the loss of the Marianas. To *Imperial General Headquarters* he reported:

> I accepted the important post of the army commander and although I exerted all-out effort, the fortune of war has not been with me. The fighting has not been in our favor since the loss of Saipan. We are continuing a desperate battle on Guam. Officers and men have been lost, weapons have been destroyed, and ammunition has been expended. We have only our bare hands to fight with. The holding of Guam has become hopeless. I will engage the enemy in the last battle with the remaining strength at Mt. Mataguac tomorrow, the 11th. My only fear is that report of death with honor (annihilation) at Guam might shock the Japanese people at home. Our souls will defend the island to the very end; we pray for the security of the Empire.
>
> I am overwhelmed with sorrow for the families of the many fallen officers and men. I request that measures be taken for government assistance to them. The remaining officers and men have high morale. Communications with the home land have been disrupted today, the 10th, after 1200 hours. I pray for the prosperity of the Empire.[29]

On the morning of the 11th, 1/306 attacked with tank support behind a heavy mortar preparation. Although the few remaining enemy troops fought back with rifles and machine guns, the overwhelming weight of American firepower gradually silenced them. Working their way down into the hollow behind a shower of white phosphorus grenades, and under the cover of direct tank fire, assault-demolition squads used pole charges and TNT to seal the caves. Sometime during the morning's battle, General Obata, true to the code of *Bushido*, took his own life in atonement for failure to fulfill his mission.

---

[29] Quoted in *GSDF Study*, p. 208.

214-881 O-67—37

Finish in the Marianas

By any rational standard, the most devastating cost of war is the lives of the men it kills and maims. In these terms, the price of Guam came high. In 21 days of battle Marine units of the Southern Troops and Landing Force lost 1,190 men killed and 377 died of wounds and suffered 5,308 wounded in action; the 77th Infantry Division casualties were 177 men killed and 662 wounded.[1] In the same period, 10,971 Japanese bodies were counted.[2] Sealed in caves and bunkers by shellfire and demolitions lay hundreds more of the enemy dead.

Perhaps as many as 10,000 Japanese were still alive in the jungles of Guam on 10 August. Except for the doomed group defending the enemy headquarters at Mt. Mataguac, there was little cohesion among the men that survived. In the months to come, when American troops in training for combat sharpened their skills in a perpetual mopping-up action, all Japanese efforts at counterattacks and guerrilla warfare floundered in a bitter struggle for survival. Starvation was a constant spectre to the men hidden in the vast stretches of jungle, and many of those that were later captured or killed were too weak to fight or hide any longer. For these survivors of the *Thirty-first Army*, Guam became a nightmare.

CHANGE OF COMMAND[3]

On 8 August, Admiral Nimitz informed Admiral Spruance of the future plans for the troops and commanders involved in the fighting on Guam. General Geiger and his staff were needed as soon as the campaign was ended to take charge of the landing force preparations for the operations pending against the Palau Islands. General Holland Smith was to be relieved as Commanding General, Expeditionary Troops and returned to Pearl Harbor to assume his duties as Commanding General, Fleet Marine Force, Pacific. General Schmidt and his VAC headquarters were to assume command of all assault troops remaining in the Marianas. In discharging this responsibility, Schmidt was to report to Admiral Spruance and later when directed, to the Commander, Forward Area, Central Pacific, Vice Admiral John H. Hoover.

[1] Army figures are derived from contemporary unit reports and those for Marine units from Headquarters Marine Corps postwar casualty statistics. A detailed unit casualty breakdown for STLF is contained in Lodge, *The Recapture of Guam*, pp. 178–180. Final official casualty totals for Marine units are listed in Appendix H.

[2] CTF 53 disp to ComFifthFlt, dtd 10Aug44 in *IIIAC Jnl*, entry of 10Aug44.

[3] Unless otherwise noted the material in this section is derived from ComFifthFlt WarD, Aug44 (OAB, NHD).

For the time being, the assault troops of IIIAC were to remain on Guam, but the 1st Provisional Marine Brigade was to depart soon for Guadalcanal, where it would join the 29th Marines and other reinforcing units to become the 6th Marine Division. Corps troops were scheduled to load out for training and staging areas in the South Pacific as shipping became available. The 77th Infantry Division, now blooded in combat and veteran in outlook, was to reorganize and refit as quickly as possible at Guam and prepare for early employment in another operation. Only the 3d Marine Division was due to remain for an extended period on the island it helped capture, but this unit, too, would be in battle again before seven months went by.

At 1030 on 10 August, shortly after the 2d Battalion, 3d Marines had accounted for the Japanese armor near Tarague, the *Indianapolis* arrived at Guam and dropped anchor in Apra Harbor. In the afternoon at 1635, Admiral Nimitz and his party, including the Commandant of the Marine Corps, Lieutenant General Alexander A. Vandegrift, landed at Orote airfield and immediately boarded the Fifth Fleet flagship for the first of a series of conferences among senior officers concerning future operations in the Pacific. On the 11th, Nimitz and Vandegrift inspected combat troops and supply installations, and before returning to Pearl, conferred with General Larsen regarding base development plans for the island.

Most of the ships that had taken part in FORAGER had already departed by the time Guam was declared secure. At noon on the 10th, Admiral Conolly turned over his responsibilities as CTF 53 and Senior Officer Present Afloat (SOPA) to Admiral Reifsnider. Conolly then flew to Pearl Harbor with key staff members to begin again the intricate task of planning an amphibious operation. Ten days later, Reifsnider in turn relinquished SOPA duties to a deputy of Admiral Hoover and sailed in the *George Clymer* for Hawaii. On his departure, Task Force 53 was dissolved.

General Geiger and his staff flew to Guadalcanal early on 12 August, turning temporary command of STLF over to General Turnage. On the same day, General Schmidt, at sea en route to Guam, reported by dispatch to assume command of all assault troops on the island. The VAC CP opened near Agana at 1430 on the 13th.

On 15 August, Admiral Nimitz' defense and development plan for the Central Pacific became effective at Guam. Admiral Turner's Joint Expeditionary Force was dissolved, and Admiral Hoover was assigned responsibility for operations at Guam as he had been for Saipan and Tinian on the 12th. On the 15th, as part of the change over, General Larsen assumed his duties as island commander.

One more step remained to be taken before the campaign for the capture of the southern Marianas became a matter of history. On 26 August, Admiral Spruance was relieved of responsibility for the Forward Area, Central Pacific and all forces under his command by Admiral William F. Halsey. For a time, Halsey's Third Fleet, using most of the ships and many of the men that had fought under Spruance, would carry on fleet operations against Japan.

As Halsey characterized the change: "Instead of using the stagecoach system of keeping the drivers and changing the horses, we changed the drivers and kept the horses. It was hard on the horses, but it was effective." [4] Spruance and his veteran staff and senior commanders would resume direction of the planning and preparations for the major amphibious campaigns aimed at the inner circle of Japanese defenses.

ISLAND COMMAND ACTIVITIES [5]

As the assault phase on Guam drew to a close, General Larsen assumed increasing responsibility for operations on the island. On 2 August, control of Orote Peninsula and Cabras Island passed to Island Command, and on the 7th, Larsen took over the operation of all extended radio circuits and a joint communications center. Supervision of all unloading activities was assigned to Island Command on 9 August. As garrison shipping arrived, the number and complexity of troops reporting to the island commander increased steadily.

General Larsen's initial task organization for base development included an advance naval base force, Lion 6, which was hard at work developing Apra Harbor as the center of a vast naval operating base. Airfield and road construction and stevedoring duties were the principal assignments of elements of the 5th Naval Construction Brigade, which included 12 Seabee battalions and 1 Marine and 4 Army battalions of aviation engineers. Supply activities were concentrated in the dumps and salvage and repair facilities developed and manned by the 5th Field Depot. For air defense, Larsen had MAG–21 and four antiaircraft battalions. V Amphibious Corps assigned him the 3d Marine Division for ground defense.

To this myriad of responsibilities for building Guam into a major staging, supply, and training base for future Pacific operations, General Larsen added the mantle of *de facto* governor of the Guamanians. The civil affairs section of Island Command had approximately 21,000 men, women, and children to care for, and to start back on the road to self-sufficiency. The cultivation of native gardens and the revival of native industries were actively fostered, and hundreds of men and women were employed as laborers and clerical workers in the burgeoning port, airfield, and supply facilities.

To protect and supervise the Guamanians, Admiral Nimitz authorized the formation of an island police, successor to the prewar Insular Patrol Force. Formed from a nucleus of former members plus military policemen from Island Command, all under a Marine officer, the new Local Security Patrol Force performed normal civilian police functions. In addition, however, these men, and many other Guamanians who volunteered as guides to American patrols, took an active part in hunting the Japanese. Isolated native villages and

---

[4] FAdm William F. Halsey and LCdr J. Bryan, III, USN, *Admiral Halsey's Story* (New York: McGraw-Hill Book Company, Inc., 1947), p. 197.

[5] Unless otherwise noted, the material in this section is derived from: VAC WarD, Aug44; Island Comd WarD, 1Apr–15Aug44, 15–31Aug44, and Sep45; *3d MarDiv Jnl; GSDF Study.*

farms were particularly vulnerable to foraging raids by the harried enemy troops, who were trying to keep alive in the jungle.

Soon after assuming responsibility for the assault troops on Guam, General Schmidt directed that the 3d and 77th Divisions each maintain an infantry regiment and an artillery battalion in the northern part of the island with a mission of killing or capturing the remaining Japanese. The 21st Marines and the 306th Infantry, which drew the initial patrol assignments, accounted for an average of 80 enemy a day between them in the last two weeks of August. On the 22d, the 3d Division passed to Island Command control for garrison duty and took over sole responsibility for the conduct of mopping-up operations; the 306th Infantry was relieved on 26 August to return to the 77th Division base camp.

While the patrol operations continued without letup, the majority of the assault troops under VAC command either shipped out from the island or settled into a rehabilitation and training routine with the emphasis on readying the men for early employment in combat again. The III Corps Headquarters and Service Battalion and the Signal Battalion left for Guadalcanal on 15 August. On the 21st, elements of the 1st Brigade began loading ship, and the veteran troops destined to form the new 6th Division sailed for the South Pacific on the 31st. In areas assigned by Island Command, the 3d Marine Division established its unit camps along the east coast road between Pago Bay and Ylig Bay, and the 77th Division encamped in the hills above Agat along Harmon Road.

The preparations of the 3d and 77th Divisions for further combat highlighted the role that Guam was to play during the remainder of the war. In addition to its development as a major troop training area, the island was transformed into a vast supply depot and a major naval base, and was eventually the site of Admiral Nimitz' advance fleet headquarters. On the plateau of northern Guam, where the final pitched battles had been fought, two huge airfields and a sprawling air depot were wrested from the jungle to house and service B–29s of the Twentieth Air Force, which struck repeatedly at Japan. A little over a year after the date that General Geiger had declared the island secure, it housed 201,718 American troops: 65,095 Army and Army Air Forces; 77,911 Navy; and 58,712 Marine Corps. Reunited on Guam for operations against the Japanese home islands were the 3d and 6th Marine Divisions, the former returned from the fighting on Iwo Jima and the latter from the battle for Okinawa.

During the period when the American forces on Guam were settling into a bustling routine of preparation for future operations, the situation of the Japanese hold-outs steadily deteriorated. Many of the men that hid out in the jungle were weaponless, few of those that were armed had much ammunition, and virtually none that had the means to fight showed any disposition to engage the Marine patrols. The overwhelming obsession of the enemy troops was food, and starvation forced many of them to risk their lives in attempts to steal rations. Gradually, as the months wore on, two officers among the survivors, Lieutenant Colonel Takeda and Major Sato, were able to establish a semblance of organization, but for the most part, the Japanese that lived did so as individuals and small groups, fending for themselves and avoiding all contact with the Americans.

In the latter stages of the war, psychological warfare teams of Island Command were increasingly successful in overcoming the Japanese reluctance to surrender. On 11 June 1945, Major Sato, convinced of the futility of holding out any longer, turned himself in and brought with him 34 men. By the end of August, records showed that 18,377 dead had been counted since W-Day and that 1,250 men had surrendered. After the Emperor had ordered all his troops to lay down their arms, the Americans were successful in convincing Lieutenant Colonel Takeda that he should come in. On 4 September, Takeda marched out of the jungle near Tarague, bringing with him 67 men. A week later he was able to persuade another group of 46 men to surrender, the last unified element of the garrison that had defended Guam. Individual Japanese continued to hide out in the jungle for years after the war was over, despite repeated efforts to convince them that Japan had surrendered.

LESSONS OF GUAM [6]

The operations leading to the recapture of Guam, as an integral part of the overall Marianas campaign, suffered and profited as did those at Saipan and Tinian from the state of progress in amphibious warfare when FORAGER was launched. In one respect, the extent of the prelanding naval bombardment, a standard was set that was never again reached during the war. In the Palaus and the Philippines, at Iwo Jima and Okinawa, gunfire support ships never again had the opportunity for prolonged, systematic fire that Admiral Conolly exploited so successfully.[7] The destruction of Japanese positions led the IIIAC naval gunfire officer to observe:

> The extended period for bombardment plus a system for keeping target damage reports accounted for practically every known Japanese gun that could seriously endanger our landings. When the morning of the landing arrived, it was known that the assault troops would meet little resistance [from enemy artillery or naval guns.] [8]

Although a few coast defense artillery pieces and antiboat guns did manage to weather this shelling and the accompanying carrier air strikes, most were knocked out as soon as they revealed themselves. The devastation wrought among the 1st Brigade assault waves by one undetected 75mm gun at Gaan Point illustrated the probable re-

[6] Unless otherwise noted, the material in this section is derived from: *IIIAC SAR; 3d MarDiv SAR; 77th InfDiv OpRpt; 1st ProvMarBrig SAR.*

[7] Admiral Spruance noted, however, that both bombers and bombardment ships began hitting Iwo Jima at the time of the attack on Saipan, a program which was kept up "whenever we could" until the actual landing in February 1945. He stated that the time schedule between Iwo Jima and Okinawa was too short for an extended bombardment program. Adm Raymond A. Spruance ltr to ACofs, G-3, HQMC, dtd 16Jun65.

[8] *IIIAC SAR*, NGF Rpt, p. 3.

sult of a less comprehensive target destruction plan. Where the enemy guns had not been destroyed, as was the case with a pair of 6-inch naval guns in the 3d Division landing zone, the murderous effect of area neutralization fires prompted crews to abandon their exposed emplacements.

The 1st Brigade, in its comments on naval gunfire, summed up the case for the assault troops—the more preparation, the better. General Shepherd recommended:

> . . . in future operations the amount of naval gunfire placed on a well-defended beach upon which troops are to be landed be no less than that fired in the Agat area of Guam. If possible, a greater amount of ammunition should be fired. The same amount of ammunition fired over a longer period of time seems to be more effective than that amount fired in a short period.[9]

Once the III Corps had landed, the use of naval fire support was continuous and generally effective. In particular, every assault unit was high in praise of the system of providing front-line battalions a ship to fire and illuminate throughout the night. Star shells were as popular with American combat troops as they were hated by the Japanese. Marine ground commanders were impressed with the need for a greater supply of illumination ammunition; General Turnage asked that "more stars be made available for future operation,"[10] and General Shepherd stated that it would be necessary to have "at least ten times the number of star shells in a future operation covering the same period of time as was allowed for the Guam operation."[11]

Carrier aircraft were equal partners with gunfire support ships in the pre-landing bombardment; they shared with land-based planes flying from Saipan the deep support missions delivered for the troops once ashore. During the operation, IIIAC noted that at least 6,432 sorties were flown, with 3,316 strafing runs made and 2,410 tons of bombs dropped. The scout, torpedo, and fighter bombers were most effective against targets that could not be reached by the flat trajectory fire of naval guns, such as the defiladed areas of Fonte Plateau from which Japanese artillery and mortars fired on the beaches and where enemy troops assembled for counterattacks. When the target area was close to the front line, opinions on the effectiveness of air support were varied and frequently critical. Admiral Turner characterized close air support at Guam as "not very good."[12]

General Shepherd noted that because most vehicular radios, the only ones capable of operating on the Support Air Direction (SAD) net, were damaged by salt water, the brigade air liaison parties directed relatively few air strikes. Those that did take place were kept beyond a bomb safety line, 1,000 yards from the Marine front lines, because of "rather severe casualties to our troops from bombing by our supporting air-

[9] *1st ProvMarBrig SAR*, p. 19.

[10] *3d MarDiv SAR*, D–3 SAR, Anx. C.

[11] *1st ProvMarBrig SAR*, p. 19. Out of a total of 106,110 shells (8,429.6 tons) fired during STEVEDORE, 5,039 were star shells. *CTF 56 OpRpt*, NGF Rpt, Anx II, App A.

[12] Adm Richmond K. Turner ltr to Maj Carl W. Hoffman, dtd 13Mar52.

craft."[13] The 77th Division had only one air strike directed by its air liaison parties, but the 3d Division made frequent use of ground-controlled strikes within 500 yards or less of its assault troops. On four occasions division troops were the target of misdirected bombing and strafing, and General Turnage recommended more accurate briefing of pilots to prevent repetition of such incidents.

The most crucial area of air support operations was communications. The SAD net was crowded at all times, and General Turnage observed that very few close support strikes were carried out on time or within limits set by requesting agencies. The method of operation worked out by Commander, Support Aircraft of TF 53 called for all requests from battalion air liaison to clear through regiments. He also frequently checked with divisions "since frontline reports from battalions were not sufficient to establish the whole front line near the target area."[14] Once the air liaison officer had shifted to the SAD frequency, he adjusted the dummy runs made by the flight leader or air coordinator until the plane was on target. Then a single bomb was dropped and if it was accurate, the entire flight would follow and attack. The time consumed in request, processing, approval, and final execution was generally 45 minutes to an hour or more. Although the Commander, Support Aircraft considered the time spent justified by the success of the missions, ground units generally asked for more immediate control of planes by air liaison officers and for a method of operations and system of communications that would ensure a faster response to the needs of assault troops. In this conclusion, that air liaison parties should have more direct contact with supporting planes, the infantrymen got firm backing from the Commander, Support Aircraft, Pacific Fleet, in his comments on air operations in the Marianas.[15] He also pointed out there was a need for greater understanding "on the one hand by the Ground Forces of the capabilities and limitations of aircraft, and on the other hand by the pilots of what they are supposed to accomplish."[16] There was undoubtedly generous room for improvement in air support techniques, and this need was sorely felt, because when planes were properly used they proved themselves invaluable in close support.

To General Geiger and many other Marines, a partial solution to air support problems lay in increased use of Marine aviation. The IIIAC air officer pointed out that Marine bombing squadrons had clearly demonstrated their capability in providing close (100 to 500 yards) support to ground troops (notably at Bougainville while working with the 3d Marine Division). He commented "that troop commanders, whether justifiably so or not, have repeatedly expressed a desire that Marine Bombing Squadrons be used for close support of their troops."[17] In reinforcing this finding with a recommendation

[13] *1st ProvMarBrig SAR*, p. 15.

[14] ComSptAirPac Rpt of Ops in Spt of the Capture of the Marianas, dtd 11Sep44 (OAB, NHD), p. 30.

[15] *Ibid.*

[16] *Ibid.*, p. 9.

[17] *IIIAC SAR*, Air Rpt, p. 4.

that specially trained Marine air support groups be placed on CVEs, the Expeditionary Troops air officer concluded:

> The troop experience of senior Marine pilots combined with the indoctrination of new pilots in infantry tactics should insure greater cooperation and coordination between air and ground units.[16]

In assessing the operations of another supporting weapon, armor working directly with the infantry, both the 3d Division and the 1st Brigade were unanimous in praising the medium tank as the most effective weapon for destroying enemy emplacements. Point-blank fire by the 75mm guns collapsed embrasures, cave defenses, and bunkers even after enemy fire drove supporting infantry to cover. The 3d Tank Battalion, which employed flamethrower tanks for the first time, was well satisfied with the new weapon, but found that attempts to mount infantry flamethrowers on tanks were generally unsatisfactory. The 3d Division recommended that in the future one tank of each platoon be equipped to spew flame. Although the brigade had no flame tanks, it did successfully employ borrowed Army tank destroyers armed with a 3-inch gun which showed great penetrating power in attacking cave positions. Operations in northern Guam demonstrated that armor and dozer blades were an effective combination against the jungle. The Marines frequently employed tanks working in conjunction with bulldozers in breaking trails; the 77th Division found that a dozer with an armored cab was the most effective vehicle for penetrating the heavy brush.

In general, infantry weapons proved reliable despite the weather and prolonged rough usage in the clutching jungle, but flamethrowers were easily damaged, with the firing mechanism a particularly sore spot. General Shepherd recommended that sufficient replacement flamethrowers be carried to the target to maintain initial allowances.

During the Guam operation, 3d Division and 1st Brigade experiments with the use of war dogs produced varied results. The dogs proved effective on night security watch and generally reliable on patrol, although they failed to alert Marines to hidden enemy troops on several occasions. Little need was found for the messenger dogs, for the SCR–300 radio provided reliable communications for isolated units. Patrols of the 3d Division found a new use for the dogs, though—investigating caves for hidden enemy before Marines entered; this technique proved best suited to the more vicious and aggressive animals.

Marine infantry battalions on Guam operated under a new table of organization, one that included in each rifle company the machine guns and mortars that had formerly been part of separate battalion weapons companies. The change worked well, gave closer support to the riflemen when needed, provided both company and battalion commanders with better control of supporting weapons, and simplified frontline supply channels. Since machine guns were prime targets for enemy fire, casualties among the crews were heavy, but replacements were found more eas-

[16] *CTF 56 OpRpt*, Air Rpt, p. 6.

ily among rifle company personnel. The other important change in infantry organization, the 13-man rifle squad with its three 4-man fire teams, proved to be a harder-hitting and more flexible fighting unit than its 11-man predecessor.

Brigade and division artillery, closely trained with the troops they supported, were an integral part of a tank-infantry-artillery team. Most ground commanders echoed General Shepherd's comment that "artillery was the most effective weapon employed during the operation." [19] Firing batteries quickly landed, promptly registered, and thundered into action early on W-Day. Whenever the front lines advanced appreciably, the artillery followed. The 12th Marines displaced five times between 1–10 August, the 77th Division artillery battalions made four moves to remain in direct support, and in northern Guam, one of the 75mm battalions of the brigade moved forward five times and the other four. In all displacements, artillery units were handicapped by the 50 percent reduction in motor transport imposed by reduced shipping space; vehicles were frequently pooled to effect rapid movement and keep the howitzers within supporting distance.

In the initial stages of the assault, the DUKW proved invaluable to the artillery units.[20] Not only did the amphibian trucks keep an adequate supply of ammunition close to the firing batteries, they also provided a satisfactory means of getting the 105mm howitzers ashore early in the fighting. Prior to the FORAGER Operation, the lack of a suitable vehicle to land the 105s in the assault had prompted the retention of the lighter and more maneuverable 75-mm in the Marine artillery regiment. Colonel John B. Wilson, commanding the 12th Marines, now recommended that the remaining 75mm pack howitzer units be replaced by 105mm battalions. This exchange would give the division more firepower and simplify ammunition handling and supply.

The key to effective fire support was rapid and efficient communications between forward observers and fire direction centers. Radios were used when necessary, but wire was employed to carry most of the traffic. The 12th Marines found the use of a forward switching central to be "extremely advantageous." [21] Artillery liaison party wire teams were required to maintain lines back only to a switching central in the vicinity of an infantry regimental CP; from there artillery battalion wiremen took care of the trunks to the fire direction center (FDC).

Centralized fire coordination was a feature of the Guam operation. The

[19] *1st ProvMarBrig SAR*, p. 18. Ammunition expended by artillery units during the campaign totaled:

Unit	75mm	105mm	155mm
III Corps Arty	.....	.....	25,346
1st ProvMarBrig	42,810	.....	.....
3d MarDiv	45,235	36,827	.....
77th InfDiv	.....	20,197	4,617
Total	88,045	57,024	29,963

[20] Infantry regiments also used DUKWs to land their 37mm guns and key command radio jeeps in the early waves, prior to the landing of artillery. *Craig 22Jun65 ltr.*

[21] *12th Mar SAR*, p. 2.

corps air and naval gunfire officers worked closely with the Corps Artillery FDC. Once his CP was functioning ashore, the Corps Artillery commander, General del Valle, was assigned operational control of all artillery on the island. This system enabled him to mass fires quickly and assign reinforcing missions as the situation required. In addition, he was able, in the light of the overall campaign picture, to make effective assignment of ammunition priorities, transportation, and firing positions.

General del Valle was not satisfied with the procedures used to get his own corps units ashore. He reported that his battalions were "prevented from entering the action ashore at an early stage with sufficient ammunition and suitable communications to render the desired support to the attack of the Corps during its critical stages." [22] In particular, he noted that the unloading was out of his control and at variance with the planned scheme of unloading and entry into action. He wrote that "as long as this control is vested in other officers, not especially concerned with, nor interested in, the operations of Corps [Artillery] satisfactory results will not be achieved." [23]

Ammunition supply was a particularly pressing problem in the first days of the operation when the heavy 155mm shells and powder began to come ashore in large quantities. Shore parties were hard put to handle the multiple transfers from boat to amphibian, vehicle to truck, and truck to dump. Large working parties of artillerymen were needed to handle their own ammunition on the beaches and in dumps ashore. The general recommended that an ammunition company and a DUKW company be assigned to Corps Artillery in the future to move ammunition directly from ship to battalion and battery dumps ashore.

Since no Japanese aircraft visited the air space over Guam, the antiaircraft batteries of Corps Artillery were not used in their primary function. The versatility of the guns and the destruction wrought by their firepower was clearly demonstrated, however, by their frequent use in support of ground troops. General del Valle drew particular attention to the employment of the 9th Defense Battalion in perimeter defense and in the patrolling in southern Guam as an illustration of the range of usefulness of antiaircraft units.

Many problems in landing troops and supplies at Guam were anticipated; others, as they occurred, were solved by combat team, brigade, division, and Corps Service Group shore parties. The effort to keep the assault troops supplied adequately required thousands of men, a force greater in strength than the 1st Brigade. The 3d Division had ship unloading details of approximately 1,200 men and shore working parties that numbered 3,300; the 1st Brigade left 1,070 men on board ship and used 1,800 on beach and reef; and the 77th Division, employing three battalions of shore party engineers plus some 270 garrison troops with low landing priorities, had 583 soldiers unloading ships and 1,828 working ashore. Almost one-fifth of the total strength of IIIAC

[22] *IIIAC Arty SAR*, Encl B.

[23] *Ibid.*, p. 18.

was engaged in the initial shore party effort.[24]

In allotting troops for the shore parties, General Geiger assigned the brigade assault forces a replacement unit. This organization, the 1st Provisional Replacement Company (11 officers, 383 enlisted men) was employed as shore party labor when the need was greatest. After the first flood of supplies was manhandled ashore, a fast-paced but orderly routine was established to unload assault and resupply shipping. The manpower requirements of the shore parties lessened, and the replacements were then fed into combat units as required. This use of replacements proved a sound concept, for it cut demands on assault troops for shore party labor and provided a ready source of trained men to fill the gaps caused by casualties. In later Marine operations in the Pacific, replacement battalions moved to the target with the assault echelon for use both as part of the shore party and as fillers in combat units.

Once the round-the-clock labor of the first 48 hours of unloading ended, the major portion of the task of handling supplies from ship to beach dumps fell to the specialists, the Army shore party engineers and the Marine pioneers. The Marine units proved adept at improvisation and in making-do with what they had plus what they could borrow, but they needed more heavy equipment. Corps reported that the pioneers:

> . . . are grossly ill-equipped when there are any beach difficulties or obstacles to overcome. The organizations attached to Corps for this operation had insufficient equipment for reef transfer of cargo, clearing beaches, and building access roads thereto. Even though an additional 25 Trackson cranes were provided, these were insufficient for reef transfer, beach, and dumps. A large number of lifts were beyond the capacity of any cranes belonging to the organizations mentioned. Some organizations totally lacked lighting equipment, others had antiquated equipment with run-down batteries which could not be used when beach operations were put on a 24-hour basis. Fortunately, Army Shore Party Battalions had sufficient equipment to meet minimum requirements for all Corps beaches.[25]

A good part of the construction work that was necessary to maintain and improve the beach areas and dumps fell to the Seabees, who operated as part of the shore parties in both beachheads. Division and brigade engineers were primarily concerned with direct support of the combat teams. Road and trail construction in forward areas, mine clearance, demolitions of obstacles and enemy defenses, and the operation of water points were all part of combat engineering tasks.

One responsibility shared by Seabees, engineers, and pioneers was the maintenance of an adequate network of roads. Under the impact of heavy traffic, the existing roads disintegrated. There was a constant struggle to repair the main arteries and to build new roads required by combat operations. The restriction on cargo space had hit the engineer units as hard as any or-

[24] Total engaged strength of IIIAC was 54,901, divided as follows: 3d Marine Division (20,338); 77th Infantry Division (17,958); 1st Provisional Marine Brigade (9,886); Corps Troops (6,719). *CTF 56 OpRpt*, Encl F.

[25] *IIIAC SAR*, ServGruRpt, Encl C, p. 5.

ganizations on the island, for much needed equipment had been left behind in the Solomon and Hawaiian Islands. Even when Corps Artillery prime movers equipped with angle dozer blades were borrowed, there were insufficient bulldozers and roadgraders to handle the tremendous road-building task. Frequent rains complicated all road operations, for mud prevented coral surfacing from binding and drainage problems caused an epidemic of floods.

Provident but temporary help was provided in this situation by the garrison force Seabee and engineer battalions, whose main mission was airfield construction. The profusion of difficulties faced by equipment-short assault units prompted the corps engineer to recommend that in future operations:

> . . . a minimum of one engineer battalion with heavy grading equipment (a Naval Construction Battalion, a Marine Separate Engineer Battalion, or an Army Aviation Engineer Battalion) be included in the assault echelon of each Marine or Army division, or fraction thereof, in the assault forces."[26]

The limitations posed by the lack of good roads and the chronic shortage of transportation hampered supply operations to some extent. Nowhere, however, was the course of combat endangered by this situation. When assault troops started moving north, units attempted to maintain a 5-day level of stocks in forward supply dumps, but there were never enough trucks available to meet this goal. The 5th Field Depot was able to supply all units and build up reserve stocks to 20-day levels in most categories. The necessity of feeding thousands of natives ate into the resupply rations, however, and the depot was never able to attain much more than a 10-day level of reserve food. Considered as a whole, logistics problems were competently handled and "the supply system on Guam worked smoothly and efficiently."[27]

One of the most heartening aspects of the operation, as indeed it was of other American assault landings, was the effectiveness of the medical treatment of casualties. If a man was hit, he knew that a Navy corpsman or an Army aidman would be at his side as soon as possible. Whatever the difficulties, evacuation was prompt; in the assault phase, the system of routing casualties from forward aid stations through beach and shore party medical sections to ships offshore brought wounded men on board specially equipped LSTs and APAs within an hour after the first wave landed.[28] Once field hospitals were set up ashore, many of the less seriously sick and wounded were treated on the island, but there was a steady flow of casualties via ship to base hospitals. Transports with specially augmented medical staffs and facilities for casualty care evacuated 2,552 men from Guam, and the hospital ships *Solace* and *Bountiful* carried 1,632 more.

The risks taken by the corpsmen, aidmen, and doctors in their concern for the wounded were great. The frequent

[26] *Ibid.*, Encl B, p. 5.

[27] Isely and Crowl, *Marines and Amphibious War*, p. 384.

[28] As the transport casualty berths began to fill up on W–plus one, however, some landing craft had to search for a ship which could take their wounded. BGen John S. Letcher ltr to CMC, dtd 12Jun65.

flurries of activity around aid stations, which were usually located on the natural routes of approach to the front lines, often drew Japanese mortar and artillery fire. Enemy small arms fire often seemed to be centered on the men that were trying to save the lives of assault troop casualties. In the course of the Guam campaign, the 3d Division had 3 medical officers and 27 corpsmen killed in action and 12 officers and 118 corpsmen wounded; the 1st Brigade had 1 officer and 9 corpsmen killed and 1 officer and 35 corpsmen wounded. The 77th Division lost 10 medical aidmen killed and had 35 wounded in action.

An analysis of the lessons learned by Americans at Guam seems incomplete without the viewpoint of the Japanese on their own operations. A postwar study of their role concludes with the judgement:

> . . . that Japanese troops on Guam took charge of the most extensive front as a division under the absolute command of sea and air by the enemy and checked the enemy from securing beachheads by organized resistance in the coastal area for the longest period, in spite of heavy enemy bombing and shelling for the longest time. In view of this, it is no exaggeration to say that this result was the best in the history of the war.[39]

A further comment based on this study by a present-day Japanese general, writing in an article authored jointly with a Marine veteran of the Guam operation, points out a principle by which the defense might have been even more effective:

> With no attempt to distract from the ability of the Japanese commanders, they were forced by Imperial General Headquarters policy to 'defeat-the-enemy-on-the-beach,' and accepted battle on two widely separated, and not mutually supporting, fronts. Their fighting strength was sapped by Col Suenaga's, and subsequent, counterattacks. These attacks, launched piecemeal, could only be indecisive. If Gen Takashina had defended the vital area of Guam, Apra Harbor, he would have seriously delayed subsequent U. S. operations. By so doing he could have delayed the devastating B-29 raids on his homeland. Instead he located his forces behind the landing areas and thus violated the cardinal rule of island defense—*defend the vital area.*[40]

CENTRAL PACIFIC PROVING GROUND

In a little more than nine months, November 1943 to August 1944, the art and science of amphibious warfare made enormous progress. The knowledge gained had been dearly won by the thousands of Americans killed and the many wounded between D-Day at Tarawa and the end of organized resistance at Guam. Each step of the way revealed weaknesses which required correction and problems which required answers. This crucial period of the war was a time when the officers and men of the Pacific Fleet and the Pacific Ocean Areas discovered—by trial and error—the most effective means of wresting a stubbornly-defended island from enemy hands.

Tarawa was the primer, and from the analytical reports of the commanders there and from their critical evalu-

[39] *GSDF Study*, p. 215.

[40] MajGen Haruo Umezawa, JGSDF and Col Louis Metzger, "The Defense of Guam," *Marine Corps Gazette*, vol. 48, no. 8 (Aug64), p. 43.

CORSAIRS OF MAG–21 *taxi down the airstrip on Orote Peninsula, 10 days after the island was secured. (USMC 92396)*

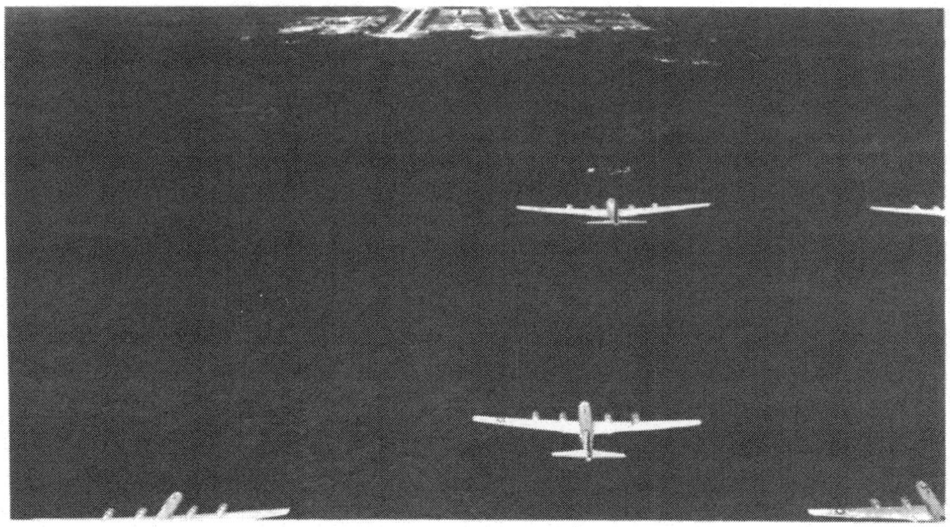

B–29s *returning from a strike on Japan approach North Field, which was wrested from the jungle battleground of northern Guam. (USAF 59056AC)*

ation of what went wrong, of what needed improvement, and of what techniques and equipment proved out in combat, came a tremendous outpouring of lessons learned. The development of the LVT(A) was expedited to provide close-in fire support for assault waves, and the value of the LVT was emphasized and its role expanded in future operations. Deficiencies in naval gunfire and aerial bombardment were pinpointed, and measures were taken to improve the delivery and effectiveness of both prelanding bombardment and fire support once the assault troops were ashore. The shortcomings of communications between ship and shore and air and ground drew particular attention, and the training and equipment of air and naval gunfire liaison teams was improved and intensified.

The performance of the fast carrier task forces in the Gilberts campaign clearly demonstrated that Americans had the power to isolate a target area, protect the amphibious forces, and permit a longer and more thorough softening-up of the objective. The carriers provided the means to keep the enemy off-balance, and with the voracious submarines that ranged the Japanese shipping lanes, choked off reinforcements and defensive supplies. From Tarawa onward, as one Japanese admiral said: "Everywhere, I think, you attacked before the defense was ready. You came more quickly than we expected." [31]

The carrier attack on Truk convinced the enemy that its vaunted naval base was vulnerable and therefore useless, and the fact that Truk was of no value to the Japanese meant "that its seizure was abandoned as a U. S. objective." [32] The momentum generated by the drive into the Marshalls at Kwajalein and the quickly planned and executed capture of Eniwetok was rewarding. The time of the attack on the Marianas was advanced by months. The swiftly rising power of Admiral Nimitz' forces, born as much of experience as of new strength, gave meaning to the principle formulated by the foremost naval historian of the war, Rear Admiral Samuel Eliot Morison, who stated that "the closer that one offensive steps on another's heels, the greater will be the enemy's loss and confusion, and the less one's own." [33]

Problems of coordination and control in the ship-to-shore movement and in operations ashore cropped up in the Marshalls as they had in the Gilberts, but the difficulties had less effect and pointed the way to better solutions. Naval gunfire was measurably more effective, artillery was used to good account from offshore islands at most objectives, and prelanding aerial strikes were better briefed and executed. Air support techniques and communications procedures remained a worrisome trouble spot in need of improvement. As the LVT had starred at Tarawa, the DUKW shone at Kwajalein, where its performance marked the growth of a family of amphibious vehicles which eased the problems posed by Pacific reefs.

A floating service squadron based in the Marshalls, which could replenish

[31] *USSBS Interrogation* No. 429, Adm Kishisaburo Nomura, IJN, II, p. 387.

[32] *Moore ltr.*

[33] Morison, *New Guinea and the Marianas*, p. 341.

and repair fleet units, vastly extended the range and duration of fast carrier operations and justified the decision to expedite the decisive thrust into the Marianas. Amphibious planners, sparked by Admirals Spruance and Turner, merged assault and base development plans into a unified whole which ensured a continued rapid advance to the ultimate objective, Japan. The spring and summer of 1944 saw the flowering of a vital skill, logistics planning, whose incredible complexity met the need to sustain massive assaults and at the same time provided a continuous flow of men, supplies, and equipment for a host of existing and future requirements.

The attack on Saipan and the following operations at Tinian and Guam demonstrated the ability of a Marine headquarters to operate above corps level and to prosecute successfully a variety of land campaigns on objectives larger than the fortress atolls. Admiral Spruance's plan, like all Fifth Fleet operations plans in amphibious campaigns, provided for action to be taken in case of attack by a major enemy naval force.[34] This foresight was in good part responsible for the favorable result of the Battle of the Philippine Sea, which Admiral King noted:

> . . . crippled Japanese naval aviation for the remainder of the war. Planes could be replaced, pilots could not, and, as was discovered later in the year at the Battle for Leyte Gulf, the Japanese no longer had the trained and seasoned aviators that were necessary for successful operations against our fleet.[35]

The fact that the attack on Saipan lured the Japanese carriers to defeat might alone be enough to call it the decisive operation of the Central Pacific campaign. The capture of the island, however, meant far more. It toppled the war party government of Premier Tojo in Japan, ensured the success of operations against Tinian and Guam, and secured the prime objective—the very long range bomber fields from which B–29s could ravage Japan.

A new pattern of Japanese defense, made possible by room to maneuver, emerged on Saipan. After beach positions fell, the enemy withdrew fighting to final defenses with the sole aim of making the battle as costly as possible to the Americans. The losses suffered by VAC were heavy but unavoidable against a determined foe. When the turn of Tinian came, every effort was bent towards improving the fire support from air and naval gunfire to limit American casualties. Artillery pounded the smaller island for days, and, under the cover of intensive supporting fires, a masterful shore-to-shore assault hit the Japanese defenses from an unexpected front. The result was a quick ending to a battle that might well have claimed the lives of many more Marines than those that did fall.

Intelligence gained at Saipan of the strength and probable defensive tactics of the *29th Division* on Guam was instrumental in lengthening and increas-

[34] Admiral Spruance did not expect "the Japanese fleet would come out to attack us, primarily because I thought the enemy would want shore based air support; and I knew that the first thing we would do in the Marianas would be to take out all of the enemy aircraft, and thereafter keep them out." Spruance 16Jun65 ltr, *op. cit.*

[35] King and Whitehill, *Naval Record*, p. 559.

ing the effectiveness of preliminary air and naval gunfire bombardment against the largest of the three Marianas target islands. Contemporary Japanese testimony amply supports the conclusion that this fire severely disrupted defensive preparations. Although the dual landings and subsequent operations in the rugged terrain ashore posed difficult problems of coordination and control, IIIAC units readily adapted their tactics to meet the enemy defense. The seizure of this island gave the Navy a base that by the end of the war was capable of supporting one-third of the Pacific fleet and provided the Army Air Forces additional B–29 bases for the aerial campaign against Japan.

In the Marianas as well as in the Gilberts and Marshalls, one aspect of the operations remained unsatisfactory—air support of ground troops. The complex and crowded communications setup caused multiple problems, inadequate pilot briefing led to inaccuracy, and, most important from the point of view of ground commanders, slow response to strike requests made air a far less effective supporting weapon than it might have been. The recognition of the need for improvement was not confined to the men that were supported, for a veteran Navy bombing squadron commander reported to CinCPac:

> In the Guam and Saipan operations, *close support* was actually almost nonexistent. Beyond tactical support by bombing before the troops landed, and some strategical bombing of rear areas and communications during the battles, little help was actually given the troops on the front lines. It is believed that the entire system must be changed and streamlined to make possible the real Close Support that we are capable of giving the troops.[86]

Marine commanders pressed hard for increased use of Marine air in close support. They wanted pilots, planes, and a control system oriented to ground needs and quickly responsive to strike requests. The winds of change were in the air in the summer of 1944 and refinements in close support techniques were coming. Operations later in the year saw planes bombing and strafing closer to frontline positions and evidenced a steady increase in the employment of Marine squadrons in this task as well as in air-to-air operations. Admiral Nimitz, in his comments on operations in the Marianas, noted:

> Four CVE's have been designated for close (troop) support and will embark Marine aircraft squadrons. It is not anticipated that Marine squadrons will furnish all close air support but they will be used with Marine divisions when the situation permits. In addition a certain number of Marine aviators are being assigned to the various amphibious force flagships to assist in the control of support aircraft.[87]

[86] CO VB–14 ltr to CinCPac, dtd 31Jul44, Encl I to CO USS *Wasp AR*, 6–30Jul44, dtd 31Jul44 (OAB, NHD). In contrast to this comment, General Shoup, chief of staff of the 2d Marine Division on Saipan, stated on 21 February 1963: "I might say openly that the finest close air support for ground troops that I experienced in World War II came from Navy squadrons at Saipan." 88th Congress, 1st Session, *Department of Defense Appropriations 1964, Hearings before a Subcommittee of the Committee on Appropriations, House of Representatives* (Washington, 1963), pt. 2, p. 383.

[87] ComInCh, *The Marianas*, p. 2–8. The operations of Marine squadrons on board CVEs and fast carriers will be covered in the fifth volume of this series.

At the conclusion of the Marianas campaign, senior commanders were generally satisfied that their forces were experts in the techniques of the amphibious assault and veterans in the flexibility of response it required. The admirals and generals were far from complacent, however, for the operations ahead promised to be even more demanding, bigger in scope, and perhaps tougher and more costly. In joint operations, despite occasional and human friction, forces of the Army, Navy, and Marine Corps had worked well together and learned from each other. There was a will to win that overrode every disagreement and setback, a pervading spirit of "let's get on with the job."

In assessing the performance of the Marines in this period, General Vandegrift, writing as Commandant to his predecessor, General Thomas Holcomb, summarized an inspection trip in the Pacific, pointing out that he had:

> . . . covered 22,000 miles in eighteen days, saw all the force, corps, and division commanders and practically all the regimental and battalion commanders in the field. I went to Saipan, Tinian, and Guam, getting to Guam just before the show was over. Our people did a superb job on all three of those islands. . . .[89]

That comment could as well apply to every man, of whatever service, that played a part in the success of GALVANIC, FLINTLOCK, CATCHPOLE, and FORAGER. Our people did a superb job.

---

[89] LtGen Alexander A. Vandegrift ltr to Gen Thomas Holcomb, dtd 5Sep44 (Vandegrift Correspondence File).

Bibliographical Notes

This history is based principally upon official Marine Corps records: the reports, diaries, journals, orders, plans, etc., of the units and commands involved in the operations described. Records of the other armed services have been consulted where they were pertinent. On matters pertaining to activities at high strategic levels, the authors consulted the records of the Joint Chiefs of Staff.

In order to cover the inevitable gaps and inadequacies that occur in the sources consulted, extensive use has been made of the knowledge of key participants in the actions described. These men, representing all services, have been generous with their time in answering specific queries, in making themselves available for interviews, and in commenting critically on draft manuscripts, not only of this volume but also of preliminary monographs. The historical offices of the Army, Navy, and Air Force have made detailed reviews of draft chapters and furnished much material of value to the history. The War History Office of the Defense Agency of Japan has read and commented upon the passages bearing on Japanese operations and provided worthwhile information that has been incorporated into the narrative.

Because this volume deals with the whole of the naval campaign in the Central Pacific, many of the records used relate to more than one of the operations. This is particularly true of the material concerning FORAGER. Such sources have been fully cited in the text and are discussed in relation to the particular operation where they have the most pertinency. All records cited, unless otherwise noted, are on file at, or obtainable through, the Archives of the Historical Branch, G-3 Division, Headquarters, U. S. Marine Corps.

A number of published works of general interest have been consulted frequently in the writing of this volume. The more important of these are listed below.

Books

Wesley Frank Craven and James Lee Cate, eds. *The Pacific: Guadalcanal to Saipan, August 1942 to July 1944—The Army Air Forces in World War II*, v. 4. Chicago: University of Chicago Press, 1950. The Air Force official history covering the period of the Central Pacific campaign. Well documented, the book is a reliable source for the actions of the Seventh Air Force and the attitudes and decisions of its commanders.

Philip A. Crowl. *Campaign in the Marianas—The War in The Pacific—United States Army in World War II*. Washington: Office of the Chief of Military History, Department of the Army, 1960. The Army official history of the operations in the Marianas with considerable detail of the actions of the 27th and 77th Infantry Divisions. It provides a well-reasoned analysis of the Smith against Smith controversy and is particularly useful for its sections on strategic background and planning.

Philip A. Crowl and Edmund G. Love. *Seizure of the Gilberts and Marshalls—The War in the Pacific—United States Army in World War II*. Washington: Office of the Chief of Military History, Department of the Army, 1955. This official Army history necessarily deals with Marine and Navy actions to a large extent with emphasis on the lessons of amphibious warfare learned in the early Central Pacific operations. It provides a good small unit narrative of Army participation in the Makin, Kwajalein, and Eniwetok fighting.

Jeter A. Isely and Philip A. Crowl. *The U. S. Marines and Amphibious War*. Princeton: Princeton University Press, 1951. An essential source for the study of the development of amphibious tactics and techniques and their application in the Pacific in World War II.

Takushiro Hattori. *Dai Toa Senso Zenshi* [The Complete History of the Greater East

Asia War]. Tokyo: Matsu Publishing Company, 1955. A manuscript translation of this excellent study is available at the Office of the Chief of Military History, Department of the Army. The author, a ranking staff officer during the war and an historian afterwards, has written a comprehensive history which contains enough detail to provide a useful strategic review from the Japanese viewpoint of every major campaign of the war.

Samuel Eliot Morison. *History of United States Naval Operations in World War II*. v. VII, VIII. Boston: Little, Brown and Company, 1951, 1953. These two volumes by Rear Admiral Morison, *Aleutians, Gilberts and Marshalls, June 1942–April 1944* and *New Guinea and the Marianas, March 1944–August 1944*, comprise a highly readable account of Navy operations in the Central Pacific. Written with considerable assistance and cooperation from the Navy, the histories are, however, very much the personalized work of the author and are most effective in their descriptions of naval actions and of Japanese operations.

Robert Sherrod. *History of Marine Corps Aviation in World War II*. Washington: Combat Forces Press, 1952. Although this is an unofficial history, it was written with substantial Marine Corps research support and contains valuable aviation unit historical data unavailable elsewhere. Much of the very readable text is based upon interviews and eyewitness accounts that were not retained for later study.

United States Strategic Bombing Survey (Pacific), Naval Analysis Division. *The Campaigns of the Pacific War* and *Interrogations of Japanese Officials*, 2 vols. Washington: Government Printing Office, 1946. Together these three volumes give an interesting account of the Japanese side of the war; however, they were prepared too soon after the event and contain many inaccuracies. The books are most useful in providing an understanding of Japanese military thinking through interviews and translations of relevant documents.

The War Reports of General of the Army George C. Marshall. Chief of Staff, General of the Army H. H. Arnold, Commanding General, Army Air Forces, Fleet Admiral Ernest J. King, Commander-in-Chief, United States Fleet and Chief of Naval Operations. Philadelphia and New York: J. B. Lippincott Company, 1947. A convenient compilation of the official reports of the chiefs of the armed services issued during and just after the war, which provides an excellent overall review of operations.

PART I

LAUNCHING THE CENTRAL PACIFIC OFFENSIVE

Official Documents

Fairly complete sets of the earlier ORANGE plans have been retained by the Operational Archives Branch, Naval Historical Division, Department of the Navy. Included in this collection are many studies and lectures which aid in tracing the development of Pacific strategy. At the Federal Records Center, Alexandria, Virginia, the World War II Records Division holds the files of the War Plans Division, War Department, which contain an accumulation of drafts of ORANGE Plans submitted between 1923 and 1928. Aside from Major Earl H. Ellis' Operation Plan 712, his "Security of Advanced Bases and Advanced Base Operations," and a collection of tactical plans drawn up during the 1930s, the Archives of the Marine Corps Historical Branch contain little material on the evolution of American strategy.

Copies of the various reports and minutes that show the development of Anglo-American wartime strategy are available in the ABC Files in the World War II Records Division in the Alexandria Federal Record Center. The more important material contained in these files, and similar ones of the Joint Chiefs of Staff, has been published and analysed in both official and unofficial histories and is more readily obtainable from these sources by private researchers.

Information regarding the status of Marine Corps units and personnel, particularly officers, during the period covered is contained in various tables of organization, station lists, and status sheets for air and ground units held in the Archives of the Historical Branch. Registers of Navy and Marine Corps regular officers and combined lineal lists of Marine Corps officers on active duty, both issued periodically during the war years, are useful sources for personal statistics. Major depository libraries should hold copies of the registers, which were printed by the Government Printing Office; the Marine Corps Archives has a complete set of the lineal lists which were printed at the Marine Corps Schools, Quantico, Virginia.

Japanese Sources

In the years immediately following the end of the war, former Japanese officials working under the auspices of General MacArthur's headquarters prepared a series of monographs detailing Japanese actions in many Pacific and Asian campaigns and at the various headquarters in the home islands. In the mid-1950s, a number of these original studies were revised and expanded, again by knowledgeable Japanese. The monographs vary considerably in their value, but, on the whole, they are honestly presented and useful in gaining an insight of Japanese planning and operations. The Office of the Chief of Military History, Department of the Army, which has a complete file of these studies, has prepared an annotated guide and index, *Guide to Japanese Monographs and Japanese Studies on Manchuria 1945-1960* (Washington, 1961), which is an excellent aid in evaluating the individual items.

Since much of the work done on these studies was in response to requests for information on campaigns in which the Army was principally involved, there is less available on the Central Pacific operations than there is on those in the South and Southwest Pacific. Almost all of the monographs of general scope, however, provide useful background information on Japanese war plans as they concerned the Central Pacific.

Books and Periodicals

The first two volumes of this series, *Pearl Harbor to Guadalcanal* and *Isolation of Rabaul*, were useful in reviewing the role of the Marine Corps in the development of amphibious doctrine and in the opening stages of the war. Among a number of books and articles concerning Pacific strategy, the following were the most useful.

FAdm Ernest J. King and Cdr Walter Muir Whitehill. *Fleet Admiral King: A Naval Record*. New York: W. W. Norton Inc., 1949. Admiral King's autobiography covers his entire naval career and gives revealing insights into the character of the man and his contributions to American strategy.

Maurice Matloff and Edwin M. Snell. *Strategic Planning for Coalition Warfare 1941–1942—The War Department—United States Army in World War II*. Washington: Office of the Chief of Military History, Department of the Army, 1953. An excellent background study of the formative stages of Allied strategy in the war.

John Miller, Jr. *CARTWHEEL: The Reduction of Rabaul—The War in the Pacific—United States Army in World War II*. Washington: Office of the Chief of Military History, Department of the Army, 1959. The book provides a companion account to operations in the Central Pacific with particular emphasis on MacArthur's planning.

John Miller, Jr. "The Casablanca Conference and Pacific Strategy," *Military Affairs*, v. 13, no. 4 (Winter 49). A concise account of the happenings at Casablanca and their effects.

Louis Morton. *Strategy and Command: The First Two Years—The War in the Pacific—United States Army in World War II*. Washington: Office of the Chief of Military History, Department of the Army, 1962. Perhaps the best account of American strategy in the Pacific with considerable coverage of its developmental stages.

Louis Morton. "American and Allied Strategy in the Far East," *Military Review*, v. 29, no. 9 (Dec49). This article contains much of the information on the ORANGE plans that was later developed in the official history cited above.

United States Army, War Department. *Handbook on Japanese Military Forces*. TM-E 30-480. Washington, 1Oct44. A basic source on the organization and equipment of Japanese land forces with useful detail on weapons characteristics and textbook tactics.

Adm Raymond A. Spruance. "The Victory in the Pacific," *Journal of the Royal United Service Institution*, v. 91, no. 564 (Nov46). An interesting but brief survey of the Pacific war with emphasis upon planning and strategy.

PART II

THE GILBERTS OPERATION

Official Documents

Although adequate material is available on the planning of Operation GALVANIC, the action reports of the units involved in the fighting both on Tarawa and Makin set forth only the general progress of the two battles. The message files and unit journals are very helpful in adding necessary detail.

All officers involved in the campaign devoted a great deal of effort to assessing the merits and defects of weapons, tactics, and amphibious techniques. The recommendations of battalion and regimental commanders have been, for the most part, condensed and included in the VAC Action Report and its many enclosures. Other important recommendations concerning naval elements appear in the reports originated by V Amphibious Force and Task Force 53.

As the Gilberts were a testing ground for many amphibious developments, comments comparing actions in later operations with those during GALVANIC occur frequently in Navy and Marine Corps reports. Any study of this operation should include reference to the action reports of higher commanders during subsequent campaigns in the Central Pacific.

Unofficial Documents

While writing the monograph used so extensively in preparing this account of the battle for Tarawa, Captain Stockman sent copies of his preliminary draft to various individuals who had taken part in the operation. Many of these men replied and their comments have been cited throughout this section. Similarly, the draft manuscript of this volume was sent to key participants and to the historical agencies of the other services, and the replies received have been used as applicable in revising the narrative. All such comments are retained in the files of the Marine Corps Historical Archives.

Of particular assistance in writing this section were extensive interviews by the authors with General Shoup, Admiral Hill, and General Julian Smith, who supplemented in this way their written comments on the draft. Admiral Spruance, and his former chief of staff, Admiral Moore, were quite helpful in developing the story of the planning background of the operation.

By no means all of the material uncovered by draft comments and interviews has been used in this book or in the Stockman monograph. The files contain much unpublished information that is of value to the student of the operation, particularly in regard to details of small unit action and the assessment of the accomplishments and character of individuals.

Japanese Sources

In addition to the Hattori manuscript mentioned previously, two of the Japanese monographs in the series held by the Office of the Chief of Military History have been useful. No. 48, *Central Pacific Operations Record, Volume I* (Dec41–Aug45), provided some data on the defensive preparations in the area and brief coverage of the operations in the Marianas, and No. 161, *Inner South Sea Islands Area Naval Operations, Part I, Gilbert Islands* (Nov41–Nov43), concerns the seizure of the Gilberts and Marshalls and provides a general review of naval operations in the Central Pacific.

In terms of pertinent captured documents, the Gilberts provided far less material than was the case in many other operations. Few knowledgeable prisoners were taken. As a consequence, it is quite difficult to reconstruct the action from the Japanese viewpoint. Much of the information available on the conduct of the Japanese defense was provided by American intelligence officers who made exhaustive investigations of the ruins of the defensive works on both Makin and Tarawa.

Books and Periodicals

In addition to the works of general interest cited above, the following have been most useful in shedding light on the Gilberts campaign.

Richard W. Johnston. *Follow Me! The Story of the Second Marine Division in World War II*. New York: Random House, 1948. This work contains some vivid impressions of the fighting on Betio and considerable information on the organization of the division.

Robert Sherrod. *Tarawa: The Story of a Battle*. New York: Duell, Sloan and Pearce,

1944. Considering the handicaps imposed by wartime security, this is perhaps the best account of the battle to be written by a journalist.

LtGen Julian C. Smith. "Tarawa," *U. S. Naval Institute Proceedings*, v. 79, no. 11 (Nov53). This story of the 2d Marine Division at Tarawa written by its commanding general is a valuable source for command decisions.

Capt James R. Stockman. *The Battle for Tarawa*. Washington: Historical Section, Division of Public Information, HQMC, 1947. The official monograph dealing with the Marine Corps role in GALVANIC, this booklet concentrates most of its narrative on the combat action ashore on Betio.

Capt Earl J. Wilson, *et al. Betio Beachhead: U. S. Marines' Own Story of the Battle for Tarawa*. New York: G. P. Putnam's Sons, 1945. The combined efforts of several Marine Corps combat correspondents have produced an uneven, though at times colorful, book.

PART III

THE MARSHALLS: QUICKENING THE PACE

Official Documents

In general, coverage of the FLINTLOCK operation is more than adequate, but fewer official reports of CATCHPOLE have survived. VAC prepared a detailed account of the Kwajalein landings as did the other major commands that participated in this phase of the Marshalls action. Tactical Group 1 and Task Group 51.11 submitted lengthly accounts of the Eniwetok venture, but the Marine battalions that did the bulk of the fighting provided reports that do not measure up in quality. Unit journals of the 4th Marine Division and Tactical Group 1 are an invaluable source of hourly and daily action and include the substance of most important orders and periodic situation reports.

Piecing together a narrative of the battles for the islands of Eniwetok Atoll is somewhat difficult, for journals and reports of the fighting often do not agree in detail with the reminiscences of those who fought there. Such disagreements have been resolved in this volume in favor of the official records. Where journals and reports are incomplete, interviews and correspondence with participants have provided the information necessary to fill the gaps.

Unofficial Documents

The numerous letters and transcripts of interviews which originated when the draft manuscript of *The Marshalls: Increasing the Tempo* was distributed for review were extremely valuable in the preparation of this account of FLINTLOCK and CATCHPOLE. In general, the monograph comment file has more detail on small unit action than the similar collection of letters and interviews gathered in the review of the draft of this section. Most of the participants queried during the writing of this volume were senior commanders and staff officers who could provide an informed and critical commentary on the treatment of the overall aspects of the campaign. Of particular use in this review were interviews and correspondence with General Schmidt and Admirals Hill and Moore.

The Office of the Chief of Military History assisted the writing of this section by making available the voluminous notes taken by Lieutenant Colonel S. L. A. Marshall during and immediately after the conquest of the southern part of Kwajalein Atoll. These notes, together with similar material gathered concerning the Makin operation, provide a significant body of detailed information regarding Army actions in the early Central Pacific landings.

Japanese Sources

Japanese strategy is set forth in the Hattori manuscript and the two monographs, Nos. 48 and 161, previously cited. Another monograph, No. 173, *Inner South Seas Islands Area Naval Operations, Part II, Marshall Islands* (Dec41–Feb44), describes the efforts to defend the Marshalls against American air attacks and contains a journal with entries from 23 November 1943 to 1 March 1944.

During the course of the fighting, JICPOA received many captured documents, most of them diaries and orders originated in the *1st Amphibious Brigade*. This material provides a revealing insight into life on a beleagured atoll as well as a knowledge of Japanese tactics. Unlike the story of the Gilberts battles, an account of the operations in the Marshalls can be fleshed out to become two-sided in terms of men who actually did the fighting.

Books

Once again Isely and Crowl, *Marines and Amphibious War*, Crowl and Love, *Gilberts and Marshalls*, and Morison, *Aleutians, Gilberts, and Marshalls* were invaluable sources. Among the other works which provided useful information were:

LtCol Robert D. Heinl, Jr. and LtCol John A. Crown. *The Marshalls: Increasing the Tempo.* Washington: Historical Branch, G-3 Division, HQMC, 1954. This official monograph, which covers the operations in small unit detail, served as the framework for the account presented here.

LtCol S. L. A. Marshall, AUS. *Island Victory.* Washington: Infantry Journal Press, 1944. This book gives a vivid and detailed account of the capture of Kwajalein Island by the 7th Infantry Division.

Carl W. Proehl, ed. *The Fourth Marine Division in World War II*. Washington: Infantry Journal Press, 1946. Like most division histories, this book concentrates on personalities and highlights of action and training and gives the reader a good grasp of the character of the unit and its men.

Part IV

SAIPAN: THE DECISIVE BATTLE

Official Records

By this stage in the war, action reports had become quite detailed and informative. The basic accounts of higher headquarters, such as Task Force 51 and Northern Troops and Landing Force, bulge large with appendices and annexes which include reports of general and special staff officers as well as important orders and journals. The records of smaller units are also more comprehensive and uniform and journals of personnel, intelligence, operations, and logistics sections provide far more information than any single volume can use, let alone a portion of such a volume. Saipan records are voluminous and sometimes contradictory, but on the whole furnish an excellent basis for an operational narrative. Where conflicts have occurred between the accounts of different reporting levels, the version presented by the unit closest to the action described has generally been the one accepted.

A valuable source of information on the Saipan fighting and the Smith against Smith controversy is the report of the Buckner Board. Included in its many annexes, designated exhibits, are firsthand accounts of the fighting by Army commanders and revealing descriptions of the combat readiness of various elements of the 27th Infantry Division. A copy of this report is available in the files of the World War II Records Division of the Alexandria Federal Record Center.

Unofficial Documents

Unfortunately, almost all of the letters and interview transcripts gathered by Major Hoffman in the preparation of his monograph have been lost. A diligent search of every possible depository where they might have strayed turned up nothing. Since the monograph contains much material based on the missing papers, and many quotes from them, it has been used frequently as a source for the information they contained. Wherever this has occurred, the footnotes clearly indicate it.

The circulation of the draft of this section produced a fair amount of comments from key participants which have been used as appropriate. Admirals Hill and Moore again furnished most useful reviews and added considerably to the authors' understanding of the naval aspects of the campaign. Many officers of the several services addressed themselves in one form or another to a discussion of the relief of General Ralph Smith; none objected to the account in this volume, which has been written after careful examination of both partisan and objective versions occurring in other works.

An interesting source of informal review of General Holland Smith's accounting of the relief is provided in his correspondence with General Vandegrift at that time. These letters, part of a file of personal correspondence with general and flag officers sent and received by the Commandant, are held in Archives of the Historical Branch. General Smith's letters and his public reports concerning the relief are consistent in all important details.

Japanese Sources

Thousands of documents were captured on Saipan: many were roughly translated there and others were later abstracted by JICPOA. A considerable body of Japanese information in fragmented form is contained in the journals and reports of unit intelligence agencies. Prisoner of war interrogations are also useful, although the majority of the information gained in this way must be checked and rechecked because the prisoner accounts conflict sharply in many instances. The JICPOA translations, and those issued by Admiral Nimitz' headquarters as CinCPac–CinCPOA documents, contain a wealth of military and human interest data which must be researched carefully since the titles of the documents often do not give an accurate clue to the contents. A complete file of these intelligence papers is held by the Operational Archives Branch of the Naval History Division.

Among the Japanese monographs in the series mentioned previously, several give coverage to the Marianas though none particularly concerns Saipan. The most useful include: No. 45, *History of the Imperial General Headquarters, Army Section* (1941–1945), which gives an overall review of the Pacific War and contains appendices of Army orders and unit designations; No. 49, *Central Pacific Operations Record* (April–November 1944), which is concerned primarily with operations in the Palau Islands, but discusses activities in the other areas and gives contemporary estimates of the enemy (Allied) situation; and No. 90, *The "A-GO" Operations* (May–June 1944), which details the buildup of the *Combined Fleet* and naval air arm prior to the Battle of the Philippine Sea.

Books and Periodicals

In addition to the overall sources, particularly the Crowl Army history and the Morison Navy account, the histories of the 2d and 4th Marine Divisions have been helpful in developing the narrative. Other works of value are listed below.

Gen Henry H. Arnold, USAF. *Global Mission.* New York: Harper and Brothers, 1949. The memoirs of the wartime leader of the Army Air Forces provide interesting background on the planning for the Marianas and the establishment of B–29 bases there.

Maj James A. Donovan. "Saipan Tank Battle," *Marine Corps Gazette*, v. 32, no. 10 (Oct48). The author, executive officer of 1/6 during the battle, gives a highly readable description of the Japanese tank attack.

Maj Carl W. Hoffman. *Saipan: The Beginning of the End.* Washington: Historical Division, HQMC, 1950. This official monograph is written in good style and considerable detail and gives adequate coverage to Navy and Army actions.

Gen George C. Kenney, USAF. *General Kenney Reports: A Personal History of the Pacific War.* New York: Duell, Sloan and Pearce, 1949. MacArthur's air commander gives an inside report of the controversy over Pacific strategy from the Southwest Pacific Area standpoint.

Edmund G. Love. *The 27th Infantry Division in World War II.* Washington: Infantry Journal Press, 1949. Longer and far more detailed than the usual division history, this book gives a good insight of the emotional jolt caused by the relief of the division commander and the disparaging remarks published about the unit in the United States.

Robert Sherrod. *On to the Westward, War in the Central Pacific.* New York: Duell, Sloan and Pearce, 1945. Carrying on from his narrative of Tarawa, the author gives a news correspondent's view of the Marianas fighting.

Gen Holland M. Smith and Percy Finch. *Coral and Brass.* New York: Charles Scribner's Sons, 1949. Reliable for the personal opinions and actions of General Smith, these memoirs are not too accurate concerning details of unit combat action.

Capt James R. Stockman. "The Taking of Mount Topatchou," *Marine Corps Gazette*, v. 32, no. 10 (Oct48). Written by an officer of 1/29, this article gives a concise and clear account of the maneuvers and fighting involved in seizing the mountain.

Tadao Yanihara. *Pacific Islands Under Japanese Mandate.* New York: Oxford University Press, 1940. A mixture of scholarship and propaganda, the book is useful in reviewing the history of Japanese presence in the islands.

PART V

THE INEVITABLE CAMPAIGN: TINIAN

Official Records

Since the command structure for Tinian was basically the same as that for Saipan, although the commanders changed in some cases, many of the reports of higher headquarters cover both operations. Much of this material is contained in separate documents, however, as these reports, particularly that of Task Force 56, were issued in multiple volumes. Like the operation itself, the records of it are models. Following the general practice of the Historical Branch in all of its histories, discrepancies between unit reports were resolved in this volume in favor of the lowest reporting unit. All commands were impressed with the unique features of the operation, particularly the landing across the White Beaches and the logistical setup, and there is much discussion of these in the various reports.

At the end of December 1944, Admiral King's headquarters issued a booklet, CominCh P–007, *Amphibious Operations: Invasion of the Marianas*, that is a valuable synthesis of reports received from major subordinate commands regarding their part in FORAGER. Equally useful for its account of all three operations, the compilation furnishes an excellent review of the unusual aspects of the assault on Tinian.

Unofficial Documents

The file of comments gathered by Major Hoffman during the writing of the monograph on Tinian have not, like those covering Saipan, disappeared. They are available in the Archives of the Historical Branch for further reference. Senior officers concerned in the planning of the operation made extensive comments on the draft manuscript so that the author could give an accurate picture of discussions leading to the selection of the beaches and also emphasize other features of the operation that elicited the almost universal praise it received. In comments on the draft of this section, many of the same men called attention to their earlier detailed remarks regarding the monograph and limited their review to a discussion of the overall aspects of the campaign as presented here. Many of the veterans of the amphibious development stages of the 1930s evaluated Tinian as a classic or a textbook example for the conduct of an amphibious operation.

Japanese Sources

The Hattori manuscript and the Japanese monographs previously cited are as useful as background information for Tinian as they are for Saipan and Guam. Many of the documents captured on Saipan furnish considerable information on the troops, weapons, and defensive dispositions on the smaller island. The 4th Marine Division published a file of representive translations of material gathered on Tinian that is an excellent source of information on Japanese operations. JICPOA and CinCPAC–CinCPOA publications of similar material keyed to the island on which it was recovered add another useful source to the body of intelligence available.

Books

Almost all the published sources listed under the Saipan section also concern themselves with Tinian. Admiral Morison's unofficial Navy history is helpful for its account of the naval aspects of the campaign, and Isely and Crowl have a good discussion of amphibious warfare developments. Among the few additional sources consulted were:

Lt John C. Chapin. *The Fourth Marine Division in World War II*. Washington: Historical Division, HQMC, Aug45. A pamphlet history, this small book highlights the actions of the division which made the assault landing on Tinian.

Major Carl W. Hoffman. *The Seizure of Tinian*. Washington: Historical Division, 1951. This official monograph drew unsolicited praise from several of officers who commented on the more generalized version of the campaign in this volume. The book provides excellent coverage of the planning phase and small unit detail, sparked by participants' comments, of the fighting ashore.

PART VI

VICTORY AT GUAM

Official Records

Although the Guam operation was a cohesive part of FORAGER, it is not particularly well represented in the reports of the Joint Expeditionary Force and Expeditionary Troops. These records of higher headquarters concern themselves largely with the campaign in the northern islands. As a result, the prime sources for a higher headquarters view of the campaign are the reports of Task Force 53 and the III Amphibious Corps. Both of these are useful but not as detailed as similar accounts by attack force and landing force headquarters at Saipan and Tinian. The 3d Marine Division action report includes concise reports of subordinate units as appendices, but the 1st Brigade report is largely a narrative at brigade level with a journal and file of orders issued attached. Only scattered examples of the war diaries and other records originated by the smaller Marine units on Guam have survived, and the reconstruction of narrative of action draws from dissimilar sources for like units, as the footnotes indicate. On the whole, however, there is enough material available to reconstruct an accurate account.

Unofficial Documents

The circulation of the draft manuscript of the monograph on Guam by Major Lodge, who made a special effort to elicit comment from officers of supporting arms and services, drew a number of detailed replies. Gaps in the small unit reports were readily filled by the information supplied by reviewers. From their comments, it was obvious that many of these men had retained copies of records that they had once originated or prepared. A file of comments concerning the Stockman-Carleton booklet on the Marianas campaign was used freely in the preparation of both the monograph and this section. Perhaps the most useful letters among the many that were received were those from the former commanding officer of the 9th Marines, General Craig, who provided detailed and extensive reviews on several occasions. The comments received on the draft manuscript of this section from senior commanders and staff officers are filed with those occasioned by earlier accounts.

Japanese Sources

The translations and interrogations of higher intelligence agencies and of the major commands on Guam were primary sources of Japanese information, but the body of information of this type was not as large as it was on Saipan and Tinian. In order to supplement this information, Lieutenant Colonel Takeda, the senior surviving member of the Guam garrison, was queried by the Historical Branch regarding many puzzling gaps in the story of the Japanese defense. His reply to these questions, together with an earlier and briefer account of the activities of the *29th Division* which he prepared after his surrender, were used frequently in the preparation of the monograph and this section. Another useful document, filed like the Takeda letters in the Historical Branch Archives, was an extensive history of the campaign prepared by officers of the Japanese Ground Self Defense Force and published in their staff school journal in a series of three articles. A translation of the text of this history written in 1962, was made available to the Historical Branch and it has been used throughout the preparation of this section.

Books and Periodicals

All the overall secondary sources relating to the Marianas campaign and to the assessment of the operations in the Central Pacific were consulted again in writing this account of Guam. In addition, the following were the most useful publications directly related to the operation.

Lt Robert A. Aurthur and Lt Kenneth Cohlmia. *The Third Marine Division*. Washington: Infantry Journal Press, 1948. More compact in format than most division histories, this book is a good source for unit background.

LtCol F. Clay Bridgewater, USA. "Reconnaissance on Guam," *The Cavalry Journal*, v. LIV, no. 3 (May–Jun45). The commanding officer of the 77th Division Reconnaissance Troop tells the story of its training and action.

MajGen Andrew D. Bruce, USA. "Administration, Supply, and Evacuation of the 77th Infantry Division on Guam," *Military Review*, v. 24, no. 10 (Dec44). The division commander reviews the combat support activities of his unit on Guam with useful comments on the reasons for various command decisions.

Bevan G. Cass, ed. *History of the Sixth Marine Division*. Washington: Infantry Journal Press, 1948. As the predecessor of the division, the 1st Brigade action is covered in some detail in this volume, but the majority of the book is devoted to the Okinawa campaign.

Kenneth W. Condit and Edwin T. Turnbladh. *Hold High the Torch, A History of the 4th Marines.* Washington: Historical Branch, G-3 Division, HQMC, 1960. An official account of the history of one of the regiments of the 1st Brigade with some detail of its actions on Guam.

1stLt Anthony A. Frances. "The Battle for Banzai Ridge," *Marine Corps Gazette*, v. 29, no. 6 (Jun45). A vivid story of the 21st Marines in the fighting for the ridges in the 3d Division beachhead.

Historical Division, War Department. *Guam, Operations of the 77th Division (21 July–10 August 1944).* Washington, 1946. A monograph covering the Army division action in considerable detail written by the division historian.

Capt Lucius W. Johnson, MC, USN. "Guam—Before December 1941," *U. S. Naval Institute Proceedings*, v. 72, no. 3 (Mar46). These recollections by a member of the prewar Navy garrison provide an interesting picture of island life from the American viewpoint.

1stLt Millard Kaufman. "Attack on Guam," *Marine Corps Gazette*, v. 29, no. 4 (Apr45). This article is a generalized account of the battle by a former member of the 1st Brigade.

Capt Edwin H. Klein. "The Handling of Supplies on Guam," *Marine Corps Gazette*, v. 29, no. 2 (Feb45). A review of supply operations on the island, the article concerns itself with shore party and field depot operations.

Maj Orlan R. Lodge. *The Recapture of Guam.* Washington: Historical Branch, G-3 Division, HQMC, 1954. This official monograph devotes a chapter to supporting arms and services and is particularly useful for the development of the Japanese side of the campaign.

LtCol Max Myers, USA, ed. *Ours to Hold it High: The History of the 77th Infantry Division in World War II.* Washington: Infantry Journal Press, 1947. The story of the Statue of Liberty Division provides a good overall view of the fighting on Guam and helpful information on its training and personnel.

Cdr H. E. Smith, CEC, USN. "I Saw the Morning Break," *U. S. Naval Institute Proceedings*, v. 72, no. 3 (Mar46). An eyewitness account of the Guam landing, this article contains vivid descriptive passages.

Laura Thompson. *Guam and its People.* Princeton: Princeton University Press, 1947. This book is primarily a sociological history of Guam with emphasis on the prewar years.

Charles O. West, *et. al.*, eds. *Second to None! The Story of the 305th Infantry in World War II.* Washington: Infantry Journal Press, 1949. This history is a useful account of the background and actions of the Army regiment that was initially attached to the 1st Brigade.

MajGen Haruo Umezawa, JGSDF, and Col Louis Metzger. "The Defense of Guam," *Marine Corps Gazette*, v. 48, no. 8 (Aug64). This article is a summary of the Japanese defense of Guam based in large part on the GSDF study used in the writing of this section.

Guide to Abbreviations

AA ---------- Antiaircraft
AAF ---------- Army Air Forces
AAR ---------- After Action Report
ABC ---------- American-British-Canadian
ACofS ---------- Assistant Chief of Staff
ADC ---------- Assistant Division Commander
Admin ---------- Administrative
AF ---------- Air Force
AFB ---------- Air Force Base
AFFE ---------- Army Forces in the Far East
AFPOA ---------- Army Forces, Pacific Ocean Areas
AGC ---------- Amphibious Command Ship
AKA ---------- Cargo ship, attack
Altn ---------- Alternate
Amtrac ---------- Amphibious Tractor
Anx ---------- Annex
APA ---------- Transport, attack
APD ---------- Transport, high speed
App ---------- Appendix
AR ---------- Action Report
Ar ---------- Army
Arty ---------- Artillery
Atk ---------- Attack
AUS ---------- Army of the United States
B–24 ---------- Army four-engine bomber, the Consolidated Liberator
B–25 ---------- Army two-engine bomber, the North American Mitchell
B-29 ---------- Army four-engine bomber, the Boeing Super-Fortress
Bat ---------- Battle
BGen ---------- Brigadier General
Bn ---------- Battalion
Bomb ---------- Bombardment
Br ---------- Branch
Brig ---------- Brigade
Btry ---------- Battery
Bu ---------- Bureau
Bul ---------- Bulletin
C ---------- Commander (units)
C–1 ---------- Corps Personnel Office(r)
C–2 ---------- Corps Intelligence Office(r)
C–3 ---------- Corps Operations and Training Office(r)
C–4 ---------- Corps Logistics Office(r)
C–47 ---------- Army twin-engine transport, the Douglas Skytrain
Capt ---------- Captain
Cbt ---------- Combat
CCS ---------- Combined Chiefs of Staff
Cdr ---------- Commander
CEC ---------- Civil Engineer Corps
Cen ---------- Central
CG ---------- Commanding General
Chap ---------- Chapter
CinC ---------- Commander in Chief
CMC ---------- Commandant of the Marine Corps
Cmt ---------- Comment
CNO ---------- Chief of Naval Operations
CO ---------- Commanding Officer
Co ---------- Company
CofS ---------- Chief of Staff
Col ---------- Colonel
Com ---------- Command
Comd ---------- Commander (units)
CominCh ---------- Commander in Chief
Con ---------- Conversation
CP ---------- Command Post
CT ---------- Combat Team
Cor ---------- Corps
CVE ---------- Escort carrier
D–1 ---------- Division Personnel Office(r)
D–2 ---------- Division Intelligence Office(r)
D–3 ---------- Division Operations and Training Office(r)
D–4 ---------- Division Logistics Office(r)
DA ---------- Department of the Army
DE ---------- Destroyer Escort
Def ---------- Defense
Dep ---------- Depot
Dept ---------- Department
Disp ---------- Dispatch
Div ---------- Division

DMS ---------- Destroyer Minesweeper
Ed ---------- Dated
DUKW ---------- Amphibious truck
Ech ---------- Echelon
Ed ---------- Editor
Est ---------- Estimate
Evac ---------- Evacuation
Expc ---------- Experience
F4U ---------- Navy-Marine single engine fighter, the Chance-Vought Corsair
FAdm ---------- Fleet Admiral
FBHL ---------- Final Beachhead Line
FE ---------- Far East
Fld ---------- Field
Flt ---------- Fleet
FMF ---------- Fleet Marine Force
FO ---------- Field Order
For ---------- Force
FRC ---------- Federal Record Center
Ftr ---------- Fighter
Fwd ---------- Forward
G-1 ---------- Division (or larger unit) Personnel Office(r)
G-2 ---------- Division (or larger unit) Intelligence Office(r)
G-3 ---------- Division (or larger unit) Operations and Training Office(r)
G-4 ---------- Division (or larger unit) Logistics Office(r)
Gar ---------- Garrison
Gen ---------- General
GHQ ---------- General Headquarters
GO ---------- General Order
GPO ---------- Government Printing Office
Gnd ---------- Ground
Gru ---------- Group
Hist ---------- History; Historical
Hq ---------- Headquarters
HQMC ---------- Headquarters, United States Marine Corps
H&S ---------- Headquarters and Service
IIIAC ---------- III Amphibious Corps
IJA ---------- Imperial Japanese Army
IJN ---------- Imperial Japanese Navy
IMAC ---------- I Marine Amphibious Corps
Inf ---------- Infantry
Incl ---------- Including
Info ---------- Information
Instl ---------- Installation
Instn ---------- Instruction
Intel ---------- Intelligence
JASCO ---------- Joint Assault Signal Company
JCS ---------- Joint Chiefs of Staff
JICPOA ---------- Joint Intelligence Center, Pacific Ocean Areas
Jnl ---------- Journal
Jnt ---------- Joint
JPS ---------- Joint Planning Staff
KIA ---------- Killed in Action
LCdr ---------- Lieutenant Commander
LCC ---------- Landing Craft, Control
LCI ---------- Landing Craft, Infantry
LCI(G) ---------- Landing Craft, Infantry (Gunboat)
LCM ---------- Landing Craft, Mechanized
LCT ---------- Landing Craft, Tank
LCVP ---------- Landing Craft, Vehicle and Personnel
Ldg ---------- Landing
Log ---------- Logistics; Logistical
LSD ---------- Landing Ship, Dock
LST ---------- Landing Ship, Tank
LT ---------- Landing Team
Lt ---------- Lieutenant
Ltr ---------- Letter
LVT ---------- Landing Vehicle, Tracked
LVT(A) ---------- Landing Vehicle, Tracked (Armored)
M4 ---------- Medium tank
M5 ---------- Light tank
Maj ---------- Major
MAG ---------- Marine Aircraft Group
Mar ---------- Marine
MAW ---------- Marine Aircraft Wing
MAWG ---------- Marine Air Warning Group
MBG ---------- Marine Bomber Group
MBDAW ---------- Marine Base Defense Aircraft Wing
MIA ---------- Missing in Action
Mil ---------- Military
MIS ---------- Military Intelligence Service
Mm ---------- Millimeter
MNFG ---------- Marine Night Fighter Group
MS ---------- Manuscript
Msg ---------- Message
Mtg ---------- Meeting
Nav ---------- Navy; Naval
NCO ---------- Noncommissioned Officer
ND ---------- Navy Department
NGF ---------- Naval Gunfire
NHD ---------- Naval History Division

No — Number
NTLF — Northern Troops and Landing Force
O — Order; Officer
OAB — Operational Archives Branch
O/B — Order of Battle
Occ — Occupation
OCMH — Office of the Chief of Military History
Off — Office
ONI — Office of Naval Intelligence
Op — Operation
OPD — Operations Division
OPlan — Operation Plan
Org — Organization
OY — Navy-Marine single-engine observation plane, the Consolidated-Vultee Sentinel
P–47 — Army single-engine fighter, the Republic Thunderbolt
Pac — Pacific; Pacific Fleet
PB4Y — Navy-Marine four-engine bomber, the Consolidated Liberator
Pers — Personnel
Phib — Amphibious
POA — Pacific Ocean Areas
POW — Prisoner of War
P(p) — Page(s)
Prelim — Preliminary
Pt — Party
Pub — Public
R4D — Navy-Marine twin-engine transport, the Douglas Skytrain
RAdm — Rear Admiral
RCT — Regimental Combat Team
Rec — Record
Recom — Recommendation
Recon — Reconnaissance
Regt — Regiment
Rel — Relations
Rep — Representative
rev — Revised
RJ — Road Junction
Rpt — Report
S–3 — Battalion or Regimental Operations and Training Office(r)
Salv — Salvage
SAR — Special Action Report
SC — Submarine Chaser
SCR — Signal Corps Radio
Sct — Scout
Sec — Section
Ser — Serial
Serv — Service
Ship — Shipping
Sho — Shore
Sig — Signal
Sit — Situation
Sked — Schedule
SMS — Supply and Maintenance Squadron
SNLF — Special Naval Landing Force
SP — Shore Party
Spl — Special
Spt — Support
Sqn — Squadron
Sta — Status
Stf — Staff
STLF — Southern Troops and Landing Force
Strat — Strategic
Subj — Subject
Subor — Subordinate
Sum — Summary
Sup — Supply
Suppl — Supplement
TBS — Talk Between Ships radio
TBX — Semi-portable low-power field radio
TBY — Portable low-power field radio
Tel — Telephone
Tg — Telegraph
TF — Task Force
TG — Task Group
Tk — Tank
TM — Technical Manual
TO — Theater of Operations
T/O — Table of Organization
TQM — Transport Quartermaster
Tr — Translated; Translator
Trac — Tractor
Trans — Transport
Trng — Training
Trp — Troop
U — Unit
UDT — Underwater Demolition Team
USA — United States Army
USAF — United States Air Force
USAFCPA — United States Army Forces, Central Pacific Area

214-881 O-67—39

USAFPOA — United States Army Forces, Pacific Ocean Areas
USMC — United States Marine Corps
USN — United States Navy
USS — United States Ship
USSBS — United States Strategic Bombing Survey
V — Volume
VAC — V Amphibious Corps
VB — Navy dive bomber squadron
Veh — Vehicle
VHF — Very High Frequency
Vic — Vicinity
VMF — Marine fighter squadron
VMF(N) — Marine night fighter squadron
VMO — Marine observation squadron
WarD — War Diary
WD — War Department
WIA — Wounded in Action
WO — Warrant Officer
Wpn — Weapons
WW — World War

Military Map Symbols

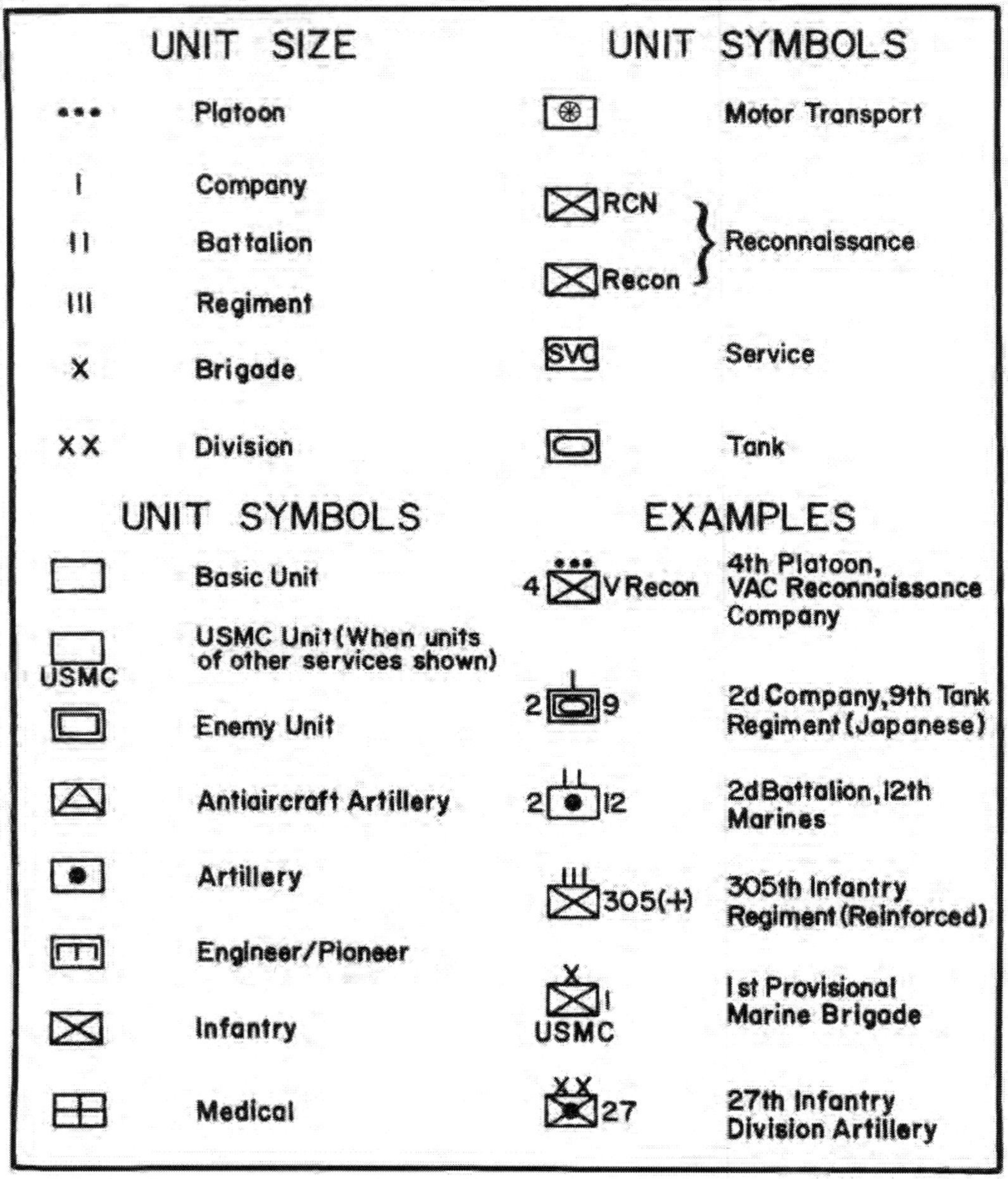

UNIT SIZE

•••	Platoon
I	Company
II	Battalion
III	Regiment
X	Brigade
XX	Division

UNIT SYMBOLS

	Motor Transport
RCN	Reconnaissance
Recon	Reconnaissance
SVC	Service
	Tank

UNIT SYMBOLS

	Basic Unit
USMC	USMC Unit (When units of other services shown)
	Enemy Unit
	Antiaircraft Artillery
	Artillery
	Engineer/Pioneer
	Infantry
	Medical

EXAMPLES

4 ••• V Recon	4th Platoon, VAC Reconnaissance Company
2 I 9	2d Company, 9th Tank Regiment (Japanese)
2 II 12	2d Battalion, 12th Marines
III 305(+)	305th Infantry Regiment (Reinforced)
X 1 USMC	1st Provisional Marine Brigade
XX 27	27th Infantry Division Artillery

T.L. RUSSELL

Chronology

The following listing of events is limited to those coming within the scope of this book, and those forecasting events to be treated in the volumes to follow.

1941

9Dec ---------- Japanese occupy Tarawa and Makin Islands in Gilberts.

10Dec -------- Guam surrenders to Japanese landing force.

23Dec -------- Wake Island surrenders to Japanese.

1942

1Feb ---------- U. S. carrier task forces raid Japanese positions in Gilberts and Marshalls.

30Mar -------- Pacific Ocean divided into Pacific Ocean Areas under Adm Nimitz, and Southwest Pacific Area under Gen MacArthur.

4–8May ------- Battle of the Coral Sea.

4–6Jun ------- Japanese are decisively defeated in main Battle of Midway.

7Aug --------- 1st MarDiv lands on Guadalcanal and Tulagi to launch first U. S. offensive of the war.

17Aug ------- 2d RdrBn lands from submarines at Makin Island. Raid is completed following day.

20Aug ------- First Marine aircraft arrive on Henderson Field, Guadalcanal.

2Oct --------- 5th DefBn occupies Funafuti, Ellice Islands.

1943

14–23Jan ----- Casablanca Conference to determine strategy for 1943. Agreement reached to advance toward Philippines through Central and Southwest Pacific, and to terminate hostilities only upon "unconditional surrender" of the enemy.

8Feb --------- Japanese complete evacuation of over 11,000 troops from Guadalcanal.

9Feb --------- Organized resistance on Guadalcanal ceases.

21Feb -------- Army troops, reinforced by Marine raiders and antiaircraft units, seize Russell Islands without opposition.

2–5Mar ------- Battle of Bismarck Sea. U.S. and Australian aircraft bomb Japanese destroyers and troop transports en route to Lae, New Guinea.

15Mar -------- Central Pacific Force redesignated Fifth Fleet; South Pacific Force becomes Third Fleet.

26Apr -------- General MacArthur issues ELKTON III, superseding previous ELKTON plans. Plan calls for mutually supporting advances in South Pacific and Southwest Pacific Area toward Rabaul, Operation CARTWHEEL.

12–25May ---- TRIDENT Conference held in Washington. General approval given to "U. S. Strategic Plan for the Defeat of Japan" calling for drive on Japan through Central Pacific.

21Jun -------- Elements of 4th RdrBn open Central Solomons campaign with landing at Segi Point, New Georgia.

14–24Aug ---- QUADRANT Conference in Quebec. CCS decide to attack Japan along both Central and Southwest Pacific routes.

4Sep ——— VAC formed under command of MajGen Holland M. Smith to train and control troops for amphibious landings in Central Pacific.

15Sep ——— 2d MarDiv is formally assigned to VAC.

19Sep ——— TF 15 and Seventh Air Force launch coordinated attacks against Tarawa.

20Sep ——— 4th MarDiv assigned to VAC.

24Sep ——— VAdm Raymond A. Spruance recommends an amphibious operation against Makin.

5Oct ——— CinCPac–CinCPOA issues plan for offensive in Central Pacific. D-Day for landings in Gilberts set for 19Nov43, later postponed to 20Nov.

13Oct ——— Photographic coverage of Makin Atoll obtained.

20Oct ——— Photographic coverage of Tarawa obtained.

25Oct ——— VAdm Spruance issues operation plan for GALVANIC, Gilberts Operation.

31Oct–7Nov ——— Northern Attack Force (TF 52) rehearses for GALVANIC off Hawaii. Southern Attack Force (TF 53) rehearses at Efate, New Hebrides.

1Nov ——— IMAC lands on Bougainville with 3d and 9th Marines and 2d RdrRegt in assault.

10Nov ——— Main body of Northern Attack Force for GALVANIC leaves Pearl Harbor.

12Nov ——— Southern Attack Force completes rehearsal and departs from New Hebrides.

19Nov ——— Final air bombardment of Gilberts, Marshalls, and Nauru in preparation for Gilberts invasion.

20Nov ——— VAC assault troops, 2d MarDiv at Tarawa and elements of 27th InfDiv at Makin, make landings in the Gilberts. Tarawa landings successful despite heavy casualties.

21Nov ——— Marines on Tarawa strengthen their hold on island. VAC ReconCo lands on Apamama Atoll.

22Nov ——— MajGen Julian Smith establishes CP on Tarawa.

22Nov–7Dec ——— SEXTANT Conference held at Cairo. Tentative timetable for offensive against Japan established.

23Nov ——— End of organized resistance on Tarawa and Makin.

1944

3Jan ——— Joint Expeditionary Force (TF 51 under RAdm Turner) issues operation plan for FLINTLOCK, assault on Marshalls.

13Jan ——— CinCPac–CinCPOA Campaign Plan GRANITE outlines tentative operations and timetable for Central Pacific offensive. Main body of TF 53 departs San Diego.

22Jan ——— Main body of FLINTLOCK attack force sail from Hawaii for Marshalls.

23Jan ——— Attack force reserve for FLINTLOCK and Majuro Attack Group sail for target.

29–30Jan ——— Carrier planes and naval vessels join in final neutralization of Marshalls.

31Jan ——— VAC assault troops seize small islands of Kwajalein Atoll as artillery positions for main landing support. VAC ReconCo secures Majuro Atoll.

1Feb ——— 4th MarDiv lands at Roi-Namur and 7th InfDiv at Kwajalein Island.

2Feb ——— 7th InfDiv troops meet increased resistance. 4th MarDiv completes mopup of Roi and capture of Namur.

MajGen Harry Schmidt announces end of organized resistance on Namur.

4Feb — 7th InfDiv completes capture of Kwajalein Island. RAdm Harry W. Hill given command of Task Group 51.11 with the mission of seizing Eniwetok Atoll.

7Feb — 4th MarDiv concludes uneventful search of islands of northern Kwajalein Atoll.

15Feb — Eniwetok Expeditionary Group (TG 51.11 under RAdm Hill) leaves Kwajalein for Eniwetok.

17Feb — Tactical Group 1, VAC, begins landing in Eniwetok Atoll (Operation CATCHPOLE).

18Feb — 22d Marines (reinforced) secure Engebi.

19Feb — Elements of 27th InfDiv and 22d Marines land on Eniwetok Island.

21Feb — Capture of Eniwetok Island completed.

22Feb — 22d Marines assault and secure Parry.

12Mar — JCS direct seizure of Southern Marianas, target date 15Jun44.

20Mar — 4th Marines seize Emirau in the Bismarcks. Adm Nimitz issues FORAGER Joint Staff Study setting forth the purpose of the Marianas operation.

26Apr — Expeditionary Troops operation order states mission ". . . to capture, occupy, and defend Saipan, Tinian, and Guam. . . ."

17–19May — Northern Troops and Landing Force maneuvers and rehearses at Maui and Kahoolawe, Hawaiian Islands.

25May — LSTs carrying assault elements of the 2d and 4th MarDivs depart Pearl Harbor.

29–30May — Portion of the Northern Troops and Landing Force not embarked in LSTs departs Pearl Harbor.

11Jun — Carrier planes of TF 58 begin preinvasion softening of Marianas. Northern Attack Force departs Eniwetok for Saipan.

13Jun — TF 58 continues aerial bombardment of Marianas and begins naval bombardment.

14Jun — VAC assault troops approach Saipan. Underwater demolition and minesweeping operations conducted along coast.

15Jun — 2d and 4th MarDivs land on Saipan.

16Jun — VAdm Spruance postpones landing on Guam because major naval battle appears imminent. Naval surface forces begin preinvasion bombardment of Guam. Elements of 27th InfDiv land on Saipan during night 16–17Jun44. Japanese launch strong, unsuccessful, tank-infantry night attack against 6th Marines.

18Jun — 4th MarDiv drives to east coast of Saipan, cutting island in two. 27th InfDiv captures Aslito airfield.

19Jun — 4th MarDiv begins clearing northern part of Saipan. 27th InfDiv troops to clear Nafutan Point and south coast of Saipan.

19–20Jun — Battle of the Philippine Sea. Carrier aircraft of TF 58 engage planes from enemy carriers and inflict crippling losses.

22Jun — VAC attacks northward on Saipan.

24Jun — 2d Marines reach outskirts of Garapan.

30Jun — Commanders' conference on Saipan decides landings on

21Jul44. Conclusion of battle for central Saipan.

2Jul ——— 2d Marines seize Garapan. Japanese fall back to final defense line in northern Saipan.

8Jul ——— Southern Attack Force begins naval bombardment of Guam.

9Jul ——— Saipan declared secure. Japanese garrison of about 22,000 is virtually destroyed.

12Jul ——— FMFPac activated with LtGen Holland M. Smith as commander.

14Jul ——— Joint Staff Study for Operation STALEMATE (invasion of the Palaus) issued.

20Jul ——— Volume of aerial attacks against Guam reaches peak. Naval bombardment continues. Preinvasion air and naval bombardment of Tinian also in progress.

21Jul ——— IIIAC assault troops land on Guam. 3d MarDiv and 1st ProvMarBrig push inland and by nightfall hold two beachheads. Elements of 77th InfDiv also go ashore.

22Jul ——— Softening up of Tinian continues. Marines repel counterattacks on Guam and continue advance inland.

24Jul ——— 4th MarDiv lands on Tinian and secures beachhead.

25Jul ——— 4th MarDiv expands beachhead on Tinian. Japanese unsuccessfully counterattack IIIAC positions on Guam.

26Jul ——— 2d MarDiv lands on Tinian. 1st ProvMarBrig opens attack to clear Orote Peninsula on Guam.

28Jul ——— Marines on Tinian continue rapid advance against light resistance from retreating enemy.

30Jul ——— Marines capture Tinian town and compress Japanese into southern tip of island. MajGen Geiger issues orders for pursuit of enemy northward on Guam.

1Aug ——— Organized enemy resistance ends on Tinian.

10Aug ——— End of organized Japanese resistance on Guam, though hundreds of Japanese remain to be mopped up.

15Aug ——— IIIAC, having completed its operations in the Marianas, is committed to invasion of the Palaus.

15Sep ——— JCS decide to invade Central rather than Southern Philippines and advance target date for invasion of Leyte from 20Dec to 20Oct44. 1st MarDiv lands on southwest shore of Peleliu Island.

17Sep ——— 81st InfDiv, as part of IIIAC, lands on Angaur.

30Sep ——— Peleliu, Angaur, Ngesebus, and Kongauru declared occupied.

9Oct ——— Third Fleet bombards Marcus Island.

14Oct ——— VAC directed to prepare plans for Iwo Jima operation.

20Oct ——— U. S. Army troops invade Leyte.

24–26Oct ——— Battle of Leyte Gulf which ends in U. S. naval victory.

25Nov ——— CinCPOA issues operation plan for invasion of Iwo Jima. Tentative date 3Feb 45.

1945

9Jan ——— Sixth Army invades Luzon.

13Feb ——— Final rehearsals for Iwo Jima operation concluded off coast of Tinian.

19Feb ——— Assault troops of VAC land on Iwo Jima.

26Mar ——— End of Japanese resistance on Iwo Jima.

1Apr ——— Tenth Army, including IIIAC, lands on Okinawa.

7Apr ——— Battle of East China Sea. Japanese fleet units heading toward Okinawa are intercepted by planes of TF 58.

25May ______ JCS direct invasion of Japan, scheduled for 1Nov45.

14Jun ______ JCS order commanders in Pacific to prepare plans for immediate occupation of Japan.

22Jun ______ End of organized resistance on Okinawa.

16Jul ______ Atomic bomb successfully tested at Los Alamos, New Mexico.

6Aug ______ Tinian-based B-29 drops atomic bomb on Hiroshima.

9Aug ______ Tinian-based B-29 drops atomic bomb on Nagasaki. Russia invades Manchuria.

12Aug ______ Soviet troops move into Korea.

15Aug ______ Offensive action against Japan halted.

30Aug ______ 4th Marines go ashore at Yokosuka. Army troops land at Atsugi airfield.

2Sep ______ Japanese sign instrument of surrender in Tokyo Bay.

Fleet Marine Force Status–31 October 1943[1]

Unit and location	Strength			
	USMC		USN	
	Off	Enl	Off	Enl
Outside U.S.A.				
Central Pacific Area				
Oahu				
Headquarters & Service Battalion, VAC	78	435	10	20
Signal Battalion, VAC	66	670	26	7
Reconnaissance Company, VAC	6	93	1	
Transient Center, VAC	3	18		
Rear Echelon, Transient Center, VAC*	31	289	14	73
Corps Motor Transport Company, VAC*	5	105		
Headquarters & Service Company, Marine Forces, 14th Naval District	31	289	2	4
6th Base Depot	23	324		2
1st Defense Battalion	63	1,460	3	27
15th Defense Battalion	61	1,284	4	21
4th Marine Division*	1,002	17,132	129	1,588
4th Radio Intelligence Platoon*	1	29		
1st Armored Amphibian Battalion*	36	855	1	9
1st Amphibian Tractor Detachment*	1	20		
1st Joint Assault Signal Company*	25	347		
1st Ammunition Company*	9	272		
12th Marine Depot Company*	4	173		
25th Replacement Battalion*	39	1,141	7	76
30th Replacement Battalion*	51	1,205	8	125
33d Replacement Battalion*	169	2,412	8	104
Headquarters Squadron, MAW, Pac	20	64	1	
Marine Fighter Squadron–114	44	189	1	8
Marine Fighter Squadron–225	56	245	1	8
Marine Scout-Bomber Squadron–245	48	280		
Marine Utility Squadron–252	32	336		
Headquarters Squadron–23, MAG–23	63	692	10	16
Service Squadron–23, MAG–23	15	249		
Marine Fighter Squadron–113, MAG–23	47	242	1	8
Area Sub-Total	2,029	30,850	227	2,096

See footnote at end of table.

Unit and location	Strength			
	USMC		USN	
	Off	Enl	Off	Enl
Midway				
6th Defense Battalion	49	1,925	5	32
Headquarters Squadron–22, MAG–22	10	133	3	12
Service Squadron–22, MAG–22	6	139	-------	-------
Marine Scout-Bomber Squadron–231, MAG–22	58	280	1	8
Marine Fighter Squadron–422, MAG–22	47	238	1	8
Area Sub-Total	170	2,715	10	60
Johnston				
16th Defense Battalion	49	913	4	19
Marine Scout-Bomber Squadron–133	33	184	1	4
Area Sub-Total	82	1,097	5	23
Palmyra				
Scout Bombing Detachment–1	14	100	1	4
Samoan Area				
Samoa				
Headquarters Squadron–31, MAG–31	26	400	9	16
Service Squadron–31, MAG–31	10	247	-------	-------
Marine Scout-Bomber Squadron–341, MAG–31	45	278	-------	-------
Area Sub-Total	81	925	9	16
Tutuila				
Headquarters, Defense Force, Samoa	31	172	6	4
Signal Company, Defense Force, Samoa	10	165	-------	-------
Harbor Defense Group	23	277	2	28
Base Depot, FMF	43	763	10	19
5th Separate Medical Company	-------	24	5	98
7th & 8th Marine Depot Companies	6	220	-------	-------
2d Defense Battalion	57	957	5	29
22d Marines (Reinforced)	175	3,229	20	437
8th Garrison Replacement Detachment*	8	250	2	50
9th Garrison Replacement Detachment*	8	250	2	50
Headquarters Squadron–4, 4th MBDAW	39	248	8	-------
Marine Utility Squadron–353, 4th MBDAW	28	283	1	8
Headquarters Squadron–13, MAG–13	44	415	7	20

See footnote at end of table.

Unit and location	Strength			
	USMC		USN	
	Off	Enl	Off	Enl
Service Squadron–13, MAG–13	13	250		
Marine Fighter Squadron–224, MAG–13	45	242	1	8
Marine Scout-Bomber Squadron–241, MAG–13	32	276	1	6
Headquarters Squadron–14, MAG–14	39	472	5	15
Area Sub-Total	601	8,493	75	772
Upolu-Savaii				
3d Marine Brigade	76	1,098	15	126
Wallis				
Headquarters Company, 8th Defense Battalion (Reinforced)	5	50		
8th Defense Battalion	75	1,100	5	35
3d Separate Medical Company		13	6	76
Marine Fighter Squadron–311, MAG–31	45	235		
Marine Scout-Bomber Squadron–331, MAG–31	46	280		8
Area Sub-Total	171	1,678	11	119
Funafuti				
5th Defense Battalion	54	1,284	47	636
Marine Fighter Squadron–111, MAG–13	45	234	1	6
Marine Scout-Bomber Squadron–151, MAG–13	44	285	1	8
Area Sub-Total	143	1,803	49	650
Nanomea				
7th Defense Battalion	49	1,104	9	26
Marine Fighter Squadron–441, MAG–13	33	235	1	8
Area Sub-Total	82	1,339	10	34
Nukufetau				
2d Airdrome Battalion (17th Defense Battalion)	35	885	5	63
South Pacific Area				
Enroute or Ordered to the Area				
Marine Fighter Squadron–216*	47	242	1	8
Marine Fighter Squadron–223*	38	237	1	7

See footnote at end of table.

Unit and location	Strength			
	USMC		USN	
	Off	Enl	Off	Enl
Marine Scout-Bomber Squadron–243*	40	282	1	9
Marine Photographic Squadron–254*	40	399	1	8
Marine Fighter Squadron–321*	44	232	-------	8
Area Sub-Total	209	1,392	4	40
Guadalcanal				
Headquarters & Service Battalion, IMAC	153	875	15	23
1st Corps Motor Transport Battalion	26	615	1	9
1st Corps Medical Battalion	1	124	40	356
1st Corps Naval Construction Battalion	-------	-------	27	806
1st Corps Signal Battalion	53	714	5	22
1st Corps Tank Battalion (Medium)	35	696	1	13
3d Marine Division	953	17,002	140	1,664
1st 155mm Artillery Battalion	36	554	1	10
2d 155mm Artillery Battalion	28	511	1	9
1st Parachute Regiment	105	2,079	17	78
2d Marine Raider Regiment	88	1,923	12	74
3d Defense Battalion	53	1,241	4	19
Marine Air Base Squadron–1, 1st MAW	13	384	1	8
Service Squadron–14, MAG–14	13	144	-------	5
Repair & Salvage Squadron–1, MAG–14	10	244	1	8
Area Sub-Total	1,567	27,106	266	3,104
New Caledonia				
Headquarters Company, Supply Service, IMAC	60	247	1	3
Transient Center, IMAC	99	1,527	34	228
14th Defense Battalion	60	809	5	20
1st Base Depot	81	1,728	1	23
1st Marine Raider Regiment	77	1,732	19	59
Barrage Balloon Group, IMAC	28	469	5	8
Marine Beach Jumper Unit No. 1	9	71	-------	-------
4th, 5th, & 6th Separate Wire Platoons*	3	135	-------	-------
9th & 10th Marine Depot Companies*	8	318	-------	-------
11th Marine Depot Company*	4	173	-------	-------
29th Replacement Battalion*	55	1,387	8	118
31st Replacement Battalion*	50	1,245	7	75
32d Replacement Battalion*	36	773	-------	-------
37th Replacement Battalion*	1	7	-------	-------
Marine Air Depot Squadron–1, 1st MAW	21	291	5	10
Headquarters Squadron–25, MAG–25	50	413	32	105

See footnote at end of table.

Unit and location	Strength USMC Off	Strength USMC Enl	Strength USN Off	Strength USN Enl
Service Squadron–25, MAG–25	20	402		
Marine Utility Squadron–152, MAG–25	44	212		
Marine Utility Squadron–153, MAG–25	47	209		
Marine Utility Squadron–253, MAG–25	44	198		
Area Sub-Total	797	12,346	117	649
Espiritu Santo				
Headquarters Squadron–1, 1st MAW	128	916	9	24
Marine Scout-Bomber Squadron–134, 1st MAW*	49	296	1	8
Marine Photographic Squadron–154, 1st MAW	32	393	3	8
Headquarters Squadron–11, MAG–11	59	580	9	16
Service Squadron–11, MAG–11	22	374		6
Marine Scout-Bomber Squadron–143, MAG–11	33	329	1	8
Marine Fighter Squadron–213, MAG–11	31	342	1	4
Marine Fighter Squadron–214, MAG–11	61	243	1	7
Marine Fighter Squadron–222, MAG–14	33	234	1	8
Marine Scout-Bomber Squadron–233, MAG–21	44	308	1	5
Marine Scout-Bomber Squadron–236, MAG–21	44	289	1	8
Area Sub-Total	536	4,304	28	102
Russells				
10th Defense Battalion	60	1,130	4	26
4th Base Depot	53	1,200	9	18
Marine Fighter Squadron–211, MAG–11	34	235	1	8
Headquarters Squadron–21, MAG–21	19	361	7	30
Service Squadron–21, MAG–21	10	218		
Marine Night Fighter Squadron–531 (Forward Echelon), MAG–21	17	179	1	6
Area Sub-Total	193	3,323	22	88
Efate				
Marine Fighter Squadron–123, MAG–21	33	258	1	4
Marine Scout-Bomber Squadron–234, MAG–21	41	291	1	8
Headquarters Squadron–24, MAG–24	20	252	6	6
Service Squadron–24, MAG–24	10	216		
Headquarters Squadron–2, 2d MAW	65	348	10	15
Headquarters Squadron–12, MAG–12	16	305	6	25

See footnote at end of table.

Unit and location	Strength			
	USMC		USN	
	Off	Enl	Off	Enl
Service Squadron–12, MAG–12	10	233		
Marine Scout-Bomber Squadron–144, MAG–21	33	295	1	4
Area Sub-Total	228	2,198	25	62
New Zealand				
2d Marine Division	978	18,821	150	1,743
1st Aviation Engineer Battalion	34	347	6	19
2d Antitank Battalion	26	372	1	12
2d Base Depot	28	500	1	14
3d Base Depot	35	347	6	19
Area Sub-Total	1,101	20,387	164	1,807
Southwest Pacific Area				
Goodenough				
1st Marine Division	956	17,468	135	1,561
27th Replacement Battalion*	60	1,360	8	124
Area Sub-Total	1,016	18,828	143	1,685
New Georgia				
9th Defense Battalion	60	1,210	3	21
Marine Scout-Bomber Squadron–232, MAG–11	36	326	1	7
Marine Scout-Bomber Squadron–244, MAG–11	45	260	1	7
Marine Scout-Bomber Squadron–144, MAG–24	43	279	1	4
Area Sub-Total	184	2,075	6	39
Rendova				
11th Defense Battalion	61	1,061	4	22
Vella Lavella				
4th Defense Battalion	59	1,148	4	22
4th Base Depot, Branch No. 3	45	591	4	15
Marine Fighter Squadron–212, MAG–11	34	234	1	8
Marine Fighter Squadron–215, MAG–14	30	242	1	8
Marine Fighter Squadron–221, MAG–24	33	251	1	8
Area Sub-Total	201	2,466	11	61

See footnote at end of table.

Unit and location	Strength			
	USMC		USN	
	Off	Enl	Off	Enl
Woodlark				
12th Defense Battalion	44	1,064	4	22
Carribean Area				
Guantanamo Bay				
13th Defense Battalion	49	683	3	15
St. Thomas, Virgin Islands				
Marine Scouting Squadron–3	34	106		
West Coast, U.S.A.				
San Diego				
Headquarters Squadron, Marine Fleet Air, West Coast	41	229		
Supply Squadron–5	19	202		
Marine Scout-Bomber Squadron–132*	4	275	1	8
Area Sub-Total	64	706	1	8
El Centro				
Headquarters Squadron–43, MBDAG–43	76	922	4	10
Service Squadron–43, MBDAG–43	10	215		
Marine Fighter Squadron–122, MBDAG–43	48	172		
Marine Observation Squadron–155, MBDAG–43	43	279	1	
Marine Fighter Squadron–217, MBDAG–43	49	251		7
Marine Torpedo Bomber Squadron–243, MBDAG–43	49	316	1	8
Area Sub-Total	275	2,155	6	25
El Toro				
Headquarters Squadron–41, MBDAG–41	23	1,505		
Service Squadron–41, MBDAG–41	13	254		
Marine Torpedo-Bomber Squadron–131, MBDAG–41	62	306	2	
Marine Scout-Bomber Squadron–141, MBDAG–41	53	258	2	13
Marine Scout-Bomber Squadron–142, MBDAG–41	8	48		
Area Sub-Total	159	2,371	4	13

See footnote at end of table.

Unit and location	Strength			
	USMC		USN	
	Off	Enl	Off	Enl
Camp Elliott				
Headquarters Company, FMF, San Diego Area	75	330	6	7
Headquarters Battalion, Training Center	50	657	30	25
Infantry Battalion, Training Center	307	3,420		
School Battalion, Training Center	54	609		
Tank Battalion, Training Center	74	724		
Field Medical School Battalion, Training Center	5	25	51	773
Base Depot, FMF	30	459	2	15
Women's Reserve Battalion, FMF, San Diego Area	25	454		
Area Sub-Total	620	6,678	89	820
Camp Gillespie				
Parachute Training School	56	689	6	32
Camp Kearney				
Headquarters Squadron–15, MAG–15	276	624	6	26
Service Squadron–15, MAG–15	13	345		
Marine Utility Squadron–952, MAG–15	30	214		
Area Sub-Total	319	1,183	6	26
Camp Pendleton				
Headquarters Battalion, Training Center	66	383	9	69
Artillery Battalion, Training Center	87	762		
Amphibian Tractor Battalion, Training Center	39	315	3	
Engineer Battalion, Training Center	74	708		
Naval Construction Battalion, Training Center			19	641
Field Signal Battalion, Training Center	17	1,029		
Raider Battalion, Training Center	24	422		1
4th Parachute Battalion	27	637	5	26
Area Sub-Total	334	4,256	36	737
Miramar				
Headquarters & Service Squadron, 2d MAWG	3	28		
Headquarters Squadron, Personnel Group	20	102	1	16
Air Regulating Squadron–1	6	906		1
Air Regulating Squadron–2	7	651		123

See footnote at end of table.

Unit and location	Strength			
	USMC		USN	
	Off	Enl	Off	Enl
Air Regulating Squadron-3	270	841	15	-------
Air Regulating Squadron-4	7	674	-------	-------
Area Sub-Total	313	3,202	16	140
Mojave				
Headquarters Squadron-44, MBDAG-44	34	457	3	6
Service Squadron-44, MBDAG-44	10	247	-------	-------
Marine Fighter Squadron-121, MBDAG-44	14	125	1	-------
Marine Fighter Squadron-124, MBDAG-44	6	35	-------	-------
Marine Fighter Squadron-218, MBDAG-44	50	268	-------	9
Marine Observation Squadron-251, MBDAG-44	50	297	-------	-------
Area Sub-Total	164	1,429	4	15
East Coast, U.S.A.				
Camp Lejeune				
Headquarters Company, Training Center	6	4	-------	-------
War Dog Training Company, Training Center	7	157	-------	-------
Amphibian Base Battalion, Training Center	23	226	86	755
Artillery Battalion, Training Center	162	1,465	-------	-------
Engineer Battalion, Training Center	143	1,019	-------	-------
Infantry Battalion, Training Center	145	622	50	573
Parachute Battalion, Training Center	2	17	-------	-------
Quartermaster Battalion, Training Center	72	882	-------	-------
Range Battalion, Training Center	5	278	-------	6
Signal Battalion, Training Center	189	3,409	-------	163
Naval Construction Battalion, Training Center	-------	-------	35	1,202
18th Defense Battalion	43	648	3	23
51st Defense Battalion	71	1,654	6	38
7th Separate Infantry Battalion	8	37	1	19
7th Separate Pack Howitzer Battery	5	62	-------	-------
2d Battalion, 13th Marines	15	65	-------	-------
Area Sub-Total	896	10,545	181	2,779
Norfolk				
Base Depot, FMF	32	460	-------	6

See footnote at end of table.

214-881 O-67—40

Unit and location	Strength			
	USMC		USN	
	Off	Enl	Off	Enl
Cherry Point				
Headquarters Squadron-3, 3d MAW	57	496	20	51
Headquarters Squadron-1, MAWG-1	47	313	1	18
Air Warning Squadron-1, MAWG-1	14	178	--------	7
Air Warning Squadron-2 MAWG-1	14	179	--------	5
Air Warning Squadron-4, MAWG-1	1	60	--------	--------
Headquarters Squadron-35, MAG-35	16	237	--------	--------
Service Squadron-35, MAG-35	10	203	--------	--------
Marine Observation Squadron-351, MAG-35	22	144	--------	--------
Marine Utility Squadron-352, MAG-35	38	179	--------	--------
Marine Photographic Squadron-354, MAG-35	34	254	--------	--------
Headquarters Squadron-53, MNFG-53	33	61	3	21
Service Squadron-53, MNFG-53	10	410	--------	--------
Marine Night Fighter Squadron-531 (Rear Echelon), MNFG-53	16	119	--------	--------
Marine Night Fighter Squadron-532, MNFG-53	25	253	--------	--------
Marine Night Fighter Squadron-555, MNFG-53	3	4	--------	--------
Marine Night Fighter Squadron-534, MNFG-53	2	5	--------	--------
Headquarters Squadron-61, MBG-61	327	493	3	20
Marine Bomber Squadron-611, MBG-61	1	2	--------	--------
Marine Bomber Squadron-612, MBG-61	1	2	--------	--------
Marine Bomber Squadron-613, MBG-61	1	1	--------	--------
Marine Bomber Squadron-614, MBG-61	1	1	--------	--------
Area Sub-Total	683	3,594	27	122
Oak Grove				
Air Warning Squadron-3, MAWG-1	11	120	--------	6
Headquarters Squadron-32, MAG-32	23	273	5	41
Service Squadron-32, MAG-32	11	204	--------	--------
Marine Fighter Squadron-314, MAG-32	20	48	--------	--------
Marine Fighter Squadron-323, MAG-32	41	227	--------	--------
Marine Fighter Squadron-324, MAG-32	15	44	--------	--------
Area Sub-Total	121	916	5	47
Parris Island				
Marine Fighter Squadron, MAG-32	44	220	2	7
Marine Fighter Squadron-322, MAG-32	43	205	1	--------
Area Sub-Total	87	425	3	7

See footnote at end of table.

Unit and location	Strength			
	USMC		USN	
	Off	Enl	Off	Enl
New Bern				
Marine Fighter Squadron-313, MAG-32	32	167	------	------
Bogue				
Headquarters Squadron-33, MAG-33	16	202	6	39
Service Squadron-33, MAG-33	9	156	------	------
Marine Scout-Bomber Squadron-332, MAG-33	43	201	------	------
Marine Scout-Bomber Squadron-333, MAG-33	43	210	------	------
Marine Scout-Bomber Squadron-334, MAG-33	42	202	------	------
Area Sub-Total	153	971	6	39
Atlantic				
Headquarters Squadron-34, MAG-34	13	271	5	39
Service Squadron-34, MAG-34	4	175	------	------
Marine Scout-Bomber Squadron-342, MAG-34	44	221	------	------
Marine Scout-Bomber Squadron-343, MAG-34	43	206	------	------
Area Sub-Total	104	873	5	39
Total FMF (Ground) Overseas	5,583	99,789	808	9,095
Total FMF (Air) Overseas	2,591	20,833	167	613
Total FMF (Ground) in U.S.A.	3,185	45,919	444	6,426
Total FMF (Air) in U.S.A.	3,095	20,830	95	531
Total FMF Overseas	8,174	120,622	975	9,708
Total FMF in U.S.A.	6,280	66,749	539	6,957
Total FMF	14,454	187,371	1,514	16,665

[1] Strength figures and unit designations were abstracted from the FMF Status Reports, Ground and Air, for October 1943 held in the Archives of the Historical Branch, G-3 Division, Headquarters Marine Corps. Units en route or ordered to the indicated areas (indicated by an asterisk *) are listed under those areas regardless of their temporary location.

Table of Organization F-100–Marine Division

5 May 1944[1]

Unit	USMC		USN		Totals	
	Off	Enl	Off	Enl	Off	Enl
Division Headquarters	(66)	(186)	(4)	(1)	(70)	(187)
Headquarters Battalion	101	883	7	13	108	896
Headquarters Company	(73)	(394)	(7)	(9)	(80)	(403)
Signal Company	(17)	(275)			(17)	(275)
Military Police Company	(6)	(96)			(6)	(96)
Reconnaissance Company	(5)	(118)		(4)	(5)	(122)
Tank Battalion	35	585	1	9	36	594
Headquarters & Service Company	(14)	(99)	(1)	(9)	(15)	(108)
3 Tank Companies (each)	(7)	(162)			(7)	(162)
Service Troops	58	1,343	66	422	124	1,765
Service Battalion	(29)	(702)	(2)	(18)	(31)	(720)
Headquarters Company	(9)	(48)	(2)	(9)	(11)	(57)
Service & Supply Company	(13)	(483)		(9)	(13)	(492)
Ordnance Company	(7)	(171)			(7)	(171)
Motor Transport Battalion	(28)	(501)	(1)	(9)	(29)	(510)
Headquarters & Service Company	(13)	(171)	(1)	(9)	(14)	(180)
3 Transport Companies (each)	(5)	(110)			(5)	(110)
Medical Battalion	(1)	(140)	(63)	(395)	(64)	(535)
Headquarters & Service Company	(1)	(15)	(28)	(45)	(29)	(60)
5 Medical Companies (each)		(25)	(7)	(70)	(7)	(95)
Engineer Battalion	41	842	1	20	42	862
Headquarters & Service Company	(23)	(263)	(1)	(20)	(24)	(283)
3 Engineer Companies (each)	(6)	(193)			(6)	(193)
Pioneer Battalion	38	672	3	32	41	704
Headquarters & Service Company	(11)	(81)	(3)	(32)	(14)	(113)
3 Pioneer Companies (each)	(9)	(197)			(9)	(197)
Artillery Regiment	159	2,415	8	57	167	2,472
Headquarters & Service Battery	(23)	(193)	(4)	(9)	(27)	(202)
2 105mm Howitzer Battalions (each)	(33)	(556)	(1)	(12)	(34)	(568)
Headquarters & Service Battery	(15)	(133)	(1)	(12)	(16)	(145)
3 Howitzer Batteries (each)	(6)	(141)			(6)	(141)
2 75mm Pack Howitzer Battalions (each)	(35)	(555)	(1)	(12)	(36)	(567)
Headquarters & Service Battery	(14)	(132)	(1)	(12)	(15)	(144)
3 Pack Howitzer Batteries (each)	(7)	(141)			(7)	(141)
3 Infantry Regiments (each)	137	2,936	11	134	148	3,070
Headquarters & Service Company	(24)	(218)	(5)	(14)	(29)	(232)
Weapons Company	(8)	(195)			(8)	(195)
3 Infantry Battalions (each)	(35)	(841)	(2)	(40)	(37)	(881)
Headquarters Company	(14)	(157)	(2)	(40)	(16)	(197)
3 Rifle Companies (each)	(7)	(228)			(7)	(228)
Division Totals	843	15,548	119	955	962	16,503

[1] All unit strength figures enclosed in parentheses are included in strength totals of parent units.

MAJOR WEAPONS AND TRANSPORTATION—MARINE DIVISION

Weapons	Number	Transportation	Number
Carbine, .30 cal., M–1	10,953	Ambulance:	
Flamethrower, portable, M2–2	243	¼-ton, 4 x 4	52
Flamethrower, mechanized, E4–5	24	½-ton, 4 x 4	12
Gun:		Car, 5-passenger	3
37mm, M3, antitank	36	Station wagon, 4 x 4	3
75mm, motor carriage, M–3, w/armament, radio-equipped (TCS)	12	Tractor: miscellaneous	71
Gun, Machine:		Trailer:	
.30 cal., M1919A4	302	¼-ton, cargo	135
.30 cal., M1917A1	162	½-ton, dump	19
.50 cal., M2	161	1-ton, cargo	155
Gun, submachine, .45 cal	49	1-ton, water	74
Howitzer:		miscellaneous	110
75mm pack	24	Truck:	
105mm	24	¼-ton, 4 x 4	323
Launcher, rocket, antitank, M1A1	172	¼-ton, 4 x 4, with radio	85
Mortar:		1-ton, 4 x 4, cargo	224
60mm	117	1-ton, 4 x 4, reconnaissance	11
81mm	36	2½-ton, 6 x 6, cargo	150
Pistol, .45 cal	399	2½-ton, 6 x 6, dump	53
Rifle, .30 cal., M–1	5,436	miscellaneous	68
Rifle, Browning, automatic	853		
Shotgun, 12 gauge	306		
Tank, Army medium, with armament	46		
Vehicle, recovery, M32B2	3		

Marine Task Organization and Command List[1]

MARINE GROUND UNITS

A. GILBERT ISLANDS[2]

(13 November–8 December 1943)

V Amphibious Corps

(20–25Nov43)

CG ______ MajGen Holland M. Smith
CofS _____ BGen Graves B. Erskine
C–1 ______ LtCol Albert F. Metze
C–2 ______ LtCol St. Julien R. Marshall
C–3 ______ Col Peter P. Schrider
C–4 ______ Col Raymond E. Knapp

V Corps Headquarters and Service Battalion

(20–25Nov43)

CO ______ Maj Thomas R. Wert

V Corps Reconnaissance Company

(18Nov–3Dec43)

CO ______ Capt James L. Jones

2d Airdrome Battalion

(21Aug–8Dec43)

CO ______ LtCol Thomas G. McFarland

[1] Unless otherwise noted, names, positions held, organization titles, and periods of service were taken from the muster rolls of the units concerned, held in the Diary Unit, Files Section, Records Branch, Personnel Department, Headquarters Marine Corps. Units are listed only for those periods, indicated by the dates below parent unit designation, for which they are entitled to campaign participation credit. This information is derived from muster rolls and U.S. Bureau of Naval Personnel, Navy and Marine Corps Awards Manual-NAVPERS 15,790 (Rev. 1953) with changes (Washington, 1953-1958). The muster rolls have been the final authority when there is a conflict in dates of unit entitlement within the overall campaign period as cited by the Awards Manual. In the case of Marine air units, many of which participated in the campaigns as flight or advance echelons only, the unit commander who was actually in the combat area is shown where muster rolls reveal this information. In order to conserve space, only units of battalion and squadron size, or larger, and sizeable separate detachments are listed for each operation, although smaller organizations may have participated also.

[2] Operations within the Gilberts include the capture of Tarawa, Makin and Apamama.

2d Defense Battalion

(24Nov–8Dec43)

CO ______ Col Norman E. True

5th Defense Battalion

(2Oct42–8Dec43)

CO ______ Col George F. Good, Jr.

8th Defense Battalion

(28Nov–8Dec43)

CO ______ Col Lloyd L. Leech

25th Replacement Battalion

(13Nov–8Dec43)

CO ______ Maj John H. McMillan

Headquarters, 2d Marine Division

(20Nov–4Dec43)

CG ______ MajGen Julian C. Smith
ADC _____ BGen Leo D. Hermle
CofS _____ Col Merritt A. Edson
D–1 ______ LtCol Cornelius P. Van Ness
D–2 ______ LtCol Thomas J. Colley
D–3 ______ LtCol James P. Riseley
D–4 ______ LtCol Jesse S. Cook, Jr.

Division Headquarters and Service Battalion

(20Nov–4Dec43)

CO ______ LtCol Lyman G. Miller

Division Special and Service Troops

(20–24Nov43)

CO ______ Col Robert C. Thaxton

2d Amphibian Tractor Battalion

(20Nov–5Dec43)

CO ______ Maj Henry C. Drewes (KIA 20-Nov43)

Maj Henry G. Lawrence, Jr. (from 20Nov43) (WIA 20Nov-43)

2d Antitank Battalion
(20–30Nov43)

CO ______ Capt Saul Glassman

2d Medical Battalion
(20–24Nov43)

CO ______ LCdr Justin J. Stein, MC, USN

2d Service Battalion
(20–24Nov43)

CO ______ Col Clarence H. Baldwin

2d Special Weapons Battalion
(20–24Nov43)

CO ______ Maj Guy E. Tannyhill

2d Tank Battalion
(20–24Nov43)

CO ______ LtCol Alexander B. Swenceski (WIA 20Nov43)

2d Marines
(20–24Nov43)

CO ______ Col David M. Shoup
ExO ______ LtCol Dixon Goen
R–3 ______ Maj Thomas A. Culhane, Jr.

1st Battalion, 2d Marines

CO ______ Maj Wood B. Kyle

2d Battalion, 2d Marines

CO ______ LtCol Herbert R. Amey, Jr. (KIA 20Nov43)
Maj Howard J. Rice (from 20-Nov43)

3d Battalion, 2d Marines

CO ______ Maj John F. Schoettel

6th Marines
(20Nov–8Dec43)

CO ______ Col Maurice G. Holmes
ExO ______ LtCol Russell Lloyd
R–3 ______ Maj Loren E. Haffner

1st Battalion, 6th Marines

CO ______ Maj William K. Jones

2d Battalion, 6th Marines

CO ______ LtCol Raymond L. Murray

3d Battalion, 6th Marines

CO ______ LtCol Kenneth F. McLeod

8th Marines
(20–24Nov43)

CO ______ Col Elmer E. Hall
ExO ______ LtCol Paul D. Sherman
R–3 ______ Maj Hewitt D. Adams

1st Battalion, 8th Marines

CO ______ Maj Lawrence C. Hays, Jr. (WIA 22Nov43)

2d Battalion, 8th Marines

CO ______ Maj Henry P. Crowe

3d Battalion, 8th Marines

CO ______ Maj Robert H. Ruud

10th Marines
(20Nov–1Dec43)

CO ______ BGen Thomas E. Bourke[3]
ExO ______ LtCol Ralph E. Forsyth
R–3 ______ LtCol Marvin H. Floom

1st Battalion, 10th Marines

CO ______ LtCol Presley M. Rixey

2d Battalion, 10th Marines

CO ______ LtCol George R. E. Shell

3d Battalion, 10th Marines

CO ______ LtCol Manly L. Curry

4th Battalion, 10th Marines

CO ______ LtCol Kenneth A. Jorgensen

5th Battalion, 10th Marines

CO ______ Maj Howard V. Hiett

18th Marines
(20–24Nov43)

CO ______ Col Cyril W. Martyr
ExO ______ LtCol Kenneth P. Corson
R–3 ______ LtCol Ewart S. Laue

1st Battalion, 18th Marines (Engineer)

CO ______ Maj August L. Vogt

2d Battalion, 18th Marines (Pioneer)

CO ______ LtCol Chester J. Salazar

---

[3] No replacement colonel was available for the artillery regiment when Bourke was promoted to brigadier general, therefore he remained in command for the Tarawa operation.

3d Battalion, 18th Marines
(18th U. S. Naval Construction Battalion)
CO ______ Cdr Lawrence Emerson Tull, CEC, USN

B. MARSHALL ISLANDS OPERATIONS[4]
(29 January–2 March 1944)

1. Kwajalein-Majuro (29Jan–8Feb44)

V Amphibious Corps[5]
(1–7Feb44)
CG ______ MajGen Holland M. Smith
Cofs ______ BGen Graves B. Erskine
C–1 ______ LtCol Albert F. Metze
C–2 ______ LtCol St. Julien R. Marshall
C–3 ______ Col John C. McQueen
C–4 ______ Col Raymond E. Knapp
C–5 ______ Col Joseph T. Smith

V Corps Headquarters and Service Battalion
(1–7Feb44)
CO ______ Maj Thomas R. Wert

V Corps Signal Battalion
(1–7Feb44)
CO ______ LtCol James H. N. Hudnall

V Corps Reconnaissance Company
(30Jan–2Feb44)
(17–23Feb44)
CO ______ Capt James L. Jones

4th Amphibian Tractor Battalion
(1–6Feb44)
CO ______ LtCol Clovis C. Coffman

10th Amphibian Tractor Battalion
(1–8Feb44)
CO ______ Maj Victor J. Croizat

1st Armored Amphibian Battalion
(1–6Feb44)
CO ______ Maj Louis Metzger

1st Defense Battalion
(1–8Feb44)
CO ______ Col Lewis A. Hohn

[4] Includes: Kwajalein-Majuro Occupation, 29Jan–8Feb44; Eniwetok Occupation, 17Feb–2Mar44.

[5] Tactical Group I, V Amphibious Corps, consisting of headquarters staff and the 22d Marines (Reinforced), as well as other supporting units, represented the Fleet Reserve during the Kwajalein-Majuro operation and is shown in detail under Eniwetok, 17Feb–2Mar44.

15th Defense Battalion
(2–8Feb44)
CO ______ LtCol Francis B. Loomis, Jr.

Headquarters, 4th Marine Division
(1–8Feb44)
CG ______ MajGen Harry Schmidt
ADC ______ BGen James L. Underhill
CofS ______ Col William W. Rogers
D–1 ______ Col Merton J. Batchelder
D–2 ______ Maj Gooderham L. McCormick
D–3 ______ Col Walter W. Wensinger
D–4 ______ Col William F. Brown

Division Headquarters and Service Battalion
(1–8Feb44)
CO ______ LtCol Melvin L. Krulewitch

Division Special and Service Troops
(1–8Feb44)
CO ______ Col Emmett W. Skinner

4th Medical Battalion
(1–8Feb44)
CO ______ LCdr Stewart W. Shimonek, MC, USN

4th Motor Transport Battalion
(1–8Feb44)
CO ______ LtCol Ralph L. Schiesswohl

4th Service Battalion
(1–8Feb44)
CO ______ Maj John L. Lamprey, Jr.

4th Special Weapons Battalion
(1–8Feb44)
CO ______ LtCol Alexander A. Vandegrift, Jr.

4th Tank Battalion
(1–8Feb44)
CO ______ Maj Richard K. Schmidt

14th Marines
(1–8Feb44)
CO ______ Col Louis G. DeHaven
ExO ______ LtCol Randall M. Victory
R–3 ______ Maj Frederick J. Karch

1st Battalion, 14th Marines
CO ______ LtCol Harry J. Zimmer

2d Battalion, 14th Marines
CO ______ LtCol George B. Wilson, Jr.

3d Battalion, 14th Marines

CO ______ LtCol Robert E. MacFarlane

4th Battalion, 14th Marines

CO ______ Maj Carl A. Youngdale

20th Marines
(1–8Feb44)

CO ______ Col Lucian W. Burnham
ExO ______ LtCol Nelson K. Brown
R–3 ______ Maj Melvin D. Henderson

1st Battalion, 20th Marines (Engineer)

CO ______ Maj Richard G. Ruby

2d Battalion, 20th Marines (Pioneer)

CO ______ LtCol Otto Lessing

3d Battalion, 20th Marines
(121st U. S. Naval Construction Battalion)

CO ______ LCdr William G. Byrne, CEC, USN

23d Marines
(1–8Feb44)

CO ______ Col Louis R. Jones
ExO ______ LtCol John R. Lanigan
R–3 ______ Maj Edward W. Wells

1st Battalion, 23d Marines

CO ______ LtCol Hewin O. Hammond

2d Battalion, 23d Marines

CO ______ LtCol Edward J. Dillon

3d Battalion, 23d Marines

CO ______ LtCol John J. Cosgrove, Jr.

24th Marines
(1–8Feb44)

CO ______ Col Franklin A. Hart
ExO ______ LtCol Homer L. Litzenberg, Jr.
R–3 ______ LtCol Charles D. Roberts

1st Battalion, 24th Marines

CO ______ LtCol Aquilla J. Dyess (KIA 2Feb44)
Maj Maynard C. Schultz (from 2Feb44)

2d Battalion, 24th Marines

CO ______ LtCol Francis H. Brink

3d Battalion, 24th Marines

CO ______ LtCol Austin R. Brunelli

25th Marines
31Jan–8Feb44)

CO ______ Col Samuel C. Cumming
ExO ______ LtCol Walter I. Jordan
R–3 ______ LtCol William F. Thyson, Jr.

1st Battalion, 25th Marines

CO ______ LtCol Clarence J. O'Donnell

2d Battalion, 25th Marines

CO ______ LtCol Lewis C. Hudson, Jr.

3d Battalion, 25th Marines

CO ______ LtCol Justice M. Chambers

2. Eniwetok (17Feb–2Mar44)
Tactical Group 1, V Amphibious Corps
(17Feb–2Mar44)

CG ______ BGen Thomas E. Watson
S–1 ______ (none shown)
S–2 ______ Maj Robert W. Shaw
S–3 ______ LtCol Wallace M. Greene, Jr.
S–4 ______ LtCol Robert J. Straub

10th Defense Battalion
(21Feb–2Mar44)

CO ______ LtCol Wallace O. Thompson

22d Marines (Reinforced)
(17–25Feb44)

CO ______ Col John T. Walker
ExO ______ Col Merlin F. Schneider
R–3 ______ LtCol Floyd R. Moore

1st Battalion, 22d Marines

CO ______ LtCol Walfried H. Fromhold

2d Battalion, 22d Marines

CO ______ LtCol Donn C. Hart

3d Battalion, 22d Marines

CO ______ Maj Clair W. Shisler

2d Separate Pack Howitzer Battalion
(17–25Feb44)

CO ______ LtCol Edwin C. Ferguson (WIA 18Feb44)

C. SAIPAN (11 June–10 August 1944)

Expeditionary Troops
(15Jun–24Jul44)

CG ______ LtGen Holland M. Smith
CofS ____ BGen Graves B. Erskine
G–1 ______ LtCol Albert F. Metze
G–2 ______ LtCol Thomas R. Yancey, USA
G–3 ______ Col Robert E. Hogaboom
G–4 ______ LtCol Joseph C. Anderson, USA

V Amphibious Corps—Northern Troops and Landing Force
(15Jun–24Jul44)

CG ______ LtGen Holland M. Smith (to 11Jul44)
MajGen Harry Schmidt (from 12Jul44)
CofS ____ BGen Graves B. Erskine
C–1 ______ LtCol Albert F. Metze
C–2 ______ Col St. Julien R. Marshall
C–3 ______ Col John C. McQueen
C–4 ______ Col Raymond E. Knapp
C–5 ______ Col Joseph T. Smith

V Corps Headquarters and Service Battalion

CO ______ Maj Thomas R. Wert (15Jun–24Jul44)

V Corps Amphibious Reconnaissance Battalion
(15Jun–24Jul44)

CO ______ Capt James L. Jones

V Corps Medical Battalion
(15Jun–24Jul44)

CO ______ LCdr William B. Clapp, MC, USN

V Corps Signal Battalion
(15Jun–24Jul44)

CO ______ Col James H. N. Hudnall

2d Amphibian Tractor Battalion
(15Jun–24Jul44)

CO ______ Maj Henry G. Lawrence, Jr. (to 2Jul44)
Maj Fenlon A. Durand (from 3Jul44)

5th Amphibian Tractor Battalion
(15Jun–24Jul44)

CO ______ Capt George L. Shead

10th Amphibian Tractor Battalion
(15Jun–24Jul44)

CO ______ Maj Victor J. Croizat

11th Amphibian Tractor Battalion
(15Jun–24Jul44)

CO ______ Maj Walter S. Haskell, Jr.

2d Armored Amphibian Battalion
(15Jun–24Jul44)

CO ______ LtCol Reed M. Fawell, Jr.

2d 155mm Howitzer Battalion
(15Jun–16Jul44)

CO ______ LtCol Marvin H. Floom

4th 105mm Howitzer Battalion
(15Jun–24Jul44)

CO ______ LtCol Douglas E. Reeve

2d Separate Engineer Battalion
(12Jul–10Aug44)

CO ______ LtCol Charles O. Clark

7th Field Depot
(15Jun–24Jul44)

CO ______ LtCol Edwin D. Partridge (to 25Jun44)
Col Earl H. Phillips (from 26Jun44)

Headquarters, 2d Marine Division
(15Jun–24Jul44)

CG ______ MajGen Thomas E. Watson
ADC ____ BGen Merritt A. Edson
CofS ____ Col David M. Shoup
D–1 ______ LtCol James T. Wilbur
D–2 ______ LtCol Thomas J. Colley
D–3 ______ LtCol Wallace M. Greene, Jr.
D–4 ______ Col Robert J. Straub

Division Headquarters and Service Battalion
(15Jun–24Jul44)

CO ______ Maj Melvin A. Smith

Division Special Troops
(15Jun–24Jul44)

CO ______ (none shown)

2d Medical Battalion
(15Jun–16Jul44)

CO ______ LCdr Claude R. Bruner, MC, USN

2d Motor Transport Battalion
(15Jun–24Jul44)

CO ______ Maj Milton J. Green

2d Service Battalion
(15Jun–24Jul44)

CO ______ Maj Edward V. Dozier

2d Tank Battalion
(15Jun–24Jul44)

CO ______ Maj Charles W. McCoy

2d Marines
(15Jun–24Jul44)

CO ______ Col Walter J. Stuart
ExO ______ LtCol John H. Griebel
R–3 ______ Maj Samuel D. Mandeville, Jr.

1st Battalion, 2d Marines

CO ______ LtCol Wood B. Kyle

2d Battalion, 2d Marines

CO ______ LtCol Richard C. Nutting

3d Battalion, 2d Marines

CO ______ LtCol Arnold F. Johnston (WIA 16 and 21Jun44)
Maj Harold "K" Throneson (from 21Jun44)
LtCol Arnold F. Johnston (from 5Jul44)
LtCol Walter F. Layer (from 11Jul44)

6th Marines
(15Jun–26Jul44)

CO ______ Col James P. Riseley
ExO ______ LtCol Kenneth F. McLeod (KIA (25Jun44)
LtCol Russell Lloyd (from (25 Jun44)
R–3 ______ Maj Loren E. Haffner

1st Battalion, 6th Marines

CO ______ LtCol William K. Jones

2d Battalion, 6th Marines

CO ______ LtCol Raymond L. Murray (WIA 15Jun44)
Maj LeRoy P. Hunt, Jr. (from 15Jun44)
LtCol Edmund B. Games (from 11Jul44)

3d Battalion, 6th Marines

CO ______ LtCol John W. Easley (WIA 15Jun44)
Maj John E. Rentsch (from 16Jun44)
LtCol John W. Easley (from 3Jul44)

8th Marines
(15Jun–24Jul44)

CO ______ Col Clarence R. Wallace
ExO ______ LtCol Jack P. Juhan
R–3 ______ Maj William H. Souder, Jr.

1st Battalion, 8th Marines

CO ______ LtCol Lawrence C. Hays, Jr. (WIA 15Jun44)

2d Battalion, 8th Marines

CO ______ LtCol Henry P. Crowe (WIA 15Jun44)

3d Battalion, 6th Marines

CO ______ LtCol John C. Miller (WIA 15Jun44)
Maj Stanley E. Larsen (WIA 15Jun44)
LtCol Gavin C. Humphrey (from 10Jul44)

10th Marines
(15Jun–24Jul44)

CO ______ Col Raphael Griffin
ExO ______ LtCol Ralph E. Forsyth (KIA 23Jun44)
Col Presley M. Rixey (from 24Jun44)
R–3 ______ LtCol Howard V. Hiett (WIA 17Jun44)
Maj Richard B. Church (WIA 24Jun44)
Maj Wade H. Hitt (from 24Jun44)

1st Battalion, 10th Marines

CO ______ Col Presley M. Rixey (to 24Jun44)
Maj Wendell H. Best (from 25Jun44)
LtCol Donovan D. Sult (from 16Jul44)

2d Battalion, 10th Marines

CO ______ LtCol George R. E. Shell (WIA 16Jun44)

Maj Kenneth C. Houston (from 16Jun44)
Maj David L. Henderson (from 16Jul44)

3d Battalion, 10th Marines

CO ______ Maj William L. Crouch (KIA 7Jul44)
Maj James O. Appleyard (from 8Jul44)
LtCol William C. Capehart (from 16Jul44)

4th Battalion, 10th Marines

CO ______ LtCol Kenneth A. Jorgensen

18th Marines
(15Jun–24Jul44)

CO ______ LtCol Russell Lloyd (to 24Jun44)
Col Cyril W. Martyr (from 25Jun44)
ExO ____ LtCol Ewart S. Laue
R–3 _____ Capt Murdoch J. McLeod

1st Battalion, 18th Marines (Engineer)

CO ______ LtCol August L. Vogt

2d Battalion, 18th Marines (Pioneer)

CO ______ LtCol Chester J. Salazar

18th U.S. Naval Construction Battalion

CO ______ Cdr Lawrence Emerson Tull, CEC, USN

Headquarters, 4th Marine Division
(15Jun–24Jul44)

CG ______ MajGen Harry Schmidt (to 11Jul44)
MajGen Clifton B. Cates (from 12Jul44)
ADC ____ BGen Samuel C. Cumming
CofS ____ Col William W. Rogers
D–1 _____ Col Walter I. Jordan
D–2 _____ LtCol Gooderham L. McCormick
D–3 _____ Col Walter W. Wensinger
D–4 _____ Col William F. Brown

Division Headquarters and Service Battalion
(15Jun–24Jul44)

CO ______ LtCol Melvin L. Krulewitch

Division Special and Service Troops
(15Jun–24Jul44)

CO ______ Col Orin H. Wheeler

4th Medical Battalion
(15Jun–24Jul44)

CO ______ LCdr George W. Mast, MC, USN

4th Motor Transport Battalion
(15Jun–24Jul44)

CO ______ LtCol Ralph L. Schiesswohl

4th Service Battalion
(15Jun–24Jul44)

CO ______ Col Richard H. Schubert

4th Tank Battalion
(15Jun–24Jul44)

CO ______ Maj Richard K. Schmidt

14th Marines
(15Jun–24Jul44)

CO ______ Col Louis G. DeHaven
ExO _____ LtCol Randall M. Victory
R–3 _____ Maj Frederick J. Karch

1st Battalion, 14th Marines

CO ______ LtCol Harry J. Zimmer

2d Battalion, 14th Marines

CO ______ LtCol George B. Wilson, Jr.

3d Battalion, 14th Marines

CO ______ LtCol Robert E. MacFarlane

4th Battalion, 14th Marines

CO ______ LtCol Carl A. Youngdale

20th Marines
(15Jun–24Jul44)

CO ______ LtCol Nelson K. Brown
ExO ____ Capt William M. Anderson
R–3 _____ Maj Melvin D. Henderson

1st Battalion, 20th Marines (Engineer)

CO ______ Maj Richard G. Ruby

2d Battalion, 20th Marines (Pioneer)

CO ______ Maj John H. Partridge

121st U. S. Naval Construction Battalion

CO ______ LCdr William G. Byrne, CEC, USN

23d Marines
(15Jun–24Jul44)

CO ______ Col Louis R. Jones

ExO ______ LtCol John R. Lanigan (WIA 12Jul44)
R-3 ______ Maj Edward W. Wells (to 16Jul44)
Capt William E. Buron (from 17Jul44)

1st Battalion, 23d Marines

CO ______ LtCol Ralph Haas

2d Battalion, 23d Marines

CO ______ LtCol Edward J. Dillon (WIA 6 and 9Jul44)

3d Battalion, 23d Marines

CO ______ LtCol John J. Cosgrove, Jr. (WIA 19Jun44)
Maj Paul S. Treitel (from 19Jun44)

24th Marines

(15Jun–24Jul44)

CO ______ Col Franklin A. Hart
ExO ______ LtCol Austin R. Brunelli
R-3 ______ LtCol Charles D. Roberts

1st Battalion, 24th Marines

CO ______ LtCol Maynard C. Schultz (KIA 16Jun44)
Maj Robert N. Fricke (16–27Jun44)
LtCol Otto Lessing (from 28Jun44)

2d Battalion, 24th Marines

CO ______ LtCol Richard Rothwell

3d Battalion, 24th Marines

CO ______ LtCol Alexander A. Vandegrift, Jr. (WIA 29Jun44)

25th Marines

(15Jun–24Jul44)

CO ______ Col Merton J. Batchelder
ExO ______ LtCol Clarence J. O'Donnell
R-3 ______ LtCol William F. Thyson, Jr.

1st Battalion, 25th Marines

CO ______ LtCol Hollis U. Mustain

2d Battalion, 25th Marines

CO ______ LtCol Lewis C. Hudson, Jr.

3d Battalion, 25th Marines

CO ______ LtCol Justice M. Chambers (WIA 22Jun44)
Maj James Taul (22–23Jun44)
LtCol Justice M. Chambers (from 24Jun44)

1st Battalion, 29th Marines

(15Jun–24Jul44)

CO ______ LtCol Guy E. Tannyhill (WIA 17Jun44)
LtCol Rathvon McC. Tompkins (WIA 2Jul44)
Maj William W. McKinley (4–15Jul44)
LtCol Orin K. Pressley (from 16Jul44)

D. TINIAN (24 July–1 August 1944)

Expeditionary Troops

(24Jul–10Aug44)

CG ______ MajGen Harry Schmidt
CofS ______ BGen Graves B. Erskine
C-1 ______ LtCol Albert F. Metze
C-2 ______ Col St. Julien R. Marshall
C-3 ______ Col John C. McQueen
C-4 ______ LtCol Raymond E. Knapp
C-5 ______ Col Joseph T. Smith

V Amphibious Corps—Northern Troops and Landing Force

(24Jul–10Aug44)

CG ______ MajGen Harry Schmidt
CofS ______ BGen Graves B. Erskine
G-1 ______ LtCol Albert F. Metze
G-2 ______ LtCol Thomas R. Yancey, USA
G-3 ______ Col Robert E. Hogaboom
G-4 ______ LtCol Joseph C. Anderson, USA

V Corps Headquarters and Service Battalion

(24Jul–7Aug44)

CO ______ Maj Thomas R. Wert

V Corps Amphibious Reconnaissance Battalion

(24Jul–10Aug44)

CO ______ Capt James L. Jones

V Corps Medical Battalion

(24Jul–7Aug44)

CO ______ LCdr William B. Clapp, MC, USN

V Corps Signal Battalion
(24Jul–7Aug44)
CO ______ Col James H. N. Hudnall

2d Amphibian Tractor Battalion
(24Jul–10Aug44)
CO ______ Maj Fenlon A. Durand

5th Amphibian Tractor Battalion
(24Jul–10Aug44)
CO ______ Maj George L. Shead

10th Amphibian Tractor Battalion (Less Company A)
(24Jul–7Aug44)
CO ______ Maj Victor J. Croizat

2d Armored Amphibian Battalion
(24Jul–10Aug44)
CO ______ LtCol Reed M. Fawell, Jr.

4th 105mm Howitzer Battalion
(24Jul–7Aug44)
CO ______ LtCol Douglas E. Reeve

7th Field Depot
(24Jul–10Aug44)
CO ______ Col Earl H. Phillips

Headquarters, 2d Marine Division
(24Jul–10Aug44)
CG ______ MajGen Thomas E. Watson
ADC ______ BGen Merritt A. Edson
CofS ______ Col David M. Shoup
D-1 ______ LtCol James T. Wilbur
D-2 ______ LtCol Thomas J. Colley
D-3 ______ LtCol Wallace M. Greene, Jr.
D-4 ______ Col Robert J. Straub

Division Headquarters and Service Battalion
(24Jul–10Aug44)
CO ______ Maj Melvin A. Smith

Division Special Troops
CO ______ (none shown)

2d Medical Battalion
(24Jul–10Aug44)
CO ______ LCdr Claude R. Bruner, MC, USN

2d Motor Transport Battalion
(24Jul–10Aug44)
CO ______ Maj Milton J. Green

2d Service Battalion
(24Jul–10Aug44)
CO ______ Maj Edward V. Dozier

2d Tank Battalion
(24Jul–10Aug44)
CO ______ Maj Charles W. McCoy

2d Marines
(24Jul–10Aug44)
CO ______ Col Walter J. Stuart
ExO ______ LtCol John H. Griebel
R-3 ______ Maj Samuel D. Mandeville, Jr.

1st Battalion, 2d Marines
CO ______ LtCol Wood B. Kyle

2d Battalion, 2d Marines
CO ______ LtCol Richard C. Nutting

3d Battalion, 2d Marines
CO ______ LtCol Walter F. Layer

6th Marines
(26Jul–10Aug44)
CO ______ Col James P. Riseley
ExO ______ LtCol Russell Lloyd
R-3 ______ Maj Loren E. Haffner

1st Battalion, 6th Marines
CO ______ LtCol William K. Jones

2d Battalion, 6th Marines
(24Jul–8Aug44)
CO ______ LtCol Edmund B. Games

3d Battalion, 6th Marines
CO ______ LtCol John W. Easley (KIA 2Aug44)
Maj John E. Rentsch (from 2Aug44)

8th Marines
(24Jul–10Aug44)
CO ______ Col Clarence R. Wallace
ExO ______ LtCol Jack P. Juhan
R-3 ______ Maj William H. Souder, Jr.

1st Battalion, 8th Marines
CO ______ LtCol Lawrence C. Hays, Jr.

2d Battalion, 8th Marines
CO ______ LtCol Lane C. Kendall

3d Battalion, 8th Marines

CO ------- LtCol Gavin C. Humphrey

10th Marines

(24Jul–10Aug44)

CO ------- Col Raphael Griffin
ExO ----- Col Presley M. Rixey
R–3 ------ Maj Wade H. Hitt

1st Battalion, 10th Marines

CO ------- LtCol Donovan D. Sult

2d Battalion, 10th Marines

CO ------- Maj David L. Henderson

3d Battalion, 10th Marines

CO ------- LtCol William C. Capehart

4th Battalion, 10th Marines

CO ------- LtCol Kenneth A. Jorgensen

18th Marines

(24Jul–10Aug44)

CO ------- Col Cyril W. Martyr
ExO ----- LtCol Ewart S. Laue
R–3 ------ Capt Murdoch J. McLeod

1st Battalion, 18th Marines (Engineer)

CO ------- Col August L. Vogt

2d Battalion, 18th Marines (Pioneer)

CO ------- LtCol Chester J. Salazar

18th U. S. Naval Construction Battalion

CO ------- Cdr Lawrence Emerson Tull, CEC, USN

Headquarters, 4th Marine Division

(24Jul–7Aug44)

CG ------- MajGen Clifton B. Cates
ADC ----- BGen Samuel C. Cumming
CofS ----- Col William W. Rogers
D–1 ------ Col Walter I. Jordan
D–2 ------ LtCol Gooderham L. McCormick
D–3 ------ Col Walter W. Wensinger
D–4 ------ Col William F. Brown

Division Headquarters and Service Battalion

(24Jul–7Aug44)

CO ------- LtCol Melvin L. Krulewitch

Division Special and Service Troops

(24Jul–7Aug44)

CO ------- Col Orin H. Wheeler

4th Medical Battalion

(24Jul–7Aug44)

CO ------- LCdr George W. Mast, MC, USN

4th Motor Transport Battalion

(24Jul–7Aug44)

CO ------- LtCol Ralph L. Schiesswohl

4th Service Battalion

(24Jul–7Aug44)

CO ------- Col Richard H. Schubert

4th Tank Battalion

(24Jul–7Aug44)

CO ------- Maj Richard K. Schmidt

14th Marines

(24Jul–7Aug44)

CO ------- Col Louis G. DeHaven
ExO ----- LtCol Randall M. Victory
R–3 ------ Maj Frederick J. Karch

1st Battalion, 14th Marines

CO ------- LtCol Harry J. Zimmer (KIA 25Jul44)
Maj Clifford B. Drake (from 26Jul44)

2d Battalion, 14th Marines

CO ------- LtCol George B. Wilson, Jr.

3d Battalion, 14th Marines

CO ------- LtCol Robert E. MacFarlane

4th Battalion, 14th Marines

CO ------- LtCol Carl A. Youngdale

20th Marines

(24Jul–10Aug44)

CO ------- LtCol Nelson K. Brown
ExO ----- Maj Richard G. Ruby
R–3 ------ Maj Melvin D. Henderson (to 2Aug44)
LtCol Otto Lessing (from 3Aug44)

1st Battalion, 20th Marines (Engineer)

CO ------- Maj Richard G. Ruby

2d Battalion, 20th Marines (Pioneer)

CO ------- Maj John H. Partridge

121st U. S. Naval Construction Battalion

CO ______ LCdr William G. Byrne, CEC, USN

23d Marines
(24Jul–10Aug44)

CO ______ Col Louis R. Jones
ExO ______ LtCol John R. Lanigan
R-3 ______ Capt William E. Buron

1st Battalion, 23d Marines

CO ______ LtCol Ralph Haas

2d Battalion, 23d Marines

CO ______ LtCol Edward J. Dillon

3d Battalion, 23d Marines

CO ______ Maj Paul S. Treitel

24th Marines
(24Jul–10Aug44)

CO ______ Col Franklin A. Hart
ExO ______ LtCol Austin R. Brunelli
R-3 ______ LtCol Charles D. Roberts

1st Battalion, 24th Marines

CO ______ LtCol Otto Lessing

2d Battalion, 24th Marines

CO ______ Maj Frank E. Garretson (to 27Jul)
LtCol Richard Rothwell (from 27Jul)

3d Battalion, 24th Marines

CO ______ LtCol Alexander A. Vandegrift, Jr.

25th Marines
(24Jul–7Aug44)

CO ______ Col Merton J. Batchelder
ExO ______ LtCol Clarence J. O'Donnell
R-3 ______ LtCol William F. Thyson, Jr.

1st Battalion, 25th Marines

CO ______ LtCol Hollis U. Mustain

2d Battalion, 25th Marines

CO ______ LtCol Lewis C. Hudson, Jr.

3d Battalion, 25th Marines

CO ______ LtCol Justice M. Chambers

1st Battalion, 29th Marines
(24Jul–10Aug44)

CO ______ LtCol Orin K. Pressley

E. GUAM (12 July–August 1944)

Expeditionary Troops
(21Jul–15Aug44)

CG ______ MajGen Harry Schmidt
CofS ______ BGen Graves B. Erskine
C-1 ______ LtCol Albert F. Metze
C-2 ______ Col St. Julien R. Marshall
C-3 ______ Col John C. McQueen
C-4 ______ LtCol Raymond E. Knapp
C-5 ______ Col Joseph T. Smith

III Amphibious Corps—Southern Troops and Landing Force
(21Jul–15Aug44)

CG ______ MajGen Roy S. Geiger
CofS ______ Col Merwin H. Silverthorn
C-1 ______ Col William J. Scheyer
C-2 ______ LtCol William F. Coleman
C-3 ______ Col Walter A. Wachtler
C-4 ______ LtCol Frederick L. Wieseman
C-5 ______ Col Dudley S. Brown

III Corps Headquarters and Service Battalion
(21Jul–15Aug44)

CO ______ LtCol Floyd A. Stephenson

III Corps Medical Battalion
(21Jul–15Aug44)

CO ______ LCdr William H. Rambo, MC, USN

III Corps Motor Transport Battalion
(21Jul–15Aug44)

CO ______ Maj Franklin H. Hayner (to 1Aug44)
Maj Kenneth E. Murphy (from 1-3Aug44)
Maj Franklin H. Hayner (from 4Aug44)

III Corps Signal Battalion
(21Jul–15Aug44)

CO ______ LtCol Robert L. Peterson

3d Amphibian Tractor Battalion
(21Jul–15Aug44)

CO ______ LtCol Sylvester L. Stephan

4th Amphibian Tractor Battalion
(21Jul–15Aug44)

CO ______ LtCol Clovis C. Coffman

1st Armored Amphibian Battalion
(21Jul–13Aug44)

CO ______ Maj Louis Metzger

1st Separate Engineer Battalion
(21Jul–15Aug44)

CO ______ LtCol Orin C. Bjornsrud

2d Separate Engineer Battalion
(21Jul–15Aug44)

CO ______ LtCol Charles O. Clark

III Corps Artillery
(21Jul–15Aug44)

CG ______ BGen Pedro A. del Valle
CofS ____ Col John Bemis
A–1 _____ Maj James H. Tatsch
A–2 _____ WO David G. Garnett
A–3 _____ LtCol Frederick P. Henderson (FA)
LtCol Edgar O. Price (AA)
A–4 _____ Maj Frederick W. Miller

1st 155mm Howitzer Battalion
(21Jul–15Aug44)

CO ______ Col James J. Keating

2d 155mm Howitzer Battalion
(21Jul–15Aug44)

CO ______ LtCol Marvin H. Floom

7th 155mm Gun Battalion
(21Jul–15Aug44)

CO ______ LtCol John S. Twitchell

9th Defense Battalion
(21Jul–15Aug44)

CO ______ LtCol Archie E. O'Neil

14th Defense Battalion
(21Jul–15Aug44)

CO ______ LtCol William F. Parks

Headquarters, 3d Marine Division
(21Jul–15Aug44)

CG ______ MajGen Allen H. Turnage
ADC ____ BGen Alfred H. Noble
CofS ____ Col Ray A. Robinson
D–1 _____ LtCol Chevey S. White (KIA 22Jul44)
Maj Irving R. Kriendler (from 22Jul44)
D–2 _____ LtCol Howard J. Turton (to 28Jul44)
LtCol Ellsworth N. Murray (from 29Jul44)
D–3 _____ Col James A. Stuart (to 28Jul44)
LtCol Howard J. Turton (from 29Jul44)
D–4 _____ LtCol Ellsworth N. Murray (to 28Jul44)
Col William C. Hall (from 29Jul44)

Division Headquarters and Service Battalion
(21Jul–15Aug44)

CO ______ LtCol Newton B. Barkley

3d Medical Battalion
(21Jul–15Aug44)

CO ______ Cdr Raymond R. Callaway, MC, USN

3d Motor Transport Battalion
21Jul–15Aug44)

CO ______ LtCol Thomas R. Stokes

3d Service Battalion
(21Jul–15Aug44)

CO ______ LtCol Durant S. Buchanan

3d Tank Battalion
(21Jul–15Aug44)

CO ______ LtCol Hartnoll J. Withers

3d Marines
(21Jul–15Aug44)

CO ______ Col William C. Hall (to 28Jul44)
Col James A. Stuart (from 29Jul44)
ExO ____ Col James Snedeker
R–3 _____ Maj John A. Scott

1st Battalion, 3d Marines

CO ______ Maj Henry Aplington II

2d Battalion, 3d Marines

CO ______ LtCol Hector de Zayas (KIA 26Jul44)
Maj William A. Culpepper (from 26Jul44)

3d Battalion, 3d Marines

CO ______ LtCol Ralph L. Houser (WIA 22Jul44)

214-881 O-67—41

Maj Royal R. Bastian, Jr. (from 24Jul44)

9th Marines

(21Jul–15Aug44)

CO ______ Col Edward A. Craig
ExO _____ LtCol Jaime Sabater (WIA 21Jul44)
LtCol Ralph M. King (from 30Jul44)
(none shown 11–15Aug44)
R-3 ______ Capt Evan E. Lips

1st Battalion, 9th Marines

CO ______ LtCol Carey A. Randall

2d Battalion, 9th Marines

CO ______ LtCol Robert E. Cushman, Jr.

3d Battalion, 9th Marines

CO ______ LtCol Walter Asmuth, Jr. (WIA 21Jul44)
Maj Donald B. Hubbard (WIA 1Aug44)
Maj Jess P. Ferrill, Jr. (from) 2Aug44)

12th Marines

CO ______ Col John B. Wilson
ExO _____ LtCol John S. Letcher
R-3 ______ LtCol William T. Fairbourn (to 13Aug44)
Maj Thomas R. Belzer (from 14-Aug44)

1st Battalion, 12th Marines

CO ______ LtCol Raymond F. Crist, Jr.

2d Battalion, 12th Marines

CO ______ LtCol Donald M. Weller (to 13-Aug44)
LtCol William T. Fairbourn (from 14Aug44)

3d Battalion, 12th Marines

CO ______ LtCol Alpha L. Bowser, Jr.

4th Battalion, 12th Marines

CO ______ LtCol Bernard H. Kirk (WIA 21Jul44)

19th Marines

(21Jul–15Aug44)

CO ______ LtCol Robert E. Fojt
ExO _____ LtCol Edmund M. Williams
R-3 ______ Maj George D. Flood, Jr.

1st Battalion, 19th Marines (Engineer)

CO ______ LtCol Walter S. Campbell

2d Battalion, 19th Marines (Pioneer)

CO ______ Maj Victor J. Simpson

25th U. S. Naval Construction Battalion

CO ______ LCdr George J. Whelan, CEC, USN

21st Marines

(21Jul–15Aug44)

CO ______ Col Arthur H. Butler
ExO _____ LtCol Ernest W. Fry, Jr.
R-3 ______ Maj James H. Tinsley

1st Battalion, 21st Marines

CO ______ LtCol Ronald R. Van Stockum

2d Battalion, 21st Marines

CO ______ LtCol Eustace R. Smoak

3d Battalion, 21st Marines

CO ______ LtCol Wendell H. Duplantis

1st Provisional Marine Brigade

(21Jul–15Aug44)

CG ______ BGen Lemuel C. Shepherd, Jr.
CofS _____ Col John T. Walker
B-1 ______ Maj Addison B. Overstreet
B-2 ______ Maj Robert W. Shaw
B-3 ______ LtCol Thomas A. Culhane, Jr.
B-4 ______ LtCol August Larson

4th Marines (Reinforced)

(21Jul–15Aug44)

CO ______ LtCol Alan Shapley
ExO _____ LtCol Samuel D. Puller (KIA 27Jul44)
Capt Charles T. Lamb (from 27-Jul44)
R-3 ______ Maj Orville V. Bergren

1st Battalion, 4th Marines

CO ______ Maj Bernard W. Green

2d Battalion, 4th Marines

CO ______ Maj John S. Messer

3d Battalion, 4th Marines

CO ______ Maj Hamilton M. Hoyler

22d Marines

(21Jul–15Aug44)

CO ______ Col Merlin F. Schneider
ExO ____ LtCol William J. Wise
R-3 _____ LtCol Horatio C. Woodhouse, Jr.

1st Battalion, 22d Marines

CO ______ LtCol Walfried H. Fromhold (to 31Jul44)
Maj Crawford B. Lawton (from 1Aug44)

2d Battalion, 22d Marines

CO ______ LtCol Donn C. Hart (to 27Jul44)
Maj John F. Schoettel (from 28-Jul44)

3d Battalion, 22d Marines

CO ______ LtCol Clair W. Shisler

1st Marine Brigade Artillery Group

(21Jul–15Aug44)

CO ______ LtCol Edwin C. Ferguson

Pack Howitzer Battalion, 4th Marines

(21Jul–15Aug44)

CO ______ Maj Robert Armstrong

Pack Howitzer Battalion, 22d Marines

(21Jul–15Aug44)

CO ______ Maj Alfred M. Mahoney

53d U. S. Naval Construction Battalion

CO ______ LCdr Edward M. Denbo, CEC, USN

1st Provisional Base Headquarters, Island Command, Guam

(26Jul–15Aug44)

CG _____ MajGen Henry L. Larsen
CofS ____ Col Robert Blake
A-1 _____ Col Lee N. Utz
A-2 _____ Col Francis H. Brink
A-3 _____ Col Benjamin W. Atkinson (to 8Aug44)
LtCol Shelton C. Zern (from 9-Aug44)
A-4 _____ Col James A. Mixson
A-5 _____ Col Charles L. Murray

Headquarters and Service Battalion, 1st Provisional Base Headquarters, Island Command, Guam

(26Jul–15Aug44)

CO ______ LtCol Victor A. Barraco

5th Field Depot

(21Jul–15Aug44)

CO ______ LtCol Walter A. Churchill

MARINE AIR UNITS

Headquarters and Detachments, 4th Marine Base Defense Aircraft Wing

(C—17Jun–10Aug44)
(E—17Jun–15Aug44) [a]

CG ______ BGen Thomas J. Cushman
CofS ____ Col Frank H. Lamson-Scribner
W-1 ____ Maj Theodore Brewster
W-2 ____ Capt Charles J. Greene, Jr.
W-3 ____ LtCol Lee C. Merrell, Jr.
W-4 ____ LtCol Harrison Brent, Jr. (to 26Jul44)
LtCol John B. Jacob (from 26-Jul44)
CO, HqSqn-4 Maj Melchior B. Trelfall

Forward Echelon, Marine Aircraft Group 21

(E—27Jul–15Aug44)

CO ______ Col Peter P. Schrider
ExO ____ Col James A. Booth, Jr.
GruOpsO _ LtCol Robert W. Clark
CO, HqSqn-21 Maj Robert F. Higley
CO, SMS-21 Maj Charleton B. Ivey

Marine Aircraft Group 22

(B2—20Feb–2Mar44)

CO ______ Col James M. Daly
ExO ____ LtCol Richard D. Hughes
GruOpsO _ LtCol Julian F. Walters
CO, HqSqn-22 1stLt John W. Hackner
CO, SMS-22 Capt John A. Hood

---

[a] Under each unit listed there will appear a letter designation for each operation in which the unit participated, and dates of involvement. Following are the campaigns and dates of entitlements:

A. Gilbert Islands ________ 13Nov–8Dec43
B. Marshall Islands
  1. Kwajalein-Majuro ________ 29Jan–8Feb44
  2. Eniwetok ________ 17Feb–2Mar44
C. Saipan ________ 11Jun–10Aug44
D. Tinian ________ 24Jul–1Aug44
E. Guam ________ 12Jul–15Aug44

Marine Aircraft Group 31

(B1—7–8Feb44)

CO ______ Col Calvin R. Freeman
ExO ______ Col Edward B. Carney
GruOpsO _ LtCol Ralph K. Rottet
CO, HqSqn–31 Capt Warren S. Adams, II
CO, SMS–31 Capt Neil A. Vestal

Marine Air Warning Squadron 1

(B2—20Feb–2Mar44)

CO ______ Capt William D. Felder, Jr.

Marine Observation Squadron 1

(E—21Jul–15Aug44)

CO ______ Maj Gordon W. Heritage

Marine Air Warning Squadron 2

(E—21Jul–15Aug44)

CO ______ Capt George T. C. Fry

Marine Observation Squadron 2

(C—17Jun–10Aug44)

CO ______ Maj Robert W. Edmondson (to 26Jun44)
Capt John A. Ambler (from 27-Jun44)

Marine Observation Squadron 4

(C—15Jun–10Aug44)

CO ______ Capt Nathan D. Blaha (to 26-Jun44)
1stLt Thomas Rozga (from 26-Jun44)

Marine Air Warning Squadron 5

(redesignated Marine Assault Air Warning Squadron 5, effective 10Jul44)

(C—15Jun–10Aug44)

CO ______ Capt Donald D. O'Neill

Ground Echelon, Marine Fighter Squadron 111

(B1—7–8Feb44)

CO ______ Maj "J" Frank Cole

Marine Fighter Squadron 112

(B1,B2—2Feb–2Mar44)

CO ______ Maj Herman Hansen, Jr.

Marine Fighter Squadron 114

(C-2–7May44)

CO ______ Capt Robert F. Stout

Marine Scout-Bomber Squadron 151

(B1,B2–29Feb–2Mar44)

CO ______ Maj Gordon H. Knott

Marine Fighter Squadron 216

Detachment, Ground Echelon (E—30Jul–15Aug44)
Remainder Squadron (E-4–15Aug44)

CO ______ Maj John Fitting, Jr.

Marine Fighter Squadron 217

Detachment, Ground Echelon (E—30Jul–15Aug44)
Remainder Squadron (E-4–15Aug44)

CO ______ Maj Max R. Read, Jr.

Forward Echelon, Marine Fighter Squadron 224

(B1—7–8Feb44)

CO ______ Maj Darrell D. Irwin

Marine Fighter Squadron 225

(E—30Jul–27Aug44)

CO ______ LtCol James A. Embry, Jr.

Marine Scout-Bomber Squadron 231

Ground Echelon (B1—3Feb–2Mar44)
Flight Echelon (B1—21Feb–2Mar44)

CO ______ Maj Elmer G. Glidden, Jr.

Marine Utility Squadron 252

Detachment, Flight Echelon
(C—20Jun–10Aug44)
(E—20Jun–15Aug44)
(D—24Jul–1Aug44)

CO ______ LtCol Neil R. MacIntyre (to 20-Jul44)
Maj Robert B. Meyersburg (21-Jul–1Aug44)
LtCol John V. Kipp (from 2Aug-44)

Marine Fighter Squadron 311

(B1—7–8Feb44)

CO ______ Maj Harry B. Hooper, Jr.

Detachment, Flight Echelon, Marine Scout-Bomber Squadron 331

(A—30Nov–8Dec43)

CO ______ Maj Paul R. Byrum, Jr.

Detachment, Flight Echelon, Marine Utility Squadron 353

(A—26Nov–8Dec43)
(C—3–10Aug44)
(E—3–10Aug44)

CO ------- LtCol Edmund L. Zonne

Marine Fighter Squadron 422

(B1,B2—2Feb–2Mar44)

CO ------- Maj Elmer A. Wrenn

Marine Night Fighter Squadron 532

Advance Echelon (C—6Jul–10Aug44)
Rear Echelon (C—12Jul–10Aug44)
Advance Echelon (E—6Jul–10Aug44)
Rear Echelon (E—12Jul–10Aug44)

CO ------- Maj Everette H. Vaughan

Marine Night Fighter Squadron 534

Advance Echelon (E—29Jul–15Aug44)
Flight Echelon (E—4–15Aug44)

CO ------- Maj Ross S. Mickey

Detachment, Flight Echelon, Marine Utility Squadron 952

(C—2–5Jul44)

CO ------- LtCol Malcolm S. Mackay

Marine Casualties[1]

Location and Date	KIA		DOW		WIA		MIAPD		Total	
	Officer	Enlisted	Officer	Enlisted	Officer	Enlisted	Officer	Enlisted	Officer	Enlisted
Marines										
Tarawa[2] (20Nov–8Dec43)	51	853	9	84	109	2,124	------	88	169	3,149
Kwajalein/Majuro (29Jan–8Feb44)	13	162	1	30	41	590	------	181	55	963
Eniwetok (17Feb–2Mar44)	4	177	1	37	27	541	2	37	34	792
Saipan (11Jun–10Jul44)	137	1,940	18	349	493	8,082	------	708	648	11,079
Guam (21Jul–15Aug44)	80	1,076	15	380	288	5,077	------	17	383	6,550
Tinian (24Jul–1Aug44)	22	278	4	61	97	1,824	1	2	124	2,165
Aviation[3]	19	18	------	6	30	90	7	15	56	129
Sea-duty[3]	------	13	------	2	1	57	------	4	1	76
Total Marines	326	4,517	48	949	1,086	18,385	10	1,052	1,470	24,903
Naval Medical Personnel Organic to Marine Units[4]										
Tarawa	2	28	------	------	2	57	------	------	4	85
Marshalls[5]	------	5	------	1	2	34	------	------	2	40
Saipan	1	70	------	6	7	330	------	------	8	406
Guam	3	43	1	4	11	195	------	------	15	242
Tinian	1	23	------	2	------	40	------	------	1	65
Marine Aviation	------	5	------	------	1	4	------	------	1	9
Total Navy	7	174	1	13	23	660	------	------	31	847
Grand Total	333	4,691	49	962	1,109	19,045	10	1,052	1,501	25,750

See footnote at end of table.

[1] These final Marine casualty figures were compiled from records furnished by Statistics Unit, Personnel Accounting Section, Records Branch, Personnel Department, HQMC. They are audited to include 26 August 1952. Naval casualties were taken from NavMed P-5021, *The History of the Medical Department of the Navy in World War II*, 2 vols (Washington: Government Printing Office, 1953), II, pp. 1-84. The key to the abbreviations used at the head of columns in the table follows: KIA, Killed in Action; DOW, Died of Wounds; WIA, Wounded in Action; MIAPD, Missing in Action, Presumed Dead. Because of the casualty reporting method used during World War II, a substantial number of DOW figures are also included in the WIA column.

[2] Includes Apamama.

[3] Includes operations in Gilberts, Marshalls, and Marianas during periods indicated above.

[4] See Footnote (1) above.

[5] Includes Kwajalein/Majuro and Eniwetok during periods indicated above.

Unit Commendations

THE SECRETARY OF THE NAVY,
Washington.

The President of the United States takes pleasure in presenting the PRESIDENTIAL UNIT CITATION to the

SECOND MARINE DIVISION (REINFORCED)

consisting of

Division Headquarters, Special Troops (including Company C, 1st Corps Medium Tank Battalion), Service Troops, 2nd, 6th, 8th, 10th and 18th Marine Regiments in the Battle of Tarawa, as set for in the following

CITATION:

"For outstanding performance in combat during the seizure and occupation of the Japanese-held Atoll of Tarawa, Gilbert Islands, November 20 to 24, 1943. Forced by treacherous coral reefs to disembark from their landing craft hundreds of yards off the beach, the Second Marine Division (Reinforced) became a highly vulnerable target for devastating Japanese fire. Dauntlessly advancing in spite of rapidly mounting losses, the Marines fought a gallant battle against crushing odds, clearing the limited beachheads of snipers and machine guns, reducing powerfully fortified enemy positions and completely annihilating the fanatically determined and strongly entrenched Japanese forces. By the successful occupation of Tarawa, the Second Marine Division (Reinforced) has provided our forces with highly strategic and important air and land bases from which to continue future operations against the enemy; by the valiant fighting spirit of these men, their heroic fortitude under punishing fire and their relentless perseverance in waging this epic battle in the Central Pacific, they have upheld the finest tradition of the United States Naval Service."

For the President.

JAMES FORRESTAL,
Secretary of the Navy.

THE SECRETARY OF THE NAVY,
Washington.

The President of the United States takes pleasure in presenting the PRESIDENTIAL UNIT CITATION to the

FOURTH MARINE DIVISION, REINFORCED

consisting of

Division Headquarters; Division Special Troops; Division Service Troops; 23rd, 24th, 25th Marines; 20th Marines (Engineers); 1st JASCO; 534th and 773rd Amphibian Tractor Battalions (Army); 10th Amphibian Tractor Battalion; Company "C" 11th Amphibian Tractor Battalion; 708th Amphibian Tank Battalion (Army); VMO-4; 2nd Amphibian Truck Company; 14th Marines (Artillery); 311th and 539th Port Companies (Army); Detachment 7th Field Depot; 1st Provisional Rocket Detachment, V Amphibious Corps; Detachment, Air Warning Squadron #5; 4th 105mm (Howitzer) Corps Artillery, V Amphibious Corps; 14th Marines (Artillery), (less 3rd and 4th Battalions); Headquarters, Provisional LVT Groups, V Amphibious Corps; 2nd Armored Amphibian Battalion; 2nd and 5th Amphibian Tractor Battalions; 715th Amphibian Tractor Battalion (Army); 1341st Engineer Battalion (Army); 1st Amphibian Truck Company; 2nd Tank Battalion; 1st and 2nd Battalions, 10th Marines (Artillery) and the 1st Provisional Rocket Detachment, for service as set forth in the following

CITATION:

"For outstanding performance in combat during the seizure of the Japanese-held islands of Saipan and Tinian in the Marianas from June 15 to August 1, 1944. Valiantly storming the mighty fortifications of Saipan on June 15, the Fourth Division, Reinforced, blasted the stubborn defenses of the enemy in an undeviating advance over the perilously rugged terrain. Unflinching despite heavy casualties, this gallant group pursued the Japanese relentlessly across the entire length of the island, pressing on against bitter opposition for twenty-five days to crush all resistance in their zone of action. With but a brief rest period in which to reorganize and re-equip, the Division hurled its full fighting power against the dangerously narrow beaches of Tinian on July 24 and rapidly expanded the beachheads for the continued landing of troops, supplies and artillery. Unchecked by either natural obstacles or hostile fire, these indomitable men spearheaded a merciless attack which swept Japanese forces before it and ravaged all opposition within eight days to add Tinian to our record of conquests in these strategically vital islands."

For the President.

JAMES FORRESTAL,
Secretary of the Navy.

THE SECRETARY OF THE NAVY,
Washington.

The President of the United States takes pleasure in presenting the PRESIDENTIAL UNIT CITATION to the

THIRD MARINES, REINFORCED,
serving as the THIRD COMBAT TEAM, THIRD MARINE DIVISION,

consisting of

Third Marine Regiment; Second Battalion, Ninth Regiment; Company "C", Third Tank Battalion; Company "C", Nineteenth Marine Regiment (Combat Engineers), and Third Band Section

for service as set forth in the following

CITATION:

"For extraordinary heroism in action against enemy Japanese forces during the invasion and recapture of Guam, Marianas Islands, from July 21 to August 10, 1944. Crossing a 400-yard reef under frontal and flanking fire from strongly defended positions on dominating terrain, the THIRD Marine Regiment (Reinforced), serving as the THIRD Combat Team, assaulted the steep slopes of the objectives and by evening had captured Adelup Point and Chonito Cliff. With no reserve available to be committed in their zone of action during the ensuing eight days, the gallant officers and men of this team fought their way forward through a maze of hostile caves and pillboxes and over rugged terrain to secure Fonte Canyon and the northeastern slopes of Fonte Ridge despite constant mortar, machine-gun, small-arms and artillery fire which blasted all echelons, shore party and lines of communication and supply. Seriously depleted by heavy casualties, including two battalion commanders, the THIRD Combat Team was continually in action as the left assault regiment until the cessation of organized resistance and the securing of the island on August 10, after twenty-one days of furious combat. By their effective teamwork, aggressive fighting spirit and individual acts of heroism and daring, the men of the THIRD Combat Team achieved an illustrious record of courage and skill, in keeping the highest traditions of the United States Naval Service."

For the President.

JAMES FORRESTAL,
Secretary of the Navy.

THE SECRETARY OF THE NAVY,
Washington.

The Secretary of the Navy takes pleasure in commending the
FIRST PROVISIONAL MARINE BRIGADE

for service as follows:

"For outstanding heroism in action against enemy Japanese forces during the invasion of Guam, Marianas Islands, from July 21 to August 10, 1944. Functioning as a combat unit for the first time, the First Provisional Marine Brigade forced a landing against strong hostile defenses and well camouflaged positions, steadily advancing inland under the relentless fury of the enemy's heavy artillery, mortar and small arms fire to secure a firm beachhead by nightfall. Executing a difficult turning movement to the north, this daring and courageous unit fought its way ahead yard by yard through mangrove swamps, dense jungles and over cliffs and, although terrifically reduced in strength under the enemy's fanatical counterattacks, hunted the Japanese in caves, pillboxes and foxholes and exterminated them. By their individual acts of gallantry and their indomitable fighting teamwork throughout this bitter and costly struggle, the men of the First Provisional Marine Brigade aided immeasurably in the restoration of Guam to our sovereignty."

All personnel serving in the First Provisional Marine Brigade, comprised of: Headquarters Company; Brigade Signal Company; Brigade Military Police Company; 4th Marines, Reinforced; 22nd Marines, Reinforced; Naval Construction Battalion Maintenance Unit 515; and 4th Platoon, 2nd Marine Ammunition Company, during the above mentioned period are hereby authorized to wear the NAVY UNIT COMMENDATION Ribbon.

JAMES FORRESTAL,
Secretary of the Navy.

THE SECRETARY OF THE NAVY,
Washington.

The Secretary of the Navy takes pleasure in commending the
TWELFTH MARINES, THIRD MARINE DIVISION

for service as follows:

"For outstanding heroism in action against enemy Japanese forces in the Empress Augusta Bay Beachhead, Bougainville, Solomon Islands, from November 1, 1943, to January 12, 1944; and in the invasion and seizure of Guam, Marianas, July 21 to August 10, 1944. Divided for landing into small elements dispersed over 5000 yards of beach at Empress Augusta Bay, the TWELFTH Marines overcame perilous surf and beach conditions and an almost inpenetrable wall of jungle and swampy terrain to land their pack howitzers, initial ammunition and equipment by hand, to occupy firing positions, emplace guns, set up all control facilities and deliver effective fire in support of the THIRD Marine Division beachhead by afternoon of D-Day. In action for 73 days while under continual Japanese air attacks, the TWELFTH Marines aided in smashing an enemy counterattack on November 7–8, silenced all hostile fire in the Battle of Cocoanut Grove on November 13, and delivered continuous effective fire in defense of the vital beachhead position. At Guam, they landed in the face of enemy mortar and artillery fire through treacherous surf and, despite extreme difficulties of communication, supply and transportation, and the necessity of shifting from one type of fire to another, rendered valuable fire support in night and day harassing fires, counterbattery fires and defensive barrages, including the disruption of an organized counterattack by seven Japanese battalions on the night of July 26–27. By their individual heroic actions and their skilled teamwork, the officers and men of the TWELFTH Marines served with courage and distinction during the THIRD Marine Division's missions to secure the Empress Augusta Bay Beachhead and to aid in the recapture of Guam, thereby enhancing the finest traditions of the United States Naval Service."

All personnel attached to and serving with the TWELFTH Marines during these periods are hereby authorized to wear the NAVY UNIT COMMENDATION Ribbon.

JAMES FORRESTAL,
Secretary of the Navy.

THE SECRETARY OF THE NAVY,
Washington.

The Secretary of the Navy takes pleasure in commending the

TWENTY-FIRST MARINES, REINFORCED, serving as the TWENTY-FIRST REGIMENTAL COMBAT TEAM, THIRD MARINE DIVISION

consisting of

the Twenty-First Marines; Company "B", Nineteenth Marines (Combat Engineers); Company "B", Third Tank Battalion; Second Band Section,

for service as follows:

"For outstanding heroism in action against enemy Japanese forces during the assault, seizure and occupation of Guam, Marianas Islands, from July 21 to August 10, 1944. Landing as the center Regimental Combat Team of the Division at Asan, the Twenty-First Marine Regiment, Reinforced, serving as the Twenty-First Regimental Combat Team, swept rapidly over enemy beach defenses toward a strategic high ridge which afforded the enemy observation of the Division landing area and enabled him to deliver accurate mortar and artillery fire on the beaches. Under heavy mortar and small-arms fire as they stormed the two narrow defiles which constituted the only approach to the vertical cliffs, these gallant Marines established two bridgeheads covering the defiles and, by midafternoon, had consolidated the Combat Team's position atop the cliffs, thus materially reducing the volume and accuracy of hostile fire and facilitating establishment of the Division artillery ashore and the landing of supplies and equipment. Halted by direct, short-range enfilade artillery fire from commanding terrain in an adjacent zone, they held tenaciously to their vital position in the face of continuous mortar fire by day, sharp nightly counterattacks and mounting casualties. When the enemy launched a full-scale counterattack with his remaining organized forces in the pre-dawn hours of July 26, wiping out one company of the Combat Team and penetrating the front lines, these officers and men waged a furious battle in the darkness; they annihilated approximately 2,000 Japanese troops in front of and within their position; and, by their individual heroism and gallant fighting spirit, dealt a crushing blow to organized enemy resistance on Guam, thereby upholding the finest traditions of the United States Naval Service."

All personnel attached to and serving with the Twenty-First Regimental Combat Team on Guam from July 21 to August 10, 1944, are authorized to wear the NAVY UNIT COMMENDATION Ribbon.

JOHN L. SULLIVAN,
Secretary of the Navy.

THE SECRETARY OF THE NAVY,
Washington.

The Secretary of the Navy takes pleasure in commending the

TWENTY-SECOND MARINES, REINFORCED, TACTICAL GROUP ONE, FIFTH AMPHIBIOUS CORPS

consisting of

Twenty-second Marines, Second Separate Pack Howitzer Company; Second Separate Tank Company; Second Separate Engineer Company; Second Separate Medical Company; Second Separate Motor Transport Company; Fifth Amphibious Corps Reconnaissance Company; Company D, Fourth Tank Battalion, Fourth Marine Division; 104th Field Artillery Battalion, U. S. Army; Company C, 766th Tank Battalion; U. S. Army; Company D, 708th Provisional Amphibian Tractor Battalion, U. S. Army; and the Provisional DUKW Battery, Seventh Infantry Division, U. S. Army.

for service as follows:

"For outstanding heroism in action against enemy Japanese forces during the assault and capture of Eniwetok Atoll, Marshall Islands, from February 17 to 22, 1944. As a unit of a Task Force, assembled only two days prior to departure for Eniwetok Atoll, the Twenty-second Marines, Reinforced, landed in whole or in part on Engebi, Eniwetok and Parry Islands in rapid succession and launched aggressive attacks in the face of heavy machine-gun and mortar fire from well camouflaged enemy dugouts and foxholes. With simultaneous landings and reconnaissance missions on numerous other small islands, they overcame all resistance within six days, destroying a known 2,665 of the Japanese and capturing 66 prisoners. By their courage and determination, despite the difficulties and hardships involved in repeated reembarkations and landings from day to day, these gallant officers and men made available to our forces in the Pacific Area an advanced base with large anchorage facilities and an established airfield, thereby contributing materially to the successful conduct of the war. Their sustained endurance, fortitude and fighting spirit throughout this operation reflect the highest credit on the Twenty-second Marines, Reinforced, and on the United States Naval Service."

All personnel attached to and serving with any of the above units during the period February 17 to 22, 1944, are authorized to wear the NAVY UNIT COMMENDATION Ribbon.

JOHN L. SULLIVAN,
Secretary of the Navy.

THE SECRETARY OF THE NAVY,
Washington.

The Secretary of the Navy takes pleasure in commending the

AMPHIBIOUS RECONNAISSANCE BATTALION
FLEET MARINE FORCE, PACIFIC

for service as follows:

"For outstanding heroism in action against enemy Japanese forces in the Gilbert Islands, from November 19 to 26, 1943; the Marshall Islands, from January 30 to February 23, 1944; Marianas Islands, from June 15 to August 4, 1944; and Ryukyu Islands, from March 26 to July 24, 1945. The only unit of its kind in the Fleet Marine Force, Pacific, the Amphibious Reconnaissance Battalion rendered unique service in executing secret reconnaissance missions on enemy-held islands. Frequently landing at night from submarines and other vessels prior to the assault, the small unit entered areas where friendly aircraft, Naval gunfire and other forms of support were unavailable and, under cover of darkness, moved about in hostile territory virtually in the presence of enemy troops. Despite hazards incident to passage through dark and unfamiliar hostile waters, often through heavy surf onto rocky shores, the Battalion persevered in its mission to reconnoiter enemy islands and obtain information vital to our assault forces and, on several occasions, succeeded in overcoming all enemy resistance without the aid of regular troops. Carrying out its difficult tasks with courage and determination, the Amphibious Reconnaissance Battalion contributed materially to the success of our offensive operations throughout four major campaigns and achieved a gallant record of service which reflects the highest credit upon its officers and men and the United States Naval Service."

All personnel attached to and serving with the Amphibious Reconnaissance Battalion during one or more of the above-mentioned periods are authorized to wear the NAVY UNIT COMMENDATION Ribbon.

JAMES FORRESTAL,
Secretary of the Navy.

THE SECRETARY OF THE NAVY,
Washington.

The Secretary of the Navy takes pleasure in commending the

FIRST SEPARATE ENGINEER BATTALION

for service as follows:

"For exceptionally meritorious service in support of military operations on Guadalcanal, December 10, 1942, to February 27, 1943; Tinian from August 20, 1944, to March 24, 1945; and Okinawa from April 14 to September 2, 1945. Faced with numerous and difficult problems in engineering throughout two major campaigns, the First Separate Engineer Battalion initiated new techniques and procedures in construction, repair and maintenance, executing its mission under adverse conditions of weather and terrain and in spite of Japanese shellings, artillery fire, bombing raids, sickness and tropical storms. Technically skilled, aggressive and unmindful of great personal danger, the officers and men of this gallant Battalion constructed, developed and maintained vital routes of communication, airfields and camp facilities; they served as combat engineer units in performing demolitions, mine detection and disposal and bomb disposal tasks in support of various units of the Fleet Marine Force; and they built bridges and repaired air-bombed air strips toward the uninterrupted operations of Allied ground and aerial forces. Undeterred by both mechanical and natural limitations, the First Separate Engineer Battalion completed with dispatch and effectiveness assigned and unanticipated duties which contributed immeasurably to the ultimate defeat of Japan and upheld the highest traditions of the United States Naval Service."

All personnel attached to the First Separate Engineer Battalion during any of the above mentioned periods are hereby authorized to wear the NAVY UNIT COMMENDATION Ribbon.

JAMES FORRESTAL,
Secretary of the Navy.

THE SECRETARY OF THE NAVY,
Washington.

The Secretary of the Navy takes pleasure in commending the

III AMPHIBIOUS CORPS SIGNAL BATTALION

for service as set forth in the following

CITATION:

"For extremely meritorious service in support of military operations, while attached to the I Marines Amphibious Corps during the amphibious assault on Bougainville, and attached to the III Amphibious Corps during operations at Guam, Palau and Okinawa, during the period from November 1, 1943 to June 21, 1945. The first American Signal Battalion to engage in amphibious landings in the Pacific Ocean Areas, the III Amphibious Corps Signal Battalion pioneered and developed techniques and procedures without benefit of established precedent, operating with limited and inadequate equipment, particularly in the earlier phase of these offensive actions, and providing its own security while participating in jungle fighting, atoll invasions and occupation of large island masses. Becoming rapidly experienced in guerrilla warfare and the handling of swiftly changing situations, this valiant group of men successfully surmounted the most difficult conditions of terrain and weather as well as unfamiliar technical problems and, working tirelessly without consideration for safety, comfort or convenience, provided the Corps with uninterrupted ship-shore and bivouac communication service continuously throughout this period. This splendid record of achievement, made possible only by the combined efforts, loyalty and courageous devotion to duty of each individual, was a decisive factor in the success of the hazardous Bougainville, Guam, Palau and Okinawa Campaigns and reflects the highest credit upon the III Amphibious Corps Signal Battalion and the United States Naval Service."

All personnel attached to the III Amphibious Corps Signal Battalion who actually participated in one or more of the Bougainville, Guam, Palau and Okinawa operations are hereby authorized to wear the NAVY UNIT COMMENDATION Ribbon.

JAMES FORRESTAL,
Secretary of the Navy.

214-881 O-67—42

THE SECRETARY OF THE NAVY,
Washington.

The Secretary of the Navy takes pleasure in commending the

THIRD BATTALION, TENTH MARINES, SECOND MARINE DIVISION
FLEET MARINE FORCE

for service as follows:

"For outstanding heroism while serving with the 2nd Marine Division in action against enemy Japanese forces on the Island of Saipan in the Marianas, July 7, 1944. When Japanese forces initiated a final concerted attack down the west coast of the island before dawn of July 7, the 3rd Battalion, 10th Marines, was occupying a newly won position astride the railway along the west coast road, with two batteries disposed on the left of the railroad and the remaining two on the right and echeloned to the rear. The mounting enemy attack penetrated the extreme left flank of our lines and moved between the coast road and the railway. Security elements to the front of the forward batteries recognized and gave battle to the oncoming force of approximately 600 Japanese supported by tanks. Battalion howitzers opened up at point-blank range, firing shells with cut fuzes; gunners employed ricochet fire when the fanatic banzai troops over-ran the forward section; and the cannoneers, command post and supply personnel in the rear positions united as one to engage the infiltrating Japanese soldiery. Under the forceful direction of skilled officers, this artillery battalion functioned effectively as an infantry unit despite the lack of specific training, the four batteries waging a furious and prolonged battle from quickly organized strongpoints and holding the line indomitably until relieved several hours later. Strengthened by fresh troops, the defending garrison continued its counter-and-thrust tactics and, recapturing the heavy guns which had fallen into hostile hands, knocked out three of the enemy tanks and annihilated approximately three hundred Japanese troops. By their valor, determination and sustained fighting spirit, the intrepid officers and men of the 3rd Battalion, 10th Marines, had succeeded in breaking the enemy's last desperate effort to oppose the seizure of Saipan, thereby hastening the conquest of this strategically important base. Their gallant defense of a vulnerable position in the face of overwhelming disparity adds new luster to the traditions of the United States Naval Service."

All personnel attached to the 3rd Battalion, 10th Marines, 2nd Marine Division, on July 7, 1944 are hereby authorized to wear the NAVY UNIT COMMENDATION Ribbon.

JAMES FORRESTAL,
Secretary of the Navy.

THE SECRETARY OF THE NAVY,
Washington.

The Secretary of the Navy takes pleasure in commending the

NINTH MARINE DEFENSE BATTALION

for service as follows:

"For outstanding heroism in action against enemy Japanese forces at Guadalcanal, November 30, 1942, to May 20, 1943; Rendova-New Georgia Area, June 30 to November 7, 1943; and at Guam, Marianas, July 21 to August 20, 1944. One of the first units of its kind to operate in the South Pacific Area, the NINTH Defense Battalion established strong seacoast and beach positions which destroyed 12 hostile planes attempting to bomb Guadalcanal, and further engaged in extensive patrolling activities. In a 21-day-and-night training period prior to the Rendova-New Georgia assault, this group calibrated and learned to handle new weapons and readily effected the conversion from a seacoast unit to a unit capable of executing field artillery missions. Joining Army Artillery units, special groups of this battalion aided in launching an attack which drove the enemy from the beaches, downed 13 of a 16-bomber plane formation during the first night ashore and denied the use of the Munda airfield to the Japanese. The NINTH Defense Battalion aided in spearheading the attack of the Army Corps operating on New Georgia and, despite heavy losses, remained in action until the enemy was routed from the island. Elements of the Battalion landed at Guam under intense fire, established beach defenses, installed antiaircraft guns and later, contributed to the rescue of civilians and to the capture or destruction of thousands of Japanese. By their skill, courage and aggressive fighting spirit, the officers and men of the NINTH Defense Battalion upheld the highest traditions of the United States Naval Service."

All personnel attached to and serving with the NINTH Defense Battalion during the above mentioned periods are authorized to wear the NAVY UNIT COMMENDATION Ribbon.

JOHN L. SULLIVAN
Secretary of the Navy.

Index

U.S. GOVERNMENT PRINTING OFFICE : 1967 O—214-881

20501511R00382

Printed in Great Britain
by Amazon